EXPERIENTIA SUPPLEMENTUM 25

Proceedings

of the

Fourth International Symposium on Polarization Phenomena in Nuclear Reactions

Zürich, 1975

Editors: W. Grüebler and V. König

1976

SPRINGER BASEL AG

Published with the financial assistance of UNESCO
(Subvention – 1975 – DG/2. 1/414/40)

CIP-Kurztitelaufnahme der Deutschen Bibliothek

International Symposium on Polarization Phenomena in Nuclear Reactions ‹04, 1975, Zürich›
Proceedings of the Fourth International Symposium on Polarization Phenomena in
Nuclear Reactions: Zürich, 1975/ed.: W. Grüebler and V. König.
  (Experientia: Suppl.; 25)
  ISBN 978-3-0348-5507-5   ISBN 978-3-0348-5506-8 (eBook)
  DOI 10.1007/978-3-0348-5506-8

NE: Grüebler, Willi [Hrsg.]

© Springer Basel AG 1976
Ursprünglich erschienen bei Birkhäuser Verlag Basel 1976
Softcover reprint of the hardcover 1st edition 1976
ISBN 978-3-0348-5507-5

Preface

The Fourth International Symposium on Polarization Phenomena in Nuclear Reactions took place from August 25 to 29, 1975, at the Swiss Federal Institute of Technology in Zürich (ETHZ). Apart from the host institution the Symposium was also supported by the International Union of Pure and Applied Physics, the Swiss National Science Foundation and the Swiss Physical Society. The program of the Symposium was set up with the advice of an International Program Committee with the following members:

Prof. I. Ja. Barit, USSR Academy of Sciences, Moscow, USSR
Prof. E. Baumgartner, University of Basel, Basel, Switzerland
Prof. H.E. Conzett, Lawrence Berkeley Laboratory, Berkeley, USA
Dr. W. Grüebler, Laboratorium für Kernphysik, ETH Zürich
Prof. W. Haeberli, University of Wisconsin, Madison, USA
Prof. S.S. Hanna, Stanford University, Stanford, USA
Prof. J. McKee, University of Manitoba, Winnipeg, Canada
Prof. G.C. Morrison, University of Birmingham, England
Dr. G.G. Ohlsen, LASL, Los Alamos, USA
Prof. J. Raynal, C.E.N. Saclay, France
Dr. M. Simonius, Laboratorium für Kernphysik, ETH Zürich

The Local Organizing Committee consisted of

Dr. R. Balzer
Dr. W. Grüebler
Dr. H. Jung
Dr. V. König
Prof. J. Lang
Dr. M. Simonius
Prof. W.G. Weitkamp (on leave from University of Washington, Seattle)

It was generally felt that the Fourth Polarization Symposium should emphasize the importance of polarization measurements in the different fields of nuclear physics and explain the physical content of polarization phenomena.

The Symposium was organized in fifteen plenary sessions and fourteen discussion sessions. The invited papers were intended to be surveys to cover advances since the Madison meeting, as well as possible future developments. Dr. M. Simonius organized extensive discussion sessions among smaller groups of experts which provided the occasion to discuss the problems in special areas. More than 200 contributed papers were submitted of which only very few were presented orally in the plenary sessions. They had been distributed at the beginning of the Symposium in a preliminary printed form as a base for the discussion sessions. Reports on the discussions have been presented in plenary sessions by rapporteurs.

In the Proceedings, the invited papers form the first part and are printed in the order in which they were given, followed by the discussion in the plenary session. The reports of the discussion sessions have been added to the corresponding survey papers, if submitted to the Symposium in written form. In the interest of a rapid publication of these Procee- dings, a reproduction by photo-offset techniques of author-prepared manuscripts was used.

Thanks are due to all sponsors and to the members of the Internatio- nal Program Committee. The Local Organizing Committee acknowledges the prompt delivery of the manuscripts by the authors. We are very grateful to Mrs. N. Doebeli for the excellent re-typing of many invited and contributed papers which have not been in accordance with the require- ments, and for her competent assistance in editing the Proceedings. I would also like to acknowledge the help and constant support of the members of the polarization group in Zürich, the employees of the Labora- torium für Kernphysik and the Department of Physics of the ETHZ. The prompt and careful completion of the Proceedings by the publisher is very much appreciated.

Zürich, November 1975   W. Grüebler

Contents

SYMMETRY PROPERTIES

TRANSFER REACTIONS

PARTICLE–γ ANGULAR CORRELATIONS AND NEW
APPLICATIONS OF POLARIZED BEAMS

Contents

OPENING ADDRESS

H. Staub, President of the Swiss National Committee of IUPAP
University of Zürich, Switzerland

As chairman of the Swiss National Committee of the International
Union of Pure and Applied Physics I have the honor to welcome you in the
name of the Union to the 4th International Symposium on Polarization
Phenomena in nuclear reactions. Like a week ago at the opening of the
cyclotron conference, I wish to express our satisfaction that the organizers of the conference have asked for the Union's sponsorship, although
this engagement encompasses always a very small financial assistance.
This shows therefore that the help of the Union in organizing and coordi-
nating scientific conferences, as it has been done now for more than
50 years, is still appreciated by the scientific community.

The first nuclear polarization symposium took place in 1960 in Basel,
and whilst we see many familiar faces of persons who attended that conference, we are missing today its organizer, Paul Huber, who died Febru-
ary 5th, 1971 at the age of sixty, shortly after the third conference in
Madison. I think it is appropriate at this occasion to say a few words
in memoriam of Paul Huber whose name is so intimately connected, not only
with the Union of which he was a vice president from 1951 to 1957 and
chairman of its nuclear physics commission, but with nuclear physics and
particularly with nuclear polarization phenomena. These problems were
some kind of guiding thread throughout his scientific career.

Almost exactly 40 years ago, Paul Huber, Ernst Baldinger and myself
investigated the neutrons from the d-d reaction. This work became Paul
Hubers Ph.D. thesis. For the detection of the neutrons we used a Helium
filled recoil ionization pulse chamber and one of our main concerns was
the completely inexplicable observation, that the Helium recoil distribution was quite anisotropic. We did at that time not realize that it was
caused by the strong spin-orbit interaction in the ^{5}He $p_{3/2}$, $p_{1/2}$ states
which makes ^{4}He an excellent polarization analyzer. The next important
contribution of Huber and his co-workers was the demonstration of the
polarization of the neutrons from the d-d reaction in 1952 simultaneously
with the same work by Ricamo at our host laboratory in Zürich. The cul-
mination for Paul Huber was the organization of the first nuclear polarization symposium in 1960 in Basel, where he and his collaborators presented one of the first working sources of polarized deuterons with
strong field magnetic separation. At this first conference, the definition of the sign of polarization was proposed and later adopted and is
now known as the Basel convention.

Beside his considerable contribution to nuclear physics, Huber was
an outstanding teacher, able to convey his enthusiasm for physics to his
students and collaborators. Another property of his character, which always surprised me, was his strong sense of responsibility as scientist
towards society. He never refused his cooperation in scientific organi-
zations or for his university. He was president of the Swiss Academy of
Science and rector of the University of Basel. He served on innumerable
advisory committees for the Swiss government.

Today at the opening of the 4th nuclear polarization symposium I
recognize many of Hubers former students and collaborators but I miss

that friendly face of the father of these conferences. I am sure he would have enjoyed it, taking place at the laboratory where he got his scientific education by Paul Scherrer.

Ladies and Gentlemen, in conclusion I wish you in the name of the International Union of Pure and Applied Physics a successful and enlightening symposium.

Survey Papers

NUCLEON-NUCLEON INTERACTION

D. Schütte

Institut für Theoretische Kernphysik der Universität, Bonn, Germany

1. Introduction

It is the purpose of this talk to give a survey on the present status of understanding of the nucleon-nucleon (N-N) interaction. The main emphasis will be put on the theoretical interpretation of the N-N interaction due to its relation to elementary particle physics phenomena at medium energies with the aim to point out that an understanding of the N-N interaction is only possible if one keeps in mind that "mesons are around in the air" if nucleons begin to interact with each other.

The organisation of this talk is as follows. In section 2 there shall be a brief discussion of the present status of the empirical knowledge of N-N scattering. In section 3 and 4 there will be given an outline on the problem of the determination of a N-N potential in a purely phenomenological way and within the one-boson-exchange (OBE) model. The attempts to relate the N-N interaction to π-π and π-N interactions are reviewed in section 5. The problem of predicting nuclear many-body properties is discussed in section 6. Hereby, special attention will be paid to the possible importance of incorporating N^* and mesonic degrees of freedom into the description of nuclei. In the last section conclusions shall be drawn with special emphasis on which kind of further experimental information is needed to obtain a clearer insight into the structure of the N-N interaction.

2. Empirical informations on the N-N interaction

Besides the properties of the deuteron, the most unambiguous experimental information on the N-N interaction is provided by the N-N scattering data, whereby any useful measurement includes polarization experiments. These N-N scattering data are traditionally analysed in terms of phase shifts $\delta_{LJ}(E)$ and $\epsilon_J(E)$ (see ref.[1]). The experimental knowledge of the phase shifts has been discussed in detail by M.H. MacGregor at the last polarization conference in Madison[2]. Since then, the situation has not essentially changed, and one can summarize it as follows.
i) The T=1 phase shifts (resulting from p-p scattering) are commonly accepted to be well known.
ii) The information on the T=0 phase shifts (being determined from p-n scattering) is still uncomplete. There are relatively large errors yielding ambiguities in the solution of the χ^2 minimum determining the phases as pointed out by Signell and Holdemann[3]. Also it is still unclear whether there is a sign change in the ϵ_1 phase for small energies[2].

As we shall see by the following analysis of the theoretical interpretation of the N-N interaction, a more accurate knowledge of the T=0 phase shifts should be of great interest.

3. Phenomenological potential model for the N-N interaction

Traditionally, the theoretical description of the N-N data is done within standard many-body theory. Hereby one assumes - disregarding all mesonic and other exotic degrees of freedom - for the description of nuclei the existence of a many-body Hamiltonian with two-body forces V.

$$H = H_o + V \tag{1}$$

(H_o represents the kinetic energy of the nucleons).

For any V one can then calculate the corresponding phase shifts $\delta_{LJ}(V)$ and $\epsilon_J(V)$. Thus the experimental phase shifts represent a clear information on V. This information is unambiguous since the two nucleon problem defined by H is rigorously solvable so that one does not get involved with problems of approximations as it is the case with any other many nucleon system.

However, the phase shifts do not determine V uniquely. For given $\delta_{LJ}(E)$ and $\epsilon_J(E)$ there exist large classes of phase shift equivalent potentials[4]. Consequently, any determination of V is based on a special ansatz for its functional form (e.g. maximum locality, separability) and there exist many N-N potentials which reproduce the N-N scattering data satisfactorily[4]

4. The OBE model for the N-N interaction

In order to improve on the problem of uniqueness of V and to get a deeper understanding of V one has invoked the classical idea of Yukawa that nuclear forces should be derivable from meson exchange. Quantitatively, the most successful model working with this assumption is the OBE model[5]. The idea hereby is to define V^{OBE} by a suitable identification with OBE Feynman diagrams

$$V^{OBE} = \underset{\text{bosons}}{\Sigma} \left| - - - \right|$$

The bosons which are exchanged are T=0 and 1 pseudoscalar (η,π), scalar (σ,δ) and vector mesons (ω,ρ). The logical steps for constructing the OBE potential can be described as follows.

i) Start with a field theoretical Hamiltonian of interacting nucleons and bosons

$$H^F = H_o + H_o^b + W \tag{2}$$

where H_o^b is the kinetic energy of the mesons and W is the interaction yielding diagramatic elements of the Type $\left| - - \right.$.

W is parametrized in terms of coupling constants, meson masses and form factors which are needed to regularize the interaction at high energies.

ii) Define an "effective" N-N potential, acting in the space of nucleons only, by eliminating the mesonic degrees of freedom. This elimination can be performed by standard methods of many-body theory[6] and yields in the OBE approximation an energy dependent "quasi-potential"[7] ($<\alpha\beta|$ are two nucleon states)

$$<\alpha\beta|V(z)|\bar{\alpha}\bar{\beta}> = <\alpha\beta|W \frac{1}{z-H_o-H_o^b} W|\bar{\alpha}\bar{\beta}> \quad \text{linked} \tag{3}$$

(V(z) corresponds to the diagram $\left| \diagdown \diagup \right|$ evaluated in the Bloch-Horowitz scheme[6]).

The OBE potentials are constructed from V(z) by eliminating artificially the z-dependence. This is not unique, and one may define a "covariant" or retarded expression[8]

$$<\alpha\beta|V^{OBE}|\bar{\alpha}\bar{\beta}> = \frac{1}{2}<\alpha\beta|V(E_\alpha+E_\beta)+V(E_{\bar{\alpha}}+E_{\bar{\beta}})|\bar{\alpha}\bar{\beta}> \tag{4}$$

or a static OBE potential[9] by setting

$$<\alpha\beta| V^{OBE}_{static} |\bar{\alpha}\bar{\beta}> = - <\alpha\beta| W \frac{1}{H^b_o} W |\bar{\alpha}\bar{\beta}> . \qquad (5)$$

This yields "OBE potentials in momentum space" which can be transformed to configuration space by neglecting higher orders in the derivatives[10,11].
iii) The parameters of V^{OBE} (contained in W) are determined, after specifying the ansatz, in the standard way by an adjustment to a best fit of N-N scattering data (including the deuteron).

An alternative way to "derive" an OBE ansatz for V consists in suitably recasting the Bethe-Salpeter equation[5] which describes the N-N scattering within covariant field theory[5]. This is convenient if one is interested in the relation of N-N scattering to the π-π and π-N interactions (see section 5). However this makes more difficult to obtain an extension of the OBE scheme to applications in the many-body case. Just as within the preceding scheme, also this prescription does not lead to a unique definition of V^{OBE}.

OBE potentials have been used to describe the N-N data with the following successes.
a) It is possible to obtain OBE potentials which give a quantitative fit to the N-N data, as good as with the best phenomenological potentials (e.g. Reid's potential)[5]. (This is true for any specification of V^{OBE}).
b) The number of parameters in V^{OBE} is considerably smaller (about 8) than in the phenomenological potentials. In addition, the coupling constants and masses have an independent physical meaning. The masses of the known mesons and the pion coupling constant can be fixed at their experimental values.
c) If one reduces V^{OBE} to a static, maximum local ansatz[10], one gets for large distances ($r \geq 1$ fm) the same expressions as with the phenomenological (maximum local) potentials (Reid or Hamada-Johnston potentials). In this way, the long range part of the N-N potential becomes clearly physically interpretable on the basis of OBE. (For large r, non-localities are expected to play a minor role so that all OBE potentials should be equivalent in this region). The different parts of the N-N interaction then have the following interpretation[10].
i) The long range part ($r > 1.4$ fm) is provided by the one-pion exchange potential and consists of a tensor and a $\vec{\sigma}_1\vec{\sigma}_2$ part.
ii) The most important part of the $\vec{L}s$ potential is generated by the vector mesons. A minor contribution comes from the scalar particles.
iii) The medium range attractive central potential is due to the exchange of scalar particles.
iv) The short range central repulsion is a consequence of ω-exchange.
This "physical" interpretation of the N-N potential is, however, limited by the following difficulties.
a) The non-uniqueness of the definition of V^{OBE} (kind of elimination of the z-dependence, choice of form factors, use of configuration space or momentum space expressions) leads to an uncertainty in the structure of V^{OBE} at small distances ($r < 1$ fm).
b) Compared to their determination from e.m. form factors of the nucleons, the coupling constants of the nucleons are too large (by factors of 2 to 3) if they are adjusted to a fit of N-N data[5]. Recently, this point could be improved by using eikonal regularization functions which were independently determined from the nucleonic e.m. form factors[12].
c) All OBE potentials have to work with a scalar particle (the 0^+ (T=0) σ-particle) which is experimentally not observed. The σ-meson is considered to be a substitute for 2π-exchange. This can be shown quantitatively

by studying the relation of N-N scattering to ππ scattering and π-N
scattering which will be outlined now.

4. Relation to π-π and π-N scattering

One of the main interests in the N-N interaction lies in its relation
to the π-π and π-N interactions so that, by considering all these interactions, one can test rather general structures predicted by field theory[13].
Schematically, these relations can be visualized as follows.

Within covariant perturbation theory, π-π scattering is given by a
class of diagrams of the type

$$T_{\pi\pi} \quad = \quad \text{---}\Box\text{---}$$

Having (in principle) determined this (off shell) π-π scattering one may
compute the π-N scattering amplitude $T_{\pi N}$ by combining all contributions
to $T_{\pi N}$ containing no final state interaction

(like ⟩|), denoted graphically by

with the π-π scattering amplitude $T_{\pi\pi}$

$$T_{\pi N} \quad = \quad \Box\text{---}\Box\text{---} \; .$$

The classes of diagrams yielding $T_{\pi\pi}$ and $T_{\pi N}$ (off shell) can now again be
combined to give the total 2π exchange contribution to N-N scattering (i.e.
the contribution to the N-N scattering of all processes which are accompanied by two pions going from one nucleon to the other) represented by
the class of diagrams

$$T^{(2\pi)}_{NN} = \Box\text{---}\Box\text{---}\Box$$

(The separation of $T^{(2\pi)}_{NN}$ into this special class of diagrams guarantees
that there is no double counting).

Thus one can calculate $T^{(2\pi)}_{NN}$ from the knowledge of the (off shell)
π-π and π-N scattering. In practice, this involves difficult on-to-off
shell extrapolations which are partly accomplished by rewriting the above
scheme using dispersion relations, unitarity and crossing symmetry[13].
Though there are still problems with the uniqueness of the extrapolations,
numerical results of different authors fairly agree[14,15].

Using a maximum local ansatz vor V, the knowledge of $T^{(2\pi)}_{NN}$ can be
transformed into a potential which is interpreted as the 2π exchange
contribution to V. Again this potential can be taken seriously only for
large r (r>1 fm) since for small r (high energies) the reliability of
the calculated $T^{(2\pi)}_{NN}$ breaks down and multi-meson exchanges come into
play. Supplemented by one-pion and one-omega exchange, the resulting
potential can be compared to phenomenological potentials showing remarkable agreement[14,15]. Thus the long range part of the N-N potential is
now understandable on completely physical grounds, leading to a picture
like in the OBE case where, however, the σ and ρ mesons are replaced by
the influence on 2π exchange. Also the ω-coupling constant can be taken
physically.

These "theoretical" potentials can be refined by fixing the long range part from the theory and adjusting the (suitably parametrized) short range part independently by fitting the N-N data. Hereby one may obtain a reproduction on the N-N data which is even better than with Reid's potential[16].

It has been tried to avoid the appearance of new parameters for the short range part of V by extrapolating the given expressions of $T_{NN}^{(2\pi)}$ and of the π and ω potentials in momentum space to high energies using eikonal regularization functions (the same for all parts of the potential) adjusted to the e.m. form factor of the nucleon. If one makes full use of the uncertainties in the experimental knowledge of π-π scattering, the ω-coupling constant and the form factor, it is possible to obtain in this way a N-N potential yielding fairly quantitative N-N data[17].

Thus the consideration of the relation to π-π and π-N scattering data has led to a clear success in the understanding of the N-N interaction. However, it should be stressed that one obtains primarily only relations to N-N scattering, the translation to a potential is only possible by specifying an ansatz for V which, in addition, becomes energy dependent which makes it unclear how to apply it to the nuclear many-body case.

5. Relation to nuclear many-body data

When inserted into the nuclear many-body problem all N-N potentials produce short range correlations (due to the short range repulsion present in any realistic N-N force). The only properties of nuclei which at present can be satisfactorily calculated microscopically in such a situation are binding energies and saturation densities of nuclei. This is due to the success of Brueckner theory[18] and its refinements[19]. In the following we shall discuss the present understanding of nuclear matter properties on the basis of Brueckner's theory.

Starting from all available phase shift equivalent N-N potentials one observes that they yield for nuclear matter different saturation energies and densities which arrange themselves on a characteristic line, the so-called Coester-line[19,20] which does not meet the experimental point. With respect to this structure phenomenological, OBE and 2π potentials yield quite equivalent structures since they all work with the traditional concept of a potential[21]. The uncertainty of the potential for small distances seems not to be important for this relation to nuclear matter properties. Though this is the result of only the Brueckner approximation, studies of higher order corrections show that the experimental energy-density value cannot be reached from the Coester line[19].

It is tempting to conjecture that this failure of standard many-body theory is due to the fact that by introducing the artificial concept of a N-N potential one has disregarded the N^* and mesonic degrees of freedom in the treatment of the nucleus. The importance of taking into account N^* (excited states of the nucleon, especially the 3/2 - 3/2 - Δ-resonance) for calculating nuclear properties has been pointed out by A.M. Green and coworkers[22]. Extending the traditional many-body scheme by adding Δ states to the Hilbert space they adjust the NΔ potential to N-N scattering and predict nuclear matter data by taking into account the Δ degree of freedom in standard Brueckner theory (which is straight forward). As a result one obtains in this way saturation energies and densities well off the Coester line making hope to explain the experimental nuclear matter properties by a suitable ansatz for the potential. However, it should be remarked that these results are only preliminary since the Δ resonance has been incorporated only in the $^1S_0$ channel. Also an extension to include mesonic degrees of freedom into Brueckner theory can be made[7]. Herefore, one works with the field theoretical Hamiltonian H^F (eq.2) throughout, avoiding in this way the definition of a N-N potential. In analogy to standard many-body theory H^F is then treated within non-covariant perturbation theory, and one includes only OBE diagrams. For the two-nucleon problem this yields the scattering equation of Kadyshevsky[23] involving the (energy dependent) quasipotential V(z) (eq.3) :

$$T(z) = V(z) + V(z)\frac{1}{z - H_o} T(z) \qquad (6)$$

For nuclear matter, one obtains in the two hole line approximation (disregarding mass renormalization corrections) the standard relation of the total binding energy to a Brueckner G-matrix

$$E = \sum_{k<k_F} <k|H_o|k> + \frac{1}{2} \sum_{kl} <kl|G(\varepsilon_k+\varepsilon_l)|kl> \qquad (7)$$

however, G obeys the modified Bethe-Goldstone equation

$$G(z) = U(z) + U(z) \frac{Q}{z-h} G(z) \qquad (8)$$

where

$$<\alpha\beta|U(z)|\bar{\alpha}\bar{\beta}> = <\alpha\beta|W \frac{1}{z-h-H_o^b} W|\bar{\alpha}\bar{\beta}> \qquad (9)$$

is now defined with the selfconsistent single particle potential h in the denominator. (h is determined by the usual selfconsistency condition). Thus U takes into account the fact that the nucleons "feel" the mean potential h also during meson exchange. The z dependence of V and U expresses the mathematically correct elimination of the mesonic degrees of freedom. Numerical calculations, obtained by adjusting H^F, i.e. W, to N-N scattering described by eq. 6, yield for nuclear matter 18 MeV per particle on the Coester line[24]. However, this result surely has to be modified by taking into account the Δ degree of freedom and mass renormalization corrections.

6. Summary

Summarizing one may say that in the last years there was a clear success in the theoretical understanding of the N-N interaction due to the investigation of the relations to elementary particle physics. The long range part of the N-N interaction can be explained in terms of one-pion, one-omega and two-pion exchange. In order to relate the N-N interaction to the structure of nuclei, it has become evident that it is necessary to include N* and mesonic degrees of freedom and to abandon the too naive picture of a standard N-N potential.

This understanding of the N-N interaction is by no means complete, thus there are used quite different methods according to whether one wants to relate the N-N interaction to pion-pion and pion-nucleon interaction or to nuclear many-body phenomena, which have to be unified. Moreover, this understanding is limited by experimental uncertainties which may mask the fact that our present interpretation of the N-N interaction is indeed much more deficient.

Thus there is a rather poor knowledge of pion-pion scattering. In connection with this conference, however, I should stress that a further experimental effort to improve the measurement of the T=0 phase shifts (including, of course, polarization experiments) would be very valuable. In fact, all OBE potentials are, due to the small number of parameters, already fixed by the T=1 phase shifts, the T=0 phases being rather well predicted within their large errors. Whether this behaviour persists when the T=0 phase shifts are known with better accuracy is an open question. The answer to this question will in any case contribute a new step to a more complete understanding of the N-N interaction.

References

1) M.H. MacGregor, R.A. Arndt and R.M. Wright, Phys. Rev. 182 (1969)1714
2) Third Polarization Symposium, page 57
3) P. Signell and J. Holdemann, Phys. Rev. Letters 27(1971)1393
4) See review article of M.K. Srinavasta and D.W.L. Sprung, "Advances in Nuclear Physics", Vol.8
5) See review of K. Erkelenz, Phys. Reports 13c(1974)193
6) C. Bloch and J. Horowitz, Nucl. Phys. 8(1958)91
7) D. Schütte, Nucl. Phys. A221(1974)450
8) G. Schierholz, Nucl. Phys. B7(1968)432
   K. Holinde, K. Erkelenz and R. Alzetta, Nucl. Phys. A194(1972)161
9) A. Gersten, R. Thompson and A.E.S. Green, Phys. Rev. D3(1971)2076
10) R.A. Bryan and B.L. Scott, Phys. Rev. 135(1964)434
11) K. Erkelenz, K. Holinde and K. Bleuler, Nucl. Phys. A139(1969)308, T. Ueda and A. Green, Phys. Rev. 174(1968)1304
12) K. Holinde and R. Machleidt, to be publ. in Nucl. Phys.
13) See review of G.E. Brown and A.D. Jackson, "The Nucleon-Nucleon Interaction", published by NORDITA
14) W.N. Cottingham, M. Lacombe, B. Loiseau, J.R. Richard and R. Vinh Mau, Phys. Rev. D8(1973)800
15) M. Chemtob, J.W. Durso and D.O. Riska, Nucl. Phys. B38(1972)141
16) M. Lacombe, B. Loiseau, J.M. Richard, R. Vinh Mau, P. Pires and R. de Tourreil, Orsay preprint IPNO/TH 75-09
17) A.D. Jackson, D.O. Riska and B. Verwest, Stony Brook preprint
18) See review of H.A. Bethe, Ann. Rev. Nucl. Sci. 21(1971)93
19) H. Kümmel and K.H. Lührmann, Nucl. Phys. A191(1972)525 J.G. Zabolitzky, Nucl. Phys. A228(1974)285
20) F. Coester, S. Cohen, B. Day and C.M. Vincent, Phys. Rev. C1(1970)769
21) K. Holinde and R. Machleid, to be published in Nucl. Phys.
22) A.M. Green and J.A. Niskanen, Helsinki preprint No 8-75
23) V.G. Kadyshevsky, Nucl. Phys. B6(1968)125
24) K. Kotthoff, Dissertation Bonn, 1975

DISCUSSION

Loiseau:
   The energy dependence of the potential of Vinh-Mau et al. is linear and can be transformed into a velocity dependence. Then this velocity dependent potential is used for nuclear calculations.

Schütte:
   If you interpret your energy dependence as being of the same origin as in the Kadyshevski-equation there is no intrinsic reason to treat this dependence in the way you propose.

Loiseau:

We have found in collaboration with W. Nutt, that the inclusion of a pion—nucleon vertex function (pion—nucleon form factor calculated from the $\pi \, \pi \rightarrow N \, \bar{N}$ interaction) in the one—pion—exchange—potential (OPEP) can reduce the tensor OPEP by about 20 % for NN distances around 1 to 1.3 fm. If this correction is included in the potential of Vinh-Mau et al. the $^3D_1$ phase is in better agreement with pheno-menology.

Schütte:

This is a very interesting result.

HIGH ENERGY EXPERIMENTS WITH A POLARIZED PROTON
BEAM AND A POLARIZED PROTON TARGET*

A.D. Krisch

Randall Laboratory of Physics, The University of Michigan

Ann Arbor, Michigan 48104 USA

During recent years there has been an increasing interest in the importance of spin in high energy strong interactions. This came from the very successful experiments using polarized proton targets at Berkeley[1], CERN[2], and Argonne[3]. This work suggested that the further studies of spin, which require a polarized beam, might yield very exciting results. Around 1970 our group began working with the Argonne ZGS staff to accelerate a beam of polarized protons and scatter them from a polarized proton target. In 1973 a polarized beam was accelerated to multi-GeV energies for the first time. The ZGS now has a beam of 5×10^9 protons at 6 GeV/c with a polarization of more than 70%. The acceleration process is described in Khoe et al.[4]. Our group's experiments using the polarized beam and a polarized target to measure the spin dependence of the total and elastic proton-proton cross sections are described in Parker et al.[5]. My collaborators deserve great credit for the success of this project. Recently other groups have been using the ZGS polarized beam to study the spin dependence of various high energy processes. Several of these experiments have surprising and probably significant results. It now appears that spin is quite important at high energy.

In this talk I will first briefly describe the process of jumping the depolarizing resonances encountered during beam acceleration; I feel this has some physics interest in itself. Next I will describe my own group's experiments in some detail. Finally, I will discuss the results obtained by some of the other groups using the ZGS polarized beam.

The polarized protons originate in a polarized ion source purchased from ANAC in New Zealand which now gives 50 μa of 20 KeV protons with a polarization of 75% ± 5%. This source was placed in the new pre-accelerator II dome shown in fig. 1 where a Cockcroft-Walton column accelerates the protons to 750 KeV. The protons are then fed into the main LINAC line by a switching magnet and accelerated to 50 MeV. The polarization at 50 MeV is measured using a polarimeter which continuously measures the left-right asymmetry in p-Carbon elastic scattering at 55° where by cross calibration we found the analyzing power to be 88% ± 5%[6]. At 50 MeV the beam polarization is 75% ± 5%. The beam is then injected into the ZGS, accelerated, and then extracted to the high energy polarimeter described below.

The main problems in accelerating polarized protons in a synchrotron are "depolarizing resonances"[7]. These occur when the Larmor precessional frequency becomes equal to the frequency with which the protons see magnetic imperfections. Then each proton gets a similar perturbation every time it passes through a horizontal fringe field. These perturbations can then add coherently and rapidly depolarize the beam. The precessional frequency and the frequency of perturbations due to vertical betatron oscillations are

$$\omega_p = \left(\frac{g}{2} - 1 \right) \gamma \frac{eB_0}{m} , \qquad \omega_k = (k \pm \nu) \frac{eB_0}{m}$$

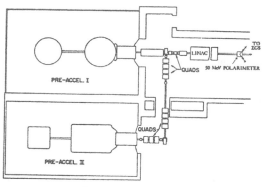

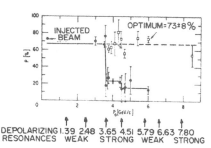

DEPOLARIZING 1.39 2.48 3.65 4.51 5.79 6.63 7.80
RESONANCES   WEAK   STRONG WEAK   STRONG

Fig. 2. Measured Polarizations
as a Function of Momentum. Open
Squares are with Quadrupoles
Compensating for Resonances and
Black Dots are with Quadrupoles
Off.

Fig. 1

The resonance condition which occurs when they are equal is $(g/2 - 1)\gamma = k t \nu$; where ν is the number of betatron oscillations per turn around the ZGS ($\sim.80$), and k is the harmonic number (8 and 16 are the strongest). The first 3 strong resonances were jumped using two pulsed quadrupoles installed in the ZGS. By pulsing these 20 μsec, ν was rapidly changed and the beam passed through each resonance in a few turns before coherent depolarization could occur. The timing and strength of these pulses was tuned by maximizing the polarization measured in the high energy polari- meter shown in fig. 3. This measures p-p elastic scattering at $P_\perp^2 = .5(GeV/c)^2$ where the asymmetry parameter is $A = .100 \pm .006$ at 6 GeV/c. As shown in fig. 2, the pulsed quadrupoles indeed overcome the resonances.

The experimental apparatus used by our group is also shown in fig. 3. Notice the 85→90% polarized proton target (PPT) and the F_{123}-B_{123} magnetic spectrometer. We studied elastic pp scattering at 6 GeV/c

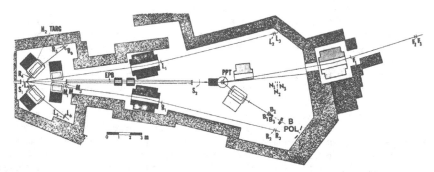

Fig. 3. Layout of the experiment. The polarized beam passes through the H_2 target and its polarization is measured by the number of elastic events seen in the L and R spectrometers of the polarimeter. The beam then scatters in the polarized target and the elastic events are counted by the F and B counters. The M and N counters are monitors.

in the range $P_\perp^2 = .5 \to 2.0$ (GeV/c)^2 in all 4 possible initial spin states
↑↑,↑↓,↓↑, and ↓↓ (the spins are perpendicular to the scattering plane).
Note that the states ↑↓ and ↓↑ are identical due to rotational invariance
of space. In fig. 4 we plotted A and C_{nn} while in fig. 5 we plotted the
ratio of the cross section in each spin state to the spin average cross
section. These were obtained from our data using the equations:

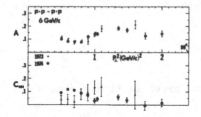

$$\frac{d\sigma}{d\Omega}(\uparrow\uparrow) = \left\langle\frac{d\sigma}{d\Omega}\right\rangle (1 + 2A + C_{nn})$$

$$\frac{d\sigma}{d\Omega}(\downarrow\downarrow) = \left\langle\frac{d\sigma}{d\Omega}\right\rangle (1 - 2A + C_{nn})$$

$$\frac{d\sigma}{d\Omega}(\uparrow\downarrow) = \frac{d\sigma}{d\Omega} \quad (\downarrow\uparrow) = \left\langle\frac{d\sigma}{d\Omega}\right\rangle (1 - C_{nn})$$

where C_{nn} is given by

$$C_{nn} = \frac{N_{\uparrow\uparrow} + N_{\downarrow\downarrow} - N_{\uparrow\downarrow} - N_{\downarrow\uparrow}}{P_B P_T \Sigma N_{ij}}$$

Fig. 4. The asymmetry parameter A
and C_{nn} are plotted against $P_\perp^2$ for
pp elastic scattering at 6 GeV/c.

and $\langle d\sigma/d\Omega \rangle$ is the spin-average cross section. The N_{ij} are the normalized event rates in each spin state. We obtained A by averaging over
either the beam or target polarization. This gave the consistency check

$$A = \frac{N_{\uparrow\uparrow} + N_{\uparrow\downarrow} - N_{\downarrow\uparrow} - N_{\downarrow\downarrow}}{P_B \Sigma N_{ij}} = \frac{N_{\uparrow\uparrow} + N_{\downarrow\uparrow} - N_{\uparrow\downarrow} - N_{\downarrow\downarrow}}{P_T \Sigma N_{ij}}$$

which held within ±1%. Notice that C_{nn} has a maximum at around
$P_\perp^2 = 0.7$(GeV/c)^2 which is where A has a minimum.

In fig. 6 we have a plot of the differential elastic cross sections
themselves against $P_\perp^2$. Notice the relation between the change in the spin
dependence and the break in the cross section. In the "diffraction peak"
region of $P_\perp^2 < 1$(GeV/c)^2 the three cross sections are parallel and $\sigma(\uparrow\uparrow)$
is about 40% larger than $\sigma(\downarrow\downarrow)$ and $\sigma(\uparrow\downarrow)$ which are about equal. In the
region after the break the cross sections are again parallel but $\sigma(\uparrow\uparrow)$ is
now about twice as big as $\sigma(\downarrow\downarrow)$; and $\sigma(\uparrow\downarrow)$ is about half-way between them.

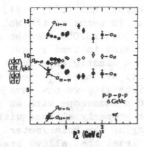

Fig.5. The ratio of the differential elastic pp cross section for each
spin state to the spin average cross section <dσ/dt> is plotted against
$P_\perp^2$. The spins are measured perpendicular to the scattering plane. The
σ_{ij} refer to the initial spins only while the $\sigma_{ij \to kl}$ indicate initial
and final spins.

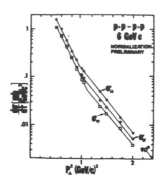

Fig. 6. The differential elastic proton proton cross section for each
initial spin state is plotted against $P_\perp^2$.

The sharp breaks and the two cases of rather different but parallel behavior are quite striking.

Next we turn to the measurement of the 3-spin cross sections using
the B-polarimeter shown in fig. 3. This measures the left-right asymmetry
in p-C scattering of the recoil protons from a 13cm carbon target. The
analyzing power of this polarimeter was calibrated using the ZGS polarized
beam accelerated to about 250 MeV. For each of the 4 initial spin states
we measured the number of recoil protons in each final spin state; we thus
obtained the cross sections for the 8 different 3-spin states such as
$\sigma(\uparrow\uparrow\rightarrow0\uparrow)$. Our notation is σ(Beam, Target $\rightarrow$ Scattered, Recoil) and 0 denotes unmeasured. If we assume parity invariance then all 8 transversity
single flip amplitudes must be zero, (e.g., $\sigma(\uparrow\uparrow \rightarrow \downarrow\uparrow)$=0. Thus we measured directly the square of the magnitude of the amplitudes for the 8 pure
transversity states but know nothing about their phases. Rotational invariance gives relations between the antiparallel non flip and double
flip cross sections $\sigma(\uparrow\downarrow \rightarrow \uparrow\downarrow)=\sigma(\downarrow\uparrow \rightarrow \downarrow\uparrow)$ and $\sigma(\downarrow\uparrow \rightarrow \uparrow\downarrow) = \sigma(\uparrow\downarrow \rightarrow \downarrow\uparrow)$.
T invariance gives a relation between the parallel double flip cross
sections $\sigma(\uparrow\uparrow \rightarrow \downarrow\downarrow) = \sigma(\downarrow\downarrow \rightarrow \uparrow\uparrow)$. We have assumed that P and T invariance
hold in obtaining the 5 independent pure transversity cross sections shown
in fig. 5 at $P_\perp^2 = .5(GeV/c)^2$. It is interesting that the double flip cross
sections are typically 10 times smaller than the non flip cross sections
and that $\sigma(\uparrow\uparrow \rightarrow \uparrow\uparrow)$ is some 80% larger than $\sigma(\uparrow\downarrow \rightarrow \uparrow\downarrow)$ and $\sigma(\downarrow\uparrow \rightarrow \downarrow\uparrow)$.
Another way to present this data is in terms of the Wolfenstein parameters.
We find that D_{nn} = 0.81±0.10, K_{nn} = 0.14±0.08. For the antiparallel non
flip states we had sufficient precision to test P invariance to 5%.

Recently we improved our 3-spin elastic experiment to obtain more
precise tests of P and T, especially at large $P_\perp^2$ where they have not been
tested. Due to the increased ZGS intensity we now have 5000 analyzed events
at $P_\perp^2$ = 1.0 $(GeV/c)^2$. We installed a hodoscope with an on-line computer to
insure that the recoil protons are properly aligned with the recoil polarimeter, and we are carefully studying the polarimeter. We hope to eliminate asymmetries and biases to a level that allows precise statements
about P and T at large $P_\perp^2$.

We also measured the spin dependence of σ_{TOT} in p-p scattering using
the polarized beam and our polarized target. It was a standard good geometry attenuation experiment and the beam spin was reversed every pulse
to eliminate errors due to beam movement and other drifts. The data are

shown in fig. 7 where the difference between σ(↑↓) and σ(↑↑) is plotted against P_{LAB}; the spins are perpendicular to the beam. The difference is clearly non zero. There are two somewhat surprising features of this data. The antiparallel cross section σ_{tot}(↑↓) is larger than σ_{tot}(↑↑);

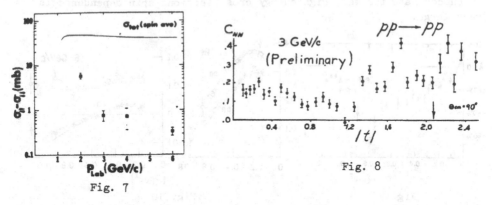

Fig. 7

Fig. 8

however, in elastic scattering C_{nn} is generally positive implying that σ_{el}(↑↑) may be larger than σ_{el}(↑↓). Next, notice the very large value of σ_{tot}(↑↓) - σ_{tot}(↑↑) of almost 6 mb at 2 GeV/c, much larger than the .76 mb at 3 GeV/c. The physical origins of this large difference are not clear.

There are several experiments with the ZGS polarized beam that I will not be able to discuss in detail[8]. The Indiana group has done some very nice experiments on A in p-p elastic scattering at large $P_\perp^2$ with a polarized target and at very small $P_\perp^2$ with the polarized beam. They have also measured D_{nn} with a polarized target. The Argonne-Chicago-Ohio State group has a very interesting new result on Λ production with the polarized beam. There was a survey exposure of 6 GeV/c polarized protons in the large bubble chamber which is being analyzed. Several groups are preparing polarized inclusive experiments. The parity test in σ_{TOT} will be discussed later in this meeting.

The Argonne-Northwestern group has been independently studying p-p elastic scattering with the polarized beam and their 85 → 90% polarized target[8]. They have concentrated somewhat on smaller t where their beam of ∿3 10⁵ gives good precision. They are very interested in the energy dependence of spin effects and have measurements at 2,3,4 and 6 GeV/c. Their C_{nn} data at 3 GeV/c is shown in fig. 8. Notice the maximum at small t, the dip, and the broad maximum near 90°. Comparing this with the 6 GeV/c C_{nn} data there is apparently considerable energy dependence. This group is planning to completely measure the 5 p-p elastic amplitudes including the phases. They have built a superconducting solenoid magnet of 120 KG-m to rotate the beam spin into the horizontal plane. They are constructing a new polarized target with the spin horizontal and recoil polarimeters to measure the recoil spin in various orientations. Measuring the phases and magnitudes of the 5 independent amplitudes is very important.

The Argonne Effective Mass Spectrometer group has several very interesting results using their general purpose spectrometer with an

unpolarized target and the polarized beam[8]. They first measured the
spin dependence of an inelastic two-body process p + p → Δ^{++} + n at
6 GeV/c. As shown in fig. 9, at large t the asymmetry becomes impress-
ively large. Thus, in at least one of the many inelastic channels
which dominate the very high energy cross sections spin dependence is
very large.

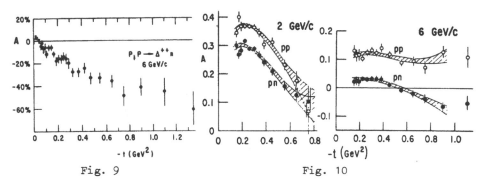

Fig. 9   Fig. 10

     This group also measured the Asymmetry in n-p and p-p elastic scat-
tering at 2 → 6 GeV/c. The data is shown in fig. 10. At 2 GeV/c the
n-p Asymmetry is similar to the p-p but somewhat smaller. At higher
energy the nature of the n-p Asymmetry changes completely and it changes
sign near -t = .5$(GeV/c)^2$. This is completely different than the π^+p
and π^-p Asymmetries which are mirror images.

     In summary there is apparently much more spin dependence than we
high energy people had anticipated. Spin may well continue to be important above 100 GeV, and people at the CERN PS and ISR, and Fermilab
are beginning to think about polarized beams and gas jets. If the different spin cross sections continue to be unequal by factors of 2 ore
more, then in a few years many high energy people may be studying spin.
In that case, you nuclear people who have been in this business for some
decades, may find you have many followers.

References

* Supported by a research grant from the USERDA
1) P. Grannis et al., Phys. Rev. 148, (1966) 1297
2) M. Borghini et al., Phys. Lett. 24B, (1967) 77; Phys. Lett. 31B,
   (1970) 405; 36B, (1971) 501;
   M.G. Albrow et al., Nucl. Phys. B23, (1970) 445
3) N.E. Booth et al., Phys. Rev. Lett. 21, (1968) 651;
   23, (1969) 192; 25, (1970) 898; Phys. Rev. D8, (1973) 45
4) T. Khoe, R.L. Kustom, R.L. Martin, E.F. Parker, C.W. Potts,
   L.G. Ratner, R.E. Timm, A.A. Krisch, J.B. Roberts, J.R. O'Fallon,
   Particle Accel, 6, (1975) 213
5) E.F. Parker, L.G. Ratner, B.C. Brown, W. DeBoer, R.C. Fernow,
   S.W. Gray, A.D. Krisch, H.E. Miettinen, T.A. Mulera, J.B. Roberts,
   K.M. Terwilliger, J.R. O'Fallon, Phys. Rev. Lett. 31, (1973) 783;
   32, (1974) 77, 34, (1975) 558; Phys. Lett. 52B, (1974) 243
6) R.M. Craig et al., 2nd International Symposium on Polarization
   Phenomena, Karlsruh (1965) pp. 322-323

7) M. Froisart and R. Stora, Nucl. Instr. and Meth. 7, (1960), 297
   D. Cohen, Rev. Sci. Inst. <u>33</u>, (1962) 161
8) Argonne Summer Study on High Energy Physics with Polarized Proton
   Beams (1974), J.B. Roberts editor

DISCUSSION

Raynal:
  Is it more difficult to look in the vertical plane than to add a
  polarimeter ? In the vertical plane you get a coefficient as interesting as C_{nn}

Krisch:
  It is very difficult to rotate the scattering plane into the vertical
  plane since the spectrometer is 30 m long and contains heavy magnets.
  It is probably easier to rotate the PPT and to use a solenoid to
  rotate the beam spin, and to leave the scattering plane horizontal.
  The ANL–Northwestern group plans to try this.

REPORT ON DISCUSSION SESSION: TWO NUCLEON SYSTEM

D. Schütte

Institut für Theoretische Kernphysik der Universität Bonn, Germany

Concerning the informations on the N-N interaction, the following most relevant points, brought up in the discussion sessions, should be emphasized.

1) The on-shell knowledge of the N-N interaction – connected with the measurement of low energy p-n scattering – has improved with respect to the fact that experimentalists now agree that the ε_1 – phase is indeed always positive (see contributed paper p.443) (as wanted by the theorists). A number of remarkable efforts to reduce the general errors in the low energy p-n scattering were reported (cf.p.437 and 439), however, without definitive success up to the present time. The value of such measurements from the theoretical point of view was stressed.

2) The hope to determine directly the off-shell N-N interaction from suitable experiments has to be damped in view of the development of the understanding of the relevant processes: N-N-bremsstrahlung – in the experimental accessible energy range – gives practically only on-shell informations so that the present discrepancy between theory and experiment (which shows up in the latest bremsstrahlung experiment) is in fact an inconsistency between experiments. Also it is questionable whether relevant off-shell informations can be extracted from the evaluation of three body reactions within the Fadde'ev theory – the sensitivity to variations in the off-shell extrapolation may be small.

3) Concerning the question of a generalization of the theoretical scheme for the description of the N-N problem by taking into account delta-resonances, it seems to be very difficult to find a unique experimental evidence for a delta-delta component in the deuteron. In view of the principal importance of such a generalization of standard nuclear physics, further experimental and theoretical effort is needed to obtain a clearer insight into the situation. Valuable experimental informations would emerge from inelastic N-N scattering experiments in the energy region where the delta-resonance dominates (cf.p.458) and from an unambiguous determination of the deuteron D-state probability (cf.p.445). Theoretically, a coupled channel approach to the N-N problem (including the delta) has to be further developed.

POLARIZATION EFFECTS IN THE THREE-NUCLEON SYSTEM

P. Doleschall[+]

Central Research Institute for Physics

H-1525 Budapest Pf. 49, Hungary

It is a difficult task to give a new summary of the polarization effects in the three-nucleon system since Prof. Conzett has done it on the Quebec Conference on Few Body Problems in Nuclear and Particle Physics. I should like to suggest everybody who is interested in this topic, to get acquainted with his detailed and competent report. As it is not allowed to repeat his talk word for word, I try to approach the problem from another direction.

First of all, I believe, a short historical introduction up to the present theoretical development in the many-body systems would not be useless. The starting point was the work of the Soviet mathematician L.D. Faddeev[1]. He derived a new equation for the non-relativistic quantum mechanical three-body problem and proved in a strict mathematical manner that the solution of his equation is unique and equivalent to the solution of the original Schrödinger equation. Following Faddeev's work some variants of his equation were composed. The best known are the Lovelace[2], the Alt-Grassberger-Sandhas (AGS)[3], and Ebenhöh[4] equations. In these equations the three-body transition operators are the unknown quantities and this is very useful in the practical calculations, at least if the two-body transition operators are separable ones. At present we have some new Faddeev-type equations by Osborn and Kowalski[5] and by Karlsson and Zeiger[6]. This last one seems to be very promising in the practical calculations if the two-body transition operators are not separable. Of course there are many more Faddeev- and non-Faddeev-type three-body equations, but it is impossible to mention all of them.

Faddeev's original work has shown the types of difficulties which arise going from two to three. Following his line Yakubovskii[7] derived an N-body equation. However, his equation is very complicated, it is an equation for non-physical quantities and the transition operators must be composed from them after the solution. If I know well, Sloan[8] was the first who derived a four-body equation where the two-cluster transition operators of the four-body system are the unknown quantities. Finally (at least at present) Bencze[9] derived an N-body equation which connects similarly to Sloan's equation only the two-cluster transition operators of the N-body system. In spite of the fact, that the uniqueness and equivalence proof of Bencze's equation seems to be not so strict as it is in Yakubovskii's work, the size of Bencze's equation (which turns out to be the Faddeev equation in the case of N=3 and Sloan's equation in the case of N=4) is considerably smaller than that of Yakubovskii. I have to remark again that there are a lot of other types of N-body equations. But at present the above mentioned Sloan's and Bencze's equations seems to me the most practical ones, in spite of the fact that there are already some four-body calculations performed on the basis of other equations.

From the above short and not too accurate survey you can see, that in principle we have already even numerically treatable N-body equations. However, at present the three-body equation has a special property: it can be solved by the present efficient computers with the necessary accuracy. We are able to control all kind of mathematical or physical approximations applied in a three-body calculation. This enables us to use only that type of approximations which do not distort the final results. This may be the most important point that I should like to emphasize.

When we are doing a theoretical calculation for a more complicated
physical system, at first we have to accept some basic physical principles.
In the next step we usually are forced to introduce some approximations to
make our equations mathematically tractable. On the basis of the non-relativistic quantum mechanics the two-body system was until the Faddeev's
work the only one which could be treated mathematically exactly in the
general case. Today, due to the Faddeev equation and the large and efficient computers, the three-body problem is solvable even for an arbitrary
nucleon-nucleon (N-N) interaction. This is a great thing in nuclear
physics because we have for the first time a mathematically correct and
well controlled method to check our basic physical principles.

To express this statement in a more concrete way: due to the Faddeev
equation the three-nucleon system may become a theoretical test of the
N-N interaction, or which seems to be even more important: it may become
a test of our basic physical principles. At present we presume that the
low energy (up to 50 MeV in C.M. system) three-nucleon system is basically non-relativistic and the possible effects of the nucleon structure are
included in the N-N interaction. The possible presence of a significant
three-nucleon force does not contradict the above mentioned basic principle, however, in the first approach it seems to be better to exclude the
existence of the three-nucleon force.

For such a non-relativistic three-particle system the Faddeev equation is the adequate one for the theoretical description. We have to find
one more N-N interaction which reproduces the low energy (up to 200-300
MeV in laboratory system) nucleon-nucleon measurements, and then to check
what results give these N-N interactions in the three-nucleon system. You
must notice that there are no free parameters to adjust by the three-
nucleon calculation. The only possibility is the use of the different on-
shell equivalent N-N interactions.

In the last ten years the theoretical investigation of the three-
nucleon system has followed this line. However, when the first calculations with the so called realistic potentials of Hamada-Johnston and Reid
give a significantly underbound triton and an incorrect charge form factor, some of the theorists have drawn the sudden conclusion that either
the existence of a significant three-nucleon force or the presence of
significant relativistic effects has been proved. Since our topic is the
polarization effects, I should not go into the details of the bound state
calculations.

The first elastic scattering and break-up calculations were performed
by Aaron, Amado and Yam and by Phillips[10]. They used the separable rank-1
S-wave Yamaguchi interactions which give identically zero polarizations.
Later Kloet and Tjon[11] using the S-wave part of some local potentials obtained similar results. The trouble was that their differential cross
sections agreed with the experiments fairly well. If the simplest S-wave
interactions reproduce more or less correctly the experimental data, what
kind of sensitivity of the three-nucleon scattering states to the N-N
interaction may be expected? However, we must not forget, that the differential cross sections are quantities averaged over the initial and final
spin states. Therefore their sensitivity to the N-N interaction must be
reduced by the averaging. The different kind of polarizations in principle
must be more and more sensitive if the averaging is less and less.

The first attempts to calculate the polarizations of the elastic n-d
scattering on the basis of the Faddeev equation were made by Aarons and
Sloan[12], by Pieper[13], and by the present author[14]. These calculations
have shown that the tensor polarizations are dominated by the tensor force,

the vector polarizations by the P-wave interactions. But both are modified
by the other interaction. The agreement with the available experimental
data was qualitatively good, however, some systematic deviation between
the calculated and measured nucleon polarization was found at 14.1 and
22.7 MeV nucleon bombarding energy. The fig. 1 illustrates the situation.

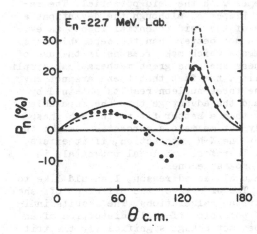

Fig. 1. The solid line is
calculated with ^1S_o, $^3S_1-^3D_1$
and P-waves. The dashed line
represents a calculation with
^1S_o, $^3S_1$ and P-waves. The
dots are experimental data
from ref. 17.

It must be mentioned that at energy 14.1 MeV Pieper's calculations
reproduce the experimental vector polarizations significantly better. However, this is a fortuitous effect, since Pieper's calculations are approximate ones. A detailed comparison by Pieper[15] shows that for the vector
polarizations his approximation gives significantly different result than
that of the exact calculation.

Some experimentalists consider these results as predictions. I should
like to emphasize that the present calculated vector polarizations are not
predictions at all. The measurements were considered to be correct. The
new measurements of the neutron polarization by Morris et al.[16] and three
contributions to this conference by Preiswerk et al. from Basel, by Steinbock et al. from South Africa, and by Brock et al. from Auckland confirm
the expected fact that the neutron and proton polarization[17] are practically the same. The measurement of the deuteron vector polarization at
45.4 MeV deuteron laboratory energy by Rad et al.[18] has shown the same
type of deviation from the theory as it was found at the nucleon polarization.

The calculated and measured[19] tensor polarizations are in good
agreement, at least much better than the vector polarizations. However,
we must be cautious, remembering the above mentioned fortuitous result at
the vector polarizations.

Summarizing all of these, we have to conclude that the disagreement
between the calculations and experiments presented on fig. 1 is a failure
of our theoretical description. Since the calculations were mathematically
exact, the only reason of the deviation could be either the incorrect N-N
interaction or incorrect basic physical principles. Let us concentrate on
the first and list the possible reasons:

   1) all calculations of Aarons-Sloan, Pieper and the present author
      were performed using separable N-N interactions,
   2) some components of the N-N interaction were not correct on the
      energy-shell,

3) even if the N-N interaction would be on-shell correct, its off-
   shell behaviour might be wrong,
4) a significant three-body force is missing.

It seems that the easiest possibility to avert the objection against the
separable N-N interaction would be to mention the calculations by Stolk
and Tjon, presented in a contribution to this conference. They performed
perturbation calculations like Pieper with the Reid potential. The results are not better than those of Pieper's, although the deviations are
not the same. However, the failure of the Reid potential is not an evidence that the local potentials are not better than the separable ones.
Nevertheless, there are some arguments which make reasonable the use of
the separable N-N interactions. These are: the great mathematical simplification coming from the separability, the fact that there are not any
significant differences between the three-nucleon results obtained by
local or separable interactions, and the advantage that the separable
interactions are more flexible, by them a better fit to the N-N phase
shifts can be achieved. And finally let me make a sarcastic remark: the
N-N interaction experts insist that the N-N interaction, if it exists,
must be at least partly non-local. Therefore the local potentials in
principle are not better than the separable ones.

Before discussing the most natural second reason, I should like to
say some words about the last two. The effects of the different off-shell
properties are well discussed in triton calculations. The results indicate that there is a wide range of possible off-shell distortion of an
original N-N interaction, which does not change significantly the triton
properties. This effect arose the opinion that the three-nucleon system
is not too sensitive to the off-shell effects. Therefore the investigation
of the possible off-shell effects in scattering does not seem too promising.

The problem of the three-nucleon force is much more difficult. The
existence of a significant three-nucleon force is always a threatening
possibility. The trouble is that the three-nucleon force appears as a
new and complicated parameter which must be defined just via the three-
nucleon calculations. At this point we find ourselves in the same situation as an N-N interaction expert who would like to extract the details
of the N-N interaction from the N-N measurements. However, our position is
even worse, because if we could reproduce the experimental data by introducing a three-nucleon force, it would not mean an unconditional success. In principle there is a possibility that we can find a correctional
three-nucleon force even if our N-N interaction were incorrect. Therefore
we have to be extremely cautious looking for a three-nucleon force without
being convinced of the correctness of our N-N interaction. My opinion is
that at present the only economical thing that can be done is: to pray
that nature does not produce a significant three-nucleon force. And besides that intensively investigate the reason for the theory-experiment
disagreement in the N-N interaction.

Of course the most natural defect could be the poor on-shell properties of the used N-N interactions. All of the earlier calculations used
the rank-1 Yamaguchi tensor force, the rank-1 singlet S-wave and P-wave
interactions. It is well known that the Yamaguchi tensor force represents
very poorly even the low energy mixing parameters and $^3D_1$ phase shifts.
Besides that, all P-wave interactions were correct more or less only up
to 100 MeV. The $^1S_0$ and $^3S_1$ phase shifts are reproduced also up to 200 MeV.

However, it became clear very soon that the poor high energy behaviour
of the $^1S_0$ and P-wave interactions did not influence significantly the
polarizations. The corrected P-wave interactions with more sophisticated

form factors[20]) practically gave the same result in the three-nucleon calculations. The same effect was found by Pieper, who compared the different,
but up to energy 50-100 MeV more or less equivalent P-wave interaction
sets. The present author performed an unpublished calculation, in which
he checked the dependence of the three-nucleon T-matrices on the $^1S_0$ interaction. A two term (rank-2) separable $^1S_0$ interaction, which reproduced
the average n-p and p-p MAW[21]) data up to 460 MeV, was used with $^3S_1$ and
P-wave interactions. The results were practically the same as with the
rank-1 $^1S_0$ interaction.

All these results indicate that the low energy three-nucleon system
does not depend on the high energy properties of the N-N interaction. In
spite of the fact, that such a type of investigation was not performed
for the tensor force, at present we suppose that this property is true
for the tensor force also. Remember now what was told about the off-shell
sensitivity of the three-nucleon system. You see, that even certain on-
shell sensitivity does not exist.

At this stage of calculations the tensor force remained the only
hope. Pieper's perturbation calculations [13,15]) have shown a very weak
dependence of the vector polarizations on the tensor force. His result
contradicts both the exact and the present Stolk's and Tjon's calculations. The fig. 1 represents an exact calculation which shows that the
presence of the tensor force modifies the vector polarization significantly. Due to the tensor force the backward maximum becomes correct, but at
middle and forward angles the tensor force pushes the results in the wrong
direction. Nevertheless, the magnitude of the effect of the tensor force
is not negligible.

Therefore the next attempt was to improve the more or less known on-
shell properties of the tensor force. It is really more or less known because the last MAW phase shift analysis[21]) did not give a unique solution
for the mixing parameters. However, all of the low energy mixing parameters are much smaller than those of the Yamaguchi tensor force. Since no
simple (rank-1 or 2) tensor force was available which fitted the low
energy MAW[21]) data, new ones were developed[20]) with complicated form factors. I should like to emphasize that these tensor forces must not be considered as new realistic tensor forces. They were made for a special purpose, they satisfy certain requirements and nothing more. Unfortunately,
any simultaneous fit to even the low energy mixing parameters and $^3D_1$
phase shifts was not achieved. Therefore two rank-2 tensor forces (the
T4M force reproduces the mixing parameters, the T4D force the $^3D_1$ phase
shifts up to 100-150 MeV) were used in three-nucleon calculations. The
fig. 2 represents the obtained neutron vector polarizations. It can be
seen immediately, that there are drastic changes at the middle and forward angles. Especially the T4M force changed the effect of the Yamaguchi
(YY4) tensor force. However, these changes are not necessarily produced by
the real tensor force. The separable tensor forces comprise the central
and spin-orbit part of the $^3S_1$ and $^3D_1$ interactions. Since the T4M interaction gives in absolute value twice larger low energy $^3D_1$ phase shifts
than those of MAW phase shifts, the drastic change might be produced by
the stronger $^3D_1$ part of T4M. If the effect of the D-wave interactions on
the vector polarizations would be additive and proportional to the strength
of the interactions, the calculations with the $^3D_2$ interaction[20]) seem to
prove that the drastic change is really produced by the too strong $^3D_1$
part of T4M interaction. However, in that case it is not completely clear
why the minimum at angles $100°-110°$ remained unchanged going from YY4 to
T4D. If all effects would be additive, then the decrease of the mixing

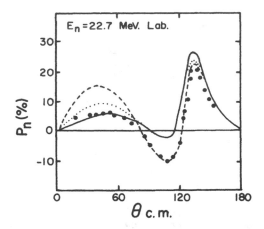

Fig.2. ——— YY4, ····· T4D,
---- T4M. The dots are
experimental data from
ref. 17.

parameters would lead to a deeper minimum as it was found in the first
calculations (see fig.1), where the appearence of the larger mixing parameters pushed up the quite good minimum produced by the pure P-wave
interactions. Therefore a simultaneous decrease of the mixing parameters
and $^3D_1$ phase shifts, which happened going from YY4 to T4D, would have
produced a deeper minimum. Instead of that the minimum remained unchanged.
Therefore we can conclude that at least at the middle angles the effect
of the different parts of the N-N interactions does not seem to be additive. Nevertheless we cannot exclude that the too strong $^3D_1$ part of T4M
produces the drastic change in the vector polarizations.

In this way we have to recognize that the present available results
do not give enough information on the reason, why we are not able to
achieve agreement with the experimental data.

Our final aim is of course to find an on-shell correct tensor force
and then to check the effect of this tensor force on the polarizations.
Some attempts were made in this direction by the Lyon group (C. Fayard,
G.H. Lamot, and E. Elbaz) [22]. They introduced different separable interactions and calculated their effect on the three-nucleon system. They
have found that a combination of the Mongan[23] and Graz potentials reproduces the vector polarizations. However, as it was pointed out by
Pieper[24], Mongan overlooked a phase and therefore his tensor forces fit
the mixing parameters very poorly. Besides this, Lyon group performed the
calculations using the Sloan's-Pieper's perturbative method which produces some deviation from the exact calculations.

In order to have more precise information a new rank-3 tensor force
was developed which fits both the mixing parameters and $^3D_1$ phase shifts
up to 100 MeV. I have not mentioned it so far, but a natural requirement
is that all tensor forces must reproduce the deuteron properties, and
they do. I attempted to perform a three-nucleon calculation at 22.7 MeV
nucleon laboratory energy just before the conference. Unfortunately the
lack of the computer time hindered the completion of the calculations.

Nevertheless, let us think about the possible results. The first one
is that the new calculation does not reproduce the experiments. In that
case we may change the P_D value, which was chosen in the present calculation to be 4% which is an arbitrary value. In the next step we may change
the tensor force and $^1P_1$ interaction by fitting the unconstrained MAW[21]

phase shifts. However, both changes must be made in the case of success, to check whether the agreement is a fortuitous result or it provides some real information about the N-N interaction.

Now a question could be asked: why have I spoken all the time about the vector polarization? This question would be justified, but I hope I have a reasonable answer. At present the vector polarizations are the quantities at 22.7 MeV nucleon laboratory energy which are measured with high accuracy. The tensor polarizations are measured[18] at 25.1 MeV deuteron laboratory energy and they must be of course reproduced too. By the way, here we can find a promising disagreement between the calculations and the experimental values of $Q=1/(2\sqrt{2})(T_{20}+\sqrt{6}T_{22})$. The deviation appears in the same region of angles (70°-110°) where the first vector polarization disagreement was found at 14.1 MeV. Since the vector polarization disagreement became more characteristic at higher energies, there is a hope that the same effect will occur at the tensor polarizations. Besides the good accuracy of the vector (and sometimes the tensor) polarization measurements, there is another cause which makes their reproduction very important. Since subsequent to the differential cross sections the most averaged quantities are these single polarizations, in principle they are the less sensitive polarizations. Therefore in the first approach they must be reproduced. This is an essential requirement for the N-N interaction. Without reproducing the single polarizations the N-N interaction cannot be considered a correct one. Therefore a possible agreement with experimental polarization transfer or spin correlation coefficients does not give a substantial information to the theory because the agreement most probably means an accidental coincidence.

Of course nobody says that it is not worth to compare the experimental and theoretical polarization transfer and spin correlation coefficients. We need all kind of accurate polarization measurements. A complete set of experimental data is especially important for the three-nucleon phase shift analysis. In a contribution to this conference Schmelzbach et al. from Zürich report about a phase shift analysis of p-d elastic scattering. They determine which kind of polarizations are sensitive to the change of phase shifts and mixing parameters. It would be probably very useful to join the informations which are obtained from Faddeev calculations and from the three-nucleon phase shift analysis. Those elastic T-matrix elements which are not or not too sensitive to the details of the N-N interaction could be used in a phase shift analysis, and the predictions of the phase shift analysis could show us which parts of the used N-N interactions are wrong.

The solutions of the Faddeev equation and the three-nucleon phase shift analyses could show: 1) which region of angles is sensitive to the N-N interaction or to the change of the three-nucleon phase shifts, 2)what accuracy is needed for the exclusion one or more model N-N interaction. Since the measurements of the more complicated polarizations with the necessary accuracy are extremely difficult, it would be useful to know which points are uncertain in the theory and then to perform the measurements in these points. However, this is a typical advice of a theorist who is convinced of the correctness of his calculations in the sense that they catch the physical essence. It would be the largest and most productive shock if an experiment produce a different result from the theory in a region of angles where no sensitivity to the N-N interaction was established.

At present we have some calculations for spin correlation coefficients by Lyon group[22]. They have found, that the theoretical spin correlation coefficients substantially depend on the N-N interaction at forward

angles. I am sorry to admit that the experimentalists made much more on
that field in spite of the fact that the theoretical part is much easier.
If once the Faddeev equation is solved, to obtain the all kind of polarizations is not a difficult theoretical task, but you need some unpleasant
computational work. The only excuse for us might be, that we concentrated
on the well established failure of the theory at the vector polarizations.
In the next period we have to take care about other types of polarizations
too.

My last duty, I believe, to talk about the break-up process. At present we learn a strong increase of polarization measurements in this
field. Besides the analysing powers even the polarization transfer coefficients are measured by Walter et al. and by Graves et al. They will report
on their results on this conference. Unfortunately the break-up calculations are under-developed. Even the differential cross section calculations
are practically on the same level as they were 3-5 years ago. The preparation of the break-up codes, which can include the higher partial wave components of the N-N interaction, is in progress. One of them, what I know
about, is made in Amsterdam, another by the present author. We need half
or one year to have some results for publication.

Finally I should like to emphasize again the main point: the three-
nucleon calculations might be really exact ones. It is a fashionable
custom to put the word 'exact' in inverted commas, probably showing that
it does not mean a really exact thing. However, in that case practically
all of our words might be supplied by inverted commas. The word 'exact'
means in the three-nucleon calculations, that accepting the non-relativistic quantum mechanics as a correct theory for the low energy three-
nucleon system, the Faddeev equation is solvable mathematically exactly.
The only uncertain one is the input N-N interaction (and the possible
three-nucleon interaction). Since we do   not know the real N-N interaction, in that sense all of the Faddeev calculations are approximate
ones. However, if we solve our Faddeev equation otherwise exactly, all
kind of deviations between the calculations and experiments are produced
by the used N-N interaction.

References

+ Present address: Department of Mathematical Physics, Lund Institute
   of Technology, P.O. Box 725, S-220 07 Lund 7, Sweden

1) L.D. Faddeev, ZhETF(USSR) 39 (1960) 1459, JETP(Sov.Phys.) 12 (1961)
   1014
2) C. Lovelace, Phys. Rev. 135 (1964) B1225
3) E.O. Alt, P. Grassberger and W. Sandhas, Nucl. Phys. B2 (1967) 167
4) W. Ebenhöh, Nucl. Phys. A191 (1972) 97
5) T.A. Osborn and K.L. Kowalski, Ann. Phys.(N.Y.) 68 (1971) 361
6) B.R. Karlsson and E.M. Zeiger, Phys. Rev. D11 (1975) 939
7) O.A. Yakubovskii, Yad. Fiz. 5 (1966) 1312 (trans.: Sov. J. Nucl. Phys.
   5 (1967) 937)
8) I.H. Sloan, Phys. Rev. C6 (1972) 1945
9) Gy. Bencze, Nucl. Phys. A210 (1973) 568
10)R. Aaron, R.D. Amado and Y.Y. Yam, Phys. Rev. 140 (1965) B1291,
   R. Aaron and R.D. Amado, Phys. Rev. 150 (1966) 857,
   A.C. Phillips, Phys. Rev. 142 (1966) 984, Phys. Lett. 20 (1966) 50
11)W.M. Kloet and J.A. Tjon, Phys. Lett. 37B (1971) 460, Ann. Phys.(N.Y.)
   79 (1973) 407
12)J.C.Aarons and I.H. Sloan, Nucl. Phys. A182 (1972) 369, A198 (1972) 321

13) S.C. Pieper, Phys. Rev. Lett. 27 (1971) 1738, Nucl. Phys. A193 (1972) 529, Phys. Rev. C6 (1972) 1157

14) P. Doleschall, Phys. Lett. 38B (1972) 298, 40B (1972) 443, Nucl. Phys. A201 (1973) 264

15) S.C. Pieper, Phys. Rev. C8 (1973) 1702

16) C.L. Morris, R. Rotter, W. Dean and S.T. Thornton, Phys. Rev. C9 (1974) 1687

17) F.N. Rad, J. Birchall, H.E. Conzett, S. Chitalapudi, R.M. Larimer and R. Roy, Phys. Rev. Lett. 33 (1974) 1227

18) A. Fiore, J. Arvieux, Nguyen Van San, G. Perrin, F. Merchez, J.C. Gondrad, C. Perrin, J.L. Durand and R. Darves-Blanc, Phys. Rev. C8 (1973) 2019

19) J.C. Faivre, D. Garreta, J. Jungerman, A. Papineau, J. Sura and A. Tarrats, Nucl. Phys. A127 (1969) 169

20) P. Doleschall, Nucl. Phys. A220 (1974) 491

21) M.H. MacGregor, R.A. Arndt and R.M. Wright, Phys. Rev. 182 (1969) 1714

22) C. Fayard, G.H. Lamot and E. Elbaz, Lett. Nuovo Cim. 7 (1973) 423, Les Observables de la Diffusion Elastique Nucleon-Deuton, preprint

23) T.R. Mongan, Phys. Rev. 175 (1968) 1260, 178 (1969) 1597

24) S.C. Pieper, Phys. Rev. C9 (1974) 883

DISCUSSION

Brady:

You compare your polarization calculations to proton ($\vec{\text{p}}$-d) data. Can you estimate how large Coulomb effects would be and what angles they would be most important?

Doleschall:

The correct theoretical treatment of the Coulomb interaction at present is impossible, since we have no solvable exact equations for p-d scattering. I tried simply to add the corresponding Coulomb T-matrix to the n-d T-matrix. Their interference in the differential cross-section exists, but the minimum and maximum at forward angles were not at the right place. This shows that the Coulomb distorted nuclear part of the p-d T-matrix is not the same as the n-d T-matrix. I do not know how this difference influences the polarizations. Fortunately, the measurements of the neutron polarization at 14-22 MeV indicate, that there are not significant differences at least at these energies.

Cramer:

Wouldn't you expect the differences between neutron and proton scattering to be most severe at forward angles, in the very region where the deviations between the calculations and the data seem to be most severe?

Doleschall:

The deviation at forward angles is produced most probably by the too large $^3D_1$ component of the T4M force. Therefore the cause of this theory-experiment disagreement is probably independent on the possible difference between the n-d and p-d polarizations. Nevertheless the usual expectation is that the p-d polarizations can be different from the n-d ones mainly at forward angles.

Ohlsen:

> Are you in a position to calculate the polarization effects in three-particle breakup channels? If so, can you give suggestions as to the kinematic conditions for which such measurements might be most useful. I am particularly interested in cases for which the projectile is polarized, either p or d, in the H($\vec{d}$,pp)n and D($\vec{p}$,pp)n reactions.

Doleschall:

> At present I prepare a breakup code and I hope it will be completed soon. Since this code includes the tensor force and higher partial wave components of the N-N interaction, it can be used for polarization calculations. However, before making any suggestion, we have to perform some calculations to see what polarization effects exist in the breakup process and what is their sensitivity to the N-N interaction.

Hackenbroich:

> Is it not possible that the interaction neglected by you (e.g. in ^{3}D states) may change the picture for tensor polarizations - so that your calculation is indeed sensitive to forces on shell?

Doleschall:

> The vector polarizations are sensitive to the D-wave interactions at 22.7 MeV. However, the published calculations have not shown the same magnitude of sensitivity of the tensor polarizations to the $^3D_2$ interactions. By the way, the vector and tensor polarizations are sensitive to the low energy on-shell properties of the N-N interaction.

THE DEUTERON D-STATE FROM THREE BODY STUDIES[*]

W. R. Gibbs
Theoretical Division
Los Alamos Scientific Laboratory, Los Alamos, New Mexico 87545

I. INTRODUCTION

The deuteron D-state has been a subject of interest to physicists for many years. Recently it has been regarded as a key to the under-standing of certain features of the nucleon-nucleon interaction. To see why this is true let us consider some of the history of the deuteron wave function along with some recent developments.

The fact that the deuteron has a quadrupole moment $(Q_D = 0.287 \pm 0.002 \text{fm}^2)$[1] implies that it has an $L = 2$ component. If we assume the discrepancy between the deuteron magnetic moment $(\mu_D = 0.857406 \ (1) \text{n.m.})$[2] and the sum of the neutron and proton magnetic moments to be due entirely to the orbital angular momentum in the D-state we obtain an estimate of ~4%.

On the other hand if one assumes a local (or semi-local) potential between nucleons and requires a simultaneous fit to the nucleon-nucleon phaseshifts, the deuteron binding energy and Q_D the resultant value of the percent D-state (P_D) is about 7%.[3]

In order to reconcile these two results one has considered rel-ativistic corrections[4] and the meson exchange current contributions to the magnetic moment.[5] These effects are very difficult to calculate with any certainty but have been estimated to resolve the discrepancy to within a factor of two.[5] However, a recent measurement of the $\rho \rightarrow \pi + \gamma$ width[6] reduces the estimate of the exchange current contribution by a factor of ~3.5, and it is not clear that the solution by invoking such effects is still valid.

Recently, with the coming of reliable three body calculations[7] for these "realistic" potentials, it has become apparent that they underbind the triton by about 1 MeV and miss the n+d doublet scattering length $(^2a = 0.65 \pm 0.04)$[8] by about a factor of three.

In the meantime it was observed that with separable potentials, by introducing extreme non-locality, one can obtain reasonable fits to both phase shifts and Q_D with a variable P_D. Different values of the triton binding energy and ^2a result for each value of P_D. These facts may be correlated in a simple way by plotting[9] the triton binding energy vs ^2a. In Figure 1 are shown the results of a recent study by Afnan and Read[10]. As can be seen the results of separable potential calcula-tions fall on a straight line (the Phillips line). The new results of this study are that the "local" potentials also fall on this line.

While it is reasonable to assume that the nucleon-nucleon potential is very non-local at short distances, it is not possible to determine the precise features from the three body system because of the complex nature of the calculations. Three body forces and/or relativistic corrections could possibly correct the three body values while leaving P_D at about 7% .

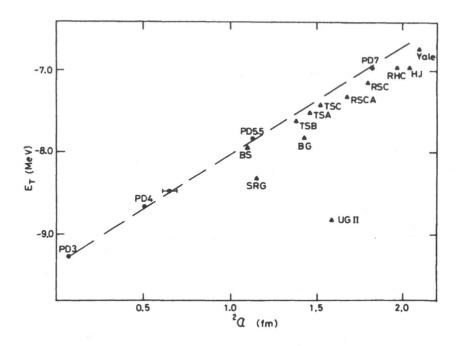

Fig. 1. Binding energy of the triton vs the n-d doublet scattering length; from Afnan and Read[10].

Thus one may see that a direct measurement of the percentage of deuteron D-state would allow us to choose between (essentially) two alternatives: a) short range non locality in the two nucleon interaction, b) three body forces.

II.   PRESENT KNOWLEDGE OF THE D-STATE

There is a theoretical lower limit for P_D of 0.44% first obtained by Levinger.[11] Values close to this limit were obtained in separable potential calculations by Mongan and others.[12] Levinger later pointed out[13] that the low value he obtained was due to the fact that a long range force was allowed. If the potential is required to have a one boson-exchange tail, Klarsfield[14,15] has shown that the limit $P_D \geqslant 3.3\%$ is more realistic. Since Mongan's potentials do not have a OBE tail, that is presumed to be the reason for the low value of P_D which comes from these calculations. While there is no firm upper limit on P_D no one has suggested a value greater than 9-10%. A precise value within these rather broad limits is difficult to obtain.

Remler and Miller[16] quote a value of ~7-9% from p-d elastic scattering, but they cannot exclude values as low as 2%. Figure 2 shows a fit that they obtain for various values of P_D. While these results indicate a large P_D the uncertainties in the calculation make it impossible to draw a definite conclusion.

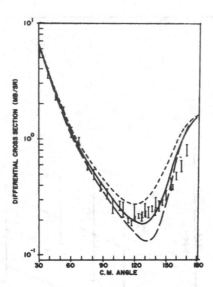

Fig. 2. P-D elastic scattering for various D-states. The solid line is
for P_D = 6.963%, the large dashed, P_D = 4%, and the small dashed
P_D = 12%; From Remler and Miller. [16]

Nasser et al[17] quote a value of P_D = 6.7% from analysis of p-d
back scattering from 100-400 MeV using the Kerman-Kisslinger[18] model.
They comment, however, that the value of P_D is uncertain since they use
a modified s-state wave function and the validity of this modified wave
function has not been checked. Rinat has recently shown[19] that large
angle p-d scattering can be fit without the assumption of the existence
of Δ's in the deuteron. It is not clear what effect Rinat's work has on
the value of P_D extracted by Nasser et al.

In elastic p-d and π-d scattering at high energies there is an
interference minimum which is filled in due to the existence of the D-
state. Attempts have been made to use this filling to estimate P_D.

For example, Michael and Wilkin[20] estimate $P_D \approx$ 9%. The difficulty with this method is that other effects also fill in this minimum
(spin flip, off-shell effects, etc.). A particular example of this can
be found in the p-^{4}He scattering at 1 GeV, where a corresponding minimum is "filled in" and we do not understand why. Modern scattering
theories are now giving us some understanding of this[21]. From this
view-point the estimate of Michael and Wilkin is probably an upper,
since once a minimum is filled by an incoherent effect it cannot be
deepened again by another effect.

A somewhat surprising sensitivity to P_D has appeared in a recent
LAMPF experiment by Preedom et al.[22] The measurement is of the angular distribution of the final protons in the reaction $\pi^+ + d \rightarrow$ p+p at
T_π = 40,50 and 60 MeV.

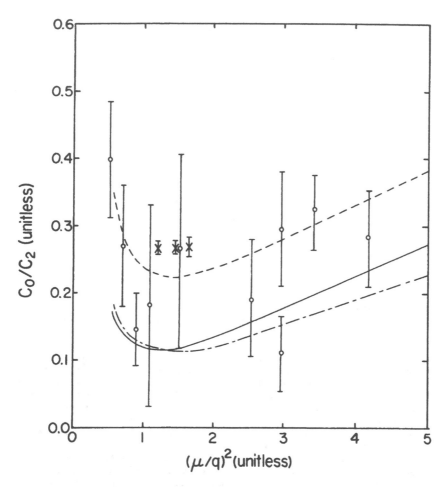

Fig. 3. Angular distribution coefficient in $\pi^+ + d \rightarrow p+p$. The solid curve is for P_D = 7% (H-J), the long dashed curve, P_D = 7.55% (BCM) and the short dashed, P_D = 4.56% (BCM). The data marked with x are from Preedom et al.[22]

This angular distribution must have the form:

$$\sigma(\theta) = c_0 + c_2\cos^2\theta + c_4\cos^4\theta + \ldots \qquad (1)$$

Figure 3 shows c_0/c_2 plotted vs the inverse pion momentum square. The curves are from Goplen et al.[23] One may judge from the error bars the improvement in the quality of the data. I wish to thank this group for permission to show their data before publication.

If one takes this plot at face value it puts the 7% D-state several standard deviations from the data points and suggests $P_D \approx 4\%$. There are problems which indicate we should not be so hasty to conclude this. The agreement between theory and experiment for the absolute cross sections is only moderate for 50 and 60 MeV and is poor for the 40 MeV point. A more general comment is that we do not yet completely understand the theory of such reactions.

In summary, our knowledge of P_D from direct measurement is rather scant and we must do considerably better to carry out the program suggested in the introduction.

III. TECHNIQUES FOR MEASURING P_D

The first method to be considered is n-d or p-d scattering. From the work of Doleschall [24], Sloan and Aarons [25], Pieper [26], and Fayard, Lamot and Elbaz [27], one might conclude that measurement of the polarizations in this process provides an excellent method for studying the N-N interaction but does not come close to giving a definitive measurement of P_D. Figure 4 is a curve from Fayard et al which shows a typical result indicating that, even though the measurements are very accurate, there is little hope of distinguishing between various values of P_D. The previous speaker has dealt with this subject so I will not discuss it any further.

One of the most promising methods for determining P_D was proposed by Levinger and co-workers [28,29,13] following a suggestion of Bertozzi. By measuring the tensor polarization of the recoiling deuteron following elastic electron scattering, information about the deuteron wave function can be obtained. At moderate moementum transfers (1-2 fm^{-1}) P_D can be measured while at larger values of the momentum transfer the short range part of the wave function can be probed [30,31]. Only one experiment (on the vector polarization) has been performed [32].

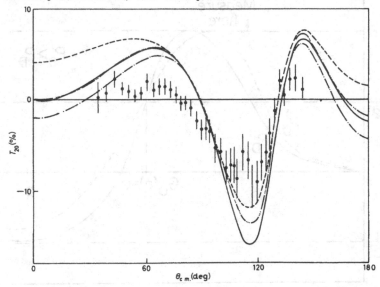

Fig. 4. T_{20} for various values of P_D from 1.1% to 7.0%. From Fayard et al [27].

We can see how this measurement works with one simple equation.
Defining $\chi \equiv G_2(q)/G_0(q)$ then:

$$T_{20} = (2\chi + \chi^2/\sqrt{2})/(1 + \chi^2) \qquad\qquad 2)$$

T_{20} has a maximum of $\sqrt{2}$ when $G_0 = \sqrt{2}\,G_2$. Thus near the zero of G_0 (if
there is a zero) the tensor polarizarion has a preordained value and is
of no use in obtaining P_D (although it is a good way to find the zero).
For small values of q the polarization vanishes. Thus intermediate momentum values are best (q ~ 1–2 fm^{-1}). I have sketched in Figure 5 a
typical comparison of two deuteron wave functions with form factors of
the same shape but with different values of P_D.

Of course there are some problems with this measurement. One of
them is that the contribution of the magnetic form factor must be removed. It is not clear how difficult this will be for a polarization
measurement. Another problem of greater complexity is that of meson
exchange currents. Two recent calculations (Blankenbecler and Gunion[33])
and Chemtob, Moniz and Rho[34]) estimate large effects for large q^2.

Chemtob et al also estimates the effect on T_{20} to be small in our
region of interest. In this calculation they assume that the ω and ρ
coupling constants are the negative of each other. This has the result
of not effecting the position of the zero in $G_0(q)$. The recent measurement which indicates that the ρ–π–γ coupling is smaller than previously
estimated would alter the position of the zero (if the corresponding ω
constant remained fixed) and increase the uncertainty. However a recent
electron scattering experiment at SLAC by B. Chertok et al [35] (fig. 6)
shows that these estimates are much too large, even with the new coupling
constant. The probable eventual result is that this effect is _not_ important for the measurement of T_{20} but the situation of the moment is
confused.

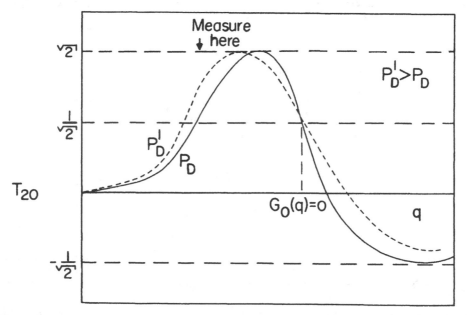

Fig. 5.   Expected dependence to T_{20} on momentum transfer.

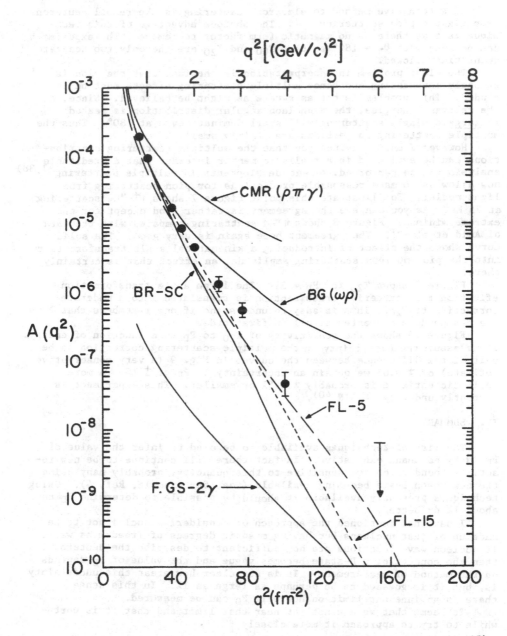

Fig. 6. Comparison of recent electron scattering results by Chertok [35] with Ref. 34 (CMR) and 33 (BG).

An alternative method to electron scaqtering is the recoil deuterons from elastic pion scattering [36]. The obvious advantage of this technique is that there is no magnetic form factor to remove. The experiment can be done with $\theta_\pi = 180°$ so that T_{00} and T_{20} are the only two non-zero quantities allowed.

The clear problem in interpretation is the fact that the pion is strongly interacting and hence multiple scattering effects must be included. This problem is not as severe as might be believed. Since, at the relevant energies, the π-nucleon angular distribution is peaked at <u>large</u> q single scattering will still dominate even at 180°. Thus the multiple scattering corrections are fairly small.

However I must convince you that the multiple scattering contributions can be evaluated in a realistic manner in order that a creditable analysis can be performed. Recent developments in multiple scattering[37,38] now allow us to make reasonable predictions for pion scattering from light nuclei. To illustrate this point Figure 7 shows π^+-^{4}He scattering at 75 MeV. As you can see the agreement is rather good except for the extreme minimum. Figure 8 shows π^+-d scattering compared with the data of Auld et al [39]. The agreement here again is very good. The solid curve shows the effect of introducing a kinematical angle transformation into the pion-nucleon scattering amplitude, an effect that is certainly there.

Figure 9 shows T_{20} for P_D = 3%. The large angle transformation effect on the scattering cross section is translated into a moderate correction to T_{20}. This is easy to understand if one remembers that the change does not enter at all in first order.

Figure 10 shows the sensitivity of T_{20} to P_D as a function of energy. If we assume the uncertainty in the multiple scattering correction to be half of the difference between the curves in Fig. 9 (a very conservative estimate) of $\pm$ 0.03 we obtain an uncertainty in P_D of $\pm$ 1%. A more realistic estimate is probably $\pm$ 0.5% or smaller. This experiment is currently underway at SIN [40].

IV. SUMMARY

The list of techniques available to be used to infer the value of P_D is by no means exhaustive. In fact there will continue to be new reactions found which are sensitive to this quantity, probably many using the new meson beams becoming available (see, for example, Ref. 4). Using techniques presently available it should be possible to determine P_D to about 1% or better.

I have not mentioned the approach of considering nuclei not to be made up of just nucleons but having mesonic degrees of freedom as well. If nucleon wave functions are not sufficient to describe the deuteron then the concept of a D-state becomes vague and the value of P_D depends on the method of measurement. It is not clear how great this uncertainty is, but it is guessed to be perhaps as large as 1%. In this sense there is an inherent limit on how well P_D can be measured.

It seems that we are not yet near that limit and that it is worthwhile to try to approach it more closely.

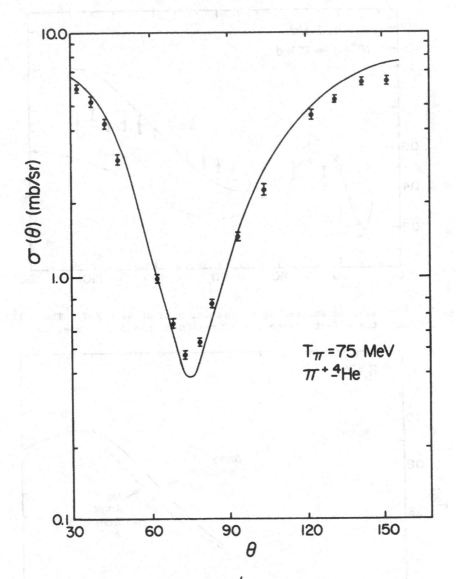

Fig. 7. Pion elastic scattering on ⁴He at 51 MeV.

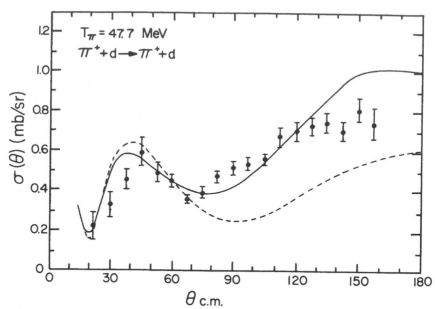

Fig. 8. Pion elastic scattering on the deuteron at 47.7 MeV. The solid line includes the effect of the angle transform; the dotted does not.

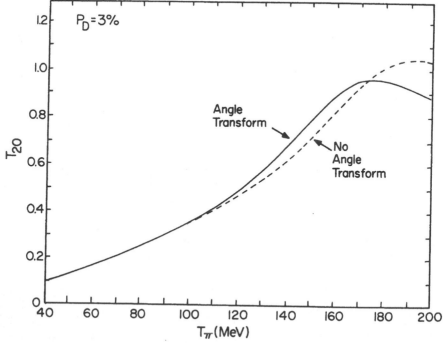

Fig. 9. T_{20} $\theta_\pi = 180°$ vs pion kinetic energy with and without the angle transform.

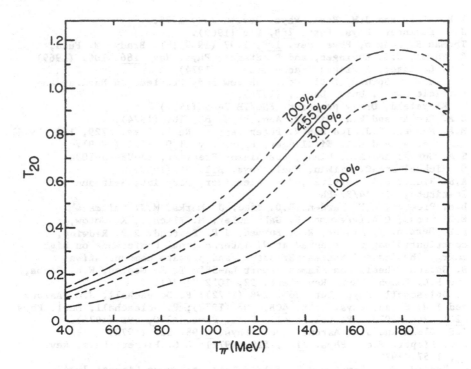

Fig. 10. T_{20} $\theta_{\pi}=180°$ vs pion kinetic energy for several values of P_D.

V. ACKNOWLEDGEMENTS

I have made extensive use of several excellent recent reviews of the two-and-three-nucleon problem and I refer the reader to them for more detail [13,14,29,42]. I wish to thank B. F. Gibson and G. J. Stephenson, Jr. for comments on the manuscript.

REFERENCES

*   Supported by the U.S. ERDA
1) R.V. Reid and M.L. Vaida, Phys. Rev. Lett. **29**, 494 (1972).
2) I. Lindgren, in K. Seigbahn's, "Alpha, Beta and Gamma Ray Spectroscopy," p. 162.
3) T. Hamada and I.D. Johnston, Nucl. Phys. **34**, 382 (1962); R.V. Reid, Ann. Phys. **50**, 411 (1968).
4) G. Breit, Phys. Rev. **71**, 400 (1947).
5) R.J. Adler, Phys. Rev. **141**, 1499 (1966).
6) B. Gobbi et al, Phys. Rev. Lett. **33**, 1450 (1974).
7) Y.E. Kim and A. Tubis, Ann. Rev. Nucl. Sci. **24**, 69 (1974).
8) W. Dilg, L. Koester and W. Nislter, Phys. Lett. **36B**, 208 (1971); E. Fermi and L. Marshall, Phys. Rev. 75, 578 (1949); D. G. Hurst and J. Alcock, Can. J. Phys. 29, 36 (1951).
9) A.C. Phillips, Nucl. Phys. A107, 209 (1968).

10) I.R. Afnan and J.M. Read, Phys. Rev. C July 1975.
11) J.S. Levinger, Phys. Lett. 29B, 216 (1969).
12) Thomas R. Mongan, Phys. Rev. 178, 1597 (1969); T. Brady, M. Fuda,
13) E. Harms, J.S. Levinger, and R. Stagat, Phys. Rev. 186, 1069 (1969).
    J.S. Levinger, Springer Tracts Vol. 71 (1974).
14) Donald W.L. Sprung, Intl. Conf. on Few Body Problems in Nuclear and
    Particle Phys., Univ. Laval, (1974).
15) S. Klarsfeld, Orsay preprint IPNO/TH 74-5 (1974).
16) E.A. Remler and R.A. Miller, Ann. Phys. 82, 189 (1974).
17) M.A. Nasser, G.J. Igo and V. Perez-Mendez, Nucl. Phys. A229, 113 (1974).
18) A.K. Kerman and L.S. Kisslinger, Phys. Rev. 180, 1483 (1969).
19. S.A. Gurvit and A.S. Rinat, Los Alamos Preprint, LA-UR-75-1034.
20) C. Michael and C. Wilkin, Nucl. Phys. B11, 99 (1969).
21) A.S. Rinat, S.A. Gurvitz and Y. Alexander, Preprint, Weizmann
    Institute, WIS-74/37-Ph.
22) B.M. Preedom, C.W. Darden, R.D. Edge, J. Marks, M.J. Saltmarsh,
    E.E. Gross, C.A. Ludemann, K. Gabethuler, M. Blecher, K. Gotow,
    P.Y. Bertin, J. Alster, R.L. Burman, J.P. Perroud, R.P. Redwine,
    contributed paper presented at VI International Conference on High
    Energy Physics and Nuclear Structure and private communication.
23) B. Goplen, Thesis Los Alamos Report LA-5854-T; B. Goplen, W.R. Gibbs,
    and E.L. Lomon, Phys. Rev. Lett. 32, 1012 (1974).
24) P. Doleschall, Phys. Lett 38B, 298 (1972); P. Doleschall, J.C. Aarons
    and I.H. Sloan, Phys. Lett. 40B, 605 (1972); P. Doleschall, Nucl. Phys.
    A201, 264 (1973); P. Doleschall, Nucl. Phys. A220, 491 (1974).
25) I.H. Sloan and J.C. Aarons, Nucl. Phys. A198, 321 (1972).
26) S.C. Pieper, Nucl. Phys. A193, 529 (1972); S.C. Pieper, Phys. Rev.
    C6, 1157 (1972).
27) C. Fayard, G.H. Lamot and E. Elbaz, Lett. al Nuovo Cimento 7, 423
    (1973).
28) T.J. Brady, E.L. Tomusiak and J.S. Levinger, Canad. J. Nucl. Phys.
    52, 1322 (1974).
29) J.S. Levinger, Acta Physica Hung. 33, 135 (1973).
30) M.J. Moravcsik and P. Ghosh, Phys. Rev. Lett. 32, 321 (1974).
31) F. Coester and A. Ostabee, Phys. Rev. C11, 1836 (1975).
32) R. Prepost, R.M. Simonds and B.H. Wiik, Phys Rev. Lett. 21, 1271 (1968).
33) R. Blankenbecler and J.F. Gunion, Phys. Rev. D, 718 (1971).
34) M. Chemtob, E.J. Moniz and M. Rho, Phys. Rev. C, 344 (1974).
35) Benson T. Chertok, et al, Bull. Am. Phys. Soc. 20, 708 (1975).
36) W.R. Gibbs, Phys. Rev. C3, 1127 (1971).
37) W. R. Gibbs, Phys. Rev. C5, 755 (1972); Ibid C10, 2166 (1974).
38) W.R. Gibbs, Summer School on Nuclear Physics, Erice, Italy, Sept. 1974.
39) E.G. Auld, et al, Intl. Conf. on Few Body Problems in Nuclear and
    Particle Phys., Univ Laval (1974).
40) W. Grüebler SIN Proposal No.R-73-01.1
41) Harold W. Fearing, Phys. Rev. C11, 1493 (1975).

DISCUSSION

Roman:

   The measurements of tensor analysing power in (d,p) reactions on
   complex nuclei (Wisconsin, Birmingham) show the need to include
   deuteron D-state contribution in DWBA calculations (Johnson and
   Santos; Robson and Delic). Is this not relevant to the problem ?

Gibbs:

Certainly it is relevant but one has, of course, the N-N final state interaction involved. Thus again, as in the three body case, we must unravel the three body forces (if they exist), short range N-N wave functions etc.

Adelberger:

You seem to think very highly of π-d scattering. Why should π-d scattering be easier to calculate reliably than n-d scattering or $D(\gamma, np)$?

Gibbs:

The π-N off-shell behavior is apparently more easily obtained than the N-N off-shell behavior. This is because reactions are available in which the pion is absorbed. If the energy is well chosen the π-N interaction is weaker than the N-N interaction.

Wäffler:

I would like to point out that a deuteron D-state probability of 3 to 4 percent as used in your calculations contradicts with 6 to 7 percent needed to explain the experimentally observed angular distribution in the deuteron photo-disintegration.

Gibbs:

I agree, there is a contradiction among various experiments.

Ohlsen:

It has always seemed to me that disintegration of a deuteron by Coulomb scattering from a very heavy nucleus should be a clean way to determine the deuteron D-state. Why is this not so ?

Gibbs:

The final state nucleon-nucleon interaction is still present. If the potential is non-local this will be different because of the different energy.

SYMMETRIES AND POLARIZATION

Markus Simonius

Laboratorium für Kernphysik, Eidg. Techn. Hochschule

8049 Zürich, Switzerland

1. Introduction

Symmetries, important in many fields of nuclear and particle physics, play a particularly important role in polarization experiments where some of them find their most important, if not their only, experimental manifestation. In this brief review we shall discuss in particular those symmetries which are relevant in nuclear reactions and their analysis: parity, time reversal, and isospin or charge symmetry. All of them are known to be broken in nature. However, in spite of that, for practical purpose in the analysis of nuclear scattering data, parity and time reversal are excellent symmetries. Only isospin and charge symmetry is appreciably broken also at this level.

In sec. 2 we list the most important consequences of these symmetries for the observables in nuclear reactions. For the rest of the paper we then turn to a discussion of the status and purpose of their tests and the role polarization measurements play in them. We also discuss briefly a recent test for second class currents in weak decays which has to do with a symmetry property of weak interactions.

2. Constraints due to symmetries

We recall here some of the important and predominantly used consequences of the relevant symmetries for nuclear reactions and their analysis. This is not meant to be a complete treatment, rather it gives examples of the types of relations one obtains for the observables. Corresponding relations for the amplitudes may be found e.g. in ref.[1]. Most relations depend on the coordinate system chosen. We use the Madison Convention[1,2] (helicity frame) - also for the notation (for others see ref.[1].

2.1. Parity conservation (reflection symmetry)

For reactions with two body final state, $a+b \rightarrow c+d$ or $b(a,c)d$ parity conservation implies

$$A_x = A_z = p_x = p_z = 0 \tag{1}$$

$$T_{kq} = (-1)^k T_{kq}^* = \begin{cases} \text{real for even } k \\ \text{pure imaginary for odd } k \end{cases} \tag{2}$$

Corresponding rules are found for coefficients involving the polarization of more than one particle, or higher tensors in the case of cartesian tensors: In spherical tensor notation all such coefficients are <u>real</u> if Σk is even and <u>pure imaginary</u> otherwise where Σk is the sum over the ranks k of all particles involved. In cartesian notation all coefficients with $n_x + n_z =$ odd vanish where $n_x(n_z)$ is the total number of x's(z's) in the coefficient. For reactions of the spin structure $1/2 + 0 \rightarrow 1/2 + 0$ one has in addition

$$p_y = \pm A_y \quad \text{and} \quad it_{11} = \pm iT_{11} \tag{3}$$

with plus sign for elastic scattering or if the product of the intrinsic parities of the two particles is unchanged in the reaction, and minus sign otherwise, i.e. if it changes.

In the decay of a nucleus or particle in a state of well defined spin and parity only _even k_ (i.e. even tensor moments) contribute to the angular distribution[1]

$$w(\theta,\phi) = \sum_{kq} t_{kq} \, B_k \, Y_{kq}^*(\theta,\phi) . \tag{4}$$

Here $Y_{kq}(\theta,\phi)$ are the usual spherical harmonics, (θ,ϕ) the measured direction of emission and t_{kq} the spherical polarization tensors of the decaying state in an arbitrary given coordinate system in the restframe of the decaying nucleus. B_k (independent of q) contains all the information on the decay. The rule that only even k occur in eq.(3) is due to parity conservation of the decay and independent of the production process(e.g. nuclear reaction in a sequential decay process). It is violated in weak decays.

2.2. Time reversal invariance

This connects a reaction to its inverse and implies for instance

$$A_y = P_y \tag{5}$$

$$T_{kq} = (-1)^{k-q} t_{kq} \tag{6}$$

detailed balance

where the two sides of the equations correspond to $b(\vec{a},c)d$ and $d(c,\vec{a})b$, respectively, and the right hand side of eq.(6) refers to the center of mass helicity frame[1]. For elastic scattering only the relations for polarization observables are nontrivial and for spin 1/2 on spin 0 scattering even these are implied already by parity conservation according to eq.(3). Of interest below are also the first rank polarization transfer coefficients for elastic scattering $b(\vec{a},\vec{a})b$. Again using the center of mass helicity frame one finds[1,3]

$$A_x^z = - A_z^x \tag{7}$$

while all others allowed by parity conservation are trivial.

2.3. Isospin or charge symmetry

The one of these better fulfilled at a fundamental level (see sec.5) is charge symmetry which for nuclear physics may be defined as invariance under replacement of all protons by neutrons and vice versa. Denoting by a', b', c', d' nuclei or particles obtained from a,b,c,d by this operation one obtains, since this does not affect space and spin degrees of freedom,

$$\text{Obs} \, [\theta;b(a,c)d] = \text{Obs} \, [\theta;b'(a',c')d'] \tag{8}$$

if charge symmetry is exactly valid. Here Obs [ ] is an arbitrary observable for the reaction specified. Of course arrows denoting polarizations are over corresponding particles and for frame of reference etc. one has to adhere to the specification of the reaction given in the brackets.

Above relation contains among other things the most important applications of the Barshay-Temmer theorem[4] and its generalizations[5] though its original formulation and proof somewhat conceals this simple content.

For reactions not connected by charge symmetry one may still get relations of this form if the initial or final configuration is such that its total isospin is unique. For the differential cross section the relation may then contain in addition the square modulus of isospin Clebsch-Gordan coefficients which, however, drop out in the usually normalized polarization observables[5]. Isospin relations between observables valid even if more than one total isospin contributes are discussed in ref.[6].

An interesting relation for (p,n) charge exchange reactions between isospin multiplets (quasi elastic scattering), mentioned also by Arnold et al.[7], has been discussed by Conzett[8] recently. Since here isospin invariance or charge symmetry exchanges initial and final state one can combine it with time reversal invariance to obtain eq.(5) or (6) also for the same (not the inverse) reaction, as in the case of elastic scattering. Tests of this relation for several reactions have been reported to this conference[9,10] and generalizations including polarization transfer (7) are discussed in a contribution by Arnold[11].

3. Parity violation in the nuclear interaction

The existence of a weak parity violating (PV) component in the nuclear interaction of the order of $10^{-6} - 10^{-7}$ in rough qualitative agreement with estimates based on the current-current form of weak interactions has been verified by now in many measurements - most of them involving γ-transitions. These experiments have been reviewed recently by Boehm[12] and will not be listed here again. Most of them are based on variants of Lobashov's[13] famous current integration technique in order to achieve the enormous statistics necessary. Unfortunately a reliable quantitative theoretical analysis of most of these data is very difficult due to nuclear structure problems, in particular for the transitions in heavy nuclei.

A comprehensive analysis of the data based mainly on measurements in light systems has recently been given by the author at the Mainz Meeting[14]. We refer the reader there for details and more references. The aim of the analysis is the determination of parity violating meson-nucleon coupling constants. Taking into account the restriction due to CP conservation, only the following (light) mesons can contribute to the PV nuclear interaction

$$
\begin{array}{cccc}
 & \pi^{\pm} & \rho & \omega \\
\Delta I = & 1 & 0,1,2 & 0,1
\end{array}
\qquad (9)
$$

The second line gives the possible ranks ΔI of isospin violation of the PV couplings of these mesons to the nucleons.

The $\pi^{\pm}$ exchange dominates in the PV analysing power (asymmetry) in thermal n-p photocapture with polarized neutrons:

$$
\vec{n} + p \rightarrow d + \gamma \qquad \text{(thermal)} \qquad (10)
$$

which is sensitive only[15] to $\Delta I=1$. This coupling depends crucially on neutral current contributions in the hadronic weak interactions whose presence would enhance the effect by a factor of about 20. Its scale can be predicted from hyperon decay measurements using SU(3) symmetry

or current algebra[16]. This and other measurements which would enable one
to isolate the $\Delta I=1$ PV interaction still wait to be done. They are obviously of great interest in view of the recent development of weak interaction theories which predict, and measurements which verified, neutral
current contributions in weak interactions involving leptons.

Now to the vector mesons, ρ and ω. Cabibbo theory in conjunction
with the so-called factorization approximation predicts a pure $\Delta I=0$ and
2 ρ-exchange interaction. This is certainly not reliable but it sets a
scale on which to base the analysis of data. We shall refer to it as the
Cabibbo-factorization prediction. It can be compared to the following experiments both of which are insensitive to $\Delta I=1$.

$^{16}O^{*}(2^{-},8.87$ MeV$) \rightarrow {}^{12}C + \alpha$ (Neubeck et al.[17]):

$$\Gamma_\alpha = (1.03 \pm .28)\text{x}10^{-10} \text{ eV} . \tag{11}$$

This is sensitive only to the $\Delta I=0$ part of the interaction,
and agrees roughly with the Cabibbo-factorization prediction.

$n + p \rightarrow d + \vec{\gamma}$, thermal (Lobashov et al.[18]):

$$P_\gamma = -(1.3 \pm .45)\text{x}10^{-6} \tag{12}$$

P_γ is the circular polarization of the emitted γ. It is mainly,
but not exclusively, sensitive to the $\Delta I=2$ part and roughly -60
times larger than the Cabibbo factorization prediction[19].

A rough analysis thus yields

  (i) The $\Delta I=0$ part is $\sim$ as predicted.
  (ii) The $\Delta I=2$ part is enhanced by a factor $\sim$ -60 (isotensor
       enhancement).

This large isotensor enhancement, if real, is certainly not understood at present. It is contrary to general prejudice which would rather
favour isotensor suppression (or isoscalar domination) based on the
hadronic $\Delta I=1/2$ rule. It seems however compatible with measurements in
heavy nuclei which all give larger values than predicted (this could,
however, also be due to $\Delta I=1$ enhancement due to neutral current contributions as discussed above). Clearly Lobashov's measurement, which was
never checked independently, should be repeated.

Now to proton-proton scattering: the quantity of interest is the
analyzing power[20]

$$A_L \equiv A_z \equiv T_{10} = \frac{\sigma_+ - \sigma_-}{\sigma_+ + \sigma_-} \tag{13}$$

where σ_+ (σ_-) are the cross sections with 100% beam polarization parallel (antiparallel) to the beam. The newest result of Potter et al.[21] at
15 MeV is

$$A_L = (-1.8 \pm 2) \times 10^{-7} \tag{14}$$

which certainly constitutes a splendid achievement in precision of a
scattering experiment. Though it is "only" an upper limit we shall see
that it is very important. It must be compared to the following theoretical predictions[14,22]:

$$A_L = 0 \quad \text{for Cabibbo + factorization} \tag{15}$$

$$A_L = .3 \times 10^{-7} \text{ same with } \Delta I=0 \text{ only} \qquad (16)$$
$$(\Delta I=2 \text{ suppression})$$

$$A_L = (1.8 \pm .6) \times 10^{-6} \text{ for isotensor enhancement (ii).} \qquad (17)$$

The zero in the first line is due to the fact that in the factorization approximation there is no neutral interaction in Cabibbo's theory. The $\Delta I=2$ part cancels the $\Delta I=0$ part exactly. If the former is suppressed according to the concept of isotensor suppression mentioned above one obtains the second line. Brown et al.[23] obtain a somewhat larger value since they modify also the isoscalar part of the interaction relative to the Cabibbo-factorization prediction.

The third line, eq.(17) finally, corresponds to the $\Delta I=2$ enhancement obtained from the analysis of Lobashov's experiment. It is clearly in conflict with the measurement (14) of Potter et al.[21]. Does this establish incompatibility between the two experiments?. There are two crucial points in the theoretical part of the comparison: the first one concerns the problem of short range correlations between the nucleons (short range part of the regular interaction). These are not well known, in particular for s-waves. But the parity violating matrix elements depend strongly on them since the PV interaction due to vector meson exchange is of very short range ($m_\rho^{-1} \approx .26$ fm). The magnitude of the uncertainty is controversial and it is not clear to what extend it can change the _relative_ magnitude of the effects which is alone relevant for this comparison. Though we think the uncertainty in the relative magnitudes due to short range correlations is smaller[14], a relative factor of 2-4, say, can not be excluded with certainty. The second point concerns the number of parameters. There are more coupling constants than measurements by which they can be determined. Already including only $\Delta I=0$ or 2 one has three couplings available (including the ω, see the list (9)). These enter in an independent way into the three measurements - quadratically in α-decay and linearly in the two others. They can certainly be adjusted to fit the three experiments. In addition there are $\Delta I=1$ couplings which can contribute to p-p scattering but not to the two others. Thus the two conflicting experiments could be reconciled. But this would involve very strong "accidental" cancellations between large couplings which is certainly very unsatisfactory. How this would affect the upper limit given by Adelberger et al.[24] has not been investigated. However, there are enough parameters to fit also this constraint - in principle.

Clearly there is still a lot of challenging work to be done for theoreticians as well as experimenters. For theoreticians this includes the problem of understanding the couplings, and in particular the enhancement discussed - if it is real - on a more fundamental basis.

What are the experiments of interest? We mentioned the asymmetry (17) and the repetition of the circular polarization (12) in thermal n-p photocapture. We now concentrate on scattering of polarized nucleons, though there are also other measurements in light nuclei which are of interest (see e.g. ref.[55]).

For $\vec{p}$-p scattering the primary objectives are the energy dependence and the enhancement of the sensitivity. What is to be expected from the energy dependence can, for energies below ~ 400 MeV, be seen in fig.1.

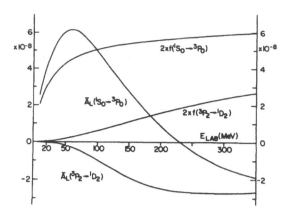

Fig. 1. Contribution of the two lowest PV transition amplitudes f to $\bar{A}_L$ (angular average)[20] in p-p scattering, Cabibbo-factorization prediction with $\Delta I=2$ suppression. Calculations done with RSC potential neglecting mixing for $f(^3P_2 \to {}^1D_2)$.

It shows the energy behaviour of the contributions of the two lowest and thus dominating parity violating p-p scattering amplitudes $f(^1S_0 \to {}^3P_0)$ and $f(^3P_2 \to {}^1D_2)$ to the angle averaged analysing power $\bar{A}_L$, which is the quantity one obtains by measuring essentially total cross sections - outside the Coulomb region - in eq.(13). At low energy A_L is angle independent anyhow as the cross section and thus $A_L(\theta) = \bar{A}_L$. Above $\sim$ 100 MeV the difference becomes, however, significant[23].

The first point to be emphasized with respect to fig. 1 is the fact that the pronounced energy dependence of the curves does not contain any information of interest for our problem since it is essentially determined by kinematics and the well known p-p scattering phase shifts. The quantities of interest are the two scales of the curves of $\bar{A}_L$ or f for the two different transitions. As given in the figure they correspond to the Cabibbo-factorization prediction with $\Delta I=2$ suppression (only $\Delta I=0$ ρ-exchange corresponding to eq.(16)). They both depend only on two parameters, the parity violating $pp\rho^0$ and $pp\omega$ coupling constants. These can be determined if the two scales are known. In fact, to a good approximation ω-exchange does not contribute to the $^3P_2 \to {}^1D_2$ transition at all[14]. The preferred energy regions for the determination of the two scales are clearly below 70 and around 230 MeV, where $\bar{A}_L$ is dominated by only one of the two contributions. Naturally the two amplitudes may in principle be disentangled also from the angular dependence of A_L. Measurements at higher, in particular at very high energy[25,21], eventually at large momentum transfer where not much is known about what will happen, are of course of great interest. However, they would not be very helpful for the analysis discussed above since not enough is known on the strong interaction part involved.

Of course, the most important thing at the moment is to enhance the sensitivity and if possible actually see parity violation. The optimal energy from the point of view of magnitude of the effect is around 50 MeV. Work toward higher sensitivity is going on in Los Alamos. Considerable improvement is hoped to be possible[21]. A program has also been

initiated in Berkeley[26]. Here at ETH in Zürich a measurement at 12 MeV
is in an advanced state of preparation, which is planned to be continued
at 50 MeV at SIN[27]. The scattering chamber and detection system[28] are
shown in fig. 2.

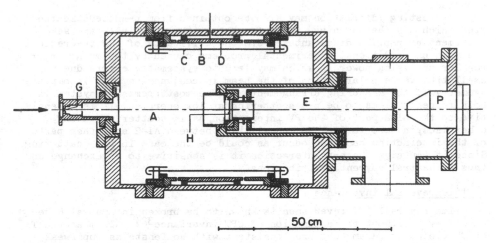

Fig.2. Scattering chamber for PV experiment:
A: Target area. B: Ionization region. C: High voltage electrodes.
D: Measuring electrode. E: Faraday cup. G and H: Windows. P: Polarimeter

Its main feature is that the same cylindrical pressure ($\sim$ 4 atm H_2)
vessel serves as target* (region A) and ionization chamber (region B) between a high voltage ($\sim$ 10 kV) electrode C and a measuring electrode D
along the mantle of the cylinder. The currents from the electrode D as a
measure of the number of scattered protons and from the Faraday cup E
are integrated and digitized in phase with fast polarization reversal·
and transmitted to an on-line computer for further evaluation. The
electronics has been developed and built by R. Balzer[29] and the special
programs for the on-line evaluation were written by F. Horber. Tests of
this equipment with unpolarized beam have been completed in May. Measurements in cooperation with W. Grüebler's group are planned to be taken up
as soon as the necessary equipment to reverse the proton polarization
is installed in the polarized ion source of the ETH tandem accelerator[30].

Besides p-p scattering, n-p scattering would be of great interest.
This contains the different meson and isospin parts of the PV interaction
in a different combination and would therefore help to disentangle them.
In particular, unlike p-p scattering it contains one π-exchange and there-
fore is sensitive to neutral current contributions as discussed above.
Brown et al.[23] have calculated A_L for n-p scattering. However, they

---

* For measurements at SIN a high pressure target will be inserted in
  the center of the chamber.

found the π-exchange contribution to be very small - even in the case of
neutral currents. Measurements of n-p scattering seem to be even harder
than those for p-p scattering. None have been done yet. Missimer[31] has
analyzed the possibility of disentangling the isospin contributions
using in addition n-n scattering. This seems to be a rather academical
exercise.

Interesting information may also be obtained from $\vec{p}$-nucleus scattering which has the advantage to be feasable with essentially the same
equipment as p-p. Its disadvantages are the difficulty of the theoretical
analysis and additional experimental problems caused by the much larger
regular analysing power A_y which may lead to systematic errors due to
small transverse polarization of the beam in conjunction with incomplete
symmetry of the measuring arrangement[20]. The most promising system in
both respects seems to be $\vec{p}$-d scattering at low energies. It is insensitive to the $\Delta I=2$ part of the PV interaction, as is scattering on any
other isospin zero target. No cancellation between $\Delta I=2$ and other parts
of the interaction can thus occur as could be the case in p-p scattering.
Since it contains the n-p interaction it is sensitive to $\pi^{\pm}$ exchange and
thus to neutral current contributions.

4. Time reversal invariance

Time reversal (T) invariance is known to be broken in neutral K decays - even without the assumption of CPT invariance[32]. All measured T
or CP violating parameters are consistent with Wolfensteins superweak
theory which describes them in terms of only one free parameter. However,
experiments are not yet able to exclude the alternative possibilities:
milliweak, electromagnetic, or millistrong T violation, where "milli"
indicates that its strength is reduced by 10^{-3} relative to the corresponding T or CP conserving interaction in accordance with the strength of
CP violation in neutral K decay.

Time reversal violation in electromagnetic or millistrong interactions would lead to <u>time reversal violating</u> effects in <u>parity conserving</u>
nuclear processes. These are the object of all nuclear physics tests of
time reversal invariance. Their magnitude is usually assumed to be of
the order of 10^{-3} relative to regular effects[12]. However, we shall see
that in the case of millistrong interactions this expectation is wrong.
Their effect would be several orders of magnitude smaller. In fact we
shall see that even a full violation of time reversal invariance in the
strong interaction could not have been detected by present nuclear
physics tests since its effects are smaller than the best upper limits
obtained up to now. We thus have the rather surprising fact that, even
if strong interaction would not be time reversal invariant, for practical
low energy nuclear physics purpose time reversal invariance would still
be a good symmetry.

Experimental tests of time reversal invariance have been reviewed
recently by Richter[33] and Boehm[12]. The best published results for
angular correlation as well as detailed balance tests yield upper limits
of about

$$\sim 3 \times 10^{-3}.$$

A current detailed balance measurement in $^{24}Mg + \alpha \leftrightarrow ^{27}Al + p$ performed
in Bochum[34] reached an upper limit of $\sim 10^{-3}$ and is hoped to still lower
it to about 3×10^{-4}.

In elastic scattering time reversal invariance can only be tested
with the help of polarization measurements - the simplest ones being
P-A-tests i.e. tests of the polarization asymmetry equality (5). As seen

in eq. (3) it cannot be violated in a reaction of the form $1/2 + 0 \to 1/2 + 0$. For tests with polarized nucleons the simplest spin structure is thus $1/2 + 1/2 \leftrightarrow 1/2 + 1/2$. In addition, in order to be sensitive the target should not contain a large inert spin zero core whose influence dominates the scattering process. Thus, apart from the most fundamental system, nucleon-nucleon scattering, p-^{3}He and p-^{3}H seem to be the most promising cases though proton scattering on high spin (odd) heavy deformed nuclei could perhaps be interesting candidates too [1,35]. $\vec{p}$ - ^{3}He and $\vec{p}$ - ^{3}H scattering have also some distinct experimental advantages since due to their (almost) equal kinematics the calibration of beam and analyzer may be eliminated by comparing the ratios of P and A for the two reactions [1,35]. The elimination of the calibration may also be achieved by testing the relation (7) for polarization transfer as done, in a particular way, by Handler et al.[36] for 425 MeV p-p scattering. Naturally one could perform tests with polarized deuterons instead of protons which could be scattered also on spin zero targets. A disadvantage seems to be that one then has to disentangle the different polarization tensors.

What are the expected violations? Unlike the case of parity violation there exists no established theory on which to base any prediction. For some models of electromagnetic T-violation, calculations of Huffman[37], and Clement and Heller[38] have shown that these could cause effects of the order of 10^{-3} in some nuclear processes, but no estimates have been given for polarization tests in elastic scattering.

A particular strong (not millistrong!) T violating A_1 exchange interaction, introduced by Sudarshan[39], has been adapted and evaluated by Brian et al.[40,41] for nucleon nucleon scattering. Their prediction for P-A is given in fig. 3 with some data cited there. There seem to be some

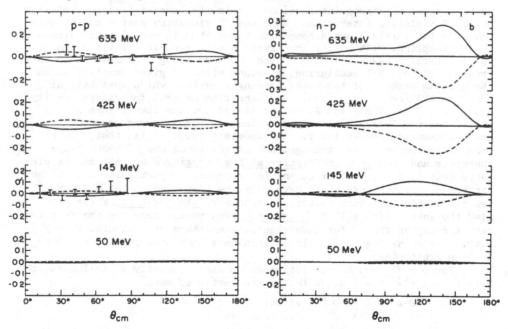

Fig.3. P-A for proton-proton and neutron-proton scattering according to Brian and Gersten[40].

unpublished data cited in ref.[41] which reduce the n-p prediction between 500 and 600 MeV by a factor of about 1/3 - still well within the
domain of strong interactions since in Sudarshan's model T conserving
and T violating couplings have essentially the same strength. Note that
the p-p predictions in fig.1 are much smaller than those for n-p and
that both are very small at low energies in spite of the fact that it
is caused by a strong T violation. We shall see that this is a general
feature not restricted to the particular model of Sudarshan.

A general analysis of the possibilities of T violation in the nucleon-nucleon interaction with main emphasis on strong T violation, was
given recently by the author[42]. It reveals the strong kinematic limitations imposed by parity conservation and crossing symmetry. The most
important features may be seen in table 1 in which $\pi\gamma$ denotes Huffman's
electromagnetic T violating interaction[37] while the other exchanges

Table 1. T-violation in the nucleon-nucleon system

Transitions	ΔI	exchanges el.magn.	strong
$^3S_1 \leftrightarrow {}^3D_1$	0	$\pi\gamma$	$A_1(1100)$
$^1P_1 \leftrightarrow {}^3P_1$	1		$\rho^\pm(770)$
$^1D_2 \leftrightarrow {}^3D_2$	1		$\rho^\pm(770)$
$^3P_2 \leftrightarrow {}^3F_2$	0,1,2	$\pi\gamma$	$A_1(1100)$

may be caused by strong interactions. ΔI is the rank of isospin violation
of the T violating interaction. Note that T violation must occur as asymmetry in the partial wave decomposition and thus in non-diagonal elements
in ℓ-s coupling. With parity conservation this restricts T violation for
the low partial waves to those listed. Corresponding requirements in the
crossed channel, $\bar{N}$-N scattering, together with CPT gives constraints on
the quantum numbers of the total exchanged systems which must have at
least spin 1 (no π, η, σ, etc.) and cannot be neutral for natural parity
1^-, 2^+ etc. [42]. The ρ and A_1 mesons listed are the lowest mass $J^\pi = 1^-$
and 1^+ mesons. They essentially represent 2π and 3π exchange with these
quantum numbers, respectively. The important property is, that, apart
from electromagnetic $\pi\gamma$ exchange, all interactions are of short range
character and thus <u>very ineffective at low energies</u> since all matrix elements contain non-zero orbital angular momenta. In particular this is so
for the p-p or n-n case since here only the last line in table 1 contributes due to the exclusion principle. Thus for <u>like nucleons</u> or for pure
$\Delta I=2$ the interaction will be <u>suppressed</u> even more. These are the features
seen already in fig. 3 for Sudarshan's A_1 exchange which contains only
$\Delta I=0$. No ρ or $\pi\gamma$ exchange calculations have been done yet for nucleon-
nucleon scattering.

For $\vec{p} - {}^3He$ scattering estimates obtained recently in collaboration
with D. Wyler[43] yield at 12 MeV in the center of mass*

$$P-A \leq 5 \times 10^{-4} \quad \text{strong T viol.}$$
$$\leq 5 \times 10^{-3} \quad \text{el. magn. } \pi\gamma$$

We would like to thank P. Heiss for providing the p-^{3}He scattering
wave functions used in the calculations.

The strong T violation effect was obtained with a ρ-exchange with the usual two couplings but completely out of phase i.e. relative imaginary instead of relative real. A_1-exchange results are smaller. For a milli-weak interaction the first line would still be 10^{-3} times smaller. These calculations are to our knowledge the first estimates of T violating effects which would result from strong interactions in nuclear physics.

It must be emphasized that the small value obtained for $\bar{p}$-^{3}He scattering is not due to a peculiarity of this system. On the contrary this system is certainly more sensitive than most heavier ones where scattering is predominantly off a spin zero core. Exceptions are perhaps some isospin-forbidden transitions where the sensitivity may be enhanced by a factor of up to 10 if the T violating interaction has $\Delta I \neq 0$ as pointed out by Nikolaev and Shmatikov[44]. No such tests have been done yet. In any case we see from this analysis that T violation due to a milli-strong interaction would be orders of magnitudes smaller than the accuracy of any scattering measurement done so far or foreseeable with present day techniques and even a strong interaction T violation would be at best at the edge of measurability. The only place in nuclear physics where such very small T violating effects could perhaps be seen is in a highly retarded γ-decay as e.g. in ^{170}Hf where the enhancement could be many orders of magnitude[45]. However, in the case of an upper limit these measurements are still not very helpful since - or as long as - one cannot _prove_ that the effect has to be enhanced.

For electromagnetic T violation the estimate given above for $\bar{p}$-^{3}He is very crude[43] and should not be taken too literally. However, it shows that an electromagnetic T violation might give effects which could be seen in accurate measurements. These are therefore certainly worth while doing in spite of our discussion above. For $\bar{p}$-^{3}He the best angular range is around 110°-125° center of mass. This can be seen in a model independent way as discussed in more detail in ref.[35].

Naturally of primary importance are also tests in nucleon-nucleon, in particular $\vec{n}$-p scattering at intermediate and higher energies where strong interaction T violation would give larger effects (see fig.3). ρ-exchange T violation could lead to different angle dependence in n-p scattering than shown in fig. 3.

5. Isospin and charge symmetry

Isospin and charge symmetry are both violated by the electromagnetic interaction between nucleons. Even if the Coulomb interaction between the protons is eliminated, isospin in the nucleon-nucleon interaction is broken due to the mass splitting of the π-mesons and higher ones, 3.3% between π^0 and $\pi^\pm$, which, of course, itself is believed to be of electromagnetic origin. This mechanism does, however, not yield a charge symmetry breaking i.e. a difference between n-n and p-p interaction after correction for the Coulomb interaction between the protons - which thus is the better symmetry. Its classic, very sensitive, test are the scattering lengths which according to the compilation of Henley and Wilkinson[46] are

$$a_{pp}^{c.c.} = -17.1 \pm .2 \text{ fm}$$
$$a_{nn} = -16.4 \pm .9 \text{ fm}$$

where $a_{pp}^{c.c.}$ means the Coulomb corrected p-p scattering length. Clearly
the two are equal within the errors. However, this comparison contains
a problem, the Coulomb correction. As pointed out by Arnold and Seyler[47]
and emphasized by Sauer[48] the Coulomb correction depends on the strong
nucleon-nucleon interaction or short distance wave function. As long as
these are not known more reliably one cannot make a significant comparison.   On the other hand, however, these authors point out that one can
postulate equality of the two scattering lengths and from this obtain
restrictions on the Coulomb corrections and thus indirectly on the short
range correlations of the two nucleon system which are hardly obtainable
otherwise.

This leads us to the general theme of isospin or charge symmetry
tests in nuclear reactions as well as in bound state problems where the
Coulomb energy anomaly still poses an unresolved problem[49]. The main
information obtainable from these tests is not on the fundamental symmetry itself but on the nuclear wave functions used for the calculation
of the Coulomb corrections. This information is in general to some extent
independent and often complementary to the information obtainable from
other sources and thus an additional test of the adequacy of models.
Though polarization measurements are not a necessary requirement for
testing charge symmetry,they can be important for the analysis of its
breaking since they yield more information on the scattering process.
A starting point in some cases might be an R-matrix type of analysis
which allows one to eliminate the "outer" Coulomb effects in the long
range part of the scattering wave functions.

Prominent deviations from charge symmetry have been found in
$d(\alpha,{}^3H){}^3He$ (or its inverse) by several groups[51-54] between threshold at
$E_d=21.5$ MeV and 41 MeV. Around 32 and 41 MeV DWBA calculations can give
effects of the right order of magnitude but do not reproduce the structure[51,53] which is not too surprising since DWBA is of doubtful validity in this domain. At lower energies isospin mixing of the ^6Li compound
levels is suggested as a possible explanation[54] - again without an
actual fit. A refined analysis has still to be done.

Deviations from the polarization - asymmetry equality (5) in the
charge exchange reaction $^3H(p,n)^3He$ have been analyzed by Arnold et al[7]
who found that it implies stronger $^3P_2 \leftrightarrow {}^3F_2$ transitions than anticipated
but no quantitative explanation is given. Two similar tests from TUNL[9,10]
are reported to this conference and a brief discussion on the intermediate
states involved is given by Arnold[11].

6. Second class currents

Second class currents are currents which could contribute to weak decays
and which have opposite transformation properties under G parity

$$G = C \, e^{i\pi I_2}$$

than the usual - first class - vector and axial vector currents. Here C
is the charge conjugation and $e^{i\pi I_2}$ the charge symmetry operation. Though
the possibility of second class currents has been pointed out 17 years
ago[56] the question of their existence is not settled yet. There are two
reasons for this: One is that second class currents contribute only to
recoil effects which are small for kinematical reasons[56,57]. And the
second that, though the dominating part of the charge symmetry breaking

effects of the Coulomb interaction are taken care of by comparing ft values rather than the decay spectra directly, it is very hard to separate the symmetry breaking due to second class currents from that originating from remaining Coulomb effects[58].

This is where polarization comes in since the separation of Coulomb effects can apparently be done more reliably from the energy dependence of the asymmetry of β emission from polarized nuclei than from the decay spectrum alone[57,59]. Of particular interest are asymmetries of analog decays, i.e. decays between mirror nuclei[57] among other reasons since here one does not have to compare between two mirror decays. The simplest and most fundamental system is neutron decay but no experiments are done yet.

Such an experiment has been done, however, for the analog decay ^{19}Ne → ^{19}Fe + e^+ + ν by a Princeton group[60]. The ^{19}Ne is produced by the ^{19}F(p,n)^{19}Ne charge exchange reaction. It is then polarized in a Stern-Gerlach type arrangement which, unlike in the usual atomic beam polarized ion sources operates directly on the nuclear magnetic moment since the total electronic spin is zero. The polarization has to be kept several seconds in a cell in order to cover a significant fraction of the halflife of 17.37 sec. of ^{19}Ne. The results of this experiment apparently require the presence of second class currents[60]. However, they still seem to be preliminary, so no definite conclusion can be drawn yet. Nevertheless this example shows how nonconventional methods in polarization measurements applied to somewhat nonconventional nuclear physics problems can yield information on fundamental problems. Other methods, where the polarized states are obtained by polarization transfer are mentioned by Hanna[61].

7. Conclusions

In conclusion we hope to have shown that the investigation of fundamental symmetries and their violation in nuclear physics offer a lot of challenging problems - many of them outside the normal nuclear physics routine. The efforts are certainly worth while; yet, only a small number of experimenters is involved in such studies. It is to be hoped that some of the pioneering spirit which goes into building better and better polarized ion sources remains to do such difficult and demanding experiments with them.

References

1) M. Simonius, Lecture Notes in Physics, Vol. 30 (Springer, Berlin, 1974) 38
2) Polarization Phenomena in Nuclear Reactions, H.H. Barshall and W. Haeberli eds. (University of Wisconsin Press, Madison, 1971) p. XXV
3) G.G. Ohlsen, Pep. Prog. Phys. 35 (1972) 717
4) S. Barshay and G.M. Temmer, Phys. Rev. Lett. 12 (1964) 728
5) M. Simonius, Phys. Lett. 37B (1971) 446
6) M.G. Doncel, L. Michel and P. Minnaert, Phys. Lett. 38B (1972) 42 and 42B (1972) 96
7) L.G. Arnold, R.G. Seyler, T.R. Donoghue, L. Brown and U. Rohrer, Phys. Rev. Lett. 32 (1974) 310
8) H.E. Conzett, Phys. Lett. 51B (1974) 445

9) P.W. Lisowski, G. Mack, R.C. Byrd, W. Tornow, S.S. Skubic,
   R.L. Walter and T.B. Clegg, contribution to this Symposium, p. 499
10) R.C. Byrd, P.W. Lisowski, S.S. Skubic, W. Tornow, R.L. Walter and
    T.B. Clegg, contribution to this Symposium, p. 501
11) L.G. Arnold, contribution to this Symposium, p. 503
12) F. Boehm, Sixth Intern. Conf. on High Energy Physics and Nuclear
    Structure, Santa Fee 1975
13) V.M. Lobashov et al., Yadern. Fiz. 13 (1971) 555, Soviet Journ.
    Nucl. Phys. 13 (1971) 313
14) M. Simonius, in Interaction Studies in Nuclei, H. Jochim and
    B. Ziegler, eds., (North Holland, 1975) p.3
15) G.S. Danilov, Phys. Lett. 18 (1965) 40
16) B.H.J. McKellar, Phys. Lett. 26B (1967) 107
17) N. Neubeck, H. Schober and H. Wäffler, Phys. Rev. C10 (1974) 320
18) V.M. Lobashov, D.M. Kaminker, G.I. Kharkevich, V.A. Kniazkov,
    N.A. Lazovoy, V.A. Nazarenko, L.F. Sayenko, L.M. Smotritski and
    A.I. Yegorov, Nucl. Phys. A197 (1972) 241
19) B. Desplanques, Nucl. Phys. A242 (1975) 423
    H.J. Pirner and M.L. Rustgi, Preprint
    K.R. Lassey and B.H.J. McKellar, Preprint
20) M. Simonius, Phys. Lett. 41B (1972) 415
21) J.M. Potter, contribution to this Symposium, p. 91
22) M. Simonius, in Few Particle Problems in the Nuclear Interaction,
    I. Slaus et al. eds. (North-Holland, Amsterdam, 1972) p. 221, and
    Nucl. Phys. A220 (1974) 269
23) V.R. Brown, E.M. Henley and F.R. Krejs, Phys. Rev. C9 (1974) 935.
    The asymmetry A defined in this paper is $2A_L$
24) E.G. Adelberger, H.E. Swanson, M.D. Cooper, J.W. Tape and T.A. Train-
    or, Phys. Rev. Lett. 34 (1975) 402
    E.G. Adelberger, contribution to this Symposium, p. 97
25) E.M. Henley and F.R. Krejs, Phys. Rev. D11 (1975) 605
26) H.E. Conzett, private Communication
27) R. Balzer, W. Grüebler, V. König, J. Lang, M. Simonius, M. Daum,
    G.H. Eaton, A. Janett and G. Heidenreich, SIN Proposal Z-75-02.1
28) L.Ph. Roesch, M. Simonius and U. Peyer, Jahresbericht Lab. f. Kern-
    physik ETHZ, 1975, p. 304
29) R. Balzer, ibid. p. 309
30) R. Risler, W. Grüebler, V. König, P.A. Schmelzbach, B. Jenny and
    W.G. Weitkamp, contribution to this Symposium, p.842
31) J. Missimer, Preprint
32) R.C. Casella, Phys. Rev. Lett. 22 (1969) 554
33) A. Richter, in Interaction Studies in Nuclei, H. Jochim and B. Zieg-
    ler eds. (North-Holland, 1975) p. 191
34) A. Richter, private communication
35) M. Simonius, in Interaction Studies in Nuclei, H. Jochim and B.Zieg-
    ler eds. (North-Holland, 1975) p. 419
36) R. Handler, S.C. Wright, L. Pondrom, P. Limon, S. Olsen and
    P. Kloeppel, Phys. Rev. Lett. 19 (1967) 933
37) A.H. Huffman, Phys. Rev. D1 (1970) 882 and 890
38) C.F. Clement and L. Heller, Phys. Rev. Lett. 27 (1971) 545
    C.F. Clement, Ann. Phys. (N.Y.) 75 (1973) 219
39) E.C.G. Sudarshan, Proc. Roy. Soc. A305 (1968) 319

40) R. Bryan and A. Gersten, Phys. Rev. Lett. 26 (1971) 1000
41) J. Binstock, R. Bryan and A. Gersten, Phys. Lett. 48B (1974) 77
42) M. Simonius, Phys. Lett. 58B (1975) 147
43) M. Simonius and D. Wyler, to be published
44) N.N. Nikolaev and M. Zh. Shmatikov, Phys. Lett. 52B (1974) 293
45) K.S. Krane, B.T. Murdoch and W.A. Steyert, Phys. Rev. C10 (1974) 840
46) E.M. Henley and D.H. Wilkinson, in Few Particle Problems in the Nuclear Interaction, I. Slaus et al. eds. (North-Holland, Amsterdam, 1972) p. 242
47) L.G. Arnold and R.G. Seyler, Phys. Rev. C7 (1973) 574
    L.G. Arnold, Phys. Rev. C10 (1974) 923
48) P.U. Sauer, Phys. Rev. Lett. 32 (1974) 626
    Phys. Rev. C11 (1975) 1786
49) J.W. Negele, Comm. Nucl. and Part. Phys. 6 (1974) 15
50) D. Robson and A. Richter, Ann. Phys. (N.Y.) 63 (1971) 261
51) E.E. Gross, E. Newman, M.B. Greenfield, R.W. Rutkowski, W.J. Roberts and A. Zucker, Phys. Rev. C5 (1972) 602
52) G.J. Wagner, G. Mairle, P. Kleinagel and R. Bilwes, in Few Particle Problems in the Nuclear Interaction, I. Slaus et al. eds. (North-Holland, Amsterdam, 1972) p. 747
53) W. Dahme, P.J.A. Buttle, H.E. Conzett, J. Arvieux, J. Birchall and R.M. Larrimer, contribution to this Symposium, p. 497
54) U. Nocken, U. Quast, A. Richter and G. Schrieder, Nucl. Phys. A213 (1973) 97
55) M. Gari, in Interaction Studies in Nuclei, H. Jochim and B. Ziegler eds. (North-Holland, 1975) p. 307
56) S. Weinberg, Phys. Rev. 112 (1958) 1375
57) B.R. Holstein and S.B. Treiman, Phys. Rev. C3 (1971) 1921
    B.R. Holstein, Revs. Mod. Phys. 46 (1974) 789
58) D.H. Wilkinson, Phys. Lett. 48B (1974) 169 and in Interaction Studies in Nuclei, H. Jochim and B. Ziegler eds. (North-Holland, 1975) p. 147
59) M. Morita and I. Tanahita, Phys. Rev. Lett. 35 (1975) 26
60) F.P. Calaprice, in Interaction Studies in Nuclei, H. Jochim and B. Ziegler eds. (North-Holland, 1975) p. 83
61) S.S. Hanna, contribution to this Symposium, p. 407

DISCUSSION

Adelberger:

Does attributing time reversal violation to the strong interaction explain the CP violating decays of the K_L^0 meson ?

Simonius:

It is expected to - that's why it has been introduced. But no actual calculations have been done to my knowledge apart from order of magnitude arguments, which in a natural way lead to a millistrong interaction.

Krisch:

You indicated that you obtained a very tight upper limit on T violation at low energy. Have you tried to extend your calculations to higher energy ?

Simonius:

I have not made calculations. The calculations of Brian et al. with
A_1 exchange I've shown, already show sizeable effects ($\sim 20\%$) for
n–p scattering around 500 MeV. At very high energy one should be sensitive to the inner region of the interaction where ρ and A_1 exchange
is important as long as one stays away from the forward direction
where it is supersedet by absorption (diffraction).

Minamisono:

I would like to make a comment about the experiment on second class
currents in weak interaction. Besides Calaprices experiment, the β
decay asymmetries of ^{12}B and ^{12}N have been studied intensively as a
function of β–energy by Sugimoto et al. in Osaka and Grenacs et al.
in Leuven. To obtain the polarized ^{12}B and ^{12}N they used (d,p) and
(^{3}He,n) reactions with unpolarized incident beams and recoil angle
selection.

TESTS OF PARITY CONSERVATION IN p-p AND p-NUCLEUS SCATTERING*

James M. Potter
Los Alamos Meson Physics Facility, University of California
Los Alamos Scientific Laboratory, Los Alamos, New Mexico 87545

Parity nonconservation in nuclear scattering implies a weak force between hadrons. This force is predicted by the current-current-type theories of weak interactions and has been observed as a mixing of nuclear levels of opposite parity[1]. Quantitative comparisons of experiment and theory have not been wholly successful. Whether the difficulties lie in the understanding of the nucleon-nucleon interaction or in the complexities of the structure of heavy nuclei is not known.

Ideally, one should study simple systems whose properties can be more easily related to the basic internucleon force. In the only experiment that finds an effect in a two-nucleon system, Lobashov et al.[2] measured the circular polarization of the photons from $np \rightarrow d\gamma$ to be $(-1.30 \pm 0.45) \times 10^{-6}$. This result is nearly two orders of magnitude greater than the Cabbibo theory prediction.

This paper describes two experiments which look for a parity violation directly, in proton-proton scattering at 15 MeV and proton-nucleus scattering at 6 GeV. Preliminary results have been published.[3,4]. The collaborators on these experiments are: J. D. Bowman, C. M. Hoffman, C. F. Hwang, J. L. McKibben†, R. E. Mischke, D. E. Nagle, and J. M. Potter from the Los Alamos Scientific Laboratory; D. M. Alde, P. G. Debrunner, H. Frauenfelder, and L. B. Sorensen from the University of Illinois; and H. L. Anderson‡ and R. Talaga‡ from the University of Chicago.

The relative magnitude of the weak nucleon-nucleon force may be estimated from the ratio of weak to strong coupling constants to be $\sim 10^{-5}$. More detailed calculations by Simonius[5], and by Brown, Henley, and Krejs[6] estimate the parity-violating effect to be $\sim 10^{-7}$.

Experimentally, a parity-violating effect is sought by looking for a dependence of the total cross section on the pseudoscalar quantity $\langle \hat{s} \cdot \hat{p} \rangle$, where s and p are the spin and momentum of the incident proton, respectively. An interference between the parity-conserving and parity-nonconserving parts in the scattering amplitude results in a total cross section that depends on whether the spin of the incident proton is parallel or anti-parallel to its momentum. The experimental results are expressed in terms of the longitudinal asymmetry, $A_L = (\sigma_+ - \sigma_-)/(\sigma_+ + \sigma_-)$ where $\sigma_+(\sigma_-)$ is the total cross section for a +(-) helicity incident proton.

Because of the high sensitivity required to measure such a small effect, the problems of obtaining adequate statistics and reducing systematic errors must be carefully considered.

Good statistics are obtained with a high counting rate and low extraneous noise. These experiments utilize integral counting techniques to permit counting rates not feasible by conventional means. Extraneous noise from slow fluctuations in the data is reduced by fast reversal of the polarization of the incident beam.

The principle sources of systematic errors are changes in the beam position and angle, in the residual transverse polarization component, and in the beam current that are correlated with the polarization reversal. The first two of these errors are removed by symmetrizing the apparatus about the beam axis, while the third is eliminated by normalizing the data to the beam current.

The 15-MeV experiment was performed at the Los Alamos Tandem Van de

Graaff accelerator. The 200-na polarized beam is produced by a Lamb-
shift ion source that has been modified to permit polarization reversal
at 1 kHz. The techniques for rapidly reversing the polarization in a
manner consistent with the systematic error requirements of this experiment are described in a separate paper.[7] Figure 1 outlines the basic
principles of the experiment. The actual configuration of the apparatus
is outlined schematically in fig. 2.

The target vessel is filled with H_2 gas at 3.5-atm pressure. Beam
scattered through angles from 5° to $\sim 90^O$ illuminates a square array of
scintillation cells, 0.38 m long and separated by 76 mm, centered on the
beam axis. The currents from 12 photomultipliers (PM's) viewing the
scintillators are summed to form the scattered signal S. The transmitted beam is measured on a gold beam stop in a vacuum chamber behind
the pressurized region. The rear detector is divided into a central
disc and outer quadrants which are used to provide beam position information. The summed signal from the beam stop is called B.

Figure 3 is a block diagram of the analog-signal-processing electronics. The gains of the inputs are adjusted so that each PM contributes equally to S and so that $S \cong B$. An analog divider forms the ratio
$Y = (S-B)/B$ to normalize the scattered beam signal. The change in Y
with polarization, corrected for the actual polarization of the beam,
corresponds to the fractional change in total cross section, i.e., the
parity-violating effect. A phase-locked amplifier (PLA), synchronized
to the polarization reversal, measures the change in Y. The output of
the PLA is integrated by a DVM and accumulated in a computer.

In addition to the usual small signal measurement problems, such as
ground loops and dc offset errors, the principle sources of systematic
error are properties of the beam generated by the source: position and
angle modulation, current modulation, and polarization misalignment.

A transversely polarized beam produces an azimuthal asymmetry in
the scattering from the nonparity-violating analyzing power of the scattering. If the S detector does not have perfect azimuthal symmetry, a
reversing residual component of transverse polarization will produce an
error in A_L. Assuming uniform scintillator quadrants parallel to the
beam, the signal S_i from the i-th quadrant is proportional to the flux
through that quadrant. The error in A_L from transverse polarization is

$$e(A_L) = A\pi^{-1}(G_y \Delta P_x/\Delta P - G_x \Delta P_y/\Delta P) .$$

In this expression $\Delta P_x/\Delta P(\Delta P_y/\Delta P)$ is the reversing component of trans-
verse polarization in the horizontal (vertical) plane, $G_x = (G_R - G_L)/(G_R + G_L)$ is the gain asymmetry for the x direction, $G_y = (G_U - G_D)/(G_U + G_D)$ is the gain asymmetry for the y direction, and A is the weighted
average of the analyzing power over the scintillator. The subscripts R,
L, U, and D refer to the quadrants right, left, up, and down, respectively. The rejection of transverse polarization is limited by the gain
stability of the PM's. The measured error for a fully transversely polarized beam is $\sim 10^{-5}$. The effective analyzing power is estimated by
measuring $(S_U - S_D)/S_D$ to be 10^{-3}, implying a gain asymmetry of $\sim 10^{-2}$.

The relative polarization misalignment is determined by comparing
the up-down and left-right asymmetries for the aligned beam with the up-
down asymmetry for a fully transversely polarized beam with a carbon
foil analyzer inserted just ahead of the detector. The effective analyzing power for the carbon foil in the parity detector is 0.17 which is
large enough that the asymmetry from residual transverse polarization
can be read directly on the PLA output meter using the rapid reversal.
Horizontal and vertical spin precessors are used to align the polariza-

tion vector by minimizing the up-down and left-right asymmetries, respectively. The relative polarization alignment was adjusted to < 0.5%
in each plane.

Any change in the beam current at the detector that is correlated
with the polarization reversal is referred to as current modulation.
Current modulation arises directly from changes in the source current
with reversal. Or, it may arise indirectly as beam losses from scraping
on apertures in the beam line vary with correlated motion (position modulation) of the beam.

Since, in principle, the normalization of the S signal removes any
dependence of A_L on the time structure of the beam current, the error
from current modulation depends on the accuracy of the normalization.
Three sources of error affecting the current modulation rejection are
considered: (a) the accuracy of the analog divider, (b) the matching of
the S and B channel amplifiers, and (c) the characteristics of the S and
B detectors. The error in A_L may be expressed

$$e(A_L) = \Delta P^{-1}(S_a + S_b + S_c)\Delta B/B_0 \ ,$$

where S_i, i = a, b, or c, is the sensitivity to current modulation from
a, b, or c, above, and $\Delta B/B_0$ is the fractional change in B with reversal.

Ideally, the analog divider output is given by y = -10z/x for x > 0,
where all quantities are expressed in volts. In practice, the accuracy
of the ratio depends on the offset voltages, nonlinearities, and frequency responses of each input. If the z input is S(t)-B(t) and the x
input is B(t), and if S(t) = kB(t), the sensitivity S_a from the divider
is estimated to be

$$S_a = (k-1)\left[\delta_x/B_0 - a_x B_0/x_0 + a_z B_0/z_0 + H_z(f_0) - H_x(f_0)\right] + \delta_z/B_0 \ ,$$

where B_0 is the average amplitude of the B signal, $\delta_x(\delta_z)$ is the offset
voltage of the x(z) input, $a_x(a_z)$ is the nonlinearity of the x(z) input
relative to the full-scale input $x_0(z_0)$, and $H_x(f_0)H_z(f_0)$ is the frequency response, normalized to unity at f = 0, of the x(z) input at the
reversal frequency f_0. If the relative gain k is set to ~ 1, the error
from the divider is dominated by the z offset term.

The combined sensitivity $S_{bc} = S_b + S_c$ from the S and B amplifiers
and detectors is

$$S_{bc} = kH_z(f_0)\left[H_S(f_0) - H_B(f_0)\right] + k\delta_S/B_0 - \delta_B/B_0 + ka_S - a_B \ ,$$

where the subscripts S(B) refer to the S(B) channel. The principle offset terms are δ_S from the PM leakage currents and δ_B from the B channel
current-to-voltage converters, which are 100 times more sensitive than
their S channel counterparts.

With carefully matched amplifiers one important error comes from
the S detector frequency response which has the form

$$H_S(f) = 1 + \Sigma_n \alpha_n (1 + i2\pi f\tau_n^S)^{-1} \ .$$

Here the α_n are the amplitudes and the τ_n^S are the time constants of the
components of the S signal due to phosphorescence of the scintillator or
sensitivity to radioactive decay products. Another important term in
the sensitivity to current modulation is the linearity of the PM's.

Changes in the beam position correlated with the reversal are referred to as position modulation. Position modulation arises in the
source because of the presence of charged particles (H⁻ ions) in the

94   *J.M. Potter*

polarization reversal region. Sensitivity to position modulation occurs
because of asymmetries of the S detector and variation of the solid angle of the S detector with beam position. For an infinite length model
of the S detector, the change in A_L with a reversal-correlated position
change δ_x, δ_y is given by

$$e(A_L) = (\Delta P\pi)^{-1}(G_x\delta_x/w + G_y\delta_y/w) \quad,$$

where w is the half-width of the scintillator array. Intuitively, the
error from angle modulation will be somewhat less than the error from a
position modulation $\delta x = \ell\delta\alpha$, where ℓ is the length of the scintillator.

Second-order terms arise from changes in the solid angle of the exit aperture for the S detector and of the entrance aperture of the B detector with beam position and angle. In a simple two-dimensional model
the error from position and angle modulation is

$$e(A_L) = \Delta P^{-1}\left[k_1(\delta x/w)(\overline{\delta x}/w) + k_2\left[(\delta x/w)\overline{\delta\alpha} + (\overline{\delta x}/w)\delta\alpha\right] + k_3\delta\alpha\overline{\delta\alpha}\right] \quad,$$

where $\delta x/w(\delta\alpha)$ is the position (angle) modulation and $\overline{\delta x}/w(\overline{\delta\alpha})$ is the
average position (angle) error. The coefficients k_1 are proportional to
w/ℓ and are of order unity for $w/\ell = 0.1$.

An important factor in the reduction of sensitivity to position
modulation is a variation of the rapid polarization reversal technique
known as alternating reversal. Because the polarization reversal depends on turning a magnetic field off and on and is independent of the
sign of the reversal field, linear error terms are reduced by two orders
of magnitude by alternating the sign of the reversal field from one period to the next. With alternating reversal the error from position and
angle modulation becomes

$$e(A_L) = 2\Delta P^{-1}\left[k_1(\delta x/w)^2 + k_2\delta\alpha(\delta x/w) + k_3(\delta\alpha)^2\right] \quad.$$

Much effort by J. L. McKibben has gone into the development of the
fast reversal technique to minimize the systematic errors originating in
the ion source. Table I lists the magnitude of the known sources of
systematic error.

With the polarization reversal technique used it is possible to reverse the phase of the polarization with respect to the fast reversal
reference signal, i.e., the system can be configured to measure either
A_L or $-A_L$. By combining data from both configurations, current modulation and ground loop errors can be made to cancel. The measured value
for A_L is $(-1.8 \pm 2.0) \times 10^{-7}$, where the error quoted is statistical.

The 6-GeV experiment was performed using the polarized external
proton beam at the Argonne National Laboratory Zero Gradient Synchrotron.
This experiment was a transmission experiment with a thick target (0.2 m
Be or 1 m water). A pair of plastic scintillators and a pair of ionization chambers were used to make two independent measurements of the incident (I) and transmitted (T) beam currents. The polarized beam arrived at the experimental area vertically polarized. The polarization
was made longitudinal by bending the beam down ~ 7.75°. The polarization was reversed every beam pulse at the ground state polarization
source. The longitudinal asymmetry was measured by correcting the
change in the ratio of transmitted to incident beam for the measured
magnitudes of the polarization and transmission.

While systematic errors were reduced by the normalization and symmetry of the detectors, additional scintillators were used to monitor

beam position upstream and downstream, beam size, and transverse polarization. Figure 4 is a pictorial representation of the experiment.

The signals from all of the scintillators integrated over each beam pulse were stored on magnetic tape for off-line analysis. The period between beam pulses was used to measure offset currents and to calibrate the gains of the PM's. Regression analysis was used to reduce the statistical noise in the measurement of transmission due to beam motion and spot size fluctuations. This analysis also served to reduce the sensitivity to motions of the beam correlated with the reversal, although no evidence for such correlations was observed.

A measurable transverse polarization component was noticed, however. This was corrected by using a calibration of the sensitivity of the transverse polarization measurement to deliberately introduce transverse polarization. This correction is a few parts in 10^6 and is determined to 1 part in 10^6. Table II lists the limits of systematic error for the most recent data.

The measured asymmetry at 6 GeV with the water target is $(15 \pm 2) \times 10^{-6}$. This nonzero result could result from a systematic error not yet considered, the parity-violating decay of polarized hyperons produced in the target. A Monte Carlo calculation estimates that the asymmetry due the hyperons, which has nothing to do with the parity-violating internucleon force, could be as large as 10^{-5}. An experiment has been proposed which will eliminate the hyperon effect and increase the sensitivity to 1 part in 10^7.

The ultimate sensitivity of the 15-MeV experiment is expected to be about 3×10^{-8}. It appears feasible to reduce systematic errors to $\sim 1 \times 10^{-8}$. Improving the statistical accuracy to the 3×10^{-8} level requires higher target gas pressure, a more intense beam, and longer running time for the experiment.

References

*Work supported by the U. S. Energy Research & Development Administration
†15-MeV experiment only
‡6-GeV experiment only

1) E. Fischbach and D. Tadic, Phys. Rep. 6 (1973) 123; M. Gari, Phys. Rep. 6 (1973) 317
2) V. M. Lobashov et al., Nucl. Phys. A197 (1972) 241
3) J. M. Potter et al., Phys. Rev. Lett. 33 (1974) 1307
4) J. D. Bowman et al., Phys. Rev. Lett. 34 (1975) 1184
5) M. Simonius, Phys. Lett. 41B (1972) 415
6) V. R. Brown et al., Phys. Rev. Lett. 30 (1973) 770
7) J. L. McKibben and J. M. Potter, this conference, p.863

TABLE I

15-MeV EXPERIMENT ERRORS

Error Source	Net Error
Position modulation	3.5×10^{-8}
Polarization misalignment	2.7×10^{-8}
Current modulation	6.1×10^{-8}
Miscellaneous	$< 4 \times 10^{-8}$

TABLE II

6-GeV EXPERIMENT

Beam Property	Error Limit (90% confidence)
Position	2.9×10^{-6}
Angle	1.3×10^{-6}
Size	0.5×10^{-6}
Intensity	1.1×10^{-6}

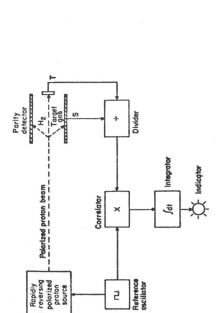

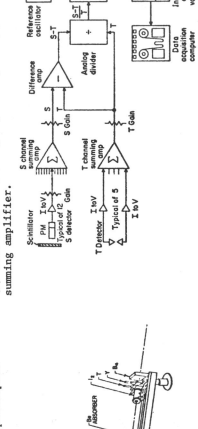

Fig. 1. Outline of experimental principles for 15-MeV experiment.

Fig. 2. System for p-p parity-conservation test: (a) steering plates, (b) steering amplifier, (c) position detector, (d) steering coil, (e) window, (f) aperture, (g) H_2 gas, (h) pressure vessel, (i) scintillator, (j) photomultiplier, (k) vacuum chamber, (l) beam stop and position detector, (m) side-detector summing amplifier, and (n) rear-detector summing amplifier.

Fig. 3. Signal-processing electronics for 15-MeV experiment.

Fig. 4. Apparatus for 6-GeV parity-conservation test. The beam position monitors are B_1–B_4; the scintillators are P, I, T, and Y; and ion chambers are I_1–I_3.

PARITY MIXING[†]

E.G. Adelberger, University of Washington, Seattle, WA 98195, USA

The field of parity mixing in light nuclei bears upon one of the exciting and active problems of physics -- the nature of the fundamental weak interaction. It is also a subject where polarization techniques play a very important role. I shall begin by reviewing enough weak interaction theory to motivate the parity mixing experiments. I shall then discuss two very attractive systems where the nuclear physics is so beautifully simple that the experimental observation of tiny effects directly measures parity violating (PV) nuclear matrix elements which are quite sensitive to the form of the basic weak interaction. Since the measurement of very small analyzing powers and polarizations may be of general interest to this conference, I will devote some discussion to experimental techniques.

The theory of the weak interaction is developing at a rapid pace. It is usually assumed that the weak Lagrangian consists of the symmetric product of two currents[1] --

$$\mathscr{L} \propto G[J_\lambda J^{\lambda^+} + J^\lambda J_\lambda^+ + Z_\lambda Z^{\lambda^+} + Z^\lambda Z_\lambda^+]$$

where J_λ is the conventional "charged" or Cabibbo current and Z_λ is the "neutral" current whose effects have recently been seen in high energy ν experiments. We know that the charged current consists of 4 pieces

$$J = J_e + J_\mu + \cos \theta_c J_0 + \sin \theta_c J_1.$$

The action of these currents are

J_e	$e^- \to \nu_e$	$\Delta Q = 1$		
J_μ	$\mu^- \to \nu_\mu$	$\Delta Q = 1$		
J_0	$n \to p$	$\Delta Q = 1$	$\Delta T = 1$	$\Delta S = 0$
J_1	$\Lambda \to p$	$\Delta Q = 1$	$\Delta T = 1/2$	$\Delta S = 1$

The two leptonic currents will not concern us in this discussion. The two hadronic currents J_0 and J_1 share the full strength of the weak interaction according to the Cabibbo angle $\theta \approx 15°$. In the Cabibbo theory the weak N-N force responsible for nuclear parity mixing is described by the "diagonal" products

$$\cos^2\theta_c[J_0{}_\lambda J_0^{\lambda^+} + J_0^\lambda J_0^+{}_\lambda] + \sin^2\theta_c[J_1{}_\lambda J_1^{\lambda^+} + J_1^\lambda J_1^+{}_\lambda].$$

The term involving the J_0 currents carries $\Delta T = 0,2$ since it is the symmetrical product of two isovectors; the term involving the J_1 currents carries $\Delta T = 1$ since it is the symmetrical product of two isospinors. Due to the smallness of $\sin^2\theta_c \approx 1/20$ the effects of the J_0 current dominate over those of the J_1 current. This dominance is somewhat reduced by the effects of the hard core in the strong N-N interaction since the $\cos^2\theta_c$ terms lead to exchanges of ρ and heavier mesons while the $\sin^2\theta_c$ terms have a one π exchange contribution[2]. Even though the weak N-N force has a strength $\sim 10^{-7}$ of the normal N-N force it can be studied because it has a unique signature -- viz. it violates parity. This occurs because each of the currents J_λ is a mixture of polar and axial vector currents $J_\lambda = V_\lambda - A_\lambda$.

† Work supported in part by U.S. Energy Research & Development Admin.

Very little is known about the neutral current Z. It is found that
Z carries $\Delta S = 0$ but not $\Delta S = 1$. However the Lorentz structure of the
current is not well determined. Is it V or A or some combination? Various theories of weak interactions propose quite different character for
Z. Some postulate a neutral current interaction which does not violate
parity and therefore would not affect nuclear parity mixing[3]). Other
theories, such as those of Weinberg[4]) and Salam[5] assume a neutral weak
current which is a V-A mixture. The V component mixes with the electromagnetic vector current and consequently has $\Delta T = 0,1$. The A component
is just the "rotated" charged current and has $\Delta T = 1$. The neutral currents in such theories yield a $\Delta T = 1$ PV N-N force which is not suppressed by the factor $\sin^2\theta_c$. Thus Weinberg-Salam theories enhance the $\Delta T = 1$
PV interaction with respect to the Cabibbo theory by roughly one order of
magnitude[6]).

A measurement of the strength of the $\Delta T = 1$ PV N-N force would
clearly be extremely interesting. The isospin structure of the PV force
is best studied in light nuclei. Experiments have studied PV in p+p[7]),
n+p[8]) and $^{16}O^{9}$) but none of these systems are sensitive to the $\Delta T = 1$
interaction. I shall discuss parity mixing in the ^{19}F and ^{18}F systems.
These cases are exceptionally interesting because the nuclear physics is
remarkably simple and the observable PV effects are sensitive to the
$\Delta T = 1$ interaction.

Let us first consider the case of ^{19}F which we have recently studied
in Seattle[10]). Because of the favorable energy splittings (see fig. 1)
the parity impurities in the ground ($1/2^-$) and 110 keV ($1/2^+$) states are
well approximated by simple two state mixing

$$|110\rangle = |-\rangle + \varepsilon|+\rangle \qquad \varepsilon = \frac{\langle -|H_{PV}|+\rangle}{110 \text{ keV}}$$

$$|g.s.\rangle = |+\rangle - \varepsilon|-\rangle$$

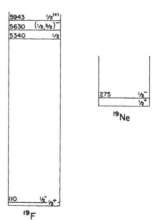

so that a measure of the parity impurity in
this case directly yields the PV matrix
element. In this case both the $\Delta T = 0$ and
$\Delta T = 1$ PV interactions are involved since the
levels have $T = 1/2$. To measure ε one needs
to observe a pseudoscalar quantity. We
chose to polarize the ^{19}F nuclei in the 110
keV state and measure the PV anisotropy of
the 110 keV γ rays $W(\theta) = 1 + \delta P_F\, \vec{\sigma}_F\cdot\hat{k}_\gamma$
where P_F is the net polarization of the
$^{19}F^*$'s. The relation between δ and ε is
found by considering the mixed M1/E1 γ-
transition between the 110 keV and ground
states

$$\langle g.s.|E1 + M1|110\rangle = \langle +|E1|-\rangle +$$
$$\varepsilon\{\langle +|M1|+\rangle - \langle -|M1|-\rangle\} + \theta(\varepsilon^2).$$

Fig. 1. Level diagram of
^{19}F showing only the low-
lying $J = 1/2$ states.

We get δ from the M1/E1 interference

$$\delta \quad 2\varepsilon\frac{\{\langle +|M1|+\rangle - \langle -|M1|-\rangle\}}{\langle +|E1|-\rangle}.$$

Since the E1 matrix element is determined by the radiative lifetime of
the 110 keV state and the M1 matrix elements are determined by the magnetic moments of the ground and 110 keV levels, a measurement of δ direct-
ly measures $\langle -|H_{PV}|+\rangle$ (Ref. 11). The experimental effect is magnified

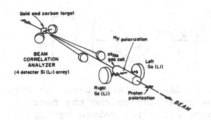

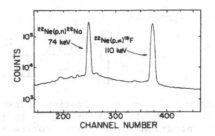

Fig. 2. Geometry of the ^{19}F experi-   Fig. 3. Gamma ray spectrum.
ment.

because the M1 matrix element is very strong (1.6 W.u.) while the E1 is
retarded ($\sim 10^{-3}$ W.u.).

The ^{19}F*'s were polarized by producing them in a reaction induced by
polarized protons. The experimental geometry is shown in fig. 2. The
reaction ^{22}Ne(p,α)^{19}F was chosen because its simple spin structure 0^+ +
$1/2^+ \to 0^+ + 1/2^-$ leads to a large polarization transfer even after one
integrates over all α angles. For example if the outgoing α's are s-
wave, the angle integrated polarization transfer coefficient K_{xx} = -1.
The ^{19}F* polarization is retained for the ~ 1 ns lifetime of the 110 keV
level by choosing the ^{22}Ne gas pressure to be low enough that the recoiling ^{19}F*'s radiate before they come to rest. The experimental geometry
is shown in fig. 2. The proton spin is switched rapidly every 0.31 sec
between left and right by reversing the axial B fields in a Lamb-shift
source operated according to the adiabatic field reduction scheme. Gamma
rays are detected in two 10 cm$^2 \times 7$ mm thin-window planar Ge(Li)'s and
digitized in two fast (7 μsec/event) pulse height analyzers. The outputs
of the ADC's are routed into different regions of memory according to
spin state. A gamma ray spectrum is shown in fig. 3.

In any experiment which measures a very small effect it is necessary
to demonstrate that systematic errors are under control. In our experiment systematic errors arise because the ion source introduces small modulations of the beam energy, intensity, angle and position which are correlated with the spin reversal. In addition the angle of the proton spin
fluctuates by 1°. Let me give some feeling for the magnitudes involved.
Suppose that instead of being clever we used a brute force approach. In
this case a position modulation of only $\pm 1.2 \times 10^{-5}$ cm would introduce a
fake asymmetry of $\delta = 10^{-5}$.

The sensitivity to many spurious modulations is greatly reduced by
design of the experiment. The most important points are the two symmetric γ-ray counters and the use of an isotropic γ-ray as a normalization which allows us to measure a relative anisotropy instead of an absolute one. This is done by choosing a bombarding energy so that in addition to the 110 keV radiation from ^{22}Ne(p,α) we also produce 74 keV radiation from ^{22}Ne(p,n). The 74 keV γ-rays come from the decay of a J = 0
level and hence are rigorously isotropic. Systematic errors are further
reduced by the following stratagem. The axial magnetic fields in the
ion source (which are switched between the + and - states) produce a
beam which is longitudinally polarized. The spin is rotated 90° by
crossed $\vec{E}$ and $\vec{B}$ fields so that it will be transversely polarized at the
target. We employ two configurations, a and b, of the crossed $\vec{E}$ and $\vec{B}$
fields. In configuration a the axial state + corresponds to spin right
at the target; in configuration b the axial state - corresponds to spin
right. We always acquire data in both configurations. We obtain δ from

the expression

$$\bar{K}\delta = \frac{1}{P_p} \left[\left(\frac{R^+_{110}L^-_{110}}{R^+_{74}L^-_{74}} \cdot \frac{L^+_{74}R^-_{74}}{L^+_{110}R^-_{110}} \right)_a \cdot \left(\frac{R^+_{110}L^-_{110}}{R^+_{74}L^-_{74}} \cdot \frac{L^+_{74}R^-_{74}}{L^+_{110}R^-_{110}} \right)_b^{-1} - 1 \right]$$

where $\bar{K}$ is the polarization transfer coefficient averaged over the lifetime of the 110 keV level, P_p is the beam polarization and the rest of
the notation is obvious. Notice that there are 4 two-valued variables
in the experiment; counter, gamma ray, axial field state, and $\vec{E} \times \vec{B}$ configuration. The true PV effect is a 4-fold correlation of these variables. However essentially all spurious effects are 3-fold or lower
order correlations and cancel exactly in our expression for δ. For example, dead time effects due to beam intensity modulation are not a function of the γ-ray variable and hence are removed in our ratio. From 5
independent measurements we obtain $\bar{K}\delta =(13.1\pm5.8)\times10^{-5}$. We studied the
possible systematic errors in the asymmetry by deliberately inducing
modulations of the beam energy,intensity, position and angle under circumstances where there could not be a PV effect (e.g., with an unpolarized beam) and measured upper limits on any possible "fake" PV anisotropies caused by these modulations of $\bar{K}\delta_{FAKE}$ < 2.2×10^{-5}.
We were able to do this because a beam correlation analyzer (shown
in fig. 1) continuously monitored the modulation of the beam angle,
position and vertical component of the spin. This device may have other
applications in precision polarized beam experiments. It consists of a
thin Au + C target viewed by two pairs of solid state detectors -- one
pair at ±3.9° and one pair at ±55°. At 3.9° the scattering from both Au
and C is Rutherford and varies extremely rapidly with angle. At 55° the
scattering from Au is still Rutherford while that from C is nuclear and
has a big analyzing power. The 55° cross sections vary slowly with angle
and the counters are placed very close to the target so that the counting
rate in these detectors is primarily a function of the horizontal displacement of the beam and the vertical component of the proton spin.
With this device we were able to monitor angle and position modulations
to an accuracy of ∿10^{-5} radians and ∿10^{-4} cm respectively. The angle of
the proton spin could be measured to an accuracy of 0.16°.
Combining our results we obtain δ = -(18±9)×10^{-5}. Recently Gari et
We determined $\bar{K}$ the net polarization transfer to the ^{19}F*'s at time
of γ-ray emission by measuring the circular polarization (CP) of the 110
keV radiation. For a J = 1/2 → J = 1/2 transition the CP of the photon
is numerical equal to the linear polarization of the initial state along
the photon propagation vector. The CP was measured using a transmission
Compton polarimeter and yields $\bar{K}$ = -0.73±0.15.
Combining our results we obtain δ = -(18±9)×10^{-5}. Recently Gari et
al.[12]) and Box and McKellar[13]) have calculated δ using correlated nuclear
wavefunctions and 2 body PV N-N forces. Although the two calculations
use different ^{19}F wavefunctions the predicted δ's are very similar.
Using the Cabibbo theory Gari et al. find the ΔT = 0 interaction gives
δ = ±(6-10)×10^{-5} while the ΔT = 1 interaction gives δ = ±1.4×10^{-5} (the
range of values for the ΔT = 0 interaction is due to uncertainties in the
short range correlations). Since δ is an interference term linear in the
PV amplitude the ΔT = 0 and ΔT = 1 contributions add algebraically δ_{TOT} =
$\delta_{\Delta T=0} + \delta_{\Delta T=1}$. The sign ambiguity in the theoretical δ's arises because one is not confident of the sign of the retarded (10^{-3} W.u.) E1
matrix element. With a Weinberg-Salam model of the weak interaction the
ΔT = 0 and ΔT = 1 contributions to δ are ±(8-13)×10^{-5} and 22×10^{-5}

respectively. Because of the large uncertainty, the
experimental result is consistent with theories
based on charge currents only as well as those which
include neutral currents. However there is no large
discrepancy here as one finds in the n+p system
where the observed effect[8]) is 100 times greater
than the theory[2]).

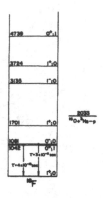

The ^{19}F system is not an ideal one to detect an
enhancement of the $\Delta T = 1$ PV interaction because the
$\Delta T = 0$ interaction is also allowed. However systems
where only the $\Delta T = 1$ interaction is involved, the
nuclear physics is simple, and the effects are expected to be big enough to be measurable are hard to
find. As first pointed out by Henley[14]) an extremely favorable case occurs in the ^{18}F system
(see fig. 4). Here a nearly degenerate $J = 0$
doublet occurs: the 1081 keV 0^-;T = 0 and the
1042 keV 0^+;T = 1 levels. The parity impurities
in these states are again well approximated by
simple two-state mixing

Fig. 4. Level dia-
gram of ^{18}F showing
only the low-lying
$J = 0$ and $J = 1$
states.

$$|1081\rangle = |-\rangle + \epsilon|+\rangle$$
$$|1042\rangle = |+\rangle - \epsilon|-\rangle \qquad \epsilon = \frac{\langle -|H_{PV}|+\rangle}{39 \text{ keV}}.$$

In this system the PV interaction is purely $\Delta T = 1$. In the case of a
$J = 0$ level one can't measure ϵ by detecting a $\vec{\sigma}_F \cdot \hat{k}_\gamma$ anisotropy in the
angular correlation. Instead one must observe the circular polarization
(CP) $\vec{\sigma}_\gamma \cdot \hat{k}_\gamma$ (the M1/E1 interference) of the gamma ray decays to the 1^+
T = 0 ground state. The CP is given by

$$CP_{1081} = 2\epsilon \frac{\langle g.s.|M1|+\rangle}{\langle g.s.|E1|-\rangle}$$

$$CP_{1042} = 2\epsilon \frac{\langle g.s.|E1|-\rangle}{\langle g.s.|M1|+\rangle}$$

where the M1 and E1 matrix elements are determined by the lifetimes of
the 1042 and 1081 keV states respectively. Now a very lovely thing occurs because of isospin selection rules on the γ-rays. The $\Delta T = 0$ E1 is
isospin forbidden, while the $\Delta T = 1$ M1 is isospin favored. The effect of
this is to cause $\tau_{1081}/\tau_{1042} \approx 10^4$. As a result $CP_{1081} \approx 200\epsilon$ while
$CP_{1042} \approx 0.02\epsilon$ so that CP_{1081} is electromagnetically enhanced by a factor
of 100. This enhancement coupled with the extraordinarily small energy
splitting of only 39 keV means that it should be possible to detect
CP_{1081}. M. Gari, J.B. McGrory and R. Offerman[15]) have calculated the
CP_{1081} expected from the Cabibbo and Weinberg-Salam models and find
3.6×10^{-4} and 5.7×10^{-3} respectively. As expected the PV neutral currents
of the Weinberg-Salam type theories give a very obvious signature.

An experiment to measure CP_{1081} with a design sensitivity of 2×10^{-3}
is currently being performed by a Seattle, Caltech, CSLA collaboration[16]).
The experimental geometry is shown in fig. 5. The ^{18}F activity is produced using the $^{16}O(^3He,p)$ reaction induced in a H_2O target by a 20 μA
beam of 3.8 MeV ^3He from the CSLA accelerator. The reaction and energy
were chosen to give good production of 1081 and 1042 keV γ rays and because essentially no neutrons are produced. Since the lifetime of the
γ-ray counters is limited by neutron damage this is an important consideration. The CP is measured with two transmission Compton

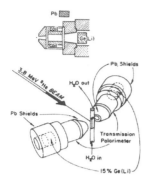

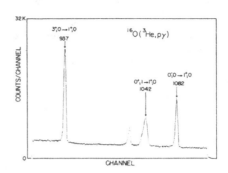

Fig. 5. Geometry of the ^{18}F   Fig. 6. Gamma ray spectrum.
experiment.

polarimeters and Ge(Li) detectors, using pulse counting techniques in
order to preserve the energy resolution. The magnetization of the polarimeters is rapidly reversed once a second. The two polarimeters are always magnetized such that if $\vec{B}$ in one magnet points toward the target,
$\vec{B}$ in the other points away. In this way true circular polarization
dependent effects are distinguished from spurious effects associated with
beam energy and position changes correlated with magnetization state. A
spectrum of γ-rays observed in a 15% Ge(Li) detector following transmission through the 8.0 cm long polarimeter magnet is shown in fig. 6.
The 937 keV gamma ray is an E2 transition from the 937 keV 3^+ state of
^{18}F. The 1042 keV gamma ray is doppler broadened. The quality of the
spectrum (due to careful γ-ray collimation) is impressive considering
that the magnet attenuates 1081 keV γ-rays by a factor of 38. The output
of each Ge(Li) (denoted by ℓ or r) is digitized in a fast ADC and routed
into different portions of memory according to the magnetization state
(denoted by + or -) of the polarimeters. The experimental design is
similar to that employed on the ^{19}F measurement. Again we have 2 detectors, 2 gamma rays and we switch repetitively between two states. We
avoid essentially all systematic errors be measuring relative CP's instead of absolute CP's, viz.,

$$
CP_{1081} \propto \frac{\left.\dfrac{\ell_{1081}}{\ell_{937}}\right| + \left.\dfrac{r_{1081}}{r_{937}}\right| -}{\left.\dfrac{\ell_{1081}}{\ell_{937}}\right| - \left.\dfrac{r_{1081}}{r_{937}}\right| +}
\qquad
CP_{1042} \propto \frac{\left.\dfrac{\ell_{1081}}{\ell_{937}}\right| + \left.\dfrac{r_{1042}}{r_{937}}\right| -}{\left.\dfrac{\ell_{1042}}{\ell_{937}}\right| - \left.\dfrac{r_{1042}}{r_{937}}\right| +}
$$

Problems were encountered in designing a target that could withstand the
20 μA beam of 3.8 MeV ^{3}He required to obtain a high enough counting rate
[∿50,000 sec^{-1} total rate in each Ge(Li)] for a practical experiment.
The present arrangement consists of a very uniform 3.2 mm φ beam incident
upon a 7.5×10^{-4} mm thick Ni foil which is backed by rapidly flowing H_2O.
The H_2O flow cools the foil, prevents ruptures and carries away the β^+
activity which otherwise would overload the Ge(Li) counters. The experiment is still in the early stages and serious data taking has just begun.
     Perhaps you are wondering why we are trying to measure the CP of the
^{18}F radiation (which is difficult because the analyzing efficiency of our

polarimeters is only ~1.5%) when we could be doing a Los Alamos style measurement of the helicity dependence of the total cross section[7]). For example we could measure the longitudinal analyzing power of the total cross sections leading to the 1042 and 1081 keV states in the $^{18}O(p,n)^{18}F$ reaction induced by longitudinally polarized protons. It would be easy to measure the total cross section with the isotropic yield of the 1042 and 1081 keV radiation. We don't try this or any other technique involving PV analyzing powers of nuclear reactions such as that suggested by Berovic[17]) because there is no reliable way to extract a PV matrix element from the PV analyzing power. It is easy to see why this occurs. Using the notation of Berovic[18]) with the z-axis along the beam the amplitudes for the $^{18}O(p,n)$ reaction leading to unmixed 0^{+} and 0^{-} states can be written

$$\begin{pmatrix} a & b \\ -b & a \end{pmatrix} \qquad \text{and} \qquad \begin{pmatrix} a' & b' \\ b' & -a' \end{pmatrix} \qquad \text{respectively.}$$

After we mix the states the longitudinal analyzing power is

$$A_L^{1081} = \frac{\sigma_R - \sigma_L}{\sigma_R + \sigma_L}\Bigg|_{1081} = \frac{2\epsilon}{\sigma_{1081}} \, \mathrm{Re}[a'a - b'b]$$

$$A_L^{1042} = - \frac{2\epsilon}{\sigma_{1042}} \, \mathrm{Re}[a'a - b'b].$$

The difficulty comes in determining the relative phases between the amplitudes of the reaction leading to the 1042 keV level and that leading to the 1081 keV level. Measurements of the normal parity allowed analyzing powers can yield the phases between a and b, but cannot directly give the relative phase of a and a'. Studies of parity violation using reaction analyzing powers can provide quantitative information only in those very few cases (such as p+p) where the reaction is well enough understood that the phases are known.

I have discussed only a few of the extremely interesting problems in weak interaction physics that one can attack using polarization techniques. Other topics such as the study of induced terms in nuclear β decay were not even mentioned. It is rare to find a field with so many good problems and so few practitioners. Hopefully conferences such as this will persuade some polarization experts to switch from DWBA to V-A!

References

1) For a good account of "Cabibbo" weak interaction theory (before the discovery of neutral weak currents, etc.) see E.D. Commins, in Weak Interactions (McGraw-Hill, New York, 1973)
2) M. Gari, Phys. Rep. 6C (1973) 319
3) For example see H. Fritzsch and P. Minkowski, Phys. Lett. to be published
4) S. Weinberg, Phys. Rev. Letts. 19 (1967) 1264; 27 (1971) 1688
5) A. Salam, in Elementary Particle Theory (Nobel Symposium No. 8)

ed. N. Svartholm (John Wiley, New York, 1969)
6)   M. Gari and J. Reid, Phys. Lett. 53B (1974) 237
7)   J.M. Potter et al., Phys. Rev. Lett. 33 (1974) 1307
8)   V.M. Lobashov et al., Nucl. Phys. A197 (1972) 241
9)   K. Neubeck, H. Schober and H. Wäffler, Phys. Rev. C 10 (1974) 320
10) E.G. Adelberger, H.E. Swanson, M.D. Cooper, J.W. Tape, and T.A.
     Trainor, Phys. Rev. Lett. 34 (1975) 402
11) E. Maqueda, Phys. Lett. 23 (1960) 571
12) M. Gari, A.H. Huffman, J.B. McGrory, and R. Offerman, Phys. Rev.
     C 11 (1975) 1485;
     M. Gari, Ruhr-universität Bochum preprint RUB TP II/115 (1975)
13) M.A. Box and B.H.J. McKellar, Phys. Rev. C 11 (1975) 1859
14) E.M. Henley, Phys. Lett. 28B (1968) 1
15) M. Gari, J.B. McGrory and R. Offerman, Phys. Lett. 55B (1975) 277
16) E.G. Adelberger, C.A. Barnes, M. Lowry, R.E. Marrs, F.B. Morinigo,
     and H. Winkler, private communication, 1975
17) N. Berovic, Phys. Lett. 35B (1971) 475
18) N. Berovic, Nucl. Phys. A157 (1970) 106

POLARIZATION SYMMETRIES IN DIRECT REACTIONS:
^{3}He$(\vec{t},d)$ ^{4}He AND ^{2}H$(\vec{d},p)$ ^{3}H

H. E. Conzett

Lawrence Berkeley Laboratory, University of California
Berkeley, California, 94720

The observable consequences of particle symmetry or charge symmetry in reactions of the type $b(a,c)c'$ are well known. Here, the final-state particles c and c' are the charge-symmetric members of an isospin doublet or are identical particles, $c = c'$. The symmetries imposed on the angular distributions of the cross section and the analyzing powers have been given by Barshay and Temmer[1] and Simonius[2]:

$$\sigma(\theta) = \sigma(\pi - \theta) \tag{1A}$$

for a state of definite isospin.

$$T_{kq}(\theta) = (-1)^q \, T_{kq}(\pi - \theta) \tag{1B}$$

These symmetries are exact for $c = c'$, but significant deviations from (1) have been observed in the ^{4}He$(d,t)^3$He reaction (or its inverse) at deuteron energies ranging from threshold (21.5 MeV) to 41 MeV. Nocken et al[3] have concluded that the lower energy cross-section data from the inverse reaction[4] are explained in terms of the compound-nucleus reaction mechanism, and they suggest that the deviations from symmetry about $\theta = \pi/2$ result from isospin mixing in the ^{6}Li intermediate nucleus. At the higher energies E_d = 32 and 41 MeV, deviations from the symmetries (1) in both the cross-section[5] and in the deuteron vector analyzing power[6] have, in large measure, been explained via DWBA calculations as resulting from the slightly different proton and neutron transfer amplitudes[5,7,8]. This explanation in terms of expected Coulomb effects thus retains the concept of basic charge-symmetry at these higher energies.

There is no symmetry condition such as (1B) imposed on the polarizations, $t_{kq}(\theta)$, of the outgoing particle in the $b(a,\vec{c})c'$ reaction; so, it follows that in a reaction $a'(\vec{a},c)d$, where the identical particles (a'=a) or isospin partners are in the initial state, the conditon (1B) on the analyzing powers does, in general, not apply. However, in this paper it is shown that if the reaction mechanism is a purely direct transfer process, condition (1B) is imposed on the analyzing powers in the $a'(\vec{a},c)d$ reaction. This symmetry then becomes, in this class of reactions, a clear signature of the direct transfer mechanism as contrasted with the compound-nucleus, or intermediate state, process.

Figure 1 shows the two direct transfer amplitudes which are added coherently; N_1 and N_2 are the transferred nucleons or nucleon clusters. As examples, in the ^{2}H$(\vec{d},p)^3$H reaction $N_1 = N_2 = n$, and in the ^{3}He$(\vec{t},d)^4$He reaction $N_1 = n$ and $N_2 = p$. As is noted in fig. 1, the M-matrix amplitude $M^{(2)}_{m_d m_c m_a m_{a'}}(\pi-\theta)$ for the transfer of N_2, producing particle d at the angle $\pi - \theta$, is described in the coordinate system with the y-axis along $\vec{k}_i \times (-\vec{k}_f)$. Rotation of the coordinate system through an angle π around the z-axis, so that the y-axis is along $\vec{k}_i \times \vec{k}_f$, gives[9] $(-1)^m M^{(2)}_{m_d m_c m_a m_{a'}}(\pi-\theta)$ for the N_2-transfer amplitude. Here $m = m_a + m_{a'} - m_c - m_d$ and m_a, $m_{a'}$, m_c, m_d are the spin magnetic substates of particles a,a',c, and d. The complete M-matrix amplitude is then

$$M_{m_c m_d m_a m_{a'}}(\theta) = M^{(1)}_{m_c m_d m_a m_{a'}}(\theta) + (-1)^m M^{(2)}_{m_d m_c m_a m_{a'}}(\pi-\theta)$$

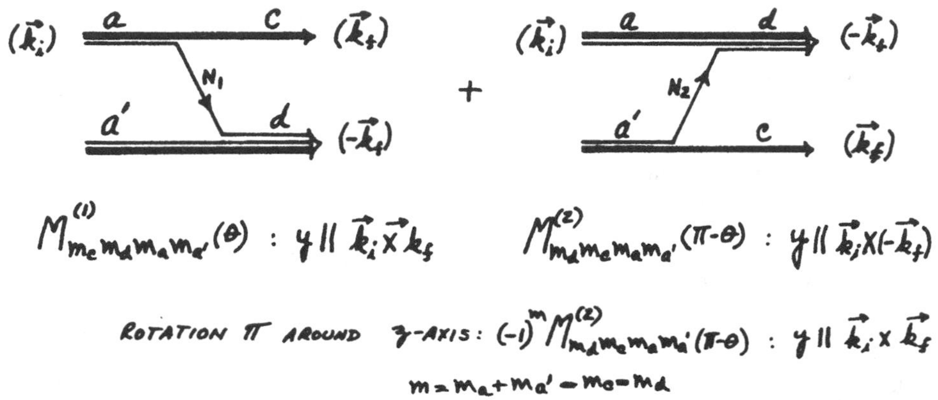

$$M^{(1)}_{m_c m_d m_a m_{a'}}(\theta) : y \parallel \vec{k}_i \times \vec{k}_f \qquad M^{(2)}_{m_d m_c m_a m_{a'}}(\pi-\theta) : y \parallel \vec{k}_i \times (-\vec{k}_f)$$

$$\text{ROTATION } \pi \text{ AROUND } z\text{-AXIS}: (-1)^m M^{(2)}_{m_d m_c m_a m_{a'}}(\pi-\theta) : y \parallel \vec{k}_i \times \vec{k}_f$$

$$m = m_a + m_{a'} - m_c - m_d$$

Fig. 1

In terms of two indices $\nu = m_a + m_{a'}$ and $\mu = m_c + m_d$ it becomes

$$M_{\mu\nu}(\theta) = M^{(1)}_{\mu\nu}(\theta) + (-1)^{\nu-\mu} M^{(2)}_{\mu\nu}(\pi-\theta). \tag{2}$$

In terms of the transition matrix M, the analyzing powers are given by the expression

$$X_{kq}(\theta) \equiv \sigma(\theta) T_{kq}(\theta) = \mathrm{Tr} M(\theta)\, \tau_{kq}\, M^{\dagger}(\theta), \tag{3}$$

where the τ_{kq} are spherical tensor operators[9,10] with elements

$$[\tau_{kq}]_{\nu\nu'} = (2S_a + 1)^{1/2} (-1)^{S_a-\nu'} <s_a \nu\, s_a - \nu'|kq>.$$

The vector addition coefficient provides that

$$[\tau_{kq}]_{\nu\nu'} \neq 0 \text{ if } q = \nu - \nu' \tag{4}$$

In terms of the elements of M, T_{kq}, and $M^{\dagger}$ (3) becomes

$$X_{kq}(\theta) = \sum_{\mu\nu\nu'} M_{\mu\nu}(\theta)\, [\tau'_{kq}]_{\nu\nu'}\, M^{*}_{\mu\nu'}(\theta), \tag{5}$$

where ν and μ span the $(2S_a + 1)(2S_{a'} + 1)$ and $(2S_c + 1)(2S_d + 1)$ dimensions of the M matrix, and $\tau'_{kq} = \tau_{kq} \times 1$ is the direct product of τ_{kq} and the unit matrix of dimension $(2S_{a'} + 1)$. Using (4), eq. (5) becomes

$$X_{kq}(\theta) = \sum_{\mu\nu} M_{\mu\nu}(\theta)\, [\tau'_{kq}]_{\nu,\nu-q}\, M^{*}_{\mu,\nu-q}(\theta), \tag{6}$$

and with eq. (2)

$$X_{kq}(\theta) = \sum_{\mu\nu} [M_{\mu\nu}^{(1)}(\theta) + (-1)^{\nu-\mu} M_{\mu\nu}^{(2)}(\pi-\theta)] [\tau'_{kq}]_{\nu,\nu-q}$$

$$[M_{\mu,\nu-q}^{(1)*}(\theta) + (-1)^{\nu-\mu-q} M_{\mu,\nu-q}^{(2)*}(\pi-\theta)]. \tag{7}$$

The particle identity or charge symmetry provides that

$$M_{m_c m_d m_a m_{a'}}^{(1)}(\theta) = M_{m_d m_c m_a m_{a'}}^{(2)}(\theta), \text{ or}$$

$$M_{\mu\nu}^{(1)}(\theta) = M_{\mu\nu}^{(2)}(\theta) \quad \text{for } a' = a. \tag{8}$$

Here and in the following, wherever an equality is written for a' = a it is an <u>approximate</u> equality for a' $\neq$ a. Using (8) in eq. (7),

$$X_{kq}(\theta) = \sum_{\mu\nu} [M_{\mu\nu}^{(2)}(\theta) + (-1)^{\nu-\mu} M_{\mu\nu}^{(1)}(\pi-\theta)] [\tau'_{kq}]_{\nu,\nu-q}$$

$$[M_{\mu,\nu-q}^{(2)*}(\theta) + (-1)^{\nu,\mu-q} M_{\mu,\nu-q}^{(1)*}(\pi-\theta)]$$

so that

$$X_{kq}(\theta) = (-1)^q X_{kq}(\pi-\theta) \quad \text{for } a' = a. \tag{9}$$

It follows ($T_{00} = 1$) that

$$T_{kq}(\theta) = (-1)^q T_{kq}(\pi-\theta) \quad \text{for } a' = a. \tag{10}$$

The symmetry (10) thus provides clear indication of a purely direct transfer process. Of course, if the reaction should proceed through a single J^π state or states of the same parity in the compound nucleus, this symmetry condition could also result. However, an examination of the energy dependence of $T_{kq}(\theta)$ should resolve any ambiguity. Also, the excitation energies of the intermediate nuclei which are accessible in this class of reactions with polarized beams would, almost certainly, be in regions of overlapping levels.

As a clear example of these considerations, fig. 2 shows angular distributions of the vector analyzing power $A_y(\theta) = (2/\sqrt{3})iT_{11}(\theta)$ in the $^2H(\vec{d},p)^3H$ reaction at energies from 11.5 to 30 MeV. The 11.5 MeV data are from Zürich[11], and the other data are from Berkeley[12]. The corresponding range of excitation energies in ^4He is 29.6 to 38.8 MeV, and one sees the obvious transition from a complete lack of symmetry in the angular distribution to one of near antisymmetry with respect to $\theta = \pi/2$, as given by eq. (10). The Zürich data range down to ^4He excitation energies of 24.2 MeV, and all the analyzing power components $T_{kq}(\theta)$ are consistent in that no such symmetry exists. Hence, analysis of these data in terms of ^4He intermediate states is required at the lower deuteron energies, whereas analysis in terms of the direct nucleon-transfer process is appropriate above 30 MeV. It should be noted from the data of fig 2. that evidence persists for some contribution from

the compound-nucleus process at energies above that of the highest
suggested excited state of ⁴He, ⁴He* (32 MeV) corresponding to E_d =
16.5 MeV[13].

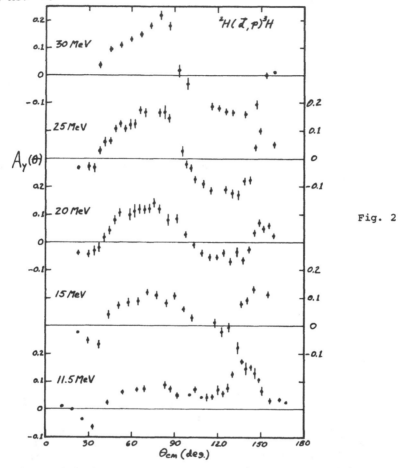

Fig. 2

Another case of interest is the ${}^3\text{He}(\vec{t},d){}^4\text{He}$ (or ${}^3\text{H}({}^3\vec{\text{He}},d){}^4\text{He}$)
reaction. As was noted before, it has been reported[3] that the lower
energy cross section data suggest isospin mixing in the compound nucleus
${}^6\text{Li}$. Support for this conclusion could be provided by measurements at
those energies of the analyzing power $iT_{11}(\theta)$ in the ${}^3\text{He}(\vec{t},d){}^4\text{He}$
reaction. The lack of symmetry of the form (10) would be confirming
evidence for the compound-nucleus reaction mechanism. With respect to
the higher energies, fig. 3 shows a DWBA calculation of the analyzing
power $A_y(\theta)$ in the ${}^3\text{He}(\vec{t},d){}^4\text{He}$ reaction at an energy equivalent to E_d =
32 MeV   in the inverse reaction. The parameter values were those used
by Dahme, Buttle et al[8] in their fits to $\sigma(\theta)$ and $iT_{11}(\theta)$ in that
inverse reaction ${}^4\text{He}(\vec{d},t){}^3\text{He}$. The near antisymmetry of $A_y(\theta)$ about
$\theta = \pi/2$ is distinct.

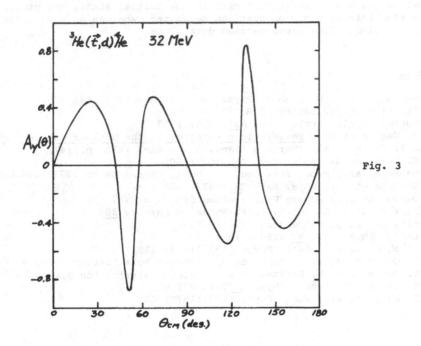

Fig. 3

Considering the presently available polarized beams, we list in Table 1 examples of reactions in which measurements of $T_{kq}(\theta)$ may provide unambiguous information on the reaction mechanism.

TABLE 1

$^2H(\vec{d},p)^3H$ $^6Li(^6\vec{Li},p)^{11}B$

$^3He(\vec{t},d)^4He$ $^6Li(^6\vec{Li},d)^{10}B$

$^3He(\vec{t},p)^5He$ $^6Li(^6\vec{Li},^3He)^9Be$

$^3He(^3\vec{He},p)^5Li$ $^6Li(^6\vec{Li},\alpha)^8Be$

                             $^6Li(^6\vec{Li},^7Li)^5Li$

                             $^6Li(^6\vec{Li},^7Be)^5He$

The $^6\vec{Li}$ + ^6Li reactions, for example, could be particularly interesting, since deviations from the symmetry (10) would provide qualitative information about states in ^{12}C which have some overlap with the appropriate cluster configurations.

In summary, we find that the analyzing powers, in reactions with identical or charge-symmetric particles in the initial state, can provide definite identification of the reaction mechanism, whereas no such information is available from cross-section data alone.

References

1) S. Barshay and G. M. Temmer, Phys. Rev. Letters 12 (1964) 728.
2) M. Simonius, Phys. Letters 37B (1971) 446.
3) U. Nocken et al., Nucl. Phys. A213 (1973) 97.
4) G. J. Wagner et al., Few Particle Problems in the Nuclear Interaction, eds. I. Slaus et al. (North Holland, Amsterdam, 1972) p. 747.
5) E. E. Gross et al, Phys. Rev. C5 (1972) 602.
6) W. Dahme et al., Proc. Int. Conf. on Nucl. Phys.(Munich 1973)Vol.1,444.
7) S. Edwards et al., Phys. Rev. C8 (1973) 456.
8) W. Dahme et al., Fourth Polarization Symp., p. 497
9) See , e.g. , M. Simonius, Lect. Notes in Physics 30 (Springer-Verlag, Heidelberg 1974) p. 38.
10) W. Lakin, Phys. Rev. 98 (1955) 139.
11) W. Grüebler et al., Nucl. Phys. A193 (1972) 129
12) H. E. Conzett, J. S. C. McKee et al., Fourth Polarization Symp, p.526
    H. E. Conzett, R. M. Larimer et al., Fourth Polarization Symp., p.524
13) E. K. Lin et al., Nucl. Phys. A179 (1972) 65.
    E. L. Haase et al., Nucl. Phys. A188 (1972) 89.

PROGRESS IN POLARIZED ION SOURCE DEVELOPMENT

Thomas B. Clegg
Department of Physics, University of North Carolina
Chapel Hill, North Carolina 27514 USA
Triangle Universities Nuclear Laboratory, Durham, North Carolina

I. INTRODUCTION

In the fifteen years since the first polarized ion source produced a useable polarized beam,[1] these devices have become important research instruments at nuclear physics accelerator installations. The number of polarized sources has grown steadily and their development has been reviewed periodically. At the Karlsruhe conference[2] this was done by Fleischmann[3] and at the Madison conference[4] by Glavish[5] and Donnally.[6] These and other reviews in the intervening period[7,8,10-12,14,15] have outlined the theory and many basic problems of operation for the two main types of polarized ion source, the "atomic beam source" and the "Lamb-shift source." The underlying physics principles for these two types of sources have not changed much in the last five years, but the hardware is slowly being refined to increase the output beam intensities, to increase the variety of polarized ion species available, to facilitate rapid changes in the beam polarization without altering the focal properties of the beam, and to adapt these ion sources to a new family of high-energy accelerators.

It will be the goal here to catalog the ion sources now being used and under development in enough detail that their relative performances can be compared. Then significant advances of the last five years will be summarized by citing examples, and possible areas for future improvements and research will be indicated.

II. PERFORMANCE OF POLARIZED SOURCES IN USE AND SOURCES UNDER DEVELOPMENT

Much of my information about polarized ion sources has been accumulated by contacting many people within the last three months. I appreciate their help and apologize if I have ommitted or mispresented any ion sources in my resulting tabulation. Tables 1 and 2 list the operational atomic beam sources and Lamb-shift sources respectively. In each table the sources are listed roughly in chronological order of their initial installation on an accelerator. Listed separately in Table 3 are the polarized sources being planned or under construction.

Both the total number of operational polarized sources and the beam intensities from the best d.c. ion sources are approximately three times greater than those reported in Madison. Atomic beam sources are more numerous and produce much more intense H^+ and D^+ beams than Lamb-shift sources. Lamb-shift sources are, however, the most intense for H^- and D^- beams. Of the atomic beam sources, the most intense sources for both positive and negative ions were purchased commercially.[61] One commercial Lamb-shift source is also operating.[62] Of the Lamb-shift sources, the spin-filter polarization scheme[63] offers a wider choice and a larger magnitude of beam polarization for deuterons than the diabatic-field-reversal scheme.[64]

Accelerated polarized proton beam energies now span completely the range 0-75 MeV. The 12 GeV polarized proton beam available at the Argonne ZGS is not likely to be surpassed in energy until a polarized beam is available at CERN. The CERN polarized beam design study is underway with a goal of building a pulsed atomic beam source with peak output beam intensity of 100-200 μA.[15]

Table 1. Operational Atomic Beam Sources

Laboratory and Installation Date	Particle	R.F. Transitions Used*	Beam Polarization** theor.	expt.	Typical Polarized Beam Current From Source (nA)	Polarized Beam Currents From Accelerator (nA unless otherwise shown)	Accelerator Type	Accelerated Beam Energy (MeV)	Polarized Beam Usage %
1. Carnegie[35] ~1962	H+	None used	1/2	~0.35	15-30	7-10	VDG	0.35-3.0	50
	D+		1/3, 1/3						
2. Saclay[19] ~1964	H+	WF, 2→4	1	0.75	2×10^3	50	cyclotron	10-28	not used since 2/75
3. Mainz[102] 1965	H+	WF	1/2	0.42	200	20	Cockroft-Walton	1.4	50
	D+	WF	1/3, 1/3						
4. Birmingham[23] ~1965	D+	WF, 3→6, 2→6, 3→5	2/3, 1	0.58, 0.90	1.0×10^3	100	cyclotron	12.3	25
5. Grenoble[22] ~1969	H+	WF	1	0.8	1.5×10^3	40-60	cyclotron	15-60	15-20
	D+	WF, 2→5, 3→5, 2→6, WF	2/3, 1	~0.58, 0.85	2.0×10^3	50-70		15-40	
6. Berkeley[18] 1969	H+	WF	1	0.8	$3 \times 10^3 - 5 \times 10^3$	200-300	cyclotron	9-55	12-15
	D+	WF, 3→5, 2→6	2/3, 1	0.54, 0.8	$3 \times 10^3 - 5 \times 10^3$	200-300		12-65	
7. Canberra[10] 1970	H-	WF, 2→4	1		80	5	EN-tandem VDG	2-12	< 5
	D-	WF, 3→5, 2→6	2/3, 1						
8. Oak Ridge[27] 1970	H+	WF, 2→4	1	~0.55	200-1000	20-60	cyclotron	20-65	not used since 4/74
	D+	WF, 2→6, 3→6, 2→5, 3→5	2/3, 1	0.35, 0.55	200-1000	20-60		12-45	
9. Stanford[31] 1970	H-	WF, 2→4	2/3, 0	0.67	200-300	40-100	FN-tandem VDG	2-18	~40
	D-	WF		0.47, 0.77					
10. Kyushu[29] 1971	D+	3→5, 2→6	1/3, 1	0.24, 0.77	300	100	Cockroft-Walton	0.05-0.15	100
11. Rutgers[30] 1971	H-	WF, 2→4	1	0.60	260	15-25	FN-tandem VDG	2-17	25
	D-	WF, 3→5, 2→6	2/3, 1	0.47, 0.70	300	15-25			
12. Ohio State[33] 1972	H+	WF	1	0.65	150-200	>60	CN-VDG	1-7	~30
	D+	3→5, WF	- , 1	- , 0.80	150-200	>30			

Table 1. Operational Atomic Beam Sources, Cont'd.

Laboratory and Installation Date	Particle	R.F. Transitions Used*	Beam Polarization** theor.	Beam Polarization** expt.	Typical Polarized Beam From Source (nA)	Polarized Beam Currents From Accelerator (nA unless otherwise shown)	Accelerator Type	Accelerated Beam Energy (MeV)	Polarized Beam Usage %
13. Argonne[16] 1973	H^+ D^+	WF, 2→4	1	0.8 ---	2.5×10^4 peak (0.4 Hz pulse)	5.9×10^9 part./pulse	Zero Gradient Synchrotron	$0.8\text{-}12 \times 10^3$	30-40
14. Auckland[103] 1973	H^+ D^+	WF, 3→5, 2→6	2/3, 1	0.8 0.52, 0.8	500 500	500 500	Cockroft-Walton	0.2	100
15. Basel[28] ~1973	H^+ D^+	WF, 2→4 WF, 3→5, 2→6	1 2/3, 1			~500	Cockroft-Walton	0.18	not used since 6/74
16. ETH-Zurich[34] 1973	H^- D^-	WF, 2→4 WF, 3→5, 2→6	1 2/3, 1	0.80 0.57, 0.87	80 120	>30 >50	EN-tandem VDG	1-13	18
17. Heidelberg[32] 1973	$^6Li^+$ $^{23}Na^+$	WF, 3→5, 2→6	2/3	>0.46	150-240 30	30-180 of $^6Li^{3+}$ 2 of $^{23}Na^{5+}$	EN-tandem VDG	4-24 6-36	20
18. Rez[24] 1973	D^+	WF	2/3, 0	0.41, 0.0	1.0×10^3	0.08	cyclotron	12	
19. Texas A & M[17] ~1973	H^+ D^+	WF WF, 3→5, 2→6	1 2/3, 1	0.74 0.53, 0.85	10^4 10^4	750 10^3	cyclotron	10-55 12-65	8
20. Eindhoven[101] 1974	H^+	WF	1	0.83	$3 \times 10^3\text{-}5 \times 10^3$	20-50	cyclotron	3-28	28
21. SIN[20,104] 1974	H^+ D^+	WF, 2→4 WF, 3→5, 2→6	1 2/3, 1	0.65	2×10^3 2×10^3	30-200,1 30-200	2-stage cyclotron	10-75,590 10-65	first used 4/75
22. Bonn[21] 1975	H^+ D^+	WF, 2→4 WF, 3→5, 2→6	1 2/3, 1	-, 0.85	$1.5 \times 10^3\text{-}2.0 \times 10^3$ $2.0 \times 10^3\text{-}2.5 \times 10^3$	~30	cyclotron	7-15 15-30	first used 7/75
23. IPCR, Saitama[25] INS, Tokyo, 1975	H^+ D^+	3→5, 2→6	1/2 1/3, 1	---, 0.55	600-1000		cyclotron	8-48 16-34	not yet used
24. Saclay-SATURNE[26] 1975	H^+ D^+	WF, 2→4 WF, 3→6, 2→6	1 2/3, 1		700		synchrotron	$0.6\text{-}3 \times 10^3$ 2.4×10^3	not yet used

* WF = weak field transition. This implies 1→3 for protons and 1→4 for deuterons. The notation is that of Ref. 8.

** For deuterium the beam polarizations listed successively are P_z, P_{zz}.

Table 2. Operational Lamb-shift Sources

Laboratory and Installation Date	Particle	Polarization Scheme	Beam Polarization* theor.	Beam Polarization* expt.	Typical Polarized Beam Currents From Source (nA)	Beam Currents From Accelerator (nA)	Accelerator Type	Accelerated Beam Energy (MeV)	Polarized Beam Usage %
1. Notre Dame[48] 1968	H⁻ / D⁻	Diabatic Field Reversal	1 / 2/3,-1	0.5 / 0.5,-0.75	15 / 10-50	3 / 5-20	FN-tandem VDG	3-17	25
2. Los Alamos[39] 1969	H⁻ / D⁻	Spin Filter	1 / 1, -2	0.89 / 0.79, 1.57	300-400 / 300-400	100-250 / 100-250	FN-tandem VDG	2-18	30
3. Wisconsin[46] 1969	H⁻ / D⁻	Diabatic Field Reversal	1 / 2/3,-1	0.78 / 0.54,-0.73	20 / 100-160	18 / 50-145	EN-tandem VDG	2-12	40
4. TUNL[42] 1970	H⁻ / D⁻	Spin Filter	1 / 1, -2	0.78 / 0.7,-1.40	200-400 / 200-400	50-110 / 50-130	FN-tandem VDG	2-16	40
5. Tsukuba[47] 1971	H⁻ / D⁻	Spin Filter	1, -2	∿0.5, 1	100	---	---	---	never installed
6. Washington[45] 1971	H⁻ / D⁻	Diabatic Field Reversal	1 / 2/3, -	0.65 / 0.50, -	90-150 / 100-200	30-50 / 40-70	FN-tandem VDG	2-18	30
7. Karlsruhe[38] 1973	D⁺	Diabatic Field Reversal	2/3, 0	0.45, 0.73	800	40	cyclotron	52	5
8. Erlangen[44] 1974	H⁻ / D⁻	Diabatic Field Reversal	1 / 2/3, 1	0.75 / 0.50, 0.75	100-150 / 150-250	30-40 / 40-60	EN-tandem VDG	2-12.5	45
9. Giessen[36] 1974	H⁻ / D⁻	Diabatic Field Reversal	1 / 2/3, 1	0.69 / 0.47, 0.52	600 / 600-1000	200-400 / 200-700	tandem	0.1-1.2	100
10. Birmingham[49] 1975	³He⁺	Adiabatic Field Reduction	1/2	0.38	2	0.15	cyclotron	33.4	40
11. Los Alamos[41] 1975	H⁻ / D⁻ / T⁻	Spin Filter	1 / 1, -2 / 1	0.85 / --- / 0.85	200-400 / --- / 150-250	100-200 / --- / 50-150	FN-tandem VDG	2-18	30
12. McMaster[43] 1975	H⁻ / D⁻	Spin Filter	1	0.80 / 0.75, 1.43	2-100 / 20-350	1-50 / 10-175	FN-tandem VDG	2-18	25
13. Munich[37] 1975	H⁻ / D⁻	Diabatic Field Reversal	1 / 2/3, 0	0.85 / 0.45, 0	700 / 800	---	MP-tandem VDG	5-24	not yet used
14. TRIUMPH[40] 1975	H⁻	Diabatic Field Reversal	1	0.80	300-400	---	cyclotron	200-500	50 expected first year

* For deuterium the beam polarization listed successively are P_z, P_{zz}.

Table 3. Polarized Sources Planned or Under Construction

Laboratory	Source Type	Particle	Accelerator Type	Accelerated Beam Energy (MeV)	Expected Data of Completion
1. Köln[50]	Lamb-shift	H⁻, D⁻	tandem VDG	2-18	1975
2. Rice/Texas A & M[51]	R.F. Discharge with Optical Pumping	³He⁺	cyclotron	25-30	1976
3. Kyoto[52]	Lamb-shift	H⁻, D⁻	tandem VDG	2-12	1976
4. LASL/LAMPF[53]	Lamb-shift	H⁻	linac	800	1976
5. RCPN Osaka[105]	Atomic Beam	H⁺, D⁺	cyclotron	8-75, 16-65	1976
6. Tsukuba[54]	Lamb-shift	H⁻, D⁻	pelletron	7-24	1976
7. Rez, Czechoslovakia[55]	Atomic Beam	H⁺, D⁺	cyclotron	40	~1976
8. Indiana[56]	Atomic Beam	H⁺, D⁺	2-stage cyclotron	50-200, 20-80	~1977
9. Manitoba[57]	Lamb-shift	H⁻, D⁻	cyclotron	22-50	1977
10. Nat. Lab. for High Energy Physics (KEK) Japan	Lamb-shift	H⁻, D⁻	synchrotron	1×10^4	1978-9
11. CERN[59]	Atomic Beam	H⁺, D⁺	synchrotron	2.8×10^4	1978-9

Polarized beams of tritons, ³He, and the alkali metals ⁶Li and ²³Na have become available for experiments within the last year. Construction of these new successful sources required many of the really new design ideas which have appeared since 1970.

III. ATOMIC BEAM SOURCES

This oldest scheme for making polarized ion beams is based on the simple Stern-Gerlach separation of atoms with electron spin in the $m_J = +\frac{1}{2}$ states from those having electron spin in the $m_J = -\frac{1}{2}$ states (see Fig. 1) by passing the atomic beam through an inhomogeneous magnetic field. In all recently constructed sources this inhomogeneous field is produced by a sextupole magnet. The atoms with $m_J = +\frac{1}{2}$ move toward the region of weaker magnetic field to minimize their energy and are focussed toward the center of the sextupole. Atoms with $m_J = -\frac{1}{2}$ sense the opposite forces and are defocussed, as shown in Fig. 2.

A schematic of a typical atomic beam source is shown in Fig. 3. This illustrates the basic components of such a source. Typically for hydrogen beams, an r.f. discharge dissociator with a carefully designed nozzle produces an intense and highly directed beam of hydrogen atoms along the sextupole axis. The $m_J = +\frac{1}{2}$ atoms focussed there pass then through one or more r.f. transition units which can enhance and change the sign of the beam polarization using the adiabatic passage technique. These neutral

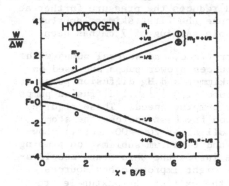

Fig. 1. Breit-Rabi diagram for the 1S$_{\frac{1}{2}}$ hyperfine states of hydrogen. (B_c=507G, ΔW=1420 MHz).

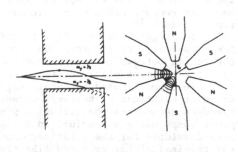

Fig. 2. The separation of $m_J = +\frac{1}{2}$ and $m_J = -\frac{1}{2}$ components of a hydrogen atomic beam in the inhomogeneous field of a sextupole magnet.

Fig. 3. Schematic diagram showing the major components of an atomic beam
polarized source for hydrogen.

polarized atoms then enter a strong axial magnetic field where a small
fraction of them are ionized by electron bombardment so they emerge as
positive, polarized ions. If negative ions are desired, subsequent acceleration and charge exchange, usually in sodium vapor, is required.
Then before injection into the accelerator, further acceleration and
spin-quantization-axis orientation may be needed, depending on the particular accelerator on which the ion source is installed.

Operation for deuterium follows closely that for hydrogen, except
that different r.f. transitions are used. For the alkali metals the
atomic beam originates from an oven, the ionizer is a hot tungsten rib-
bon covered with a surface layer of oxygen, and charge exchange to produce negative polarized ions may require another medium, typically
potassium vapor.

In the sections below design principles for recently constructed
atomic beam sources and suggested improvements are discussed briefly.
Earlier reviews[5,10,14] should be consulted for more detail.

Atomic beam formation--Typically a $\sim$25 MHz discharge is struck inside a
water- or air-cooled pyrex or quartz tube to dissociate the H_2 or D_2 gas
used. The discharge power is 0.2 to 1.0 kW, the pressure in the discharge is in the range 0.2 to 6 Torr. Lower pressure operation (<2 Torr)
is typically necessary with the inductively coupled discharge tubes than
with capacitively-coupled hairpin-type tubes.[61,65] The resulting atomic
beam escapes through a single-aperture nozzle approximately 2.5 mm in
diameter in the end of the tube. Glavish has discussed the details of
nozzle design.[5,10] The goal is to produce supersonic flow at the tip
of the nozzle. A specially designed conical skimmer aperture, $\sim$0.5 mm
larger in diameter than the nozzle diameter and often movable along the
beam axis to optimize the beam intensity, is placed within 4 to 8 mm of
the nozzle tip near the point where the atomic beam transforms into free
flow. This aperture removes the most divergent part of the atomic beam
and acts simultaneously as a differential-pumping aperture to reduce the
pressure beyond this point. A second similarly shaped but slightly
larger aperture collimates the beam and reduces the pressure further at
the entrance to the sextupole. At Argonne the first skimmer aperture
has been removed and a nozzle-to-sextupole spacing of 7 cm has been
found optimum.[16]

The total gas flow rate is 2 to 6 Torr-ℓ/sec, and several pumps with
large capacity are required. On some sources blower pumps are used on
the region between the nozzle and the skimmer and Hg diffusion pumps are
used for the next stage before the sextupole.[19,101] Oil diffusion pumps
are more commonly used to obtain higher pumping speeds. These require
frequent maintenance, however, because on the best sources the atomic
hydrogen decomposes the diffusion pump oil after 200-400 hours.[16] Low-
power pumps using a combination of ion and titanium sublimation pumping
were developed for the polarized source inside the Ohio State accelerator terminal.[66] Larger pumps like these might improve other sources .

Atomic beam intensities obtained at the exit of the sextupole are
10^{17}, 3×10^{16}, and 2.4×10^{16} atoms/cm^2-sec at Saclay,[10] Stanford,[10] and
Zurich,[65] respectively. These are a factor of 3 (ref. 10) to 40 (ref.
65) higher than obtained from dissociators with multiple capillary bun-
dles. Further improvement in intensity of up to a factor of five[65]

might be expected by cooling the dissociator to 77°K with liquid nitrogen. This is done at Bonn[21] and has
been tried at Argonne.[16] In the latter source temperatures only as low
as 250°K were reached and the beam
improved only 10%. Apparently there
was insufficient cooling to remove
the power below this temperature.
For the CERN source a low-power
microwave dissociator is planned
and liquid-nitrogen cooling should
be easier.[59] It is not known whether the dissociation degree is reduced at low temperature.[10]

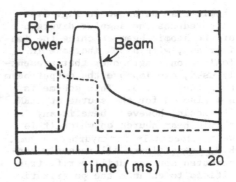

Fig. 4. Sketch of oscilloscope trace
showing dissociator power pulse and
output polarized beam pulse for the
Argonne polarized source.

For pulsed, low-duty-cycle accelerators a large increase in
polarized beam can be obtained by
pulsing the dissociator power and
the gas flow.[16] In the best performance at Argonne, pulsing the disso-
iator power at 25 Hz increased the peak beam current from 12 to 25 μA,
presumably because of the lower dissociator temperature of ∼310°K rather
than the ∼500°K typical of d.c. operation.[14] Pulsing the gas to the dissociator lowered the background pressure and raised the peak current further to 40 μA. This has the other advantage that periods between maintenance on the diffusion pumps are greatly extended. Typical r.f. and
beam pulses for this source are shown in Fig. 4. The time delay between
the beginning of the r.f. pulse and the beginning of the beam pulse increases from zero at 0.8 Torr to 2 msec at 2 Torr. At the end of the
pulse the beam shows first a fast and then a slower decay as the dissociator clears, demonstrating a larger fraction of dissociation near
the nozzle tip.

For the alkali metals an atomic beam is formed by heating the desired
metal in an oven, evaporating atoms through a heated 0.5 mm diameter
Laval nozzle and then through a 1 mm diameter heated skimmer electrode
immediately in front of the sextupole. Beams of ^{6}Li and ^{23}Na have been
obtained at Heidelberg using this scheme, and ^{7}Li and ^{39}K operation is
planned.[67] This group has also considered producing polarized beams of
the halogens.[68]

Inhomogeneous fields to polarize the beam--All recently constructed
sources use a sextupole to separate the $m_J=+\frac{1}{2}$ and $m_J=-\frac{1}{2}$ components of the
atomic beam. These range in length from 13.3 cm, which is too short for
complete separation,[33] to 50 cm.[18,34] The largest possible pole-tip
field is desirable to increase the solid angle for acceptance of the
atomic beam. Fields of 6 to 8 kG are common, but higher fields are possible with special coil windings,[59] and 10 kG fields have been realized
using a special alloy for the pole pieces.[25] Most sextupoles are tapered to reduce the axial magnetic field at the exit and thus the angular divergence of the emerging atomic beam.[10] Entrance apertures range
between 0.3 and 0.7 cm between opposite pole tips and exit apertures are
approximately twice as large. This usually approximately doubles the
output beam intensity over that obtained with an untapered sextupole.

In a contribution to this conference[69] Glavish points out that adding
a short sextupole unit between the r.f. transition unit and the ionizer
makes the two-sextupole combination an achromatic lens for the selected

atomic beam as shown in Fig. 5.
This reduces the angular divergence
of the atomic beam reaching the ionizer and, with a 50% increase in
ionization length over that presently used, can improve the output beam
by a factor of 3. This scheme is
now planned for the source at CERN.[59]
It should, however, benefit any
atomic beam source on which it is
used, especially for hydrogen beams.

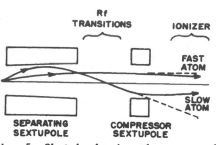

Fig. 5. Sketch showing the proposed
location of the compressor sextupole
and its achromatic focussing action.

R.F. transitions--All recently constructed sources utilize r.f. transitions to enhance the polarization
of the atomic beam emerging from
the sextupole. Other r.f. transitions provide the easiest method for
reversing the sign of the vector polarization of the beam without causing
the beam to steer differently for the two spin directions. The best in-
troductions to the adiabatic passage techniques[70] for atomic beam sources
are those of Beurtey[71] and Haeberli,[8] and the latest survey of the specific hardware used is that of Glavish.[10]

The transitions produced are typically of two basic types: low-fre-
quency transitions which are accomplished in a weak magnetic field where
m_F is a good description of the total angular momentum projection, and
high-frequency, strong-field transitions, where m_I and m_J are the good
quantum numbers. The weak-field (WF) transitions usually interchange
the populations of states with opposite sign for m_F. The strong-field
units require the use of tuned cavities to cause transitions between a
particular pair of hyperfine states.

The choice and number of transitions and the particular hardware designs vary considerably for different sources. All transition units
produce a magnetic field transverse to the beam direction which varies
linearly between $B_0-\delta_0$ and $B_0+\delta_0$ (see ref. 8) over the region where the
r.f. field is applied. It has been suggested that this field should be
increasing in the beam direction, rather than decreasing, for weak-field
transitions, but it was later pointed out that it does not matter which
slope is chosen if the r.f. field is large enough.[5] The r.f. field is
chosen parallel or transverse to the beam direction depending on the
particular transitions desired. Some transitions are more difficult to
achieve than others, and are sensitive to frequency instability and power
level for the r.f. field.[8,27a] Both weak-field and strong-field transition units have been designed for the alkali-metal-ion, polarized source
at Heidelberg [67] following the scheme outlined by Glavish (Table 4 of
ref. 10). These are designed for operation with ^{6}Li, ^{7}Li, ^{23}Na, and ^{39}K.

Strong-field ionization--All but two of the operational atomic beam
sources listed in Table 1 use a strong-field ionizer for the atomic beam
to achieve the higher values of beam polarization made possible by the
r.f. transitions. The basic design of these ionizers has remained unchanged for the last seven years.[5,72] The polarized atomic beam enters
down the axis of a solenoid. Inside the solenoid a high electron density
is maintained by an electron gun at one end and a negative, reflecting
grid at the other. The positive ions produced in electron-atom collisions are extracted by the negative grid into a gridded lens system
which accelerates and focusses them. Initially the active ionization
volume was 15 cm long, but recently constructed ionizers have been 20 to
30 cm long for increased efficiency.[34,69] It is important in these

stretched versions that the background pressure in the ionizer be lower
than 5×10^{-7} Torr.

All these ionizers are woefully inefficient for hydrogen operation,
the best reported efficiency being 6×10^{-3} for the ionizer at Texas A &
M.[10] Glavish has suggested[14] that a factor of ~ 1.4 improvement in efficiency might be obtained by doubling the electron energy to 200 eV. This
would, however, increase the energy spread of the extracted ions. He is
also considering a shaped solenoidal magnetic field for the source at
CERN in order to improve the electron density distribution.[59]

It has been suggested that increased ionization efficiency would be
possible in charge-exchange reactions with ion beams.[73,74] A test of
these ideas is reported at this conference by the Bonn group.[75] They
find that if the polarized atomic beam enters the 75 kG magnetic field
of a superconducting solenoid and interacts there with a 10 to 20 keV
H^+ beam entering the other end of the solenoid from a duoplasmatron,
they are able to obtain an ionization efficiency in their first try of
10^{-3}. They report an extracted H^{++} current after the charge-exchange process of 1 μA and suggest that improvement is possible. It is important
to realize that any ion beam emerging after charge-exchange in such a
large solenoidal field will increase significantly in emittance.[76] It
is possible that if these ions emerge at low enough energy, the increased
emittance of the beam will not be detrimental after it is accelerated.
Nevertheless, it is tempting to ask whether comparable ionization efficiencies can be obtained with much smaller fields than 75 kG. Haeberli
points out, however, that in the design of crossed-beam ionizers using
a charged incident beam and no confining magnetic field, one must pay
very careful attention to space-charge forces.[77] These potential problems with charge-exchange reactions using charged incident beams are not
likely to be as bad if one uses Haeberli's suggestion[73] of ionization
with a fast, neutral cesium beam to obtain negative polarized ions. Of
course, there are technical difficulties in handling the cesium beam.

These problems of ionization efficiency disappear in the process used
for the alkali metal source at Heidelberg.[32] Here, for example, the
polarized ^{6}Li atomic beam is incident on a tungsten strip heated to
1000°C and covered with a surface oxygen layer. These atoms remain on the
strip 2 to 100 msec before ionization by the process $^6Li + WO_2 \rightarrow {}^6Li^+ + WO_2^-$.
The ionization efficiency is nearly 100%. If the ionization takes
place in ~ 80 G magnetic field, the nuclear polarization is not destroyed
in the ionization process.

Acceleration and charge exchange--In all atomic beam sources now being
used to produce H^- or D^- beams, the H^{++} or D^- beam from the ionizer is accelerated to ~ 5 keV where it is focussed into a sodium-vapor charge-exchange canal placed in a strong $(B \gg B_c)$ solenoidal magnetic field to prevent depolarization during the charge-exchange process. It is important
that the lens system which extracts the beam from the ionizer be well-
matched to the acceptance of the canal-solenoid system. This can in
practice have a significant effect on the amount of output beam obtained.[10]

The major problem with the charge-exchange system is the necessity to
clean the spent sodium from the neighboring components of the ion source,
a process which may be required as often as every 200 hours of operation.[30] A wick-type charge-exchange cell (see Sec. IV below) has been
tried in an effort to reduce the amount of sodium escaping from the
canal, but so far it has proved unsuccessful.[59]

IV. LAMB-SHIFT SOURCES

The Lamb-shift technique produces a polarized beam from one-electron atoms in the metastable $2S_{1/2}$ excited state. In Fig. 6 a level diagram is shown for the $2S_{1/2}$ and $2P_{1/2}$ states in hydrogen. At zero magnetic field these levels are split by an energy equal to the Lamb Shift. Because of this splitting, in a magnetic field near 575 G, the 2S substates with $m_J=-\frac{1}{2}$ are nearly degenerate with the 2P $m_J=+\frac{1}{2}$ substates, giving the possibility of introducing some P-state admixture selectively into various 2S substates with applied electric fields. This shortens the lifetimes of atoms in these states. The remaining $2S_{1/2}$ metastable atoms are relatively long-lived and can be used to produce a polarized beam of hydrogen ions.

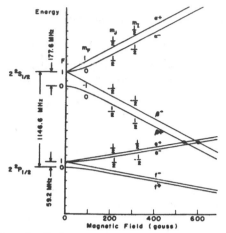

Fig. 6. Hydrogen energy-level diagram showing the Zeeman splitting of the $2S_{1/2}$ and $2P_{1/2}$ states in an external magnetic field. The α, β, e, and f designations are those of Lamb and Retherford.[78]

A schematic diagram of the familiar Lamb-shift source for hydrogen is shown in Fig. 7. A positive ion source produces an intense H^+ beam which enters a cesium canal where it undergoes charge exchange to produce a beam of neutral metastable H(2S) atoms accompanied by a H(1S) background beam. There are additional H^+ and H^- background beam components which must be removed by deflection. Then the metastable beam enters a region of electric and magnetic fields chosen to quench H(2S) atoms in the undesired hyperfine states, leaving an H(2S) beam with nuclear polarization accompanied by an unpolarized H(1S) background. From here the beam passes into a second charge-exchange medium chosen to discriminate against the H(1S) atoms. To produce polarized positive or negative ions, respectively, from the H(2S) atoms selectively, charge exchange in iodine or argon gas is chosen. Subsequent focussing, acceleration, and spin-axis orientation is required in the same manner as for atomic beam sources.

Lamb-shift sources designed as outlined in Fig. 7 will also work for deuterium and tritium with minor modifications of parameters and the necessary precautions for handling the radioactive tritium gas. Tritium operation would, in fact, be very difficult with an atomic beam source because of its excessive gas requirements. The one-electron, $^3He^+$ (2S) metastable ion is also a candidate for polarization by the Lamb-shift technique. The $^3He^{++}$ Lamb-shift source development at Birmingham[44] has required extensive research to learn that air is conveniently the best charge-exchange medium instead of cesium and argon or iodine. The most impressive accomplishment of that development project is the successful design and operation of a velocity filter to separate the emerging

Fig. 7. Schematic diagram showing the main components of the Lamb-shift polarized source for hydrogen.

polarized $^3\vec{He}^{++}$ from the intense unpolarized He^{++} beam transmitted
through the system from the positive ion source.

In the following sections explicit developments are discussed for the
major components of the Lamb-shift source.

<u>Positive ion source</u>--Eleven of the 13 hydrogen sources listed in Table 2
utilize a duoplasmatron as the plasma source from which positive ions are
extracted. At Karlsruhe[38] and Munich[37] r.f. sources are used. The magnitudes of the output polarized beams seem to indicate that for present
beam intensities it is not so important which plasma source is used as
long as the plasma has relatively low ion-energy spread.[48] Probably
more important is the configuration of the electrodes which extract the
beam and focus it through the cesium canal.

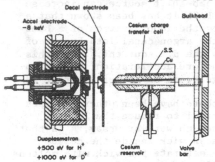

Fig. 8. Schematic of the Los Alamos
positive ion duoplasmatron and
closely spaced electrodes which
focus the beam into the cesium
canal.

Here two different philosophies
are apparent. At most laboratories
the beam is extracted from a duoplasmatron and decelerated immediately into the cesium canal in a
scheme similar to that originated
by McKibben et al.[39] at LASL and
shown in Fig. 8. At some laboratories[42,46] this system is modified
slightly by the addition of a magnetic lens between the duoplasmatron
and the cesium canal, but the distance between the deceleration electrode and the cesium canal is still
less than 10 cm. In either case
there is a beam of 1 to 5 mA entering the cesium canal which is as
small as 5 mm in diameter. Unless
almost completely neutralized by
electrons produced in collisions with cesium, the beam will quickly become highly radially divergent under the influence of space-charge
forces. In addition there is charged beam emerging from the cesium
canal and the space-charge electric fields from this beam can quench a
large fraction of the metastable beam before the charged particles are
removed from the beam by the sweeping fields.[42] It is clear that the
beam should be larger in diameter to reduce the problem of space-charge
electric fields, but it is also clear that a larger diameter beam which
is also divergent makes deflection of the charged background beam components very difficult.

In an alternative system at
Giessen,[36] Munich,[37] Karlsruhe,[38]
and Erlangen[44] the positive
beam travels through a 10-to-
50 cm long converging electrostatic lens system which focusses the beam through the
cesium canal toward an aperture ∿30 cm beyond. The Munich
system is shown in Fig. 9. In
this case the beam is allowed to
expand radially before being re-
focussed. It can be made large
enough in diameter emerging from
the cesium canal that space-

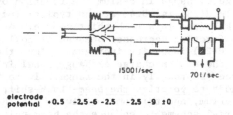

Fig. 9. Schematic of the Munich r.f.
source showing the extended lens system for beam extraction and focussing.

charge forces are reduced. Since the beam is converging here, deflection
of the unwanted charged-beam components requires only small transverse
electric fields and only a small fraction of the metastable beam is
quenched. Further improvements on this latter scheme are probably possible by using a magnetic rather than an electrostatic lens system to focus
the beam into the cesium canal[53a] since it is known that electrostatic
lenses inhibit electron trapping needed for space-charge neutraliza-
tion.[79] More intense, multi-aperture duoplasmatrons,[80] when used in this
configuration, may also provide increased current.

For application to cyclotrons or linacs[53] or for tandem accelerators
when a pulsed beam is required for time-of-flight measurements,[46] pulsing
and pre-bunching the positive beam in a Lamb-shift source can increase
the output polarized beam intensity. At LASL it has been shown that a
3 MHz, $\sim$50 V ramp signal can be applied to the duoplasmatron of a Lamb-
shift source to prebunch the H^- beam at the argon canal into bursts of
$\sim$70 nsec FWHM with peak currents $\sim$4 times the d.c. current.[81] If this
were synchronized with further bunching before acceleration, pulsed operation could be accomplished with average beams on target as great as
50% of the d.c. beams.

Cesium charge-exchange region--Although there have been other attempts
at designing the cesium charge-exchange canal to reduce the loss of
cesium, the major recent development, based on work at Orsay,[82] is that
of using a cylindrical, stainless steel mesh wick inside the cesium
canal.[41] If the canal is first heated to saturate the wick with molten
cesium and then the ends are cooled to $\sim$30°C while the temperature of the
center of the canal is automatically regulated to maintain the required
vapor density, continuous operation for many hours is possible, even in
the vertical orientation, with negligibly small cesium loss. If this is
coupled with valves before and after the cesium region for isolation
during maintenance on other parts of the source, trouble-free cesium
operation for several months is easily possible. One should be careful
to use oil in any cesium-region diffusion pump that does not deteriorate
in the presence of cesium.[83]

Removal of the charged ions emerging from the cesium canal is accomplished in the most successful sources with deflection systems which
fill up all useable space between the cesium canal and the point where
H(2S) atoms are ionized.[41] These deflectors ideally surround the beam
with multiple electrodes and allow the source operator to choose the
direction and magnitude of each deflecting field.[41,83,37]

New, accurate cross section measurements[85] have shown that after
charge exchange in cesium the fraction of neutral beam which is metastable H(2S) is $\sim$0.43 at the typical incident proton beam energy of
0.5 keV used in a Lamb-shift source. As the cesium density is raised,
this fraction decreases because of collisional quenching, but the actual H(2S) intensity increases. Thus the fraction of H(2S) emerging
from the cesium charge-exchange canal in an operational Lamb-shift
source is probably in the range 0.1 to 0.2.

Methods to polarize the beam--Lamb-shift polarized sources for hydrogen,
deuterium, and tritium now always use the spin-filter or diabatic-field-
reversal scheme to enhance the beam polarization.

The spin filter was first developed by Ohlsen and McKibben[63,86] following ideas of Lamb.[87] It allows the production of polarized beam from
H(2S) atoms in a particular α-hyperfine state by quenching atoms in the
other hyperfine states to the ground state. It requires a cylindrical
r.f. cavity[88] open at the ends to allow the beam to pass down the cylinder axis. The cavity is split into quadrants to allow simultaneous

Fig. 10. The spin-filter cavity for the polarized triton source. The beam enters from the left. Electrodes to deflect H$^+$ and H$^-$ ions can be seen to the left of the cavity. R.f. power enters through the bottom quadrant of the cavity and a small fraction is picked up through the top quadrant for control purposes. The front and rear quadrants are d.c. electrodes.

application of transverse d.c. and axial r.f. electric fields of 10-15 V/cm in an approximate 575G axial magnetic field, uniform to better than 0.5G over the length of the cavity. The 14.6 cm diameter of the cavity is determined by the desired ∿1.6 GHz frequency. The cavity length and the shape of the end pipes are chosen to provide for adiabatic entry and exit of the beam. The original Los Alamos cavity was 34 cm long, but Hardekopf reports[41] success for proton and triton operation with a cavity 27 cm long. He has no results yet for deuterium operation. A photograph of his cavity is shown in Fig. 10. In Fig. 11 data are shown which demonstrate the operation of the spin filter at TUNL.[42] One must only change the value of the spin-filter magnetic field to change the hyperfine state and therefore the beam polarization.

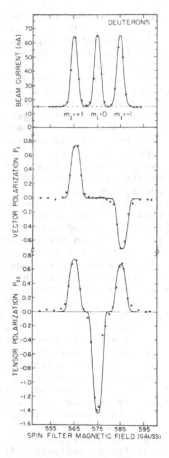

Fig. 11. Measured values of current, p_z, and p_{zz} for the polarized deuteron beam from the TUNL source as a function of the spin-filter magnetic field. The curves are fits using spin-filter parameters in the approximate calculation of Trainor.[42c] Ionization occurred in a ∿40G magnetic field.

A significant advantage of spin-filter operation is that with care the beam polarization can be measured to ±0.005 by the quench-ratio method. This has been tested both for proton[89] and triton operation.[90] This simple, quick measurement involves determining the ratio Q of the beam current in one of the peaks in Fig. 11 to the current remaining when a large electric field is applied to quench the H(2S) beam (dotted line in Fig. 11). If the current in the peak comes from a single state and the quenched beam is completely unpolarized, then the quench-ratio beam polarization $p_B = p_Q = 1 - 1/Q$ for "pure" states like the $m_I = +1$ state in Fig. 11. If Q is measured at an experimental target, one knows the fraction of the beam which is polarized but does not determine the direction of the spin quantization axis.

A possible disadvantage of the spin-filter scheme, that of complexity and cost of the r.f. system, has been reduced significantly with the newly designed simple system for the triton source.[91]

Ohlsen et al.[92] have suggested that two coaxial spin-filter cavities be tuned to different frequencies to select different states. The r.f. power to these could be alternated rapidly to switch the beam polarization without beam motion. This scheme has not been tried but would provide a very effective technique for making deuteron tensor polarization measurements by the ratio method[93] if the cavities were tuned to the $m_I = +1$ and $m_I = 0$ states.

More common than the spin-filter method is the diabatic field reversal method, or "Sona scheme,"[64] for enhancing the beam polarization. Details of the method were discussed by Donnally[6] and Clegg.[11] The technique is illustrated for spin-$\frac{1}{2}$ and spin-1 particles in Fig. 12. The important feature for each particle is that metastable atoms in state 1 make a non-adiabatic transition as the magnetic field direction is reversed quickly, causing them to become antiparallel to the new field direction. This is possible if the magnetic field direction reverses fast compared with one Larmour precession period in the field, a restriction which is satisfied only for beam particles near the axis and for small axial magnetic field gradients in the crossover region. Sources using this technique usually have crossover axial field gradients $\leq 5 G/cm$ and focus the beam through a ~ 2 cm aperture at the field crossover to limit the beam diameter.

The advantages of this system over a spin-filter system are reduced cost and the fact that only static fields are used. One disadvantage which has been experienced at several laboratories[38,46] arises because of the short lifetime of the β-state atoms created from state-1 atoms in the crossover. If a vector-polarized deuteron beam is desired, one must be careful not to quench these atoms before ionization to avoid an unwanted tensor polarization in the beam. This is accomplished by reducing the magnitude of the field after the crossover to reduce

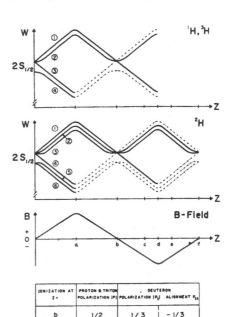

IONIZATION AT Z=	PROTON & TRITON POLARIZATION $|P|$	DEUTERON POLARIZATION $|P_z|$	ALIGNMENT P_{33}
b	1/2	1/3	-1/3
c	1	2/3	0
e	—	1/2	-1/2
f	—	0	-1

Fig. 12. Shown plotted vs Z are the Zeeman splitting for ^1H, ^2H, and ^3H and the qualitative axial magnetic field. Dotted lines show states quenched at a and d. Ionizing at different positions Z gives the beam polarizations shown.

motional electric fields experienced by the beam. The Karlsruhe group points out, however, that if only tensor polarized beams are desired, an extremely short coil configuration can be achieved because the quenching of these β-state atoms which occurs is exactly what is desired.[38]

The Sona-scheme sources do not produce beams with polarizations which can be as easily determined as with spin-filter sources. Nevertheless, Roy at TRIUMPF[40] is trying to develop a sequence of measurements which, like the quench ratio for spin filter sources, would let him measure his beam polarization.

Fast changes in polarization are possible in sources using the Sona scheme. Donnally proposed[6] that one could use transverse r.f. electric fields in the regions before and after the crossover to quench either α- or β-state metastable atoms and produce beams with pure vector or tensor polarization. This scheme has not been implemented, but the switching could be done rapidly in a manner that would not affect the beam position. In another scheme at Seattle[45a] the entire coil system is made from non-magnetic material and is wound on coil forms designed to reduce eddy currents. Programmable power supplies reverse the 575 G coil currents in 10 msec. During operation they often reverse polarity at a 10 Hz rate. Finally at Giessen[36] and TRIUMPF a small coil in the crossover region produces a transverse field of approximately 20 G, destroys the non-adiabatic transition, and reduces the proton polarization to zero. It can be switched on and off quickly to change the beam polarization.

At Los Alamos McKibben has successfully operated the source for protons with the magnetic field in the argon gas ionization region opposite to that in the spin-filter region. If the spin-filter selects H(2S) atoms in state 1, then they undergo a non-adiabatic transition upon entering the argon field which points their spin direction opposite to the argon field direction. If, in addition, a small transverse field coil is placed at the crossover, turning it on spoils the non-adiabatic transition and reverses the spin direction. This system has been carefully designed to produce minimum beam motion associated with spin reversal.[94]

Selective ionization--For negative polarized ion production argon is thought to be the best choice for a charge-exchange gas which selectively ionizes the polarized H(2S) atomic beam instead of the unpolarized H(1S) background beam.[95] Recent calculations indicate, however, that krypton might be a better choice[96] with a cross section ∿5 times greater than argon for producing H⁻ from H(2S). It is known that its selectivity against producing H⁻ from the unpolarized H(1S) is ∿2 times poorer than argon.[97] For positive polarized ion production, the best selective ionizer is iodine.[98] Cross sections show that iodine is a better selective ionizer than argon for all hydrogen atom beam energies from 150 to 750 eV. Both gases exhibit better selectivity as the beam energy is lowered over this range.[99] It is common that the negative beam after charge exchange in argon has a polarized component with better emittance than the unpolarized background component. An accelerating lens which focusses the negative beam immediately through a ∿2 mm aperture enhances the beam polarization by "scraping" off the higher emittance unpolarized beam.[89,6] This enhancement is harder to realize for T⁻ beams because the triton is more massive than hydrogen or deuterium and the emittance difference produced in the ionizing collisions is less.[91]

Some of the newest Lamb-shift sources[41,43,54] utilize helium-cycle refrigerators to cryopump the argon on ∿20°K baffles placed immediately at the ends of the charge-exchange canal. This has the advantage of producing an extremely clean vacuum system. One cannot rely completely on the cryopump in this region, however, because hydrogen is not pumped well and the partial pressure of hydrogen can build up enough to allow non-selective ionization by collisions with hydrogen, reducing the output beam polarization. The main disadvantages of cryopumping seem to be that the minimum size $\sim$2W refrigerator required costs $\sim$\$8,000, and that routine maintenance to the refrigerator is required once or twice yearly.

At Karlsruhe turbopumps are used to pump the iodine and liquid nitrogen traps keep iodine from reaching the forepumps.[38]

V. FUTURE PROBLEMS

One problem facing all polarized source builders is that of matching the optical properties of the polarized beam to the requirements of a particular accelerator. The best performance has been invariably obtained where this has been done well.[17,39] Compromises must be made occasionally which sacrifice ion optic quality for greater flexibility of spin-axis orientation. With careful thought about the type and location of spin precession devices, their optical effects, however, can be minimized.

Installation of polarized sources on high-energy, strong focussing machines requires in addition a thorough understanding of depolarizing resonances which can occur as the beam is being accelerated.[100] These arise when the Larmour precession frequency in the accelerator magnetic field becomes an integer multiple of the betatron oscillation frequency and the polarized particle feels the same depolarizing force repeatedly. At CERN these resonances are much more severe for H^+ than for D^+ beams, so initial polarized-beam operation may accelerate D^+ and strip them to produce polarized nucleons.[58]

REFERENCES

1. Proc. Int. Symp. on Polarization Phenomena of Nucleons, Basel, (Eds. P. Huber and K. P. Meyer, Birkhäuser Verlag, Basel, 1961).
2. Proc. 2nd Int. Symp. on Polarization Phenomena of Nucleons, Karlsruhe, (Eds. P. Huber and H. Schopper, Birkhäuser Verlag, Basel, 1966).
3. R. Fleischmann, Ref. 2, p. 21.
4. Proc. 3rd Int. Symp. on Polarization Phenomena in Nuclear Reactions, Madison, (Eds. H. H. Barschall and W. Haeberli, Univ. of Wisconsin Press, Madison, Wisc., 1971).
5. H. F. Glavish, Ref. 4, p. 267. (Also references therein.)
6. B. L. Donnally, Ref. 4, p. 295. (Also references therein.)
7. J. M. Dickson, Progr. Nucl. Tech. Instr. 1 (1965) 105.
8. W. Haeberli, Ann. Rev. Nuc. Sci. 17 (1967) 373.
9. Proc. Symp. on Ion Sources and Formation of Ion Beams, Brookhaven National Laboratory Report, 1971 - BNL50310.
10. H. F. Glavish, Ref. 9, p. 207. (Also references therein.)
11. T. B. Clegg, Ref. 9, p. 223. (Also references therein.)
12. J. L. McKibben, Proc. 2nd Int. Conf. on Polarized Targets (Ed. G. Shapiro, Lawrence Berkeley Laboratory Report, 1971, LBL-500, UC-34 Physics, TID-4500 (58th Ed.)) p. 307.
13. Proc. 2nd Symp. on Ion Sources and Formation of Ion Beams, Lawrence Berkeley Laboratory Report, 1974, LBL-3399.
14. H. F. Glavish, Ref. 13, p. IV-1-1.
15. E. F. Parker, IEEE Trans. Nuc. Sci., NS-22 (1975) 1466.
16. E. F. Parker, N. Q. Sesol, and R. E. Timm, IEEE Trans. Nuc. Sci., NS-22 (1975) 1718; E. F. Parker, private communication.

17. E. P. Chamberlain, R. A. Kenefick,and W. H. Peeler, Texas A. and M. Cyclotron Laboratory Progress Report, 1974, p. 87; E. P. Chamberlain, R. A. Kenefick, and H. Peeler, Bull. Am. Phys. Soc. 20 (1975) 563.

18. A. U. Luccio, D. J. Clark, D.Elo, P. Frazier, H. Meiner, D. Morris, and M. Reukas, IEEE Trans. Nuc. Sci., NS-16 (1969) 140; D. J. Clark, A. U. Luccio, F. Resmini, and H. Meiner, Proc. 5th Int. Cyclotron Conf., (Ed. R. W. McElroy,Butterworths and Co., London, 1971) p. 610; H. J. Conzett, private communication.

19. R. Beurtey, Ph.D. Thesis, U. of Paris (1963); Report CEA-R2366, CEA France (1964); R. Beurtey, et al., Saclay Progress Report CEA-N621 (1966) 81.

20. Th. Stammbach, private communication.

21. F. Barz, E. Dreesen, W. Hammon, H. H. Hansen, S. Penselin, A. Scholzen, W. Schumacher, contribution this conference; W. Schumacher, et al. Nucl. Instr. Meth. to be published.

22. P. Guy, Inst. des Sciences Nucléaires, B.P. 257, F.38044, Grenoble, France, private communication; J. L. Belmont, G. Bagieu, F. Ripouteau, D. Bouteloup, R. V. Tripier, and J. Arvieux, Ref. 4, p. 815.

23. W. B. Powell, Ref. 2, p. 47; also private communication.

24. V. Bejšovec, P. Bém, J. Mareš, and Z. Trejbal, Nucl, Instr. Meth. 87 (1970) 229; 87 (1970) 233; Proc. 3rd Symp. on Isochronous Cyclotron U-120M and Its Use for the Resolution of Physical Problems, Česke Budějovice 14-18 May, 1973, Czechoslovakia, Report JINR, P-9-7339, Dubna USSR 1973, p. 163; J. Mareš, private communication.

25. S. Motonaga, T. Fujisawa, M. Hemmi, H. Takebe, K. Ikegami, and Y. Yamazaki, contribution this conference, p. 837

26. P. A. Chamouard et al., Ref. 60; P. A. Chamouard, private communication.

27. Oak Ridge National Laboratory, Electronuclear Division Annual Reports, a) ORNL-4217 (1967) p. 106; b) ORNL-4404 (1968) p. 96; c) ORNL-4534 (1969) p. 99; d) R. S. Lord, private communication.

28. F. Seiler, private communication.

29. Y. Wakuta, H. Hasuyama, K. Tsuji, M. Sonoda, and A. Katase, Japan J. Appl. Phys. 10 (1971) 622; M. Sonoda, Y. Wakuta, H. Hasuyama, K. Tsuji, and F. Aramaki, Ref. 60; Y. Wakuta, private communication.

30. B. A. McKinnon, J. P. Ruffell, I. J. Walker, A. B. Robbins, R. N. Boyd, and H. F. Glavish, Ref. 9, p. 245; F. T. Baker, G. A. Bissinger, S. Davis, C. Glashausser, and A. B. Robbins, Progress Report of the 16 MeV Tandem Van de Graaff Accelerator, Rutgers University (June 30, 1973) p. 11; A. B. Robbins, private communication.

31. Progress Reports on Nuclear Structure Investigations with an FN Tandem Accelerator, Stanford University (Feb. 28, 1970) p. 126; (Feb. 28, 1971) p. 165; (Feb. 29, 1972) p. 170; (Feb. 28, 1973) p. 148; H. F. Glavish, private communication.

32. E. Steffens, H. Ebinghaus, F. Fiedler, K. Bethge, G. Engelhardt, R. Schäfer, W. Weiss, and D. Fick, Nucl. Instr. Meth. 124 (1975) 601; E. Steffens, et al., contributions this conference and references therein.

33. T. R. Donoghue, W. S. McEver, H. Paetz gen. Schieck, J. C. Volkers, C. E. Busch, Sr. M. A. Doyle, L. Dries, and J. L. Regner, contribution this conference and references therein; H. Paetz gen. Schieck, C. E. Busch, J. A. Keane, and T. R. Donoghue, Ref. 9, p. 263; Annual Report of Nuclear Physics Research at Ohio State Univ. Van de Graaff Accelerator Laboratory (1972-3) p. 5.

34. R. Risler, W. Grüebler, V. König, P. A. Schmelzbach, B. Jenny, and W. G. Weitkamp, contribution this conference; R. Risler, Dissertation Nr. 5395, ETH Zurich (1974).
35. L. Brown, private communication.
36. W. Arnold, H. Berg, G. Clausnitzer, H. H. Krause, and J. Ulbricht, contribution this conference; W. Arnold and G. Clausnitzer, Ref. 13, p. IV-9-1; G. Clausnitzer, private communication.
37. D. Ehrlich, R. Frick, and P. Schiemenz, contribution this conference; Jahresbericht, Beschleunigerlaboratorium der Universität und der Technischen Universität, München (1974) p. 166.
38. V. Bechtold, L. Friedrich, D. Finken., G. Strassner, and P. Ziegler, contribution this conference and references therein; L. Friedrich, private communication.
39. J. L. McKibben and James M. Potter, contribution this conf. p.863; J. L. McKibben, Symp. of No. Eastern Accelerator Personnel, Tallahassee, (March 1972); J. L. McKibben, Proc. Int. Conf. on Technology of Electrostatic Accelerators (Eds. T. W. Aitken and N. R. S. Tait, Daresburg, 1973) p. 379; see also Ref. 11.
40. G. Roy, J. Beveridge, and P. Bosman, contribution this conf. p.862; G. Roy, private communication.
41. R. A. Hardekopf, J. L. McKibben, and T. B. Clegg, Bull. Am. Phys. Soc. 18 (1973) 618; R. A. Hardekopf, et al., contributions this conference, p. 865
42. a) T. B. Clegg, G. A. Bissinger, and T. A. Trainor, Nucl. Instr. Meth. 120 (1974) 445; b) T. A. Trainor, and T. B. Clegg, Ref. 13, p. IV-5-1; c) T. A. Trainor, Thesis, Univ. of North Carolina (1973), available from University Microfilms, 300 N. Zeeb Road, Ann Arbor, Michigan 48106, USA.
43. L. D. Lund, P. R. Hanley, and K. H. Purser, Ref. 13, p. IV-3-1; J. W. McKay, Ref. 13, p. IV-4-1; J. W. McKay, private communication.
44. J. H. Feist, B. Granz, G. Graw, H. Löh, H. Schultz, H. Treiber, contribution this conference; Jahresberichts des Physikalischen Institutes Erlangen-Nürnberg (1974, 1972, and 1970/71).
45. a) T. A. Trainor and W. B. Ingalls, contribution this conf. p.858; University of Washington Nuclear Physics Laboratory Annual Reports; b) (1975) p. 18; c) (1974) p. 13, 73; d) (1973) p. 11; e) (1972) p. 6; f) (1971) p. 6; g) (1970) p. 7.
46. T. B. Clegg, G. A. Bissinger, W. Haeberli, and P. A. Quin, Ref. 4, p. 835; E. J. Stevenson, private communication.
47. Y. Tagishi and J. Sanada, Ref. 9, p. 269.
48. G. Michel, K. Corrigan, H. Meiner, R. M. Prior, and S. E. Darden, Nucl. Instr. Meth. 78 (1970) 261 ; 62 (1968) 203.
49. K. Allenby, G. H. Guest, W. C. Hardy, O. Karban, and W. B. Powell, contribution this conference; O. Karban, S. Oh, W. B. Powell, contribution this conference and references therein; W. B. Powell, private communication.
50. L. Friedrich and H. Paetz gen. Schieck, private communication.
51. D. O. Findley, S. D. Baker, E. B. Carter, and N. D. Stockwell, Nucl. Instr. Meth. 71 (1969) 125; S. D. Baker, private communication.
52. H. Sakaguchi and S. Kobayashi, contribution this conference; Kyoto University Tandem Van de Graaff Laboratory Annual Report (1975).
53. a) J. L. McKibben, R. R. Stevens, Jr., P. Allison, and R. A. Hardekopf, Ref. 13, p. IV-2-1; b) G. P. Lawrence, Ref. 9, p. 257.
54. Y. Tagishi and J. Sanada, contribution this conference; V. Tagishi, private communication.

55. J. Mareš, private communication.
56. P. Schwandt, private communication.
57. S. Oh and A. McIlwain, Ref. 13, p. IV-6-1.
58. E. F. Parker, private communication.
59. H. F. Glavish, private communication.
60. Proc. 2nd Int. Conf. on Ion Sources (Eds. F. Viehbock, H. Winter, and M. Bruck, SGAE, Vienna, Austria, 1972).
61. ANAC, Ltd., P. O. Box 16066, Auckland 3, New Zealand.
62. The source at McMaster was built by General Ionex Corp., 25 Hayward St., Ipswitch, Massachusetts 01938, U.S.A.
63. G. G. Ohlsen and J. L. McKibben, Los Alamos Scientific Laboratory Report LA-3725 (1967).
64. P. G. Sona, Energia Nucleare 14 (1967) 295.
65. R. Risler, W. Grüebler, V. König, and P. A. Schmelzbach, Nucl. Instr. Meth. 121 (1974) 425.
66. W. S. McEver, J. C. Volkers, and T. R. Donoghue, Nucl. Instr. Meth. 127 (1975) (in press).
67. E. Steffens, private communication.
68. D. Fick, private communication.
69. H. F. Glavish, contribution this conference. (p.844)
70. A. Abragam and J. M. Winter, Phys. Rev. Letters 1 (1958) 374; A. Abragam and J. M. Winter, Comp. Rend. Acad. Sci. 255 (1962) 1099.
71. R. Beurtey, Ref. 2, p. 33.
72. H. F. Glavish, Nucl. Instr. Meth. 65 (1968) 1.
73. W. Haeberli, Nucl. Instr. Meth. 62 (1968) 355.
74. R. Beurtey and M. Borghini, Jour. de Physique C2 (1969) 56.
75. W. Hammon and A. Weinig, contribution this conference. (p.846)
76. G. G. Ohlsen, J. L. McKibben, R. R. Stevens, Jr., and G. P. Lawrence, Nucl. Instr. Meth. 73 (1969) 45.
77. W. Haeberli, private communication.
78. W. E. Lamb, Jr., and R. C. Retherford, Phys. Rev. 79 (1950) 549.
79. O. B. Morgan, Thesis, Univ. of Wisconsin (1970), available from University Microfilms, Inc., 300 N. Zeeb Rd., Ann Arbor, Mich. 48106 U.S.A.
80. J. E. Osher, Ref. 9, p. 137; J. E. Osher and J. W. Hamilton, Ref. 9, p. 157.
81. G. P. Lawrence, A. R. Koelle, J. L. McKibben, T. B. Clegg, and G. Roy, Ref. 60, p. 505.
82. M. Bacall, N. Olier, and W. Reichelt, Ref. 9, post-deadline paper.
83. CONVALEX, Bendix Corp., Rochester, N.Y., U.S.A.; SANTOVAC 5, Varian Vacuum Division, Newton, Mass., U.S.A.
84. H. Brückmann, Ref. 4, p. 823.
85. Vu Ngoc Tuan, G. Gautherin, and A. S. Schlachter, Phys. Rev. A9 (1974) 1242.
86. J. L. McKibben, G. P. Lawrence, and G. G. Ohlsen, Phys. Rev. Letters 20 (1968) 1180.
87. W. E. Lamb, Jr. and R. C. Retherford, Phys. Rev. 81 (1951) 222; W. E. Lamb, Jr., Phys. Rev. 85 (1952) 259.
88. G. G. Ohlsen, J. A. Jackson, J. L. McKibben, R. R. Stevens, G. P. Lawrence, and N. A. Lindsay, Los Alamos Scientific Laboratory Report (preprint).
89. G. G. Ohlsen, J. L. McKibben, G. P. Lawrence, P. W. Keaton, Jr., and D. D. Armstrong, Phys. Rev. Letters 27 (1971) 599.
90. R. A. Hardekopf, G. G. Ohlsen, R. V. Poore, and N. Jarmie, contribution this conference. (p. 903)
91. R. A. Hardekopf, private communication.

92. G.G. Ohlsen, K.R. Crosthwaite, and P.A. Lovoi, Bull. Am. Phys. Soc. 19 (1974) 550.
93. G.G. Ohlsen and P.W. Keaton, Jr., Nucl. Instr. Meth. 109 (1973) 41.
94. J.L. McKibben and J.M. Potter, contribution this conference, p.863
95. B.L. Donnally and W. Sawyer, Phys. Rev. Letters 15 (1965) 439.
96. R.E. Olson, Nucl. Instr. Meth. 126 (1975) 467.
97. B.L. Donnally and W. Sawyer, ref. 2 p. 71.
98. L.D. Knutson, Phys. Rev. A2 (1970) 1878.
99. V. Betchold, H. Brückmann, D. Finken, L. Friedrich, K. Hamdi and G. Strassner, Ref. 60, p. 498.
100. M. Froissart, R. Stora, Nucl. Inst. Meth. 7 (1960) 297 ; J. Faure, A. Hilaire and R. Vienet, Particle Accelerators 3 (1972) 225 ; Y. Cho, C. Potts, L. Ratner, A.D. Kirsh, C. Johnson, P. Lefevre and D. Möhl, ZGS Machine Experimental Report, ANL/ARF AE 15-3 (1975) and CERN MPS-DL Note 75-8 (1975).
101. J.A. Van der Heide, Ph. D. Thesis, Eindhoven (1972).
102. H. Jochim, private communication.
103. R. Garrett, private communication
104. G. Heidenreich, M. Daum, G.H. Eaton, U. Rohrer, E. Seiner, contribution this conference, p.847
105. H. Kamitsubo, private communication.
106. P. Feldman and R. Novick, Congr. Int. Phys. Nucl. Paris (1968) p.785.
107. M. Kaminsky, Phys. Rev. Lett. 23 (1969) 819; M. Kaminsky, ref. 4 p.803.
108. L.C. Feldman, Rad. Effects. 13 (1972) 145.
109. W. Brandt and R. Sizmann, Phys. Lett. 37A (1971) 115.
110. C. Rau and R.S. Sizmann, Phys. Lett. 43A (1973) 317.
111. U. Fano, Phys. Rev. 178 (1969) 131.
112. U. Heinzmann, J. Kessler and J. Lorenz, Phys. Rev. Lett. 25 (1970) 1325 Zeitschr f. Physik 240 (1970) 42.
113. R. Shakeshaft and J. Macek, Phys. Rev. Lett. 27 (1971) 1487 ; Phys. ReV. A6 (1972) 1876.
114. J. Arianer, Institut de Physique Nucléaire (Orsay 1975) Report IPNO-PHN-75-01 ; Ch. Goldstein and M. Ulrich, Institut de Physique Nucléaire (Orsay 1975) Report IPNO-75-08; J. Arianer, Institut de Physique Nucléaire (Orsay, May 1975) preprint.
115. E. Donetz and M. Pipkin, JINR (Dubna, 1974) Report P7-7999.

APPENDIX

During discussions at the conference, Dr Clausnitzer suggested other possible schemes for obtaining beams of polarized particles. These included :

1. Optical pumping of ^{3}He (ref. 51).
2. Autoionization of ^{3}He$^-$ in magnetic fields[106].
3. Charge-exchange during channelling through or reflection from thin magnetic crystals[107,110].
4. Photoionization near threshold utilizing the Fano effect [111,112]
5. Laser-light ionization.
6. Ion-atom collisions[73,74,113]
7. Spin-exchange processes with optically-pumped substances.

Research is also begining[114] to determine whether the electron bombardment ionizers in atomic-beam polarized sources can be improved in efficiency for pulsed operation utilizing the techniques developed by Donetz.[115] Basically this scheme involves the storage of the positive polarized particles after ionization. This utilizes a very intense elec-

tron gun and magnetic compression to obtain electron current densities
of $\sim 10^3 A/cm^2$, a much better vacuum in the ionizer region ($< 10^{-10}$ Torr),
and mirror electrodes in the ionizer to confine the polarized ions. The
voltage on these electrodes are then pulsed to allow the polarized ion
beam to emerge when desired. During storage the potential well formed by
the intense electron beam inside the solenoid traps the positive ions.

DISCUSSION

Clausnitzer:

R.E. Olson (Nucl. Instr. and Meth. 126 (1975) 467) published a proposal based on multichannel Landau-Zener calculation of the charge
exchange cross sections $\sigma_{m,-1}$ for different target gases used in
Lambshift polarized sources. He quoted a ratio of approximately 5
for Krypton with respect to Argon, the selectivity still remaining
12 (Argon: 16). We have measured the negative ion yield ratio for
Krypton vs. Argon and obtained 0.5 for equal target thicknesses; the
optimum target thickness for Krypton is smaller (0.6) but the optimum
yields still lead to a ratio of only 0.6.

An explanation of this discrepancy would rest on the different reduction cross sections in the entrance and exit channel. The known
values of $\sigma_{-1,0}$ lead to a Krypton-Argon ratio of approximately 6/5
and can not explain the discrepancy. Quenching cross sections have
not been measured at the relevant energy; a crude extrapolation of
this values would lead to a ratio of 3/2 and can again not explain the
measured yield ratio. As a consequence,we presently think, that the
calculations are not reliable enough, to give absolute cross section
values'.

The energy dependence of the negative ion yield ratios for Krypton
and Argon adder gases were measured in Karlsruhe by Friedrich and
Bechthold. The same result (0.5) was obtained for 1 keV metastable
deuterium atoms; the ratio remains almost constant up to 1.3 keV and
decreases for higher energies similarly as for energies smaller than
1 keV.

STRUCTURE AND REACTIONS OF LIGHT NUCLEI

H.H. Hackenbroich

Institut für Theoretische Physik
der Universität zu Köln, 5 Köln 41, Germany

Within the period following the Madison Polarization Conference the most important contribution of polarization measurements to Few Nucleon Physics does not result from the collection of new or better data on reactions which were chosen randomly or according to the facilities of a laboratory. On the contrary: a better understanding of relevant phenomena has emerged which allows for a more fundamental theoretical description of few nucleon structures and their reactions and leads to quantitative or semiquantitative predictions for scattering matrices which can be compared to experiment. Consequently, we can say that a carefully selected experiment will result in very specific information about physical processes, and not only in numbers.

Due to this state of affairs, I will present my talk in three parts: first, I will deal very briefly with some experimental results; second, I will sketch current theoretical ideas and, third, I will devote the main part of my talk to the discussion of important phenomena.

I. Experimental Data

First I wish to mention the reports by Fiarman and Meyerhof[1] on data concerning $A = 4$ nuclear systems and by Ajzenberg-Selove and Lauritsen on systems with $A = 5 - 10$[2]. Several papers are given in the reports of other conferences, e.g. in the proceedings of the Conference on Few Particle Problems in the Nuclear Interaction, Los Angeles 1972[3] and those of the Conference on Clustering Phenomena in Nuclei, Maryland 1975[4]. There are also review papers on reactions with polarized deuterons by F. Seiler[5], D. Fick[6] and by the ETH Zürich[6a].

Second, I should report that the general formalism for polarizations given for uncharged particles by Welton[7] has been extended to the case of charged particles by P. Heiss[8]. A quite general computer program has been worked out by F. Seiler[9] based on the Welton formalism, and the extension for the Heiss formulation has been programmed by H. Aulenkamp[10].

Although there is an overwhelming wealth of new data on the $A = 4$ systems, I will deal only with those which concern two fragment reactions, keeping in mind that three particle break-up also provides data valuable to theoreticians.

The elastic scattering of polarized deuterons by deuterons has been studied by Grüebler et al.[11], Schulte et al.[12], and by Conzett et al. in a contribution to this conference; the latter data seem to be in conflict with earlier data which were approximately smaller by a factor of 2. The reactions $^2H(d,p)^3H$ have been measured by Jeltsch et al.[13], Gückel and Fick[14], Ad'yasvich et al.[15], Grüebler et al.[16], Ying et al.[17], Hardekopf et al.[18], Nemets et al.[19], Zaika et al.[20], Berovic and Clews[21], in addition to contributions to this conference, by Grüebler et al. and Conzett et al. The charge symmetrical reaction $^2\vec{H}(d,n)^3He$ has been given attention by some of the authors just mentioned, by Salzman et al.[22], and by Blyth et al.[23]. For the same reaction we also have new data on polarization transfer measurements:

^{2}H($\vec{d}$,$\vec{p}$)^{3}H by Grüebler et al.[24], Clegg et al.[25], and
^{2}H($\vec{d}$,n)^{3}He by Lisowski et al.[26]. Proton polarizations in the ^{2}H(d,$\vec{p}$)^{3}H
reaction have been measured by Hardekopf et al.[27]and the neutron polari-
zation in the ^{2}H(d,$\vec{n}$)^{3}He reaction by Spalek et al.[28], Hardekopf et
al.[29], Maayouf and Galloway[30], Galloway et al.[31], and by Sikkema and
Steendam[32].

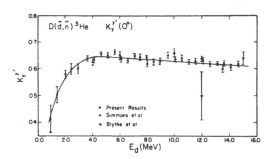

Fig.1. Results for $K_y^{y'}$ (0°) for D($\vec{d}$,$\vec{n}$)^{3}He from ref. 47

The main results from these measurements seem to be that 1) the deuteron
polarization in elastic d + d scattering is not qualitatively different
in magnitude from deuteron polarizations found in deuteron scattering on
nucleons, indicating that for small energies the deuterons cannot
approach each other very closely; consequently the short-ranged spin-
orbit forces cannot act strongly, 2) the deuteron vector analyzing power
in the two mirror break-up reactions seem to come out quite similarly
now[23] (cf. fig.2), and 3) there is some irregularity in the vicinity of

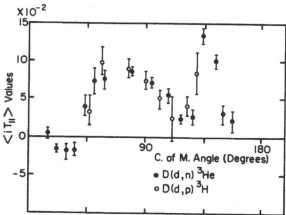

Fig.2. Comparison of iT_{11} values for ^{2}H($\vec{d}$,n)^{3}He and ^{2}H($\vec{d}$,p)^{3}H at
12.3 MeV, from ref. 23

the d - d threshold[14] which, however, cannot be interpreted as a 2$^+$
state, contrary to the previous discussion (cf. also sec.III,3) as
shown in fig. 3.

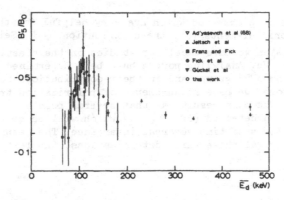

Fig. 3. Expansion coefficient B'_5 of $P_2(\theta)$ for the tensor analyzing power divided by the coefficient B_0 of the angle-independent part of the cross section for the $^2H(d,p)^3H$ reaction; from ref. 14.

Other reactions in the A = 4 systems which have been considered involve either a triton or a ^{3}He both in incoming and outgoing channels. Polarizations or analyzing powers have been given by Hardekopf et al.[33], Darvis-Blanc[34], Baker et al.[35], McSherry and Baker[36], Brown and Rolwer[37], Jarmie and Jeff[38], Jarmer et al.[39], Seagrave et al.[40], Szaloki et al.[41], in addition to contributions to this conference. Phase shifts have been presented by Hardekopf et al.[33], Kilian et al.[44] and by Fick et al. to this conference. The results of these measurements also seem to show that there is no significant difference between mirror reactions. Nevertheless, to my knowledge there is no experimental determination of the complete set of $^3H(p,p)^3H$, $^3H(p,n)^3He$ and $^3He(n,n)^3He$ matrix elements which would allow the construction of S-matrix eigenvectors which also should be eigenvectors of isospin approximately.

From the measurements on A = 5 systems, I should mention the polarization or analyzing power measurements made by Holl et al.[45], Hickey et al.[46], Lisowski et al.[47] and the contribution by Bond and Fick to this conference. Measurements of this kind lead to several phase shift analyses, e.g. those by Stammbach and Walter[48] and by Plattner et al.[49]. An R-matrix analysis for p·^{4}He scattering has been presented by the Los Alamos group[50].

There are many measurements which are concerned with the elastic scattering of deuterons by ^{3}H or ^{3}He. The deuteron analyzing power has been given by König et al.[51], Debenham et al.[52], Tanifuji and Yazaki[53], Donaghue et al.[54] and in contributions to this conference. Spin correlation measurements have been performed by Ohlsen et al.[55] and Lavoi et al.[56]. Deuteron analyzing powers for the reactions $^3H(\vec{d},n)^4He$ or $^3He(\vec{d},p)^4He$ have been determined by the Zürich group[57], Simon et al.[58], Traimov et al.[59], Garrett and Lindstrom[60] and are reported here. ^{3}He analyzing powers have been determined by Rohrer et al.[61] whereas proton polarizations are given by Leemann et al.[62] and Clare[63]. A polarization transfer measurement has been performed by Hardekopf et al.[64]. The main results of these studies seem to be that the discrepancy seen before between p-^{4}He and n-^{4}He scattering (especially in the vicinity of the 3/2$^+$ state) is negligible now and that points of maximal

analyzing power have been observed which are very helpful for the determination of S-matrix elements (cf. the contribution by F. Seiler).

In the A = 6 system we have excellent studies on the elastic scattering of deuterons by ^{4}He. Analyzing powers have been determined by the
Zürich[65] and Los Alamos[66] groups and in their contributions to this
conference. Furthermore, we have measurements of polarization transfer
for this process[67]. The main result was that several points of maximal
polarizations have been detected (cf. fig. 4) which will be very helpful
in providing better tests of time reversal invariance. The results have
been summarized in several phase shift determinations, the last being

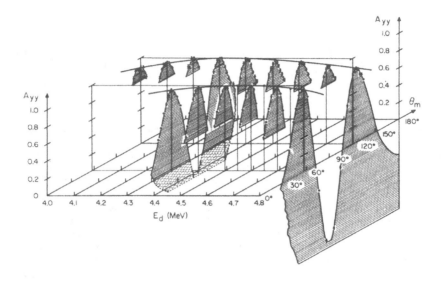

Fig. 4. Analyzing power A_{yy} as function of angle and energy for α + d
elastic scattering; from ref. 68

performed by the Zürich group[68]. The reaction ^{4}He(d,^{3}He)^{3}H has attracted
much attention because, close to the threshold, we find a large violation
of the Barshey-Temmer theorem (cf. sec.III); the deuteron analyzing power
has been given by Dahme et al.[69]. I have been especially pleased to see
the contribution by Lisowski, Byrd, Walter and Clegg which shows that
also in 3 body break-up measurements polarization transfers are measured;
this will provide new and important tests for theoretical approaches.

Unfortunately, I cannot mention all the measurements of larger
systems. In addition to the contributions to this conference, I will only
quote the ^{6}Li($\vec{n}$,n) work by Holt et al.[70], the ^{6}Li($\vec{d}$,α)^{4}He measurements
by the Zürich[71] and Hamburg[72] groups and the ^{7}Li($\vec{p}$,n)^{7}Be experiment by
Rohrer and Brown[73].

Besides the fact that I - as a theoretician - may have overlooked
important measurements, I have presented a somewhat biased selection of
papers: In my opinion, it is the most important goal of present experiments to provide as much data as possible for a set of nuclear systems in

order to present tests which can be used to check theories <u>in detail</u>. The
papers presented here seem to show that this is the goal of experimental
physicists, too.

II. Theoretical Approaches

Within recent years the main effort of theoreticians has been devoted
to the problem of reactions involving light nuclear systems and not to the
study of bound state structure. It is impossible to describe all important
contributions to this field here; I will limit myself to the main ideas
which are currently in the field.

First I wish to mention that some progress has been made to extend
the ideas which have been successfully used in three body research since
Faddeev presented his equations. The most promising approach has been recently formulated by Sandhas who was able to give exact integral equations
for the N-body case built in a similar spirit to the AGS equations[74].
The main advantage of all these equations seemed to be that one had to
solve integral equations for reaction processes which showed one pair of
particles bound and one particle free in incoming and outgoing channels.
The 3 body break-up reaction matrix elements could be computed once the
simpler ones had been obtained from the solution of the integral equations. The equivalent statement is true for the extended theory. This
demands, however, that, e.g., for the solution of the 4 body system the
solution of the AGS 3 body equations for the ^{3}H and ^{3}He ground states has
to be used and therefore to be computed first. Thus we must expect that
it will take some time before realistic calculations using this method
can be provided, - but these will be very interesting.

Second, several attempts have been made to use the continuum shell
model (cf. Mahaux and Weidenmüller[75]). I should like to mention here
some papers which were concerned with the ^{3}He + n and ^{3}H + p channels in
the ^{4}He scattering system by Londergan and Shakin[76] and by Ramavataram
et al.[77]. These authors give a reasonable description for the reactions
^{4}He(γ,p) and ^{4}He(γ,n), the latter ones also presenting fair values for
the polarizations between ^{3}H + p and ^{3}He + n. Up till now only simplified
potentials have been used in this work.

Third, many theoreticians are trying to treat reactions between
light nuclei by assuming that there are substructures present which may
be regarded as "elementary", e.g. α-clusters; consequently not only
distortion effects due to the mutual interaction of the reaction partners
are ignored but also wave functions are used which are not antisymmetrized
correctly with respect to nucleon coordinates. This procedure may be justified for the work of D.S. Chuu et al.[78] who treat the scattering states
of the ^{6}Li system by using p,n and α as "elementary", nevertheless they
also introduce the α^*, the first excited state of the α. The interaction
between these "elementary" particles is taken from suitable experimental
results; the 3 particle problem is treated by solving a Faddeev equation.
The agreement between theory and experiment is satisfactory. However, in
addition, one tries to use the basic idea mentioned before for heavier
systems and for such "elementary" clusters like ^{3}He and ^{3}H. Of course
one would run into serious difficulties if the Pauli principle would not
be taken into account in some approximation. Thus one employs the so-
called "orthogonality condition model" proposed originally by Saito[79];
i.e., one requires that all wave functions used should be orthogonal on

the Pauli-forbidden wave functions in some suitable harmonic oscillator. These orthogonality conditions can be evaluated in a simple way, however, only if all fragment wave functions are taken to be simplest shell model functions with the same width constant throughout. This procedure may be questionable; in any case one is forced to use simple potential models to be consistent with the oversimplifications made concerning the wave functions. Nevertheless, the results obtained so far for the α-like nuclear structures (e.g., the review by Brink in ref. 4 are promising; therefore several groups have started to apply these methods to other systems[80,81]).

Fourth, methods similar in spirit to field theoretical ones have been used in few nucleon research. I should mention the approach attributed to Plattner which he will discuss himself later. Partial wave dispersion relations solved by N/D methods for the coupled ^{4}He + n and ^{3}H + d channels of the ^{5}He system have been treated by Stingl and Rinat[82]. The "driving" terms of these equations have been constructed from an analysis of A = 4 subsystem experiments. The agreement between their calculations and experiment is only qualitative but the method seems to be promising.

Fifth, cluster model ideas have been used in several distinct ways. The Minnesota group (cf. the review by Tang and Thompson in ref. 4) mainly wants to study the structure of equations in order to obtain an improved understanding of such quantities as the optical potential. Consequently these authors concentrate on the study of elastic scattering or on reactions between two types of fragmentations only; other reaction processes are accounted for by the introduction of an imaginary potential (e.g., for n + ^{3}H elastic scattering[83], for A = 5[84], and for A = 7 systems[85]). A very careful analysis is presented for the structure of the interaction between the fragments, especially of the difference between these interactions in even and odd states. The parameters of the internuclear interaction used in these studies are determined in a way that cross sections are reproduced. Similar in spirit are calculations performed by the Tübingen group (cf. Wildermuth in ref. 4). The calculations were aimed to provide examples from few nucleon physics which could clarify the role of specific reaction mechanisms – as in the work by Teufel[86] on the two coupled 5/2⁻ channels in the ^{7}Li or ^{7}Be system. Some other groups also work along parallel lines, e.g., Shakin and Weiss on the γ-break up of ^{6}Li[87] and Leclerc-Willain and Libert-Heinemann on ^{5}He resonant states[88]. The main goals of our Cologne group have been the elucidation of different parts of the internucleon force in nuclear reactions and the interplay of different channels, especially those with unstable but resonant substructures. Thus we were not allowed to use very simple wave functions or to neglect important channels – accounting for absorption by introducing imaginary potentials. This demands that many coupled channels have to be considered. As a result new mathematical and computational techniques had to be developed – from efficient handling of antisymmetrisation and quadrature problems to a better handling of many coupled reaction channels. These methods have been described in a review by myself[89]. We were therefore able to treat several light nuclear systems, taking into account various fragmentations. The model for the internucleon forces used in our calculations has been derived from two body data and has been used consequently – i.e., all fragment wave functions were determined according to the same assumptions about the forces. Our calculations described cross sections and polarizations reasonably well; moreover, for some nuclear systems even the three body break up could be considered in an approximate but consistent way[90]. A review of some results has been recently given by myself in ref. 4.

III. 1. The Nucleon-Nucleon Force

Most physicists believe that the three nucleon system is most suited to study details of the internucleon forces: It is the simplest system which shows off-shell effects and possibly many-body interactions. Obviously, the theoretical analysis of this system is, though complicated, less involved than work on heavier systems; therefore most recent theoretical work could be based on Faddeev or Faddeev-like exact equations. Nevertheless, I doubt that in the present status of knowledge about the internuclear force three body studies will lead to the solution of current problems, e.g., the determination of the off-shell behaviour or to a better determination of spin flipping forces. My reasoning is based on two facts: first, the three body systems are bound very loosely, and thus the main part of the interaction is a two-body force taken to be on shell, and second, the deuteron-nucleon elastic scattering does not lead to any appreciable large polarizations. The phase shift analysis shows that we do not see any resonance-like structures; the polarizations are due to interference effects on many S-matrix elements which are unsplit at first glance. The interpretation of the - extremely accurate - measurements is therefore complicated and extremely doubtful at the moment. Nevertheless, the three body data will be very helpful for the interpretation of finer details of the forces in the future.

For the present time we should study other systems which allow for a simpler analysis of the data. The goal should be the selection of processes which are very sensitive on parts of the interaction; thus we might hope to check assumptions made in currently used potential models.

I came across one of these interesting situations by reading the contributed papers to this conference, specifically the paper by D. Fick et al. which gives a phase shift analysis for elastic p-^{3}H scattering. Here the authors compare their results with a calculation made by P. Heiss and myself and claim a difference between experimental and theoretical p-wave phases, especially in the 0^- and 2^- cases for very low energies. We repeated this calculation - which had been performed for enough energies that our results were technically correct - and added also the "small" components of the force, namely the odd central and tensor and the even spin-orbit strength compared to the previous calculation. The change of results is dramatic as seen from fig. 5. The addition of the "small" forces changes the ordering of levels from 2^-, 0^- to 0^-, 2^- ; i.e., now the tensor forces are able to determine completely the level structures for these T = 0 levels. The 2^- level is pushed up into the region of the T = 1 level with 2^- which also decreases the coupling between the p-^{3}H and n-^{3}He channels. This is even more drastically seen from the calculation in which the spin-orbit strength was reduced by a factor .5. These different choices of the intranucleon potential almost do not affect the T = 1 resonances which are governed by the spin-orbit forces - whereas the negative parity T = 0 resonances are due to a complicated balancing of tensor and spin orbit interactions which is sensitive to the central potential also due to the difference in range of these spin flipping interactions. Of course, these results will force us to deal with these reactions again.

Another interesting check for the assumptions about spin dependent forces can be obtained from the comparison of elastic deuteron scattering by nucleons, deuterons, tritons (or ^{3}He) and α-particles. For such a study deuterons are better suited than nucleons - despite the larger complexity of the analysis - due to the fact that in all cases we see the effect of the tensor forces even when only the most simple parts of the target

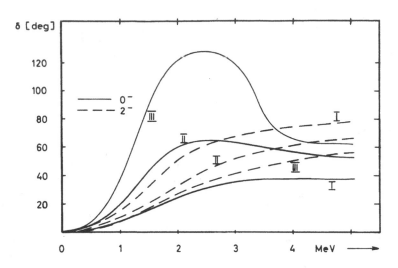

Fig. 5. Real parts of L = 1 p-^{3}H elastic phases for 0$^-$(——) and 2$^-$(---)
resulting from various assumptions about internucleon forces: I previous calc. II odd central, even spin orbit, odd tensor forces added;
III only half of spin orbit strength compared to II.

nucleus wave functions are considered. The comparison of different processes is necessary in order to clarify the role of range effects of the
interaction and of nuclear structure. In deuteron-nucleon scattering the
deuteron experiences only a relatively weak interaction which does not
allow a low energy nucleon to penetrate the centrifugal barrier; thus
the main spin-flipping effects are due to the long ranged tensor force.
On the other hand, for deuteron-α elastic scattering the force is
strongly attractive in the L = 2 case; consequently, resonances appear
which are split mainly due to the short ranged spin orbit force (cf.
fig. 2-4 of my article in ref. 4). The situation is less clear for the
d-d and d-^{3}H cases because we have more channel spins and competing
reaction processes here. We do not have a convincing phase shift analysis
for these scattering processes yet, and the theoretical description of
these processes still poses problems. For example, there are good resonating group calculations for d-d elastic scattering[91] but the complete
S-matrix (including e.g. all two fragment reactions) has never been worked out. In the five nucleon case a calculation exists which includes
all two fragment processes and - in some approximation - even the most
important three-body break-up process[92]; but this calculation shows a
sequence of D wave resonances ordered as 7/2$^+$, 5/2$^+$, 3/2$^+$, 1/2$^+$ with
phases which have phase shifts differing not too much. The experimental
situation is still unclear[5] but it is believed that these resonances are
not strictly spin-orbit ordered.
    I should also mention the interesting result reported by H.E. Conzett
et al. to this conference, that the vector analyzing power in the reaction
^{2}H(d,p)^{3}H is approximately antisymmetric around 90° for higher energies.
This was interpreted as the consequence of a direct nucleon-transfer process. One must, however, be very careful in using model considerations
like these. The antisymmetry relation is valid providing that the channel spin - or, practically, the spin - be conserved in the reaction;

thus only the central and some "diagonal" parts of the spin flipping forces seem to be essentially responsible for the reaction.

In this context I wish to express my hope that in the future data will be supplied on polarizations of fragments which are produced in γ-break up reactions (ref.9). At first glance it seems unnecessary to take these data because one might argue that the geometry of spins and orbital angular momentum of the outgoing fragments is fixed by the type of radiation responsible for the reaction which can be determined by less sophisticated measurements (provided that the target nucleus wave function is assumed to be sufficiently simple). But this argument is wrong if the scattering state wave function shows many strongly coupled channels (cf. e.g. ref. 94). Even in those cases in which the transition matrix elements from the ground state are large to parts of the scattering state wave functions only, we do not have outgoing fragments only in this channel due to the effect of the strong interactions. Thus, e.g., in the γ-break up of the α particle (which proceeds at low energies mainly via E1 radiation) we expect to see not only $n + {}^3He$ or $p + {}^3H$ with channel spin 0. The matrix elements for these transitions seem to be sensitively dependent on the assumptions made for the wave functions; therefore, we might expect that measurements of the type mentioned will provide new tests for the theory and especially for the basic assumptions about the internucleon force.

III. 2. The Importance of Substructures

S-matrices are unitary matrices; thus, the coupling between channels must have consequences for elastic scattering too. This well-known fact has been clarified in many model calculations (cf. e.g. ref. 95). Few nucleon physics has provided some examples in which complete S-matrices have been calculated; resonances in one channel have been proved to be the reason for rapid variations of S-matrix elements describing elastic scattering in other channels[96]. Therefore the possible structure of S-matrices seemed to be well understood; on the other hand, this knowledge was regarded to be of quite limited advantage because in most scattering systems even at low energies, three- (and more) fragment break-up processes can occur which correspond to S-matrices of infinite rank.

However, this difficulty in the application of well established S-matrix considerations is very often an academic one because in light nuclear systems the break-up processes seem to proceed – at least for low energies – via intermediate long living substructures, which are known as resonances in corresponding lighter systems. This dominance of two-step processes can be easily checked if one compares the known lifetimes of these resonances with a typical time for staying in the region of interaction. Due to these considerations we understand the success of the Watson-Migdal-model of three body break-up processes[97] also and especially for scattering systems with A>3; here one divides the reaction into two steps – first the formation of the unstable substructure, then its decay occuring outside the region of interaction with the third particle. In the framework of the resonating group theory an equivalent approach has been formulated by P. Heiss[98]. But, if the three body break-up (at least outside the spectator peaks) can be approximated by a step process model, we can use the methods suited for two fragment reaction processes to describe the first step; we will particularly have an S-matrix with only finite (and hopefully) small number of coupled channels for which well-known considerations may be applied.

This idea ist most important for a theoretical description which

wants to treat <u>all</u> possible reaction processes in a given scattering
system in a managable way. An example might be the calculation made for
the ^{6}Li system[90] for relatively small energies. Here one could see that
the results depended sensitively on the inclusion of the channels which
accounted for the three body break-up. The results have been compared
with the deuteron elastic polarizations measured by the Zürich group (cf.
also ref.4). For the break-up only preliminary experimental data are
available up to now, but their comparison with theoretical predictions
seems encouraging.

Thus the inclusion of resonant substructures is important for a theoretical analysis; at first glance, however,, one might think that this
will lead in essence to a microscopic foundation of the imaginary potential - interesting but not exciting. There are, however, other consequences of the idea which cannot be re-expressed in such an orthodox framework.

The first result from the fact that these unstable substructures very
often have large angular momenta; thus in the first step of the reaction
polarizations of these outgoing substructures of high rank can occur -
e.g., in ^{4}He + ^{3}He → ^{6}Li* + p the ^{6}Li* can have polarizations up to rank
6. If the orbital momentum coupling scheme for the substructure wave
functions is unique one can measure these polarizations by measuring the
three body break-up cross section[99]. Measurements of this kind will be
very helpful in determining observables which are sensitive on spin-
flipping internucleon forces as shown in ref. 90.

Second, this interpretation very often provides an easy understanding
of the quantum numbers of resonances; this is partly due to the fact that
one of the most important substructures - the α^* - has exactly the same
quantum numbers as the α particle itself. We have in the ^{5}Li and ^{5}He
scattering systems not only the two low-lying 3/2$^-$and 1/2$^-$ α-nucleon
resonances, but also 3/2$^-$ and 1/2$^-$ α^* - nucleon resonances some 20 MeV
above the first doublet (cf. fig. 6).

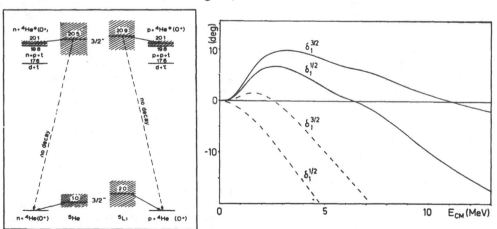

Fig. 6. Level scheme of ^{5}He and ^{5}Li, compared to results for d + ^{3}H
phase shifts for L = 1, S = 1/2, calculated with (———) and without
(---) coupling to α^* + p channels; from ref. 92.

Third, one knows that the Barshay-Temmer[100] theorem is heavily violated in the case of the α + d→^{3}He + ^{3}H reaction in an energy region

several MeV above threshold, as measured by G.J. Wagner et al.[101] and by
A. Richter and co-workers[102] (cf. fig. 7). This effect cannot be connected with a three body break-up leading to α + p + n, as expected. But the
threshold structures in the ^{6}Li system[2] may tempt us to say that the
coupling of these channels to the n + ^{5}Li[**] and p + ^{5}He[**] channels -
^{5}Li[**] and ^{5}He[**] denote the 3/2$^+$ resonances - may cause this isospin violation due to the difference in energy of the two 3/2$^+$ states; thus one
should search for ^{3}He + d + n or ^{3}H + d + p break-up processes. Strong

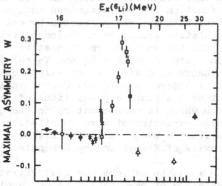

Fig. 7. Maximal asymmetry W(θ) for the reaction ^{3}H(^{3}He,d)^{4}H as function
of excitation energy in the ^{6}Li system; from ref. 102.

indications result from observations by the Freiburg group[103] and by
results of a calculation which is currently being performed by our
Cologne group (but not yet finished). Fig. 8 shows a comparison of absolute values for some α + d→^{3}He + ^{3}H reaction S-matrix elements due to
our results; the figure includes one S-matrix element which should vanish

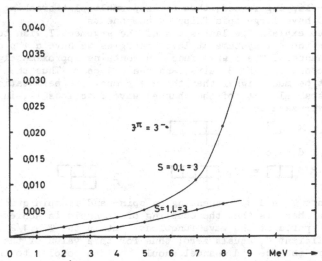

Fig.8. Calculated absolute values for two reaction matrix elements in the
J^π = 3$^-$ state for the ^{4}He(d,^{3}He)^{3}H reaction; Symbols denote final channel
quantum numbers; Upper curve: "allowed", lower curve: "forbidden" transition.

exactly if the isospin would be conserved in this reaction. In the theoretical analysis this is mainly a consequence of the presence of the
above mentioned substructure channels.

III.3. Tests for the Supermultiplet Theory

The general success of the Barshey-Temmer theorem which I mentioned
above shows that even the extreme version of the supermultiplet theory
is close to reality. In this restricted theory one not only assumes that
every bound state wave function can be factorized into a spatial and a
spin-isospin part and that both are characterized by one irreducible representation of the permutation group; in addition to this the spatial
parts of wave functions for the same supermultiplet (i.e. for wave functions which differ in the quantum numbers S_z and T_3 only) are taken to
be equal. This extreme theory also gives an excellent qualitative description for ft values for β-decay processes. Consequently one is
tempted to search for other qualitative predictions of this basic assumption - a line of research which is complementary to the quantitative
work in which a systematic expansion of wave functions according to the
complexity of Young diagrams is used - this quantitative work being summarized, e.g., by the review of Kramer and by Schenzle and Kramer in
ref. 4.

We need a few simple ideas to achieve our goal. First, we note that
the irreducible representation of permutation groups would be a good
quantum number if the internucleon force would be of pure Serber type.
Therefore the supermultiplet theory is not restricted to bound states
but may be applied to scattering states as well[104] - there is no connection to boundary conditions whatsoever. Second, the main component of
the force can be approximated by a Serber force, such that the interaction favours even pairs of nucleon and thus the simplest Young diagrams
allowed by the Pauli principle. Third, the main part of the difference
between the true force and the approximating Serber interaction results
from the spin-flipping forces; thus the supermultiplet-breaking interactions should have large spin-flipping components.

In order to explain the last steps of the argument I wish to present
an example: In the ^{5}Li systems at lower energies we have α + p channels
and ^{3}He + d channels. The α wave function contains approximately 6 even
pairs of nucleons, the ^{3}He 3 pairs, and the d 1 pair. Thus the ^{3}He + d
threshold must be much higher than the α + p one. The permutational symmetry of the spatial parts of the channel wave functions is, in the
α + p case, expressed as

$$\boxed{\square\square\square\square} \times \square = \begin{array}{l}\boxed{\square\square\square\square}\\ \boxed{\square}\end{array}$$

and in the ^{3}He + d case

$$\boxed{\square\square\square} \times \boxed{\square\square} = C_1 \begin{array}{l}\boxed{\square\square\square\square}\\ \boxed{\square}\end{array} + C_2 \begin{array}{l}\boxed{\square\square\square}\\ \boxed{\square\square}\end{array}$$

the coefficients C_1 and C_2 depending on spin- and isospin quantum numbers. The idea here is that the coupling of channels is achieved mainly
by the $\begin{smallmatrix}\square\square\square\square\\ \square\end{smallmatrix}$ parts of the wave functions. But in the T = 1/2, S = 3/2
case the coefficient C_1 equals zero; thus for this value of the channel
quantum numbers the ^{3}He + d channel should be less coupled to the α + p
channels.

The last part of the argument is that for these quantum numbers the
number of even pairs in the interaction region is much larger than in
the asymptotic region. Thus we expect a resonance close to threshold for

relative orbital momentum zero; the $3/2^+$ state in ^{5}Li. The same reasoning holds for a family of other states, as explained by Seligman and myself [105]; our reasoning was based on counting the number of even triplet pairs. Niewisch and Fick have pointed out since that one should better count the number of even pairs [106].

Much experimental evidence for these threshold states has been collected in recent years [5], even in cases which were not listed in the original paper. Thus I think that the supermultiplet scheme - used as an expansion scheme in quantitative work - seems to be an excellent basis for qualitative reasoning in the field of reactions between light nuclei.

I wish to express my appreciation to Drs. P. Heiss and T.H. Seligman for their continuous good collaboration in these studies.

References

1) S. Fiarmann and W.E. Meyerhof, Nucl. Phys. A206 (1972) 1
2) F. Ajzenberg-Selove and T. Lauritzen, Nucl. Phys. A227 (1974) 1
3) I. Slaus, S.A. Moszkowski, R.P. Haddock and W.T.H. van Oers, editors, Few Particle Problems in the Nuclear Interaction (North-Holland Publ. Comp. Amsterdam 1972)
3a) Proceedings of the Munich Nuclear Physics Conference (North-Holland Publ. Comp. Amsterdam 1973)
4) Clustering Phenomena in Nuclei (University of Maryland Press, College Park, Md. 1975)
5) F. Seiler, Nucl. Phys. A187 (1972) 379; A244 (1975) 236
6) D. Fick, Kernphysik mit polarisierten Deuteronen (Bibliographisches Institut, Mannheim 1971)
6a) Results of Measurements and Analyses of Nuclear Reactions Induced by Polarized and Unpolarized Deuterons (ETH Zürich internal report 1973)
7) T.A. Welton, in Fast Neutron Physics II (Interscience, New York-London 1963), p.1317
8) P. Heiss, Z. Physik 251 (1972) 159
9) F. Seiler, Comp. Phys. Comm. 5 (1973) 229
10) H. Aulenkamp, Diplomarbeit Cologne 1973 (unpublished)
11) W. Grüebler et al., ref. 3, p. 636; Nucl. Phys. A193 (1972) 149
12) R.L. Schulte et al., Nucl. Phys. A192 (1972) 609
13) K. Jeltsch et al., Helv. Phys. Acta 43 (1970) 279
14) F.A. Gückel and D. Fick, Z. Phys. 271 (1974) 39
15) B.P. Ad'yasevich et al., Sov. J. Nucl. Phys. 11 (1970) 411
16) W. Grüebler et al., Phys. Lett. 36B (1971) 337; Nucl. Phys. A193 (1972) 129
17) N. Ying et al., Nucl. Phys. A206 (1973) 481
18) R.A. Hardekopf et al., Phys. Rev. Lett. 28 (1972) 760
19) O.F. Nemets et al., Izv. Akad. Nak SSSR Ser. Fiz. 38 (1974) 746
20) N.I. Zaika et al., Ukr. Fiz. Zh. (SSSR) 19 (1974) 926
21) N. Berovic and C. Clews, to be published
22) G.C. Salzman et al., Nucl. Phys. A222 (1974) 512
23) C.O. Blyth et al., to be published
24) W. Grüebler et al., Nucl. Phys. A230 (1974) 353
25) T.B. Clegg et al., Phys. Rev. C8 (1974) 922
26) W. Lisowski et al., Nucl. Phys. A242 (1975) 292
27) R.A. Hardekopf et al., Nucl. Phys. A191 (1972) 468; 481
28) G. Spalek et al., Nucl. Phys. A191 (1972) 449
29) R.A. Hardekopf et al., Nucl. Phys. A191 (1972) 460
30) R.M.A. Maaouf and R.B. Galloway, Nucl. Inst. Meth. 118 (1974) 343

31) R.B. Galloway et al., Nucl. Phys. A242 (1975) 122
32) C.P. Sikkema and S.P. Steendam, Nucl. Phys. A245 (1975) 1
33) R.A. Hardekopf et al., Nucl. Phys. A191 (1972) 481
34) R. Darvis-Blanc et al., Lett. al Nuovo Cim. 4 (1972) 16; ref.3),
    p. 617
35) S.D. Baker et al., Phys. Rev. 178 (1969) 1616
36) D.H. McSherry and S.D. Baker, Phys. Rev. C1 (1970) 888
37) L. Brown and U. Rohrer, Nucl. Phys. A221 (1974) 325
38) N. Jarmie and J.H. Jeff, Phys. Rev. C10 (1974) 57
39) J.J. Jarmer et al., Phys. Rev. C9 (1974) 1292
40) J.D. Seagrave et al., Ann. Phys. (USA) 74 (1972) 250
41) G. Szaloki et al., Helv. Phys. Acta 47 (1974) 91
42) P. Schwandt et al., Nucl. Phys. A163 (1971) 432
43) J. J. Jarmer et al., Phys. Rev. C10 (1974) 494
44) K. Kilian et al., ref. 3), p.667
45) R.J. Holt et al., Nucl. Phys. A213 (1973) 147
46) G.T. Hickey et al., Nucl. Phys. A225 (1974) 470
47) P.W. Lisowski et al., Nucl. Phys. A242 (1975) 298
48) Th. Stammbach and R.L. Walter, Nucl. Phys. A180 (1972) 225
49) G.R. Plattner et al., Phys. Rev. C5 (1972) 1158
50) N. Jarmie et al., ref. 3a), p.132
51) V. König et al., Nucl. Phys. A185 (1972) 263
52) A.A. Debenham et al., Nucl. Phys. A216 (1973) 342
53) M. Tanifuji and K. Yamazaki, ref. 3), p. 739
54) T.R. Donoghue et al., Phys. Rev. C10 (1974) 571
55) G.G. Ohlsen et al., Nucl. Phys. A233 (1974) 1
56) P.H. Lavoi, Phys. Rev. C9 (1974) 1336
57) W. Grüebler et al., Nucl. Phys. A165 (1971) 505; V. König et al.,
    Nucl. Phys. A166 (1971) 393; W. Grüebler et al., Nucl. Phys. A176
    (1971) 631; V. König et al., Nucl. Phys. A185 (1972) 263
58) W.G. Simon et al., ref. 3), p. 735
59) T.A. Trainor et al., Nucl. Phys. A220 (1974) 533
60) R. Garrett and W.W. Lindstrom, Nucl. Phys. A220 (1974) 187
61) V. Rohrer et al., Helv. Phys. Acta 44 (1971) 846
62) Ch. Leemann et al., Helv. Phys. Acta 44 (1971) 141
63) J.F. Clare, Nucl. Phys. A217 (1973) 342
64) R.A. Hardekopf et al., Phys. Rev. C8 (1974) 1629
65) V. König et al., Nucl. Phys. A148 (1970) 380; W. Grüebler et al.,
    Nucl. Phys. A148 (1970) 391; V. König et al., Nucl. Phys. A166 (1971)
    393
66) G.G. Ohlsen et al., Phys. Rev. C8 (1973) 1262
67) G.G. Ohlsen et al., Phys. Rev. C8 (1973) 1639
68) W. Grüebler et al., Nucl. Phys. A242 (1975) 285
69) W. Dahme et al., ref. 3a), p. 444
70) R.J. Holt et al., Nucl. Phys. A237 (1975) 111
71) V. König et al., Helv. Phys. Acta 44 (1971) 699; R. Neff et al., Helv.
    Phys. Acta 45 (1972) 934
72) A. Lindner and H. Ebinghaus, Nucl. Phys. A230 (1974) 487
73) V. Rohrer and L. Brown, Nucl. Phys. A217 (1973) 525
74) E.O. Alt, G. Grassberger and W. Sandhas, Nucl. Phys. B2 (1967) 181
75) C. Mahaux and H.A. Weidenmüller, Shell Model of Nuclear Reactions
    (North-Holland Publ. Comp., Amsterdam 1969)
76) J.T. Londergan and C.M. Shakin, Phys. Rev. Lett. 28 (1972) 1729
77) S. Ramavataram et al., Nucl. Phys. A226 (1974) 173 and to be
    published in Nucl. Phys.

78) D.S. Chuu et al., Phys. Rev. C7 (1973) 1329
79) S. Saito, Prog. Theor. Phys. 41 (1969) 705
80) V.G. Neudatchin et al., Phys. Rev. C11 (1975) 128
81) B. Buck et al., Phys. Rev. C11 (1975) 1803
82) M. Stingl and A.S. Rinat (Reiner), Phys. Rev. C10 (1974) 1253
83) M. LeMere et al., to be published
84) F.S. Chwieroth et al., Phys. Rev. C8 (1973) 938; Phys. Rev. C9 (1974) 56
85) J.H. Koepke et al., Phys. Rev. C9 (1974) 823
86) A.G. Teufel, Nucl. Phys. A235 (1974) 19
87) C.M. Shakin and M.W. Weiss, Phys. Rev. C9 (1974) 1679
88) Ch. Leclerc-Willain and M. Libert-Heinemann, Nucl. Phys. A229 (1974) 15
89) H.H. Hackenbroich, in: The Nuclear Many-Body Problem (Editrice Compositori, Bologna 1973)
90) H.H. Hackenbroich et al., Nucl. Phys. A221 (1974) 461; P. Heiss et al., Phys. Lett. 52B (1974) 411
91) D.R. Thompson, Nucl. Phys. A151 (1970) 442
92) P. Heiss and H.H. Hackenbroich, Nucl. Phys. A162 (1971) 530
93) H.F. Glavish et al., ref. 4) p. 632
94) H. Stöve and H.H. Hackenbroich, Z. Physik 267 (1974) 403
95) P. von Brentano et al., Phys. Lett. 46B (1973) 177
96) P. Heiss and H.H. Hackenbroich, Phys. Lett. 30B (1970) 373
97) M.L. Goldberger and K.M. Watson: Collision Theory, (Wiley, New York, 1964)
98) P. Heiss, Z. Physik A272 (1975) 267
99) H. Eichner et al., Nucl. Phys. A205 (1973) 249
100) S. Barshey and G. Temmer, Phys. Rev. Lett. 12 (1964) 728
101) G.J. Wagner et al., ref. 3), p. 747
102) V. Nocken et al., Nucl. Phys. A213 (1973) 97
103) Th. Fischer et al., Verhndl. DPG (VI) 10 (1975) C2.4
104) G. John and T.H. Seligman, Nucl. Phys. A236 (1974) 397
105) H.H. Hackenbroich and T.H. Seligman, Phys. Lett. 41B (1972) 102
106) J. Niwisch and D. Fick, to be published

DISCUSSION

Schwager:

You mentioned that in the integral equation approach to few-particle scattering one has to put in the properties of the subsystems. Of course, one can develop approximation schemes for this. I have considered the connection of that approach with more conventional reaction theories and it has appeared that they are not very different. Probably the most important feature of the integral equations is that they can give an exact mathematical basis to calculations like yours, for instance.

Hackenbroich:

My hope is that in some future the "exact"- i.e. the best available numerical- solutions to subsystem properties are plugged in the equations for larger systems.

Slobodrian:

    I would like to know what type of calculations did you perform on
the d-α System, were they three body calculations or did you do a
full six nucleon calculation ?

Hackenbroich:

    We did a calculation using the methods which are described e.g. in
ref. 89. This is a six nucleon calculation which takes into account
α-d, ^{3}He- ^{3}H, ^{5}Li + n, ^{5}Li^x+ n, ... and "distortion" channels
describing the mutual distortion of fragments in the region of interaction. The internuclear potential given by Eikenmeier and myself
has been used; also the fragment wave functions have been determined
from a variational calculation starting from this potential.

ANALYTICAL METHODS IN THE STUDY OF LIGHT NUCLEI

G.R. Plattner, University of Basel,
CH-4056 Basel, Switzerland

1. Introduction

It is my aim in this talk to demonstrate that the (conjectured) analyticity of the S-matrix - or, equivalently, of scattering amplitudes -
provides us with a useful tool to extract spectroscopic information about
the ground states of light nuclei from nuclear scattering data. While
such considerations of analyticity have been extensively used in elementary particle physics, their application to nuclear physics has only recently become the subject of a sustained and relatively successful effort.

The material presented in this extremely short and cursory review
should be considered as another example of the many uses to which good
nuclear scattering and reaction data can be put. At this Polarization
Symposium I should like to stress that most of the work presented here
is based heavily on the excellent polarization data available, without
which the scattering amplitudes in question would have been too poorly
determined to be of much value.

For the sake of brevity and clarity I shall restrict the discussion
to the nonrelativistic domain and shall deal mainly with elastic scattering, where the major effort of the last few years has been concentrated.

2. Scattering Poles and Bound States

A scattering amplitude can be specified as a function of the two independent variables cm energy E (or momentum k) and cm scattering angle
Θ. The methods reviewed in this talk are based on the use of the analytic
properties of the scattering amplitude in either variable, keeping the
other fixed.

It is well known that the process shown in fig.1 (chosen to represent p-d elastic scattering as an example)
corresponds to a pole in the scattering
amplitude if the intermediate state has a
specific mass and energy (the famous Breit-
Wigner formula for such a resonant excitation of width Γ is just the mathematical
equivalent of a pole with residue $R = -i\Gamma$).
Clearly this pole is located at a p-d cm
energy $E = E_x-B$, where B is the removal
energy of a proton in ^{3}He. If the intermediate state is the bound ground state of

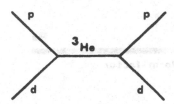

Fig.1: ^{3}He pole in p+d
elastic scattering

^{3}He, the pole corresponding to fig.1 lies at a negative energy

$$E = -B.$$ (1)

Such direct bound state poles exist only in the variable E. (In cos Θ
they are located at infinity).

The topologically very similar diagram (fig.2) describing particle
exchange also corresponds to a pole in the scattering amplitude. This
becomes evident when one considers that this process - like the one just
discussed - also consists of formation of a particle at one vertex and
its decay at the other, except that their time sequence is inverted with

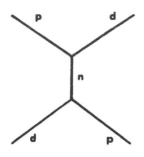

Fig.2: n-pole in p+d
       elastic scattering

respect to diagram 1. In the exchange process, the deuteron first undergoes a virtual
decay and is then formed again. The corresponding exchange pole in the scattering amplitude occurs when the (kinematically determined) energy of the scattered proton
corresponds to the proton energy from the
(virtual) decay of the deuteron:

$$E\left[(1-\cos\Theta)\frac{2m_p}{m_p+m_d} - \frac{m_p+m_d}{m_d}\right] =$$

$$\frac{m_d-m_p}{m_d}\, \mathcal{B}' \quad , \quad (2)$$

where B' is the deuteron binding energy. For fixed physical scattering
angles this yields a pole at a negative energy, and for fixed positive
energies, a pole at a virtual scattering angle beyond 180°.

As an example we see in fig.3 the positions of the poles in p-d
scattering, corresponding to virtual formation of ^{3}He and to virtual

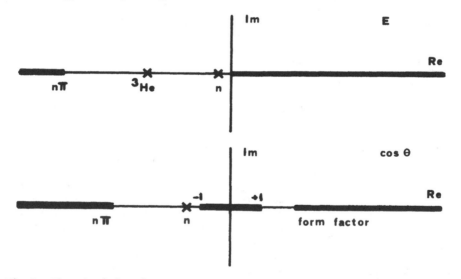

Fig.3: Singularities in the E and cos Θ planes for p+d scattering

neutron exchange. Additional singularities (cuts) are also shown, which
even in this simple case considerably complicate the analytical structure. Such left- and right-hand cuts are invariably present. Those nearest to the poles of interest usually represent the continuum states of
the constituents within the exchanged particle and are thus a manifestation of its composite structure. Here the mesons present in nuclei will
eventually enter into the picture, though to date they are still largely
ignored in this type of consideration, since their contribution is very
small.

3. Pole Residues and Wave Functions

The influence of a particular pole on the scattering amplitude is determined by its position and by its residue. Loosely speaking, a pole contributes a term

$$f_p(z) \sim \frac{R}{z - z_p} \tag{3}$$

to the amplitude, where z is the variable in which the pole is considered. The residue R determines the strength of the particle interaction at the vertices of figs.1 and 2. In the physics of elementary particles it is known as a coupling constant. It can be shown[1]) that in nuclear physics these coupling constants are related to the asymptotic wave functions of the nuclei involved. The n-exchange residue or "d-p-n coupling constant" for the process of fig.2 determines the magnitude of the asymptotic d wave function:

$$\Psi(\vec{r}) \xrightarrow{r \to \infty} const. \sqrt{R} \frac{e^{-\varkappa r}}{r} Y_{oo}(\vec{r}) , \tag{4}$$

while the ^{3}He-pole residue or ^{3}He-d-p coupling constant of fig.1 similarly determines the asymptotic d-p cluster wave function of ^{3}He.

It is evident that here we enter the realm of nuclear models, since only they can tell us what components are present in the wave functions. For instance, both ^{3}He and d have a D-state admixture to their ground states, which has a slightly different form than that given in eq.(4). Thus, while the coupling constants are defined without reference to specific models, this is not so for the corresponding magnitudes of the asymptotic wave functions. However, the whole interest in the subject of this talk stems just from the fact that information about the ground-state wave functions can be obtained directly from experimental data without the need for a microscopic theory of nuclear reactions.

4. Analytic Continuation in cos Θ

From eq.(3) it is clear that in principle the pole residue R can be determined by taking the limit

$$\lim_{z \to z_p} F(z) \equiv \lim_{z \to z_p} (z - z_p) f(z) \sim R , \tag{5}$$

since here all those contributions to f(z) which are not singular at $z = z_p$ will vanish as $(z - z_p) \to 0$, and only the pole residue R will remain. Evidently this procedure can only be used if f(z) is known near $z = z_p$. To this end, the amplitude must be analytically continued to the pole from the physical region[2]), where it has been measured by experiment. This can be achieved by fitting the amplitude F(z) (or even directly the cross-section) over the physical region to a polynomial expansion. All the investigations based on analyticity of the scattering amplitude as a function of the scattering angle Θ are carried out in this way.

A generally accepted and mathematically sound extrapolation technique was introduced by Cutkosky and Deo[3]) in 1968. Rather than in cos Θ the extrapolation is done in a new variable u obtained by a conformal mapping u=U(cos Θ) of the cos Θ plane onto a unifocal ellipse. In fig.4 we show the result of this conformal mapping as applied to fig.3b. One notices that the real axis between the left and right hand cuts has been

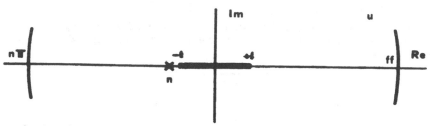

Fig.4: Singularities in the u plane for p̄+d scattering

mapped onto the major axis of the ellipse (which in this case is almost
circular). The n-exchange pole - which we are interested in - has stayed
more or less in place, while the cuts have been mapped onto the circum-
ference of the ellipse and are now much further from the physical region
or from the pole than before. It can be shown that a polynomial expansion
of F(u) processes a maximum rate of convergence inside this ellipse[3]).
Thus, in practice F(u) is represented by a minimum number of parameters
and can be reliably extrapolated to the pole. There its value is proportional to the pole residue or coupling constant which can thus be determined from the experimental data.

Quite a number of investigations of nuclear coupling constants have
been carried out in roughly this manner. I shall review the results in
section 7. In all of these studies the authors have directly extrapolated cross-section data rather than the amplitudes, though in many cases
these are well known and would represent much more experimental information than just cross sections. Not only could one take advantage of the
large body of polarization data by using the amplitudes, but also of the
fact that amplitudes for different spin states will in general contain
the coupling constants (if more than one is involved) with different relative weights, so that the various contributions can be separated. The
use of derived amplitudes rather than directly measured data points certainly also has its disadvantages, but on the whole I feel that one
should try to move in this direction.

5. Forward Dispersion Relations

After discussing analyticity in the scattering angle Θ we now turn
to analyticity in the energy E or, equivalently, in the momentum k for
a fixed value of cos Θ = 1 (for this choice of cos Θ the theory assumes
a particularly simple form). In principle, the forward scattering amplitude f(k) could be extrapolated to the pole in much the same way as described for f(Θ). However, the properties of f(k) are such that a different approach is even more worthwhile.

The bound state and exchange singularities of a particular scattering amplitude are without exception situated on the positive imaginary
k-axis as shown in fig.5. It is well known[1]) that no other singularities
exist in the whole upper half-plane Im k>0. However, there are singularities situated in the lower half-plane, which represent the contributions
from the continuum states of the scattering system, as e.g. resonances.
They are shown as dots in fig.5.

The analyticity of the forward scattering amplitude f(k) permits us
to apply Cauchy's Integral Formula, which relates the values of an analytical function on an arbitrary closed contour Γ to the poles of f(k) en-

closed by Γ. If we choose Γ as shown in fig.5, Cauchy's formula gives (apart from constants)

$$f(k) - \frac{1}{\pi i} \oint_\Gamma \frac{f(k')dk'}{k'-k} = 2\pi i \sum_j \frac{R_j}{k_j - k} \quad , \tag{6}$$

where the j-th pole is situated at k_j and has a residue R_j.

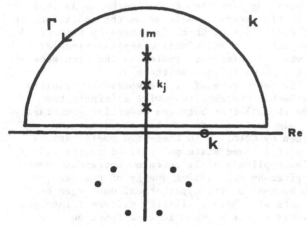

Fig.5: The complex k plane

The integral over f along the infinite semi-circle can be taken into account by making the usual and reasonable assumption that for infinite momenta f(k) approaches a constant value. It is then possible to modify Cauchy's formula in such a way that this part of the integral along Γ disappears ("subtraction")[1]. On the positive real axis f(k) is known from experiments. On the negative real axis it can be calculated using time reversal invariance[1], which relates the amplitudes for positive and negative values of k. Thus the integral can be evaluated along all parts of Γ.

If this is done, eq.(6) takes on the following form (which we can again write in the energy variable E):

$$\mathrm{Re}\, f(E) - \mathrm{Re}\, f(E_0) - \frac{(E-E_0)}{\pi} \int_0^\infty \frac{\mathrm{Im}\, f(E')dE'}{(E'-E)(E'-E_0)} = \sum_j \frac{(E-E_0)R_j}{(E_j-E)(E_j-E_0)} \equiv \Delta(E) \tag{7}$$

This equation is called a "Forward Dispersion Relation" (FDR) "subtracted at the energy E_0". It permits us to calculate the discrepancy function $\Delta(E)$ if the amplitude is known over a sufficient part of the physical region to make the dispersion integral converge. It is important to note that from the forward scattering amplitude f(E) we have now gained a new function $\Delta(E)$, from which all the contributions of the lower momentum half-plane, or, equivalently, of the second sheet of the energy plane have been eliminated. These parts of the amplitude – which are not related to the coupling constants of interest! – are usually very large. It is the strength of the FDR approach as contrasted with other methods, that it allows us to get rid of these unwanted contributions in a model-independent manner.

It is now straightforward to extrapolate the discrepancy function $\Delta(E)$ to the poles of interest. Again the conformal mapping techniques discussed earlier can be used in much the same way to obtain rapid convergence of a polynomial expansion and to determine the coupling constants. This combination of conformal mapping and dispersion relations

has been introduced only recently[4],[5]), while the older applications of an
FDR to nuclear scattering data have still used less satisfactory extrapolation methods. It is clear that all future work should take full advantage of the power of conformal mapping techniques.

6. Coulomb Problems

So far we have completely neglected a tedious problem: the influence of the charges of the scattering particles on our methods. As is immediately clear the electromagnetic interaction leads to the appearance of
a Coulomb pole at cos Θ = 1 and/or E = 0, which is caused by Rutherford
scattering. In the energy variable, an additional essential singularity
at E = 0 and an infinite number of poles are created by the occurence of
the Coulomb penetration factor $C_0^2(k)$ in the amplitude[6]).

These two problems can be taken care of if one works with amplitudes. The subtraction of the Rutherford amplitude will eliminate the
Coulomb pole. It can also be shown[6]) that both the essential singularity
and the infinite number of poles will disappear when the remaining
"nuclear" amplitude is divided by $C_0^2(\eta)$. With these two modifications
the Coulomb problem is solved for bound state poles[6]). Of course, the
poles of the Coulomb-modified amplitude still contain charge-dependent
effects. Their position is given by the binding energy of the real,
charged nucleus rather than by that of its hypothetical uncharged coun-
terpart. The coupling constants also have a slightly different interpretation in that they are related to the asymptotic wave function of a
charged system[6])

$$\Psi(\vec{r}) \xrightarrow{r \to \infty} const. \sqrt{R} \ \Gamma^2(1+\eta) \ \frac{W_{-\eta, \ell+\frac{1}{2}}(2\varkappa r)}{r} \ Y_{\ell m}(\hat{r}), \tag{8}$$

where η is the Sommerfeld parameter, k the momentum, ℓ and m the radial
quantum numbers of the bound state in question. Thus, even though these
coupling constants are not purely hadronic, they are the quantities
which one is naturally led to consider and which correspond exactly to
the actual mixture of hadronic and electromagnetic effects in the bound
state of a real charged nucleus.

If the methods outlined in section 4 are used to extrapolate elastic
scattering cross-sections rather than the amplitude, it is impossible to
eliminate the Coulomb pole at cos Θ = 1 by subtracting the Rutherford
amplitude, since Coulomb-nuclear interference terms are present. In
practice it has been tried to alleviate the problem by choosing data at
sufficiently high energies and by mapping the Coulomb pole onto the
ellipse[7]). However, the problem is still basically unsolved and may be
unsolvable.

Recently it has been rediscovered that for exchange poles the above
mentioned Coulomb modifications are not sufficient as they are for bound-
state poles[8],[9]). It was in fact realized at least eight years ago[10]) that
the electromagnetic interaction leads to the appearance of a logarithmic
cut for each exchange process, with a branch point at the pole position.
To my knowledge, this fact has been ignored with one exception[8]) in the
published literature. A thorough investigation of this effect is now
under way. From approximate calculations (DWBA) of its nature one can
already conclude that the discontinuity across this cut is strongly con-

centrated near the location of the exchange pole, so that its net effect is almost that of a multiplicative factor in the effective residue[9]). This factor is near unity for distant poles, but becomes progressively larger for exchange poles nearer to the physical region. For the d-exchange pole in p-^{3}He scattering it reaches values near 1.5! This additional Coulomb effect must thus be thoroughly understood.

7. Results

I now want to review with extreme brevity the results which have been obtained so far. They have been published by five groups: Locher and coworkers at the ETH here in Zürich, Dubnicka and Dumbrais at Dubna, Kisslinger at Pittsburgh, Borbely at Dubna, and last but - I hope! - not least the Basel group. The processes studied include the scattering of protons from p, d, ^{3}He, ^{4}He and ^{12}C, of neutrons from d, ^{3}H, ^{4}He, ^{12}C and ^{14}N, of deuterons from ^{4}He, of ^{4}He from ^{4}He, and the reactions D(d,p)^{3}H and D(d,n)^{3}He. From these investigations the following coupling constants have been deduced [11-14]) (sometimes with both FDR and cos Θ-extrapolation):

p-p-π_0, d-p-n, ^{3}H-d-n, ^{3}He-d-p, ^{4}He-d-d, ^{4}He^{3}H-p, ^{4}He-^{3}He-n, ^{6}Li-^{4}He-d, ^{8}Be-^{4}He-^{4}He, ^{12}C-^{11}C-n, ^{13}N-^{12}C-p, ^{14}N-^{13}N-n.

Remember that e.g. the ^{6}Li-^{4}He-d coupling constant determines the asymptotic α-d cluster wave function of the ground state of ^{6}Li.

The numerical values of these coupling constants agree well with each other in those cases where they have been determined from different processes or by using different methods. Where a comparison with independent experimental information is possible, the results obtained using "analytical" methods lie well within the range of published values. In the very light nuclei, which are particularly well known, the agreement between the coupling constants reviewed here and the values derived from other sources is excellent. The pp-π_0 coupling constant obtained using an FDR on the pp scattering amplitude[12]) is 15.3, at most a few percent off the accepted value. The dpn coupling constant from an FDR analysis of both n-d and p-d scattering[13,15]) is in perfect agreement with the values derived from low-energy n-p scattering or from standard deuteron wave functions. The ^{4}He-^{3}He-p coupling constant has been compared[16]) to the ^{4}He charge density distribution at large radii as measured by electron scattering, and the agreement is excellent within the quoted errors of 10 %. The values obtained for the ^{3}He-d-p coupling constant[8,13]) and those[13]) for ^{6}Li-^{4}He-d match very closely the best theoretical models for the ^{3}He and ^{6}Li ground state cluster wave functions[17,18]).

Thus it seems that no serious shortcoming has yet been uncovered, and that one can trust the coupling constants derived in this manner. Even the slightly exotic claim that an FDR analysis of p-^{4}He spin-flip scattering yields evidence[14]) for a few percent D-state admixture to the ^{4}He ground state, has thus a fair chance of being correct.

8. Conclusion

It should be pointed out, that apart from the problems already mentioned on which further work is still needed, there are other questions which to my knowledge have not yet been tackled, at least not by the nuclear physicists in the field. The question of how the left-hand cut

discontinuities in an exchange process are related to the scattering amplitudes of the constituent particles can probably be answered also in
nonrelativistic nuclear physics[19]. If so, this would constitute a major
success, enabling us to relate scattering amplitudes for different processes. Another point concerns the inelasticity of the scattering amplitudes. It is not obvious that the methods which we have described should
not be partly modified if inelastic processes are important, since e.g.
the crossing relation (time reversal invariance) on which the derivation
of FDR's is based, seems to differ in the presence of absorption from
the usual one[19]. This point is particularly disturbing, as it touches
upon the fundamental aspects of the method.

I conclude by admitting that the successes of the methods described
in this review are certainly not spectacular. However, they have provided us with valuable, largely model-independent information about the
ground states of many of the light nuclei. In addition, these approaches
have given us a new angle for regarding nuclear scattering and reactions.
Already a few surprising discoveries have been made, and I feel that one
can look forward with optimism.

References

[1]) M.L. Goldberger and K.M. Watson, Collision Theory (Wiley, New York,
     1964).
[2]) R.D. Amado, Phys. Rev. Lett. 2 (1959) 399.
[3]) R.E. Cutkosky and B.B. Deo, Phys. Rev. 174 (1968) 1859.
[4]) R.D. Viollier, G.R. Plattner and K. Alder, Proc. Int. Conf. Few Body
     Problem, Quebec 1974, in J.J. DeSwart's rapporteur talk (ed. R. Slo-
     bodrian, to be published).
[5]) M.P. Locher, SIN preprint PR-75-003.
[6]) R.D. Viollier, G.R. Plattner, D. Trautmann and K. Alder, Nucl. Phys.
     A206 (1973) 498.
[7]) S. Dubnička and O. Dumbrais, Phys. Lett 57B (1975) 327.
[8]) I. Borbely, to be published in Nuovo Cim. Lett. (1975).
[9]) M. Bornand and D. Trautmann, private communication.
[10]) W.K. Bertram and L.J. Tassie, Phys. Rev. 166 (1968) 1029.
[11]) S. Dubnička and O. Dumbrais, Phys. Reports 19C (1975), and refs.
      therein.
[12]) R.D. Viollier, G.R. Plattner and K. Alder, Phys. Lett. 48B (1974) 99.
[13]) M. Bornand, G.R. Plattner and K. Alder, to be published.
[14]) G.R. Plattner, R.D. Viollier and K. Alder, Phys. Rev. Lett. 34 (1975)
      830.
[15]) M.P. Locher, Nucl. Phys. B23 (1970) 116.
[16]) G.R. Plattner, R.D. Viollier, D. Trautmann and K. Alder, Nucl. Phys.
      A206 (1973) 513.
[17]) T.K. Lim, private communication.
[18]) T.K. Lim, Phys. Lett. 56B (1975) 321.
[19]) R.D. Viollier, private communication.

For a recent compilation of references please use ref.11.

DISCUSSION

Borbély:

   To comment the methods you have surveyed, it is clear that the FDR
   shows a very good performance. It is because a very large piece of
   information from phase-shift analysis in a wide range of energy is
   used. But it is limited to elastic scattering processes. On the
   other hand, the continuation methods use less information, therefore
   they cannot always be applied because of the limited accuracy of the
   available data, but they have the advantage that they can be applied
   to arbitrary reactions.
   I should like to add to the list of vertex constant extracted:
   I have applied the "singularity substraction method" to (d,p) reactions on light nuclei (like ^{16}O and ^{31}P) with a complete success.
   My results have been submitted to Journal of Physics G : Nuclear
   Physics. I should like to remark that not all the published results
   are correct: particularly, when d exchange is analysed by the
   continuation method, in most cases the analysis is incorrect. So
   one should be very careful in the case of two nucleon exchange. But
   in the more favourable case of one nucleon exchange, the structure
   information can be extracted with a surprisingly small error.
   Finally, I want to add that the continuation procedure could also be
   applied to polarizations (not only to the differential cross section),
   if one removes the kinematical singularities on the edge of the
   physical region.

ANALYSIS OF REACTIONS INDUCED BY POLARIZED PARTICLES[*]

F. Seiler[+]

Lawrence Berkeley Laboratory, University of California
Berkeley, California 94720

The analysis of reactions between particles with spin is usually rather difficult due to the large number of transition matrix elements with different orientations of the particle spins. In terms of these amplitudes the observables form a set of bilinear equations. For an analysis, one important aspect is to determine those subsets that allow a solution. This question has been investigated in detail by Simonius[1] who showed that, as long as experimental errors are neglected, polarization experiments of no higher than second order are needed. Consequently, the data base consists of the cross section $\sigma_o(\theta)$ and of measurements involving one and two particle polarizations. Since all of these data have to be taken at the same energy and angle, this method is generally difficult and probably restricted to a few reactions with a favorable combination of particles.

At relatively low energies and especially in reactions with resonances in the intermediate system, an analysis in terms of (ℓ,s,J) reaction matrix elements is more promising. Penetrability considerations limit the number of these elements, and an R-matrix or S-matrix approach can be used to describe their energy dependence. It is therefore possible to base the analysis on measurements of all types, taken at various energies and angles. Initially, the most important problem is to find those amplitudes $R_i \equiv \; < \ell_i' s_i' J_i^\pi | \; | \ell_i s_i J_i^\pi >$ which are responsible for the major structures of the observables as a function of energy and angle. In the last few years a set of criteria has been derived for such a preliminary analysis[2-4]. Although derived for polarized deuterons, these methods are readily generalized to particles with spin i. I shall devote my talk almost exclusively to this question, leaving the discussion of the full-scale analysis to the next speaker[5].

The natural basis for a preliminary analysis is the set of first order polarization data. Since these measurements involve only one particle polarization, they usually are both the most numerous and the most precise. An effective way to obtain information from the first order observable T_{kq} is to expand the product $\sigma_o(\theta) \cdot T_{kq}(\theta)$ in terms of Legendre polynomials $P_{L,q}(\theta)$, resulting in the expansion coefficients $a_{kq}(L)$ [ref. 4]. An immediate consequence is the separation of element combinations $R_1 R_2^*$ with equal and opposite parities into different coefficients $a_{kq}(L)$, according to the parity selection rule $\ell_1 + \ell_2 + L =$ even.

Another important aspect is the fact that a single reaction matrix element R_i can give rise to nonzero coefficients $a_{kq}(L)$ of even rank k and even degree L. For polarized particles with spin $i > 1/2$ both the cross-section $\sigma_o(\theta)$ and the 3 tensor polarization quantities of rank 2 are proportional to $|R_i|^2$. The coefficients $a_{kq}(L)$ are therefore linearly dependent. Dividing them by $a_{oo}(0)$, which is essentially the total cross section, yields the parameters $d_{kq}(L)$ [refs. 2,4]. These are independent of the magnitude of the element R_i and depend only on its spin space. Together with the linear relations they are therefore an important tool for the identification of major amplitudes. In an analysis this is a distinct advantage of rank-2 polarizations over vector polarization data.

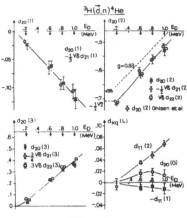

Fig. 1

The approximation that only one or two amplitudes are nonzero can be successful in identifying major elements even in the presence of other sizeable amplitudes[2,4]. Using this assumption, simple explicit criteria can be derived in many cases. Alternately, computer codes can be used, which calculate the formulae for the observables in terms of a given set of matrix elements[6,7]. Useful criteria can thus be derived 1) from the numerical values of the parameters $d_{oo}(L)$ and $d_{2q}(L)$ for isolated resonances, 2) from linear relations between the quantities $a_{kq}(L)$ for the same polarized particle by varying one of the parameters k,q and L, 3) from linear relations between the same observable, measured for different reaction partners. An example for the first two cases are levels induced by S-wave particles. The coefficient $d_{k0}(L)$ is given by

$$d_{k0}(L) = \delta(k,L)\,\delta(k,even)\,\hat{i}\hat{L}(-)^{i+I+2J+s'}(2J+1)(2\ell'+1)\begin{pmatrix}\ell' & \ell' & L \\ 0 & 0 & 0\end{pmatrix}\begin{Bmatrix}\ell' & \ell' & L \\ J & J & s\end{Bmatrix}\begin{Bmatrix}i & i & L \\ J & J & I\end{Bmatrix}$$

with $\hat{i} \equiv (2i+1)^{1/2}$. For orders $q > 0$ the linear relation is then $\qquad$ (1)

$$d_{k0}(L) = \frac{(-)^q}{2}\sqrt{\frac{(L+q)!}{(L-q)!}}\,d_{kq}(L). \qquad (2)$$

The most probable case, $k = 2$, yields[2]

$$d_{20}(2) = -\frac{1}{2}\sqrt{6}\,d_{21}(2) = \sqrt{6}\,d_{22}(2). \qquad (3)$$

Figure 1 shows data for the $^3H(\vec{d},n)^4He$ reaction[8] above the $3/2^+$ level at 107 keV. Excellent agreement is found for eq. (3) and other relations [4], while the ideal value $d_{20}(2) = -1/2\sqrt{2}$ is approached closely.

For resonances induced by particles with angular momentum $\ell > 0$, the magnitudes and relations between the coefficients $d_{kq}(L)$ are best derived using a computer code. A simple indicator for the spin space of the dominant element R_i is the coefficient

$$d_{k0}(0) = \delta(k,even)(-)^{i+I+J+2s}(2\ell+1)(2s+1)\hat{i}\hat{k}\begin{pmatrix}\ell & \ell & k \\ 0 & 0 & 0\end{pmatrix}\begin{Bmatrix}\ell & \ell & k \\ s & s & J\end{Bmatrix}\begin{Bmatrix}i & i & k \\ s & s & I\end{Bmatrix} \qquad (4)$$

In fig. 2 some of the $^3He(\vec{d},p)^4He$ data of Grüebler et al.[9] are shown. The value $d_{20}(0) = -1/7\sqrt{2}$ derived from eq. (4) for the proposed $7/2^+$ d-wave level[2,4] is in good agreement with the data near 9 MeV. Other coefficients also agree well with computer generated values, indicated by horizontal lines.

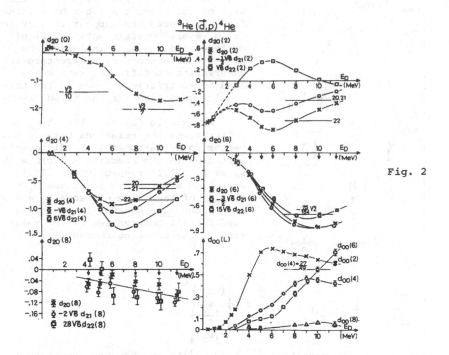

Fig. 2

For isolated resonances with only small competing amplitudes,

$$T_{k0}(0°) = \delta(k,\text{even})(-)^{i+I+\rho}(2s+1)\hat{i}\hat{k}\begin{pmatrix}ssk\\ \rho-\rho0\end{pmatrix}\begin{pmatrix}ssk\\iiI\end{pmatrix} \qquad (5)$$

also provides a good criterion for reactions in which the parameter
$\rho \leq s, s', J$ can assume only one value[2]
An important example for the second case pertains to the overlap
of two elements $R_1 R_2^*$, that satisfy the conditions

$$(\ell_1'+\ell_2'), (J_1+J_2) \geqslant \Lambda = \ell_1+\ell_2+k . \qquad (6)$$

For $q > 0$ the coefficients then obey the relation

$$d_{k0}(\Lambda) = \frac{(-)^q}{2} \frac{\Lambda!}{(\Lambda-q)!}\sqrt{\frac{(k+q)!}{k!}\frac{(k-q)!}{k!}} d_{kq}(\Lambda) . \qquad (7)$$

Conditions (6) select preferentially elements with $J > \ell$. Thus eqs. (7)
assist primarily in the identification of levels such as the $7/2^+$ d-
wave state in fig. 2 or of the overlaps shown in fig. 3 [ref. 4].
Relations (7) also provide a quantitative means to determine the highest
significant orbital angular momentum ℓ_{max}. It is precisely the set of
elements that satisfy conditions (6) which gives rise to the coefficients
$a_{kq}(\Lambda)$ of highest degree $\Lambda = 2\ell_{max}+k$. Equation (7) can thus be used
effectively to limit the set of matrix elements necessary. The absence
of coefficients with L=9 and the agreement found for L=7 and 8 (figs.
2 and 3) limit the angular momenta in ^{3}He($\vec{d}$,p)^{4}He below 12 MeV to $\ell \leqslant 3$.

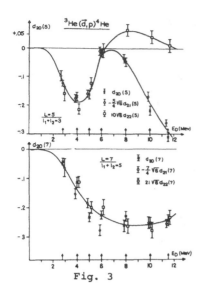

Fig. 3

Other useful restrictions on the size of the element set can be obtained from structures that appear in only one or in several coefficients $a_{kq}(L)$ of different degree L. Similarly, the presence or absence of a structure in observables of different rank k may lead to restrictions[4]) due to the triangular condition for the triad (s_1, s_2, k).

An example for the third type of criteria are the relations between the vector polarizations of the different particles in the reaction A(b,c)D with spins I,i,i',I' respectively. If in each case a structure is caused by the overlap of the large elements R_1 and R_2, and if the <u>same</u> coordinate system is used for all observables[4]), the relation in the incoming channel is[3])

$$\frac{d_{11}(L)}{d_{0011}(L)} = \frac{\hat{i}}{\hat{I}} \, \delta(s_1', s_2') \, (-)^{s_1 - s_2} \frac{\begin{Bmatrix} i & i & 1 \\ s_1 & s_2 & I \end{Bmatrix}}{\begin{Bmatrix} I & I & 1 \\ s_1 & s_2 & i \end{Bmatrix}} \, . \tag{8}$$

The formula for the quantities $d^{11}(L)/d^{0011}(L)$ in the outgoing channel is obtained by interchanging primed and unprimed quantities. Figure 4 shows that the L=2 coefficients in the ^{3}He(d,p)^{4}He reaction below 700 keV are clearly due to the overlap of two elements with channel spins $s_i = 3/2$. The other combinations would require a different relative sign.

If the same structure appears in polarizations of both the incoming and outgoing channel, the element combination responsible must obey both conditions $s_1 = s_2$ and $s_1' = s_2'$. The ratio $a_p(L)/a_{p'}(L)$ is then given by[10])

$a_p(L) =$	$a_{11}(L)$	$a_{0011}(L)$
$\dfrac{a_{p'}(L) =}{a^{11}(L)}$	$-2 \dfrac{K(i,I,s)}{K(i',I',s')} A$	$-2 \dfrac{K(I,i,s)}{K(i',I',s')} A$
$a^{0011}(L)$	$-2 \dfrac{K(i,I,s)}{K(I',i',s')} A$	$-2 \dfrac{K(I,i,s)}{K(I',i',s')} A$

with

$$K(j,J,s) = \frac{j(j+1) + s(s+1) - J(J+1)}{s(s+1)\sqrt{j(j+1)}} \tag{9}$$

and

$$A = \frac{\ell_1(\ell_1+1) - \ell_2(\ell_2+1) - J_1(J_1+1) + J_2(J_2+1)}{\ell_1'(\ell_1'+1) - \ell_2'(\ell_2'+1) - J_1(J_1+1) + J_2(J_2+1)} \tag{10}$$

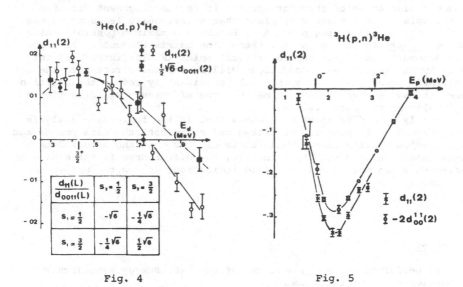

Fig. 4   Fig. 5

Since these ratios depend on the detailed spin structure of both elements, they are best suited to confirm assignments already made. In fig. 5 data for the ^{3}H(p,n)^{3}He reaction[11] show that the large overlapping 0^- and 2^- P-wave elements must belong to the class for which A=1 and thus $\ell_i = \ell_i'$.

With the methods outlined here, one or more tentative sets of major elements can be derived with a modest effort. Such sets can then be used as a first guess in an analysis. A careful preliminary analysis, accounting for all major structures, should come reasonably close to a solution. Indeed a preliminary evaluation[4] of ^{3}He(d,p)^{4}He data up to 12 MeV agrees very well with the R-matrix fit by Dodder and Hale[12]. A similar effort by the Zürich group[13] on the ^{6}Li(d,α)^{4}He reaction also shows good agreement with the level structure proposed for ^{8}Be.

A recent discussion of experimental data has shown that it is often the less conspicuous features of an angular distribution that give the least ambiguous information[4]. In order to exhaust the information content of an angular distribution, measurements should therefore be taken at as many angles as possible, even at the cost of less statistics for the individual point. This holds especially for first order polarization experiments, since a successful analysis depends largely on the constraints imposed by these data. The more difficult second order experiments are usually neither numerous nor precise enough to restrict the solution space sufficiently. Together with exhaustive first order data, however, they may provide the independent information needed to arrive at a solution.

In view of the difficulties encountered in an analysis of reactions, an effort should be made to investigate every useful indicator. One possibility is the evaluation of the data in the transverse coordinate system S^T, with the z-axis perpendicular to the scattering plane. The conditions of parity conservation give a simple geometric interpretation to the physical limits of spin-1 polarizations, the somewhat neglected Lakin cone[14,15]. Other interesting quantities also <u>have</u> a simple form. Thus in the system S^T, Johnson's[16] parameter $T_{22} - \sqrt{3/2}\, T_{20}$, which is

sensitive to tensor interaction, corresponds to the efficiency $(A_{xx}-A_{yy})^T$
It should also be noted that for spin 1, it is the alignment direction
perpendicular to the scattering plane that occasionally leads to values
of unity for the analyzing power A_{yy}. Reaching a maximum possible value
imposes <u>linear</u> conditions on some transition matrix elements. If a set
of experimental data near such a critical point (E,θ) is introduced into
the analysis, these linear conditions will be imposed directly on the set
of bilinear equations. Since this will considerably reduce the solution
space, it may be advantageous to measure the efficiency A_{yy} directly and
use it also in the analysis.

Finally, the discussion here shows that in the full-scale analysis
it may be useful to compare the fitted values to both the data points and
the expansion coefficients. While it is obvious that the actual data
points should be used to obtain the fit, the differences in the expansion
coefficients are more likely to yield indications on how to improve the
solution.

References

*   Work performed under the auspices of the U.S. Energy Reasearch and
    Development Administration.
+   On leave of absence from the University of Basel, Switzerland

1) M. Simonius, Polarization Phenomena in Nuclear Reactions,
    H. H. Barschall and W. Haeberli, eds. (University of Wisconsin Press,
    1971) p. 401
2) F. Seiler, Nucl. Phys. <u>A187</u> (1972) 379
3) F. Seiler, Helv. Phys. Acta <u>46</u> (1973) 56
4) F. Seiler, Nucl. Phys. <u>A244</u> (1975) 236
5) D. C. Dodder, these proceedings , p. 167
6) F. Seiler, Comp. Phys. Commun. <u>6</u> (1974) 229
7) R. D. McCulloch et al., Oak Ridge Natl. Lab. report ORNL-4121
8) H. A. Grunder et al., Helv. Phys. Acta <u>44</u> (1971) 622
9) W. Grüebler et al., Nucl. Phys <u>A176</u> (1971) 631
10) F. Seiler, to be published
11) L. G. Arnold et al., Phys. Rev. Lett. <u>32</u> (1974) 310
12) D. C. Dodder and G. M. Hale, private communication
13) R. Risler et al., these proceedings, p.585
14) W. Lakin, Phys. Rev. <u>98</u> (1955) 139
15) F. Seiler et al., these proceedings, p.897
16) R. C. Johnson, Polarization Phenomena in Nuclear Reactions,
    H. H. Barschall and W. Haeberli, eds. (University of Wisconsin Press,
    1971) p. 143

DISCUSSION

Lindner:
   Would'nt you agree that your linear stem relations from the fact
   that one does not use a spherical invariant basis? Using such a
   basis (triple correlation functions, e.g.) reduces the number of
   fitting parameters and of the linear relations between them. In addition there will be no need to bother about the coordinate system.

Seiler:
   I agree that your triple correlation functions describe the data with
   less parameters. It is however just the redundancies that yield information you need. Once you have diagnosed the situation, your
   parameters are of course more simple. You should however, write down
   your triple correlation coefficients in terms of (ℓ, s, J) matrix
   elements and then one could see how much diagnostic value remains.

R-MATRIX ANALYSIS*

D. C. Dodder
Theoretical Division
Los Alamos Scientific Laboratory, University of California
Los Alamos, New Mexico 87545, U.S.A.

Experiments involving polarization techniques have nowhere unearthed a richer structure of observed phenomena than in those scattering and reaction processes involving very few nucleons. This paper is concerned with a program, carried out with colleagues at Los Alamos, with its first goal as a phenomenological understanding of these processes. It is the belief of those of us working on this program that our approach is essentially dictated by the fact that the overwhelming amount of information already collected seems to be still only marginally sufficient to describe the complex behavior of these systems. This somewhat paradoxical situation, that an almost unmanageable quantity of data is still perhaps just on the threshold of encompassing all aspects of the behavior of such a system, seems to call for an approach with two main features. First, all channels and all complementary experiments should be described simultaneously. Second, an energy dependent parameterization should be used, if only because the data are not sufficiently complete at single energies. The simultaneous description of all channels allows us to make use of the very important principle of unitarity of the collision matrix S. The need for an energy dependent parameterization has led us to choose the R-matrix formalism. In this context that formalism is to be regarded as a practical tool. This paper does not discuss the role of the R-matrix in general theories of nuclear reactions. It may be, however, that the results of the present work will have a place in such discussions.

Conservation of angular momentum means that the S matrix can be written as a block-diagonal matrix with a block for each J that is filled out with complex matrix elements. Parity conservation further breaks this down to a separate block for each J^π. And of course time reversal invariance means the matrix can be made symmetric. Unitarity implies that these complex matrix elements are not all independent. The $N(N + 1)/2$ complex numbers for each $N \times N$ block of the general matrix are equivalent to half this number of independent parameters for one that is unitary. This reduction is often conveniently made by using the real reactance matrix K, defined by $iK = (S + 1)(S - 1)^{-1}$, which completely describes the collision process.

The significance of this can perhaps most easily be seen by examining the structure of one J^π submatrix in a typical case. In the d + ^{3}He, p + ^{4}He system the $3/2^+$ submatrix has the following four states:

$$d + {}^3He: \quad {}^4S, {}^4D, {}^2D$$

$$p + {}^4He: \quad {}^2D \quad .$$

A description of the p + ^{4}He part alone, freed from the constraint of unitarity, would require 2 parameters, i.e., one complex phase shift. A corresponding description of the d + ^{3}He part alone would seem to require 12. Even the common incomplete description in terms of 3 complex phase shifts and 3 mixing parameters taken to be real requires 9 real parameters. The K matrix for both channels combined, however, describing

both elastic scatterings and the reaction, has just 10 real parameters
for this state. The relaxation of the unitarity condition and the separate parameterization of the different processes gives far more freedom
than is permissible.

It is worth noting that the unitarity condition is important even
when the channels are so loosely coupled that the reaction cross section
is appreciably less than the elastic scattering cross sections, and the
loss of flux from the elastic channels due to the reaction is negligible.
This can be seen in a simple case from a common parameterization of the
general symmetric unitary 2 × 2 matrix:

$$\begin{pmatrix} \cos \epsilon \ e^{2i\delta_1} & i \sin \epsilon \ e^{i(\delta_1 + \delta_2)} \\ i \sin \epsilon \ e^{i(\delta_1 + \delta_2)} & \cos \epsilon \ e^{2i\delta_2} \end{pmatrix}$$

It follows that the phase of the off-diagonal (reaction) element is completely determined by the phases of the diagonal (elastic scattering)
elements. (If one of the elastic phases is zero, as near its channel
threshold energy, the relationship is known as Watson's Theorem.) Therefore if more than one J^π state, and hence more than one submatrix is involved in a reaction, the relative phases of the reaction matrix elements
are completely determined by the elastic phases no matter how small the
mixing parameters ϵ are. This is all obvious and well known but it is
sometimes ignored.

A problem that is faced is just how to parameterize the energy dependence of a unitary description of a collision system in a way that is
both natural and simple. Simplicity is important because we want to use
an iterative procedure for data fitting with as little computation per
iteration as possible. By being natural means being appropriate to the
physics of the system. Success in meeting this criterion can, in the last
analysis, only be judged a posteriori. A physical property of the light
nuclear systems very pertinent to the choice of parameterization is the
presence of resonances or resonance-like behavior. It turns out that
these resonances are quite usually found with reduced widths (The use of
this concept anticipates the conclusions to be reached) about equal to
single particle widths for particles of reduced mass M and interaction
radius a. This means that in some channel the behavior at a resonance is
roughly that expected from elastic scattering from a potential well of
radius comparable to a. Fortunately the description of a resonance, in
lowest approximation, is very simple in terms of a resonance energy and
partial widths as in a Breit-Wigner equation. Of course it turns out that
these quantities are really energy dependent, but this is usually a slow
dependence, and the relatively rapidly varying resonance behavior of
physical observables is described in terms of some slowly varying parameters. It also turns out that the energy dependence of the widths and
resonant energy can be mostly accounted for by the behavior of the 2-
particle wave functions in the external coulomb and centrifugal potentials. The invention which utilizes this circumstance most directly and
completely is the R-matrix formalism of Wigner and Eisenbud.[1])

A complete description of the derivation of the R-matrix theory or
even a precise statement of the assumptions underlying it are impossible
here. The starting point is the division of configuration space into
two parts, an inner region where everything is going on, including the
nuclear interaction, and an outer region where it is assumed only two-
body states are present, and where there are only coulomb or other long
range interactions between the pairs. A condition that is essential for

the method to be at all realistic, but which is easily satisfied, is that the boundary between the two regions be at small enough distances between particle pairs that the coulomb barrier and phase space effects are mainly taken care of in the external region. The crucial consequence of the theory is that the derivative matrix R has a partial fraction expansion:

$$R_{c'c} = \frac{\gamma_{\lambda c'} \, \gamma_{\lambda c}}{E_\lambda - E} \quad .$$

The $\gamma_{\lambda c}$ and E_λ are real, energy independent partial reduced widths and resonant energies, respectively. The relationship between the K matrix and the R matrix involves external wave functions evaluated at the boundary between the regions, and it is the energy dependence of those wave functions (expressed often as penetration factors and level shifts) that gives the further energy dependence of the K matrix.

There are several derivations of the partial fraction expansion. One, for a single channel, relies on a causality principle.[2] The original one of Wigner and Eisenbud identified the E_λ as eigenvalues of the true interior Hamiltonian with an Hermitian boundary condition and the $\gamma_{\lambda c}$ as the overlaps, on the boundary, of the internal eigenfunctions with external two-body functions of the appropriate energy. In this interpretation one hopes to identify, in some way, the E_λ with physical resonances of the system. This correspondence is not exact, as will be shown below. For our purposes, however, the facts that the energy dependence is simple and still can have resonance properties make the use of the method attractive.

In actual practice, that is in fitting data with R-matrix parameters, one encounters levels that correspond in varying degree with physical resonances. An interesting example is in the $^3D_3$ state in ^{4}Li, where the R-matrix parameters have been found in an analysis of p-^{3}He elastic scattering from 0 to 19.7 MeV proton energies. The parameters found in this case are E = 17.5 MeV, γ^2 = 4.62 MeV, with an interaction radius of 4.93×10^{-13} cm. At first glance this seems to be a resonance slightly above the energy range of the analysis. However the phase shift in this state never gets much above 2°. What is happening is that this one level is a good approximation over this energy range, to that infinite series of levels that would just cancel the hard sphere phase shift for a radius of 4.93×10^{-13} cm and give a resultant 0° phase shift. The E_λ's in that series are just those values of the energy where the free particle wave functions satisfy the boundary condition defining the R matrix. This circumstance does not trouble the phenomenological analysis as long as the "level" is recognized for what it is. And at the same time the parameterization is sufficient to have allowed for the presence of a true physical resonance had one existed.

A quite different situation is afforded by the 5/2⁻ state in ^{7}Be, where the following R-matrix parameters are from an analysis[3] of ^{3}He + ^{4}He and p + ^{6}Li scattering and reactions:

γ	E_λ	6.4	5.7	100.0 (fixed)	100.0 (fixed)
^{3}He + ^{4}He: $^2F_{5/2}$	.69	.13	6.0		0.0 (fixed)
p + ^{6}Li: $^4P_{5/2}$	-.12	.65	0.0 (fixed)		7.9

The corresponding phase shifts in the "nuclear bar" or Stapp[4] representation are shown in Fig. 1. These first two levels are obviously two honest resonances, which of course were already well known.[5]

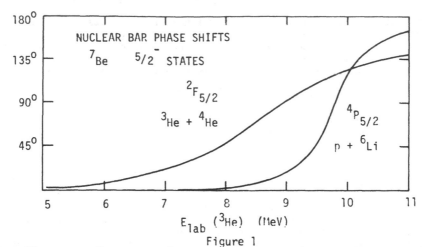

Figure 1

    There are also cases that are ot an intermediate nature. Figure 2
shows the phases of the collision matrix elements for the quartet D
states in D + ^{3}He from an analysis of the ^{5}Li system.

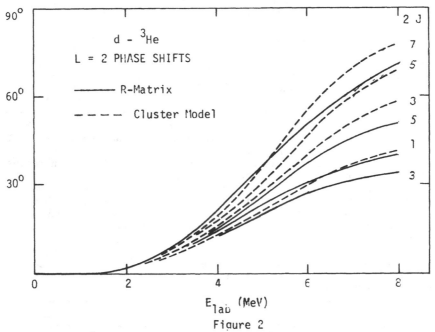

Figure 2

The dotted lines also show results of some calculations of Heiss and
Hackenbroich using a "refined cluster model." [6] It is clear that there
is a marked similarity, and that the R-matrix parameterization is suitable for a description of the model physical process. We have found
level parameters for all four J states, and our final results will probably find their way into "The Level Structure of Light Nuclei," but

there may be some who would prefer to discuss such slowly varying phase shifts without referring to levels at all. These examples should make it clear that the R-matrix formalism is quite versatile in describing a wide range of physical phenomena.

There are further consequences of the formalism that are very convenient. The energy dependent effects of the external wave functions extend right down to include a description of threshold behavior. Indeed Wigner[7] first gave a systematic treatment of threshold behavior after he was in possession of his R-matrix machinery. This threshold behavior even includes the effects below threshold of a new channel opening up. This is true because, even though the K-matrix element for a reaction must vanish below threshold, the R matrix depends in no way on any threshold, and the relationship between the two matrices, involving matrix inversions, makes open channel K-matrix elements depend on both open and closed channel R-matrix elements. Physically, this is because near a threshold the closed channels will have appreciable decaying two-body wave functions extending beyond the boundary into the external region.

It is near threshold, too, that the most dramatic effects of the coulomb interaction occur, and where there are the biggest differences between systems that, aside from their charge states, are identical under the isospin conservation hypothesis. Because the boundary radii can be chosen to include most of the coulomb effects in the external region the R-matrix parameters should depend only slightly on the charge state, and a perturbative approach should be reasonable. Indeed, in a simultaneous analysis of the $p + {}^4He$ and $n + {}^4He$ systems an assumption that there is only a constant shift of E_λ's between the two charge states is sufficient to fit both sets of data to within experimental uncertainties. The constant E_λ shift, an empirically determined number, has the appropriate magnitude for a coulomb energy perturbation in the internal regions.

A more interesting case occurs in the ^4He system with $p + {}^3H$, $n + {}^3He$, and $d + d$ channels occurring simultaneously, where the isospin conservation condition can be imposed directly on the γ_λ's.[8] Under this assumption charge independence is strict in the internal region and the rather considerable coulomb differences in the mirror channels are all accounted for in the external region.

And finally another example will reemphasize the importance of the global approach in interpreting these data. In Fig. 3 is shown the interesting energy dependence of the deuteron vector analyzing power in $^3He(\vec{d},d)^3He$ at 115° cm. This is a prediction from an analysis of ^5He in which the data set involved no deuteron vector polarizations. The bump in the curve has been tentatively ascribed to a resonance $n + {}^4He^*(0^+)$ channel by Kilian et al.,[9] who have obtained experimental results that show the same general behavior. However, in the analysis producing this curve there was no provision for an $n + {}^4He^*$ channel. The bump here is produced by interference between the resonating D waves and the negative phase shift P waves in the $D + {}^3He$ channel. This illustrates the difficulty in attempting to identify individual features of the observables with effects in particular states.

The examples shown of the R-matrix analysis demonstrate that the program suggested in this paper has been put into action. A large computer program, known as EDA, is capable of making global searches on R-matrix parameters on data covering a system's whole observable space over a large energy range. The program is written to handle any possible observable in spin space for any number of 2-particle channels for particles of any spin. The limitations on any of these quantities, as well as limitations on total number of data points, maximum ℓ value, etc.,

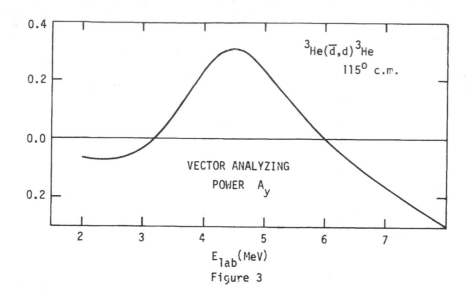

Figure 3

are those imposed by total storage capacity of the CDC computer as installed at Los Alamos. Work is under way analyzing most of the light
nuclear systems, and we are proceeding gradually to make the results
known.

    Among many colleagues in this work K. Witte and G. Hale deserve
special mention. K. Witte has done all of the detailed programming. G.
Hale has made several contributions to the design of the program and has
shared in most of the analyses. Both have contributed indispensably in
acquiring the expertise in using the computer program which we are gradually developing. And finally almost everyone at this symposium has
contributed either directly in providing experimental results or indirectly in developing experimental techniques that have made these results
possible.

References

* Work performed under the auspices of the United States Energy Research
    and Development Administration.
1) E. P. Wigner and L. Eisenbud, Phys. Rev. 72 (1947) 29
2) N. G. van Kampen, Rev. mex. fis. 2 (1953) 233
3) This analysis is a collaboration with E. K. Biegert and S. Baker of
    Rice University.
4) H. P. Stapp, T. J. Ypsilantis, and N. Metropolis, Phys. Rev. 105
    (1957) 302
5) e.g., R. J. Spiger and T. A. Tombrello, Phys. Rev. 163 (1970) 964
6) P. Heiss and H. H. Hackenbroich, Nucl. Phys. A162 (1971) 530
7) E. P. Wigner, Phys. Rev. 73 (1948) 1002
8) C. Werntz and W. E. Meyerhof, Nucl. Phys. A121 (1968) 38
9) K. Kilian, H. Treiber, R. Strausz, and D. Fick, Phys. Lett. 34B
    (1971) 283

OPTICAL MODEL POTENTIALS FOR DEUTERONS, TRITONS AND HELIONS

P. W. Keaton, Jr.
Los Alamos Scientific Laboratory, University of California[+]
P. O. Box 1663, Los Alamos, NM 87545, USA

INTRODUCTION

The recent developments of two new polarized ion sources have completely changed the scope of future reports on the elastic scattering of composite particles. Precise measurements with polarized tritons and helions* can now be used to test current models and future proposals devised to describe differential cross section and polarization values. Polarized beams of helions have been accelerated to 33 MeV at Birmingham as will be discussed by Powell at this symposium, and polarized beams of tritons have been accelerated to 15 MeV at Los Alamos as will be discussed by Hardekopf at this symposium. These data are too new and preliminary to have received extensive studies at this time. However, the systematic measurements of cross section and polarization data for intermediate and heavy weight nuclei that are undoubtedly forthcoming from these laboratories will be of considerable value.

Polarized deuteron beams are used to investigate the nature of the deuteron-nucleus tensor potential term. This term is most sensitive to the tensor analyzing power (negative of the tensor polarization value) A_{xz} or T_{21}.[‡] In the past few years several laboratories have used polarized ion sources to measure with precision A_{xz} for deuterons elastically scattered from intermediate weight nuclei. The values are small (generally less than 0.1) and 10-30% relative uncertainties are not uncommon. Therefore, a convincing statement has not been made as to the size or existence of the tensor potential term.

Furthermore, all attempts to describe deuteron cross section and four analyzing powers for various bombarding energies with a potential have failed. It turns out to be a very stringent test of any model to predict the differential cross section $\sigma(\theta)$, the vector analyzing power $A_y(\theta)$ (or iT_{11}), and the three second-rank analyzing tensors $A_{xz}(\theta)$ (or T_{21}), $\frac{1}{2}[A_{xx}(\theta) - A_{yy}(\theta)]$ (or T_{22}), and $A_{zz}(\theta)$ (or T_{20}). Finding a global fit appears hopeless at this time. In fact, for bombarding energies above the coulomb barrier, no one has found an optical model potential to fit all five observables at any energy for any target nucleus. Perhaps what is needed to solve this problem is a totally new idea.

Searches for optical model potentials that will describe $\sigma(\theta)$ for the elastic scattering of composite particles have been successful when polarization data are ignored. When doing so, the usual discrete and continuous ambiguities often arise and it is difficult not to utilize inadvertently the spin-dependent parts of the optical model potential as an ad hoc addition that improves agreement between data and calculations. As a guide to the potential shape describing deuteron scattering, Watanabe[1] suggested that the optical potential for a proton and neutron be averaged over the internal motion of the deuteron. This concept has been generalized to atomic mass-3[2] mass-4[3] and mass-6[4] projectiles and is commonly referred to as the folding model.

The folding model assumes that the composite particle-nucleus interaction can be simulated by an optical model potential, $U(\vec{R})$, where $\vec{R}$ is the displacement between the centers of mass of the bombarding particle and the target nucleus. It is an attempt to derive $U(\vec{R})$ from a fundamental point of view. For the deuteron, the coordinates $\vec{R} = \frac{1}{2}(\vec{r}_p + \vec{r}_n)$ and

$\vec{r} = (\vec{r}_p - \vec{r}_n)$ and the nuclear potentials $V_p(\vec{R} + \frac{1}{2}\vec{r})$ and $V_n(\vec{R} - \frac{1}{2}\vec{r})$ for the proton and neutron respectively are introduced. The effective potential is taken to be $V(\vec{R},\vec{r}) = V_p(\vec{R} + \frac{1}{2}\vec{r}) + V_n(\vec{R} - \frac{1}{2}\vec{r})$ averaged over the internal motion of the deuteron and $U_o(\vec{R}) = \int \chi_o^\dagger(\vec{r}) V(\vec{R},\vec{r}) \chi_o(\vec{r}) d\vec{r}$ is the first approximation to $U(\vec{R})$, where $\chi_o(\vec{r})$ is the one bound state of the deuteron. This results in the deuteron-nuclear optical model potential

$$U_o(\vec{R}) = U_c(R) + U_S(R) \; \vec{S}\cdot\vec{L} + U_T(R)[(\vec{S}\cdot\hat{R})^2 - 2/3], \tag{1}$$

where $\vec{S}$ is the deuteron spin and $\vec{S}^2 = S(S+1)$. Satchler has shown that the only three types of tensor terms that can appear in the potential are[5]

$$
\begin{aligned}
T_R &= \left[(\vec{S}\cdot\hat{R})^2 - \frac{S(S+1)}{3} \right] \\
T_p &= \left[(\vec{S}\cdot\vec{p})^2 - \frac{S(S+1)}{3}\vec{p}^2 \right] \\
T_L &= \left[(\vec{S}\cdot\vec{L})^2 + \frac{1}{2}(\vec{S}\cdot\vec{L}) - \frac{S(S+1)}{3}\frac{\ell(\ell+1)}{} \right]
\end{aligned}
\tag{2}
$$

where $\vec{p}$ is the deuteron linear momentum operator, $\vec{L}$ is the angular momentum operator $\vec{R} \times \vec{p}$, and $\vec{L}^2 = \ell(\ell+1)$. Equation 1 predicts that of these three possibilities, only T_R contributes to $U_o(\vec{R})$. Equation 1 contains a real and imaginary central term, a real spin-orbit term, and a real and imaginary tensor term. Generalized to a composite particle of N nucleons, the central term U_C is nominally N times as deep as that for a single nucleon ($\sim$ 50 MeV) and the spin-orbit term U_S is 1/N times as deep as that for a single nucleon ($\sim$ 6 MeV). For spin-1/2 particles such as the triton and the helion, the tensor term in Eq. 1, of course, vanishes. One of the primary motivations for triton and helion polarization measurements has been to test the predictions that the spin-orbit well depth is $\sim$ 2 MeV. Preliminary investigations of the newly available $\vec{t}$ and $\vec{h}^\ddagger$ data call for V_{so} = 3 to 8 MeV. However, these studies are not adequate to resolve the point at this time.

The position that has been consciously adopted here is that global optical model potentials for composite particles would be acceptable only if they correctly predict all polarization data as well as cross section data. However, in view of the adequate description of distorted waves without correct spin-dependence for some applications such as Distorted Wave Born Approximation, this position is itself worthy of future investigations. The point will not be considered further here except to explain why a number of papers on optical model potentials for deuterons, tritons, and helions are not cited in this review.

A few other points may be helpful to the reader. An excellent summary of the status of this subject up to 1971 has been published by Hodgson.[6] The work that follows is intended to trace developments in this field since that writing. A thorough compilation, with bibliography, of optical model parameters determined by fitting elastic scattering angular distributions for nucleons, deuterons, tritons, helions, and heavier particles, has been published by Perey and Perey.[7] Only parameters published in the period 1969-1972 are included but they have now incorporated earlier work and expanded the compilation up to 1975. They are in the process of publishing the updated version.[8] A set of notes presenting the details of folding model calculations for deuterons and tritons are incorporated in an informal report.[9] This report has been useful to students interested in writing computer programs to study the subject.

DEUTERONS

The Folding Model

The original folding model calculations[1] disregarded the deuteron
D-state (and thereby predicted that the tensor term is zero), ignored
distortions of the internal wave functions, neglected effects of the
Pauli principle, assumed the validity of a local nucleon potential, and
did not allow for breakup channels known to exist. Much of the subsequent work has consisted of attempts to address one or two of these de-
fects. Baumgartner[10] has estimated that for deuterons the Pauli principle produces corrections of only a few percent to the effective interaction. Kunz[11] and Johnson[35] have shown that the range of non-locality
for deuterons is one-half that for nucleons. The neglect of these two
effects would therefore appear to be justified. The Los Alamos Group[12]
extended the original model to simultaneously address all the other effects listed above. The coupled-channel approach included the deuteron
spin and D-state and incorporated deuteron internal wave-function distortions in the proximity of the nucleus. Electric and nuclear breakup
terms were calculated explicitly. Each of these effects had been explored prior to the Los Alamos work by others and are referenced therein.

The folding model predictions for the cross section and analyzing
powers of elastically scattered 15 MeV deuterons incident on ^{60}Ni and
^{90}Zr are shown as solid curves in Figs. 1 and 2. The dashed curves are
calculated with the breakup term set equal to zero. The nucleon potentials that were folded in the calculation were taken from Becchetti and
Greenlees.[13] The deuteron optical model potential thus generated was
used with a modified version of the computer code OPTIX,[14] which features coupled differential equations that allow and include the non-zero
off-diagonal elements of T_R. The comparisons between predictions and
data in Figs. 1 and 2 are quite acceptable for a calculation such as this
that starts from basic principles in which no free parameters are allowed.
The agreement is not acceptable when compared with global fits to nucleon cross section and polarization data or to certain single energy phe-

nomenological fits to deuteron cross
section and vector polarization data.
This leads to an investigation of
whether small changes in the folding
model potential can be made to bring
better agreement between predictions
and data.

The folding model potential $U(\vec{R})$
$= U_0(\vec{R}) + U_1(\vec{R})$, where $U_0(\vec{R})$ from Eq.
1 is made up of contributions from
the distorted deuteron wave function
and deuteron breakup. $U_0(\vec{R})$ is much
larger than $U_1(\vec{R})$. Real nuclear well
depths in Eq. 1 for ^{60}Ni are nominally 99 MeV for the central part, 6
MeV for the spin-orbit part, and 8
MeV (times a small form factor) for
the tensor part. Investigations of
$U_1(\vec{R})$ indicate a small imaginary
spin-orbit term less than 0.05 MeV[12]
and a small T_L term less than 0.04
MeV.[15] These terms are normally
neglected in folding model calculations. As a practical matter,

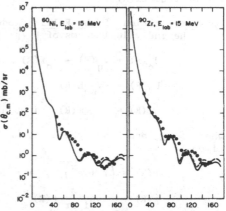

Fig. 1. Predictions of the folding model[12] with no free parameters for the differential cross
section of the deuterons on ^{60}Ni
and ^{90}Zr (solid curve). The dashed curves are calculated with the
breakup term set equal to zero.

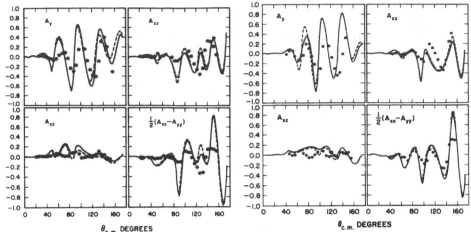

Fig. 2.   The four analyzing powers for ^{60}Ni (left) and ^{90}Zr (right) (See Fig. 1.)

$U(\vec{R})$ is dominated by $U_o(\vec{R})$, and $U_1(\vec{R})$ can be neglected entirely with only minor changes in predictions as can be seen by the dashed curves in Figs. 1 and 2. It can be expected that small adjustments to $U_o(\vec{R})$ will compensate for setting $U_1(\vec{R}) = 0$. The potential $U_o(\vec{R})$ is a close approximation to the data. The specific terms and well shapes which it predicts are often used as a starting point for devising an optical model phenomenological potential. A parameterization of the general features of $U_o(\vec{R})$ allows optical model computer programs to do least-squares searches for appropriate shapes. A phenomenological potential that can approximate $U_o(\vec{R})$ quite well has been developed and is the subject of the next section.

Parameterization of the Folding Model Potential
        The individual terms of $U_o(\vec{R})$ in Eq. 1 can be closely simulated by

$$U_c(R) = V_q(R) - V_o f(R) - i W_i g(R)$$

$$U_S(R) = -V_{so} h(R)$$

$$U_T(R) = -U_{Tr} m(R) - i U_{Ti} n(R) \tag{3}$$

where f, g, and h are the usual Woods-Saxon form factors for nucleons given by

$$f(R) = \left(1 + e^x\right)^{-1}; \quad x = (R - r_o A^{1/3})/a_o$$

$$g(R) = -4a_i \frac{d}{dr} f(R)$$

$$h(R) = -\left(\lambda_\pi^2/R\right)\frac{d}{dr} f(R); \quad \lambda_\pi^2 = 2.0 \text{ fm}^2 \tag{4}$$

A surface term has been used for g to be specific. A volume term which replaces g(R) by f(R) is sometimes used. The coulomb term $V_q(R)$ is the same for deuterons as for protons, namely a uniform charge distribution of radius $r_q A^{1/3}$ where $r_q = 1.3$ fm and A is the target atomic number. The tensor well shapes are the same sign as the corresponding central shapes for large R. The tensor form factors in Eq. 3 are approximated by[9,12]

$$m(R) = \lambda_\pi^2 \, R \, \frac{d}{dr} \left(\frac{1}{R} \frac{df}{dR} \right)$$

$$n(R) = \lambda_\pi^2 \, R \, \frac{d}{dr} \left(\frac{1}{R} \frac{dg}{dR} \right) \quad (5)$$

A pion Compton wavelength $\lambda_\pi = (\hbar/m_\pi c) = \sqrt{2}$ fm is inserted to make the form factors dimensionless. The folding model potential has been approximated according to Eq. 3 for 15 MeV deuterons on ^{60}Ni and the results are given in Table 1.

Heuristic arguments have recently been made which show that the particular set of real well-shapes given in Eqs. 3-5 are internally consistent when deduced by much more general albeit far less rigorous procedures.[16] Fermi[17] justified the Thomas form factor for the $\vec{S} \cdot \vec{L}$ term by assuming that the spin-nuclear interaction is predominantly a surface effect. No

TABLE 1

OPTICAL-MODEL POTENTIAL PARAMETERS IN EQ. 3 FOR 15 MeV DEUTERONS ON ^{60}NI

PARAMETERS[†]	LA[a]	TUNL[b]	ETH[c]	UW[d]
V_o	98.60	102.30	90.80	106.86
r_o	1.14	1.14	1.20	1.05
a_o	0.97	0.98	0.72	0.80
W_i	13.70	(N.R.)	15.50	14.22
r_i	1.28	--	1.29	1.43
a_i	0.87	--	0.77	0.70
V_{so}	5.60	5.52	5.35	7.00
r_{so}	0.97	0.98	0.79	0.75
a_{so}	0.97	0.97	0.34	0.50
U_{To}	7.60	8.17	6.00	0
r_{To}	1.14	1.12	1.20	--
a_{To}	1.02	0.98	0.82	--
U_{Ti}	1.00	(N.R.)	0	0
r_{Ti}	1.29	--	--	--
a_{Ti}	0.87	--	--	--
U_{LS}	0	0	0.15	0
r_{LS}	--	--	0.83	--
a_{LS}	--	--	0.10	--

[†]Well depths are expressed in MeV and geometry parameters in fm.
[a]Derived from the folding model in Ref. 12.
[b]Derived from the folding model in Ref. 16.
[c]Phenomenological parameters[24] beginning with LA set.
[d]Global phenomenological parameters of Eqs. 9 and 10 extrapolated to 15 MeV.[28]

assumptions are made as to the magnitude of $\vec{S}$ and therefore h(R) in Eq. 4 is an equally valid form factor for any spin. Similarly, it is suggested that m(R) of Eq. 5 is an equally valid form factor for any spin. The Thomas form is deduced by forming the scalar product of the first rank tensors (pseudovectors), $\vec{S}$ and $\vec{\nabla}f(R) \times \vec{p}$, to construct a scalar. The generalization consists of forming the scalar product of second rank tensors (i.e., the invariant contraction $\sum_q (-1)^q T_{kq} T_{kq}$ for $k = 2$) to construct a scalar. Tensors can be constructed from vectors and to do so the scalar $[\vec{A}\vec{B}]:[\vec{C}\vec{D}] = \sum_q (-1)^q T_{2q}(\vec{A},\vec{B}) \, T_{2q}(\vec{C},\vec{D})$ is a useful notation. It follows from Brink and Satchler[18] that except for an overall multiplicative constant the scalar can be expressed in terms of simple vector relations by the identity

$$[\vec{A}\vec{B}]:[\vec{C}\vec{D}] = (\vec{A} \cdot \vec{C})(\vec{B} \cdot \vec{D}) - \frac{1}{3}(\vec{A} \cdot \vec{B})(\vec{C} \cdot \vec{D}) - \frac{1}{2}(\vec{A} \times \vec{B}) \cdot (\vec{C} \times \vec{D}) \quad (6)$$

providing $\vec{B}$ and $\vec{C}$ commute. This identity yields the three tensor terms in Eq. 2 by defining $[\hat{R}\hat{R}]:[\vec{S}\vec{S}] = T_R$, $[\vec{p}\vec{p}]:[\vec{S}\vec{S}] = T_p$, and $[\vec{L}\vec{L}]:[\vec{S}\vec{S}] = T_L$. Having recalled the necessary points, the basic idea is now sketched.

Fermi argued that the potential of a projectile immersed in a uniform distribution of nucleons cannot depend on $\vec{L} = \vec{R} \times \vec{p}$ because it cannot locate the center of the distribution to determine the sign and

magnitude of $\vec{L}$. Near the surface,
however, the gradient of the nucleon density does supply a preferred direction (see Fig. 3).
For the short range nuclear forces,
we take this as $\vec{\nabla}f(R)$ where the
nucleus is assumed to be spherically symmetric. At the surface
three vectors occur that might be
used as building blocks to construct the $\vec{L}$ dependent potential
term which must be a scalar. They
are projectile spin $\vec{S}$, the projectile momentum $\vec{p}$, and the gradient of the nucleon density $\vec{\nabla}f(R)$.
The simplest combination is an
inner product between $\vec{S}$ and
$\vec{\nabla}f(R) \times \vec{p}$ which results in

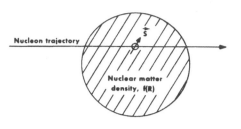

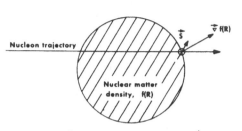

$$\vec{S} \cdot \vec{\nabla}f(R) \times \vec{p} = \frac{1}{R}\frac{df(R)}{dR}\vec{S} \cdot \vec{L} \quad (7)$$

and which is considered a justif-
ication for selecting the Thomas
form factor as the spin–orbit well
shape.

Fig. 3. A projectile of spin $\vec{S}$ immersed in a uniform nuclear density
cannot locate the center of the nucleus and therefore cannot deter-
mine angular momentum $\vec{L} = \vec{R} \times \vec{p}$.
However, at the surface the gradient
of the nucleon density does supply a
preferred direction.[17,16]

    There are several ways to com-
bine $\vec{S}$, $\vec{p}$, and $\vec{\nabla}f$ to deduce the
tensor terms in Eq. 6. The sim-
plest ones which obey basic symmetry requirements and do not re-
sult in a non-local potential are easily demonstrated to be

$$[\vec{\nabla}f \ \vec{\nabla}f]:[\vec{S}\vec{S}] = \left(\frac{df}{dR}\right)^2 T_R$$

$$[\vec{\nabla} \ \vec{\nabla} \ f]:[\vec{S}\vec{S}] = R\frac{d}{dR}\left(\frac{1}{R}\frac{df}{dR}\right)T_R$$

$$[\vec{\nabla}f \times \vec{p} \ \vec{\nabla}f \times \vec{p}]:[\vec{S}\vec{S}] = \left(\frac{1}{R}\frac{df}{dR}\right)^2 T_L \quad (8)$$

This prescription has suggested two candidates as well shapes for the T_R
term, the second one in Eq. 8 being the one originally proposed by the
Los Alamos Group[9,12] as an excellent empirical description of the folding model prediction. In addition, a new well shape is suggested for the
T_L term. Again, to the extent that this is a valid approach, the form
factors in Eqs. 4, 5, and 8 are valid for arbitrary spin. This generalization can be useful because whenever a local phenomenological potential
containing nuclear spins is being devised without a specific theory, at
least these simplicity arguments that will lead to form factors for the
various spin contributions can be invoked.

    Several comments are in order for Eq. 3 as a parameterization of the
folding model potential. It is well known that the real well depth is
not actually of the Woods-Saxon shape and deviations as large as a factor
of 2 in $\sigma(\theta)$ at back angles have been demonstrated by Perey and
Satchler.[19] Therefore, the parameters in Table I cannot be expected to
yield exactly the same results as the folding model calculations from
which they were derived. Raynal[20] has approximated the tensor form factor simply as d^2f/dR^2 and incorporated that into the Saclay computer code

MAGALI.[21] Irshad and Robson[22] have used Raynal's form factor multiplied by R^2. These terms are all similar outside the nuclear surface where they are relevant and can be adjusted toward one another by changing well depth. The T_L term is predicted to be small and a form factor is suggested in Eq. 8 that has not been tested. The non-local T_p term has not been investigated. Ramirez and Thompson[23] have investigated the effects on the folded potentials of using different, realistic deuteron wave functions. They have used Reid's soft-core, Reid's hard-core, and Erkelenz, et al., OBEP model. However, all give the same deuteron scattering potentials to within 1%. For convenience, they have also calculated the parameters appearing in the second column of Table I as a function of A for $6 \leqslant A \leqslant 216$.

Phenomenological Potentials

Deuteron polarization data were not available when the first attempts to find optical potential global fits were made. As polarization data became available, a set of parameters that brought predictions and all data into agreement were difficult to find.

One of the more successful studies of the problem is the recent work of the ETH Group[24] on ^{40}A, ^{40}Ca, and ^{60}Ni. They started with the LA parameters in Table I for ^{60}Ni and arrived at the ETH parameters in Table I. They were not able to get fits with $r_0 = 1.05$ fm as has been commonly used elsewhere.[22,25-30] Their results are given in Fig. 4. The backward angles of all analyzing powers, especially T_{20}, indicate a strong sensitivity for the tensor terms. Their work brings attention to the need for analyzing power measurements at scattering angles between 160° and 180°. These results are typical of the near success but not complete success of many attempts to describe the elastic scattering of deuterons. (See also Ref. 31-66.)

The difficulties with nuclear well shapes can be minimized if the bombarding energy is decreased until the coulomb force dominates. The Wisconsin Group[67] has bombarded ^{90}Zr with 5.5 MeV deuterons and measured the cross section and four analyzing tensors as shown in Fig. 5. They show good agreement between all data and a phenomenological optical potential calculation with a folding model T_R term added (solid lines). This is the first time all five observables have been fit and is apparently successful because the coulomb term is dominating. Even though the bombarding energy is in the subcoulomb region it is most encouraging to see that the interactions can be adequately described.

Some systematic trends of a phenomenological deuteron potential have been noted by Lohr and Haeberli[28] by relaxing the requirement that the second rank analyzing powers be fit. They have studied deuteron elastic scattering for 22 elements from ^{27}Al to ^{208}Pb in the energy region 5 to 13 MeV. The cross section and vector analyzing powers are fit satisfactorily with the values for ^{60}Ni labeled UW in Table I. Three of the nine parameters were shown to be $A^{1/3}$ dependent. Setting $C = A^{1/3}$, they give

$$V_o = 91.13 + 2.20 \ Z/C \text{ MeV}, \quad W_i = 218 \ C^{-2} \text{ MeV}, \quad a_i = 0.50 + 0.013 \ C^2 \text{ fm}. \quad (9)$$

The other seven parameters for the Lohr and Haeberli potential are:

$$
\begin{array}{llll}
r_o = 1.05 & r_i = 1.43 & r_{so} = 0.75 & r_q = 1.3 \\
a_o = 0.80 & a_{so} = 0.50 & V_{so} = 7.0 &
\end{array} \quad (10)
$$

The results are compared with data in Fig. 6.

180   P.W. Keaton, Jr.

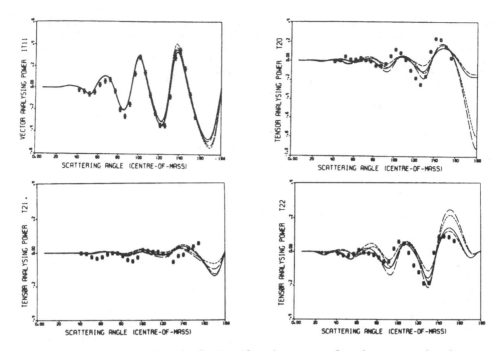

Fig. 4. Optical model calculation for deuteron elastic scattering by
^{60}Ni at 15 MeV.[24)] The curves show the following results:
- - - - - - - Without tensor terms.   — — — With T_R term.
- - - - - With T_L term.   ———— With T_R and T_L term.

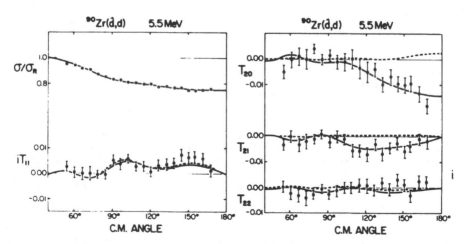

Fig. 5. Angular distribution of the differential cross section and the
four analyzing powers for deuteron elastic scattering from ^{90}Zr at 5.5
MeV.[67)] The result of optical model calculations with no tensor term is
the dashed curve and with a T_R term is the solid curve.

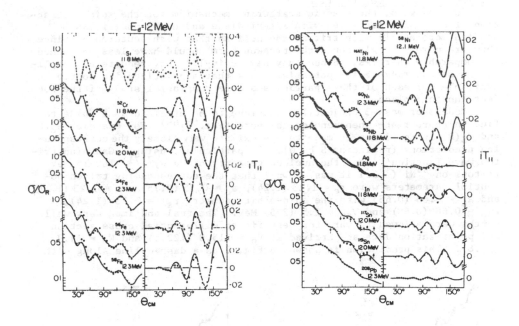

Fig. 6. Comparison between measurements and optical model calculations
for deuteron energies of 12 MeV.[28)] The dotted curves are based on Eqs.
9 and 10. The solid curves were obtained by adjusting W_i, r_i, and a_i.
The dashed curves for Si were obtained by adjusting V_o as well as the
three absorptive parameters.

Conclusions

      We leave this subject emphasizing that questions have been raised
and that answers are not forthcoming. The folding model reproduces the
general features of the differential cross section and vector analyzing
power reasonably well as can be seen in Fig. 1, but by developing the a-
bility to make precision measurements with polarized deuterons, experimentalists have explored more details of the elastic scattering mechanism.
The folding model is reviewed thoroughly here because it has been the
only viable tool to guide the experimentalists in describing their polarization data. The three tensor analyzing powers for intermediate weight
nuclei have not been described in terms of our present understanding of
the spin-nuclear interaction mechanism. Even more detail can be investigated experimentally. Polarization transfer experiments for deuterons
on intermediate weight nuclei can be measured with present technology.
The spin-spin interaction between deuterons and nuclei can also be investigated experimentally. Although these experiments are not to be dis-
couraged, it is unlikely that they can be understood until the more fundamental ones are understood.

TRITONS AND HELIONS

Folding Model Parameterization

In many ways, the elastic scattering mechanism for the spin interaction of composite particles with intermediate and heavy nuclei can be studied more easily with tritons and helions than with deuterons. Since they are more strongly bound the breakup term should have less influence at comparable energies. Since they have spin-1/2 they require no complicating second rank tensor potential. Now that a polarized ion source is available for each of these particles a more meaningful study of the subject can take place.

Models of mass-3 projectiles continue to be utilized as a guide to find an appropriate phenomenological potential.[68,69] For example, $U_c(R)$ and $U_S(R)$ in Eq. 3 can be made to approximate very closely the corresponding predictions (replacing $\vec{S}$ by $\vec{\sigma}$). In particular, using a Gaussian triton wave function[70] and the nucleon geometry of Perey[7,71] for 15 MeV tritons on ^{60}Ni (^{90}Zr) it was shown[9] that the corresponding triton potential parameters were $r_0 = 1.23$ (1.24), $a_0 = 0.75$ (0.75), $r_i = 1.23$ (1.24) and $a_i = 0.68$ (0.68) with the spin-orbit geometry $r_{so} = 1.23$ (1.24), $a_{so} = 0.68$ (0.68) and $V_{so} = 2.5$ (2.5) MeV. The real and imaginary well depths were allowed to vary for best fit to differential cross section and polarization data and arrived at $V_0 = 115.0$ (122.5) and $W_0 = 34.9$ (32.8). This prediction is shown in Fig. 7 as a dashed line along with

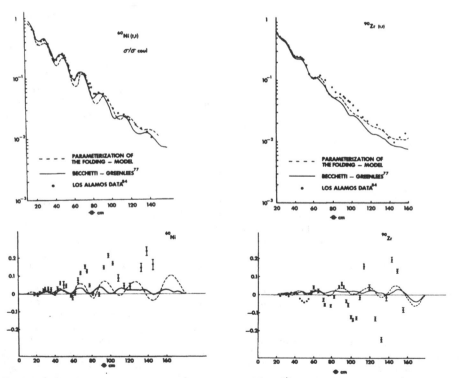

Fig. 7. Folding model predictions for ^{60}Ni$(\vec{t},t)^{60}$Ni and ^{90}Zr$(\vec{t},t)^{90}$Zr at 15 MeV (dashed). Becchetti-Greenless[77] predictions for ^{60}Ni and ^{90}Zr at 15 MeV (solid)

data from Fig. 8. The striking
feature in these graphs is the
large discrepancy between the
folding model polarization predictions and the data. This geometry
was used as starting parameters
for a phenomenological potential
which is the next subject. Unfortunately, in the preliminary
studies this technique did not
yield a systematic set of parameters for the five nuclei having
polarization data.

Phenomenological Potentials

There have been earlier attempts to measure mass-3 polarization with double scattering techniques.[72-76] This was successful
only for lighter nuclei. The combination of low count rate double
scattering experiments and of the
rapid fall-off of $\sigma(\theta)$ with scattering angle for intermediate
weight nuclei precluded any meaningful data that could address
this problem.[75,76] Consequently
there have been no global optical-
model analyses including triton
and helion polarization data.

Becchetti and Greenlees[77]
suggest an average set of parameters for tritons and helions by
analyzing differential cross section data and some reaction cross
section data. For tritons, they
propose

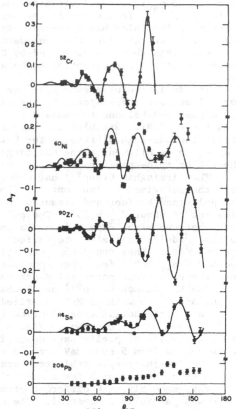

Fig. 8. Data[84] and optical model
calculation (line) to analyzing power A_y (or polarization) for 15 MeV
tritons.

$$V_o = 165.0 - 0.17\ E - 6.4\ \xi \qquad r_o = 1.20, \qquad a_o = 0.72$$
$$W_v = 46.0 - 0.33\ E - 110\ \xi \qquad r_i = 1.40, \qquad a_i = 0.84 \qquad (11)$$

and for helions they propose

$$V_o = 151.9 - 0.17\ E + 50\ \xi \qquad r_o = 1.20, \qquad a_o = 0.72$$
$$W_v = 41.7 - 0.33\ E + 44\ \xi \qquad r_i = 1.40, \qquad a_i = 0.88 \qquad (12)$$

and for both particles, they set $V_{so} = 2.5$ MeV, $r_{so} = 1.20$, $a_{so} = 0.72$,
and $r_q = 1.30$. In these formulae there is no surface imaginery term and
the volume imaginary term requires that $g(R)$ be replaced by $f(R)$ in Eq.
3. Also $\xi = (N-Z)/A$ and E is the incident LAB energy. These parameters
gave an χ^2 per point of about 16 and they did not require a spin-orbit
term to obtain this kind of agreement with data but it was included in
the potential. The predictions of these parameters are shown in Fig. 7
for ^{60}Ni and ^{90}Zr as solid lines.

The importance of some other terms that might be included in a phenomenological potential has been explored. A nuclear spin-orbit ($\vec{I}\cdot\vec{L}$)

term in the optical model potential has been investigated by the Florida
State Group.[78] They carefully measured $\sigma(\theta)$ for 14 MeV alpha particles
bombarding ^{9}Be which has spin-3/2. Agreement between the optical model
calculation and the data was considerably improved with an $\vec{I} \cdot \vec{L}$ term of
2.4 MeV. This term was estimated by Rawitscher[79] on the basis of the
nucleon-α spin-orbit potential to be an order of magnitude smaller than
reported.

The influence of a spin-spin term on the differential cross section
for helions was investigated by the Oak Ridge Group.[80,81] They used 50
MeV alphas and helions to bombard the spin-7/2 target ^{59}Co and the spinless target ^{60}Ni. Satchler and Fulmer[82] have explained somewhat empirically the odd-even difference for helions as due to the quadrupole moment which is allowed in the odd target but not in the even. Neither
the $\vec{I} \cdot \vec{L}$ or spin-spin data have been explained in a fundamental way.

The Birmingham Group[83] has begun to address the present problem
with the polarized helion source. They bombarded a ^{58}Ni target with 33
MeV polarized helions and measured the analyzing power from forward angles out to about θ_{cm} = 75°. The polarization is large when compared
with what one might expect from the folding model. The phenomenological
optical model potentials were varied to fit the cross section and polarization data. They found that the spin-orbit term tended toward $r_o = r_{so}$
and $0.2 \leqslant a_{so} \leqslant 0.3$ fm. They conclude that much more data is needed before the spin-orbit potential depth can be deduced with confidence.

The Los Alamos Group[84] has bombarded targets of ^{52}Cr, ^{60}Ni, ^{90}Zr,
^{116}Sn, and ^{208}Pb with 15 MeV polarized tritons and measured the analyzing
power and differential cross section from forward angles to about θ_{cm} =
160°. Those data are shown in Fig. 8 along with the solid curves which
are the results of preliminary best-fit searches. The spin-orbit well
depth varied from 5 to 8 MeV for the various elements. Attempts to include an imaginary spin-orbit term did not indicate improvement.

A subsequent search for a purely phenomenological set of parameters
that vary smoothly with different atomic numbers for ^{52}Cr, ^{90}Zr, and
^{116}Sn was partially successful. The central real geometry was fixed at
$r_q = 1.30$, $r_o = 1.20$, and $a_o = 0.65$, and the imaginary radius was fixed
at $r_i = 1.6$. The spin-orbit parameters were fixed at $r_{so} = 1.15$ and
$V_{so} = 6.0$. The four parameter search arrived at the following values
for ^{52}Cr, ^{90}Zr, and ^{116}Sn respectively:

$$V_o = 165.4, \ 168.0, \ 153.5 \qquad W_i = 16.7, \ 13.5, \ 13.7$$
$$a_i = 0.79, \ 0.87, \ 0.90 \qquad a_{so} = 0.63, \ 0.51, \ 0.82$$
$$\chi^2_\sigma = 1, \ 4, \ 2 \qquad \chi^2_p = 2, \ 3, \ 6$$

where all well depths are in MeV and all geometry parameters are in fm.
The values for χ^2_σ and χ^2_p are the approximate chi-squared per point for
the differential cross section and polarization values respectively.
These data indicate a spin-orbit well depth of about 6 MeV but a careful
study of the subject has not yet been made.

Conclusions
The new polarization data for tritons[84] and helions[83] are both exciting and challenging because they test our knowledge of the elastic
scattering mechanism for composite particles. The data have been fit
with phenomenological optical model potentials, but not yet in a systematic way or in a way that is compatible with the folding model. The polarization values are unexpectedly large which raises the question of

how the spin-orbit term of a composite particle could be larger than the
sum of the spin-orbit terms of its constituents. The climate seems ideal
for a vigorous joint effort of experimentalists and theoreticians.

ACKNOWLEDGEMENTS

It is a pleasure to acknowledge the help of Lynn Veeser in the optical model analyses and of Patricia Kelley in the preparation of the manuscript.

REFERENCES

†   Work performed under the auspices of the U.S. Energy Research and De-
    velopment Administration.
*   The ^{3}He ion will be referred to as helion with symbol h.
‡   The Cartesian tensors are labeled with an A and double subscripts and
    spherical tensors with a T and double subscripts. They will be used
    interchangeably and are related by a constant.
⧧   Beams of polarized tritons (helions) are written as vectors such as
    $\vec{t}(\vec{h})$.
1) S. Watanabe, Nucl. Phys. 8, 484 (1958).
2) A. Y. Abul-Magd and M. El-Nadi, Prog. Theor. Phys. 35, 798 (1966).
3) S. G. Kadmenskii, Yad. Fiz. 8, 486 (1969) [Transl.: Sov. J. Nucl.
    Phys. 8, 284 (1969)].
4) J. W. Watson, Nucl. Phys. A198, 129-143 (1972).
5) G. R. Satchler, Nucl. Phys. 21, 116 (1960); "The Scattering and
    Polarization of Spin One Particle," Oak Ridge National Laboratory
    Report No. ORNL-2861 (unpublished).
6) P. E. Hodgson, Nuclear Reactions and Nuclear Study (Clarendon Press,
    Oxford, 1971).
7) C. M. Perey and F. G. Perey, Atomic Data and Nuclear Data Tables 13,
    293-337 (1974).
8) C. M. Perey, private communication.
9) P. W. Keaton, Jr., E. Aufdembrink, and L. R. Veeser, "A Model for the
    Optical Potential of Composite Particles," Los Alamos Scientific
    Laboratory Report No. LA-4379-MS.
10) G. Baumgartner, Z. Phys. 204, 17 (1967).
11) P. D. Kunz, Phys. Lett. 35B, 16 (1971)
12) P. W. Keaton, Jr. and D. D. Armstrong, Phys. Rev. C8, 1692 (1973).
    The data have been tabulated by R. A. Hardekopf, D. D. Armstrong,
    L. L. Catlin, P. W. Keaton, Jr., and G. P. Lawrence, "Cross Section
    and Analyzing Powers in Deuteron Elastic Scattering," Los Alamos
    Scientific Laboratory Report No. LA-5050 (unpublished).
13) F. D. Becchetti, Jr. and G. W. Greenlees, Phys. Rev. 182, 1190 (1969).
14) B. A. Robson, "A Fortran Program for Elastic Deuteron Scattering,"
    Oak Ridge National Laboratory Report No. ORNL-TM-1831 (unpublished).
15) A. P. Stamp, Nucl. Phys. A159, 399-408 (1970).
16) P. W. Keaton, Jr., "Potential Well Shapes for Tensor Terms Involving
    Spin," Los Alamos Scientific Laboratory Report No. LA-6002
    (unpublished).
17) E. Fermi, Nuovo Cimento Suppl. II, Series 10, 18 (1955).
18) D. M. Brink and G. R. Satchler, Angular Momentum (Clarendon Press,
    Oxford, 1962), p. 124.
19) F. G. Perey and G. R. Satchler, Nucl. Phys. A97, 515 (1967).
20) J. Raynal, Phys. Lett. 7, 281 (1963).
21) J. Raynal, Technical Report Dph-T/69-42, Saclay (unpublished) 1969.

22) M. Irshad and B. A. Robson, Nucl. Phys. A218, 504-508 (1974).
23) J. A. Ramirez and W. J. Thompson, Fourth Polarization Symposium.
24) H. R. Bürgi, W. Grüebler, P. A. Schmelzbach, V. König, R. Risler
    Nucl. Phys. A247 (1975) 322
25) J. A. R. Griffith, M. Irshad, O. Karban, and S. Roman, Nucl. Phys.
    A146, 193-214 (1970).
26) R. C. Brown, A. A. Debenham, J. A. R. Griffith, O. Karban, D. C.
    Kocher, and S. Roman, Nucl. Phys. A208, 589-610 (1973).
27) F. T. Baker, S. Davis, C. Glashausser, and A. B. Robbins, Nucl.
    Phys. A233, 409-424 (1974).
28) J. M. Lohr and W. Haeberli, Nucl. Phys. A232, 381-397 (1974).
29) C. E. Busch, T. B. Clegg, S. K. Datta and E. J. Ludwig, Nucl. Phys.
    A223, 183-194 (1974).
30) O. Karban, A. K. Basak, J. A. R. Griffith, S. Roman and G. Tungate,
    Fourth Polarization Symposium, p. 616
31) J. L. Gammel, B. J. Hill, and R. M. Thaler, Phys. Rev. 119, 267
    (1960).
32) J. Raynal, Centre d'Etudes Nucleaires de Saclay Report No.
    CEA-R2511, Thesis (unpublished) 1965.
33) J. Testoni and L. C. Gomez, Nucl. Phys. 89, 288 (1966).
34) F. G. Perey and G. R. Satchler, Nucl. Phys. A97, 515 (1967).
35) R. C. Johnson and P. J. R. Soper, Phys. Rev. C1, 976 (1970).
36) K. Veta and K. Hara, Z. Phys. 248, 311 (1971).
37) D. J. Hooten and R. C. Johnson, Nucl. Phys. A175, 583 (1971).
38) J. R. Rook, Nucl. Phys. 61, 219 (1965).
39) F. J. Bloore, Nucl. Phys. 68, 298 (1965).
40) E. Couffan and L. J. B. Golfarb, Nucl. Phys. A94, 241 (1967).
41) G. R. Satchler, Nucl. Phys. 85, 273 (1966).
42) P. E. Hodgson, Adv. Phys. 17, 202 (1968), 15, 329 (1966).
43) S. K. Samaddar and S. Mukherjee, Nucl. Phys. A177, 598 (1971).
44) S. Mukherjee, Nucl. Phys. A118, 423 (1968).
45) F. Hinterberger, Nucl. Phys. A111, 265 (1968).
46) W. F. Junkin and F. Villars, Ann. Phys. (N.Y.) 45, 93 (1967).
47) H. P. Stapp, Phys. Rev. 107, 605 (1957).
48) C. A. Engelbrecht and H. Fiedeldey, Ann. Phys. (N.Y.) 42, 262 (1967).
49) Gy. Bencze and I. Szentpetery, Phys. Lett. 30B, 446 (1969).
50) T. H. Rihan and M. A. Sharaf, Nucl. Phys. A134, 369 (1969).
51) R. K. Satpathy, S. K. Samaddar, and S. Mukherjee, Nucl. Phys. A132,
    276 (1969).
52) G. H. Rawitscher and S. N. Mukherjee, Ann. Phys. (N.Y.) 68, 57
    (1971).
53) M. Reeves, III, Union Carbide Report No. CTC-32 (unpublished) (1970).
54) S. K. Samaddar, R. K. Satpathy, and S. Mukherjee, Nucl. Phys. A150,
    655 (1970).
55) M. Bauer and C. Block, Phys. Lett. 33B, 155 (1970).
56) G. Perrin, Nguyen Van Sen, J. Arvieux, C. Perrin, R. Darves-Blanc,
    J. L. Durand, A. Fiore, J. C. Gondraud, and F. Merchez, Nucl. Phys.
    A206, 623-632 (1973).
57) R. Roche, Nguyen Van Sen, G. Perrin, J. C. Gondraud, A. Fiore, and
    H. Müller, Nucl. Phys. A220, 381-390 (1974).
58) F. T. Baker, S. Davis, C. Glashausser, and A. B. Robbins, Nucl. Phys.
    A233, 409-424 (1974).
59) Mervat H. Simbel, Phys. Rev. C8, 75 (1973).
60) G. H. Rawitscher, Phys. Rev. C9, 2210 (1974).
61) J. D. Childs, W. W. Daehnick, M. J. Spisak, Phys. Rev. C10, 217
    (1974).

62) University of Birmingham, Department of Physics, Birmingham, England, Nuclear Structure Group Annual Progress Report, June 1975 (unpublished).

63) Eidgenössische Technische Hochschule, Zürich, Laboratorium für Kernphysik, Jahresbericht (unpublished) (1974).

64) O. Kalban, A. K. Bosak, J. A. R. Griffith, S. Roman and G. Tungate, Fourth Polarization Symposium, p.616

65) R. J. Eastgate, T. B. Clegg, R. F. Haglund, and W. J. Thompson, ibid.

66) H. R. Bürgi, W. Grüebler, P. A. Schmelzbach, V. König, B. Jenny, R. Risler, and W. G. Weitkamp, ibid. p.622

67) L. D. Knutson and W. Haeberli, Phys. Rev. Lett. $\underline{30}$, 986 (1973).

68) Bikash Sinha, Feraze Duggan, and Richard J. Griffiths, Nucl. Phys. $\underline{A243}$, 170-180 (1975).

69) M. H. Simbel, Phys. Rev. $\underline{C10}$, 1083 (1974).

70) L. I. Schiff, Phys. Rev. $\underline{133}$, B802 (1964).

71) F. G. Perey, Phys. Rev. $\underline{131}$, 745 (1963).

72) C. D. Bond, L. S. August, P. Shapiro and W. I. McGarry, Nucl. Phys. $\underline{A241}$, 29-35 (1975).

73) W. S. McEver, T. B. Clegg, J. M. Joyce, and E. J. Ludwig, Nucl. Phys. $\underline{A178}$, 529 (1972).

74) J. B. A. England, R. G. Harris, L. H. Watson and D. H. Worledge, Nucl. Phys. $\underline{A165}$, 277 (1971).

75) D. D. Armstrong and P. W. Keaton, Jr., "Polarization of Elastically Scattered Tritons and ^{3}He," Los Alamos Scientific Laboratory Report No. LA-4539 (unpublished).

76) E. J. Ludwig, T. B. Clegg, and R. L. Walter, Nucl. Phys. $\underline{A211}$, 559 (1973).

77) F. D. Becchetti, Jr., and G. W. Greenlees, Polarization Phenomena in Nuclear Reactions, edited by H. H. Barschall and W. Haeberli (The University of Wisconsin Press, Madison, 1971), p. 682.

78) R. B. Taylor, N. R. Fletcher, and R. H. Davis, Nucl. Phys. $\underline{65}$, 318 (1965).

79) G. H. Rawitscher, Phys. Rev. $\underline{C6}$, 1212 (1972).

80) J. C. Hafele, C. B. Fulmer, and F. G. Kingston, Phys. Lett. $\underline{31B}$, 17 (1970).

81) C. B. Fulmer and J. C. Hafele, Phys. Rev. $\underline{C7}$, 631 (1973).

82) G. R. Satchler and C. B. Fulmer, Phys. Lett. $\underline{B50}$, 309 (1974).

83) S. Roman, A. K. Basak, J. B. A. England, O. Karban, G. C. Morrison, J. M. Nelson, and G. G. Shute, Fourth Polarization Symposium, p.626

84) R. A. Hardekopf, L. R. Veeser, and P. W. Keaton, Jr., Fourth Polarization Symposium p.624

DISCUSSION

Cramer:

I would like to comment on a remark you made at the very beginning of your talk, concerning the importance of non-locality for composite particles: It's true that folding models indicate a 1/A dependence for the non-local range of composite particles, but the relevant parameter is $k \cdot \beta$ not β. Since the wave length are shorter and the energies tend to be higher for measurements involving composite particles, k will be larger, so the size of the non-local effects is about the same as for a nucleon.

Keaton:

Yes, thank you for clarifying that point. I do not want to imply that the non-local effects are less important for deuterons than nucleons. But rather, I would like to leave the impression that since they are about the same and since we have done so well with a local potential for nucleons, it seems reasonable to expect success with a local potential for deuterons.

Bleuler:

I was interested about your remark on the strength of the spin-orbit interaction in composite systems: Is it definitely larger than the value obtained from summing up the corresponding nucleon-nucleon interactions ?

Keaton:

Based on analyses so far, the answer is "yes" for tritons and "no" for deuterons. I don't expect that answer for deuterons to change because careful analyses have been done. However, the answer for tritons could change when more careful analyses are done.

Johnson:

I would like to point out that in the presence of the Pauli Exclusion Principle and a tensor force in the neutron-proton interaction, a tensor force of the $(\vec{p} \cdot \vec{s})^2$ type is expected in the deuteron optical potential. Calculations by Ioanides indicate that the T_p potential generated in this way has a strength similar to the T_r force obtained by you and Knutson and Haeberli using the Watanabe model.

Keaton:

That is very exciting. To my knowledge it is the first time anyone has thought of what a source of a $(\vec{p} \cdot \vec{s})^2$ force would be. We look forward to future developments of this line of thinking.

Johnson:

You referred to the result of Soper and me and generalized by Jackson and me that in a folding model the range of non-locality of the deuteron optical potential is predicted to be 1/2 of the range of non-locality of the nucleon optical potential. It may be helpful to point out that this just means that to a first approximation the equivalent local deuteron potential can be obtained by folding nucleon optical potential corresponding to 1/2 the deuteron kinetic energy into the deuteron ground state wave function.

THE SPIN-SPIN INTERACTION IN NUCLEON-NUCLEUS SCATTERING[*]

H.S. Sherif

Department of Physics, University of Alberta, Edmonton, Canada

1. Introduction

The existence of spin-spin terms in the optical potential for nucleons scattered on targets with non-zero spin has been the subject of investigation for many years. The presence of spin-spin and tensor terms in the nucleon-nucleon interaction leads one to suspect that these forces may still reveal themselves in the scattering on odd mass targets in the form of an interaction that couples the projectile and target spins. It was Feshbach[1]) who first suggested that such terms may be present in the optical model potential. Two of the possible invariant forms that may appear in the potential are: a spherical spin-spin term

$$U_{SS}(r) = - V_{SS} \, F_0(r) \, \vec{\sigma} \cdot \vec{I} \tag{1}$$

and a tensor term

$$U_{ST}(r) = - \frac{1}{2} V_{ST} \, F_T(r) [3(\vec{\sigma} \cdot \hat{r})(\vec{I} \cdot \hat{r}) - \vec{\sigma} \cdot \vec{I}] \tag{2}$$

where $\vec{\sigma}$ is the Pauli spin operator for the scattered nucleon and $\vec{I}$ is the spin operator of the target nucleus. V_{SS} and V_{ST} are the strengths of the spherical and tensor terms, respectively.

2. Microscopic Calculations of the Potentials

Many attempts have been made[2-5]) to evaluate the spherical spin-spin term (1) starting from the appropriate terms in the nucleon-nucleon potential. Satchler[4]) used a folding procedure in which the potential is constructed by averaging an effective interaction v_{pt} between the projectile nucleon p and a target nucleon t over the distribution of nucleons in the target nucleus. Starting with the two body central spin-spin potential

$$v_{pt} = -V_c \, g(r_{pt}) \, \vec{\sigma}_p \cdot \vec{\sigma}_t \tag{3}$$

One then writes

$$U_{SS}(r_p) = \langle \Phi_0 | v_{pt} | \Phi_0 \rangle \tag{4}$$

where Φ_0 is the target ground state wave function.

In the extreme single-particle model where the target is made up of an inert core plus one valence nucleon in a shell model orbit with quantum numbers ℓ and $j = I$, Eqs. (3) and (4) give

$$U_{SS}(r_p) = V_c \, (-)^{j-\ell+\frac{1}{2}} \, (\ell + \frac{1}{2})^{-1} \, F_0(r_p) \vec{\sigma} \cdot \vec{I} \tag{5}$$

where

$$F_0(r_p) = \frac{1}{2} \iint u_{\ell j}^2 \, (r_t) \, g \, (r_{pt}) r_t^2 \, dr_t \, d(\cos \theta_{pt}) \tag{6}$$

$u_{\ell j}(r)$ is the single-particle radial wave function.

In this simple model the volume integral of $U_{SS}(r)$ is related to the volume integral $J_{\sigma\sigma}$ of the spin-spin part (3) of the nucleon-nucleon interaction

$$4\pi \ V_{SS} \int F_0(r) \ r^2 \ dr = (-)^{j-\ell+\frac{1}{2}} \ (\ell + \tfrac{1}{2})^{-1} \ J_{\sigma\sigma} \qquad (7)$$

which may be used to estimate the strength V_{SS}. Since $J_{\sigma\sigma}$ is constant it follows that the depth of $U_{SS}(r)$ will tend to decrease like A^{-1} as the target mass A increases.

The above model was applied to the calculation of $U_{SS}(r)$ for ^{59}Co. The ground state was taken as a single proton hole in the $1f_{\frac{7}{2}}$ shell.

The effective interaction v_{pt} was a Serber mixture with a Yakawa shape of a range of an OPEP. The results for proton scattering are shown by the curve labelled C in Fig. 1. (The function $U_{SS}(r)$ in the Figure is

$U_{SS}(r) = -V_{SS} \ F_0(r)$ and not the full expression in Eq. (1)). The potential has a peak value of about +100 keV (at $r \approx 3$ fm).

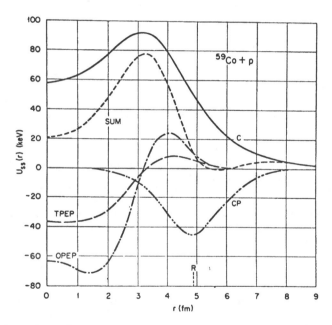

Fig. 1. The spherical spin-spin potential calculated using the central (C) and tensor (OPEP and TPEP) forces and the core polarization (CP) model. The sum is C+CP+TPEP. (Ref. 4)

Core polarization effects are expected to reduce the above single particle estimate for $U_{SS}(r)$ in much the same way as they tend to quench the contributions to the magnetic moments of odd mass nuclei from the spin of the valence nucleon. A simple collective model was adopted by Satchler; the core is presumed to undergo a collective 1^+ spin oscillation. The coupling to this of the valence and projectile nucleons is achieved through the assumption of a vibrating potential in which these nucleons move. When treated to second order these core excitations contribute an additional term

$$- (\frac{y}{4\pi}) k_p(r_p) \; k_t(r_t) \vec{\sigma}_p \cdot \vec{\sigma}_t \qquad (8)$$

to effective interaction v_{pt}. In the above expression y is a coupling parameter and the functions $k(r)$ are proportional to the derivatives of the appropriate potentials.

The contribution of this core polarization to $U_{SS}(r)$ for ^{59}Co is shown in Fig. 1 (curve labelled CP). It has a sign opposite to that of the single particle contribution, resulting in a strong cancellation between these two contributions.

The two-body tensor force will also contribute to $U_{SS}(r)$. Satchler has estimated this contribution using the OPEP and a "regularized OPEP" (labelled TPEP).[6] The corresponding curves are shown in Fig. 1. These contributions tend to be more confined to the nuclear interior and both have a weakly repulsive tail. The curve labelled SUM in Fig. 1 is the sum of the single particle, core polarization, and the TPEP contributions.

It was also shown that for neutrons on ^{59}Co, $U_{SS}(r)$ is weaker than for protons and has opposite sign. Moreover, preliminary estimates for the tensor interaction (Eq. (2)) showed that this interaction appears to be weak, perhaps with a strength of a few tens of keV.

It may be useful to show here what the single particle strengths V_{SS} are for some lighter targets. To obtain these estimates one uses Eq. (7). The quantity $J_{\sigma\sigma}$ has been determined[7] from analyses of inelastic proton scattering and charge exchange (p,n) reactions. Woods-Saxon form factors $F_0(r)$ were first fitted to the shapes calculated by Batty[8] using Eq. (6) for ^{9}Be and ^{27}Al. These were then used on the left hand side of Eq. (7) and the strength V_{SS} was deduced. For ^{10}B a form factor similar to that of ^{9}Be was used and the strength was calculated by adding the contributions from a proton-neutron pair in the $1p_{\frac{3}{2}}$ shell. The results are shown in Table 1.

TABLE 1

Single-Particle Estimates for the Shape and Strength of $U_{SS}(r)$ for Protons

Nucleus	$F_0(r)$	V_{SS}(MeV)
^{9}Be	$[1 + \exp(\frac{r-3.6}{0.75})]^{-1}$	0.1
^{10}B	$[1 + \exp(\frac{r-3.73}{0.75})]^{-1}$	-0.23[a]
^{27}Al	$[1 + \exp(\frac{r-3.54}{0.7})]^{-1}$	-0.4

[a] The value given in Ref. (14) is in error by a factor of 2.

Calculations of the spin-spin potential for a particle in nuclear matter were performed by Dabrowski and Haensel[5]. These involved the derivation of the potential for a nucleon at the top of the Fermi sea in terms of the K-matrix. Using the Reid soft core nucleon-nucleon interaction, the predicted strengths of $U_{SS}(r)$ are generally larger than those obtained from the folding single-particle model discussed above. For example, they get a depth V_{SS} = 424 keV for p + ^{59}Co (Satchler gets $\approx$ 100 keV) and $V_{SS} \approx$ 1.3 MeV for ^{27}Al which is three times stronger than the value given in Table 1. It is, of course, expected that nuclear matter predictions for the potential will be higher than those relevant to energies above the Fermi energy. The authors have essentially shown this by performing calculations for the potentials in the phase shift approximation; the calculated potentials were all smaller than those obtained from the K-matrix. Another point that emerged from these calculations is that the tensor potential is much weaker than the spherical one, confirming the conclusion made earlier by Satchler.

The question of the possible presence of an imaginary part in the spin-spin potentials has recently been taken up by Haensel[9]). His calculations for ^{59}Co indicate that the depth of the imaginary part of $U_{SS}(r)$ for protons is about 60 keV at 40 MeV and about 20 keV at 100 MeV. The neutron potential is weaker, of opposite sign, and almost energy independent.

3. Effect of the Spin-Spin Potential on the Scattering Process

Several optical model calculations which take into account the effects of spin-spin interactions have been carried out. The first of these are coupled channel calculations such as those performed by Stamp[2]) and by Drigo and Picent[10]). The derivation of the scattering amplitude is handled in the channel spin framework. These calculations take into account only the spherical spin-spin potential which is diagonal in this representation. The spin-orbit potential then serves as the coupling interaction. Theoretical predictions were given by Stamp for the "triple" scattering parameters for neutron scattering. The main result was that the most sensitive among these is the depolarization parameter D (D is equivalent to the polarization transfer coefficient $K_y^{y'}$), and that measurements of D near the differential cross section minima should yield information about the spin-spin potential. Tamura[11]) has also included the spherical spin-spin potential in his coupled channel code JUPITER.

The other approach takes advantage of the apparent weakness of the spin-spin potentials and treats these to first order in the framework of the distorted wave Born approximation. Davies and Satchler[12]) have derived a general DWBA expression for the spin-spin amplitude. They then carried out detailed calculations for the spherical spin-spin potential. These calculations indicated that one might detect the effects of the spin-spin interaction through measurements of the cross sections for polarized nucleons on polarized targets. DWBA calculations in which both the spherical and tensor terms are taken into consideration have been carried out recently[13,14]). These indicated that[14]), among the "triple" scattering parameters, the depolarization parameter seems to be the one most sensitive to the spin-spin potentials. This confirms the assertion made earlier by Stamp.

Other types of calculations[15,16]) with the spherical spin-spin potential have also been carried out. These will be briefly mentioned in the course of the following discussion of various experiments and/or analyses of experimental data that were performed in the hope of verifying the existence of the spin-spin potentials.

4. Analysis of Data on Neutron S-Wave Strength Function

An attempt has been made by Rahman Khan[15]) to place an upper limit on the strength of the spherical spin-spin potential by considering the observed differences in neutron S-wave strength functions for even and odd mass nuclei. In the mass range $45 \leq A \leq 125$ the resonance in the strength function for even nuclei occurs at $\overline{A} \approx 65$ whereas for odd mass nuclei it occurs at $A \approx 50$. A Woods-Saxon form factor for the spin-spin term was used and all the odd mass nuclei were assumed to have spin 3/2. A value $V_{SS} = 3.33$ MeV was used in the calculation, resulting in a qualitative difference between the position of the resonance for even and odd nuclei. However, one serious shortcoming in these calculations was the appearance of a small secondary maximum in the strength function near $A = 95$ in disagreement with the experimental data in this region.

More recently Newstead *et al.*[16]) have analyzed measured strength functions for several odd mass nuclei in the range $A = 143 - 177$. They used Tamura's coupled channel code[11]) which includes the spherical spin-spin interaction. Again a Woods-Saxon form factor was used. They found that the experimental results were well described by a spin-spin strength $|V_{SS}| \leq 0.5$ MeV.

5. Isobaric Analog Resonances

Some evidence for the existence of a $\vec{\sigma} \cdot \vec{I}$ type of interaction has been found in a study of isobaric analog resonances for nonzero spin targets. Spencer[17]) has determined the strength of the potential from the observed splitting of the ground state doublet in the $p + {}^{89}Y$ elastic scattering. This term was subsequently used by Spencer and Kerman[18]) in their calculations of isobaric analog resonances for a number of nuclei in the mass region A=90. They used a depth $V_{SS} = -0.7$ MeV with Woods-Saxon derivative for the shape of the potential. Nuclear matter calculations[5]) (which imply a volume type radial shape) predict a strength $V_{SS} = 0.65$ MeV. Calculations based on the single particle model (see Eq.(7)) give a much weaker strength $V_{SS} \simeq 0.16$ MeV. Note that both types of microscopic calculations give an attractive potential while the one determined by Spencer and Kerman is repulsive. It should be pointed out, however, that while they used the same potential for ${}^{87}Rb$ and ${}^{89}Y$, the microscopic models will predict potentials of opposite sign and different strengths for these two targets.

6. Spin-Correlation Experiments

Several experiments on the transmission of polarized neutrons through polarized targets have been carried out in the hope of gaining information about the spin-spin interaction. An excellent review of these has been given by Postma[19]) at the Madison Symposium, and we shall only give a brief summary here. The relevant quantity in these measurements is the spin-spin cross section defined by:

$$\sigma_{SS} = (\sigma_{\uparrow\uparrow} - \sigma_{\uparrow\downarrow})/2P_n P_t \qquad (9)$$

where $\sigma_{\uparrow\uparrow}$ and $\sigma_{\uparrow\downarrow}$ are total cross sections for parallel and antiparallel spin orientations and P_n and P_t are the neutron and target polarizations, respectively. The most commonly used targets are ${}^{165}Ho$ and ${}^{59}Co$. The experiments were carried out at several neutron energies in the range $0.3 - 8$ MeV. For ${}^{165}Ho$ all measurements were analyzed using either coupled channel or DWBA calculations in an attempt to put an upper limit

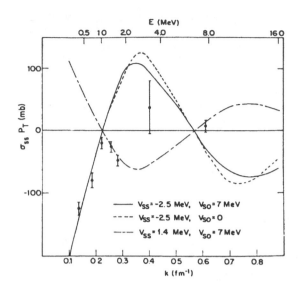

Fig. 2. "Spin-spin" effect in the total cross section of polarized
^{59}Co. The curves are DWBA predictions for the effect due
to the spherical spin-spin potential. The values quoted
for V_{SS} should be divided by the target spin $I = \frac{7}{2}$.
(Ref. 20)

on the strength V_{SS} of the spherical spin-spin interaction. All measurements were consistent with no spin-spin effect. The latest analysis by
Fisher *et al*[20] for the ^{59}Co data failed to reproduce the energy dependence of σ_{ss}. Values of V_{SS} of -0.71 and +0.4 MeV (Woods-Saxon form
factor) could be obtained by fitting the data in different energy regions. This is shown in Fig. 2. (The values quoted in the Figure for
V_{SS} should be divided by the target spin $I = 7/2$.) Inclusion of deformation effects did little in the way of improving the agreement with
experiment. No significant improvement in the fit was obtained when the
tensor spin-spin term was included. It was concluded that measurements
at higher neutron energies (~ 13 MeV) are necessary in order to reduce
the uncertainties in the analysis.

7.   Depolarization Experiments

     Before we consider the various measurements of the Wolfenstein
depolarization parameter D, let us briefly discuss the relationship
between D and the spin-flip probability S in the elastic scattering[21].
The axis of the spin-flip referred to here is along the normal to the
scattering plane.
     According to the Bohr theorem[22] spin-flip is forbidden for scat-
tering on targets with spin zero. However, for targets with non-zero
spin, the nucleons are allowed to flip their spin during the scattering
process. The spin flip probability is given by

$$S = \frac{\sigma_{+-} + \sigma_{-+}}{\sigma_{++} + \sigma_{+-} + \sigma_{-+} + \sigma_{--}} \tag{10}$$

where the σ's are differential cross sections and the subscripts denote the initial and final spin projections of the scattered nucleon.

On the other hand, the depolarization parameter is given by[21]

$$D = \frac{\sigma_{++} - \sigma_{+-} - \sigma_{-+} + \sigma_{--}}{\sigma_{++} + \sigma_{+-} + \sigma_{-+} + \sigma_{--}} \tag{11}$$

hence,

$$D = 1 - 2S \tag{12}$$

A finite spin-flip probability will thus lead to a deviation of D from unity. It is hoped that this deviation will probe the spin-spin interaction between the nucleon and the target. Depolarization experiments on proton scattering from nuclei were started about 20 years ago.

In 1956 Chamberlain et al.[23] measured the depolarization of 310 MeV protons elastically scattered by ^{12}C and $^{27}A\ell$. For both targets the values measured for D were consistent with unity. A later measurement[24] using 14.6 MeV protons scattered by ^9Be again was consistent with D = 1. A similar result was obtained[25] for 2.6 MeV neutrons scattered on ^{139}La. Batty and Tschalar[26] measured D for 50 MeV protons scattered by several targets. Unfortunately the measured values have large error bars and can only be used to place an upper limit on the strength of the spin-spin interaction.[8]

Deviations of D from unity in proton scattering were first observed by the Saclay group[27] for ~20 MeV protons scattered by ^9Be, ^{10}B, and $^{27}A\ell$. Their results are shown in the third column of Table 2.

The measurements mentioned above were all at one angle in each case. The first experiment in which an angular distribution of the D parameter was measured was performed by the Tokyo group[21]. They measured D for 1.36 MeV neutrons scattered by various targets. Large deviations of D from unity were observed. A sample from their data is shown in Fig. 3.

More recently, Birshall et al.[28] have measured D at four angles in the range $45° < \theta_{cm} < 95°$ for the elastic scattering of 25 MeV protons from ^9Be. These show definite deviations of D from unity (See Fig. 4).

Three contributions to this Symposium report new measurements of D which again show large deviations from unity. Birchall et al.[29] have performed measurements for 26 MeV protons on ^{10}B. Complete angular distributions for D are reported for 16 MeV protons on ^{14}N by Clegg et al.[30] and for 17 MeV protons on ^9Be by Baker et al.[31] In the latter measurement very large deviations of D from unity at backward angles have been observed (D = 0.64 $\pm$ 0.06 at θ ~ 150°).

At low energies the large deviations of D from unity may not indicate a large spin-spin component in the nucleon-nucleus potential. The reason for this is that small contributions from the compound nucleus mechanism to the scattering process could lead to substantial deviations of D from unity. This has been shown by Katori et al.[21] in the analysis

of their 1.36 MeV neutron data mentioned above. Their argument goes as
follows: When the scattering process involves statistical compound
contributions the cross-section is given by

$$\sigma = \sigma^{se} + \sigma^{ce} \tag{13}$$

The shape elastic part σ^{se} is calculated using the usual optical
model potential and the compound elastic part σ^{ce} is obtained from a
Hauser – Feshbach calculation. When CN contributions are taken into
account, the depolarization parameter may be written as

$$D = \frac{\sigma^{se}}{\sigma} D^{se} + 2 \frac{\sigma^{ce}_{++} - \sigma^{ce}_{+-}}{\sigma} \tag{14}$$

where D^{se} is to be calculated from the optical potential and the subscripts ++ and +- are spin indices as explained earlier. In the deriva-
tion of the above relation it has been assumed that $\sigma^{ce}_{++} = \sigma^{ce}_{--}$ and time
reversal invariance (viz. $\sigma_{+-} = \sigma_{-+}$) has been taken into account.

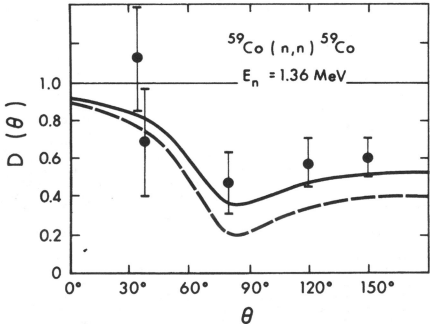

Fig. 3. Depolarization in the elastic scattering of 1.36 MeV
         neutrons by ^{59}Co. The curves are calculated from Eq. (14)
         with $D^{se} = 1$. The dashed curve is obtained using only the
         first term while the solid curve is obtained when the full
         expression is used. (Ref. 21, 32)

It is seen from Eq. (14) that even if $D^{se} = 1$, one would generally
obtain deviations of D from unity. Leaving out the second term in
Eq. (14) and taking $D^{se} = 1$, the authors obtain good qualitative fits

to the observed depolarizations. This is shown by the dashed curve in Fig. 3 for ^{59}Co. The fit to the data is further improved when the second term in Eq. (14) is taken into account[32]. This is shown by the solid curve in Fig. 3. It is, therefore, evident that the observed depolarizations for 1.36 MeV neutrons, though substantially different from unity, do not necessarily indicate the presence of a spin-spin component in the optical potential.

We now turn to the proton depolarization data. The two methods of calculation discussed earlier, namely the coupled channel method[2] which takes into account only the spherical spin-spin term and the DWBA calculations[14] which include both spherical and tensor terms, have been used in the analysis of the data. Batty[8] has used the former method to analyze the Saclay measurements[27] for ^{9}Be and ^{27}Al (^{10}B was not analyzed because Stamp's CC code cannot handle targets with integer spin). Using Woods-Saxon form factors for the spherical spin-spin term he obtained V_{SS} = 1.7 MeV for ^{9}Be and V_{SS} = 0.44 MeV for ^{27}Al . For the latter the result is consistent with the microscopic single particle estimate (See Table 1), but the value for the former seems unreasonably large (the geometry used for ^{9}Be has a much smaller radius than that in Table 1; however, for this geometry the microscopic model gives a depth V_{SS} = 0.38 MeV).

DWBA calculations have also been carried out for these data. The procedure followed is to use the microscopic estimates given in Table 1 for the spherical spin-spin interaction and vary the strength of the tensor term in an attempt to fit the data. The radial shape of the tensor interaction is taken to be that of the real part of the central optical potential. The results of this analysis are given in Table 2.

TABLE 2

Measured and DWBA Calculated Values of D for ~ 20 MeV Protons

Target	$\theta^o_{c.m.}$	$D^{(a)}_{exp}$	$D_{calc.}$	V_{SS} (MeV)	$V^{(b)}_{ST}$ (MeV)
^{9}Be	63.5	0.940 ± 0.016	0.998	0.1	0.0
			0.945	0.6	0.0
			0.942	0.1	3.3
^{10}B	65	0.926 ± 0.012	0.967	−0.23	0.0
			0.927	−0.35	0.0
			0.924	−0.23	−0.5
			0.921	−0.23	1.4
^{27}Al	44	$0.964 + 0.021$	0.971	−0.4	0.0

(a) Ref. 27.

(b) A Woods-Saxon form factor $F_T(r)$ with the same parameters as the real well has been used.

We note that the result obtained for ^{27}Al is consistent with the single particle estimate for V_{SS} and is also in agreement with the value obtained by Batty. We must keep in mind, however, that because of core polarization effects (See section 2) the single particle estimate is an upper limit. For ^{9}Be and ^{10}B we also show in Table 2 the values of V_{SS} that are required to fit the data in the absence of tensor contributions; in both cases the depth has to be increased. If V_{SS} is held fixed at its single-particle value, we must add tensor contributions in order to fit the data points. The tensor strengths required seem to be large. This is rather unsettling particularly for ^{10}B, an isospin zero nucleus for which the direct contribution to the tensor term from a pure OPEP two-nucleon tensor interaction vanishes since the latter is proportional to the isospin operator of a target nucleon.

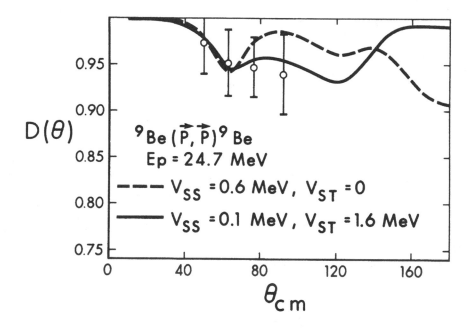

Fig. 4.   Depolarization of 24.7 MeV protons scattered by ^{9}Be
            (Ref. 28). The curves are DWBA calculations for the
            spin-spin potentials. The spherical and tensor terms
            have the same radial shape. (See Table 1).

The Berkeley[28] data for 25 MeV protons on ^{9}Be have also been ana-lyzed using DWBA calculations. Without the tensor term, the depth V_{SS} had to be increased from 0.1 MeV to 0.6 MeV in order to get a reasonable fit. This is shown by a dashed curve in Fig. 4. The solid curve shows the fit obtained with $V_{SS} = 0.1$ MeV and a tensor term of strength $V_{ST} = 1.6$ MeV. The geometry used for the tensor term is the same as that of the spherical term (thus the value V_{ST} = 1.6 MeV is not inconsistent with that shown in Table 2, V_{ST} = 3.3 MeV, since the latter was obtained using a much smaller geometry). Although the fit is quite good, one is still suspicious that the tensor term obtained may be too strong. This feeling is supported by the analysis of the data[29] on ^{10}B at 26 MeV where it turns out that no fit could be obtained with the spherical term alone and that the best one could do is by adding a tensor term with V_{ST} = 1.4 MeV and even then the fit is not that good.

The above considerations have prompted us to look for other possible factors that may contribute to the deviations of D from unity. If one assumes that at 25 MeV compound nuclear contributions are negligible, then it may happen that for the particular nuclei we are dealing with, some other mechanism is playing a role in the spin-flip process and hence making it difficult to extract reasonable values for the spin-spin terms. One such mechanism will be discussed next.

8. The Quadrupole Spin-Flip Effect[33]

If J is the angular momentum transferred to the nucleus during the elastic scattering process, we can split the scattering amplitude into pieces corresponding to different values of J. The principal monopole part (J=0) is the one usually calculated using the central optical potential. The dipole J=1 part arises, for example, from the spherical and tensor spin-spin interactions when calculated in the DWBA. For a target with spin I, values of J up to 2I are possible and hence for I > $\frac{1}{2}$ the next important contribution is the quadrupole J=2 part. The interaction responsible for this is the same as the one that produces the strong J=2 inelastic excitations of low-lying levels in nuclei. In most cases J=2 contribution to elastic scattering is masked by the dominant monopole contribution. However in some experiments these J=2 contributions have been identified[34,35]. Our interest in these as far as depolarization is concerned is that we expect them to be associated with large spin-flip probabilities. This is evidenced by the fact that the probability for spin-flip in the inelastic excitation of the first 2$^+$ levels in even nuclei by nucleons is found to be appreciable[36-38]. If the intrinsic J=2 *elastic* spin-flip probability mirrors the inelastic one and if the J=2 contribution to the scattering amplitude is sizeable one may expect measurable deviations of D from unity.

The scattering amplitude is expanded in J-multipoles as follows[12]

$$f_{M'M,\mu'\mu} = \sum_{J=0}^{2I} \sum_m (IJ,M',-m|IM) \ (2J+1)^{\frac{1}{2}} f_{m\mu'\mu}^{(J)} \tag{15}$$

Choosing the quantization axis along the normal to the scattering plane, then according to the Bohr threorem[22], spin-flip will occur for odd values of m (hence we may delete the spin projection μ').

The elastic differential cross section may be written

$$\sigma_{el} = \frac{1}{2} \sum_{J\mu m} |f_{m\mu}^{(J)}|^2 \equiv \sum_J \sigma_{el}^{(J)} \tag{16}$$

and the elastic spin flip probability S is given by

$$S_{el} = (2\sigma_{el})^{-1} \sum_{J\mu, m \text{ odd}} |f_{m\mu}^{(J)}|^2 \equiv \sum_{J=1}^{2I} S_{el}^{(J)} \qquad (17)$$

The following procedure has been suggested for evaluating the quadrupole spin-flip probability $S_{el}^{(2)}$. If we denote the inelastic differential cross section for the lowest strongly excited state in the target nucleus by σ_{in}, we may write the following exact expression for $S_{el}^{(2)}$

$$S_{el}^{(2)} = \frac{\sigma_{in}}{\sigma_{el}} \frac{\sigma_{el}^{(2)}}{\sigma_{in}} \frac{\sum_{\mu, m \text{ odd}} |f_{m\mu}^{(2)}|^2}{2 \sigma_{el}^{(2)}} \qquad (18)$$

The first ratio on the right hand side of Eq. (18) is readily obtained from experiment. For direct excitation of a low-lying level the second ratio may be calculated by adopting a specific nuclear model or may be deduced from experiment. In the rotational model, for example when one uses the sudden approximation this quantity is given by ratios of squares of the appropriate Clebsch-Gordan coefficients.

The third ratio which is the intrinsic quadrupole spin-flip probability $\hat{S}_{el}^{(2)}$ may be calculated using DWBA or coupled channel calculations or one may substitute for it the measured inelastic spin-flip probability $S_{in}^{(2)}$ for the lowest 2^+ levels of neighboring even nuclei. We have applied both procedures to the analysis of the Saclay measurements for ^{9}Be, ^{10}B and ^{27}Aℓ. Table 3 shows the comparison between their results and the calculated $D^{(2)} = 1 - 2S_{el}^{(2)}$ which is the depolarization resulting from the quadrupole spin-flip alone. The rotational model was adopted for both ^{9}Be and ^{10}B while for ^{27}Aℓ the second ratio was obtained by comparing cross sections for scattering of α-particles by ^{27}Aℓ and ^{26}Mg. For ^{9}Be and ^{10}B, the intrinsic spin-flip probility $\hat{S}_{el}^{(2)}$ was replaced by the measured values[36,37] of $S_{in}^{(2)}$ for the lowest 2^+ state in ^{12}C, while for ^{27}Aℓ the measured values[38] for the lowest 2^+ state in ^{32}S were used. We see from Table 3 that the calculated values of $D^{(2)}$ are on the average not too far off compared to the measured ones. The values predicted by DWBA or CC calculations are somewhat larger than those shown in the table.

TABLE 3

Quadrupole Spin-Flip Calculations for the Depolarization of ~ 20 MeV protons

Target	θ_{cm}^o	$D_{exp}^{(a)}$	$D^{(2)(b)}$
^{9}Be	63.5	0.940±0.016	0.924
^{10}B	65	0.926±0.012	0.94
^{27}Aℓ	44	0.964±0.021	0.974

(a) Ref. 27   (b) Ref. 33

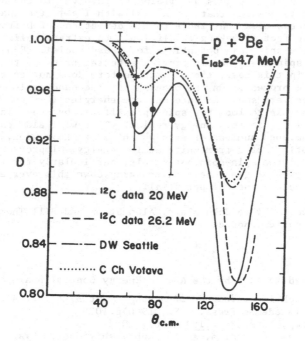

Fig. 5. Depolarization of protons on ^{9}Be at 24.7 MeV (Ref. 28).
The curves show the predicted values of D due to the quadrupole spin-flip effect. The spin-flip probabilities are
based on measurements for ^{12}C (solid and dashed curves),
on DWBA calculations (dash-dot curve), and on coupled
channel calculations (dotted curve).

Similar calculations have been carried out for the 25 MeV data on
^{9}Be. The results are shown in Fig. 5. The solid and dashed curves
were obtained using the measured inelastic spin-flip probabilities
mentioned above. The dash-dot curve is the result of a calculation
in which $S_{e\ell}^{(2)}$ was obtained from DWBA calculations while the dotted
curve gives the results obtained using CC calculations. On the whole,
it appears that a good part of the observed depolarizations at this
energy is due to the quadrupole spin-flip effect. We note however that
the data are restricted to the forward hemisphere.

At lower energies the results are not as good. Calculations for
the 17 MeV proton data[31] on ^{9}Be show that the quadrupole spin-flip
calculations fail to reproduce the observed large angle depolarizations.
These are much too small compared to the calculated values. The same
situation holds for 16 MeV proton data[30] on ^{14}N. In the latter case it
is shown by the authors that compound nuclear spin-flip effects are
responsible for most of the observed deviations of D from unity. It is
possible that the same is true for the ^{9}Be data.

Inspite of the various uncertainties involved in the above calculations, it is evident that for nuclei with $I > \frac{1}{2}$, the quadrupole
spin-flip process does contribute to the deviations of D from unity.
This has the effect of making it difficult to extract reliable information about the spin-spin potentials for these nuclei. Experiments on
targets with spin $I = \frac{1}{2}$ should provide a better probe of these potentials since, in this case, the quadrupole term does not contribute to
the scattering process. On the other hand, the quadrupole spin-flip
effect becomes very small at intermediate energies (E ~ 200 MeV), hence
it may be possible to look for spin-spin effects by measuring D for
targets with $I > \frac{1}{2}$ in this energy region. This will also ensure the
absence of compound nuclear effects which, as we have seen, can cause
drastic deviations of D from unity at low energy. Improved microscopic
calculations of the spin-spin potentials, particularly the tensor term,
will certainly be of great help in tracking down this ever evasive
part of the nuclear optical potential.

I am greateful to John Blair, Alan Stamp and Bill Thompson for
many helpful correspondences.

References

*Supported in part by the Atomic Energy Control Board of Canada

1) H. Feshbach in *Nuclear Spectroscopy* part B, ed. F. Ajzenberg-
    Selove, (Academic Press, N.Y., 1960)p.1033
2) A.P. Stamp, Phys. Rev. 153 (1967) 1052
3) K. Nagamine, A. Uchida, and S. Kobayashi, Nucl. Phys. A145 (1970)
    203
4) G.R. Satchler, Phys. Lett. 34B (1971) 37; Particles and Nuclei 1
    (1971) 397
5) J. Dabrowski and P. Haensel, Phys. Lett. 42B (1972) 163; Can. J.
    Phys. 52 (1974) 1768
6) W.G. Love, L.J. Parish, and A. Richter, Phys. Lett. 31B (1970) 167
7) W.G. Love and G.R. Satchler, Nucl. Phys. A159 (1970) 1
8) C.J. Batty, Nucl. Phys. A178 (1971) 17
9) P. Haensel, Nucl. Phys. A245 (1975) 29
10) L. Drigo and G. Pisent, Nuovo Cim. 70A (1970) 592
11) T. Tamura, Rev. Mod. Phys. 37 (1965) 679
12) K.T.R. Davies and G.R. Satchler, Nucl. Phys. 53 (1964) 1
13) T.R. Fisher, Phys. Lett. 35B (1971) 573
14) H.S. Sherif and A.H. Hussein, Phys. Lett. 41B (1972) 465; A.H.
    Hussein and H.S. Sherif, Phys. Rev. C8 (1973) 518. The numerical
    calculations reported in this reference are slightly in error. A
    discrepancy between our results for D and those reported by Stamp
    (Ref. 2) and Batty (Ref. 8) turned out to be due to an error in
    our computer code "SPINSOR". This error lead us to underestimate
    the deviations of D from unity mainly near the first cross section
    minimum. It has been corrected and our results now agree with
    those of Stamp and Batty.
15) M.Z. Rahman Khan, Nucl. Phys. 76 (1966) 475
16) C.M. Newstead, J. Delaroche, and B. Cauvin, in *Statistical Prop-
    erties of Nuclei*, ed. J.B. Garg (Planum Press, 1972)p. 367
17) J.E. Spencer, Bull. Am. Phys. Soc. 16 (1971) 558

18) J.E. Spencer and A.K. Kerman, Phys. Lett. 38B (1972) 289
19) H. Postma, in *Polarization Phenomena in Nuclear Reactions*, ed.
    H.H. Barschall and W. Haeberli, (Univ. of Wisconsin Press, 1970)
    p. 373
20) T.R. Fisher, H.A. Grench, D.C. Healey, J.S. McCarthy, D. Parks and
    R. Whitney, Nucl. Phys. A179 (1972) 241
21) K. Katori, T. Nagata, A. Uchida, and S. Kobayashi, J. Phys. Soc.
    (Japan) 28 (1970) 1116
22) A. Bohr, Nucl. Phys. 10 (1959) 486
23) O. Chamberlain, E. Segre, R.D. Tripp, C. Wiegand and T. Ypsilantis,
    Phys. Rev. 102 (1956) 1659
24) L. Rosen, in *Proc. 2nd Int. Symp. on Polarization Phenomena of
    Nucleons*, ed. P. Huber and H. Schopper, (Birkhauser Verlag 1965)
    p. 253
25) L. A. Schaller, R.L. Walter and F.O. Purser, *ibid.* p. 314
26) C.J. Batty and C. Tschalar, Nucl. Phys. A143 (1970) 151
27) R. Beurtey, P. Catillon and P. Schnabel, J. Phys. (Paris) 31
    Supp. No. 5-6, C2 (1970) 96; P. Catillon, in *Polarization
    Phenomena in Nuclear Reactions*, ed. H.H. Barschall and W. Haeberli,
    (Univ. of Wisconsin Press, 1970) p. 657
28) J. Birchall, H.E. Conzett, J. Arvieux, W. Dahme and R.M. Larimer,
    Phys. Lett. 53B (1974) 165
29) J. Birchall, H.E. Conzett, F.N. Rad, S. Chintalapudi and R.M.
    Larimer, Contribution to this Symposium, p.630
30) T.B. Clegg, W.J. Thompson, R.A. Hardekopf and G.G. Ohlsen, Contri-
    bution to this Symposium, p.634
31) M.P. Baker, J.S. Blair, J.G. Cramer, T. Trainor and W. Weitkamp,
    Contribution to this Symposium, p.628
32) K. Katori, private communication
33) J.S. Blair, M.P. Baker and H.S. Sherif, Contribution to this
    Symposium, and to be published
34) J.S. Blair and I.M. Naqib, Phys. Rev. C1 (1970) 569
35) G.R. Satchler and C.B. Fulmer, Phys. Lett. 50B (1974) 309
36) W.A. Kolasinski, J. Eenmaa, F.H. Schmidt, H. Sherif and J.R. Tesmer
    Phys. Rev. 180 (1969) 1006
37) J.J. Kolata and A. Galonsky, Phys. Rev. 182 (1969) 1073
38) M.A.D. Wilson and L. Schecter, Phys. Rev. C4 (1971) 1103

DISCUSSION

Cramer:

I would like to comment on the Seattle depolarization data. My former
student, Mike Baker, whose PhD research this measurement was, has
calculated that in spite of the spectacular back angle dip in the
depolarization, the data is consistent with an isotropic spin-flip
cross section. Thus the large depolarization we have observed may
very well be the result of compound nucleus contamination of the reaction mechanism.

Sherif:

I am inclined to agree with you that the dip is mostly due to compound nucleus effects.

Clegg:

You stated that the spin-spin effects were small on $D(\theta)$ in proton scattering from spin-1/2 targets at about 25 MeV. If this is true how good must new measurements be to be able to make a significant contribution ?

Sherif:

Single particle model estimates for the spherical spin-spin interaction for spin 1/2 nuclei in which the valence nucleon is in a $P_{1/2}$ state show that V_{ss} is comparatively weak. It is therefore unlikely that one would see measurable deviations of D from unity in the case of targets such as ^{13}C or ^{15}N, unless of course the tensor interaction turned out to be fairly strong. For spin 1/2 targets in which the valence nuclon is in an $s_{1/2}$ orbit (e.g. ^{31}P) the single particle estimate for V_{ss} is moderately strong. However, one must keep in mind that core polarization effects will reduce this strength. Preliminary calculations for protons on ^{31}P indicate that one would need 1 % accuracy in measuring D in order to be able to say something about the spin-spin interactions.

POLARIZATION EFFECTS IN DWBA CALCULATIONS

by

F. D. SANTOS

University of Lisbon, Portugal

1. Introduction

For more than two decades the distorted wave Born approximation (DWBA) has been a very useful model for the study of direct
reaction phenomena. It enables us to understand much of the physics
of a direct reaction and is a good starting point for the investigation of higher order processes[1]. The DWBA has had considerable
success in the description of polarization data in transfer and charge exchange reactions where the first order process is dominant. In
particular it is of great value to make spin assignments using, for
instance, deuteron stripping reactions[2]. On the other hand the time seems to be coming where the reliability of DWBA calculations is
sufficient to extract information on the detailed structure of light
ions used as projectiles in transfer reactions. This type of study
is likely to be promising in the future and may eventually be extended to polarization phenomena in heavy ion induced reactions.

The availability of intense beams of polarized particles has
opened the possibility of very interesting experiments from the
point of view of reaction and nuclear structure studies. Polarization
transfer experiments on light nuclei have just started. They should
present a considerable challenge to the DWBA and improve our understanding of the nature of the optical model and of higher order direct reaction processes.

It would be impossible to attempt in this talk a complete
review of DWBA calculations of polarization phenomena. I have chosen
to give particular emphasis to the presentation of a unified formalism for the discussion of polarization in direct reactions. This
formalism, suggested by the DWBA, is based on the concepts of angular momentum and spin transfer in the reaction and is useful in the
physical interpretation of polarization experiments.

2. Polarization Experiments and Description of Polarization

We shall consider the following types of polarization experi-
ments in the $A(a,b)B$ reaction: $A(a,\vec{b})B$ polarization; $A(\vec{a},b)B$ analy-
zing power; $A(\vec{a},\vec{b})B$, $A(\vec{a},b)\vec{B}$, $\vec{A}(a,\vec{b})B$ and $\vec{A}(a,b)\vec{B}$ polarization transfer; $\vec{A}(\vec{a},b)B$ and $A(\vec{a},\vec{b})\vec{B}$ spin correlation. It is convenient to describe the polarization of particles A,a,b and B with spins J_A, s_a, s_b and
J_B with spherical tensor operators[3] $\tau_{kq}(s)$ due to their simple

transformation properties under rotations. For arbitrary spin s they are irreducible tensors with matrix elements in the spin representation given by

$$\langle s\sigma' | T_{kq}(s) | s\sigma \rangle = \hat{s} (-1)^{s-\sigma} (s\sigma' s-\sigma | kq) , \quad (1)$$

where $\hat{s} = (2s + 1)^{1/2}$.

It can be easily verified that all information on the reaction which can be derived from the above polarization experiments is contained in the quantities

$$X^{k_b q_b k_B q_B}_{k_a q_a k_A q_A} = Trace \left[T T^{\dagger}_{k_a q_a}(s_a) T^{\dagger}_{k_A q_A}(J_A) T^{\dagger} T_{k_b q_b}(s_b) T_{k_B q_B}(J_B) \right] , \quad (2)$$

where T is the reaction transition matrix. In fact the X's are more general than such experiments since we only consider two polarized particles at a time. As shown by Simonius [4]) this is sufficient to determine all the transition matrix elements at a given energy and angle. If it is preferable to have a Cartesian description of a polarization experiment we have to relate, in the relevant equations, the spherical tensors $T_{kq}(s)$ with the Cartesian tensors for spin s. The basic general formalism for polarization transfer and spin correlation experiments can be found in the extensive review given by Ohlsen [5]).

3. Polarization Tensors

In a direct reaction the transfer of angular momentum j to the target nucleus is physically very significant. This transfer is comprised of an orbital angular momentum transfer ℓ and a spin transfer s defined as

$$\vec{j} = \vec{J_B} - \vec{J_A} , \quad \vec{s} = \vec{s_a} - \vec{s_b} , \quad \vec{\ell} = \vec{j} - \vec{s} . \quad (3)$$

Very often only one value of ℓ,s and j is allowed or important in a given transition that proceeds by a direct reaction mechanism. In compound nucleus reactions it is of course more convenient to consider states of the system with definite total angular momentum and channel spin. This leads to a different angular momentum coupling scheme which it is customary to adopt for instance in the R - matrix theory [6]). In direct reaction processes it is convenient to use the coupling scheme indicated by eq.(3). We note that this coupling scheme may be useful in the analysis of few nucleon problems [7]) and therefore that there is no a priori limitation in its range of applicability as regards target mass.

The exact transition amplitude T can always be expanded into terms with definite j,ℓ,s;

$$\langle J_B M_B s_b m_b | T | J_A M_A s_a m_a \rangle = \sum_{j\ell s} (J_A M_A j \mathfrak{z} | J_B M_B)(\ell \lambda s\sigma | j \mathfrak{z}) \hat{\ell}$$
$$\times (-1)^{s_b-\sigma_b} (s_a \sigma_a s_b - \sigma_b | s\sigma) B^{\ell\lambda}_{sj} \quad (4)$$

where the amplitudes [8] B are defined by inversion of this equation. We may proceed with the analysis by extending this type of expansion to the X coefficients. Inserting eq(4) into eq(2) and using eq.(1) we obtain

$$
X_{k_a q_a k_A q_A}^{k_b q_b k_B q_B} = \sum_{kk'k''jj'ss'} \hat{k}_a \hat{k}_b \hat{k}_A \hat{k}_B \, \hat{s} \, \hat{s}' \hat{j} \hat{j}' \, (-1)^{k_A - q_A} \, (k_B q_B \, k_A - q_A | k'' q'')
$$
$$
\times (-1)^{k_a - q_a} (k_b q_b \, k_a - q_a | k'q')(k'q' \, k''q'' | kq)
$$
$$
\times \begin{Bmatrix} J_A & J_A & k_A \\ J_B & J_B & k_B \\ j & j' & k'' \end{Bmatrix} \begin{Bmatrix} s_a & s_a & k_a \\ s_b & s_b & k_b \\ s & s' & k' \end{Bmatrix} M_{k''k'kq}^{ss'jj'} . \tag{5}
$$

The polarization tensors [9] $M_{k''k'kq}^{ss'jj'}$ defined through this equation have the important property of carrying definite total angular momentum transfer with quantum numbers j,j' and definite spin transfer with quantum numbers s,s'. In terms of the amplitudes B they are given by

$$
M_{k''k'kq}^{ss'jj'} = \hat{s}_a \hat{s}_b \hat{J}_A \hat{J}_B^{\,3} (-1)^{s-s'} \hat{k} \hat{k}' \hat{k}'' \sum_{\ell \lambda \ell' \lambda'} \begin{Bmatrix} j & j' & k'' \\ s & s' & k' \\ \ell & \ell' & k \end{Bmatrix}
$$
$$
\times \hat{\ell} \, (\ell \lambda \, kq | \ell' \lambda') \, B_{sj}^{\ell \lambda} \, B_{s'j'}^{\ell' \lambda' *} . \tag{6}
$$

Note therefore that in general they may involve a summation over different orbital angular momentum quantum numbers.

In eq.(5) we were able to separate to a large extent the particular "geometry" of each X coefficient from the information on reaction dynamics now entirely contained in the polarization tensors. They were defined in such a way that the eq.(5) retains its form upon inversion. In fact using orthogonality theorems[10] we obtain

$$
M_{k''k'kq}^{ss'jj'} = \sum_{k_a q_a k_A q_A k_b q_b k_B q_B} \hat{k}_a \hat{k}_b \hat{k}_A \hat{k}_B \, \hat{s} \, \hat{s}' \hat{j} \hat{j}' \, (-1)^{k_A - q_A} \, (k_B q_B \, k_A - q_A | k'' q'')
$$
$$
\times (-1)^{k_a - q_a} (k_b q_b \, k_a - q_a | k'q')(k'q' \, k''q'' | kq)
$$
$$
\times \begin{Bmatrix} J_A & J_A & k_A \\ J_B & J_B & k_B \\ j & j' & k'' \end{Bmatrix} \begin{Bmatrix} s_a & s_a & k_a \\ s_b & s_b & k_b \\ s & s' & k' \end{Bmatrix} X_{k_a q_a k_A q_A}^{k_b q_b k_B q_B} \tag{7}
$$

In Sec. 5 we emphasize that the quantum numbers j and s are physically very significant due to their close connection with the angular momentum structure of the nuclear bound states involved in the transition and with spin dependent interactions in the entrance and exit channels. In this context we point out that:

a) the linear eqs.(5) give information on the relative importance of contributions to a given polarization observable from different j and s values;

b) the linear eqs.(7) indicate which polarization observables we should measure in case we are interested in a particular value of j or/and s.

3.1 Coordinate Systems

In order to bring out the rotational properties of the M's it is convenient to use the same coordinate system for each of the operators $\mathcal{T}_{kq}$ in eq.(2). We choose the projectile helicity frame [5] S with z -axis along $\vec{k}_{in}$ and y -axis along $\vec{k}_{in} \times \vec{k}_{out}$. In polarization and polarization transfer experiments it is however more suitable for experimental purposes to describe the outgoing particle polarization in the corresponding helicity frame [5] S' with z' -axis along $\vec{k}_{out}$ and y' ≡ y. This means that usually in the experimental data the operators $\mathcal{T}_{k_a q_a}(s_a)$ and $\mathcal{T}_{k_A q_A}(J_A)$ are referred to the coordinate system S while $\mathcal{T}_{k_b q_b}(s_b)$ and $\mathcal{T}_{k_B q_B}(J_B)$ are referred to the coordinate system S'. The passage from S' to S is of course obtained by a rotation about the y -axis through the center of mass scattering angle θ. Using the same coordinate system for all the operators in eq.(2) it is clear from the angular momentum coupling [11] in eq.(7) that the $M_{k''k'kq}^{ss'jj'}$ are irreducible tensors of rank k and therefore transform as a spherical harmonic Y_{kq} under rotations. Referred to the coordinate system S the $M_{k''k'kq}^{ss'jj'}$ have the useful property that as functions of the reaction scattering angle θ they have the parity [9] of q and at $\theta = 0$ and π only those with q = 0 may not vanish.

3.2 Parity Conservation and Time Reversal Invariance

Let us now consider the symmetry properties of the M's . Just from their definition and noting that $\mathcal{T}_{kq}^{\dagger} = (-1)^q \mathcal{T}_{k-q}$ we obtain

$$M_{k''k'kq}^{ss'jj'} = (-1)^{s'-s+j'-j+k''+k'+k+q} M_{k''k'k-q}^{s'sj'j*} .$$ (8)

If we further assume parity conservation in the reaction and use a coordinate system with y -axis along $\vec{k}_{in} \times \vec{k}_{out}$, such as for instance S and S', we get

$$M_{k''k'kq}^{ss'jj'} = (-1)^{s'-s+j'-j+k''+k'} M_{k''k'k-q}^{s'sj'j*} .$$ (9)

Combination of eqs.(8) and (9) leads to the very simple result

$$M_{k''k'kq}^{ss'jj'} = (-1)^{k+q} M_{k''k'k-q}^{ss'jj'} .$$ (10)

The assumption of invariance of the interactions under time reversal allows us to relate the polarization observables in the A(a,b)B reaction to those in the inverse B(b,a)A reaction. The X coefficients for the latter reaction, analogous to those defined in eq.(2), are

$$\overline{X}^{k_a q_a k_A q_A}_{k_b q_b k_B q_B} = \text{Trace}\left[\overline{T}\, \tau^{\dagger}_{k_b q_b}(s_b)\, \tau^{\dagger}_{k_B q_B}(J_B)\, \overline{T}^{\dagger}\, \tau_{k_a q_a}(s_a)\, \tau_{k_A q_A}(J_A)\right], \quad (11)$$

where $\overline{T}$ is the reaction transition matrix for the B(b,a)A process. Time reversal invariance implies that [9]

$$\overline{X}^{k_a q_a k_A q_A}_{k_b q_b k_B q_B}(-\vec{k}_{out} \to -\vec{k}_{in}) = (-1)^{k_b + q_b + k_B + q_B + k_a + q_a + k_A + q_A}\, \overline{X}^{k_b - q_b\, k_B\, q_B}_{k_a - q_a k_A - q_A}(\vec{k}_{in} \to \vec{k}_{out}), \quad (12)$$

where the same coordinate system is used on both sides of this relation. We can also define polarization tensors $\overline{M}^{ss'jj'}_{k''k'kq}$ for the B(b,a)A reaction through an equation analogous to eq.(7) but with X replaced by $\overline{X}$ and A,a interchanged everywhere with B,b. From eqs. (7) and (12) we get

$$\overline{M}^{ss'jj'}_{k''k'kq}(-\vec{k}_{out} \to -\vec{k}_{in}) = (-1)^{j'-j+s'-s}\, M^{ss'jj'}_{k''k'kq}(\vec{k}_{in} \to \vec{k}_{out}). \quad (13)$$

Note that by using polarization tensors parity conservation and time reversal invariance in nuclear reactions are expressed through very concise and convenient relations as regards data analysis, namely eqs.(10) and (13) respectively.

4. Choice of Experiments

In a direct reaction mechanism there is usually a dominant set of values for j,ℓ,s in a given transition. However there are often contributions from other values of j,ℓ,s which are particularly interesting since they contain specific information on less well known interactions or/and less well known components of nuclear bound states. To study these effects we should focus attention on those observables which involve interference between the "strong" and "weak" angular momentum transfer quantum numbers. The selection of these observables and therefore of the most suitable experiments can be done with the help of eqs.(5) and (7). For instance as an example we note that the differential cross section, which is proportional to $\sigma = X^{0000}_{0000}$, is clearly not the most convenient observable to study interference effects since it is always an incoherent sum in j,ℓ and s.

4.1 Mixture of j,ℓ and s Values

From the angular momentum coupling in eq.(7) we find that a necessary condition for the interference between different total angular momentum transfer (J_A and J_B are taken to be non - zero) is k_A or/and $k_B \neq 0$ and $k'' \neq 0$. Assuming, as a mere simplification, that $k_a = k_b = 0$ this means that we should consider the following types of experiments: $\vec{A}(a,b)B$ analyzing power; $A(a,b)\vec{B}$ polarization and $\vec{A}(a,b)\vec{B}$ polarization transfer. These experiments in transfer and charge exchange reactions are specially sensitive to two - step processes [12]. In fact the spin orientation of the target or/and resi-

dual nucleus contains specific information on these processes particularly those that proceed through inelastic core excitation effects [13]. The above experiments are also useful in nuclear spectroscopy studies as a means for the identification of different j -values
[14]. Finally note that they should also be sensitive to spin - spin
interactions.

A necessary condition for the interference between different
spin transfer (s_a and s_b are taken to be non - zero) is k_a or/and
$k_b \neq 0$ and $k' \neq 0$. These conditions may be satisfied in the following
experiments: $A(a,\vec{b})B$ polarization; $A(\vec{a},b)B$ analyzing power; $A(\vec{a},\vec{b})B$
polarization transfer. Generally they are useful to study the intera-
ctions in the entrance and exit channels that depend on the spins $\vec{s}_a$
and $\vec{s}_b$ and also the internal structure of nuclei a and b, as discussed in Sec. 5.4. Furthermore these experiments are sensitive to two-
step processes [15,16,17].

Unlike the two previous cases the interference between different orbital angular momentum does not characterize a class of polarization experiments. In fact when more than one ℓ -value is allowed
all polarization tensors with $k \neq 0$ are likely to mix ℓ . This is
clearly a weak condition for the selection of polarization experiments.

In the above ennumeration we have left out experiments such
as: $A(\vec{a},b)\vec{B}$ polarization transfer and $\vec{A}(\vec{a},b)B$ spin correlation. These
are less specific as regards the interference in j and s. They corre-
spond generally to k' and $k'' \neq 0$ in eq.(7) and therefore tend to mix
the 2 types of interference effects.

4.2 Examples

Let us consider two examples of the above formalism.
1) $\underline{A(\vec{a},\vec{b})B \text{ polarization transfer}}$. With $k_a = k_b = 0$ the eq.(5) sim-
plifies to;

$$X_{k_a q_a 00}^{k_b q_b 00} = \sum_{ss'k} \hat{s}\,\hat{s}'\hat{k}_a\hat{k}_b(-1)^{k_a - q_a}(k_b q_b k_a - q_a | kq) \begin{Bmatrix} s_a & s_a & k_a \\ s_b & s_b & k_b \\ s & s' & k \end{Bmatrix} M_{kq}^{ss'} \,, \quad (14)$$

where the incoherent sum over j was conveniently incorporated in the
reduced polarization tensors

$$M_{kq}^{ss'} = \frac{1}{\hat{J}_A \hat{J}_B} \sum_j \hat{j}\, M_{0kkq}^{ss'jj} \,. \quad (15)$$

These tensors have been fully discussed in ref. 9 in connection with
(d,p) and (d,n) reactions.
2) $\underline{A(\vec{a},b)\vec{B} \text{ polarization transfer}}$. With $k_A = k_b = 0$ in eq.(5) we get;

$$X_{k_a q_a 00}^{00 k_B q_B} = \sum_{kjj'ss'} \frac{\hat{s}\hat{s}'\hat{j}\hat{j}'}{\hat{s}_b\,\hat{J}_A}(-1)^{k_a - q_a}(k_a - q_a k_B q_B | kq) W(k_B j J_B J_A ; j' J_B)$$

$$\times W(k_a s_a s' s_b ; s_a s)\, M_{k_B k_a kq}^{ss'jj'} \,. \quad (16)$$

These equations can be used as the starting point for a unified treatment of experiments involving the spin orientation of nuclei a and B. As examples we mention spin - flip probability measurements in inelastic scattering [18], particle - γ angular correlation $A(\vec{a},b\gamma)\vec{B}$ using a polarized incident beam, which is a very promising type of experiment as regards the study of the residual nucleus [19] and finally $A(\vec{a},b)\vec{B}$ polarization transfer used as a method of polarizing[20] the nuclei B.

5. The Physics of Spin Transfer in the DWBA

The quantum numbers j, ℓ and s as defined in eq.(4) are the overall total angular momentum transfer, orbital angular momentum transfer and spin transfer in the reaction. It is therefore important to establish the relation between these quantum numbers and the apparently similar, although quite different, quanta $\bar{j}, \bar{\ell}$ and $\bar{s}$ which describe only the changes occuring during the inelastic event [21] and are subject to the selection rules imposed by nuclear structure considerations. The weak coupling form of the distorted wave theory is a very useful model to discuss direct reactions in part because it makes the above relation particularly simple and clear. In this Sec. we shall follow closely the notation of ref. 8 .

5.1 Inelastic and Rearrangement Collisions

It is well known that the strongest spin interactions acting during the elastic scattering involve the projectile spin. The weaker interactions between the spins of projectile and target will not be considered here [22]. By further assuming weak coupling the DWBA[8] transition amplitude for the A(a,b)B reaction has the general form

$$\langle J_B M_B s_b \sigma_b | T | J_A M_A s_a \sigma_a \rangle = J \sum_{\sigma'_a \sigma'_b} \int d\vec{r}_a \, d\vec{r}_b \; \chi^{(-)*}_{\sigma'_b \sigma_b} (\vec{k}_{out}, \vec{r}_b)$$

$$\times \langle M_B \sigma'_b | V | M_A \sigma'_a \rangle \, \chi^{(+)}_{\sigma'_a \sigma_a} (\vec{k}_{in}, \vec{r}_a), \quad (17)$$

where J is the Jacobian of the transformation to the relative coordinates $\vec{r}_a$ and $\vec{r}_b$, $\chi^{(\pm)}_{\sigma'\sigma}$ are distorted waves spin matrices defined through the overlap

$$\chi^{(\pm)}_{\sigma'\sigma} (\vec{r}) = \langle \phi_{\sigma'} | \chi^{(\pm)}_{\sigma} \rangle , \quad (18)$$

where $\chi^{(\pm)}_{\sigma}$ is the full distorted wave corresponding to an asymptotic spin projection σ and $\phi_{\sigma'}$ is the particle internal wave function with spin projection σ' . The eq.(17) is also valid in the adiabatic model of deuteron stripping [23] which is an important extension of the DWBA that takes into account deuteron break - up contributions and has been successfully applied to (d,p) and (p,d) reactions in recent years. The interaction matrix element of eq.(17) is now expanded as [8]

$$J <M_B \sigma'_b | V | M_A \sigma_a> = \sum_{\bar{\ell}\,\bar{s}\,\bar{j}} (J_A M_A \bar{j}\,\bar{\mp} | J_B M_B)(-1)^{S_b - \sigma'_b} (s_a \sigma'_a s_b - \sigma'_b | \bar{s}\,\bar{\sigma})$$
$$\times (\bar{\ell}\,\bar{\lambda}\,\bar{s}\,\bar{\sigma} | \bar{j}\,\bar{\mp})\, i^{-\bar{\ell}}\, G_{\bar{\ell}\,\bar{s}\,\bar{j},\bar{\lambda}}(\vec{r}_b, \vec{r}_a). \quad (19)$$

The function G, discussed in ref. 8 , involves essentially the interaction and bound state wave functions and depends exclusively on
the inelastic transfer quantum numbers $\bar{j}$, $\bar{\ell}$ and $\bar{s}$. Note that no approximations are assumed regarding the treatment of the effects due to
the interaction finite range.

5.2 Distorted Wave Tensor Components

In order to make explicit the relation between j, ℓ, s and
$\bar{j}, \bar{\ell}, \bar{s}$ it is convenient to write the above spin matrix(18) $<\phi' | \chi^{(\pm)}>$ at
each point $\vec{r}$ as a linear combination of the $\tau_{kq}(s)$ tensors for spin s;

$$<\phi' | \chi^{(\pm)}> = \sum_{d\delta} \tau^\dagger_{d\delta}(s) \chi^{(\pm)}_{d\delta}(\vec{r}) . \quad (20)$$

Upon inversion the coefficients of this expansion are given by

$$\chi^{(\pm)}_{d\delta}(\vec{r}) = \frac{1}{2s+1} Trace\left[\tau_{d\delta}(s) <\phi' | \chi^{(\pm)}>\right] . \quad (21)$$

The $\chi^{(\pm)}_{d\delta}$, apart from the advantage of simple transformation properties under rotations, are well suited for the perturbation theory
treatment of spin   dependent forces [24].Writing the optical potential as

$$V = V_c + V_s , \quad (22)$$

where V_c and V_s are respectively the central and spin dependent
parts we have [25]

$$|\chi^{(\pm)}_\sigma> = (1 + G^{(\pm)} V_s)|\chi^{(\pm)}_c>|\phi_\sigma> , \quad (23)$$

where $G^{(\pm)}$ is the full Hamiltonian Green operator and $\chi^{(\pm)}_c$ the distorted wave corresponding to the central potential V_c. Inserting eq.(23)
into eq.(20) we get

$$\chi^{(\pm)}_{d\delta}(\vec{r}) = \delta_{do}\,\delta_{\delta o}\, \chi^{(\pm)}_c(\vec{r}) + \frac{1}{2s+1} Trace\left[\tau_{d\delta}(s) <\phi' | G^{(\pm)} V_s | \chi^{(\pm)}_c>|\phi>\right]. (24)$$

The interaction V can itself be expanded in terms of the operators

$$\tau_{kq}\quad V = \sum_{d\delta} V_{d\delta}\, \tau^\dagger_{d\delta}(s) ,\quad V_c = V_{oo} ,\quad V_s = \sum_{d>0} V_{d\delta}\, \tau^\dagger_{d\delta}(s) , \quad (25)$$

where of course the $d>0$ terms contain all the spin dependence of
the interaction. In first order perturbation theory of V_s we have
from eq.(21);

$$\chi^{(\pm)}_{oo}(\vec{r}) = \chi^{(\pm)}_c(\vec{r}) ,\qquad \chi^{(\pm)}_{d\delta}(\vec{r}) = G^{(\pm)}_c V_{d\delta} \chi^{(\pm)}_c(\vec{r}) \quad d>0 , \quad (26)$$

showing that $\chi^{(\pm)}_{d\delta}$ for $d>0$ is proportional to the corresponding component $V_{d\delta}$ of V_s. The above treatment could be easily generalized to
include higher order corrections in V_s. $G^{(\pm)}_c$ does not contain V_s.

5.3 Relation Between j,ℓ,s and $\bar{j},\bar{\ell},\bar{s}$

By inversion of eq.(4) using eqs.(17),(19) and (20) and performing the appropriate angular momentum couplings we get

$$B_{sj}^{\ell\lambda} = \delta_{j\bar{j}} \sum_{d_a \delta_a d_b \delta_b \bar{s} f} \hat{s}_a \hat{s}_b \hat{d}_a \hat{d}_b \hat{s} \hat{\bar{s}} \begin{Bmatrix} s_a & s_a & d_a \\ s_b & s_b & d_b \\ s & \bar{s} & f \end{Bmatrix} W(\ell \bar{\ell} s \bar{s}; f\bar{j})$$

$$\times (-1)^{\bar{j}-\bar{s}-\bar{\ell}+\ell-\lambda} (\bar{\ell} \lambda \ell -\lambda | f\phi)(-1)^{d_b-\delta_b}(d_a\delta_a d_b -\delta_b | f\phi)$$

$$\times i^{-\bar{\ell}} \hat{\ell} \int d\vec{r}_a d\vec{r}_b \chi_{d_b\delta_b}^{(-)*}(\vec{r}_b) G_{\bar{\ell}\bar{s}\bar{j},\lambda}^{-}(\vec{r}_a,\vec{r}_b)\chi_{d_a\delta_a}^{(+)}(\vec{r}_a), \quad (27)$$

where $\chi_{d_a\delta_a}^{(+)}$ and $\chi_{d_b\delta_b}^{(-)}$ are the distorted wave tensor components in the entrance and exit channels. The first thing we notice about this equation is the identification between j and $\bar{j}$. This results from our initial assumptions of weak coupling (neglect of two step processes) and target (residual nucleus) spin independent optical potential in the entrance (exit) channel. The difference between s and the inelastic spin transfer $\bar{s}$ involves necessarily distorted wave tensor components with d > 0. This difference is balanced through the quantum number f by a difference between ℓ and the inelastic orbital angular momentum transfer $\bar{\ell}$. The amplitudes B result in general from a complicated summation over d_a and d_b. Clearly they are not the most suitable quantities for the specific study of spin dependent distortion. Even in cases were one of the light particles is spinless, for instance b, there is still a summation over $f = d_a$. With spin independent distortion $d_a = d_b = 0$ and eq.(27) reduces to

$$B_{sj}^{\ell\lambda} = \delta_{j\bar{j}} \delta_{s\bar{s}} \delta_{\ell\bar{\ell}} \frac{i^{-\bar{\ell}}}{\hat{\ell}} \int d\vec{r}_a d\vec{r}_b \chi_{bc}^{(-)*}(\vec{r}_b) G_{\bar{\ell}\bar{s}\bar{j},\lambda}^{-}(\vec{r}_a,\vec{r}_b)\chi_{ae}^{(+)}(\vec{r}_a), \quad (28)$$

leading therefore to a complete identification between the quantum numbers j,ℓ,s and $\bar{j},\bar{\ell},\bar{s}$. Since the spin dependent forces are known to be weak relative to the central forces this means that the amplitudes $B_{sj}^{\ell\lambda}$ are mostly determined by the inelastic quantum numbers $\bar{j},\bar{\ell},\bar{s}$.

An important consequence is that when there are $\bar{j}, \bar{\ell}$ and $\bar{s}$ mixtures involved in the population of a given final state the DWBA calculations may be useful to determine the percentage admixtures. Various authors have recently pointed out [26 -28] that the vector analyzing power of one - and two - nucleon transfer reactions is strongly sensitive to mixtures in $\bar{\ell}$ and $\bar{j}$ and may be used to calculate the spectroscopic amplitudes. Note however that for the quantitative description of the data it is essential to include spin dependent distortion as Nelson et al. [29] have shown for two - nucleon transfer reactions.

Transitions with $\bar{\ell} = 0$ are characterized by a considerable reduction in the complexity of the angular momentum couplings in eq.(27) which results in a greater sensitivity of the polarization observables to spin dependent interactions 30). It would be of particular interest to analyze this simplification in connection with the prediction of the WBP model[31] for stripping reactions of a relation between polarization (or analyzing power) in elastic and transfer reactions 32).

5.4 The DWBA in Few Nucleon Structure Studies

The above analysis emphasizes that polarization measurements in inelastic and rearrangement collisions can be most useful to study specific nuclear structure problems of nuclei A and B. On the other hand we may ask what are the possibilities of studying the lighter nuclei a and b using a direct reaction A(a,b)B. Although our DWBA description of a direct reaction process is far from satisfactory there are favourable cases where nuclear structure information on very light nuclei can be derived from DWBA calculations.

An example of this situation is given by polarization phenomena involving the spins of a and b in the (d,p),(d,n),(d,t) and $(d,{}^3He)$ reactions. They all have in common $s_a = 1$ and $s_b = 1/2$ and therefore the spin transfer s can be either 1/2 or 3/2. The inelastic spin transfer $\bar{s}$ in the DWBA is also either 1/2 or 3/2. Moreover there is a one - to - one correspondance between $\bar{s}$ and the orbital angular momentum[33,34] L_a of the relative motion in nucleus a between the transferred nucleon and the outgoing particle b: $\bar{s} = 1/2$ and 3/2 are associated respectively with contributions from $L_a = 0$ and $L_a = 2$. The latter value corresponds in (d,p) and (d,n) reactions to the deuteron D-state and in (d,t) and $(d,{}^3He)$ reactions to the D-state of the overlap integrals[35,36] $\langle d|{}^3He \rangle$ and $\langle d|t \rangle$. We can obtain information on these low probability components by looking at reduced polarization tensors of the form $M_{kq}^{1/2\,3/2}$. The polarization observables which in lowest order of spin transfer involve $M_{kq}^{1/2\,3/2}$ are 9) the tensor analyzing powers T_{20}, T_{21}, T_{22} and the polarization transfer coefficients $K_z^x, K_{yz}^x, K_{xz}^y, K_x^x$ and K_{xy}^z. A still better procedure would be to measure the particular linear combinations 9) of these observables which give the polarization tensors $M_{2q}^{1/2\,3/2}$ for q = 0,1,2. The parameter D_2 which characterizes the $\bar{s} = 3/2$ contribution[33] in low and medium energy reactions is of great value in (d,t) and $(d,{}^3He)$ since it is the only direct measure we have of the asymmetry in the three - nucleon bound state wave function. Good agreement was obtained between an experimental[35] and theoretical[36] estimate of D_2 for (d,t) reactions leading to a triton intrinsic quadrupole moment of $|Q| = 0.11 \text{ fm}^2$.

There is a considerable variety of transfer reactions where an analysis analogous to that outlined above can be pursued. Restricting ourselves to reactions induced by polarized deuterons the analyzing powers of $(d,{}^6Li)$ and $(d,{}^7Li)$ reactions 37) should give information on tensor force effects in the s and p shells. Strohbusch et al.[38] have recently reported effects in $({}^6Li,d)$ cross sections which might be due to the D-state of the relative motion α-d.

5.5 Elastic Scattering

The application of the formalism developed in Secs. 3 and 4 to the elastic scattering of nuclei by nuclei in so far as it can be described by an optical potential is particularly simple. The transition matrix element for the scattering of a spin s_o particle can now be directly written as a linear combination [24] of the $\mathcal{T}_{kq}(s_o)$ tensors. Thus we have

$$\langle s_o \sigma_o' | T | s_o \sigma_o \rangle = \sum_{s\sigma} Q_{s\sigma} \langle s_o \sigma_o' | \mathcal{T}_{s\sigma}^\dagger (s_o) | s_o \sigma_o \rangle . \qquad (29)$$

As the total angular momentum transfer j is zero it is immediatly apparent from eqs.(1) and (4) that $s\sigma$ are spin transfer quantum numbers. Upon inversion we obtain

$$Q_{s\sigma} = \frac{1}{2s_o+1} \, Trace \left[T \, \mathcal{T}_{s\sigma}(s_o) \right] . \qquad (30)$$

These tensor coefficients have been studied by Hooton and Johnson [24] in connection with a perturbation theory treatment of the spin dependence of the deuteron - nucleus interaction (our $Q_{s\sigma}$ are equal to the $M_{s\sigma}$ of ref. 24 . By separating the optical potential as in eq.(22) we get

$$\langle s_o \sigma_o' | T | s_o \sigma_o \rangle = \delta_{\sigma_o' \sigma_o} T_c + \langle \chi_{c\sigma_o'}^{(-)} | V_s (1 + G^{(+)} V_s) | \chi_{c\sigma_o}^{(+)} \rangle , \qquad (31)$$

where T_c is the transition operator corresponding to the central potential V_c,

$$| \chi_{c\sigma_o}^{(\pm)} \rangle = | \chi_c^{(\pm)} \rangle | \phi_{\sigma_o} \rangle , \qquad (32)$$

ϕ_{σ_o} is the projectile internal wave function, $\chi_c^{(\pm)}$ the distorted wave corresponding to V_c and $G^{(+)}$ the full Hamiltonian Green operator. Using the tensor expansion(25) for V we obtain from eqs.(30),(31) and (32) in first order of V_s

$$Q_{oo} = T_c \quad , \quad Q_{s\sigma} = \langle \chi_c^{(-)} | V_{s\sigma} | \chi_c^{(+)} \rangle \qquad s > 0 . \qquad (33)$$

Thus Q_s is proportional to the $V_{s\sigma}$ component of V_s. In other words spin transfer s in first order of V_s arises exclusively from the component of tensorial rank s of the optical potential spin dependent part. The $Q_{s\sigma}$ also have the important property that their relation with the reduced polarization tensors $M_{kq}^{ss'}$ is very simple. In fact from eqs.(2),(14) and (29) we obtain

$$M_{kq}^{ss'} = (2s_o+1)^2 \sum_{\sigma\sigma'} Q_{s\sigma} Q_{s\sigma'}^* (-1)^{k+s'-\sigma'} (s\sigma s'-\sigma' | kq) . \qquad (34)$$

Thus in first order of V_s the $M_{kq}^{ss'}$ are just proportional to the tensor product of $V_{s\sigma}$ by $V_{s'\sigma'}$ of rank k. Higher order corrections to this statement can be easily obtained from the Born series expansion of eq.(31). Naturally the most direct sources of information on spin dependent interactions are polarization experiments in elastic

scattering. In contrast with the simple physical interpretation of
eq.(34) note that in inelastic and rearrangement collisions the relation between spin transfer and the tensor components of the optical potential spin dependent part, described by eqs.(6),(26) and (27)
is considerably complex.

The consequences of time reversal invariance in elastic scattering take a particularly simple form in a coordinate system [24])
$\bar{S}$ with $\bar{y}$ - axis along $\vec{k}_{in} \times \vec{k}_{out}$ and $\bar{z}$ - axis along the bisector of
the angle between $\vec{k}_{in}$ and $\vec{k}_{out}$. In this system time reversal invariance and parity conservation imply

$$M_{kq}^{ss'} = (-1)^{s-s'+q} M_{kq}^{ss'} \quad , \quad M_{kq}^{ss'} = (-1)^{k+q} M_{k-q}^{ss'} \, , \quad (35)$$

which considerably reduces the number of independent tensor components.

5.6 Deuteron - Nucleus Tensor Interaction

There is a considerable amount of data on the analyzing powers of deuteron elastic scattering and more recently measurements of
polarization transfer coefficients have been reported [39]. The
agreement between optical model calculations and the vector and tensor analyzing power data is still rather poor [40]. This might be
due in part to our incomplete knowledge of the tensor force terms in
the deuteron optical potential [40]. It is therefore important to find
out which polarization observables are most sensitive to these forces. Clearly the polarization tensors likely to show stronger tensor
force effects are M_{20}^{02} and M_{22}^{02} (M_{21}^{02} vanishes in the $\bar{S}$ system).
From eq.(5) we can determine the polarization observables that in
lowest order of spin transfer depend on M_{20}^{02} and M_{22}^{02}. Referred to
the coordinate system $\bar{S}$ these are the tensor analyzing powers T_{20},
T_{21}, T_{22} and the polarization transfer coefficients K_z^x, K_z^y, K_{xy}^z,
K_{yz}^{yz}, K_{xz}^{xz}, K_{xy}^{xy}. There are other polarization observables involving M_{20}^{02} and M_{22}^{02} but they also depend on polarization tensor components of lowest order in spin transfer which are probably
dominant. Using eq.(7) we get in the coordinate system $\bar{S}$ [41,42] ;

$$Re \, M_{20}^{02} = \sigma\left[\frac{2}{3}T_{20} + \frac{1}{2\sqrt{2}}\left[K_z^z - \frac{1}{2}\left(K_x^x + K_y^y\right)\right] - \frac{1}{9\sqrt{2}}\left(2K_{zz}^{zz} - K_{xx}^{xx} - K_{yy}^{yy}\right. \right.$$
$$\left.\left. -2K_{xy}^{xy} + K_{xz}^{xz} + K_{yz}^{yz}\right)\right] \, , \quad (36a)$$

$$Im \, M_{20}^{02} = \frac{\sigma}{2\sqrt{2}}\left(K_x^{yz} - K_{yz}^x + K_{xz}^y - K_y^{xz}\right) \, , \quad (36b)$$

$$Re \, M_{22}^{02} = \sigma\left[\frac{2}{3}T_{22} - \frac{\sqrt{3}}{8}\left(K_y^y - K_x^x\right) + \frac{1}{6\sqrt{3}}\left(K_{zz}^{xx} - K_{zz}^{yy} + K_{yz}^{yz} - K_{xz}^{xz}\right)\right] , \quad (36c)$$

$$Im \, M_{22}^{02} = -\frac{\sigma}{\sqrt{3}}\left[K_z^{xy} + \frac{1}{2}\left(K_{yz}^x + K_{xz}^y\right)\right] \, . \quad (36d)$$

Each of these linear combinations of tensor analysing powers and

polarization transfer coefficients separates completely the interference between spin transfer 0 and 2 from other spin transfer contributions. They are therefore very sensitive to tensor forces. Their
measurement may decide if our poor description of the existing data
is in fact due to uncertainties in the tensor interaction. They are
complementary to the measurement of the linear combination [24]
$T_{22} - \sqrt{3/2}\ T_{20}$ which has interference between spin transfer 0 and 2
but also double spin transfer 2 contributions.

6. J - Dependence in Transfer Reactions

Consider a transfer reaction A(a,b)B with a = b + x where x
is the transferred particle or cluster with spin s_x. We assume that x
is bound in nuclei a and B with angular momentum quantum numbers in
j-j coupling respectively [34] $\bar{s}, L_a$ and $\bar{j}, L_B$. In transitions induced
by spinless targets $\bar{j} = J_B$ and in the weak coupling distorted wave
theory $j = J_B$. Thus the polarization observables have contributions
from just one value of $j = J_B$. If s_x is non - zero the reaction may
lead to final states of the residual nucleus with the same L_B but different J_B. The reaction observables angular distributions often have
a characteristic dependence on j which in the above circumstances may
lead to the identification of the residual nucleus spin J_B.

From the point of view of the DWBA we may classify the j - dependent effects into two broad categories:

I) j - dependence predicted, at least qualitatively, by the DWBA theory
in its simplest form, namely in the weak coupling approximation
and with spin independent optical potentials;

II) j - dependence which may only be predicted in a distorted wave
type calculation without one or both of the above approximations.

This classification pretends to distinguish the j - dependent effects
with a simple relation to the first order process involved in a given
transition from those which only manifest itself through higher order
processes or/and spin dependent couplings in the entrance and exit
channels. The j_3 - dependence in the vector analysing power of (d,p),
(d,n),(d,t),(d,^{3}He),(^{3}He,d) and (^{3}He,α) reactions angular distributions [2, 43] in the angular region of the cross section main
peak are typical examples of type I j - dependence. The spin orbit
terms in the optical potentials considerably improve the description
of the data but they are not essential to reproduce this j - dependence. Another example of type I j - dependence is that induced by the
deuteron D-state on the tensor analyzing powers of (d,p) reactions
[44,45]. A characteristic example of type II j - dependence is the large angle j - dependence observed in $L_B = 1$ (d,p) angular distributions [46] and more recently also in (d,t) angular distributions [47].
This j - dependence cannot be explained, even qualitatively, by the
DWBA in its simplest form referred above. It is apparently due to

spin dependent distortion in the deuteron channel [48]. On the other hand the small angle j - dependence in $L_B = 3$ (d,p) angular distributions [49] is likely to be of a mixed type since it is in part due to deuteron D-state effects [33] and possibly also to higher order processes [50].

6.1 Geometric and Dynamical J - Dependence

We restrict the discussion here to experiments involving the spins of particles a and b and also to type I j - dependence. Thus we may assume spin independent distortion, which of course implies $s = \bar{s}$. Also we consider L_B fixed but allow for contributions from more than one L_a and $\bar{s}$. With these specifications we obtain [51]

$$\sum_{J_B} M_{kq}^{\bar{s}\bar{s}'}(J_B) = \hat{k}\,\hat{s}_a^3\,\hat{s}_b\,(-1)^{\bar{s}-\bar{s}'} \sum_{\ell\ell' L_a L_a'} W(L_B L_a' \ell\, k; \ell' L_a)$$
$$\times W(s_x L_a' \bar{s}\, k; \bar{s}' L_a)(-1)^\ell\,\hat{\ell}\,\hat{\ell}'\, D_{\bar{s}\ell,\bar{s}'\ell'}^{kq} \quad , \qquad (37)$$

where the sum extends over all allowed values of J_B and we made explicit in the notation that the reduced polarization tensors depend on $j = J_B$. The reaction dynamics is now entirely described by the quantities [9] $D_{s\ell,s\ell'}^{kq}$ which in general depend weakly on j. The interesting thing about eq.(37) is that we may have

$$\sum_{J_B} M_{kq}^{\bar{s}\bar{s}'}(J_B) = 0 \quad . \qquad (38)$$

although each polarization tensor does not vanish. This situation occurs when the triangular condition $\Delta(L_a, L_a', k)$, implicit in the Racah coefficient of eq.(37), is not satisfied. Most frequently there is just one L_a value and the relevant triangular relation becomes $\Delta(L_a, L_a, k)$. In other words the eq.(38) is valid whenever $k > 2L_a$. Therefore in transfer reactions induced by S - state projectiles ($L_a = 0$) all polarization tensors of rank $k > 0$ are likely to exhibit strong j - dependence. As an example consider the (d,p) reaction, neglecting deuteron D-state effects. In this case $\bar{s} = s_x = 1/2$, $J_B = L_B \pm 1/2$ and eq.(38) takes the form

$$M_{11}^{\frac{1}{2}\frac{1}{2}}(L_B + \tfrac{1}{2}) + M_{11}^{\frac{1}{2}\frac{1}{2}}(L_B - \tfrac{1}{2}) = 0 \quad . \qquad (39)$$

It is easy to show that this relation is equivalent to the well known sign rule j - dependence

$$(L_B + 1)\, T_{11}(L_B + \tfrac{1}{2}) + L_B\, T_{11}(L_B - \tfrac{1}{2}) = 0 , \qquad (40)$$

now extensively used to make spin assignments from iT_{11} angular distributions in the region of the stripping main peak [52].Outside this angular region spin dependent distortion effects are more pronounced and tend to obliterate the particular j - dependence described by eq.(40). Away from the main peak the DWBA predictions may

still be strongly j - dependent but less reliable for spin assignments. With the assumed approximations the proton polarization for
an unpolarized beam $(P_y)_0$ is proportional [21] to iT_{11} and therefore
should also satisfy eq.(40). The reason why $(P_y)_0$ does not exhibit
the sign rule j - dependence at least so clearly as iT_{11} is that
$(P_y)_0$ is more sensitive [9] to spin transfer 3/2 contributions than
iT_{11}. Spin dependent distortion and deuteron D-state effects being
stronger they tend to obscure the sign rule j - dependence in $(P_y)_0$.
On the other hand a particular linear combination [9] of polarization
transfer coefficients is predicted to have at the stripping main
peak a more marked sign rule j - dependence than iT_{11}.

In (d, α) reactions $\bar{s} = 1$ and $L_a = 0$. Thus eq.(38) leads to

$$M_{kq}^{11}(L_B + 1) + M_{kq}^{11}(L_B) + M_{kq}^{11}(L_B - 1) = 0 , \qquad (41)$$

where k can be 1 and 2. Expressed in terms of the analyzing powers
this equation takes the form

$$(2L_B - 1)T_{kq}(L_B - 1) + (2L_B + 1)T_{kq}(L_B) + (2L_B + 3)T_{kq}(L_B + 1) = 0 , (42)$$

where k = 1 and 2. Notice that the j - dependence is now common to
vector and tensor analyzing powers. In contrast with this the tensor
analyzing powers of (d,p) reactions do not have the simple sign rule
j - dependence of eq.(38) even when deuteron D-state effects are included [44]. Recent DWBA calculations of vector and tensor analyzing
powers of $(\vec{d}, \alpha)$ reactions do in fact show a strong sensitivity to
the value of J_B [29,53,54]. This j - dependence which is clearly of the
type described by eq.(41) in the region of the cross section main
peak, is well illustrated by the vector analyzing power DWBA calculations of Ludwig et al. [53] for the $^{28}Si(\vec{d}, \alpha)^{26}Al$ reaction shown in
fig.1. Note that away from this angular region deuteron spin depen-

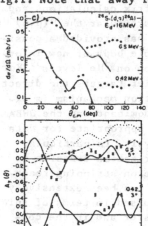

dent distortion effects are more pronounced and the simple sign rule j - dependence is no longer present.

Transfer reactions induced by non -
S - state projectiles, as it frequently occurs in heavy ion reactions, do not have
the sign rule j - dependence in vector polarization observables. If, for instance,
$L_a = 1$ the triangular condition $\Delta(L_a, L_a, k)$
is satisfied for k = 1 and 2 which implies
that eq.(38) is not valid.

The j - dependence characterized by

Fig.1 DWBA calculations with $L_B = 4$: the
solid curve assumes $J_B = 5$, the dashed curve
$J_B = 4$ and the dotted curve $J_B = 3$. The data
and DWBA calculations are from ref. 53.

eq.(38) can be described as a"geometric"type of j - dependence since
it arises exclusively from a particular angular momentum coupling in
certain polarization tensors. In fact the derivation of eq.(38) is
independent of any detailed dynamical considerations in the context
of the DWBA. Those j - dependent effects which for their interpretation involve quantitative predictions for the D's of eq.(37) may be
described as"dynamical". An example of a dynamical form of j - dependence is that observed in the T_{2q} angular distributions of (d,p) reactions and known to be induced by deuteron D-state effects[44,45]. In
particular the T_{22} angular distributions in the small angle region
show a marked ℓ and j-dependence which is apparently due to a surface type reaction mechanism [45]. As another example note that the prediction of the iT_{11} sign at the stripping main peak as a function of
j in (d,p) and (d,t) reactions requires necessarily a discussion of
reaction dynamics [55]. The dynamical forms and aspects of j - dependence are of great interest from the point of view of reaction models.

7. Polarization Transfer

        Measurements of polarization transfer coefficients on very
light nuclei have been performed for various types of reactions.Tho-
se on the mirror reactions[56,57] $^2H(\vec{d},\vec{n})^3He$ and $^2H(\vec{d},\vec{p})^3H$ involving
several angular distributions are particularly complete. For heavier[58]
nuclei a single experiment was reported at our earlier Symposium at

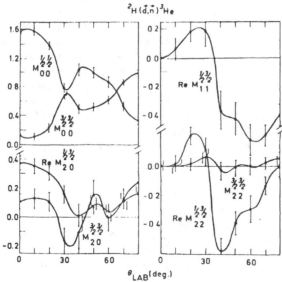

$^2H(\vec{d},\vec{n})^3He$

Madison and a substancial
amount of data on (d,p)
and (d,n) reactions is now
reported. The measurements
on very light nuclei have
been analyzed in terms of
R matrix calculations and
in particular the $^3H(\vec{p},\vec{n})^3He$
data has provided a consi-
derable body of new information on ^4He levels[59]. For
heavier targets the direct
reaction mechanism is of-
ten dominant and the DWBA
more appropriate for data
analysis. In particular
polarization transfer in
deuteron stripping react-
ions has been extensively
discussed in terms of spin
transfer and in the con-
text of the DWBA in ref.9.
        From   the

Fig.2 Polarization tensors for the
$^2H(\vec{d},\vec{n})^3He$ reaction at 10 Mev in units of
σ calculated from the data of ref. *56* .The
curves are meant to guide the eye.

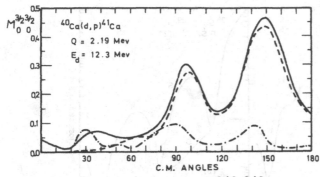

Fig. 3 DWBA calculations for $M_{0\ 0}^{3/2\ 3/2}$ (in units of σ):the dashed curve assumes spin--orbit distortion in the deuteron and proton channels but no deuteron D-state effects; the point - dash curve assumes D-state effects but no spin-orbit distortion; the full curve assumes spin-orbit distortion plus D-state effects.

angular distributions of $K_x^{x'}$, $K_x^{z'}$, K_z^x, K_y^y, $K_{yy}^{y'}$, $K_z^{z'}$ A_y, A_{xx}, A_{yy}, A_{zz} and $(P_y)_0$ given in ref. 56 for the $^2H(\vec{d},n)^3He$ reaction we can calculate a few polarization tensor components and therefore completely separate contributions from different spin transfer. The results, shown in fig. 2 referred to the S system, indicate that at forward angles the spin transfer 1/2 is distinctively more probable than spin transfer 3/2. Thus in spite of the very low target mass the spin dependent interactions and deuteron D-state effects play a relatively minor role in the small angle region of the neighborhood of the cross section main peak[60]. The marked increase with θ of $M_{0\ 0}^{3/2\ 3/2}$ reveals strong spin dependent effects at large angles and is analogous to that of the spin - flip probability in inelastic proton scattering [18]. The existing data on $(\vec{d},p)$ and $(\vec{d},n)$ polarization transfer on heavier targets is not sufficient to obtain any polarization tensor component. They should have considerable value in the study of direct reaction processes and in particular to obtain information on spin dependent distortion. For instance the scalar $M_{0\ 0}^{3/2\ 3/2}$, which gives the spin transfer 3/2 contributions to the cross section [9] can be obtained by measuring just K_x^x, K_y^y and K_z^z (at $\theta = 0°$ we need only to measure K_y^y and K_z^z). Fig. 3 shows the results of DWBA predictions for $M_{0\ 0}^{3/2\ 3/2}$ including deuteron D-state effects calculated in the local energy approximation [33]. As expected we find that this quantity is very sensitive to spin dependent distortion and that such effects have a pronounced increase with angle. The D-state effects are generally smaller and with a weaker angle dependence. The measurement should for instance be useful for a more complete understanding of the large angle $L_B = 1,j$ - dependence [46].

DWBA calculations are in very good agreement with K_y^y measurements for the $^{28}Si(\vec{d},\vec{p})^{29}Si$ ground state transition at two scattering angles [61]. Fig. 4 shows that the D-state effects are very large at 20° because the double spin transfer 1/2,namely $M_{0\ 0}^{1/2\ 1/2}$ in this case, turns out to be particularly small at that angle, as indicated by the cross section calculations. These results show that deuteron D-state effects are important in polarization transfer and should be included

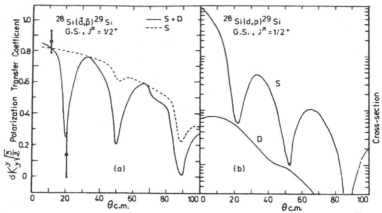

Fig. 4   (a) Polarization transfer coefficient $\sqrt{3/2}\,\sigma K_y^y$ and
DWBA predictions; (b) DWBA predictions of the S- and D-
state contributions to the cross section from ref.61 .

in DWBA calculations so as to obtain a reliable description of the data. In particular D-state effects are predicted to be larger in K_z^z
than in either K_x^x and K_y^y since the latter are relatively less dependent [9] on spin transfer 3/2.

The deviation of K_x^x, K_y^y, K_z^z from   the   value of 2/3
is exclusively due to spin transfer 3/2 contributions . These arise
in the DWBA from D-state effects and spin dependent distortion. Since
the latter is known to be small at forward angles (see fig. 3) the
difference between the $K_y^y(0°)$ measurements [58,61-64] and the value
2/3 is probably due almost entirely to D-state effects. These do not
provide direct information on the deuteron D-state probability but rather on the low momentum components of the D-state wave function [34].
The fitting of low energy $K_y^y(0°)$ measurements with DWBA calculations
would therefore provide complementary information on D_2 to that already obtained from the analysis of tensor analyzing power data [44].
At sufficient high energy the local energy approximation breaks down,
the data cannot be described just in terms of D_2 and the experiments
may turn out to be an interesting source of information on the details
of the deuteron D-state wave function.

7.1 Transparency to Polarization Transfer

Besides the contribution to the understanding of reaction dynamics the polarization transfer experiments also have the interest
of revealing reactions where the polarization transfer is suffiently
large to provide an attractive source of polarized particles. This
was pointed out for the $^2H(\vec{d},\vec{n})^3He$ reaction [65] where $K_y^y(0°)$ is quite
large. From this point of view we note that under certain conditions
discussed in ref.66 deuteron stripping reactions are transparent to
vector polarization transfer. In general transparency gives the possi-

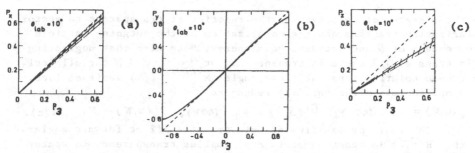

Fig. 5 Transparency to vector polarization transfer in the $^2\text{H}(\vec{d},\vec{n})^3\text{He}$ reaction[56] at 10 Mev. (a) and (c) correspond to a pure vector polarized deuteron beam with symmetry axis $\vec{n}$ respectively along the x - and z - axis; (b) $\vec{n}$ along the y - axis and p_3 and p_{33} satisfying eq.(45b). The broken lines correspond to ideal transparency and the shaded area to the error bars.The errors in (b) are to small to be represented.

bility of using polarization transfer in transfer and charge exchange reactions as a means of producing beams of polarized particles with known polarization. This may be of particular interest for particles which are otherwise difficult to polarize.

Transparency to polarization transfer in reactions with the spin structure $A(\overrightarrow{1/2},\overrightarrow{1/2})B$ and $A(\vec{1},\vec{1})B$ has been discussed in ref.41,42. We shall restrict our considerations here to transfer reactions with the spin structure $A(\vec{1},\overrightarrow{1/2})B$. This specification covers a wide range of reactions such as (d,p),(d,n),(d,t) and (d,^{3}He).Neglecting spin transfer 3/2 the Cartesian components of the outgoing particle polarization referred to the system S are given by[68,9]

$$P_x = \frac{p_x + \frac{2}{\sqrt{3}}\, p_{xy}\, A}{1 + \sqrt{3}\, p_y\, A}\ \ (43a), \quad P_y = \frac{p_y + \frac{1}{\sqrt{3}}\,(1+2p_{yy})A}{1 + \sqrt{3}\, p_y\, A}\ \ (43b), \quad P_z = \frac{p_z + \frac{2}{\sqrt{3}}\, p_{yz}\, A}{1 + \sqrt{3}\, p_y\, A}\ \ (43c),$$

where

$$A(\theta) = -i\,\frac{\sqrt{2}}{3}\ M_{11}^{\frac{1}{2}\frac{1}{2}}(\theta)\big/G(\theta) \tag{44}$$

and $p_y, p_{yy}, p_x, p_{xy}, p_z$ and p_{yz} are the Cartesian moments of the incident particle polarization in system S. Note that these equations apply to all reactions with the spin structure $A(\vec{1},\overrightarrow{1/2})B$ when we assume pure spin transfer 1/2. The degree of validity of this approximation depends on the particular reaction in question, the incident energy and scattering angle[9].The form of eqs.(43) is such that the outgoing particle polarization components are independent of $A(\theta)$ in certain incident beam polarizations[68]. In fact for

$$p_{xy} = \frac{3}{2}\, p_x\, p_y \ \ (45a), \quad p_{yy} = \frac{1}{2}\,(3p_y^2 - 1) \ \ (45b), \quad p_{yz} = \frac{3}{2}\, p_y\, p_z \ \ (45c),$$

eqs.(43) give respectively

$$P_x(\theta) = p_x \quad (46a), \qquad P_y(\theta) = p_y \qquad (46b), \qquad P_z(\theta) = p_z \quad (46c).$$

In these states of polarization the reaction is transparent to vector
polarization transfer and the polarization of the outgoing particle
independent of θ and incident beam energy. Note also that neglecting
spin transfer 3/2 there is transparency at $\theta = 0^\circ$ and π for all incident beam polarizations. At these angles $M_1^{1/2}{}_1^{1/2}(\theta)$ vanishes (Sec.
3.1) and therefore the eqs.(43) reduce to

$$P_x(0,\pi) = p_x \quad (47a), \quad P_y(0,\pi) = p_y \quad (47b), \quad P_z(0,\pi) = p_z \quad (47c).$$

The small probability of spin transfer 3/2 at forward angles
in the $^2H(\vec{d},n)^3He$ reaction at 10 Mev implies transparency to vector
polarization transfer in this angular region. This can be tested using the data of ref.56 for deuteron beams with polarization symmetry
axis $\vec{n}$ along the x,y or z axis of the S system. We represent by p_3
and p_{33} the deuteron vector and tensor polarization with respect to $\vec{n}$.
With $\vec{n}$ along the y - axis p_3 and p_{33} must satisfy eq.(45b) in order
to obtain transparency. The fig. 5 shows that at $\theta_{lab.} = 10^\circ$ the reaction is remarkably transparent along the y - axis. With $\vec{n}$ along the
x-and z- axis we get respectively eqs.(46a) and (46c) for all values
of p_3 and p_{33}. This results from the fact that[67] in these $\vec{n}$ orientations $p_y=p_{xy}=p_{yz}=0$. As expected the reaction is less transparent along the z - axis than along the x - axis since K_z^z is relatively more
dependent on spin transfer 3/2 than K_x^x. As a source of polarized neutrons the polarization transfer along the x - axis is more convenient
than along the y- axis because it does not impose conditions on the
deuteron source parameters p_3 and p_{33}.

The measurement of $K_y^y(0^\circ)$ in deuteron stripping reactions
is sufficient to determine the reaction transparency along the y -
axis. For a pure vector polarized beam, assuming parity conservation,
we get $P_y(0^\circ) = \frac{3}{2}p_y K_y^y(0^\circ)$. In the absence of spin transfer 3/2 in the
reaction 9) $K_y^y = 2/3$ and therefore we get eq.(47 b).

Transparency to vector polarization occurs also in other for-
ms of polarization transfer such as $A(\vec{d},p)\vec{B}$ when $J_A = 0$ and the neutron is transferred into an S-state[68]. This can be used as a method
for the polarization of the residual nuclei B.

7.2 Reaction Mechanisms and Spin Transfer

The $K_y^y(0^\circ)$ measurements in deuteron stripping reactions[61-65]
show that spin transfer 3/2 is very weak in the forward direction. In
fact $K_y^y(0^\circ)$ is systematically near to 2/3 and often has a structureless energy dependence[64]. This constancy as a function of incident
energy may be only indicative of very small spin transfer 3/2. It
does not necessarily imply that compound nucleus processes do not contribute to the reaction. With no spin transfer 3/2 the cross section
$\sigma = \sqrt{\frac{1}{3}} M_0^{1/2}{}_0^{1/2}$ can be quite strongly energy dependent at a given scattering angle while K_y^y at that angle is independent of this quantity
and equal to 2/3. A very striking example is given by the $K_y^y(0^\circ)$ measurements of Walter et al.[64] for the $^{12}C(\vec{d},\vec{n})^{13}N$ ground state reaction.

While the cross section varies by a factor of nearly 50%, $K_y^y(0°)$ does not deviate appreciably from 2/3 in the same energy region(see figs. 1 and 2 of ref.64). It is also significant that the spin transfer 3/2 at forward angles in deuteron stripping reactions is consistently small over a wide range of target mass and deuteron incident energy 61-65,69)in which different reaction mechanisms can be recognized.

These results show that the analysis of polarization data in terms of spin transfer reveals aspects of the reaction dynamics which do not seem to be specific of either the compound nucleus or the direct reaction mechanisms. It would be of great interest to elucidate the relations between the three models from the experimental and theoretical point of view.

References

1) G.R.Satchler, in Proc. of the Int. Conf. on nucl. phys.,ed. J. de Boer and H.J.Mang(North - Holland,1973) p.570
2) T.J.Yule and W.Haeberli, Nucl. Phys. A117 (1968)1
3) Madison Convention, Proc. 3rd Int. Symp. on polarization phenomena in nuclear reactions, ed. H.H.Barschall and W.Haeberli (Univ. of Wisconsin Press,Madison,1971) p.25
4) M.Simonius,in ibid.,p. 401; in Proc. of a Meeting on pol. nucl. phys.(Springer Verlag, 1974) p.38
5) G.G.Ohlsen, Rep. Prog. Phys. 35(1972)717
6) A.M.Lane and R.G.Thomas, Rev. Mod. Phys. 30(1958)257
7) R.M.Devries,J.L.Perrenoud and I.Slaus,Nucl. Phys. A188(1972)449
8) G.R.Satchler, Nucl. Phys. 55(1964)1
9) F.D.Santos, Nucl. Phys. A236(1974)90
10) A.R. Edmonds, Angular momentum in quantum mechanics (Princeton Univ. Press.,1957)
11) M.E. Rose, Elementary theory of angular momentum (Wiley,1965)
12) R.N. Boyd et al., in Proc. of the Int. Conf. on nucl. struct. and spect.,ed. H.P.Block and A.E.L.Dieperink(Scholar's Press,1974)178
13) F.S. Levin and H. Feshbach, Reaction dynamics(Gordon Breach,1973)
14) G. Thornton et al. Nucl. Phys. A198 (1972) 397
15) R. Smith and K. Amos, Phys. Lett. 55B(1975)162
16) J.A. Macdonald et al., Phys. Rev. 9C (1974) 1694
17) G. Berg et al., contribution to this meeting, p. 659
18) F.H.Schmidt et al. Nucl. Phys. 52 (1964) 353
19) C.R.Gould et al. Phys. Rev. Lett. 30 (1973) 298
20) E.G. Adelberger et al., Bull. Amer. Phys. Soc. 19 (1974) 478
21) N. Austern, Direct reaction theories (Wiley,1970)
22) A.H. Hussein and H.S. Sherif, Phys. Rev. C8 (1973) 518
23) R.C.Johnson and P.J.R. Soper, Phys. Rev. C3 (1970) 976; J.D. Harvey and R.C. Johnson, Phys. Rev. C3 (1971) 636
24) D.J. Hooton and R.C. Johnson, Nucl. Phys. A175 (1971) 583
25) A. Messiah, Quantum mechanics, (North - Holland,1962)Vol.2,Ch.19
26) E.J. Ludwig et al. Nucl. Phys. A230 (1974) 271

27) R.W. Babcock and P.A. Quin, contribution to this meeting, p.669
28) J.M. Nelson, O. Karban, E.J. Ludwig and S. Roman, contributions to this meeting, p.691.
29) J.M. Nelson and W.R. Falk, Nucl. Phys. A218 (1974) 441
30) P. Krammer et al., contribution to this meeting, p.651
31) C.A. Pearson and M. Coz, Nucl. Phys. 82 (1966) 533,545
32) S.K. Datta et al.,Phys. Rev. Lett. 31 (1973) 949; W.W. Jacobs et al., contribution to this meeting, p.697
33) R.C. Johnson and F.D. Santos, Particles and Nuclei 2 (1971) 285
34) F.D. Santos, Nucl. Phys. A212 (1973) 341
35) L.D. Knutson and B.P. Hichwa, contribution to this meeting, p.683
36) A.M.Eiro,A.Barroso and F.D.Santos, contribution to this meeting
37) A.E. Ceballos et al., Nucl. Phys. A208 (1973) 617; P. Martin et al., Nucl. Phys. A212 (1973) 304
38) U. Strohbusch et al., Phys. Rev. Lett. 34 (1975) 968
39) G.G. Ohlsen et al., Phys. Rev. C8 (1973) 1639; P.A. Lovoi et al., Phys. Rev. 9C (1974) 1336
40) P.W. Keaton et al., Phys. Rev. 8C (1973) 1692; M. Irshad et al., Nucl. Phys. A218 (1974) 504; R. Roche et al. Nucl. Phys. A220 (1974) 381
41) M.H. Lopes,Faculdade de Ciencias de Lisboa, Estagio, 1974, unpublished report
42) M.H. Lopes and F.D. Santos, to be published
43) B.P. Hichwa et al., contribution to this meeting; A.K. Basak et al. Nucl. Phys., A229 (1974) 219; B. Mayer et al., Phys; Rev. Lett. 32 (1974)1452; A.K. Basak, J.B.A.England, O.Karban, G.C.Morrison, J.M.Nelson,S.Roman and G.G.Shute, contribution to this meeting p.711
44) L.D. Knutson et al., Nucl. Phys. A241 (1975) 36 and references therein
45) F.D. Santos, contribution to this meeting, p.675
46) L.L. Lee, Jr., and J.P. Schiffer, Phys. Rev. 136 (1964) B405
47) M. Borsaru et al., Nucl. Phys. A237 (1975) 93
48) D.M. Rosalky et al., Nucl. Phys. A132 (1970) 469
49) R.Sherr,E.Rost and M.E.Rickey, Phys. Rev. Lett. 12 (1964) 420
50) V.K. Luk'yanov et al., Sov. Jour. Nucl. Phys. 19 (1974) 295
51) In eq.(46) of ref.9 $\hat{\jmath}^2$ and $\hat{\jmath}^2_B$ should be deleted. The following eqs.(47) and (48) must also be corrected accordingly
52) See for instance G.A. Hutlin et al., Nucl. Phys. A227 (1974) 389; J.A. Thompson, Nucl. A227 (1974) 485; S.Sen et al. Phys. Rev. 10C (1974) 1050
53) E.J. Ludwig,T.B. Clegg, P.G. Ikossi and W.W. Jacobs, contribution to this meeting, p.687
54) O.F. Nemets, Yu.S. Stryuk and A.M. Yasnogorodsky, contribution to this meeting, p.713
55) S.E. Vigdor et al., Nucl. Phys. 210 (1973) 93
56) G.C. Salzman et al. Nucl. Phys. A222 (1974) 512
57) W. Gruebler et al. Nucl. Phys. A230 (1974) 353

58) R.C. Brown et al., in Proc. 3rd Int. Symp. on pol. phen. in nucl. reac.,ed. H.H.Barschall and W.Haeberli (Univ. of Wisconsin Press, Madison,1971) p.782

59) J.J. Jarmer et al., Phys. Rev. 10C (1974) 494 and references therein

60) S.T. Thornton, Nucl. Phys. A136 (1969) 25

61) A.K. Basak et al., contribution to this meeting, p.661

62) R.K. TenHaken et al., contribution to this meeting, p.647

63) P.W. Lisowski et al., contribution to this meeting, p.653

64) R.L. Walter et al., contribution to this meeting, p.649

65) J.E. Simmons et al., Phys. Rev. Lett. 27 (1971) 113

66) F.D. Santos, Phys. Lett. 51B (1974) 14

67) S.E. Darden, in Proc. 3rd Int. Symp. on pol. phen. in nucl. reac. ed. H.H. Barschall and W. Haeberli (Univ. of Wisconsin Press, Madison, 1971) p.39

68) F.D. Santos, to be published

69) T.B. Clegg et al., Phys. Rev. 8C (1973) 922; R.A. Hardekopf et al., Phys; Rev. 8C (1973) 1629 and references therein

DISCUSSION

Slobodrian:

When looking at small interference effects should one not also consider reaction mechanisms other than those of a direct type ?

Santos:

The interference between different values of the total angular momentum transfer j and spin transfer s as defined in this talk is model independent. Those observables which in lowest order of angular momentum transfer contain this interference should be of particular interest to the study of reaction mechanisms and spin dependent interactions.

Boyd:

It is worth noting that some of the moments discussed are strongly affected by multistep processes. For example, in a (d,p) calculation, tensor analysing powers of the same magnitude as those resulting from inclusion of the deuteron d-state were produced by multistep processes alone. The (one step) spectroscopic factor in this case was about 0.15.

Santos:

We may expect the polarization experiments involving the spins of particles A and B to be sensitive to multistep processes, as for instance (d,pγ) angular correlation measurements. However other polarization observables may also turn out to be interesting from this point of view especially when the first order transition is very weak.

McKee:

Are you in a position to predict which reactions might make the best sources of secondary polarized tritons and ^{3}He beams by the transparency mechanism which you discussed in your talk ?

Santos:

I think that we now have sufficient information, in particular from the analogy with (d,p) and (d,n) reactions, to predict that (d,t) and (d,^{3}He) are transparent to vector polarization transfer at forward angles. One should choose (d,t) and (d,^{3}He) reactions where the first order direct process is strong. Measurements of K_y^y in (d,t) and (d,^{3}He) would be of great value in this respect.

POLARIZATION EFFECTS IN TRANSFER REACTIONS

S.E. Darden
Department of Physics, University of Notre Dame
Notre Dame, Indiana 46556[†]

and

W. Haeberli
Department of Physics, University of Wisconsin
Madison, Wisconsin 53706[††]

In the discussion[1] of this topic presented at the Madison conference five years ago, a table listing the transfer reactions for which
polarization effects had been measured up to that time was given. The
great majority of the experiments involved measurement of the vector analyzing power, and a total of around 200 transitions had been studied.
In the five years since the Madison meeting, measurements on over 400
additional transitions have been carried out. While most of these
studies involve (d,p) and (d,t) reactions, results on several other
transfer reactions have now been reported. The present discussion attempts to give a representative sample of recent work, but does not treat
the large amount of work which has been done on few-nucleon systems.
Measurements of deuteron tensor analyzing powers are omitted from this
summary, since they are discussed in more detail in the following paper.

The technique of using vector analyzing power (VAP) measurements to
determine the total angular momentum j of the transferred nucleon has become an essentially routine procedure in the case of (d,p) and (d,t) reactions. Such measurements have been carried out in most regions of the
periodic table. It is interesting to note that for many of the light
and intermediate odd-mass nuclei, the majority of the firm spin-parity
assignments are based on this technique. Determinations of j have been
made for values of the transferred orbital angular momentum as high as
seven, and in no case has any conflict with firm spin assignments obtained by other methods been encountered. The apparent conflict which
existed between VAP measurements and particle-γ correlation data in the
case of the (d,p) transition to the 3.61 MeV level in ^{41}Ca was resolved
when it was shown that a doublet[2-4] is involved.

One aspect of single-nucleon transfer reactions which has been
clarified in recent years concerns the sign of the polarization or vector analyzing power. It has become clear that there are no simple sign
rules which hold in all cases, but rather that the sign and angular dependence of the polarization effects depend on Q-value, target mass and
bombarding energy. Two simple models which illustrate the situation are
the semiclassical picture proposed by Newns[6] and extended by Verhaar,[7]
and the subcoulomb stripping model developed by Vigdor et al.,[8] following a suggestion by Quin and Goldfarb (fig. 1). The Newns model makes
it clear how polarization effects can arise in simple transfer reactions,
and how their sign, for angles near the first stripping maximum, will be
opposite for the two j values $\ell \pm 1/2$. Assume we are dealing with a
(d,p) reaction. The momentum vector $\vec{k}_n$ has been drawn in fig. 1 assum-
ing $|k_d| \simeq |k_p|$. Because the strong absorption of the deuterons causes
most of the reactions to occur in the half of the nucleus away from the
detector, the orbital angular momentum of the captured neutrons will
point predominantly up (with respect to $\vec{k}_d \times \vec{k}_p$). If the reaction proceeds to a state for which $j = \ell_n + 1/2$, then the cross section for an

incident beam with posi-
tive polarization, as
shown, will be greater
than for an unpolarized
beam, i.e. $A_y > 0$. Correspondingly, $A_y < 0$
for stripping to a state
for which $j = \ell_n - 1/2$.
This model does not re-
liably predict the sign
of the analyzing power,
except under rather re-
stricted conditions
which are not normally
realistic for actual
transfer reactions.
The subcoulomb strip-
ping model is much
better, and appears
to be valid whenever

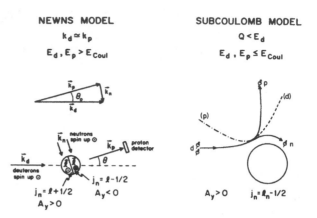

Fig. 1. Simple models of polarization effects
in (d,p) stripping (Ref. 5).

the energies of the incident and outgoing particles are below or near
the Coulomb barrier. In this case the incident and outgoing particles
follow hyperbolic trajectories and the reaction occurs with greatest
probability when the turning points of the trajectories coincide (fig.1).
For a (d,p) reaction with Q-value $<E_d$, the linear (and angular) momentum of the deuteron is larger than the linear (and angular) momentum of
the proton. Conservation of orbital angular momentum then requires $\vec{\ell}_n$
to be opposite to $\vec{k}_d \times \vec{k}_p$, i.e. opposite to the neutron spin if the incident deuteron polarization is positive. Thus the cross section is enhanced ($A_y > 0$) for $j_n = \ell - 1/2$. The sign of the $\vec{\ell}_n$ and thus the sign
of the analyzing power will reverse if the momentum of the outgoing particle is larger than that of the incident particle [e.g. (d,p) reactions
with $Q > E_d$ or (d,t) reactions with $Q \sim 0$]. The model has been shown[8]
to predict correctly the Q-dependence of the sign of A_y as a function
of angle.

Fig. 2 shows an
example of the drastic
change in the appearance
of the analyzing power
which occurs as the out-
going particle energy
changes from a value
well above the Coulomb
barrier, in which case
nuclear distortions
predominate, to a value
well below the Coulomb
barrier. The deuteron
energy has been chosen
to be about equal to
the Coulomb barrier.
These curves were cal-
culated by the DWBA
for hypothetical
$5/2^+$ transitions
in ^{80}Se.

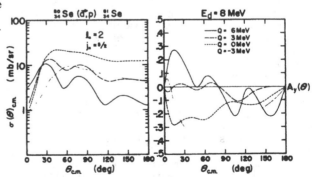

Fig. 2. DWBA calculations showing the Q dependence of $\sigma(\theta)$ and $A_y(\theta)$.

This suggests that in planning experiments to determine j-values, it may be a good idea to perform calculations of this sort, in order to avoid situations such as that depicted by the curve for Q = 3 MeV, in which the magnitude of the analyzing power is less than 0.03 for angles less than 90°. The curves shown for Q = 0 and Q = -3 MeV are characteristic of subcoulomb transitions when $Q < E_d$. The cross section is relatively featureless and drops off at forward angles, while the analyzing power is fairly large and has the same sign over practically the entire angular range. Another attractive feature of sub-coulomb transfer is the lack of sensitivity to the details of the nuclear distortions. A nice example of this is provided by the data of Vigdor and Haeberli[9] on the $^{118}Sn(\vec{d},t)$ reaction (fig. 3). The results for these $\ell = 2$ transitions exhibit the j-dependence strikingly, and are in excellent agreement with the DWBA predictions as well as with the expectation of the subcoulomb model.

An example of data on (d,p) transitions in a region of the periodic table in which few results have been available are the measurements of Yoh et al.[10] (fig. 4) on $Ge(\vec{d},p)$. The data shown are typical of the results for $\ell = 2$ and $\ell = 4$ transitions at a bombarding energy of 12 MeV. Although rather good agreement with the data is achieved in these cases by the DWBA calculations, there seem to be some discrepancies not attributable to the choice of potential parameters. Sizeable values of the analyzing power for small angles are evident from the predictions shown in fig. 4. Aymar et al.[11] have used a magnetic spectrograph to

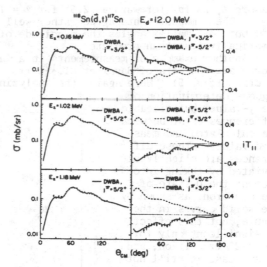

Fig. 3.   Data for $^{118}Sn(\vec{d},t)^{117}Sn$ (Ref. 9).

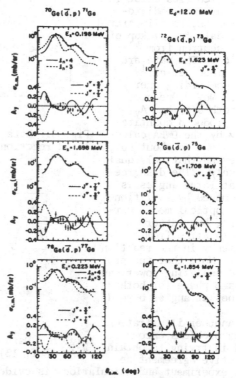

Fig. 4.   Examples of data for $Ge(\vec{d},p)$ (Ref. 10).

measure A_y as far foreward as 2.5° for $\ell = 1$ transitions on ^{58}Ni and ^{60}Ni. Their measurements agree rather well with DWBA for angles down to 10°, but for smaller angles the magnitude of the analyzing power is systematically less than predicted.

Another example of measurements in a mass region previously unexplored are the data of Basak et al.[12] from Birmingham on the ^{142}Nd($\vec{d}$,p) reaction (fig. 5). Here again, the analyzing power provides an unambiguous determination of the j-value, but sizeable differences exist between the calculated and measured values. These differences are slightly alleviated by taking into account the D-state of the deuteron, as shown by the solid curves in the figure, but this effect is significant only for the 7/2⁻ transition, and in any case is still not enough to produce satisfactory agreement. Similar difficulties exist in the light intermediate-mass region, as illustrated by the data for pickup of a $d_{3/2}$ neutron from ^{40}Ar (fig. 6). The data are those of Sen et al.[13] for a bombarding energy of 14.8 MeV. One sees that both $\sigma(\theta)$ and $A_y(\theta)$ are reproduced quantitatively by the DWBA calculations out to about 60°, and beyond that only qualitatively. The disagreement at large angles is not removed by variation in the optical model parameters, but rather appears to arise from a deficiency in the reaction model. Note that the predicted analyzing power has the same phase for both j-values at angles beyond 60°.

Another illustration of the problems encountered in trying to obtain quantitive agreement be-

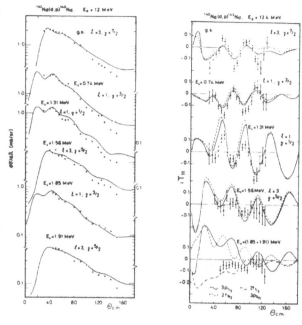

Fig. 5.   Data of Ref. 12 for the ^{142}Nd($\vec{d}$,p) ^{143}Nd reaction.

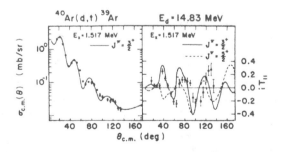

Fig. 6.   Data for a 3/2 ($\vec{d}$,t) transition in ^{40}Ar (Ref. 13).

tween experiment and calculations is evident in the data of the Saclay group[14] on ^{18}O($\vec{p}$,d)^{17}O. Although a bombarding energy of 24.5 MeV was used for these measurements, the data were generally not too well accounted for by DWBA calculations, and the authors attributed this to shortcomings in the treatment of the reaction mechanism (fig. 7).

An explanation sometimes offered for discrepancies of the sort noted in the examples just given is that the reaction proceeds to an appreciable extent by compound nucleus formation. That such is not the case, at least for nuclei of mass number greater than about 50, has been shown rather conclusively by Stephenson and Haeberli[15]. They determined the magnitude of the compound nucleus contribution to the cross section and vector analyzing power in $^{52}Cr(\vec{d},p)$ and $^{46}Ti(\vec{d},p)$ at 6 MeV by measuring the Ericson fluctuations in these quantities. Neither the inclusion of energy-averaged compound nucleus contribution (solid curves in fig. 8) nor the effect of undamped Ericson fluctuations (shaded areas in fig. 8) are able to remove the substantial discrepancies between the measurements and the DWBA calculations (dashed curves in fig. 8). For bombarding energies of 10 MeV and above, compound nucleus effects should be at least an order of magnitude smaller than those shown in fig. 8.

One effect which is undoubtedly responsible in some cases for the difficulties encountered in fitting data with the DWBA is the contribution of two-step processes to the reaction. Such processes have been shown[16,17] to be able to contribute substantially to transfer cross sections not only in transitions which proceed weakly by a single step mechanism, but in strong transitions as well. In a contribution to this symposium, Boyd and collaborators[18] have reported an investigation

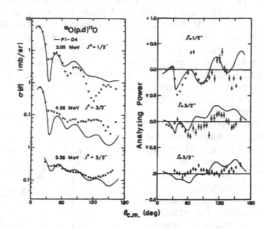

Fig. 7. Cross section and vector analyzing power for $^{18}O(\vec{p},d)^{17}O$ at E_p = 24.5 MeV. The data shown are for ℓ^p = 1 transitions (Ref. 14).

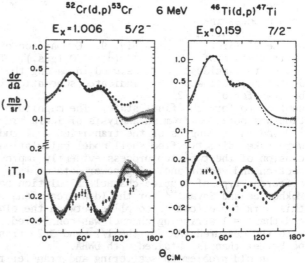

Fig. 8. Effect of compound nucleus processes in $^{52}Cr(\vec{d},p)^{53}Cr$ and $^{46}Ti(\vec{d},p)^{47}Ti$ (Ref. 15).

of the (d,p) transitions to 7/2 and 9/2 levels of ^{41}Ca at excitation energies near 3 MeV. They have carried out CCBA calculations in which the
transition amplitudes include two-step components in addition to a relatively weak single-step component. The CC calculations are able to reproduce the analyzing-power data somewhat better than the usual DWBA,
particularly for the 9/2$^+$ state. The predicted analyzing power can depend quite strongly on the details of the wave function assumed for the
final state, so that measurements of this kind should provide a sensitive test of final state configurations.

In another inves-
tigation, Hichwa and
Stephenson[19] from Wis-
consin have studied
the effect of a two-step
process in the
^{18}O($\vec{d}$,p)^{19}O transition, which is
known to proceed
strongly by single-
step neutron trans-
fer. The results
are shown in fig. 9
which also shows
schematically the
two-step process
assumed in making
the CC calculation.
This process in-
volves excitation
of the 2$^+$ first-
excited state of
^{18}O by the incident
deuterons, followed
by transfer of a 5/2$^+$
neutron to form the final state. The magnitude of β used in the calcula-

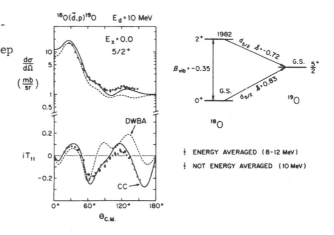

Fig. 9. Comparison of DWBA and CCBA for
^{18}O($\vec{d}$,p)^{19}O (g.s.). Thw two-step process
assumed in making the CC calculation is
indicated schematically.

tion was obtained from an analysis of inelastic deuteron scattering on
^{18}O and the strengths of the transition amplitudes to the final state
were taken directly from shell-model calculations of Wildenthal. Inclusion of the two-step process evidently improves the fit to both cross-
section and vector analyzing power data at large angles. A somewhat
different type of coupled-channel calculation has been reported by
Mukherjee and Shyam[20] for the (d,p) reactions on ^{40}Ca and ^{48}Ca. In
this work the effect of coupling between the elastic deuteron channel
and the ℓ = 1 stripping channel was considered. The overall agreement
with the measured VAP for the 1/2$^-$ and 3/2$^-$ transitions seemed to be
no better than is obtained with DWBA.

An old problem in scattering and transfer reactions is that of the
so-called derivative rule[21-23], according to which the polarization or
analyzing power is proportional to the logarithmic derivative with respect to angle of the differential cross section. For transfer reactions, the rule should apply approximately for ℓ = 0 transitions. In
practice, however, the rule seems to work only in certain cases. Vigdor[24] has recently discussed the conditions under which the derivative
rule should be valid and has presented a case in which it is particularly well satisfied. His conclusion is that the rule should hold for

$\ell = 0$ transitions if (i) the analyzing power arises from the spin-dependent distortion in only one channel and (ii) if in that channel the spatial overlap of the radial wave function with the spin-orbit potential is large and approximately constant for the partial waves which are dominant in the reaction. These conditions apply to the $^{118}\mathrm{Sn}(d,t)^{117}\mathrm{Sn}$ (g.s.) transition shown in fig. 10. The prediction of the derivative rule fits the data much better than the DWBA. This investigation has also confirmed that the triton-nucleus spin-orbit potential does not play much role in determining the analyzing power, even in $\ell_n = 0$ transitions. In addition, a correlation between $(\vec{d},d)$ and $(\vec{d},t)$ analyzing powers similar to that noted by Datta et al.[25] was observed here also, but no detailed explanation of this correlation has been given.

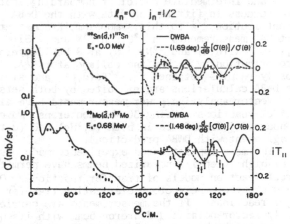

Fig. 10. Application of the derivative rule to $\ell_n = 0$ transitions in $^{118}\mathrm{Sn}(\vec{d},t)^{117}\mathrm{Sn}$ and $^{98}\mathrm{Mo}(\vec{d},t)^{97}\mathrm{Mo}$ (Ref. 24).

In contrast to the large number of even-odd nuclei for which VAP measurements have permitted spin assignments, the number of investigations involving non-zero spin targets is rather limited. In the most general case, mixtures of both ℓ and j can occur, but with this added complexity comes the possibility of obtaining useful nuclear structure information. The most reliable procedure adopted so far in studying these transitions seems to be to use calibration curves of cross section and analyzing power which have been measured on neighboring even-even nuclei. The ℓ-mixing can be determined from the cross section, and the j-mixing from the analyzing power. Applications of this technique to $(\vec{d},p)$ reactions have been reported by Kocher and Haeberli[26] and in an abstract submitted to this conference by Babcock and Quin[27]. They present data for $(\vec{d},p)$ transitions to states in $^{26}\mathrm{Mg}$ at 1.81, 3.94 and 4.84 MeV. In all three cases there appears to be mixing of $\ell_n = 0$ and $\ell_n = 2$ (the ground state spin of $^{25}\mathrm{Mg}$ is $5/2^+$). Calibration curves from transitions measured on $^{20,22}\mathrm{Ne}$ and $^{24}\mathrm{Mg}$ were used to fit the data for these mixed transitions in the neighborhood of the $\ell_n = 2$ stripping maximum. Although the fits are reasonably good in this case, the uncertainties in the method will remain high until the reaction theory is in better shape and more extensive calibration data are available.

A number of experiments[28-33] to measure polarization effects in (d,n) and $(^3\mathrm{He},n)$ reactions have been carried out since the Madison meeting in various laboratories. Many of these experiments have involved measurements of the polarization of neutrons emitted from reactions on light target nuclei utilizing relatively low incident

particle energies. As a consequence, these measurements are not usually
expected to be well reproduced by direct reaction model calculations,
and indeed they rarely are. VAP data have been reported on several
light and intermediate nuclei at bombarding energies high enough that
some success in fitting the data with the DWBA can be anticipated. In
a contribution [32] to this symposium, Krämmer et al. from Erlangen re-
port VAP measurements on $^{12}C(\vec{d},n)^{13}N$ and $^{28}Si(\vec{d},n)^{29}P$. Measurements of
iT_{11} in $^{58}Ni(\vec{d},n)$ and $^{60}Ni(\vec{d},n)$ at 8 MeV deuteron energy are reported in
a contribution by Hichwa and collaborators[31]. The data resemble very
closely results obtained for $^{54}Fe(\vec{d},p)$, and similar discrepancies with
the DWBA calculations are exhibited by both sets of data.

Vector analyzing power measurements have also been carried out for
$(\vec{d},^3He)$ reactions. Using 30 MeV deuterons, Mayer et al.[34] observed a
pronounced j-dependence in both $(\vec{d},^3He)$ and $(\vec{d},t)$ reactions on ^{208}Pb,
in agreement with DWBA predictions.

The more sophisticated experiments made possible by the availability of high intensity polarized beams have permitted additional tests of
direct reaction models of transfer reactions. One such test is the
measurement of the vector polarization transfer coefficient in $(\vec{d},\vec{p})$ and
$(\vec{d},\vec{n})$ reactions. If the measurements are carried out at 0° using a
purely vector-polarized deuteron beam with its quantization axis perpendicular to the beam direction, and if the emitted nucleon is simply a
spectator in the transfer reaction, one expects the polarization of the
outgoing nucleons to be the same as the polarization of the incoming
deuterons, provided the D-state of the deuteron is ignored. Papers[35-37]
reporting measurements for $(\vec{d},\vec{n})$ reactions at 0° have been submitted to
this symposium. Lisowski et al.[37] from the Triangle Universities Lab-
oratory present data for $(\vec{d},\vec{n})$ reactions on ^{14}N, ^{16}O and ^{28}Si for bombarding energies between 4 and 14 MeV. In all three reactions, the dom-
inant contribution to the 0° yield comes from $\ell_p = 0$ transitions and in
all three cases the neutron polarization is almost constant and about
10% less than the polarization of the incident deuterons.

In a contribution from Birmingham, a silicon polarimeter was used
to study $^{28}Si(\vec{d},\vec{p})^{29}Si$ for 12.4 MeV deuterons at angles of 10° and 20°.
While the polarization at 10° was found to be large, as in the measurements described above, at 20° a small value was observed. This is in
agreement with DWBA predictions only if the D-state of the deuteron is
taken into account. The D-state makes an important contribution to the
cross section at 20° where the first minimum in the S-state cross section occurs.

Very recently, the first beams of highly polarized tritons and
^3He-ions have been achieved and already interesting results[39-41] on
transfer reactions with these mass-three projectiles have been obtained.
The reactions induced by ^3He are discussed in an invited paper by Roman.
The first measurements on transfer reactions induced by the polarized
triton beam from the Los Alamos source are being reported at this
conference. Coffin et al.[41] have measured the analyzing power for
the $(\vec{t},\alpha)$ reaction leading to $j = \ell \pm 1/2$ single hole states in ^{89}Y
and ^{207}Tl. The magnitude of the analyzing powers was found to be fairly large, and a pronounced j-dependence is present.

The remainder of this discussion will be devoted to two- three-
and four-nucleon transfer reactions. For $(\vec{p},^3He)$ and $(\vec{p},t)$ reactions,
most of the work reported since the last symposium has been for light
target nuclei[14, 42-44], although some experiments on heavy target
nuclei have been performed[45,46]. Measurements were carried out at

Berkeley[46]) on the ^{208}Pb($\vec{p}$,t)^{206}Pb reaction leading to 0$^+$, 2$^+$, 4$^+$ and 6$^+$ states, using 40 MeV protons. Rather good fits were obtained to the cross-section data using the zero-range DWBA. For the analyzing-power data, the agreement was good only for the L = 0 transition. For the higher L values, only qualitative agreement could be obtained.

In another Berkeley experiment, Macdonald et al.[44]) made an extensive study of ($\vec{p}$,t) and ($\vec{p}$,^{3}He) transitions on ^{16}O, ^{15}N and ^{13}C. Since for the (p,t) reaction the transferred spin angular momentum S = 0, the transitions for these targets are characterized by a single value of L, whereas the (p,^{3}He) reaction can proceed by either S = 0 or S = 1. As a result, transitions for the latter reaction may involve coherent contributions of several combinations of L and S, as well as incoherent contributions from different values of J. In view of this

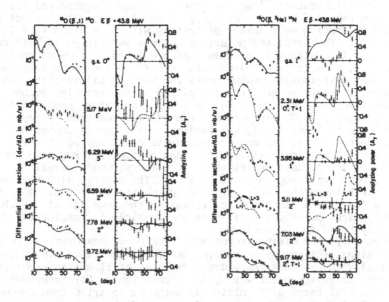

Fig. 11. Data for ($\vec{p}$,t) and ($\vec{p}$,^{3}He) on ^{16}O. The curves show the predictions of DWBA calculations (Ref. 44).

complexity it is not surprising that the agreement between the data and zero-range DWBA calculations was found to be good only for the L = 0 transitions and some of the L = 2 transitions (fig. 11). In most cases, the DWBA could not account for the analyzing power data, particularly for the (p,^{3}He) transitions. Clearly, considerable progress in the theoretical treatment of these reactions will be required before they can be fully exploited as a reliable spectroscopic tool.

Several contributions[47-50]) reporting polarization measurements in (^{3}He,$\vec{p}$) reactions on very light nuclei have been submitted to this conference from Laval University. The analysis of these data is made difficult by the multiplicity of allowed J, L and S values referred to above. In the case of ^{3}He(^{3}He,$\vec{p}$)^{5}Li, however, there is evidence[48]) that the reaction proceeds by simple transfer of a deuteron.

There have been several investigations[51-55]) of polarization effects in $(\vec{d},\alpha)$ reactions, many of them involving nuclei in the 1d-2s shell. Calculations by Nemets and collaborators[51] at the Institute for Nuclear Research in Kiev illustrate the pronounced j-dependence expected in both vector and tensor analyzing power for a reaction assumed to proceed by deuteron pickup. For the $^{27}Al(\vec{d},\alpha)^{25}Mg$ reaction initiated by 14 MeV deuterons, both iT_{11} and T_{20} show large differences between the values of transferred J = 1, 2 and 3 for L = 2. Such effects have in fact been observed by Ludwig and collaborators[52] at the Triangle Universities Laboratory in $(\vec{d},\alpha)$ measurements on ^{14}N, ^{16}O and ^{28}Si. They obtained fairly good agreement with the data for most transitions investigated using calculations based on simple deuteron pickup. A similar investigation at Birmingham[53] has been reported for $^{19}F(\vec{d},\alpha)$ and $^{40}Ca(\vec{d},\alpha)$.

Little is known about three-nucleon transfer reactions. However, Nemets and collaborators[56] at Kiev have calculated proton polarizations to be expected in the $^{28}Si(\alpha,\vec{p}_0)^{31}P$ and $^{63}Cu(\alpha,\vec{p}_0)^{66}Zn$ reactions under the assumption of transfer of tritons with $\ell_t = 0$ and $\ell_t = 1$, respectively. The predicted polarizations are very large in both cases, and for $\ell_t = 1$, a strong j-dependence is predicted at forward angles. In this case, a large j-dependence also occurs in the cross section.

Angular distribution measurements for the $(d,^6Li)$ reaction are, in many cases, strongly suggestive of α-pickup as the reaction mechanism. In a submitted contribution[57] Jacobs and collaborators present the first analyzing power measurements for a $(\vec{d},^6Li)$ reaction. They have studied $^{12}C(\vec{d},^6Li)$ and $^{16}O(\vec{d},^6Li)$ leading to the ground states of ^8Be and ^{12}C, respectively, for an incident deuteron energy of 16 MeV. Preliminary calculations assuming an alpha particle pickup show some similarities to the data, particularly for the strongly oscillatory analyzing power observed in $^{12}C(\vec{d},^6Li)^8Be$. The authors also note some resemblance between the analyzing power predicted for $(\vec{d},^6Li)$ and that predicted for elastic deuteron scattering on the same target. Such a similarity might be expected if the alpha particle is considered to be very weakly bound to the deuteron.

In summary, the use of vector analyzing power measurements to determine values of j-transfer in (d,p) and (d,t) reactions has increased substantially in the last few years. Problems still exist in obtaining good quantitative agreement with DWBA calculations. For intermediate and heavy nuclei these difficulties do not seem to arise from compound-nucleus effects but appear rather to arise from shortcomings in the DWBA theory as it is usually applied. One question which has not been resolved is whether the additional constraint imposed on the DWBA calculations by the vector analyzing power measurements improves the reliability of the spectroscopic factor determinations. It has been demonstrated that two-step processes are important in many cases, and the availability of coupled-channels programs should permit the analysis of many transitions which can not be understood in terms of simple stripping. It can be anticipated that the use of vector analyzing power measurements as a tool in nuclear spectroscopy will continue to expand since the j-dependence has now been observed in virtually all single-nucleon transfer reactions. Progress in the study of multinucleon transfer has been much slower, in great part because many amplitudes can contribute in such processes.

Substantial progress has been achieved in the understanding of polarization effects in transfer reactions since the last symposium. Further progress in this area will require the well-planned use of all the sophisticated theoretical and experimental techniques which have been presented in the contributions to this symposium.

References

† Work supported in part by the National Science Foundation
†† Work supported in part by the U.S. ERDA

1) W. Haeberli, Proc. Third Int. Symposium on Polarization Phenomena (Univ. of Wisconsin Press, 1971) p. 235
2) R.R. Cadmus and W. Haeberli, Bull. Amer. Phys. Soc. 18 (1973) 1406
3) B.P. Hichwa and R.R. Cadmus, Bull. Amer. Phys. Soc. 18 (1973) 1406
4) S.L. Tabor, R.W. Zurmühle and D.P. Balamuth, Phys. Rev. C8, (1973) 2200
5) W. Haeberli, Lecture Notes in Physics (Ed. by D. Fick) Springer Verlag, 1974
6) H.C. Newns, Proc. Phys. Soc. (London) A66 (1953) 477
7) B.J. Verhaar, Phys. Rev. Lett. 22 (1969) 609
8) S.E. Vigdor, R.D. Rathmell and W. Haeberli, Nucl. Phys. A210 (1973) 93
9) S.E. Vigdor and W. Haeberli, to be published
10) W.A. Yoh, S.E. Darden and S. Sen, to be published
11) J.A. Aymar, H.R. Hiddleston, S.E. Darden and A.A. Rollefson, Nucl. Phys. A207 (1973) 596
12) A.K. Basak, J.A.R. Griffith, M. Irshad, O. Karban, E.J. Ludwig, J.M. Nelson and S. Roman, Nucl. Phys. A229 (1974) 219
13) S. Sen, S.E. Darden, W.A. Yoh and E.D. Berners, Nucl. Phys., to be published
14) M. Pignanelli, J. Gosset, F. Resmini, B. Mayer and J.L. Escudié, Phys. Rev. C8 (1973) 2120
15) E.J. Stephenson and W. Haeberli, Fourth Polarization Symp. and to be published
16) F.A. Gareev, R.M. Jamalejev, H. Schulz and J. Bang, Nucl. Phys. A215 (1973) 570
17) A.K. Abdallah, T. Udagawa and T. Tamura, Phys. Rev. C8 (1973) 1855
18) R.N. Boyd, D. Elmore, H. Clement, W.P. Alford, J.A. Kuehner and G. Jones, Fourth Polarization Symp., p.671
19) B. Hichwa and E.J. Stephenson, Bull. Am. Phys. Soc. 20 (1975) 597
20) S. Mukherjee and R. Shyam, Phys. Rev. C11 (1975) 476
21) L.S. Rodberg, Nucl. Phys. 15 (1959) 72
22) L.C. Biedenharn and G.R. Satchler, Helv. Phys. Acta., Supplement VI (1960) 372
23) R.C. Johnson, Nucl. Phys. 35 (1962) 654
24) S.E. Vigdor, Nucl. Phys., to be published
25) S.K. Datta, C.E. Busch, T.B. Clegg, E.J. Ludwig and W.J. Thompson, Phys. Rev. Lett. 31 (1973) 949
26) D.C. Kocher and W. Haeberli, Nucl. Phys., to be published
27) R.W. Babcock and P.A. Quin, Fourth Polarization Symp. p.669
28) D. Hilscher, J.C. Davis and P.A. Quin, Nucl. Phys. A174 (1971) 417
29) P.A. Quin, J.C. Davis and D. Hilscher, Nucl. Phys. A183 (1972) 173
30) P.A. Quin and J.A. Thomson, to be published
31) B.P. Hichwa, L.D. Knutson, J.A. Thomson, W.H. Wong and P.A. Quin, Fourth International Polarization Symp. p.655
32) P. Krämmer, W. Drenckhahn, E. Finckh, and J. Niewisch, Fourth International Polarization Symp. p.651
33) W.H. Wong and P.A. Quin, Fourth International Polarization Symp. p.657

34) B. Mayer, H.E. Conzett, W. Dahme, D.G. Kovar, R.M. Larimer, and
    Ch. Lehmann, Phys. Rev. Lett. 32 (]974)]452
35) R.K. TenHaken and P.A. Quin, Fourth International Polarization Symp.
36) R.L. Walter, R.O. Byrd, P.W. Lisowski, G. Mack and T. Clegg,
    Fourth International Polarization Symp. p.649
37) F.W. Lisowski, R.O. Byrd, W. Tornow, R.L. Walter and T. Clegg,
    Fourth International Polarization Symp. p.653
38) A.K. Basak, J.A.R. Griffith, O. Karban, J.M. Nelson, S. Roman and
    G. Tungate, Fourth International Polarization Symp. p.661
39) S. Roman, A.K. Basak, J.B.A. England, O. Karban, G.C. Morrison,
    J.M. Nelson and G.G. Shute, Fourth International Polarization Symp.
40) A.K. Basak, J.B.A. England, O. Karban, G.C. Morrison, J.M. Nelson,
    S. Roman and G.G. Shute, Fourth International Polarization Symp.
41) J.P. Coffin, E.R. Flynn, R.A. Hardekopf, J.D. Sherman and
    J.W. Sunier, Fourth International Polarization Symp. p.699
42) P.W. Keaton, Jr., D.D. Armstrong, L.R. Veeser, H.T. Fortune
    and N.R. Roberson, Nucl. Phys. A179 (1972) 561
43) J.M. Nelson and W.R. Falk, Nucl. Phys. A218 (1974) 441
44) J.A. Macdonald, J. Cerny, J.C. Hardy, H.L. Harney, A.D. Bacher
    and G.R. Plattner, Phys. Rev. C9 (1974) 1694
45) G. Igo, J.C.S. Chai, R.F. Casten, T. Udagawa and T. Tamura,
    Nucl. Phys. A207 (1973) 289
46) J.A. Macdonald, N.A. Jelley and J. Cerny, Phys. Lett. 47B (1973)
    237
47) R. Pigeon, M. Irshad, S. Sen, J. Asai and R. Slobodrian, Fourth
    International Polarization Symp. p.701
48) J. Asai and R.J. Slobodrian, Fourth International Polarization
    Symp. p.703
49) M. Irshad, J. Asai, S. Sen, R. Pigeon and R.J. Slobodrian, Fourth
    International Polarization Symp. p.705
50) C.R. Lamontagne, M. Irshad, R. Pigeon, R.J. Slobodrian and
    J.M. Nelson, Fourth International Polarization Symp. p.707
51) O.F. Nemets, Yu. S. Stryuk and A.M. Yasnogorodsky, Fourth
    International Polarization Symp. p.689
52) E.J. Ludwig, T.B. Clegg, P.G. Ikossi and W.W. Jacobs, Fourth
    International Polarization Symp. p.687
53) J.M. Nelson, O. Karban, E. Ludwig and S. Roman, Fourth Inter-
    national Polarization Symp. p.679
54) R.W. Babcock and P.A. Quin, Bull. Am. Phys. Soc. 18 (1973) 1394
55) Y. Takeuchi, J.A.R. Griffith, O. Karban and S. Roman, Nucl.
    Phys. A220 (1974) 589
56) O.F. Nemets, Yu. S. Stryuk and A.M. Yasnogorodsky, Fourth
    International Polarization Symp. p.713
57) W.W. Jacobs, T.B. Clegg, P.G. Ikossi and E.J. Ludwig, Fourth
    International Polarization Symp. p.697

DISCUSSION

Boyd:
   In the $^{40}Ca(\vec{d},p)^{41}Ca$ ($11/2^+$, 3.37 MeV) study, the CCBA prediction
   fits the analysing power data well, while the DWBA gives a completely wrong answer. Thus one must use caution, based on the structure of
   the level, in applying DWBA j assignments.

Pignanelli:
   For what concerns the discussion on $^{18}O(d,p)$ two step calculations,
   one can remark that in the incoming channel the correct experimental
   β value has been used and that the inelastic effect in the final
   channel should be not so important because the weak coupling between
   the final channels in a odd mass residual nucleus. A completely different situation is found instead for a (p,t) reaction.

D-STATE EFFECTS IN TRANSFER REACTIONS

L.D. Knutson[†]⋅
University of Washington, Seattle, Washington 98195 USA

1. Introduction

For decades the deuteron has been the object of a great deal of interest, as physicists have attempted to learn more about the structure of this simple nuclear system. In spite of this, our understanding of the two-nucleon bound state is far from complete. In particular, attempts to learn about the D-state component in the deuteron wave function have not been very productive.

Recently there have been a number of papers dealing with the effects of the D-state in transfer reactions. Two reasons for studying these effects come to mind. First of all, these studies will increase our understanding of the reaction process, and will provide an additional means for testing the direct-reaction theories. But in addition, there is the hope that by studying the effects of the D-state one might find that it is possible to obtain new information about the internal structure of the deuteron.

Since the last polarization symposium, we have learned that the D-state can have large measurable effects in $(\vec{d},p)$ reactions. In this paper I will review what is known about these D-state effects, and will try to point out those areas where additional research would be valuable. The scope of this review will be limited to transfer reactions on nuclei for which the conventional distorted-wave theories are applicable. Specifically, I will limit my comments to reactions on targets with $A \geq 12$.

Since the D-state probability is a small number, one might think that the effects of the D-state are necessarily small. However, the effects which one observes in $(\vec{d},p)$ reactions arise from interference between the S-state amplitude and the D-state amplitude. These coherent effects can be quite large.

2. D-state effects in (d,p) reactions

2.1. CROSS SECTION, VECTOR ANALYZING POWER AND PROTON POLARIZATION

If one neglects the spin dependence of the optical model potentials in a DWBA calculation, the S- and D-states contribute incoherently to the differential cross section[1]). As a result, the D-state normally has very little effect on the cross section. However, there are some interesting exceptions to this rule. Fig. 1 contains measurements and DWBA calculations[2]) for two $\ell_n = 3$ transitions in the reaction $^{56}Fe(p,d)^{55}Fe$ at $E_p = 18.5$ MeV. When the D-state is neglected the calculated cross sections for the two transitions are nearly identical. Including the D-state has almost no effect on the 5/2⁻ transition, but for the 7/2⁻ transition, the effect is substantial, particularly near the minimum at 60°. It is clear that the calculations agree more closely with the data when the D-state is included.

In fig. 2 we present some results for a more typical reaction. The cross section, vector analyzing power and proton polarization data are from ref. 3.   From the DWBA calculations we see that the D-state effect is small for the cross section, somewhat larger for the vector analyzing power and still larger for the proton polarization. This trend is observed quite generally in DWBA calculations[4,5]).

Normally, the effects of the D-state are small compared to the changes which can be produced, for example, by using different optical model potentials. However, the magnitude of the D-state effects can vary considerably from one transition to another. The following guidelines may be of some use in judging whether the effects will be important in any particular case: (i) The D-state effects increase as the (d,p) Q-value increases. (ii) The effects are larger for $j_n=\ell_n+1/2$ than for $j_n=\ell_n-1/2$. (iii) The effects increase with increasing ℓ_n.

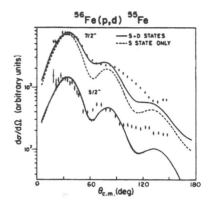

2.2. TENSOR ANALYZING POWERS

The effects of the deuteron D-state can be quite large for the tensor analyzing powers. This was first pointed out by Johnson[1], who showed that in a DWBA calculation, the S- and D-states act coherently to produce non-zero tensor analyzing powers. More recently, the impor-

Fig. 1. Angular distributions of the differential cross section for $^{56}Fe(p,d)^{55}Fe$ at $E_p=18.5$ MeV. The solid curves are DWBA calculations which include the deuteron D-state. For the dashed curves, the D-state was not included. The figure is from ref. 2.

tance of the D-state has been demonstrated in the finite-range DWBA calculations of Delic and Robson[5-8].

At the time of the last polarization symposium, there had been very few tensor analyzing power measurements for transfer reactions[9]. Since that time a number of experiments have been carried out[10-21]. In table 1, I have presented a compilation of the existing measurements for single-nucleon transfer reactions on nuclei with $A \geq 12$.

Figure 3 contains tensor analyzing power measurements[13] and DWBA calculations for some $(\vec{d},p)$ transitions on ^{52}Cr and ^{90}Zr at $E_d=10$ MeV. One notes that the measured analyzing powers are reasonably large in magnitude and show a rather complicated angular dependence. The dashed

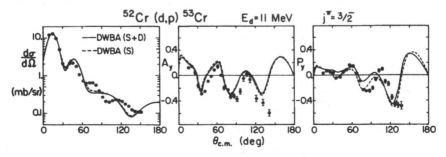

Fig. 2. Angular distributions of the cross section, vector analyzing power and proton polarization for $^{52}Cr(d,p)^{53}Cr$ at $E_d=11$ MeV. The measurements are from ref. 3. The solid curves are DWBA calculations which include the deuteron D-state. For the dashed curves, the D-state was not included.

Table 1

Tensor analyzing power measurements
for single-nucleon transfer reactions on targets with A ≥ 12

REACTION	TARGETS	ENERGIES (MeV)	QUANTITIES MEASURED	REF.
$(\vec{d},p)$	^{12}C ^{16}O ^{19}F ^{25}Mg ^{28}Si ^{40}Ca	12.3	T_{20} T_{22}	10,14
	^{12}C ^{28}Si ^{40}Ca	5.0-12.0	T_{20}	16
	^{16}O	9.3 13.3	T_{20} T_{21} T_{22}	11
	^{46}Ti ^{52}Cr	6.0 10.0	T_{20} T_{21} T_{22}	15,20
	^{52}Cr ^{54}Fe ^{90}Zr	10.0	T_{20} T_{21} T_{22}	13,15
	^{58}Ni ^{60}Ni	10.0	T_{20} + $\sqrt{6}$ T_{22}	19
	^{90}Zr ^{208}Pb	5.5 9.0	T_{20} T_{21} T_{22}	15,17
	^{117}Sn ^{119}Sn	12.0	T_{20} T_{21} T_{22}	18
	^{208}Pb	12.3 15.0	T_{20} + $\sqrt{6}$ T_{22}	12
$(\vec{d},t)$	^{118}Sn ^{208}Pb	12.0 12.3	T_{20} T_{21} T_{22}	21

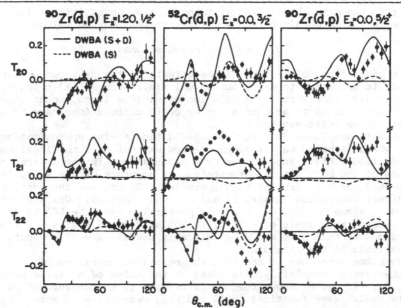

Fig. 3. Angular distributions of the three tensor analyzing powers for
$(\vec{d},p)$ reactions on ^{52}Cr and ^{90}Zr at 10 MeV. The measurements are from
ref. 13. The solid curves include the effects of the deuteron D-state
and the dashed curves do not.

curves in the figure were obtained from the standard DWBA theory in which
the deuteron D-state is neglected. These calculations show no resemblance to the data. In particular, for the $\ell_n=0$ transition, and for T_{21},
the predictions are an order of magnitude smaller than the data at forward angles. The solid curves show the results of DWBA calculations
which include the D-state. These calculations provide good qualitative
agreement with the data. For T_{21} and T_{22} at forward angles, the agreement is surprisingly good. We can conclude that near the stripping peak
the deuteron D-state is primarily responsible for the large tensor analyzing powers.

Brown et al.[10]) have pointed out that the effects of the deuteron
D-state are particularly large for T_{20} at forward angles, and have measured $T_{20}(0°)$ for a large number of transitions. The measured values of
$T_{20}(0°)$ are large in magnitude and negative (except for the $j^\pi=5/2^+$
transitions). The DWBA calculations reproduce this sort of behavior
only when one includes the deuteron D-state. Unfortunately, the measurements were carried out in a region of strong compound nucleus fluctuations[16]), and as a result, a quantitative comparison with DWBA is not
meaningful.

A particularly interesting feature of the measurements is the
behavior of T_{22} at forward angles[14,15,20,22]). For transitions with
$j_n=\ell_n+1/2$, T_{22} invariably shows a sharp negative dip in the region of
the stripping peak (see fig. 3), while for $j_n=\ell_n-1/2$ transitions, T_{22} is
positive. In one of the contributed papers, Santos[22]) has offered an
explanation of this phenomenon which is based on a simple model of the
reaction. The j_n-dependence of T_{22} may turn out to be a useful spectroscopic tool. For most (d,p) transitions, one can easily determine the
j_n value from a measurement of the vector analyzing power[9]). However,
in certain cases (such as mixed j_n transitions) the additional information which can be obtained from a T_{22} measurement may be valuable.

3. The DWBA calculations

In order to properly include the effects of the deuteron D-state in
DWBA, it is necessary to carry out a full finite-range (FR) calculation.
Unfortunately, this sort of calculation requires a large amount of computing time. In spite of this a number of FR calculations have been
carried out by Delic and Robson[5-8]).

Johnson and Santos[2]) have developed a time saving approximation
method which is based on a straightforward extension of the local-energy
approximation (LEA). Many of the DWBA calculations in the literature
make use of the LEA, but unfortunately, the accuracy of the approximation method has not been thoroughly studied. Of course, the ideal test
is a direct comparison between FR and LEA calculations. One such comparison is shown in fig. 4. For this particular case the LEA tends to
overestimate the D-state effects by 10-20% at forward angles. Further
comparisons of this sort would be extremely useful, and at present, none
are available in the literature.

When one makes use of the LEA, the relative importance of the deuteron D-state is completely determined by the value of a single parameter, D_2. It can be shown that D_2 is related to the deuteron S- and
D-state radial wave functions [$u(\rho)$ and $w(\rho)$ respectively] according to

$$D_2 = \frac{1}{15} \int_0^\infty \rho^3 w(\rho)d\rho / \int_0^\infty \rho\, u(\rho)d\rho. \tag{1}$$

From this expression we see that D_2 is primarily sensitive to the

behavior of w(ρ) at large ρ. Since the
deuteron quadrupole moment is also sensitive to w(ρ) at large ρ, the range of allowed D_2 values is somewhat restricted[2]).
Wave functions which correctly reproduce
the deuteron quadrupole moment generally
have values of D_2 in the range[17]) 0.45-
0.55 fm^2. This explains why the DWBA
calculations accurately reproduce the
magnitude of the observed D-state effects
in spite of the fact that the deuteron
D-state probability is not well known.

In general, I think it is fair to say
that the DWBA theory does a good job of
reproducing the observed deuteron D-
state effects. In fact, the calculations
and measurements usually agree more
closely for the tensor analyzing powers
than for the vector analyzing power[20]).
Normally the fits to T_{21} and T_{22} are
excellent at forward angles. At back
angles the fits are less accurate, but
it is quite possible that the addition
of tensor terms in the deuteron optical
model potential would produce some improvement[23]). This point certainly
bears further investigation. In general,
the calculations tend to be less
accurate for T_{20} than for either T_{21}
or T_{22}[18,24]).

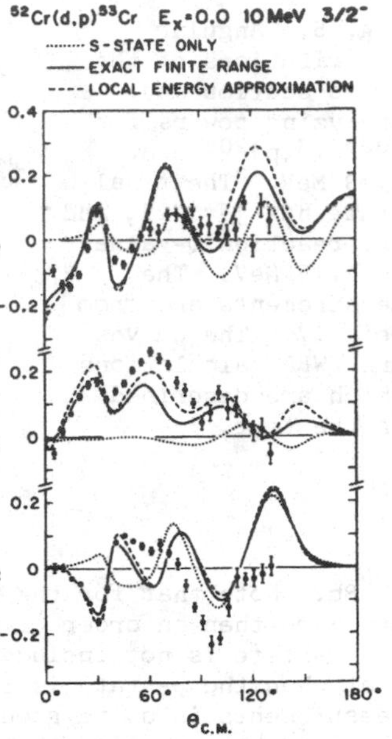

The tensor analyzing power measurements provide a means for testing the
reliability of any direct-reaction
theory. Recently, Pearson et al.[25])

Fig. 4. Tensor analyzing
powers for ^{52}Cr($\vec{d}$,p)^{53}Cr. The
solid curves are an exact FR
calculation taken from ref. 8.
The dashed curves were calculated with the LEA, using the
same optical model potentials.
The data are from ref. 13.

have extended the weakly-bound projectile
model to include the effects of the deuteron D-state, and have presented
calculations of cross sections, vector analyzing powers and proton
polarizations for (d,p) reactions. It would be quite interesting to see
how well this theory reproduces the tensor analyzing power measurements.

4. Sub-Coulomb (d,p) reactions

Reactions carried out at energies below the Coulomb barrier have
several special properties which make the deuteron D-state effects particularly interesting. Goldfarb[26]) has demonstrated that for sub-Coulomb
(d,p) reactions which have Q-values near zero the neutron capture occurs
far outside the nucleus, near the turning point of the classical Coulomb
trajectories. As a result, one expects that the DWBA calculations for
sub-Coulomb reactions should be exceptionally reliable, since the reactions are insensitive to the nuclear interactions with the target.
Furthermore, because the deuteron and proton are not subjected to any
spin-dependent nuclear forces, the calculated tensor analyzing powers are
extremely small when one neglects the deuteron D-state.

These points are illustrated in fig. 5, which contains measurements[17]) and DWBA calculations for a sub-Coulomb ($\vec{d}$,p) transition on

Fig. 5. Angular
distributions of the
cross section and
analyzing powers for
208Pb($\vec{d}$,p)209Pb at
E_d=9 MeV. The final
state has j^π=5/2$^+$, and
the reaction Q-value
is 0.15 MeV. The
measurements are from
ref. 17. The curves
are DWBA calculations
which are described
in the text.

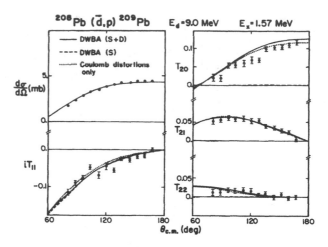

208Pb. Note that for the tensor analyzing powers, the DWBA predictions
are more than an order of magnitude smaller than the measurements when
the D-state is not included in the calculations (dashed curves), and
that when the D-state is included (solid curves), the agreement with the
measurements is quite good. The dotted curves in the figure were
obtained by repeating the D-state calculation with the nuclear optical
model potentials set to zero, so that the deuteron and proton are subject to Coulomb distortions only. The close agreement between the solid
and dotted curves demonstrates that the reaction is insensitive to
nuclear distortions.

The angular dependence of the tensor analyzing powers is quite
similar for all ($\vec{d}$,p) reactions near and below the Coulomb barrier[17].
In all cases, T_{20} is positive at back angles and decreases in magnitude
at more forward angles, T_{21} is primarily positive and peaks in magnitude
near 90°, and T_{22} is relatively small. However, the overall magnitude
of the analyzing powers varies substantially from one transition to
another[17].

Knutson et al.[27] have used a simple classical model to explain how
the deuteron D-state affects the analyzing powers for sub-Coulomb ($\vec{d}$,p)
reactions. In this model it is assumed that the deuteron and proton
follow classical Coulomb trajectories, and that the neutron transfer
occurs at the point of closest approach. This is illustrated in fig. 6.
Because the deuteron has a D-state its wave function is not spherically
symmetric, and as a result, the probability that a neutron capture will
take place depends on the orientation of the deuteron spin. It is
possible to argue[27] that the effect of the D-state is to enhance the
($\vec{d}$,p) cross section when the deuteron spins are parallel or antiparallel to the vector $\vec{q}$ in fig. 6, and to decrease the cross section
for deuterons whose spins are perpendicular to $\vec{q}$.

One can easily test this hypothesis. We consider a "completely
aligned" beam, in which all of the deuterons populate the m=±1 magnetic
substates and use σ_a to designate the cross section for such a beam.
In fig. 7 we present values of the ratio σ_a/σ_o where σ_o is the unpolarized cross section. These data were calculated directly from the tensor
analyzing power measurements shown in fig. 5. The crosses show the
values of σ_a/σ_o which correspond to the spin orientation for which the

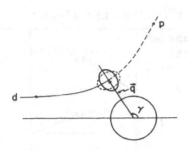

Fig. 6. Schematic representation of a sub-Coulomb (d,p)
reaction. The neutron transfer occurs at the point $\vec{q}$.
The figure is from ref. 27.

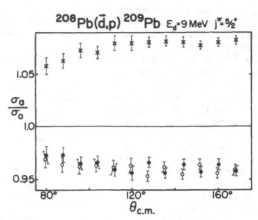

Fig. 7. Values of σ_a/σ_0 as a function of
reaction angle for three orientations of
the spin alignment axis. For the crosses
$\hat{s}$ is parallel to $\vec{q}$, and for the open and
closed circles, $\hat{s}$ is perpendicular to $\vec{q}$.

spin alignment axis ($\hat{s}$) is
parallel to $\vec{q}$ at the point
of transfer. The open and
closed circles are for two
mutually perpendicular spin
orientations for which $\hat{s}$ is perpendicular to $\vec{q}$. The results seen in fig.
7 are precisely what one would expect from the classical model.

For reactions with $\ell_n=0$ it is possible to extend the classical model
and calculate the magnitude of the D-state effects[27]. When this is done,
one comes up with an extremely simple formula for the tensor analyzing
powers. The predictions of this simple model agree surprisingly well
with the analyzing powers obtained from DWBA calculations (see ref. 27).

In the introduction I pointed out that studies of the deuteron D-
state effects are motivated, in part, by the hope that one might obtain
new information about the internal structure of the deuteron. Recently,
it has been proposed[28] that tensor analyzing power measurements for sub-
Coulomb ($\vec{d}$,p) reactions provide a means for experimentally determining
the value of the parameter D_2. This proposal is based on the idea that
DWBA calculations are very accurate for sub-Coulomb ($\vec{d}$,p) reactions
which have Q-values near zero.

The results of a preliminary analysis[28] are shown in fig. 8. The
tensor analyzing power measurements are for two transitions in
^{208}Pb($\vec{d}$,p)^{209}Pb which have Q-values near zero. The solid curves show the
DWBA calculations for $D_2=0.484$ fm^2, which is the value predicted by the
Reid soft-core deuteron wave function[29]. The dotted curves were ob-
tained with $D_2=0.432$ fm^2, which provides the best fit to the measurements.

This sort of analysis raises several issues which must be dealt with
before the method can be used in a serious attempt to measure D_2. First
of all, one must insure that the data set is not subject to an overall
normalization error. Second, one must demonstrate that the calculated
analyzing powers are insensitive to reasonable changes in the optical
model potentials. Finally, one must find a way to assess the accuracy
of the DWBA calculations. In particular, it is important to show that
two step processes, which are neglected in DWBA, do not produce large
changes in the analyzing powers. It is not difficult to argue from a
qualitative basis that the approximations used in DWBA are reasonable
for sub-Coulomb reactions[26,28]. However, what is required is a quantitative estimate of the effects of higher order processes. To my

Fig. 8. Angular distribu-
tions of the tensor
analyzing powers for two
transitions in the reac-
tion ^{208}Pb($\vec{d}$,p)^{209}Pb at
9 MeV. The solid curves
are DWBA calculations for
D_2=0.484 fm^2 which is the
value predicted by the Reid
soft-core deuteron wave
function[29]). The dotted
curves show the best fit
to the measurements which
was obtained with D_2 =
0.432 fm^2. The figure is
from ref. 28.

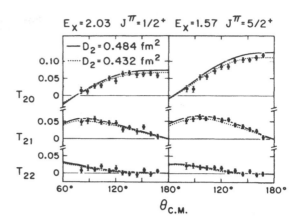

knowledge, the accuracy of the DWBA theory for sub-Coulomb reactions has
not been studied from a quantitative point of view. Such an investigation would present a challenging and interesting theoretical problem.

5. Recent developments

5.1. POLARIZATION TRANSFER

In one of the contributed papers to this symposium, Basak et al.[30])
have reported the measurement of a vector polarization transfer coefficient for the reaction ^{28}Si($\vec{d}$,p)^{29}Si. DWBA calculations show that the
deuteron D-state has a large effect on the polarization transfer, particularly at angles where the cross section minima occur. The measured
polarization transfer coefficients are in good agreement with the calculations which include the D-state. Further discussion of the D-state
effects on polarization transfer coefficients is given in a paper by
Santos[31]).

5.2. (d,t) REACTIONS

Recently it has been pointed out[21,32]) that D-state effects should
also be present in (d,t) reactions. As in the case of a (d,p) reaction,
the D-state effects should be small for the cross section and vector
analyzing power and should be relatively large for the tensor analyzing
powers.

In a (d,t) reaction, the deuteron plays the role of a spectator.
Consequently, the measurements are not sensitive to the internal structure of the deuteron, but rather, to the internal structure of the
triton. In particular, the tensor analyzing powers for ($\vec{d}$,t) reactions
arise primarily from the D-state components in the triton wave function.

In one of the contributed papers, Knutson and Hichwa[21]) have reported
a measurement of the tensor analyzing powers for ($\vec{d}$,t) reactions on
^{118}Sn and ^{208}Pb. The observed tensor analyzing powers for the ($\vec{d}$,t)
reactions are similar to typical ($\vec{d}$,p) tensor analyzing powers except
that the signs are reversed. Measurements for ^{118}Sn($\vec{d}$,t)^{117}Sn at E_d =
12 MeV are shown in fig. 9. The dashed curves are DWBA calculations in
which the D-state effects have been neglected. For the solid curves,
the D-state effects were included by making use of the LEA. A priori,

it is not clear what value of D_2 should be used for the (d,t) calculations. Knutson and Hichwa[21] have carried out several DWBA calculations in which D_2 is treated as an adjustable parameter. They find that the value $D_2 = -0.24$ fm^2 gives the best fit to the measurements, and this value was used for the calculations shown in fig. 9.

In one of the contributed papers, Barroso et al.[32] have reported a calculation of D_2 for (d,t) reactions. They make use of a simple triton wave function and obtain the value $D_2 = -0.20$ fm^2, which agrees rather well with the value deduced from the tensor analyzing power measurements. In my opinion, this is an extremely interesting result.

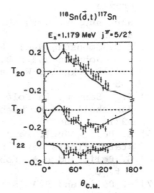

Fig. 9. Angular distributions of the tensor analyzing powers for ^{118}Sn($\vec{d}$,t)^{117}Sn at $E_d = 12$ MeV. The solid curves are DWBA calculations which include the D-state effects. For the dashed curves, the D-state effects were not included. The figure is from ref. 21.

6. Summary

Since the last polarization symposium our understanding of how the deuteron D-state affects (d,p) reactions has increased greatly. We have learned that the D-state has large effects on the tensor analyzing powers, and have studied these effects with some degree of thoroughness (although the existing data cover a very limited range of energies). However, the recent polarization transfer measurement by the Birmingham group[30] illustrates very nicely that there are still interesting things to be learned about the D-state effects.

It now appears that the tensor analyzing power measurements may be of some practical importance. For example, we have seen that T_{22} measurements can provide spectroscopic information about nuclear states. In addition, it may turn out that tensor analyzing power measurements for sub-Coulomb reactions can be used to determine the value of D_2. If this is to be done, we will need accurate new experiments at low energies, and on the theoretical side, we will need to advance our understanding of the DWBA theory and its limitations.

Little is known about the D-state effects in transfer reactions other than (d,p). The first results for (d,t) reactions have been reported at this symposium, and of course, there are many unanswered questions. For example, it is not clear whether tensor analyzing power measurements for ($\vec{d}$,t) reactions can provide information about the D-state components in the triton wave function which is not available from other sources. This point should be thoroughly investigated. Finally, the possibility that some sort of D-state effect may be present in more complex reactions [such as (d,α)] has not been studied at all.

The author would like to express his thanks to W. Haeberli and E.J. Stephenson for many stimulating discussions.

References

† Work supported in part by the U.S. Energy Research and Development Administration

1) R.C. Johnson, Nucl. Phys. A90 (1967) 289
2) R.C. Johnson and F.D. Santos, Particles and Nuclei 2 (1971) 285
3) P.J. Bjorkholm, W. Haeberli and B. Mayer, Phys. Rev. Lett. 22 (1969) 955
4) F.D. Santos, in Polarization phenomena in nuclear reactions, ed. H.H. Barschall and W. Haeberli (University of Wisconsin Press, Madison, 1971) p. 758
5) G. Delic, Nucl. Phys. A158 (1971) 117. In these calculations the D-state effects were underestimated as the result of a computing error
6) G. Delic and B.A. Robson, Nucl. Phys. A156 (1970) 97. See note in ref. 5
7) G. Delic and B.A. Robson, Nucl. Phys. A193 (1972) 510. See note in ref. 5
8) G. Delic and B.A. Robson, Nucl. Phys. A232 (1974) 493
9) W. Haeberli, in Polarization phenomena in nuclear reactions, ref. cit., p. 235
10) R.C. Brown, A.A. Debenham, G.W. Greenlees, J.A.R. Griffith, O. Karban, D.C. Kocher and S. Roman, Phys. Rev. Lett. 27 (1971) 1446
11) K.W. Corrigan, R.M. Prior, S.E. Darden and B.A. Robson, Nucl. Phys. A188 (1972) 164
12) R.F. Casten, E. Cosman, E.R. Flynn, O. Hansen, P.W. Keaton, N. Stein and R. Stock, Nucl. Phys. A202 (1973) 161
13) N. Rohrig and W. Haeberli, Nucl. Phys. A206 (1973) 225
14) R.C. Johnson, F.D. Santos, R.C. Brown, A.A. Debenham, G.W. Greenlees, J.A.R. Griffith, O. Karban, D.C. Kocher and S. Roman, Nucl. Phys. A208 (1973) 221
15) L.D. Knutson, E.J. Stephenson, N. Rohrig and W. Haeberli, Phys. Rev. Lett. 31 (1973) 392
16) H.O. Meyer and J.A. Thomson, Phys. Rev. C 8 (1973) 1215
17) L.D. Knutson, Ph.D. Thesis, University of Wisconsin (1973) available from University Microfilms, Ann Arbor, Michigan, and to be published
18) L.D. Knutson, J.A. Thomson and H.O. Meyer, Nucl. Phys. A241 (1975) 36
19) J.A. Aymar, M. Miki, A. Mora, S.E. Darden, H.R. Hiddleston and A.A. Rollefson, Bull. Am. Phys. Soc. 20 (1975) 694
20) E.J. Stephenson and W. Haeberli, Fourth polarization symposium, p.663
21) L.D. Knutson and B.P. Hichwa, Fourth polarization symposium, p.683
22) F.D. Santos, Fourth polarization symposium, p.675
23) G. Delic and B.A. Robson, Nucl. Phys. A127 (1969) 234
24) H.O. Meyer and W.A. Friedman, Phys. Lett. 45B (1973) 441
25) C.A. Pearson, D. Rickel and D. Zissermann, Nucl. Phys. A148 (1970) 273
26) L.J.B. Goldfarb, in Lectures in theoretical physics, ed. P.D. Kunz, D.A. Lind and W.E. Brittin (University of Colorado Press, Boulder, 1966) Vol. VIII C, p. 445
27) L.D. Knutson, E.J. Stephenson and W. Haeberli, Phys. Rev. Lett. 32 (1974) 690
28) L.D. Knutson and W. Haeberli, to be published
29) R.V. Reid, Ann. of Phys. 50 (1968) 411
30) A.K. Basak, J.A.R. Griffith, O. Karban, J.M. Nelson, S. Roman and G. Tungate, Fourth polarization symposium, p. 661
31) F.D. Santos, Nucl. Phys. A236 (1974) 90
32) A. Barroso, A.M. Eiro and F.D. Santos, Fourth polarization symp. p.685

DISCUSSION

Cramer:

Sub-Coulomb (d,p) reactions are currently regarded as the most promising way of obtaining relatively model-independent spectroscopic informations, particularly single-nucleon spectroscopic factors. Would you comment on the relevance of D-state effects and tensor analysing power determinations on the accurate extraction of spectroscopic factors ?

Knutson:

Cross section measurements for sub-Coulomb (d,p) reliably determine the asymptotic normalization of the bound neutron wave function. The tensor analysing powers are completely insensitive to the wave function normalization. Also I should point out that the D-state makes very little contribution to the unpolarized cross section for the sub-Coulomb case.

Clegg:

You suggested that the forward angle dependence of T_{22} showed reliable sensitivity to the angular momentum J of the transfered nucleon, being one sign for $J = \ell + \frac{1}{2}$ and the opposite for $J = \ell - \frac{1}{2}$. The cases you showed were for spin 0 targets. Do you know of any cases where this measured quantity has been used, for non-spin-zero targets when more than one J-transfer is possible, to establish the fraction of each J-transfer which is active in the reaction process?

Knutson:

This has not yet been done, but is certainly an interesting possibility. With the vector analysing power, one normally determines the J-mixing by using measurements of iT_{11} for pure J transitions on nearby nuclei. For T_{22} it would be better to make use of DWBA calculations, because T_{22} changes considerably as a function of Q-value, and because DWBA apparently reproduces T_{22} at forward angles with a high degree of reliability.

INTERACTIONS OF POLARIZED ^{3}He WITH NUCLEI

S. Roman

The University of Birmingham, England

1. INTRODUCTION

Three main areas of ^{3}He polarization studies are of considerable current interest: i) The elastic scattering, due to the information it provides concerning the spin dependence of the optical potentials. ii) Transfer reactions, where polarization measurements reveal the spin dependence of the reaction mechanisms. This may be utilized to derive unique spectroscopic information. iii) Few-nucleon systems studies. The latter area is not discussed here.

The progress of polarization measurements using unpolarized beams has been very limited due to the formidable difficulties of the double scattering experiments. Since the first acceleration of a polarized ^{3}He beam in 1974, using the recently developed ion source [1,2], a number of experimental results has become available. The observed magnitude of polarization effects has exceeded expectations not only for the elastic scattering but also for one-nucleon transfer reactions.

A review of the results obtained to date is given in table 1. The measurements at the top of the table have been obtained by double scattering techniques. The only useful angular distributions for the elastic scattering were obtained by McEver et al.[7] and Bond et al.[9]. Measurements of proton polarizations in the (^{3}He,p) two-nucleon transfer reactions on light targets are reported by the Laval group [11,12]. The list of results obtained with the polarized ^{3}He beam at Birmingham includes some preliminary data and the analyses are incomplete.

2. ELASTIC SCATTERING

The available evidence suggests that the elastic scattering of ^{3}He at energies above about 25 MeV is a phenomenon which can be described by a local optical potential [19,20] of a form similar to that which has proved successful in describing the nucleon-nucleus and deuteron-nucleus scattering. In the case of nucleons which have been studied most extensively, optical-model analyses have shown that it is not in general possible to determine even the central potential parameters uniquely from the differential cross-section data alone. In the case of ^{3}He, the interaction is mainly determined by partial waves of high l-values with very little contribution from the nuclear interior. Due to the surface character of the interaction, the ambiguities of the ^{3}He optical potential are even more pronounced than for nucleons. The additional constraint provided by the polarization data could facilitate resolution of these ambiguities.

The theoretical estimates of the spin-orbit potential for composite particles by Abul-Magd and El-Nadi[21] based on a development[22] of the Watanabe model[23] for deuterons, predict the ^{3}He spin-orbit potential to be

Table 1.   Polarization measurements in ^{3}He interactions

Reaction	E_x MeV	E_b MeV	Ref.	Remarks
^{12}C	elastic	31.5	3,4)	25°
^{12}C , ^{27}Al	elastic	30	5)	25°
^{12}C,	elastic	36	6)	24°
^{9}Be, ^{12}C, ^{27}Al, ^{47}Ti, Cu	elastic	31.6	7)	25°
^{9}Be, ^{12}C, ^{16}O	elastic	18 (20)	8)	
^{12}C	elastic	28	9)	
^{7}Li(^{3}He,$\vec{p}$)^{9}Be	g.s.	14	10)	
^{10}B(^{3}He,$\vec{p}$)^{12}C	4.43	10	11)	
^{12}C(^{3}He,$\vec{p}$)^{14}N	g.s.	2.4 - 3.6	12)	
^{9}Be($^3\vec{\mathrm{He}}$,^{3}He)	elastic		13)	*), **)
($^3\vec{\mathrm{He}}$,d)^{10}B	4 states		14)	
($^3\vec{\mathrm{He}}$,a)^{8}Be	16.9			
^{12}C($^3\vec{\mathrm{He}}$,^{3}He)	elastic		13)	
($^3\vec{\mathrm{He}}$,d)^{13}N	g.s.;(3.51+3.56)		14)	
($^3\vec{\mathrm{He}}$,a)^{11}C	g.s.;1.995			
^{26}Mg($^3\vec{\mathrm{He}}$,^{3}He)	elastic		15)	
($^3\vec{\mathrm{He}}$,a)^{25}Mg	3 states		15)	
^{27}Al($^3\vec{\mathrm{He}}$,^{3}He)	elastic		15)	
($^3\vec{\mathrm{He}}$,a)^{26}Al	2 states		15)	
^{28}Si($^3\vec{\mathrm{He}}$,^{3}He)	elastic		16)	
($^3\vec{\mathrm{He}}$,d)^{29}P	4 states		16)	
($^3\vec{\mathrm{He}}$,a)^{27}Si	g.s.		16)	
^{58}Ni($^3\vec{\mathrm{He}}$,^{3}He)	elastic		17)	
($^3\vec{\mathrm{He}}$,d)^{59}Cu	3 states		18)	
($^3\vec{\mathrm{He}}$,a)^{57}Ni	6 states		16)	

*) The measurements with polarized ^{3}He beam were obtained at E_b = 33.3 MeV.

**) Inelastic ^{3}He scattering data are available for most nuclei.

about 2 MeV, by folding the potentials of constituent nucleons. Johnson discussed the situation at the Third Polarization Symposium [24], pointing out that these folding models would suggest $P(^3He) \simeq 1/9\ P(nucleon)$. This result was even then known to be misleading, since it disagreed with the measurements of McEver et al. [8]. Before these measurements became available, earlier experimental attempts to evaluate the spin-orbit potential have not produced conclusive results.

At the same time extensive elastic scattering cross-section data for a wide range of targets have been accumulating and, although insensitive to the choice of the spin-orbit well depth, systematic optical model searches have yielded V_{so} values consistently larger than those predicted by the constituent folding model [21], e.g. refs 25-27. Equally, the cross section data were insensitive to the choice of the spin-orbit geometry which was kept equal to the geometry of the central potential. In contrast, the most obvious finding from the measurements for ⁹Be and ¹²C [13] was the extreme sensitivity of the optical potential to the spin-orbit geometry parameters.

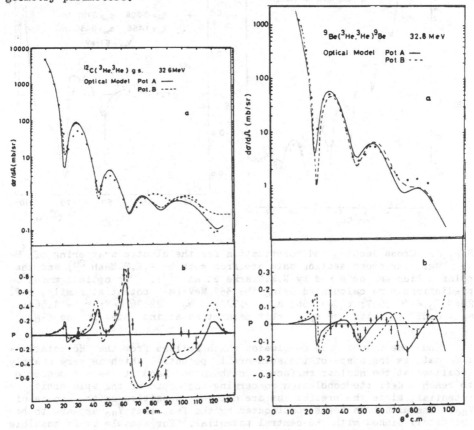

Fig. 1. Cross sections (a) and polarizations (b) for the elastic scattering of ³He by ⁹Be and ¹²C. The curves are optical model predictions. Work of ref. 13.

The subsequent measurements of polarization in the elastic scatter -
ing of ^{3}He by heavier nuclei have established that this sensitivity is
not a property peculiar to the very light targets. The conclusions concerning the ^{3}He spin-orbit term from preliminary optical model analyses
of the polarization data available to date for targets listed in table 1,
essentially confirm the findings of the analysis of ^{3}He scattering by
the light nuclei, ^{9}Be and ^{12}C ^{13}). The preferred spin-orbit radius is
one very nearly equal to the radius of the central potential with a diffuseness parameter very much smaller than the diffuseness of the central
terms.

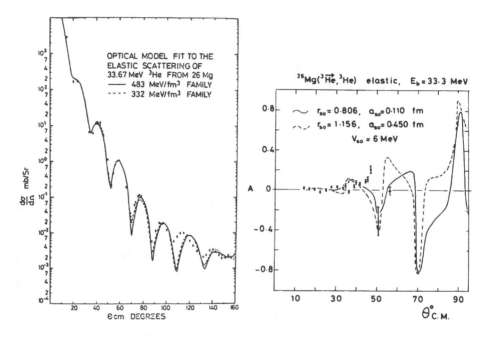

Fig. 2. Cross section and polarization for the elastic scattering of ^{3}He
by ^{26}Mg. The cross section data are from work by A.M.S.Meah ^{28}) and the
polarization was obtained by N.M.Clarke et al. ^{15}). The optical model
predictions were calculated using the 332 MeV·fm^3 potential family, ref.
28: V_O = 107.657; r_O =1.156; a_O = 0.722; W_V = 22.96 MeV; r_w = 1.564;
a_w = 0.872 fm, with the spin orbit parameters as indicated in the fig.

Thus one unambiguous conclusion which follows from the ^{3}He polarization data is the shape of the spin-orbit potential which is very sharply
localized at the nuclear surface. On the other hand it was not possible
to reach a definite conclusion concerning the depth of the spin-orbit
potential, since the predictions are not very sensitive to the choice of
V_{so}. The situation seems complicated by the fact that V_{so} appears to be
intimately linked with the central potential. For example it is possible
to obtain very different V_{so} values as a result of parameter searches
within different potential families for the same data set. Most of the

calculations, however, showed preference for a spin-orbit potential somewhat deeper than the prediction of the folding model [21]).

The calculations showed that polarization data help in selection of a suitable discrete potential family in the sense that some potential families are rejected when polarization data are included in the optical model analysis: notably the very deep potential families are excluded.[26]) e.g. the family represented by a volume integral J_R = 566 MeV·fm^3 for ^{58}Ni could not reproduce the polarization data.

Two features of the ^{3}He elastic scattering have been noted and are not well understood: The first one concerns large-angle cross-section data. Some angular distributions have been measured out to very large scattering angles ($\sim$ 170°) [26]) and the standard optical-model calculations have been unable to account fully for the data in this region. The situation becomes much worse as soon as an attempt is made to fit the cross-section and polarization data simultaneously. The limited searches carried out so far suggest that the cross-section data above about 140° ought to be excluded in order to reproduce the polarization.

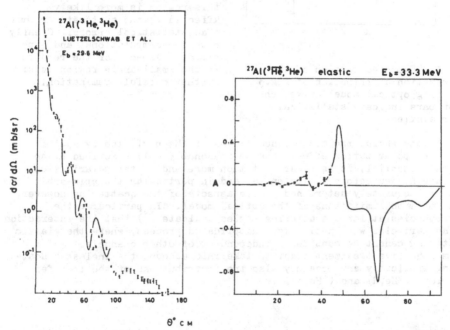

Fig. 3. Cross section and polarization for the elastic scattering of ^{3}He by ^{27}Al. The cross section data are from ref.29 and the polarization was obtained by N.M.Clarke et al. [15]). The curve is a result of a 5 parameter optical model search with the following parameters: V_O = 108.43; r_O =1.37 a_O = 0.645; W_D = 18.830; r_W = 1.0603; a_O = 0.980; V_{SO} = 2.497 MeV; r_{SO} = 1.37; a_{SO} = 0.2382 fm.

The second region causing some difficulties is at small scattering angles, where a small amplitude oscillation of the polarization may be

observed if the measurements are carried out to a sufficient accuracy.
This feature is clearly seen in fig. 4, where the forward angle portion

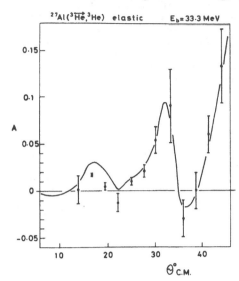

of the polarization distribution
for the elastic scattering of ^{3}He
by ^{27}Al of fig. 3 has been drawn
on an expanded scale. It is worth
noting that the polarization
curves resulting from optical model searches tend to miss the
small-angle oscillation, unless
the search is deliberately biased
to reproduce this region.

   At the small scattering
angles the Coulomb contribution
is exceedingly large and it is
possible that the interference
effects are not adequately accounted for by the calculations. On
the other hand the experimental
accuracy of the measurements in
this region is more likely
affected by the systematic rather
than statistical errors. Clearly
both the computational and experimental aspects of the results
in the small-angle region require
further careful examination.

Fig. 4. Expanded portion of the
polarization distribution of fig.3
including optical model curve. The
error bars indicate statistical
uncertainties.

   On the whole, the elastic scattering of ^{3}He particles by nuclei
including polarization effects can be reasonably well described by an
optical potential. It is clear that much more and better polarization
data are needed before the potential and in particular the spin-orbit
part, is accurately determined. Some aspects of the scattering however,
point to the to limitations of the optical-model. In particular the
discrepancies at large scattering angles indicate [13] that the interaction
of ^{3}He particles with nuclei is a complicated process, where the elastic
scattering cannot be considered independent of other channels.
A comprehensive treatment ought to take into account the inelastic scattering explicitly and possibly also the contributions of the transfer
reactions (^{3}He,d) and (^{3}He,α).

3. REACTIONS

   Investigation of polarization effects in nuclear reactions induced
by ^{3}He particles has two distinct objectives: Firstly to improve the
understanding of the reaction mechanisms leading to the development of
appropriate theories and models and secondly to apply the established
spin-sensitivity of the analysing power to nuclear spectroscopy.
   The distorted-wave theory, which with some refinements has been
successful in describing deuteron induced reactions, can be used for

describing reactions involving helium ions, provided that the conditions
for its applicability are satisfied. With ^{3}He projectiles however, this
is not always so, especially when a large momentum transfer takes place
or when one particle is much more strongly absorbed than the other.
These limitations are not well enough understood for reliable extraction
of spectroscopic factors from distorted-wave analyses.

The one-nucleon transfer reactions (^{3}He,d) and (^{3}He,α) have been
extensively exploited in spectroscopy since they lead to the same final
nuclei as (d,n) and (p,d) reactions, respectively. The use of ^{3}He particles is advantageous in the formation of excited states of the residual
nucleus. Because the kinetic energy of the projectile is associated with
its higher mass, the momentum transfer tends to be also large favouring
large orbital momentum transfers.

Austern [30]) discussed the conditions of applicability of the distorted-wave theory to these reactions, stressing the need to use the correct
^{3}He wave function and to include finite-range corrections. The theoretical aspects of spectroscopic applications of (^{3}He,d) reactions have been
studied by Bassel [31]), who made a comparison between the theoretically
derived form factors with experimental normalizations. Apart from the
normalization, the differential cross-sections show very little structure
and there is no significant j-dependence. Measurements of the analysing
powers, which are expected to be more sensitive to spin dependent
effects than are the usually featureless cross-sections, were
undertaken hoping that they may help to define the limits of
applicability of theory more clearly.

Consideration of the ^{3}He reactions in parallel with deuteron
reactions leads directly to an expectation of strong j-dependent
effects. Since the j-dependence of the analysing power of (d,p) and
(d,t) reactions is associated with the spin-orbit forces acting on a
nucleon in a bound state, it is reasonable to expect effects of
similar magnitude in other one-nucleon transfer, direct reactions.
Furthermore, the incident- and out-going channel distortions are of
secondary importance. This leads to expectation of strong j-dependent
effects in the analysing power of (^{3}He,d) and (^{3}He,α) reactions.

The measurements for the (^{3}He,d) reactions are first
considered. The results obtained for ^{12}C show large analysing powers.
The experimental results shown in figs. 5 and 6 are not in a good

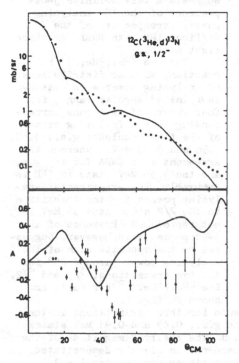

Fig. 5. Differential cross section
and analysing power of the reaction
^{12}C(^{3}He,d)^{13}N g.s., DWBA prediction
for j=$\frac{1}{2}$, finite range parameter
FR = 0.77.

agreement with distorted-wave calculations including finite range corrections, especially the ground state transition. For the transition in the
^{12}C(^{3}He,d)^{13}N reaction leading to the unresolved doublet at (3.51 + 3.56)
MeV shown in fig. 6, a comparison with a DWBA prediction for j = 3/2 ·
suggests. that the 3.51 MeV state makes the dominant contribution to the
doublet.

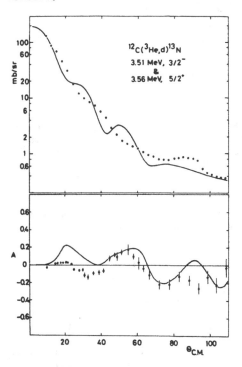

For the ^{9}Be(^{3}He,d)^{10}B reaction
^{3}He analysing powers were determined for the following states of
the final nucleus: g.s., 0.717,
1.74 and 2.15 MeV, discussed in
the contribution by Basak et al.
14).

The pattern of the oscillations of the analysing powers
for the (^{3}He,d) reactions enables
useful comparisons between distributions for transitions with
equal values of the transferred
orbital angular momentum l
(e.g. fig. 2, ref. 14). This
suggests a very definite j-dependence of the ^{3}He analysing
power, irrespective of the
difficulties with DWBA for these
light nuclei.

For the ^{28}Si(^{3}He,d)^{29}P
reaction, angular distributions
of analysing power were obtained
in a limited angular range, for
four groups of deuterons corresponding to the following states
of the final nucleus: g.s., 1.38,
1.96 and 3.45 MeV. Whereas the
agreement with DWBA for the g.s.
and the 1.38 MeV state in ^{29}P is
reasonable, the experimental analysing powers for the transition
to the 5/2 state at 1.96 MeV is
not reproduced. Presence of the
j- dependence is, however, suggested by the opposite sign of the
analysing power around 25° for
the two transitions, 3/2 and 5/2.
The ^{28}Si(^{3}He,d)^{29}P results are
shown in fig. 7.

Fig. 6. Differential cross section
and analysing power of the reaction
^{12}C(^{3}He,d)^{13}N, E_x = (3.51 + 3.56)
MeV, unresolved doublet. DWBA
prediction for j = 3/2.

The ^{3}He analysing power measurements for three transitions in the
^{58}Ni(^{3}He,d)^{59}Cu reaction leading to the g.s., 0.49 and 0.91 MeV states
of the final nucleus are shown in fig. 8. The observed magnitude of the
analysing power is large and the j-dependence is clearly demonstrated.
The j-dependent behaviour is especially striking for the two l = 1
transitions having the opposite j = l + ½ and j = l - ½ combinations 18).
since, due to the close proximity of these states, there is no Q-value
effect which would significantly shift the pattern of oscillations.

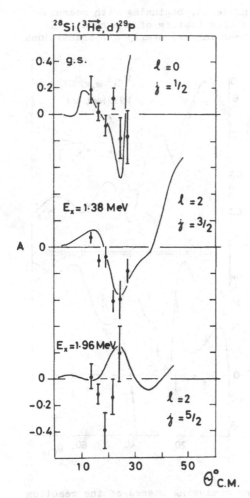

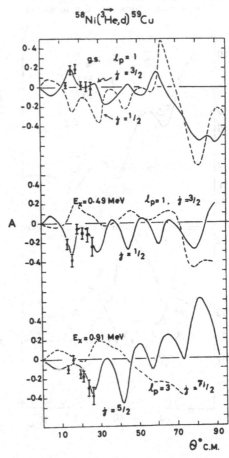

Fig. 7. Analysing power of the reaction ^{28}Si(^{3}He,d)^{29}P for the g.s., 1.38 and 1.96 MeV states of the final nucleus. The curves are DWBA predictions for the spin values indicated.

Fig. 8. Analysing power of the reaction ^{58}Ni(^{3}He,d)^{59}Cu for the g.s., 0.49 and 0.91 MeV states of the final nucleus. The curves are DWBA predictions in zero range for the spin values indicated.

This strong j-dependence of the ^{3}He analysing power is in contrast with the differential cross-section measurements, which have been obtained for a number of transitions in the ^{58}Ni(^{3}He,d)^{59}Cu reaction at various bombarding energies [32-34], showing very little structure. The zero range DWBA calculations are in a good agreement with the measurements of the ^{3}He analysing power for ^{58}Ni(^{3}He,d)^{59}Cu, accounting fully for the observed j-dependence.

The (³He,α) reactions are next considered, beginning with measurements on the light target nuclei. A striking feature of the results is
the large magnitude of the analysing power in very nearly all transitions.

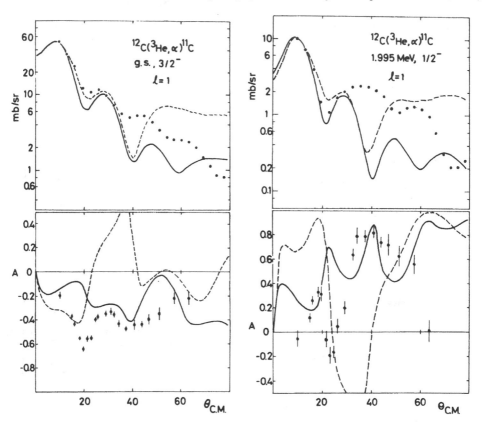

Fig. 9. Differential cross sections and analysing powers of the reaction
¹²C(³He,α)¹¹C for the g.s. and the 1.995 MeV excited state of the final
nucleus. DWBA predictions, including zero range (dashed lines) and finite
range (continuous line) calculations are shown.

The measurements for the ¹²C(³He,α)¹¹C reaction [14] are shown in
fig. 9. The j-dependent behaviour is very strong, not only in the first
peak of the analysing power around 20°, corresponding to a cross section
minimum, where the analysing powers are of an opposite sign - but also
in the remaining regions of the angular distribution. The DWBA calculations are not in agreement with the data, but it is very interesting to
note that the finite-range predictions are much better than the zero-
-range, confirming the suggestion made by Austern [30].
    The j-dependence for the l = 1 transitions established in the case
of the ¹²C(³He,α)¹¹C reaction, namely the fact that the analysing power
for the l = 1, j = 3/2 transition is predomonantly negative whereas

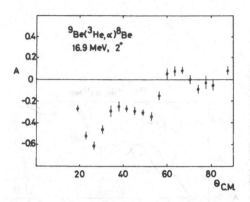

Fig. 10. Analysing power of the reaction ^{9}Be(^{3}He,α)^{8}Be to the 16.9 MeV state in ^{8}Be.

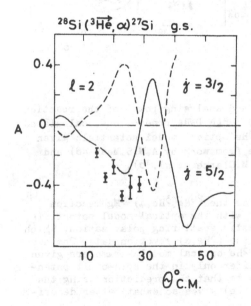

Fig. 11. Analysing power of the ^{28}Si(^{3}He,α)^{27}Si g.s. reaction. The curves are DWBA predictions for the spin values indicated.

the l = 1, j = ½ transition is positive, can be used to examine other l = 1 transitions in order to find the j-value even if the DWBA predictions are not reliable. Using this feature, it can be concluded that the 16.9 MeV, 2$^+$ state in ^{8}Be is populated predominantly by j = 3/2 transfer. This result is in agreement with the theoretical prediction of Cohen and Kurath [35] that the dominant configuration of this state in ^{8}Be is 1p$_{3/2}$. This has been achieved by comparing the analysing power of the ^{9}Be(^{3}He,α) ^{8}Be reaction shown in fig. 10, with that for the ^{12}C(^{3}He,α)^{11}C g.s. transition which is l = 1, j = 3/2, fig. 9. Since the two experimental distributions are virtually identical, it is concluded that the transferred spin values must be the same.

The large magnitude of the observed polarization effects in the (^{3}He,α) reactions on the light nuclei and the strong j-dependence suggested that the predominant mechanism of the interaction must be a simple one, occurring at the nuclear surface and with very little being contributed by the incoming particle distortion. This leads to an expectation of a better agreement with DWBA for reactions on heavier target nuclei, borne out by the results for the (^{3}He,α) reaction on ^{28}Si, ^{26}Mg and ^{58}Ni.

The measurements for the ^{28}Si(^{3}He,α)^{27}Si g.s. and the ^{26}Mg(^{3}He,α)^{25}Mg g.s. reactions are shown in figs. 11 and 12, respectively. Both are l = 2, j = 5/2 transitions and the agreement with DWBA is good. The results of the calculations for the l = 2, j = 3/2 combination are also presented, suggesting a very strong j-dependent effect.

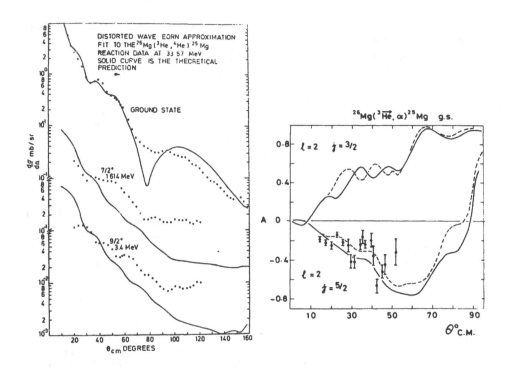

Fig. 12. Differential cross sections and analysing power of the reaction $^{26}Mg(^3He,\alpha)^{25}Mg$. The curves are zero range DWBA predictions, calculated for the spin values indicated, using ^3He optical model potentials given in fig. 2. The cross section data are from work by A.M.S.Meah 28) and the analysing power was obtained by N.M.Clarke et al.15).

The distorted-wave calculations for the $^{26}Mg(^3He,\alpha)^{25}Mg$ reaction shown in fig. 12 have been carried out with the optical model potentials used in the description of the ^3He elastic scattering polarization, which has been measured simultaneously with the $(^3He,\alpha)$ reaction data. The two sets of predictions correspond to the ^3He optical model parameters given in fig. 2. These two parameter sets differ only in the spin-orbit potential geometry. It is interesting to note that the prediction using the reduced difuseness parameter (dashed line) gives somewhat better description of the data.

Finally, the measurements of the analysing power for the $(^3He,\alpha)$ reaction on ^{58}Ni are presented in fig. 13. The ground state transition is l = 1, j = 3/2. Five other alpha particle groups have been observed, leading to the following states in ^{57}Ni: 0.76, 2.56, 3.24, 4.26 and 5.28 MeV. Except for the g.s. and the 0.76 MeV state all the remaining transitions are l =3, j = 7/2 of which only the state at E_x = 2.56 MeV is shown.

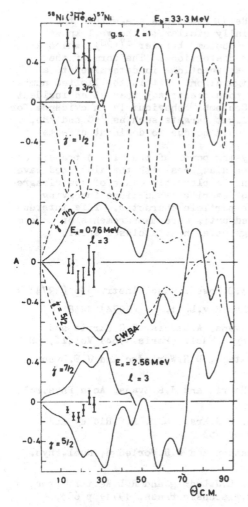

Fig. 13. Analysing power of the reaction ^{58}Ni(^{3}He,α)^{57}Ni leading to the g.s., 0.76 and 2.56 MeV states of the final nucleus. The curves are zero range DWBA predictions calculated for the spin values indicated. For the 0.76 MeV state results of calculation with no nuclear distortion (CWBA) are included.

Two striking features of the results for the ^{58}Ni(^{3}He,α)^{57}Ni reaction are apparent: The first one concerns the ground state l = 1, j = 3/2 transition. The observed analysing power is large and positive, in agreement with the DWBA prediction calculated for the appropriate relative orientation of the orbital and spin angular momenta. But the sign of the analysing power, both of the experimental points and the DWBA prediction, is opposite to the sign rule established for the l = 1 transitions for the light targets, fig. 9, i.e. for l = 1, j = 3/2 the analysing po-wer was predominantly negative. The second finding concerns the magnitude of the analysing power for the l = 3 transitions, which is much smaller than predicted. The assignments of these transitions are well known [36] (j = 5/2 for both the 0.76 and 2.56 MeV states) and the sign of the analysing power seems to be correctly predicted by the DWBA, but not the magnitude.

The calculations showed that $(^3He, \alpha)$ reactions are, within the distorted wave approximation, essentially similar to the well known deuteron stripping reactions near the Coulomb barrier [37,38], which are very little influenced by the nuclear distortions. The form of the analysing power distributions for these reactions is determined to a large extent by the bound state configuration. Within the limited number of cases studied to date, it is only possible to say that at an incident ^3He beam energy of 33 MeV this 'sub-Coulomb' behaviour is in evidence for ^{26}Mg and ^{28}Si, but not for the very light targets such as ^{12}C and ^9Be, where the reversal of the sign rule could be due to dominant effects of nuclear distortions.

The small magnitude of the analysing power of the $l = 3$ transitions in the $(^3He, \alpha)$ reactions on ^{58}Ni which disagrees with the distorted wave calculations, requires an explanation. A plausible cause of the disagreement could be the failure of DWBA to describe transitions involving states which are not of a good single-particle character. This interesting question cannot be fully answered until more measurements of $(^3He, \alpha)$ reactions become available, including nuclei near a closed shell.

References

1) W.E.Burcham, O.Karban, S.Oh and W.B.Powell, Nucl.Instr. **116** (1974) 1

2) O.Karban, S.Oh and W.B.Powell, Phys.Rev.Lett. **33** (1974) 1438

3) W.E.Burcham, J.B.A.England, J.E.Evans, A.Garcia, R.G.Harris and C.Wilne, Proc. Congres Int. de Phys. Nucl. (Paris, 1964) Vol.II, p877

4) J.B.A.England, R.G.Harris, L.H.Watson, D.H.Worledge and J.E.Evans, Phys.Lett. **30B** (1969) 476

5) W.E.Burcham, J.B.A.England, R.G.Harris and J.E.Evans, Acta Phys.Pol. **A38** (1970) 521

6) R.L.Hutson, S.Hayakawa, M.Chabre, J.J.Kraushaar, B.W.Ridley and E.T.Boschitz, Phys.Lett. **27B** (1968) 153

7) J.B.A.England, R.G.Harris, L.H.Watson and D.H.Worledge, Nucl.Phys. **A165** (1971) 277

8) W.S.McEver, T.B.Clegg, J.M.Joyce, E.J.Ludwig and R.L.Walter, Proc. 3rd Polarization Symposium (Univ.Wisconsin Press, 1971) p 603; Nucl.Phys. **A178** (1972) 529

9) C.D.Bond, L.S.August, P.Shapiro and W.I.McGarry, Nucl.Phys. **A241** (1975) 29

10) M.Irshad, J.Asai, S.Sen, R.Pigeon and R.J.Slobodrian, Proc. 4th Polarization Symposium (Zurich, 1975) p.705

11) C.R.Lamontagne, M.Irshad, R.Pigeon, R.J.Slobodrian and J.M.Nelson, Proc. 4th Polarization Symposium (Zurich, 1975) p. 707

12) H.Oehler, M.I.Krivopustov, H.I.Vibike, F.Asfour, I.V.Sizov and G.Schirmer, Proc. 3rd Polarization Symposium (Univ.Wisconsin Press, 1971) p 619

13) W.E.Burcham, J.B.A.England, R.G.Harris, O.Karban and S.Roman, Nucl.Phys. **A246** (1975) 269

14) A.K.Basak, J.B.A.England, O.Karban, G.C.Morrison, J.M.Nelson, S.Roman and G.G.Shute, Proc. 4th Polarization Symposium (Zurich, 1975)p. 711

15) N.M.Clarke et al., Kings College - Birmingham measurements, unpublished

16) Birmingham measurements, unpublished

17) S. Roman, A.K.Basak, J.B.A.England, O.Karban, G.C.Morrison, J.M.Nelson and G.G.Shute, Proc. 4th Polarization Symposium (Zurich, 1975), p.626

18) S.Roman, A.K.Basak, J.B.A.England, O.Karban, G.C.Morrison, J.M.Nelson and G.G.Shute, Proc. 4th Polarization Symposium (Zurich, 1975) p.709

19) P.E.Hodgson, Adv. in Physics 17 (1968) 563

20) P.E.Hodgson, Nuclear Reactions and Structure (Clarendon Press, Oxford, 1971) p 253

21) A.Y.Abul-Magd and M.El-Nadi, Progr.Theor.Phys. 35 (1966) 798

22) J.R.Rook, Nucl.Phys. 61 (1965) 219

23) S.Watanabe, Nucl.Phys. 8 (1958) 484

24) R.C.Johnson, Proc. 3rd Polarization Symposium (Univ.Wisconsin Press, 1971) p 603

25) C.B.Fulmer and J.C.Hafele, Phys.Rev. C7 (1973) 631

26) M.E.Cage, D.L.Clough, A.J.Cole, J.B.A.England, G.J.Pyle, P.M.Rolph, L.H.Watson and D.H.Worledge, Nucl.Phys. A183 (1972) 449

27) C.J.Marchese, R.J.Griffiths, N.M.Clarke, G.J.Pyle, G.T.A.Squier and M.E.Cage, Nucl.Phys. A191 (1972) 627

28) A.M.Shahabuddin Meah, Ph.D. Thesis, Univ. of London, 1973, unpublished

29) J.W.Luetzelschwab and J.C.Hafele, Phys.Rev. 180 (1969) 1023

30) N.Austern, Direct Nuclear Reaction Theories (Wiley, 1970) p 185

31) R.H.Bassel, Phys.Rev. 149 (1966) 791

32) G.C.Morrison and J.P.Schiffer, Isobaric Spin in Nucl.Phys., Eds. J.D.Fox and D.Robson (Acad.Press, N.Y., 1966) p 748

33) D.J.Pullen and B.Rosner, Phys.Rev. 170 (1968) 1034

34) M.E.Cage et al. Nucl.Phys. to be published

35) S.Cohen and D.Kurath, Nucl. Phys. A101 (1967) 1

36) D.E.Rundquist, M.K.Brussel and A.I.Yavin, Phys.Rev. 168 (1968) 1296

37) J.A.R.Griffith and S.Roman, Phys.Rev.Lett. 26 (1970)1496

38) A.A.Debenham, J.A.R.Griffith, M.Irshad and S.Roman, Nucl.Phys. A151 (1970) 81.

DISCUSSION

Cramer:
    There are well known ambiguities in the ^{3}He optical potential, such
that elastic cross section data can be fitted using 50 MeV deep real
central potentials or potentials hundreds of MeV deep. Therefore, in
considering the ^{3}He spin-orbit potential, the same ambiguities apply?
In other words, is the V_{SO} depth really the relevant parameter, or
should one perhaps look instead at the ratio $V_{SO}(r_c)/V_{central}\,(r_c)$,
where r_c is some critical radius like the turning point of the dominant partial wave?

Roman:
    The optical model calculations showed that the preference for a small
spin-orbit diffuseness parameter is irrespective of the discrete
potential family. The fits are less sensitive to the spin-orbit potential strength V_{SO}, but it is true that V_{SO} is intimately linked
with the central potential. However, more systematic data are needed
to study these ambiguities.

INELASTIC SCATTERING (COUPLED CHANNELS)

J. Raynal

Service de Physique Théorique
Centre d'Etudes Nucléaires de Saclay
BP N°2 - 91190 Gif-sur-Yvette
France

I.- INTRODUCTION

Dealing with polarization phenomena in coupled channel computations of inelastic scattering, we are mainly interested in the spin-orbit interaction. In fact, the subject of coupled channel calculations cannot be separated clearly from the DWBA because the latter is an approximation which shows very often the main results. However, some results cannot be obtained without coupled channels : the multistep inelastic scattering is a trivial example. Less expected are the possible effects of the quadrupole moment of the excited state on the analysing power of protons and of a strong coupling on the elastic polarization of deuterons at low energy. This last point appears in the communication to this conference by Basak et al, and, perhaps, by Baker et al. These effects are independent of the spin-orbit interaction.

To present the main successes (and failures) of the coupled channel calculations, we shall consider the macroscopic description of the nucleus by the rotational model which is a "generalized optical model" describing the target in its ground state and excited states by a deformed optical potential. Experimental evidences show that, strictly speaking, such a model does not exist because the spin-orbit deformation must depend on the nuclear structure of the excited state. This can be qualitatively understood in a microscopic description using the nucleon-nucleon interaction.

II.- EXPERIMENTAL RESULTS

One of the earliest experiments with a polarized beam of protons was performed in 1965 on ^{54}Fe at 18.6 MeV [1]. This energy was chosen because the cross-section was already known with precision [2]. Since that time, a large number of other experiments have been performed. However, the results obtained on ^{54}Fe exhibit the main features of the analysing power in inelastic scattering because they were obtained for the first 2^+ at 1.41 MeV and the second one at 2.96 MeV. The angular distributions of the cross-sections and the analysing powers are quite similar, but the absolute values of the analysing power are very large for the first 2^+ and definitely smaller for the second one, chiefly forwards.

Measurements done with targets ranging from Ti to Sn were classified by the experimentalists into two groups :

1) large analysing powers which are the first 2^+ of ^{54}Fe, ^{52}Cr, ^{50}Ti, ^{92}Mo, ^{90}Zr and ^{88}Sr,

2) small analysing powers : the first 2^+ of ^{64}Ni, ^{62}Ni, ^{60}Ni, ^{58}Ni and the second 2^+ of ^{54}Fe and ^{52}Cr.

Results for ^{56}Fe, ^{92}Zr and ^{94}Mo are intermediate. Note that the nuclei of which the first 2+ shows a large analysing power have 28 or 50 neutrons and an open shell of protons; on the contrary, a small analysing power is related to an open shell of neutrons. However, the division into two groups was essentially due to the difficulties encountered when trying to explain the experimental results of the first group [3,4,5]. Besides the striking differences between the analysing powers on ^{90}Zr and ^{92}Zr or the two 2+ of ^{54}Fe, the details of the angular distribution must be explained. This was achieved by Blair and Sherif [6,9] using DWBA and we shall extend their calculation to any spin.

III.- GENERALIZED OPTICAL MODEL AND COUPLED CHANNELS

In the rotational model, the interaction between the particle and the target is an optical potential [10].

$$V(r) = - Vf(x_r) - iW_v f(x_v) + 4iW_s \frac{d}{dx_s} f(x_s)$$

$$+ \left(\frac{\hbar}{m_\pi c}\right)^2 V_{ls} \frac{1}{r} \frac{d}{dr} f(x_{ls}) (2 \vec{l} \ \vec{s}) + V_{coulomb} \tag{1}$$

where the Wood Saxon form factors $f(x)$ are :

$$f(x_j) = \left[1 + \exp\left(\frac{r - R_j A^{1/3}}{a_j}\right) \right]^{-1} \tag{2}$$

and the reduced radii R_j depend on the angle with the intrinsic axis of the nucleus $\hat{r}'$ and is parametrized by quadrupole and hexadecupole deformations β_2 and β_4 :

$$R_j(\Theta) = R_j^o \left(1 + \beta_2 Y_2^o(\Theta) + \beta_4 Y_4^o(\Theta)\right) \tag{3}$$

The central part of the potential (1) can be expanded into multipoles

$$V(r, \hat{r}') = 4\pi \sum_\lambda V_\lambda(r) \ Y_\lambda^\mu(\hat{r}) \ Y_\lambda^{\mu *}(\hat{r}') \tag{4}$$

by numerical integration on ϵ in the rotational model (in the vibrational model, one uses a Taylor expansion with respect to the deformations which are phonon creation and annihilation operators).

A state of the target, member of a rotational band starting with a 0^+, is described by :

$$|\psi_{IM}> = \sqrt{\frac{2I+1}{8\pi^2}} \ R_{Mo}^{(I)*}(\Omega) \chi(r')$$

where $\chi(r')$ is the intrinsic wave function. All the radial functions y_i related to an ingoing or outgoing wave $|\ell_i j_i m_i>$ which can be coupled with the target to the same J,M lead to a set of coupled equations

$$y_i'' + \sum_j V_{ij} y_j = E_i y_i$$

with

$$V_{ij} = \sum_\lambda G_{ij}^\lambda V_\lambda(r) \tag{5}$$

The geometrical coefficients

$$G^{\lambda}_{i\mathfrak{f}} = (-)^{J+\lambda+s} (2s+1) \sqrt{(2I_i+1)(2I_f+1)(2\ell_i+1)(2\ell_f+1)(2j_i+1)(2j_f+1)}$$

$$\begin{pmatrix} \ell_i & \ell_f & \lambda \\ o & o & o \end{pmatrix} \begin{Bmatrix} j_i & j_f & \lambda \\ \ell_f & \ell_i & s \end{Bmatrix} \begin{Bmatrix} j_i & j_f & \lambda \\ I_f & I_i & J \end{Bmatrix} \begin{pmatrix} I_i & I_f & \lambda \\ o & o & o \end{pmatrix} \tag{6}$$

includes a nuclear matrix element which is the last 3-j symbol. The monopole
part (λ=o) is the optical model.

For a 0^+-2^+ excitation, there are 6 coupled equations : one for 0^+
and 5 for the 2^+. There are only two nuclear matrix elements : between 0^+
and 2^+, and between two 2^+ functions. They are related respectively to E2
transition probability and quadrupole moment of the excited state. DWBA
takes into account only the first of them. The sign of the quadrupole moment
is related to the sign of the 2^+-2^+ nuclear matrix element and will affect
the results of the calculation. In the vibrational model the 2^+-2^+ nuclear
matrix element vanishes and the sign of the 0^+-2^+ nuclear matrix element
does not change the results.

IV.- SPIN-ORBIT DEFORMATION

The optical model (1) includes a spin-orbit potential

$$\frac{1}{r} \frac{d}{dr} \left\{ V(r) \right\} \quad (2\vec{L}.\vec{s}) \tag{7}$$

where $V(r)$ is a potential of which the form factor is similar to the one
of the real potential. For a state with given ℓ and j values, the operator
$(2\vec{L}.\vec{s})$ has the eigenvalue which we shall note by γ. For a spin 1/2.

$$\begin{aligned} \text{if} \quad j &= \ell+1/2 \qquad \gamma = \ell \\ \text{if} \quad j &= \ell-1/2 \qquad \gamma = -\ell-1 \end{aligned} \tag{8}$$

For a spin 1 :

$$\begin{aligned} \text{if} \quad j &= \ell+1 \qquad = 2\ell \\ \text{if} \quad j &= \ell \qquad = -2 \\ \text{if} \quad j &= \ell-1 \qquad = -2\ell-2 \end{aligned} \tag{9}$$

In the first analyses, people wanted to deform the spin-orbit interaction as they had already done for the real and the imaginary potential. But
the coupling potentials (5) are not symmetric unless the spin-orbit potential
is written as

$$\frac{1}{2} \left[\frac{1}{r} \frac{d}{dr} \{V(\vec{r})\} (\vec{L}.\vec{\sigma}) + (\vec{L}.\vec{\sigma}) \frac{1}{r} \frac{d}{dr} \{V(r)\} \right] = \frac{1}{r} \frac{d}{dr} \{V(r)\} \frac{\gamma_i + \gamma_f}{2} \tag{10}$$

With this prescription, various form factors [4] were tried. However Blair
and Sherif obtained very good results [6-9] in DWBA computations using :

$$\vec{\nabla} \{V(\vec{r})\} \wedge \frac{\vec{\nabla}}{i} \vec{\sigma} \tag{11}$$

which reduced to (7) when $V(\vec{r})$ is isotropic. This expression can be used
for any spin because it is conserved by convolution with the intrinsic wave
function of a composite particle if there are only S states in this intrinsic
wave function. For example, the deuteron D state introduces correction terms
but, when only S states are taken into account, $V(\vec{r})$ is replaced by the
result of the convolution in (11).

Let us consider a multipole $V_\lambda(r) Y_\lambda^\mu(r)$ of the form factor in (11) and $2 \vec{s} = \sum_{n=1}^{2|s|} \vec{\sigma}_n$, with $2|s|$ Pauli matrices coupled to maximum angular momentum. Performing on each matrix the transformations used [11-12] for spin 1/2, we get

$$2 \left[\vec{\nabla} V_\lambda(r) Y_\lambda^\mu(r) \right] \wedge \frac{\vec{\nabla}}{i} \vec{s} = \sum_{n=1}^{2|s|} - \frac{V_\lambda(r)}{r^2} \left[\vec{L} \, Y_\lambda^\mu(r) \cdot \vec{L} \right]$$

$$+ \frac{1}{r} \left[\frac{d}{dr} V_\lambda(r) \right] Y_\lambda^\mu(r) (\vec{L} \cdot \vec{\sigma}_n) - \frac{V_\lambda(r)}{r} \vec{\sigma}_n \cdot \left[\vec{L} \, Y_\lambda^\mu(r) \right] \frac{1}{r} \frac{d}{dr}$$

$$+ \frac{V_\lambda(r)}{r^2} Y_\lambda^\mu(r) (\vec{L} \cdot \vec{\sigma}_n) + \frac{V_\lambda(r)}{r^2} (\vec{L} \cdot \vec{\sigma}_n) Y_\lambda^\mu(r) (\vec{L} \cdot \vec{\sigma}_n) \qquad (12)$$

The summation on **n** is straightforward, except for the last term for which

$$\sum_{n=1}^{2|s|} < \ell_f \, s \, j_f \mid (\vec{L} \cdot \vec{\sigma}_n) \, Y_\lambda^\mu(r) \, (\vec{L} \, \vec{\sigma}_n) \mid \ell_i \, s \, j_i >$$

$$= \sum_{n=1}^{2|s|} \sum_{s'} < \ell_f \, s \, j_f \mid (\vec{L} \cdot \vec{\sigma}_n) \mid \ell_f s' \, j_f > < \ell_f \, s' \, j_f \mid Y_\lambda^\mu(r) \mid \ell_i \, s' \, j_i >$$

$$< \ell_i \, s' \, j_i \mid (\vec{L} \cdot \vec{\sigma}_n) \mid \ell_i \, s \, j_i >$$

where summation on s' includes only s'=s and $\bar{s}'$=s-1. For $\bar{s}'$=s

$$< \ell \, s \, j \mid \vec{L} \cdot \vec{\sigma} \mid \ell \, s \, j > = \frac{1}{2|s|} < \ell \, s \, j \mid 2 \, \vec{L} \cdot \vec{s} \mid \ell \, s \, j >$$

which can be expressed with the eigenvalues of $2\vec{L}\vec{s}$ and introduces no modification in the geometrical coefficient (6). On the contrary, for s'=s-1,

$$f_{\ell s j} = < \ell \, s \, j \mid \vec{L} \, \vec{\sigma} \mid \ell \, s \, -1 \, 1 > = - \frac{\{ [\ell(\ell+1) - (j-s)(j-s+1)] [(j+s)(j+s+1) - \ell(\ell+1)] \}^{1/2}}{2|s|}$$

$$\tag{13}$$

introduces a 6j symbol with s-1 instead of s.

Gathering all the terms

$$2 \vec{\nabla} V_\lambda(r) \, Y_\lambda^\mu(r) \wedge \frac{\vec{\nabla}}{i} \vec{s} \to \frac{1}{r} \frac{d}{dr} V_\lambda(r) \, \gamma_i + (\gamma_i - \gamma_f) \frac{V_\lambda(r)}{r} \frac{d}{dr}$$

$$+ \frac{V_\lambda(r)}{r^2} \left\{ |s| \left[\lambda(\lambda+1) - \ell_i(\ell_i+1) - \ell_f(\ell_f+1) \right] + \gamma_f + \frac{\gamma_i \gamma_f}{2|s|} + F \right\} \qquad (14)$$

with

$$F = - 2|s| f_{\ell_i s \, j_i} f_{\ell_f s j_f} \begin{Bmatrix} \ell_i & j_i & s-1 \\ j_f & \ell_f & \lambda \end{Bmatrix} \begin{Bmatrix} \ell_i & j_i & s \\ j_f & \ell_f & \lambda \end{Bmatrix}^{-1}$$

where → indicates that this expression must replace $V_\lambda(r)$ in (5). For s=1

$$F = \frac{\ell_i \ell_f (\ell_i+1)(\ell_f+1)}{\ell_i(\ell_i+1)+\ell_f(\ell_f+1)-\lambda(\lambda+1)} \delta_{\ell_i j_i} \delta_{\ell_f j_f}$$

and vanishes for $s=\frac{1}{2}$. In this last case :

$$\vec{\nabla} \, V_\lambda(r) Y_\lambda^\mu(r) \wedge \frac{\vec{\nabla}}{i} \vec{\sigma} \to \frac{1}{r}\frac{d}{dr} V_\lambda(r)\gamma_i + (\gamma_i-\gamma_f)\frac{V_\lambda(r)}{r^2}\frac{d}{dr}$$

$$+ \frac{V_\lambda(r)}{2r^2}\left\{ \lambda(\lambda+1) - (\gamma_f-\gamma_i)(\gamma_f-\gamma_i+1)\right\} \tag{15}$$

In (14) and (15) the radial wave function is assumed multiplied by r.

V.- DISCUSSION OF THE EFFECTS OF SPIN-ORBIT DEFORMATION

The interaction (15) is more difficult to handle than the interaction (10). In particular, it introduces first derivatives in the coupled equations, and numerical methods are less efficient in principle . However, these terms have been introduced in the program ECIS which uses an iteration procedure (Equations Couplées en Iterations Sequentielles) of which results are presented by different authors to this conference. A parametrization of the deformed spin-orbit allows direct comparisons of the interactions (10) and (15).

At first look, (10) is linear in $\gamma_i+\gamma_f$ and (15) quadratic in $(\gamma_i-\gamma_f)$. For large angular momentum ℓ, γ_i and γ_f are of the order of ℓ, and have the same sign if $j_i-\ell_i=j_f-\ell_f$ and opposite signs if $j_i-\ell_i = \ell_f-j_f$. However, for $s = 1/2$ the geometrical factor

$$\bar{G}_{if}^\lambda = \sqrt{(2j_i+1)(2j_f+1)(2\ell_i+1)(2\ell_f+1)} \begin{pmatrix} \ell_i & \ell_f & \lambda \\ o & o & o \end{pmatrix} \begin{Bmatrix} j_i & j_f & J \\ \ell_j & \ell_i & 1/2 \end{Bmatrix} \tag{16}$$

$$= -\sqrt{(2j_i+1)(2j_f+1)} \begin{pmatrix} j_f & \lambda & j_i \\ -1/2 & o & 1/2 \end{pmatrix}$$

which appears in the equations, is of the order of $\sqrt{\ell}$ when $j_f+j_i+\lambda$ is odd and $1/\sqrt{\ell}$ when $j_f+j_i+\lambda$ is even. So, for large values of ℓ and with respect to the deformed central potential for $j_i-\ell_i=j_f-\ell_f$ as a unit :

- the deformed central potential is of the order $1/\ell$ when $j_i-\ell_i \neq j_f-\ell_f$

- the deformed spin-orbit potential (10) is of the order ℓ when $j_i-\ell_i = j_f-\ell_f$ and $1/\ell$ in the other case

- the deformed spin-orbit potential (15) is of the order ℓ when $j_i-\ell_i \neq j_f-\ell_f$ and remains constant in the other case.

The behaviour of expression (15) is that of a vector : the difference $\gamma_i-\gamma_f$ is also found to be the ratio of the geometries for a transfer of spin and a scalar interaction:

$$\begin{Bmatrix} \ell_i & \ell_f & J \\ 1/2 & 1/2 & 1 \\ j_i & j_f & J \end{Bmatrix} = \frac{\gamma_i - \gamma_f}{\sqrt{3J(J+1)}} \begin{Bmatrix} \ell_i & \ell_f & J \\ 1/2 & 1/2 & 0 \\ j_i & j_f & J \end{Bmatrix}$$

However, the transfer of spin does not increase linearly with the angular momentum as the expression (15) does.

These two deformed spin-orbit potential are expected to give different results, mainly in the angular region where coherence effects can occur, namely the forward direction.

However, results of numerical calculations are not as clear as one would expect from the theory. A first example is shown on Fig. 1 : the dashed curve is the analysing power of 18.6 MeV protons on the 2^+ at 1.17 MeV of ^{62}Ni without spin-orbit deformation; the continuous curve is obtained with spin-orbit deformation and dotted curve with the interaction (10) multiplied by 4. In this case, deformed spin-orbit does not seem to be necessary, or, at least its deformation must be smaller than the deformation of the central potential. Interaction (10) gives very similar results when multiplied by some enhancement factor.

Figure 2 and 3 show that an enhancement factor 2 to 3 is needed to fit the first 2^+ of ^{54}Fe but the effect of the spin-orbit deformation must be damped for the second 2^+. But Figure 4 shows that the enhancement of the spin-orbit deformation does not reproduce the experimental data with another set of optical model parameters. These curves [12] were obtained with parameters used in ref. 3) where the second set of optical parameters gave large values of the analysing power when used with the interaction (10) multiplied by 4. Necessity of enhancement for the spin-orbit deformation is also seen directly on the experimental results for the first 2^+ of ^{90}Zr and ^{92}Zr which are very similar but with smaller absolute values for ^{92}Zr. So each level must be defined by a central deformation and an inde-pendant spin-orbit deformation and optical model parameters play an impor-tant role and can give quite misleading results.

Since the Madison conférence, coupled channel calculations on inelastic proton scattering involving analysing powers are not very numerous. Except for special topics which will be explained below, they are chiefly limited to the s-d shell [13-14]. Among the communication to this conference, there is only the "study of ^{112}Cd by inelastic scattering of 30 MeV polarized protons" by R. de Swiniarski et al which includes coupled channels effects.

VI.- MICROSCOPIC JUSTIFICATION OF SPIN-ORBIT DEFORMATION

The two body spin-orbit interaction behaves like the interaction (15) with respect to the incident and outgoing particles. All the nucleon-nucleon interactions agree for a smaller range for the spin-orbit interaction than for the central potential. The zero-range limit is

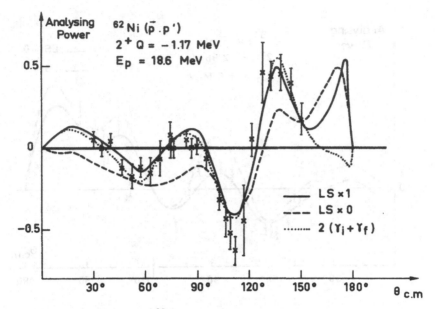

Fig.1. Analysing power for ^{62}Ni($\vec{p}$,p') (vibrational model)

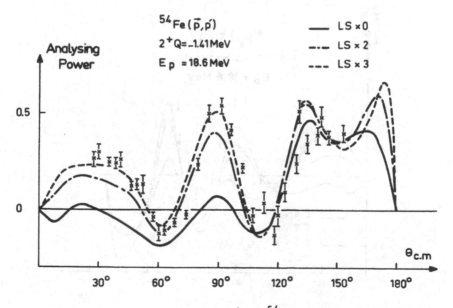

Fig.2. Analysing power for the first 2^+ of ^{54}Fe (vibrational model)

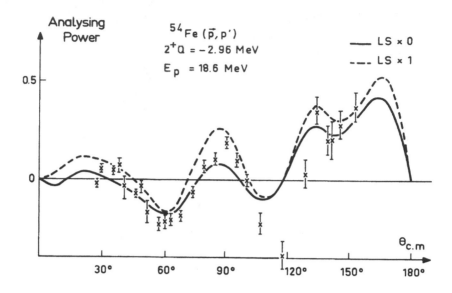

Fig.3. Analysing power for the second 2$^+$ of ^{54}Fe (vibrational model)

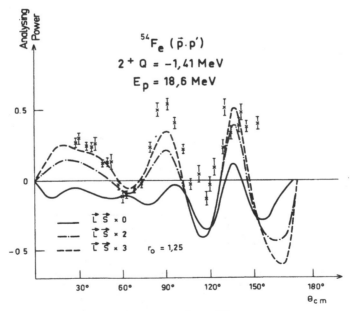

Fig.4. Analysing power for the first 2$^+$ of ^{54}Fe with an other optical
       potential

$$\frac{1}{r^2} \, \overline{G}_{if}^{\lambda} \, \overline{G}_{ph}^{\lambda} \left\{ \left[\lambda(\lambda+1)(\alpha_{if}^{\lambda^2} + \alpha_{hp}^{\lambda^2} - 2) - 1 \right] V_{\lambda}(r) + \right.$$

$$\left[1-(-)^{\ell_p+\ell_h+\lambda} \, \alpha_{if}^{\lambda}\alpha_{hp}^{\lambda} \right] \left[-(\gamma_i+\gamma_f+1)V_{\lambda}(r) + (\gamma_p+\gamma_h \right.$$

$$\left. -\gamma_i-\gamma_f)r\frac{d}{dr}\{V_{\lambda}(r)\} + \left[1+(-)^{\ell_p+\ell_h+\lambda} \right] \left[(\gamma_i+\gamma_p+1)V_{\lambda}(r) \right.$$

$$\left. \left. +(\gamma_h-\gamma_p-\gamma_i+\gamma_f) \left(V_{\lambda}(r)r\frac{d}{dr} +r \, \phi_h \frac{d}{dr} \, (\phi_p) \right) \right] \right\} \quad \quad (17)$$

for a particle hole excitation. Here $\overline{G}_{ph}^{\lambda}$ is the geometrical coefficient (16) for the bound particles; $\alpha_{ij}^{\lambda}.= (\gamma-\gamma')/\sqrt{J(J+1)}$ for a natural parity state and $\alpha_{ij}^{\lambda}.=(\gamma+\gamma'+2)/\sqrt{J(J+1)}$ for an unnatural one; $V_{\lambda}(r)$ is the product of particle and hole functions, that is the transition density. This matrix element is antisymmetric: permutation of i and h and summation on λ with the adequate 6-j symbol gives only a change of sign. This interaction is T=1: when a proton configuration is excited by the incident proton, its value is twice the one for a neutron configuration, whereas the ratio for the central interaction are inverse.

If the central interaction is written as $V_o+V_1\vec{\sigma}_1.\vec{\sigma}_2$ and the zero range limit is used, V_1 introduces terms similar to the spin-orbit interaction for a single particle-hole excitation, but vanishes for a recoupling in the same shell. Fig. 5 shows the important difference of the spin-orbit interaction when the first 2^+ of ^{54}Fe is assumed to be $(f_{7/2})^2$ or $2p_{3/2}-f_{7/2}^{-1}$ (central interaction : $-10(1-\vec{\sigma}_1 \vec{\sigma}_2)$ with an Yukawa form-factor of range 1.4 fm).

However, microscopic calculations fail to reproduce experimental results. Communication by Escudie et al shows better results without antisymmetrization, which is an incorrect calculation because the nucleon-nucleon interaction parameters are what they claim to be only when exchange is taken into account. Furthermore, the nucleon-nucleon spin-orbit interaction is too small with respect to nuclear spin-orbit potential or other phenomenological parameters as the Skyrme force which uses [15] exactly the expression (17): in the same units, the spin-orbit strength of Escudie et al is about 30, the one used by D.L. Pham et al is 25 instead of an usual value of 100 MeV. This discrepancy was not found in earlier calculations[12] due to an error by a factor 4.

The macroscopic limit of the expression (17), obtained by assuming mean values 1 for α_{ph}^{λ} and 0 for γ_p,γ_h and $\phi_h \frac{d}{dr}(\phi_p)$ and taking hermitian part of the whole, is exactly the interaction (15). One can understand qualitatively that the spin-orbit deformation must be modulated from less than the central deformation (for Ni) to up to 3 times (for ^{54}Fe or ^{90}Zr). In fact, the best values are often 50% larger than the central deformation, because many measurements are done in the s-d shell were there can be recoupling in the same shells.

Attemps has been made to explain this increase of spin-orbit deformation on the basis of the folding of a short range nucleon-nucleon interaction[16]. This doesnot explain the modulation with nuclear structure. However, a communication to this conference by R. de Swiniarski et al analyses new measurements on ^{90}Zr at 30 MeV with $\beta_{LS}=1.5$ β instead of $\beta_{LS}=3$ β which was needed at 20 MeV. Unhappilly these authors do not show us if the same experimental relation found at 20 MeV between ^{90}Zr and ^{92}Zr remains at 30 MeV, as it was already known [17] for ^{54}Fe and ^{56}Fe.

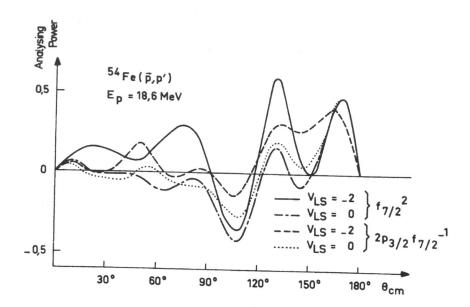

Fig.5. Results obtained with a single configuration in a microscopic
model for ^{54}Fe(p,p')

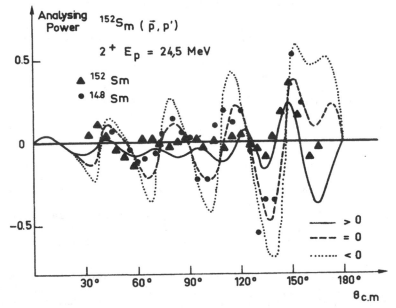

Fig.6. Effect of the quadrupole moment of the 2^+ in ^{148,152}Sm ($\vec{p}$,p')

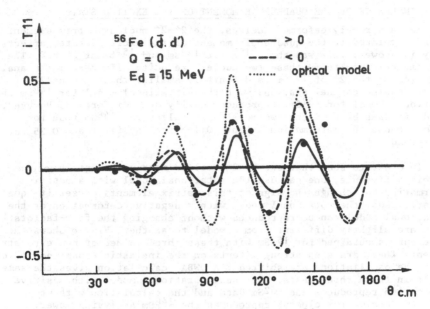

Fig.7. Effect of the quadrupole moment of the 2^+ in ^{56}Fe $(\vec{d},d)$

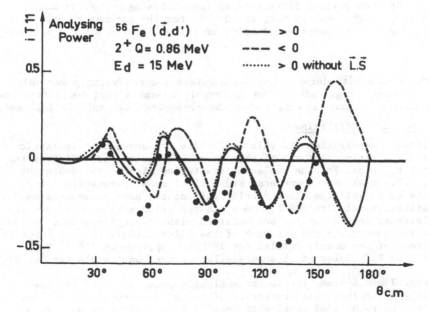

Fig.8. Effect of the quadrupole moment of the 2^+ in ^{56}Fe $(\vec{d},d')$

VII.- INFLUENCE OF THE QUADRUPOLE MOMENT OF THE EXCITED STATE

For a strongly deformed nucleus, the 2^+-2^+ nuclear matrix element, which is related to the quadrupole moment of the excited state, affects the analysing power. This was shown [18] for ^{148}Sm and ^{152}Sm at 25 MeV. The experimental results for these two nuclei are quite different : the analysing pwer on the 2^+ of ^{152}Sm is definitely smaller than for the 2^+ of ^{148}Sm. Coupled channel analyses with the vibrational model for ^{148}Sm and rotational model for ^{152}Sm reproduce easily this difference. However, the difference between the two models is smaller for ^{148}Sm than for ^{152}Sm because the deformation is only 0.115 for the first and 0.25 for the second.

In order to show that the 2^+-2^+ nuclear matrix element is responsible for this difference, one can use the rotational model with a positive deformation but change the sign of this matrix elements or set it equal to zero. Thus, the rotational model with a negative deformation or the vibrational model can be reproduced without changing the form-factors which are slightly different from a model to another. Fig. 6 shows the three curves obtained for ^{152}Sm with these three values of nuclear matrix element. There are also strong effects on the inelastic cross-section of which the oscillations are shifted . A DWBA calculation gives the same result in these three cases. The usual rotational model with positive deformation reproduces the ^{152}Sm data and the calculation without 2^+-2^+ nuclear matrix element reproduces the ^{148}Sm analysing power.

This effect is present only for a large deformation. It is quite general as it can be observed in the s-d shell in the same energy region, but without such evident difference on the analysing power. It must disappear at higher energy but, at 40 MeV, for the parameters used here for Sm, there is an important damping of the oscillations for a positive deformation.

The spin orbit deformation can modulate the analysing power only when it assumes large values. The quadrupole moment effect comes from the real potential because it exists when the real potential only is deformed .

VIII.- DEUTERON SCATTERING

Three communications to this conference are devoted to inelastic deuteron scattering, but two of them at such a low energy that they run into new troubles. For the inelastic scattering of 12,4 MeV deuterons on Cd, A.K. Basak and coworkers used optical model parameters which describe quite well the elastic polarization; the same parameters used in a rotational model with β=0.2 gives an elastic polarization with very weak oscillations, inverse of the one obtained without coupling. This is due to the low energy and the strength of the deformation without a critical importance of the model. Results for 15 MeV deuterons on ^{148,152}Sm reported by F.J. Baker et al show similar effects on the elastic curve.

Fig. 7 and 8 show the vector analysing power obtained for the elastic and the inelastic scattering of 15 MeV deuterons [19] on ^{56}Fe, assuming the rotational model with β=.225. As already done on Fig. 6, the 2^+-2^+ nuclear matrix element was also inversed (dashed line). Fig. 8 shows the small influence of deformed spin-orbit in this case and Fig. 7

the tremendous effect of the coupled channels in the elastic scattering.
No conclusion can be drawn from these calculations : if these nuclei are
so deformed, the optical model parameters which fitted so well the elastic
scattering have no meaning and should be rejected. But we do not know
where to find good optical model parameters if they exist.

IX.- CONCLUSION

The microscopic analyses of the inelastic scattering do not succeed
yet to reproduce analysing powers but macroscopic model succeeds quite
easily. The spin-orbit interaction must be deformed, more or less, to
describe data. Even the n $.C^{12}$ calculation reported by R.Casparis
et al would be improved with deformed spin-orbit. Besides the problems
relevant to coupled-channels as scattering on a rotational band or two-
phonons excitation, the coupled channel calculation showed unexpected
effects on the analysing power for which the spin-orbit deformation does
not matter. The effect of quadrupole moment suggested new experiments,
in particular with low energy deuteron scattering, which led to problems
not yet solved.

REFERENCES

1) A. Garin, C. Glasshauser, A. Papineau, R. de Swiniarski and J. Thirion -
   Physics Letters 21 (1966) 73.
2) S.F. Eccles, H.F. Lutz and V.A. Madsen - Phys. Rev. 141 (1966) 1067
3) C. Glasshauser, R. de Swiniarski and J. Thirion - Phys. Rev. 164
   (1967) 1437.
4) C. Glasshauser and J. Thirion - Advances in Nuclear Physics Vol 2 -
   Plenum Press New-York (1969) 79
5) C. Glasshauser, R. de Swiniarski et al - Phys. Rev. 184 (1969) 1217
6) H. Sherif and J.S. Blair - Physics Letters 26B (1968) 489
7) H. Sherif - Spin dependent effects in proton inelastic scattering -
   Thesis University of Washington (1968)
8) H. Sherif and R. de Swiniarski - Physics Letters 28B (1968) 96
9) H. Sherif - Nuclear Phys. A131 (1969) 532
10)T. Tamura - Rev. of Mod. Phys. 37 (1965) 679
11)J. Raynal - Symposium sur les Mécanismes de Réactions Nucléaires et
   Phénomènes de Polarisation Université Leval , Quebec (1969).
12)J. Raynal in the Structure of Nuclei (IAEA, Vienna) (1972) p. 75
13)R. de Swiniarski, A.D. Bacher et al - Phys. Rev. Letters 28 (1972) 1139
14)R.M. Lombard and J. Raynal - Phys. Rev. Letters 31 (1973) 1015
15)J.P. Blaizot and J. Raynal - Lett. al N.C. 12 (1975) 508
16)B.J. Verhaar - Phys. Rev. Lett. 32 (1974) 307
17)O. Karban, P.D. Greaves et al - Nuc. Phys. A147 (1970) 461
18)A.B. Kurepin, R.M. Lombard and J. Raynal - Phys. Lett. 45B (1973) 184
19)F.J. Baker, S. Davis et al - Nucl. Phys. A 223 (1974) 409

DISCUSSION

Glashausser:
    Baker and Love at the University of Georgia have recently found the
    very interesting result that the introduction of the quadrupole mo-
    ment of the ground state into a coupled-channel calculation for ^{63}Cu
    explains the apparent J dependence previously observed for inelastic
    scattering of protons and, particularly, deuterons.

Raynal:
    I received their preprint recently, but the details of their calcula-
    tions are not very clear for me, so I have no comment.

REPORT ON DISCUSSION SESSION:
SPIN-SPIN INTERACTION AND COUPLED CHANNELS

J. Raynal

Centre d'Etudes Nucléaires de Saclay, Gif-sur-Yvette, France

The experimental data relevant to these two topics can be handled in the scope of a microscopic or a macroscopic approach. For the first of them, "spin-spin interaction" means microscopic interpretation, whereas the same experiments could be interpreted as effects of the quadrupole moment of the target ground state in the macroscopic approach. Discussion on this subject was chiefly on the energies and the targets to be used in future experiments; in particular, the target must have spin 1/2 to eliminate quadrupole moment effects.

Five contributions on inelastic proton scattering on Fe and Ni isotopes were included in another session, so I could not quote them in my talk. These experiments, performed at three or four energies between 20 and 30 MeV, are analysed by coupled channels in the framework of Greenlees' optical model. The authors conclude on a ratio between deformed spin-orbit and deformed central potential decreasing with energy and becoming almost the same for ^{54}Fe and ^{56}Fe. However, the energy dependence of the calculated curves is stronger than the one of the experimental curves and the fit to the analysing power worsens at small angles. Anyway, at a given energy, the analysing powers on different Ni isotopes look similar but are different between ^{54}Fe and ^{56}Fe: nuclear structure effects can be seen on these experimental curves even if they do not show up in their analysis.

In the microscopic approach to inelastic scattering, which is limited to DWBA, fits are not so good as with the collective models, but there are some hopes for better results. Until now, calculations were made with the even part of the nucleon-nucleon force, because the odd part was thought to be less important and was not so well known. In many cases, direct calculations give good fits to the data, but the agreement is lost when exchange contributions are included. But exchange is needed to cancel spurious odd interactions when even interactions are used. This suggest the need of an odd interaction in such calculations. As improvements to a communication to this Conference, Resmini reported preliminary calculations with even+odd interactions for which the direct+ exchange remains good. The microscopic description of analysing powers will perhaps been improved in the near future.

TECHNIQUES OF POLARIZATION MEASUREMENT – A SURVEY*

Gerald G. Ohlsen

Los Alamos Scientific Laboratory, University of California

Los Alamos, NM 87545 USA

In this paper, I plan to discuss experimental methods for measuring analyzing powers, spin correlation functions, and polarization transfer functions, with an emphasis on sources of error and techniques or precautions which minimize them. I consider this to be an extension of the excellent paper by R. Hanna[1], presented at the Karlsruhe conference, in the directions of a) higher spin, b) more detailed concern with the nature and magnitude of various systematic errors, and c) more attention to "higher order" experiments, i.e., to experiments in which two particles are polarized.

In the following talk[2], Dr. Grüebler will speak about absolute polarization standards. In this talk, I will assume that either the absolute magnitude of the polarization of the beam is known, or that a calibrated polarimeter is available.

I. Spin 1/2 Analyzing Power Measurements

Polarized protons, tritons, and ^{3}He particles are now available from ion sources[3]. For spin 1/2 particles such as these, there are several unique circumstances which allow extremely accurate analyzing power measurements to be made. Two proton experiments at the ± 0.0002 level have been reported[4,5]. While the general principles for the measurement of spin-1/2 analyzing powers will be well known to this audience, it is such a basic point of departure for our discussion that I wish to review the ideas and make a few observations.

In accordance with the usual conventions, we assume a Cartesian coordinate system with z along the incident beam momentum, $\vec{k}_{in}$, y along $\vec{k}_{in} \times \vec{k}_{out}$, where $\vec{k}_{out}$ is the scattered particle momentum, and x such as to define a right-handed coordinate system. This will be referred to as the "projectile helicity frame" (see fig. 1a). Thus, if the spin is perpendicular to $\vec{k}_{in}$ and pointing "up" in space, the cross section for scattering to the left, $I_L(\theta)$, is

$$I_L(\theta) = I_0(\theta)[1 + p_y A_y(\theta)]. \tag{1}$$

For scattering to the right, the y axis points "down," so that p_y is negative; the corresponding cross section, $I_R(\theta)$, is therefore

$$I_R(\theta) = I_0(\theta)[1 - p_y A_y(\theta)]. \tag{2}$$

In both expressions, $I_0(\theta)$ is the cross section for an unpolarized beam, p_y is the beam polarization, and $A_y(\theta)$ is the analyzing power.

The actual number of counts recorded in detectors located at the left (detector 1) and right (detector 2) are,

$$L_1 = nN\Omega_1 E_1 I_0(\theta)\,[1 + p_y A_y(\theta)]$$

$$R_2 = nN\Omega_2 E_2 I_0(\theta)\,[1 - p_y A_y(\theta)] \tag{3}$$

where n is the number of incident particles, N is the target thickness (atoms/cm^2), and Ω_1, E_1 and Ω_2, E_2 are the solid angles and efficiencies for detector 1 and detector 2, respectively (see fig. 2). For gas target

G. G. Ohlsen

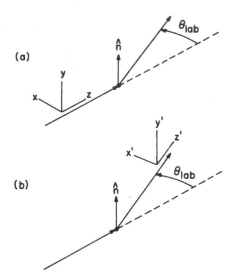

Fig. 1 a) Projectile helicity frame
        b) Outgoing reactant laboratory helicity frame

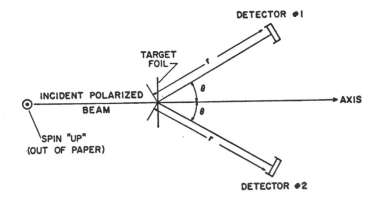

Fig. 2. Schematic diagram of a symmetric detector pair
         (solid target geometry)

geometry, N represents the density of the target (atoms/cm^3) and Ω the gas
target G factor.

If we now flip the polarization, i.e., let $p_y \rightarrow -p_y$, we obtain

$$R_1 = n'N'\Omega_1 E_1 I_0(\theta) \ [1 - p_y A_y(\theta)]$$

$$L_2 = n'N'\Omega_2 E_2 I_0(\theta) \ [1 + p_y A_y(\theta)],$$

(4)

where the primes are used to indicate that the integrated charge and the
effective target thickness may not be the same for the two counting
periods. Notice that in the present convention, detector 2 is now called
a left detector and detector 1 a right detector.

There are four unknowns in the problem, which may be taken to be $p_y A_y$, I_0, $\Omega_1 E_1 / \Omega_2 E_2$, and $nN/n'N'$. There are four observed numbers, and in this case all are independent so that the four unknowns may be determined. In particular, if we form the geometric means

$$L \equiv (L_1 L_2)^{\frac{1}{2}} = [nn'NN'\Omega_1\Omega_2 E_1 E_2]^{\frac{1}{2}} I_0 (1 + p_y A_y),$$

$$R \equiv (R_1 R_2)^{\frac{1}{2}} = [nn'NN'\Omega_1\Omega_2 E_1 E_2]^{\frac{1}{2}} I_0 (1 - p_y A_y),$$

$$(5)$$

we find the left-right asymmetry, ε, to be

$$\varepsilon \equiv (L-R)/(L+R) = p_y A_y ,$$

$$(6)$$

which is independent of relative detector efficiencies and solid angles, of relative integrated charge, and of target thickness variations. Those quantities common to the two channels, i.e., n and N, are averaged over the data accumulation period so that time fluctuations in the beam current or target density are of no consequence. On the other hand, those quantities which are different in the two channels, E and Ω, must not vary with time. If we define the geometric mean of the number of particles detected by detector 1 (2) in the two intervals as $N_1 (N_2)$, we have

$$N_1 \equiv (L_1 R_1)^{\frac{1}{2}} = \{nn'NN'[1 - (p_y A_y)^2]\}^{\frac{1}{2}}\Omega_1 E_1 I_0,$$

$$N_2 \equiv (L_2 R_2)^{\frac{1}{2}} = \{nn'NN'[1 - (p_y A_y)^2]\}^{\frac{1}{2}}\Omega_2 E_2 I_0.$$

$$(7)$$

The ratio of these is

$$N_1/N_2 = (\Omega_1 E_1)/(\Omega_2 E_2)$$

$$(8)$$

That is, this ratio provides a check on the performance of the apparatus; it is just the quantity which must be constant in time if the asymmetry is to be accurate.

A. Proper and Nonproper Flips

In the above discussion, it was assumed that the beam polarization was exactly reversed and that no other change in the beam characteristics occurred. We refer to this as a "proper" flip. Under these conditions, there is exact cancellation of any apparatus asymmetries. The only possible effect of misalignment is a shift in the angle at which the analyzing power is determined. This same result can be achieved with fixed beam polarization direction if the apparatus is rotated about the beam axis in such a way that the beam position remains invariant with respect to the detectors. This can be achieved experimentally, for example, by requiring the beam current readings on rotating sets of slits preceding and following the scattering apparatus to be the same in the "ordinary" and "flipped" configuration.

A proper flip can be compared to a "nonproper" flip, in which the apparatus is rotated about an axis which does not coincide with the beam axis (see fig. 3) and in which the beam position in space remains fixed. Suppose that the beam alignment is such that, to first order, the yield to the left is enhanced by the factor $1 + \varepsilon'$ and to the right by $1 - \varepsilon'$.

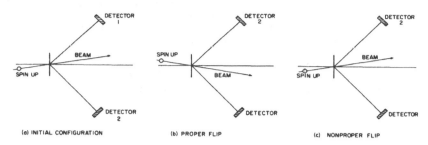

Fig. 3.   Diagram showing a proper and a nonproper flip

Then, for the initial configuration, the observed counts will be:

$$L_1 = nN\Omega_1 E_1 I_0(\theta) [1 + p_y A_y(\theta)](1 + \epsilon')$$

$$R_2 = nN\Omega_2 E_2 I_0(\theta) [1 - p_y A_y(\theta)](1 - \epsilon') \ . \tag{9}$$

If a proper flip were executed, for the second counting interval we would obtain

$$L_2 = n'N'\Omega_2 E_2 I_0(\theta) [1 + p_y A_y(\theta)](1 - \epsilon')$$

$$R_1 = n'N'\Omega_1 E_1 I (\theta) [1 - p_y A_y(\theta)](1 + \epsilon') \ . \tag{10}$$

If instead a nonproper flip were executed, we would obtain

$$L_2 = n'N'\Omega_2 E_2 I_0(\theta) [1 + p_y A_y(\theta)](1 + \epsilon')$$

$$R_1 = n'N'\Omega_1 E_1 I_0(\theta) [1 - p_y A_y(\theta)](1 - \epsilon') \ . \tag{11}$$

Thus, forming the geometric means, we find

$$\epsilon = p_y A_y \qquad\qquad \text{(proper flip)}$$

$$\epsilon = [p_y A_y + \epsilon']/[1 + \epsilon' p_y A_y] \quad \text{(nonproper flip)} \tag{12}$$

Note that ϵ' is exactly the asymmetry which would be observed if the incident beam were unpolarized or if the analyzing power, A_y, were zero. Expressions for ϵ' for gas and solid target geometry as a function of misalignment parameters are given in ref. 6. Note also that the larger the value of $p_y A_y$, the smaller the error in A_y that ϵ' induces.

If electronic windows are narrow, a nonproper flip can introduce error through shifts in the scattered particle energies as well as through the pure geometrical arguments given above.

Even though one makes what is intended to be a proper flip, there will likely be a component of nonproper flip present. For example, reversing the spin direction at the ion source might actually change the beam position slightly. Rotating a scattering chamber, even though slit balances were precisely maintained, might result in a shift of the beam centroid because of an inhomogeneous beam profile. Perhaps most significant is the difference in the average beam position for the two counting intervals which arises from random variations in the beam trajectory. "Rapid reversal" of the spin vector, such as reported by McKibben and Potter[7] can be used to combat the random movement problem. One must be sure that correlated beam displacement effects are not introduced, of course. A suitable beam position stabilizer can be used to combat random or correlated position variations.

The use of symmetric detectors for the measurement of spin-1/2 analyzing powers has become almost universal. In cases where the detector is complicated or expensive, for example, one may have to make do with a single detector, and it is of interest to ask what desirable experimental features are thereby lost. First, one loses a factor of two in counting rate. Second, the accuracy of the measurements then depends in first order on the accuracy of beam current integration and on variations in the target thickness. Third, an asymmetry measured with symmetric geometry depends, up to second order terms, only on the mean value of the polarization over both counting intervals. In contrast, with one detector, there is a first order dependence on the difference between average polarization in the two counting intervals. Fourth, one loses the ability to cancel the effects of dead-time in the counting equipment.

B. Sources of Error - Summary

The following is a summary of the possible systematic errors in a spin-1/2 analyzing power measurement.

1. Beam polarization uncertainty or fluctuation. Most results will depend linearly on the beam polarization. However, as noted, the use of two symmetric detector pairs can permit the measurement of one analyzing power relative to another without a first order dependence on the beam polarization. A caution - if the beam polarization is inhomogeneous, as is known to be the case for the beam produced by a Lamb-shift ion source, for example, an error can be introduced if the two detector sets view targets illuminated with different portions of the beam.

2. Nonproper flip component. This is one of the most severe problems, since the sensitivity can be substantial. The experimental geometry should be made large to reduce the sensitivity, and beam position monitoring or control can be used to reduce the nonproper flip component.

3. Target impurities, background, and slit edge effects. Polarization experiments have an advantage over cross section measurements in regard to these problems, since one is unconcerned with absolute counting efficiency. This allows windows to be narrow, so that dilution of the observed asymmetry can be minimized. Good energy resolution can reduce or eliminate these problems.

4. Uncertainty in the scattering angle. Although a proper flip eliminates false asymmetries exactly, the exact angle to which the measurement corresponds needs to be determined. If the actual scattering angles for the two arms are θ_1 and θ_2, the analyzing power measured corresponds, in first order, to $\frac{1}{2}(\theta_1 + \theta_2)$.

5. Dead time corrections. Even when the slower electronic components are common to the two channels, for higher rates, the dead time difference arising from the faster but separate components may need to be accounted for.

6. Instability of electronics. This is not usually a severe problem with modern counting equipment. In special cases, such as in the current measurements used in parity violation experiments[8] in lieu of single event counting, it is a crucial problem. If unstable electronics must be tolerated, rapid alternation of the beam polarization can be used to combat it.

II.   Spin-1 Analyzing Power Measurements

Polarized deuterons are now widely available from ion-sources,[3] and one source of polarized ^{6}Li has been built[9]. The cross sections for a reaction induced by a spin-1 projectile for left, right, up, and down azimuthal directions are

$$I_L = I_0[1 + \frac{3}{2} p_z \sin\beta \, A_y + \frac{1}{2} p_{zz}(\sin^2\beta \, A_{yy} + \cos^2\beta \, A_{zz})],$$

$$I_R = I_0[1 - \frac{3}{2} p_z \sin\beta \, A_y + \frac{1}{2} p_{zz}(\sin^2\beta \, A_{yy} + \cos^2\beta \, A_{zz})], \qquad (13)$$

$$I_U = I_0[1 + p_{zz}\sin\beta\cos\beta \, A_{xz} + \frac{1}{2} p_{zz}(\sin^2\beta \, A_{xx} + \cos^2\beta \, A_{zz})],$$

$$I_D = I_0[1 - p_{zz}\sin\beta\cos\beta \, A_{xz} + \frac{1}{2} p_{zz}(\sin^2\beta \, A_{xx} + \cos^2\beta \, A_{zz})].$$

where β is the angle between the beam quantization axis and the beam direction, k_{in}, and where left, right, up, and down are as would be defined by an observer "aligned" with the transverse component of the quantization axis looking along k_{in}. p_z and p_{zz} are the vector and tensor polarization of the incident spin-1 beam referred to its own quantization axis, where $|p_z| < 1$ and $-2 < p_{zz} < 1$. A_{xx}, A_{yy}, and A_{zz} are related in that their sum must vanish. The Cartesian analyzing powers are related to the spherical analyzing powers by

$$A_{xx} = \sqrt{3} \, T_{22} - \frac{1}{\sqrt{2}} T_{20} \qquad\qquad A_{yy} = -\sqrt{3} \, T_{22} - \frac{1}{\sqrt{2}} T_{20}$$

$$A_{zz} = \sqrt{2} \, T_{20} \qquad\qquad\qquad A_{xz} = -\sqrt{3} \, T_{21} \qquad (14)$$

$$A_y = \frac{2}{\sqrt{3}} iT_{11} \qquad \frac{1}{2}(A_{xx} - A_{yy}) = \sqrt{3} \, T_{22}$$

and are bounded by the limits

$$|A_y| < 1, \quad |A_{xz}| < \frac{3}{2}, \quad -2 < A_{jj} < 1 \text{ for } j = x, y, \text{ or } z, \text{ and } |\frac{1}{2}(A_{xx} - A_{yy})| < \frac{3}{2}.$$

Each of the Cartesian analyzing powers is an even or an odd function of the scattering angle, θ, and thus is readily measured with a symmetric two detector system. Because of the relation $A_{xx} + A_{yy} + A_{zz} = 0$, only two of these particular three variables need be measured. Thus, there is great flexibility in experimental design. For $\beta = 90°$, A_{xx} and A_{yy} are the measurable quantities, while for $\beta = 0°$ A_{zz} is measurable. Many authors elect to report the spherical tensor T_{22}, even though they actually measure A_{yy} and A_{xx} or A_{zz}. This practice is not desirable, as the systematic errors in the two measurements tend to be dissimilar, and mixing them removes the opportunity for later detection. T_{22} can be directly measured only with $\beta = 54.7°$, as discussed below. Except for these points the distinction between spherical and Cartesian forms is a matter of normalization.

A.   General Principles of Measurement

Measurement of the four independent analyzing tensors may be approached through either the azimuthal dependence of the cross section, or through its dependence on the vector and tensor beam polarization, or both. In seeking an optimum method for given experimental constraints one should observe several basic principles, as follows.

First, the effect being measured should be maximized. For example, a method in which p_{ZZ} is varied between +1 and -1 is to be preferred over one in which it is varied between +1 and 0. The latter condition will require a counting interval four times longer to achieve a given statistical accuracy than will the former. In addition, since most systematic effects will cause a fixed error in the observed ratio or asymmetry, the systematic error in the observed analyzing power will be twice as great in the latter case than in the former.

Second, a method in which as many factors as possible are cancelled should be sought. For most spin-1 methods, it is not possible to remove dependence on current integration and the like. However, detector efficiency and solid angle effects can always be cancelled by taking appropriate geometric means or ratios of counts for a given detector.

Third, methods which do not depend on first order errors in the spin angle, β, or in the azimuthal angle, φ, are to be preferred. Errors in β and φ can induce errors by introducing "crosstalk" between the observed analyzing powers. In addition, for certain methods, an error in β can produce scale errors in the analyzing powers.

Fourth, proper flips should be used. A proper flip, once again, is one in which the beam polarization is altered with no other changes in the character of the beam, e.g., with no beam steering. Changes in the state selection process at the ion source will generally quality, although those changes which involve reversal of magnetic fields need to be examined carefully. In the spin-1 case, one might use more than two states of polarization in an experimental sequence, so the word "flip" no longer adequately describes the concept; nevertheless, we will retain it. We will also use the term "proper flip" to describe apparatus rotations of 90°, 180°, or 270° in which the trajectory of the beam is held fixed with respect to each of the detectors.

Fifth, symmetric detector pairs should be used. In the spin one case, crosstalk between the observables can often be eliminated by symmetric geometry. An example is presented in the following.

Consider the measurement of A_{zz}. Since it gives rise to no azimuthal variation, one <u>must</u> utilize the dependence of the yield on p_{zz} to determine it. For β = 0°, and for a detector at any azimuthal position, the yield, T, is

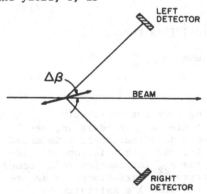

$$T = nN\Omega EI_o (1+\tfrac{1}{2} \, p_{ZZ} A_{zz}) \qquad (15)$$

where the factors n, N, Ω, and E are as previously defined. If the yields T, T' are observed for the beam polarization p_{ZZ}, p'_{ZZ}, respectively, assuming the factor nN to be the same for the two cases, we find

$$A_{zz} = \frac{2(T - T')}{p_{ZZ}T' - p'_{ZZ}T} . \qquad (16)$$

Now suppose we consider a left-right detector pair, and that the spin quantization axis has a small component Δβ in the plane of the scattering (see Fig. 4). A contribution to the left detector yield proportional to sinΔβ cosΔβ A_{xz} (≈ Δβ A_{xz}) will then arise. For example, if A_{xz} = 0.5

Fig. 4. Spin misalignment pertinent to measurements of A_{zz}. The two headed arrow denotes a tensor polarization.

and $\Delta\beta = 0.05$ (corresponding to $\sim 3°$), an error term of 0.025 would be generated. Since this would be combined with $\frac{1}{2}A_{zz}$, an error in A_{zz} of 0.05 would result. However, such an error in the right detector has the opposite sign, so if two detectors are used the effect is cancelled.

A particularly serious case of cross-talk that cannot be "cured" by symmetric geometry occurs between A_{xz} and A_y. A_y tives rise to a left-right asymmetry while A_{xz} gives rise to an up-down asymmetry. Because of the numerical factors intrinsic to the formalism, a given value of A_{xz} produces an asymmetry three times smaller than does the same value of A_y. Thus, if the azimuthal angles deviate from the ideal values, the mixing of left-right and up-down asymmetries may severely distort the A_{xz} measurement while the A_y measurement may be little affected. (Because of the small intrinsic asymmetry produced, A_{xz} measurements are also very susceptible to false asymmetries.) The use of a beam with pure tensor polarization can be used to measure A_{xz} with no error induced by cross talk from A_y. Polarization-dependent misalignment effects of this type are considered in great detail in ref. 6.

The most immediate generalization of the standard spin 1/2 methods involves the use of $\beta = 54.7°$, together with detectors in each of the four half-planes. This can be achieved in an apparatus such as the LASL supercube[10]. This geometry has not found wide acceptance because of its complexity, particularly with regard to the design of suitable targets, but it does offer the advantage of complete flexibility regarding the various spin-1 measuring schemes one might propose. To cancel detector solid angles and efficiencies, one can use four counting intervals and appropriate "four-fold" geometric means, e.g.:

$$L = (L_1 \, L_2 \, L_3 \, L_4)^{1/4} \tag{17}$$

where L_1 is the yield obtained during the counting interval when detector 1 is in the left position, L_2 the yield when detector 2 is in the left position, etc. Four-way slits at the entrance and exit of the chamber are used to guarantee that proper flips are executed. The following asymmetries can be formed:

$$\varepsilon_3 = \frac{3}{2} \, P_Z A_y = 2[L-R]/[L+R+U+D]$$

$$\varepsilon_4 = \frac{1}{3}\sqrt{2} \, P_{ZZ} A_{xz} = 2[U-D]/[L+R+U+D] \tag{18}$$

$$\varepsilon_5 = -\frac{1}{6} \, P_{ZZ}(A_{xx}-A_{yy}) = [(L+R) - (U+D)]/[L+R+U+D].$$

This separation of observables occurs because

$$L+R+U+D = 1 + \frac{1}{2}P_{ZZ} \, (3\cos^2\beta - 1) \, A_{zz} \tag{19}$$

which becomes unity for $\beta = 54.7°$.

In this method, independence of the measurement on relative current integration, dead time, etc., is achieved. In addition, dependence on first order variation in the beam polarization is eliminated in the sense discussed above for the spin 1/2 case; that is, one set of four detectors can be used to transfer p_z and p_{zz} information to a second set with no first order dependence on polarization variations. A severe disadvantage arises, however, from the (first order) sensitivity to errors in β. For example, for the $\sin^2\beta$ dependence of ε_5, one calculates that a $1°$ error in β results in a 2.5% error in $\frac{1}{2}(A_{xx}-A_{yy})$. However, no alteration of the beam polarization is required in the course of the measurements, so it is particularly suitable for cases in which only one

spin state is available or for intercalibration of the polarization between several spin states.

Consider next the measurement of A_y and A_{yy} with a left and right
detector. If two counting intervals separated by a proper flip are used,
and L and R represent the geometric means, as in eq. 5,

$$\epsilon_1 = \frac{L-R}{L+R} = [\tfrac{3}{2} p_Z \sin\beta \, A_y]/[1 + \tfrac{1}{2} p_{ZZ}(\sin^2\beta \, A_{yy} + \cos^2\beta \, A_{zz})] \qquad (20)$$

which, for $\beta = 90°$, becomes

$$\epsilon_1 = [\tfrac{3}{2} p_Z A_y]/[1 + \tfrac{1}{2} p_{ZZ} A_{yy}] \qquad (21)$$

If a beam with zero tensor polarization is used (ideally $p_Z = 2/3$,
$p_{ZZ} = 0$), A_y is determined just as was the case for spin 1/2. If p_{ZZ} is
small, a crude knowledge of A_{yy} suffices to correct the measurement of
A_y. This is, for example, the approach of Cadmus and Haeberli[11], and
one of the methods used by König et al.[12] although König et al. do not
use a method which cancels detector efficiency and solid angle effects.

For fixed values of p_Z and p_{ZZ}, ϵ_1 is a quantity which can be compared to calculations. Since it can be measured with high accuracy with
no dependence on current integration, etc., this possibility is sometimes worth considering. However, one normally desires to know A_y and
A_{yy} separately. To extract A_{yy}, a method in which the magnitude of p_{ZZ}
is varied must be used. König et al.[12] alternate between states with
$p_Z = 1/3$, $p_{ZZ} = -1$ and $p_Z = -1/3$, $p_{ZZ} = 1$. This allows the extraction
of both A_{yy} and A_y, but in this mode of operation the sensitivity to A_y
is relatively small. At Los Alamos,[6] we use counting intervals with
each of the three pure states $m_I = 1$, 0, and -1 (whose ideal polarization
values are $p_Z = 1$, $p_{ZZ} = 1$; $p_Z = 0$, $p_{ZZ} = -2$; and $p_Z = -1$, $p_{ZZ} = 1$,
respectively). For the left detector, one observes:

$$L^{(1)} = n^{(1)} N^{(1)} \Omega E I_o (1 + \tfrac{3}{2} pA_y + \tfrac{1}{2} pA_{yy})$$

$$L^{(0)} = n^{(0)} N^{(0)} \Omega E I_o (1 - pA_{yy}) \qquad (22)$$

$$L^{(-1)} = n^{(-1)} N^{(-1)} \Omega E I_o (1 - \tfrac{3}{2} pA_y + \tfrac{1}{2} pA_{yy}) .$$

where p is the fractional polarization of the beam. Thus, one has three
observed quantities, and five unknowns:
$n^{(1)} N^{(1)}/n^{(0)} N^{(0)}$, $n^{(-1)} N^{(-1)}/n^{(0)} N^{(0)}$, I_o, pA_y, and pA_{yy}. If current
integration, etc., is reliable, the first two unknowns become known and
the equations can be solved. Precise values of the relative vector and
tensor beam polarizations for the three states can be incorporated into
the calculations.[6] Inclusion of a right detector, although important
for systematic error cancellation, does not change the relation between
knowns and unknowns, since one new unknown (Ω_L/Ω_R) and one independent
new observation is introduced.

    B. Sources of Error - Summary

    At this point, I am sure that I have given the (correct) impression
that there are a great many ways to make spin-1 analyzing power measurements. I will conclude the discussion by summarizing the sources of first
order error which must be considered for any given method that are in
addition to the earlier list for spin 1/2 analyzing powers.

    1. Current integration, dead time, etc. Most spin-1 schemes rely
on accurate control of these factors.

2. Crosstalk between observables. A way can usually be found to measure each of the analyzing powers which is free of crosstalk. Symmetric detectors and proper flips suffice to eliminate some types; others may be eliminated only by use of a pure vector or pure tensor polarized beam.

3. Spin–alignment errors. In addition to the crosstalk problem, spin angle errors can lead to scale errors in observables. For each observable, there exists a method to reduce the error induced by faulty spin-alignment to second order.

4. Intercalibration of spin states. The values of p_Z and p_{ZZ} for beams with at least two different states of polarization must be known if all of the spin-1 analyzing powers are to be measured. (One of the "states" of polarization may be "unpolarized.")

A much greater effort should be made to accurately evaluate experimental errors. As a guide, it is quite difficult to make A_y measurements with errors less than ± 0.01, A_{xz} measurements with errors less than ± 0.03, and A_{xx}, A_{yy}, A_{zz} or $\frac{1}{2}(A_{xx} - A_{yy})$ measurements with errors less than ± 0.01 even without considering any uncertainty in p_Z and p_{ZZ}. In the measurement of deuteron analyzing powers, with the large beam currents available nowadays, the statistical errors are rarely the most significant ones.

III. Polarization Transfer Experiments

The role of experiments in which two particles are polarized is, generally, to furnish a new "view" of the amplitudes, just as one might photograph an object from a new angle to obtain more information about it. Thus, such measurements are neither more nor less significant than, say, analyzing powers. In relatively simple cases, such as p–α and d–α elastic scattering, it is likely that these extra "views" of the amplitudes are not needed, whereas in higher spin cases they probably are. In a given model, however, a particular observable might offer a direct measure of a given parameter. For example, since the spin transfer coefficient K_y^y ($\equiv$ the Wolfenstein D parameter) must be unity for spin-1/2 on spin-0 scattering, deviations of K_y^y from 1 in the case of proton-heavy nucleus scattering have been interpreted as direct evidence for a spin-spin term in the corresponding optical potential[13]. However, it appears in more recent work [14] that other effects may also reduce K_y^y from unity. (In all such effects, of course, the target spin must play _some_ role.)

In a polarization transfer experiment, it is convenient to express the polarization of the projectile in its helicity frame (z along $\vec{k}_{in}$, y along $\vec{k}_{in} \times \vec{k}_{out}$) and the polarization of the outgoing particle in its laboratory helicity frame (z' along k_{out}(lab), y' along $\vec{k}_{in} \times \vec{k}_{out}$). These frames are shown in fig. 1. The y and y' axes are identical, but are usually distinguished for the sake of notational consistency[15].

We consider, as an example, the measurement of $K_y^{y'}$ for a spin 1/2 projectile and outgoing particle. The pertinent expressions for the observables are[15]

$$I(\theta) = I_0(\theta)[1 + p_y A_y(\theta)]$$

$$p_{y'}I(\theta) = I_0(\theta)[P_{y'}(\theta) + p_y K_y^{y'}(\theta)]$$

(23)

where $p_{y'}$ is the outgoing particle polarization, $P_{y'}(\theta)$ is the polarization function, and other quantities are as previously defined. Since
the polarization of the projectile is along the y-axis, there are no
other outgoing particle polarization components present. The polarization after the first scattering to the left may be written (see fig. 5)

$$P_{y'} = [P_{y'}(\theta) + p_y K_y^{y'}(\theta)]/[1 + p_y A_y(\theta)] \tag{24}$$

and the number of outgoing particles, n', may be written

$$n' = nN\,\Delta\Omega\,I_0(\theta)[1 + p_y A_y(\theta)] \tag{25}$$

where $\Delta\Omega$ is the solid angle of acceptance of the analyzing device. The
yield, denoted by L, in a second scattering to the left will be

$$L = n'N^{(2)}\,\Omega_L^{(2)} I_0^{(2)}(\theta)\,[1 + P_{y'}A_y^{(2)}(\theta)], \tag{26}$$

where quantities pertaining to the second scattering are indicated by a
superscript "2". Substituting eqs. (24) and (25) into eq. (26), we have

$$L = nNN^{(2)}\,\Delta\Omega\,\Omega_L^{(2)} I_0 I_0^{(2)}\,[1 + p_y A_y + (P_{y'} + p_y K_y^{y'})A_y^{(2)}] \tag{27}$$

and correspondingly, for a second scattering to the right

$$R = nNN^{(2)}\,\Delta\Omega\,\Omega_R^{(2)} I_0 I_0^{(2)}\,[1 + p_y A_y - (P_{y'} + p_y K_y^{y'})A_y^{(2)}]. \tag{28}$$

It suffices for our discussion to condense the various factors into one
proportional to the number of projectiles, n_{++}; into one that is proportional to the solid angle of each of the two detectors, Ω_L or Ω_R; and
into an overall scale factor, I_0, which actually includes the differential cross sections, $I_0(\theta)$ and $I_0^{(2)}(\theta)$, so that

$$L_{++} = n_{++}\Omega_L I_0\,[1 + p_y A_y + (P_{y'} + p_y K_y^{y'})A_y^{(2)}]\,(1 + \varepsilon')$$
$$R_{++} = n_{++}\Omega_R I_0\,[1 + p_y A_y - (P_{y'} + p_y K_y^{y'})A_y^{(2)}]\,(1 - \varepsilon') \tag{29}$$

We have included the terms $1 + \varepsilon'$ and $1 - \varepsilon'$ to allow for a possible intrinsic instrumental asymmetry which may arise from non-uniform illumina-
tion of the second scatterer.* Variations of energy across the scatterer

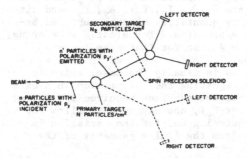

Fig. 5. Schematic diagram of
a double scattering experiment.
The particles emitted from the
first target may have their
spins reversed either by the
precession device or by switching from left to right scattering (dotted portion of diagram)
while simultaneously reversing
the projectile spin.

---

*In defining a single false asymmetry parameter, we are assuming that the
cross section variation with energy and angle is independent of beam polarization. This is clearly incorrect in calculations of accuracy higher
than first order.

would also result in such an effect. We denote by the first subscript
the sense of the polarization of the beam and by the second subscript
whether or not the sense of the polarization has been reversed at the
analyzer. Notice that, contrary to the convention in the previous sections, we now use left and right to denote particular detectors without
regard to the state of polarization of the beam.

If the beam polarization is reversed, two more measurements can be
made:

$$L_{-+} = n_{-+}\Omega_L I_0 [1 - p_y A_y + (P_{y'} - p_y K_y^{y'}) A_y^{(2)}](1 + \varepsilon')$$

$$R_{-+} = n_{-+}\Omega_R I_0 [1 - p_y A_y - (P_{y'} - p_y K_y^{y'}) A_y^{(2)}](1 - \varepsilon')$$

$$(30)$$

This is assumed to be a proper flip, so the factors $1 + \varepsilon'$ and $1 - \varepsilon'$ are
unchanged. If the outgoing particles are precessed $180°$, we can observe
four more quantities:

$$L_{+-} = n_{+-}\Omega_L I_0 [1 + p_y A_y - (P_{y'} + p_y K_y^{y'}) A_y^{(2)}](1 \pm \varepsilon')$$

$$R_{+-} = n_{+-}\Omega_R I_0 [1 + p_y A_y + (P_{y'} + p_y K_y^{y'}) A_y^{(2)}](1 \mp \varepsilon')$$

$$(31)$$

and

$$L_{--} = n_{--}\Omega_L I_0 [1 - p_y A_y - (P_{y'} - p_y K_y^{y'}) A_y^{(2)}](1 \pm \varepsilon')$$

$$R_{--} = n_{--}\Omega_R I_0 [1 - p_y A_y + (P_{y'} - p_y K_y^{y'}) A_y^{(2)}](1 \mp \varepsilon')$$

$$(32)$$

where the upper sign on ε' applies for a proper flip of the outgoing particle and the lower for a nonproper flip.

We now consider several schemes whereby $K_y^{y'}$ might be extracted. First
consider making a proper flip of the polarization of the outgoing particles. For example, this can be done for neutrons by means of a solenoidal magnetic field. (In practice, the spins are usually precessed $\pm 90°$
rather than $0°$ and $180°$.) By the usual geometric mean, the quantity
$P_{y'} \equiv [P_{y'} + p_y K_y^{y'}]/[1 + p_y A_y]$ can be extracted from eqs. (29) and (31);
a similar quantity with p_y negative can be extracted from eqs. (30) and
(32). In these cases, the unknowns are Ω_L/Ω_R, I_0, $P_{y'}$, A_y, $K_y^{y'}$, and either
n_{++}/n_{+-}, or n_{-+}/n_{--}. Thus two auxiliary quantities must be supplied before $K_y^{y'}$ can be extracted from either expression. For elastic scattering,
time reversal invariance implies $P_{y'} = A_y$, so for that case, only A_y
would need to be supplied. A caution: if A_y is measured in good geometry and with good energy resolution, as is usually the case, suitable
averages should be calculated before inputting the information in a case
such as this. In particular, note that the averages required for A_y and
for $P_{y'}$ may be different. That is, for $P_{y'}$, one may need to allow for
the finite geometry of the second scattering and for the variation in
polarization direction which results from the finite geometry of the
first scattering.

Instead of precessing the outgoing particles with a magnetic field,
one might flip their spins by switching from left to right in the first
scattering, while simultaneously reversing the projectile spin. This is
a procedure which can be used for outgoing charged particles. However,
the instrumental asymmetry, ε', then does not cancel so one obtains an
observed asymmetry, ε,

$$\varepsilon = [p_y \cdot A_y^{(2)} + \varepsilon']/[1 + p_y \cdot A_y^{(2)} \varepsilon'] \qquad (33)$$

just as in the second of eqs. (12). That is, one has executed a nonproper flip, and therefore one additional unknown has been introduced. In order to use this result, ε' would have to be estimated. Veeser et al.[16] have described a useful means for determining ε', as follows. The false asymmetry arises from a displacement of the centroid (x_0 in fig. 6) of the scattered beam with respect to the axis of the polarimeter, and from a shift in the mean angle of the scattered particles with respect to this axis (k_x in fig. 6). Two pairs of detectors at small angles where the polarimeter analyzing power is zero are required to determine both x_0 and k_x; once these are known, ε' can be calculated for the conditions of interest, as in eqs. (28) or (30) of ref. 6. An alternative method for determining ε' might make use of elastic scattering of a polarized beam from a spin-0 target, for which $K_y^y = 1$. Then, if A_y were known, A_y, P_y', and $K_y^y{}'$ could be input and ε' extracted. This value of ε' could then be used for correcting measurements for which ε' cannot be eliminated, so long as the geometrical and kinematic conditions were sufficiently similar.

If we consider a proper flip of the beam polarization rather than the outgoing particle polarization, we require the extraction of K_y^y from either eqs. (29) and (30) or eqs. (31) and (32). This is the method used, for example, by Clegg et al.[17] and by Baker[18]. ε' is cancelled, since a proper flip is used, so once again there are six unknowns and four observations. Thus, supplying P_y' and A_y, or A_y alone for elastic scattering, suffices to determine K_y^y. Alternatively, one could input the charge ratio, n_{++}/n_{+-} or n_{+-}/n_{--}, together with the statement that $P_y' = A_y$.

If all eight of eqs. (29) through (31) are used, one can show that only seven of the eight observations are independent, as follows. The (nine) unknowns are n_{++}/n_{+-}, n_{++}/n_{-+}, n_{++}/n_{--}, Ω_L/Ω_R, I_0, P_y', A_y, K_y^y, and ε'. Formation of the geometric means eliminates two charge ratios and the factor Ω_L/Ω_R; however, only four observed quantities remain. Thus, we again conclude that two pieces of auxiliary information must be supplied. One advantage in this case is that the effective values of P_y', A_y, and K_y^y are obtained directly, and one does not need to perform any averaging over measured A_y values prior to obtaining a result for K_y^y. One then depends, however, on current integration, etc.

If only one detector is used, this has the consequence that reliable charge integration <u>must</u> be one of the factors which is controlled. In some cases this is difficult to achieve, and for such cases two detectors should definitely be used. For example, if fast electronic coincidences are used, as between a passing counter and a side counter, it is diffi-

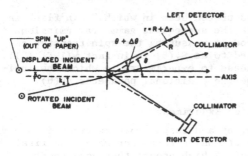

Fig. 6. Diagram showing the displacement and angular shift contribution to a false asymmetry in a nonproper flip.

cult to eliminate absolute efficiency variations with time, whereas the
relative efficiency between left and right systems can be controlled
relatively well.

We now turn briefly to a more complex case – namely polarization
transfer in a $(\vec{d},\vec{n})$ reaction, with the initial spin oriented along the
y axis[19]. The equations for the observables are[15]

$$I = I_o(1 + \frac{3}{2} P_z A_y + \frac{1}{2} P_{zz} A_{yy})$$

$$P_{y'} I = I_o(P_{y'} + \frac{3}{2} P_z K_y^{y'} + \frac{1}{2} P_{zz} K_y^{y'}) \ . \tag{34}$$

One notices that, compared to the spin-1/2 case, there are two new unknowns: A_{yy} and the tensor to vector spin transfer coefficient $K_y^{y'}$. For
the left detector we may obtain six numbers; e.g. counts with $m_I = 1$, 0,
and –1 deuterons incident for left and then right scattering (nonproper
flip sequence) or with the once scattered particles precessed 180° or
not precessed (proper flip sequence). The unknowns are: I_o, A_y, A_{yy},
$P_{y'}$, K_y^y, K_{yy}^y, ϵ', and five projectile charge ratios, or 12 variables in
all. The addition of a right detector and the calculation of geometric
means eliminates three of the charge ratios, so we are left with six numbers from which we seek to determine nine unknowns. Thus, three pieces
of auxiliary information must be supplied. If a proper flip scheme is
used, so that ϵ' may be neglected, and if charge integration is good, as
was assumed, for example, in ref. 19, the calculation can be made. Clearly
other auxiliary input variables may be chosen; in particular, choosing
A_y, A_{yy} and ϵ' would be useful in some cases.

Even more complex cases can be handled. For example, $^4\mathrm{He}(\vec{d},\vec{d})^4\mathrm{He}$
scattering has been studied in a current integration free fashion. In
many cases the algebra becomes quite complex, and in that particular case
computer methods for solving various quintic equations was required[20].

In some cases, the nature of the experiment permits a particularly
simple measurement to be made. For example, for a spin rotation parameter measurement for a spin-1/2 on spin-0 scattering, one can precess the
incident spin until the outgoing polarization is longitudinal, as detected by a polarimeter[21]. Excellent sensitivity can be achieved in this
way. However, the accuracy of such a measurement is limited by the accuracy with which the direction of the beam polarization is known. Measurement of K_x^x and K_z^x with the beam quantization axis at $\beta = 90°$ and $\beta =
0°$, respectively, would yield a result for the spin rotation parameter
which is independent, in first order, of the beam polarization direction.
In this case, the accuracy of the measurement would be limited by the accuracy with which the relative beam polarization for the two measurements
is known.

Experiments with an unpolarized proton beam in which "spin flip" in
inelastic scattering is detected via the subsequent gamma-ray emissions[22]
along the y axis constitutes a method of measuring the spin transfer co-
efficient K_y^y. The quantity referred to as $S_1(\theta)$ is in fact $\frac{1}{2}(1-K_y^y(\theta))$.
Boyd et al.[23] have applied these ideas to experiments in which the inci-
dent beam is polarized; in that case the additional parameter $\Delta S(\theta)$
($\equiv \frac{1}{2}(A_y - P_{y'})$) can be determined.

A. High Resolution Spin Transfer Experiments

Spin transfer experiments have, by and large, been restricted to
elastic scattering or to nuclear reactions in cases for which the first
excited state of the residual nucleus is high enough that good energy

resolution is not needed. For neutron work, one can use time-of-flight
methods to obtain better energy resolution. For high resolution charged
particle work, one can use either a magnetic spectrometer, so that a polarimeter with poor energy resolution can be tolerated, or a silicon polarimeter, as discussed below.

In fig. 7, we show a schematic diagram of a proposed "spin transfer
spectrometer." The desired features are 1) zero energy dispersion at the
output, so that all of the particles may enter a polarimeter with a small
acceptance; 2) zero net spin rotation, so that it could be used, for
example, for K_x^x measurements as well as for K_y^y measurements; 3) large
dispersion and a radial focus at some intermediate point where slits can
be used to define the energy spread that is accepted; and 4) double focussing, so that a large acceptance solid angle is possible. The design
in fig. 7 uses bends of $50.2°$, since this corresponds to the angle
through which a proton beam must be bent to transform longitudinal polarization into transverse polarization. Thus, by putting a polarimeter
just beyond the energy-defining slits, one could measure polarization
transfer coefficients such as K_x^z and K_z^z. Since there would be energy
dispersion at this point, one would have to use a thin primary target, so
that overall rates would be considerably lower than would be obtained
with the device in its primary configuration. This device is purely
hypothetical at the moment, and presumably a sophisticated design with
much large solid angle, etc., could be developed. The illustrated geometry achieves an acceptance of about 2.5 msr.

Existing QDDD spectrometers[24] have an extremely large acceptance
(~ 14 msr) and therefore might be useful for spin transfer measurements.
Their dispersion is large, so thin (~ 10 to 20 keV) targets would be required. With a solid target used in transmission mode, however, the limit would be imposed by energy straggling, and thicknesses of perhaps 100
keV could be tolerated.

Silicon polarimeters have been used where good energy resolution is
required. In these devices, a silicon detector is used as a (thick) scatterer. The pulse in the scatterer and in a side detector are added, so
that in principle all events lead to the same summed pulse height. Resolution of ~ 100–150 keV is achieved in practice, and count rates of ~ 100
kHz are tolerated.
Two basic designs
have been used; "re-
flection geometry,"
as used by Baker[18]
following the design
of Martin et al.[25]
(see fig. 8), and
"transmission ge-
ometry" with a scat-
tering angle around
$30°$, as described
most recently by
Birchall et al.[26]
These devices have
efficiencies of a
few parts in 10^5,
but their analyzing
powers are low
(~ 0.25 for trans-

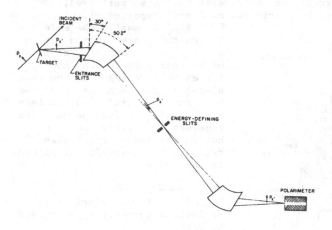

Fig. 7. Proposed spin-transfer spectrometer.

mission geometry, ~ 0.40 for reflection geometry) and energy dependent. They are useful only at energies above about 10 MeV. This is to be compared to an efficiency of ~ 10^{-4} with an analyzing power of 0.6 which has been obtained with a low resolution vane-type high pressure helium polarimeter[27].

IV. Spin Correlation Experiments

The following discussion is restricted to experiments in which polarized beams and targets are used; i.e., to "initial channel" spin correlation experiments. The structure of a spin correlation experiment is quite similar to that for an experiment in which the analogous polarization transfer coefficients are to be determined. For the spin-1/2 on spin-1/2 case, with beam and target polarization along the y-axis, the cross section for scattering to the left is

$$I_L(\theta) = I_0(\theta)[1 + p_y A_y(\theta) + p_y^T A_y^T(\theta) + p_y p_y^T C_{yy}(\theta)] \tag{35}$$

where the superscript "T" denotes a target quantity and we use $C_{yy}(\theta)$ for the spin correlation coefficient. As has been customary to date, both beam and target polarization are specified in the <u>projectile</u> helicity frame. The yields into a left and right counter, L and R, are

$$L = nN\Omega_L I_0[1 + p_y A_y + p_y^T A_y^T + p_y p_y^T C_{yy}]$$

$$R = nN\Omega_R I_0[1 - p_y A_y - p_y^T A_y^T + p_y p_y^T C_{yy}] \ , \tag{36}$$

since C_{yy} is an even function while A_y and A_y^T are odd functions of the scattering angle. Thus eight numbers may be obtained using four counting intervals with the projectile and target polarization in the ++, +-, -+, and -- configurations. Arguments concerning numbers of unknowns which are very similar to those given above for $K_y^{y'}$ measurements can be given. However, since both beam and target spin reversals are usually proper flips, ε' usually need not be considered.

The forms of the cross section for various particle spins and for various orientations of the polarizations are tabulated in ref. 15. A great many p-p spin correlation experiments have been performed, and several spin-1 on spin-1/2 experiments have been reported[28].

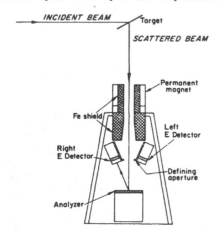

If p_y^T is small, the effect of C_{yy} on the cross section can be quite small. Thus, C_{yy} effects might easily be overwhelmed by, for example, changes in the (possibly) larger term $p_y A_y$ that can be induced by drifts in the beam polarization during the counting sequence. Crosstalk between the horizontal and vertical planes can also be a significant problem in this case.

Fig. 8. Silicon polarimeter of reflection geometry type as used by Baker.

V. Conventions

Some conventions for the reporting of spin transfer and spin correlation coefficients may be adopted at this conference. The following
comments may be considered in this connection.

1. All spin-1/2 and spin-1 observables reported should be in terms
of the standard operators defined in the "Madison Convention."

2. Coordinate systems for the definitions of the polarizations of
the various particles should be adopted. A specific case in point is
whether to use the projectile helicity frame or the target helicity frame
to describe the polarization of a target.

3. All definitions should be framed to relate closely to experiments as they are actually carried out, so that no combining of raw observations is required. This serves to a) avoid confusion and errors
in transformations and b) avoid mixing of (perhaps dissimilar) systematic errors. In this connection it might be noted that different observables suggest different experiments. Thus, for example, in a spin-1
on spin-1/2 correlation experiment, the measurement of C_{xx} and C_{yy} suggests experimental configurations in which the beam and target quantization axes are parallel to each other and parallel or perpendicular to
the scattering plane[15]. The direct measurement of the spherical observables $C_{11,11}$ and $C_{11,1-1}$ suggests configurations in which the quantization axes are inclined at 45° with respect to the detector planes and in
which both parallel and perpendicular beam vs. target polarization are
needed. In practice, the Cartesian observables are the ones that have
been measured.

4. Specific symbols for various quantities should be specified.

VI. Conclusion

In closing, there are a few points which I would like to re-emphasize.
First, the distinction between proper and nonproper flips is a crucial
one. The concept is applicable to the analysis of all types of polarization experiments, and a proper flip should be used whenever possible.
Secondly, I believe that we experimentalists should do a more complete
job of estimating and reporting our systematic errors. I hope that the
lists of possible sources of error given in the above might prove useful
in this regard. Thirdly, in a counting sequence, one should use the
most widely separated states of polarization available. This results in
maximizing the physical effect that one observes while leaving the systematic error effects nominally unchanged. Lastly, one should take maximum advantage of the fact that, in all polarization experiments, one
measures ratios, so that many factors may be made to cancel.

VII. Acknowledgments

I would like to thank R. Hardekopf, M. Baker, R. Walter, and N.
Jarmie for several helpful discussions.

References

* Work performed under the auspices of the U.S. Energy Research
  Development Administration

1) R. C. Hanna, Proc. 2nd Int. Symp. on Polarization Phenomena of
   Nucleons, Karlsruhe, Germany, 1965, edited by P. Huber and
   H. Schopper (Birkhauser Verlag) Basel and Stuttgart, 1966, p. 280
2) W. Grüebler, these proceedings, p.307
3) T. B. Clegg, these proceedings, p.111
4) P. A. Lovoi, G. G. Ohlsen, N. Jarmie, C. E. Moss, and D. M. Stupin,
   these proceedings, p.450
5) J. D. Hutton, W. Haeberli, and L. D. Knutson, Bull. Am. Phys. Soc.
   20 (1975) 576
6) G. G. Ohlsen and P. W. Keaton, Jr., Nucl. Instr. and Meth. 109 (1973)
   41. In eq. (30) of this reference, $\cos \theta$ should be replaced by
   $\csc \theta$
7) J. L. McKibben and J. M. Potter, these proceedings, p.863
8) J. M. Potter, these proceedings, p.91
9) U. Holm and H. Ebinghaus, Nucl. Instr. and Meth. 95 (1971) 39
10) G. G. Ohlsen and P. A. Lovoi, these proceedings, p.907
11) R. R. Cadmus, Jr. and W. Haeberli, these proceedings, p.901
12) V. König, W. Grüebler, and P. A. Schmelzbach, these proc. p.895, 897
13) A. P. Stamp, Phys. Rev. 153 (1967) 1052
14) J. S. Blair and H. S. Sherif, Bull. Am. Phys. Soc. 19 (1974) 1010;
    A. H. Hussein and H. S. Sherif, Phys. Rev. C8 (1973) 518
15) G. G. Ohlsen, Reports on Prog. in Physics 35 (1972) 717
16) L. R. Veeser, D. D. Armstrong, and P. W. Keaton, Jr., Nucl. Phys.
    A140 (1970) 177
17) T. B. Clegg, W. J. Thompson, R. A. Hardekopf, and G. G. Ohlsen,
    these proceedings, p.634
18) M. Baker, thesis, Univ. of Washington (1975)
19) G. C. Salzman, G. G. Ohlsen, J. C. Martin, J. J. Jarmer, and T. R.
    Donoghue, Nucl. Phys. A222 (1974) 512
20) G. G. Ohlsen, G. C. Salzman, C. K. Mitchell, W. G. Simon, and
    W. Grüebler, Phys. Rev. C8 (1973) 1639; G. C. Salzman, C. K.
    Mitchell, and G. G. Ohlsen, Nucl. Instr. and Meth. 109 (1973) 61
21) W. G. Weitkamp, W. Grüebler, V. König, P. A. Schmelzbach, R. Risler,
    and B. Jenny, these proceedings, p.514
22) F. H. Schmidt, R. E. Brown, J. B. Gerhard, and W. A. Kolasinski,
    Nucl. Phys. 52 (1964) 353
23) R. N. Boyd et al., Phys. Rev. Letters 29 (1972) 955
24) C. A. Weidner, M. Goldschmidt, D. Rieck, H. A. Enge, and S. B.
    Kowalski, Nucl. Instr. and Meth. 105 (1972) 205
25) J. P. Martin, J. L. Foster, H. R. Hiddleston, and S. E. Darden,
    Nucl. Instr. and Meth. 113 (1973) 477
26) J. Birchall et al., Nucl Instr. and Meth. 123 (1975) 105
27) R. A. Hardekopf, D. D. Armstrong, and P. W. Keaton, Jr., Nucl. Instr.
    and Meth. 114 (1974) 17
28) G. G. Ohlsen, R. A. Hardekopf, D. P. May, S. D. Baker, and W. T.
    Armstrong, Nucl. Phys. A233 (1974) 1; J. Chauvin, M. Fruneau, D.
    Garreta, Proc. Second Int. Conf. on Polarized Targets, Berkeley,
    1971, ed. G. Shapiro, Lawrence Berkeley Laboratory, LBL-500, p. 359;
    N. Berovic, W. E. Burcham, C. J. Clews, R. L. Maughan, and P. M.
    Rolph, Proc. Second Int. Conf. on Polarized Targets, Berkeley, 1971,
    ed. G. Shapiro, Lawrence Berkeley Laboratory, LBL-500, p. 355

DISCUSSION

König:

I would like to make two remarks about the so-called "supercube system" for tensor analysing power measurements.

1) Calibrating the beam alignment axis is a tedious and time consuming process. In contrast to your 54.7° measurements our measuring system is nearly independent of small deviations of the spin axis from the correct value.

2) In principle it is not necessary to have 4 detectors in 2 planes. With our system with 2 detectors right and left in the plane containing the spin axis one can avoid the complicated shape for a gas target you need in the supercube. Further it is easier to measure at extremely forward and backward angles or to use detectors for more scattering angles θ.

Ohlsen:

I agree. The 54.7° method depends on first order errors in the angle β. It is primarily of value for intercalibrating the relative p_{zz} of various spin states and for cases in which current integration for some reason is not possible or not sufficiently accurate. If better than a few percent accuracy is desired, this method is not a good one to use. On the other hand, spin angle accuracies of $\pm 1/2^\circ$ are not difficult to obtain in our particular setup. The $\beta=54.7^\circ$ method was perhaps overemphasized in my talk because it is conceptually most similar to standard spin-1/2 methods.
I also agree that two plane detection is not necessary in principle. In fact, our present favorite method, using $\beta = 90^\circ$ with $m_I=1$, $m_I=0$ and $m_I=-1$ beams, decouples the horizontal and vertical plane observations. On the other hand, A_{xx} and A_{yy} measurements simultaneously made are likely to be more accurate relative to each other than in cases when they are measured in separate counting intervals.
The philosophy of the supercube is that all schemes for spin 1 measurements can be implemented. Other than cost, there is no disadvantage in having these extra degrees of freedom available. If one choose to use a one-plane system, he can certainly do so in the supercube.

Trainor:

In one of the polarimeter schemes you described it is assumed that $P_y' = A_y$ for the measurement. Please comment on the effect which a finite polarimeter geometry has on this assumption.

Ohlsen:

Of course $P_y' = A_y$ is true only point-by-point. One performs finite-geometry calculations which assume $P_y' = A_y$ for each ray only.

ABSOLUTE CALIBRATION OF POLARIZATION MEASUREMENTS

W. Grüebler

Laboratorium für Kernphysik, Eidg. Techn. Hochschule
8049 Zürich, Switzerland

1. Introduction

The absolute determination of the polarization of particle beams has
been a difficult experimental problem for a long time. Apparently one
needs an analyser, the analysing power of which is known to a high precision. In the past, the double scattering technique has been used for
vector polarized particles. The precision however was limited, due to the
well known disavantages of this method, to an accuracy of about ± 0.02.
The p-α scattering is in this respect best known scattering over a wide
energy range[1,2,3].

For the measurement of the polarization of deuteron beams from polarized ion sources prior to the injection into an accelerator, the reaction
$^3H(d,n)^4He$ has provided the conventional method of calibrating the tensor
polarization at energies up to about 100 keV. At this low energy only incoming s waves contributes to the reaction cross section and therefore
the vector analysing power is zero. The tensor analysing power can be calculated assuming a pure $J = 3/2^+$ reaction; that means there is no $J = 1/2^+$
contribution[4]. Because of this simplified assumption, there is some doubt
about the correctness of the analysing power value which is generally
used. The same remark also applies to the mirror reaction $^3He(d,p)^4He$ if
it is used with the same assumption. Particular care has to be taken if
this reaction is used for measuring the polarization of beams degraded
from higher energies. In this case, the contribution of p waves in the entrance channel and charge-exchange depolarization effects of the deuterons[5] make this method less reliable as an absolute calibration.

The situation has been drastically changed since the last Polarization Symposium. Modern methods use the fact that model independent linear
and quadratic relations exist between the polarization observables of two
interacting particles. Similar relations also occur between the scattering
amplitudes of the corresponding reactions. Investigation of these relations reveals regions where the analysing power has to reach a theoretical
maximum value. The precise angle and energy of these calibration points
must be determined experimentally.

In another method, reactions with particular spin and parity configurations are used. This has the unique feature that the analysing power is
independent of energy and angle and can be calculated without any assumption about the reaction mechanism. The cross sections of these interesting
reactions however are quite small and show large fluctuations as a function
of energy and angle. Therefore this method is only suitable as a primary
standard for a few calibration points. For routine use convenient
reactions should be calibrated as secondary standards over large energy
region from these high precision points.

2. Spin - 1/2 particles

Shortly after the Madison Symposium, Plattner and Bacher[6] showed
that for spin -1/2 particles scattered from particles without spin the
analysing power A_y must reach a value ± 1 at some specific angle θ and
energy E, if the scattering amplitudes fulfil certain conditions. The
method is best understood by normalizing the non-spin-flip amplitude f
and the spin flip amplitude g by an overall phase factor such that f

becomes real and equal to unity for all scattering angles and energies. The necessary and sufficient relation to obtain values $A_y = \pm 1$ is then

$$g(\theta_1, E_1) = \pm i \ .$$

From experimentally determined phase shifts one can compute the amplitude g and plot the trajectory it describes in the complex plane as a function of scattering angle θ at a fixed energy E. Such g-trajectories are shown for p-α scattering at five energies in ref.6. Each time the g-trajectory passes through one of the points $\pm$ i a value $A_y = \pm 1$ occurs. In order to prove the existence of such a point, it is sufficient to show that for two different energies, one of the points $\pm$ i switch from inside to the outside of the closed g-trajectories. This is true, because the amplitude g is a continuous smooth function of energy and angle. To find the calibration point in question one can experimentally search for the highest value A_y as a function of angle and energy in that energy interval. In the particular case of p-α scattering three such points have been found for $A_y = 1$. By interpolation between phase shifts one obtains the results given in table 1[6].

Table 1. $A_y = 1$ points in p-α scattering

E_{lab} (MeV)	θ_{cm} (deg.)
1.90 ± 0.02	88.0 ± 0.25
6.35 ± 0.04	128.8 ± 0.1
12.30 ± 0.04	125.5 ± 0.1

Corresponding point in n-α scattering and ^{3}He - ^{4}He scattering are predicted in ref.6.

In general one will have to start with some reasonably calibrated analyser, so that the phase shifts used for the calculation of the g-trajectories can be reliably determined from the experiment. If no reliable phase shifts exist, the relations between polarization observables can be used directly to find fairly precise calibration points. For example a combination of particles which have the spin configuration $1/2 + 0 \rightarrow 0 + 1/2$ obey the quadratic relation

$$(A_y)^2 + (K_x^{x'})^2 + (K_z^{x'})^2 = 1 \ . \tag{1}$$

$K_x^{x'}$ and $K_z^{x'}$ are identical with the Wolfenstein parameters R and A respectively. When one of the observables is found to approach very nearly its maximum allowable value of unity the other terms in the quadratic relation must be close to zero. At appropriate energies and angles where it is known that the analysing power of the spin 0 target approaches its maximum value $|A_y| = 1$, relatively crude measurements of the two other observables determine the difference of $(A_y)^2$ from unity fairly precisely because of the quadratic nature of the relation. Examples are the following reactions: ^{4}He(p,p)^{4}He, ^{4}He(n,n)^{4}He, ^{4}He(t,t)^{4}He, ^{12}C(t,p)^{14}C and ^{1}H(π,π)^{1}H.

Experimentally, this method was first used by the Los Alamos group for p-α scattering near 12 MeV[7]. The same null technique has recently been used for the calibration of the triton beam from the new polarized triton source at Los Alamos[8]. In this case the polarization transfer coefficients of the t-α scattering are predicted from a R-matrix analysis and $K_z^{x'}$ plotted versus $K_x^{x'}$. Calculations have been made for contours at 11.0 and 12.0 MeV, which indicate that between these two energies a corresponding contour must pass through the origin, resulting in a $|A_y| = 1$. A more sophisticated method based on the same principles for π-p scattering is described in ref.9.

3. Spin-1 particles

For spin-1 particles the spin and parity configuration

$$1^+ + 0^+ \rightarrow 0^+ + 0^+ \tag{2}$$

is particular interesting for the absolute calibration of the tensor polarization. It has been shown[10] that as a consequence of parity invariance such a reaction has the remarkable and unique property that the analysing power is independent of bombarding energy and reaction angle. The analysing powers can be calculated without any assumptions about the reaction mechanism. The results are:

$$T_{20} = 1/\sqrt{2} \qquad T_{22} = \frac{1}{2}\sqrt{3} \qquad iT_{11} = T_{21} = 0$$

For practical application this type of reaction requires reasonable level spacing of the 0^+ state in the final nucleus. Since many of these reactions do not conserve isobaric spin, the cross section for this transition is quite small. As a result of the much stronger neighboring transitions, a fairly high background is observed in the corresponding spectra. Taking convenient level spacing into account, the following reactions are candidates for calibration experiments:

$$^{12}C(d,\alpha_2) \quad {}^{10}B^*(1.74 \text{ MeV}) \quad \Delta T = 1$$
$$^{16}O(d,\alpha_1) \quad {}^{14}N^*(2.31 \text{ MeV}) \quad \Delta T = 1$$
$$^{34}S(d,\alpha_2) \quad {}^{32}P^*(0.51 \text{ MeV}) \quad \Delta T = 0$$
$$^{42}Ca(d,\alpha_4) \quad {}^{40}K^*(1.64 \text{ MeV}) \quad \Delta T = 0$$

Whereas the first two reactions are isospin forbidden, the last two conserve isobaric spin. The accuracy of this method is in practical application mainly limited by the background subtraction. The technique is suitable as a primary standard for absolute calibration at a few energies but is to tedious for routine use.

The $^{16}O(d,\alpha_1)^{14}N^*(2.31 \text{ MeV})$ reaction has been used by several groups[11,12,13,14] to establish the calibration of other reactions as secondary standards.

In the spin configuration (2) any two parities can be changed. The relevant quantity is the product π_{tot} of all intrinsic parities, which in this case should be +1.

A reaction with a transition

$$1^+ + 0^+ \rightarrow 0^+ + 0^-$$

to an odd parity state in the final nucleus is discussed in ref.15. For such a $\pi_{tot} = -1$ reaction the following relations hold:

1) $T_{20} + \sqrt{6}T_{22} = -\sqrt{2}$

2) $|T_{11}|^2 + \frac{1}{2}|T_{20}|^2 + |T_{21}|^2 + |T_{22}|^2 = 1$

or in Cartesian description[7]:

1) $A_{yy} = 1$

2) $A_y^2 + (\frac{2}{3}A_{xz})^2 + [\frac{1}{3}(A_{zz} - A_{xx})] = 1$

The first relation can then also be used to calibrate the tensor polarisation. It has been proposed[7,15] to measure in addition T_{21} and determine with the aid of the second relation an absolute value of iT_{11}. This calibration procedure does not give the sign of iT_{11}. Reactions of the above type however occur but rarely in nuclear physics. One possibility is the $^{16}O(d,\alpha_3)^{14}N^*$ reaction to the 0^- final state with 4.91 MeV excitation energy. This 0^--state has also convenient spacing to its neighbor states.

The method of ref.6 using relations between scattering amplitudes can also be transposed to spin-one particles. Based on experimental determined phase shifts, one can calculate all M-matrix elements. Then relations between these matrix elements can be found for which a component of the analysing power reaches its theoretical maximum value. Because of the simple spin structure, d-α elastic scattering has been investigated extensively in this respect. The conditions on the matrix elements for obtaining maxima of the different analysing power has recently been published[16]. For the maxima of most analysing powers, the time reversal invariance gives additional conditions which are only valid at the corresponding maxima. These conditions can be used to test time reversal conservation.

Assuming only parity conservation, the simplest relation is found for the tensor component maximum $A_{yy} = 1$. For this case the spin non-flip amplitude M_{11} is equal to minus the spin flip amplitude M_{1-1}. For a proof of the existence of a value $A_{yy} = 1$ one has to study the function

$$M_{11}(E,\theta) + M_{1-1}(E,\theta) \tag{3}$$

in order to find the scattering angle and the corresponding energy where the function is equal to zero. Since the M-matrix elements are complex functions, it is again convenient to study the behaviour of the matrix elements or a combination of matrix elements in the complex plane. In order that $A_{yy} = 1$, the trajectory of this function has to pass through the origin. Calculations of such trajectories for energies between 3.0 MeV and 17.0 MeV are shown in ref.16. These trajectories are different from the polarization function of the spin-1/2 case in as much as they are not closed curves. This behaviour happens because the function under investigation is symmetric around $\theta = 180^\circ$, i.e. the values of the function run on the same trajectory from 180° back to 360°.

From these trajectories it is immediately clear that between 4.30 and 4.81 there has to exist not only one but two points where $A_{yy} = 1$[16]. This follows from the fact that the curve of 4.30 MeV does not enter the first quadrant, while the 4.81 trajectory however run through all four quadrants since the trajectory displacement is a smooth function of energy and angle, the curve passes with two branches through the origin. The crossing happens at roughly $\theta_{cm} \simeq 60^\circ$ and $\theta_{cm} \simeq 120^\circ$. The existence of another such point is predicted the same way near 12 MeV[16].

In order to have a survey over the whole energy and angular region, which is investigated experimentally a contour plot of the quantities Re $(M_{11} + M_{1-1}) = 0$ and $I_m (M_{11} + M_{1-1}) = 0$ can be drawn. The crossing points of the two curves are the points in question. It can really be seen from ref.16 that the three points are the only existing $A_{yy} = 1$ points in the energy region between 3.0 and 17 MeV and no other point has escaped the investigation.

Although this procedure reveals the energies and angles of the maxima fairly well, one has to keep in mind that the calculations are based on experimental phase shifts with statistical and instrumental errors. Furthermore the experimental data used to determine the phase shifts are not measured absolutely but have a certain scale factor. This should make it clear that in order to find the exact energy and angle of

A_{yy} = 1 points, an experimental mapping of the corresponding region is
required. The maximum itself is determined by an interpolation method.
The maximum value is known and can then serve to calibrate absolutely the
beam polarization or a secondary standard reaction. Examples of the
corresponding experimental data and the fitted curves for d-α scattering
can be found in ref.16. The energies and angle found for maximum A_{yy}
points for this scattering between 3 and 17 MeV are listed in table 2.

Table 2.   Energies and angles for A_{yy}= 1

$E_{d,lab}$(MeV)	θ_{cm} (deg.)
4.30	120.7
4.57	58.0
11.88	55.3

A_{yy}= 1 points are also suspected for reaction with light nuclei.
Seiler[17,18] found by an analysis of the experimental data from ref. 19
and 20 the following possible reactions

1) ^{3}He(d,p)^{4}He E_d ~ 9.0MeV θ ≃ 27°

2) ^{6}Li(d,α)^{4}He E_d ~ 5.5MeV θ ≃ 30°

E_d ~ 9.0MeV θ ≃ 90°

For the existence of a vector analysing power value A_y = ± 1 of d-α
scattering the corresponding linear relations among the M-matrix elements
show[16] that A_{yy} = 1 is a necessary but not sufficient condition. Two
other conditions have to be fulfilled besides M_{11} + M_{1-1} = 0. In such a
case one calculates for the remaining analysing powers the values
A_{xz} = 0, A_{xx} = A_{zz} = - 1/2. This means that a value A_y = ± 1 implies
that all other analysing powers are fixed. The existence of such a point
cannot be proven in the same way as for A_{yy}. The additional relations show
that near 12 MeV and θ_{cm} = 55° a possible A_y = -1 point may exist in accordance with A_{yy} = 1 [16]. Since the contours representing all three relations
show very small angles at the crossing points, there is a large uncertainty
in the precise scattering angle and energy of the crossing. Therefore an
excitation curve at θ_{cm} = 55.3°, the angle found for A_{yy} = 1, has been
measured. Although A_y reaches a value fairly near to -1 this maximum is
shifted to higher energy[21]. From this excitation curve it is obvious
that in this energy region no point exists where A_y is exactly equal to
-1.

Three contributed papers to this conference, two from Berkeley[20,23]
and one describing the first measurements with the polarized deuteron beam
at SIN[24] report a possible double maximum A_y = 1 and A_{yy} = 1 of d-α
scattering near 28 MeV.

An alternative method to calibrate vector polarization absolutely is
to take advantage of the well known ratio of vector to tensor polarization
in a polarized beam from a polarized ion source. This allows one to calibrate the vector analysing power at an A_{yy} = 1 point to a high precision
too.

4. Secondary standards

In order to monitor the beam polarization for routine measurements at arbitrary energies, it is convenient to use secondary standards, which must be calibrated with methods discussed above. Lamb shift sources incorporating a nuclear spin filter can use the quenching ratio method[25]. For tensor polarized deuterons, the ^{3}He(d,p)^{4}He reaction at $0°$ has been proposed previously[26] as secondary standard. The analysing power T_{20} of this reaction has been found to be very large between 2 and 16 MeV[14], [19]. A new calibration in this energy range has been reported to this Symposium[27]. The calibration uses the ^{16}O(d,α_1)^{14}N* reaction at 6.64 MeV as primary standard for the energy range between 0 and 12 MeV[19] and a point at 9.80 MeV for the 6.6 to 15.8 MeV energy range[14]. The $A_{yy} = 1$ points of the d-α elastic scattering have been used as a independent consistency check. The results[27] show clearly the attractive features of this reaction, namely a high analysing power coupled with a smooth behaviour without sharp resonances over a wide energy range. Parts of the excitation function can be very well reproduced by second order expansions[27].

In conclusion one can say that we have now powerful tools to calibrate the polarization much more precisely than with the double scattering method. By careful measurement an experimental error of 0.001 should be obtainable.

References

1) W.G. Weitkamp and W. Haeberli, Nucl. Phys. <u>83</u> (1966) 46
2) R.C. Hanna, Proceedings of the 2nd International Symposium on Polarization Phenomena of Nucleons (P. Huber, H. Schopper Eds., Birkhäuser, 1966) p. 280
3) H.H. Barschall, Proceedings of the 2nd International Symposium on Polarization Phenomena of Nucleons (P. Huber, H. Schopper Eds., Birkhäuser, 1966) p. 393
4) A. Galonsky, H.B. Willard and T.A. Welton, Phys. Rev. Lett. <u>2</u> (1959) 349
5) W.W. Lindstrom, R. Garett and U.v.Möllendorf, Nucl. Instr. and Meth. (1971) 385
6) G.R. Plattner and A.D. Bacher, Phys. Lett. <u>36B</u> (1971) 211
7) P.W. Keaton, Jr., D.D. Armstrong, R.A. Hardekopf, P.M. Kurjan and Y.K. Lee, Phys. Rev. Lett. <u>29</u> (1972) 880
8) R.A. Hardekopf, G.G. Ohlsen, R.V. Poore and N. Jarmie, contribution to this Symposium, p.903
9) C. Daum and P.W. Keaton, Jr., submitted to Nuclear Instruments and Methods
10) B.A. Jacobsohn and R.M. Ryndin, Nucl. Phys. <u>24</u> (1961) 505
11) S.E. Darden, R.M. Prior and W.K. Corrigan, Phys. Rev. Lett. <u>25</u> (1970) 1673
12) P.W. Keaton, Jr., D.D. Armstrong, G.P. Lawrence, J.L. McKibben and G.G. Ohlsen, Proceedings Third International Symposium on Polarization Phenomena in Nuclear Reactions (Eds. H.H. Barschall and W. Haeberli) p. 849
13) V. König, W. Grüebler, H. Ruh, R.E. White, P.A. Schmelzbach, R. Risler and P. Marmier, Nucl. Phys. <u>A166</u> (1971) 393
14) T.A. Trainor, T.B. Clegg and P.W. Lisowski, Nucl. Phys. <u>A220</u> (1974) 533
15) M. Simonius in Polarization Nuclear Physics (Lecture Notes in Physics, Vol. 30, ed. D. Fick) Springer Verlag 1974, p. 74

16) W. Grüebler, P.A. Schmelzbach, V. König, R. Risler, B. Jenny and
    D. Boerma, Nucl. Phys. A242 (1975) 285
17) F. Seiler, contribution to this Symposium, p.552
18) F. Seiler, F.N. Rad, H.E. Conzett and R. Roy, contribution to this
    Symposium, p.587
19) W. Grüebler, V. König, A. Ruh, P.A. Schmelzbach, R.E. White and
    P. Marmier, Nucl. Phys. A176 (1971) 631
20) W. Grüebler, V. König, P.A. Schmelzbach, Results of Measurements
    and Analyses of Nuclear Reactions Induced by Polarized and Unpola-
    rized Deuterons, ETH Zürich, 1973
21) W. Grüebler, P.A. Schmelzbach, V. König, R. Risler, B. Jenny,
    D.O. Boerma and W.G. Weitkamp, contribution to this Symposium, p.564
22) H.E. Conzett, W. Dahme, Ch. Leemann, J.A. Macdonald and J.P. Meulders,
    contribution to this Symposium, p.566
23) H.E. Conzett, F. Seiler, F.N. Rad, R. Roy and R.M. Larimer,
    contribution to this Symposium, p.568
24) G. Heidenreich, J. Birchall, V. König, W. Grüebler, P.A. Schmelzbach,
    R. Risler, B. Jenny and W.G. Weitkamp, contribution to this Symp. p.570
25) G.G. Ohlsen, J.L. McKibben, G.P. Lawrence, P.W. Keaton, Jr., and
    D.D. Armstrong, Phys. Rev. Lett. 27 (1971) 599
26) W. Grüebler, V. König, A. Ruh, R.E. White, P.A. Schmelzbach, R. Risler
    and P. Marmier, Nucl. Phys. A165 (1971) 505
27) P.A. Schmelzbach, W. Grüebler, V. König, R. Risler, D.O. Boerma and
    B. Jenny, contribution to this Symposium, p.899

DISCUSSION

Clegg:
    For a Lamb-shift polarized ion source equipped with a spin-filter,
    quench-ratio measurements of the beam polarization are possible. The
    Los Alamos group has shown for proton, deuteron and triton beams that
    the beam polarization determination from the quench ratio measured
    near the experimental target agrees to better than 0.005 with the
    beam polarization measured simultaneously with one of the true abso-
    lute calibration standards you mentioned. The quench ratio can then
    be used with confidence to monitor the beam polarization at energies
    where absolute standards are not available. Also the quench-ratio
    has the extremely attractive feature that measurements of the beam
    polarization are possible in many different experimental beam legs
    without the necessity for constructing a nuclear reaction beam polar-
    ization monitor for each beam leg. The quench ratio tells, of course,
    only the fraction of the beam which is polarized, and the experiment-
    er must be careful that the spin-filter parameters are set correctly
    and that the spin quantization axis direction is accurately known at
    the scattering chamber.

Ohlsen:
    I believe it is fair to regard the quench ratio as a primary standard.
    There are various corrections with magnitudes of tenths of one per-
    cent which must be applied. However, independent techniques for
    evaluating these corrections have been found. We think this will be
    particularly valuable for LAMPF. On the other hand, without some
    nuclear standards against which to check, one would be worried that
    all of the corrections which needed to be applied had not been thought
    of.

Grüebler:

The quench-ratio method is a very nice secondary standard if calibrated near the experimental target with one of the mentioned primary standards at a few energies. However, one relies on the stability of the beam polarization during the experiment since in the
quench-ratio method only a sample of the beam is taken before and
after each run. A simultaneously monitored beam polarization, which
averages over the same period as the experimental running time, is
safer in this respect. On the other hand a nuclear reaction beam
polarization monitor is a fairly simple device which can be moved
from one experiment to the other.

Donoghue:

In your ^{3}He(d,p) results for $A_{zz}(0^{\circ})$, you calibrate the polarization
of the deuteron beam by d-α scattering above 4 MeV, and then assume
the beam polarization is constant as you go to lower energy. Can you
comment on the problem of beam depolarization effects in the tandem
terminal which arise because of charge exchange in the residual gas ?
I'm particularly interested in how this affects your lower energy
data.

Grüebler:

The ^{3}He(d,p) results are calibrated by the ^{16}O$(d,\alpha_1)^{14}$N* reaction at
6.64 MeV. The d-α scattering at the $A_{yy} = 1$ points was used as a
consistency check in order to eliminate such depolarization effects
as you mentioned at other energies. Within the accuracy of the measurements, no such depolarization was observed. This, however, does
not exclude depolarization at lower energies.

Ohlsen:

In response to the remark of Dr. Donoghue regarding depolarization in
a tandem terminal, I would like to say that we now believe these
effects to be, at worst, only a few tenths of a percent rather than
up to several percent. A Nuclear Instruments and Methods paper on
these matters is in preparation. The basic change in our understanding has to do with our earlier neglect of the possibility of direct
double charge exchange collisions on a residual gas atom. For very
low pressure, these phenomena actually dominate. I would like to
acknowledge Dr. Clausnitzer who first pointed out to us that we
should consider these phenomena.

REPORT ON DISCUSSION SESSION:
TECHNIQUES OF POLARIZATION MEASUREMENTS; STANDARDS

G.G. Ohlsen
Los Alamos Scientific Laboratory, Los Alamos

Several interesting points were brought out in this discussion session, as follows. Most of these relate to the various contributed papers and details can usually be found there.

1) Mott-Schwinger scattering has finally been effectively used for neutron polarization detection by Galloway and Lugo. It appears, however, that the method does not compete with n-^{4}He scattering for practical applications, except in that it is in principle absolute.

2) C.P. Sikkema has constructed a ^{4}He wire chamber which has an output which distinguishes left from right scattering. The resolution is ~5%, which is comparable to ^{4}He gas scintillation counter performance. This might permit a complete and accurate mapping of the n-^{4}He analyzing power in a relatively small amount of running time. Since gas pressures would be low, presumably no multiple scattering correction would be needed.

3) Lisowski et al. have generated neutrons with known polarization by means of T($\vec{d}$,$\vec{n}$)^{4}He longitudinal polarization transfer at 0°. These neutrons were used to calibrate n-α scattering at several energies in the 20-30 MeV range. It was pointed out that D($\vec{t}$,$\vec{n}$)^{4}He transverse polarization transfer at 0° could also produce neutrons of known polarization. These phenomena occur because of angular momentum conservation and the simple spin structure of these reactions.

4) W. Haeberli emphasized that one detector polarization measurements are sometimes essential, in spite of the advantages of two detector methods. For example, if a thick target must be used, but good resolution is required, only a detector which views the target in transmission can be used. In such cases, a two or more detector monitor placed downstream can be used to control or compensate for spin direction errors, etc.

5) The spin -1 analyzing power T_{22}, as actually measured by most workers, is a linear combination of two independently measured quantities. Thus, at least in principle, an error matrix is required for a complete presentation of such measurements.

POLARIZATION EFFECTS IN GIANT ELECTRIC DIPOLE AND QUADRUPOLE RESONANCES[†]

H. F. Glavish

Department of Physics, Stanford University, Stanford, Ca. 94305 U.S.A.

1. Introduction

Most nuclei can be excited into strong electromagnetic multipole modes such as E1, M1, E2 and so forth. These modes are thought to be fundamental in character and have been studied intensively throughout the past two decades. They are usually observed as giant resonances in the following types of reactions:

(i)   Photonuclear reactions (γ,p), (γ,n), $(\gamma,2n)$

(ii)   Capture reactions (p,γ), (d,γ), (α,γ)

(iii)   Electro-excitation (e,e')

(iv)   Heavy particle excitation (p,p'), (d,d'), (α,α')

In photonuclear and capture reactions, if the particles involved are the same in each case, then the two reactions are related in a fundamental way by time reversal symmetry. In either case a γ-ray carries the multipole signature of the excitation and as a consequence the processes provide a very direct method for study of the giant resonances. Analysis of experimental data is often facilitated by the rapid convergence of the multipole expansion for the γ-rays, although at the same time, this means it is difficult to observe the higher order multipoles if they are superimposed on ones of lower order.

Electro-excitation (e,e') of a nucleus shares with photo-excitation the important feature that the electromagnetic part of the reaction mechanism is in principle fairly well understood so that unknowns in the problem may be related to nuclear structure considerations. Electro-excitation extends beyond photo-excitation in one important aspect. Whereas a photon producing an excitation energy of $\hbar\omega$ always transfers momentum $\hbar\omega/c$, an electron producing the same excitation energy may transfer momentum beginning near $\hbar\omega/c$ and ranging on upwards. By varying the momentum transfer it is often possible to selectively enhance the various multipole excitations.

Heavy particle excitation is a method that has evolved in the last few years for observing giant resonances. In this case there is no longer a simple electromagnetic probe and the giant resonances are superimposed on a background of complicated and unrelated excitations. To obtain quantitative information it is necessary to delineate the electromagnetic multipole resonance from the so-called background, and from other neighboring multipole resonances.

In recent times important developments have occurred in each of the above methods. The purpose of this talk is to focus our attention on one of these developments, namely the application of polarized beams to capture reactions and, inversely, polarization measurements in photonuclear reactions. This new work has enabled definitive and essentially model independent measurements of E2 strength to be made for various light nuclei. Also it has enabled us to explore more completely the validity of the shell model description of the giant E1 resonance.

These introductory remarks will be completed by showing examples of representative experimental data for each of the methods used to observe the giant resonances. In fig. 1 the results of Tanner et al.[1] are shown for proton capture on ^{15}N i.e. $^{15}N(p,\gamma)^{16}O$. The giant E1 resonance in

[†] Supported in part by the National Science Foundation.

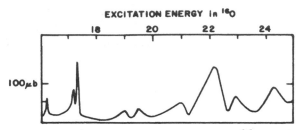

Fig. 1.   The cross section as a function of the ^{16}O excitation energy
          for the reaction $^{15}N(p,\gamma)^{16}O$.   Ref. [1]).

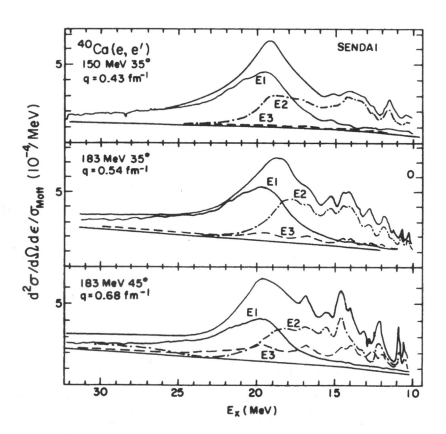

Fig. 2.   Electro-excitation of ^{40}Ca showing the linear background and
          empirical fits to isolate various multipole resonances.
          Ref. [4]).

[16]O, often called the "Giant Dipole Resonance" (GDR), is discernible at excitations in the range 21 to 27 MeV by the large cross-section and by the large value for the energy integrated cross section $\int \sigma dE$. Angular distribution measurements[2]) on the outgoing γ-ray show that the GDR is predominantly dipole in character. The large value of $\int \sigma dE$ supports an El assignment and rules out a dominant Ml assignment. Measurements[3]) of the γ-ray polarization have also confirmed an El assignment for the giant resonance.

In fig. 2 the results for electro-excitation[4]) on [40]Ca are shown. A linear background subtraction and a semi-empirical fit are used to isolate the various multipoles. Actually, it turns out not to be possible to distinguish between E2 and E0, a common problem with (e,e') experiments. The results from (α,α') experiments[5]) are shown in fig. 3. The vertical arrows mark the energy location $60/A^{1/3}$ where a compact iso-scalar E2 resonance has been predicted on quite general grounds by Bohr and Mottelson[6]). A resonance is evident in the data at this location except for the lightest nucleus [27]Al.

2. Polarization effects

2.1. GENERAL

In a capture or photonucleon reaction it is generally true that two (or more) partial waves in the nucleon channel contribute to the formation of an El excitation. The partial waves contribute to the angular distribution independently and by mutual interference. The reaction matrix elements associated with the partial waves cannot be delineated because their phases are unknown.

It is in this respect that polarization measurements have proved to be valuable. Interference between the partial waves can produce a polarization effect which if detected provides physically independent information. In many instances this enables the reaction matrix elements to be extracted uniquely, with respect to both their phases and amplitudes.

Nucleon polarization effects associated with the GDR were first detected by Bertozzi et al.[7]) with the reaction [16]O($\gamma,\vec{n}$)[15]O. They found that neutrons emerging at an angle of 45° were quite strongly polarized throughout most of the GDR. This type of measurement has been continued at Yale[8,9]). Also, photoneutron polarization studies of the giant Ml resonance in [208]Pb are reported[10]) at this conference. The ($\gamma,\vec{n}$) experiments are difficult. In particular one must distinguish neutrons which leave the residual nucleus in a ground state from other neutrons which may be present.

The first polarization effect associated with the proton channel of a GDR was measured at Stanford[11]) with the reaction [11]B($\vec{p},\gamma$)[12]C. Here an asymmetry was observed in the angular distribution of the γ-rays when the reaction was initiated with polarized protons. The polarized protons were obtained from the Stanford FN Van de Graaff accelerator which is fitted with a polarized ion source. Meanwhile these ($\vec{p},\gamma$) studies have been extended to include the GDR of [4]He, ^{16}O, [20]Ne, [28]Si, [32]S, [90]Zr and also measurements of E2 strength. Measurements of the ($\vec{p},\gamma$) type have also been made by Weller et al.[12,13]). Their results for the GDR of [59,57,55]Co are reported at this conference[13]).

2.2. ANGULAR DEPENDENCE OF ANALYZING POWER AND DIFFERENTIAL CROSS SECTION

The differential cross section for an (x,γ) reaction has an angular dependence of the form

$$\sigma(\theta,E) = A_o(E)\{1 + \sum_{k=1} a_k(E) P_k(\cos\theta)\} \tag{1}$$

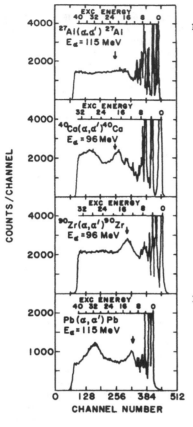

Fig. 3. (Left) Results for (α,α') excitations of ^{27}Al, ^{40}Ca, ^{90}Zr and ^{208}Pb. Ref. [5].

Fig. 4. (Below) Angular distributions $\sigma(\theta)$ and analyzing powers $A(\theta)$ (plotted as $\sigma(\theta)A(\theta)$) for the reaction ^{15}N$(\vec{p},\gamma)^{16}$O at three proton bombarding energies. Ref. [21].

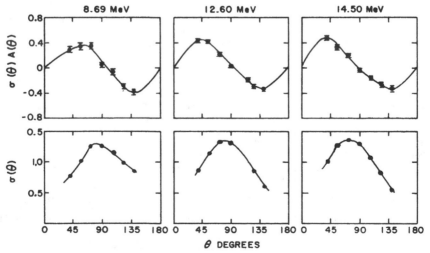

while the analyzing power $A(\theta,E)$ is expandable in terms of associated Legendre Polynomial functions:

$$A(\theta,E)\sigma(\theta,E) = A_o(E) \sum_{k=1} b_k(E) P_k^1(\cos\theta) \qquad (2)$$

The overall normalization factor $A_o(E) = \int \sigma(\theta,E) d\Omega/4\pi$ contains the energy dependence of the cross section $\int \sigma(\theta,E) d\Omega$.

The angular complexity of $\sigma(\theta,E)$ and $A(\theta,E)$ depends on the number of non-vanishing coefficients a_k and b_k which are present. The rules which govern this are the same for b_k as they are for a_k:

(i) For pure γ-radiation of multipolarity L; $k \leqslant 2L$ and k is even.

(ii) For interfering radiations of multipolarity L,L'; $k \leqslant L + L'$ and k is $\{\begin{smallmatrix}\text{even}\\\text{odd}\end{smallmatrix}\}$ when the radiations have the $\{\begin{smallmatrix}\text{same}\\\text{opposite}\end{smallmatrix}\}$ parity.

Table 1

Radiation Type	Non-vanishing coefficients	
	$\sigma(\theta,E)$	$A(\theta,E)$
pure E1 or M1	a_2	b_2
pure E2	a_2,a_4	b_2,b_4
(E1,M1)	a_1	b_1
(E2,E1)	a_1,a_3	b_1,b_3
(E1,M2) and (E2, M1)	a_2	b_2

The coefficients a_k and b_k can be expressed as a sum of bilinear products of the reaction matrix elements. A specific example, for the reaction $^{15}N(p,\gamma)^{16}O$, is considered in section 4.

The results in table 1 show that pure dipole radiation will be characterized by a_2 and b_2 coefficients alone. The presence of $k = 3$ and $k = 4$ coefficients will be a signature for quadrupole radiation. The $k = 1$ coefficients can result from either (E1,M1) interference or (E1,E2) interference.

3. Electromagnetic sum rules

Angular distribution and polarization measurements alone are not able to distinguish between E and M radiation. To make this distinction requires a measurement of the γ-ray polarization, which has been made in isolated cases[3]), or a consideration of the electromagnetic sum rules[14]) which define upper limits for the energy weighted, integrated, photonuclear, cross sections: i.e. $\int \sigma_L(E) f_L(E) dE \leqslant S_L$ where $f_L(E)$ is the weighting function, L is the radiation multipolarity and S_L is the sum rule limit.

There are three sum rules of particular interest to us here:

(i) Isovector $(\Delta T = 1)$ E1: $\int \sigma_{E1}(E) dE \cong 60 \dfrac{NZ}{A}$ mb-MeV

(ii) Isoscalar $(\Delta T = 0)$ E2: $\int \dfrac{\sigma_{E2}(E)}{E^2} dE \cong 2.4 \times 10^{-4} \dfrac{Z^2}{A^{1/3}}$ mb-MeV^{-1}

(iii) Isovector $(\Delta T = 1)$ E2: $\int \dfrac{\sigma_{E2}(E)}{E^2} dE \cong 2.4 \times 10^{-4} \dfrac{NZ}{A^{1/3}}$ mb-MeV^{-1} (3)

The first, which applies to El radiation, is the same as the Thomas-Reiche-Kuhn sum rule of atomic physics. Its validity rests on having a well defined number of particles $A = N + Z$ and that these particles experience an interaction potential which commutes with the El electromagnetic transition operator. For nuclei with $A > 40$ one universally observes a GDR centered at $\cong 76$ $A^{-1/3}$ MeV, of width $\cong 5$ MeV, and integrated cross section about 20% higher than the El sum rule[15]. For lighter nuclei the GDR is also observed but there are small departures from the above systematics[15]. The large values for the integrated cross section rules out assignments other than El for the bulk of the photonuclear strength observed in the GDR.

The isoscalar E2 sum rule was derived by Gellman and Telegdi[16] and its validity depends only on a knowledge of the value of $<r^2>$ for the nuclear charge distribution. The isovector E2 sum rule, on the other hand, is subject to more uncertainty since it depends on the commutator of the nuclear interaction with the E2 electromagnetic transition operator[14]. The existence or otherwise of a giant quadrupole resonance (GQR) is far less clear than the GDR and is currently a subject of intensive study[17]. A compact isoscalar E2 resonance has been predicted[6] on quite general grounds at a location of about $60/A^{1/3}$ and seems to have been identified[5,18,19] in nuclei with $A \geqslant 40$ in heavy-particle excitation experiments such as (p,p'), (d,d') and (α,α').

4. ^{16}O

The ^{16}O nucleus has been the subject of a great deal of study. On the experimental side the GDR has been observed in (γ,n) and (p,γ) reactions and these have included polarization measurements[7,20-23]. On the theoretical side the nucleus has a doubly closed shell and many calculations have been made in the framework of the shell model[24-28].

4.1. REACTION MATRIX ELEMENTS FOR THE NUCLEON CHANNELS

If we consider the spins and parities involved in the capture reaction, leading to a ground state ^{16}O nucleus

$$p + {}^{15}N \rightarrow {}^{16}O + \gamma$$
$$1/2^+ \quad 1/2^- \qquad 0^+$$

we deduce the following proton partial wave restrictions for the various γ-radiations: El: $s_{1/2}$, $d_{3/2}$; E2: $p_{3/2}f_{5/2}$; Ml: $p_{1/2}$; $p_{3/2}$. In the region of the GDR of ^{16}O it follows that $(Ml)^2 << (El)^2$ and so the contribution to a_2 and b_2 from Ml radiation can be safely neglected. To the extent that higher order multipoles M2, E3, etc. can also be neglected, the unpolarized angular distribution coefficients a_2, a_3, a_4 and the analyzing power coefficients b_2, b_3, b_4 will depend only on the El and E2 reaction matrix elements. The explicit dependence can be found from the usual Racah Algebra:

$$A_o = 0.75(s_{1/2}^2 + d_{3/2}^2) + 1.25(p_{3/2}^2 + f_{5/2}^2)$$

$$A_o a_2 = 1.061 s_{1/2} d_{3/2} \cos(s,d) - 0.375 d_{3/2}^2 + 0.625 p_{3/2}^2 + 0.7143 f_{5/2}^2$$
$$- 0.4374 p_{3/2} f_{5/2} \cos(p,f)$$

$$A_o a_3 = 1.936 s_{1/2} f_{5/2} \cos(s,f) + 2.012 d_{3/2} p_{3/2} \cos(d,p) - 1.095 d_{3/2} f_{5/2} \cos(d,f)$$

$$A_o a_4 = 3.499 p_{3/2} f_{5/2} \cos(p,f) - 0.7143 f_{5/2}^2$$

$$A_o b_2 = -0.5303 s_{1/2} d_{3/2} \sin(s,d) + 0.3645 p_{3/2} f_{5/2} \sin(p,f)$$

$$A_o b_3 = -0.6455 s_{1/2} f_{5/2} \sin(s,f) + 0.6708 d_{3/2} p_{3/2} \sin(d,p)$$

$$+ 0.0913 d_{3/2} f_{5/2} \sin(d,f)$$

$$A_o b_4 = -0.8748 p_{3/2} f_{5/2} \sin(p,f) \tag{4}$$

Since one of the phases can be arbitrarily chosen (we choose the $s_{1/2}$ phase = 0) there are seven real quantities needed to define the E1 and E2 reaction matrix elements, at a particular energy. The six experimental coefficients a_2, a_3, a_4, b_2, b_3, b_4 plus A_o are sufficient for this purpose. Note that without polarization measurements the individual reaction matrix elements could not be determined and definitive values for the E1 and E2 strength could not be obtained.

4.2. EXPERIMENTAL RESULTS FROM THE $(\vec{p},\gamma)$ REACTION

The first experiments[20] measured only the a_2 and b_2 coefficients which enabled the E1 matrix elements to be found with the assumption $(E2)^2 \ll (E1)^2$. Later, the experiments[21] were extended to include measurements of all of the coefficients up to $k = 4$.

Typical data from the later experiments are shown in fig. 4. The γ-rays were detected with a 24 cm $\times$ 24 cm NaI spectrometer[29]. Instrumental asymmetries were eliminated by fast flipping of the proton polarization using rf transitions in the polarized ion source. The target was ^{15}N gas contained in a gas cell at a pressure of 0.6 atm.

The experimental data in fig. 4 shows the strong dipole character of the γ-radiation in that the angular dependence of $\sigma(\theta)$ is derived mainly from $P_2(\cos\theta)$ while that for $\sigma(\theta)A(\theta)$ is derived mainly from $P_2^1(\cos\theta)$. However, least squares fits to the data establish as well, the presence of non-zero $k = 1$, 3 and 4 coefficients. The values obtained for a_2 and b_2 are shown in fig. 5.

A least squares search routine was used to extract the E1 and E2 reaction matrix elements and phases using eq. (4). The results for E1 are shown in fig. 5 and for E2 in fig. 6. It should be noted that the amplitudes of the matrix elements are plotted as relative amplitudes: i.e. at each energy their normalization is adjusted such that $A_o = 1$ in eq. (4).

Because of the quadratic nature of eq. (4) there are actually two solutions at each energy. These are labeled I and II in fig. 5. Presumably one of the solutions is physically correct and the other is merely a mathematical solution. On theoretical grounds (see section 4.4) solution I is thought to be the correct physical solution. The E2 amplitudes and the extracted E2 cross section plotted in fig. 6 corresponds to solution I. Actually it turns out that solution II produces essentially the same E2 cross section as does solution I.

The E2 cross section exhausts 50% of the E2 sum rule expressed in eq. (3). One might suppose that a similar E2 strength would be observed in the neutron channel on the basis that the unpolarized (p,γ) and (n,γ) data show remarkable similarity[1,30,31]. In any case, the $(\vec{p},\gamma)$ measurements have established a large amount of E2 strength spread over an energy of a few MeV, and centered one or two MeV above the center of the GDR. As to the iso-character of the E2 strength one cannot be sure except

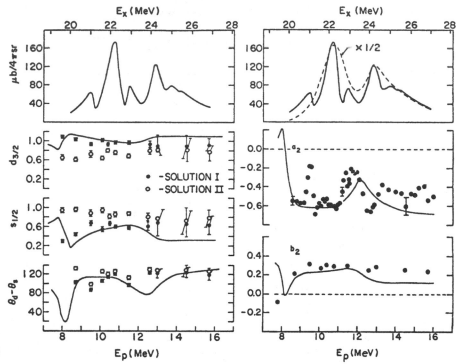

Fig. 5. Left: The amplitudes and phases of the El reaction matrix elements
obtained by fitting to the data of fig. 4 using eq. (4). For
reference the cross section for the $^{15}N(p,\gamma)^{16}O$ reaction is shown
in the upper part of the figure. Right: Values obtained for the
a_2 and b_2 coefficients of eqs. (1) and (2) by fitting to the data
of fig. 4. The solid curves in the lower parts of the figure are
the results of the theoretical calculations discussed in section
4.4. The broken curve located at the upper right is the calculated cross section. Ref. [34]).

that (α,γ) measurements[32]) have shown the presence of isoscalar E2
strength below the GDR exhausting 55% of the isoscalar E2 sum rule. Thus
it would seem that at least some of the E2 strength observed in the $(\vec{p},\gamma)$
reaction is isovector in nature.

4.3. EXPERIMENTAL RESULTS FROM THE $(\gamma,\vec{n})$ REACTION

    The first such measurements were made by Bertozzi et al.[7]). They
are in overall agreement with, although not nearly as precise as, the
$(\vec{p},\gamma)$ measurements. The $(\gamma,\vec{n})$ reaction has also been studied at Yale and
their results presented at Asilomar[9]) were not very different from the
$(\vec{p},\gamma)$ results. However, more recent work by the same group shows a very
different behaviour in the $(\gamma,\vec{n})$ reaction. At present there does not
seem to be an explanation for the dramatic differences.

4.4. THEORETICAL WORK

    The strong El excitations in ^{16}O are interpreted in the framework

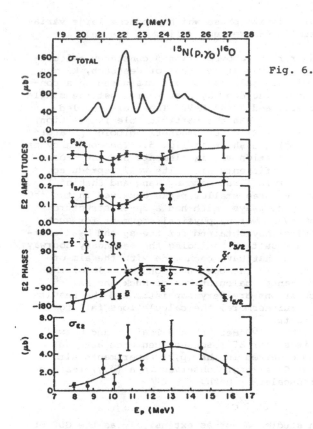

Fig. 6. The amplitudes and phases of the E2 reaction matrix elements obtained by fitting to the data of fig. 4 using eq. (4). The total cross section (E1 + E2) is shown in the upper part of the figure and the extracted E2 cross section is in the lower part of figure.

of the shell model as single particle excitations out of a closed shell to a higher-lying unfilled shell[24]. Such states can be coupled strongly to the closed shell configuration by E1 radiation, and at the same time have sufficient energy to decay by nucleon emission. Calculations must take into account the interaction between the excited particle and the hole left in the lower shell[25]. This particle-hole interaction mixes the unperturbed single particle excitations in such a way as to push one or a few of the levels a long way from the unperturbed levels and endow them with most of the E1 strength[26].

The GDR of ^{16}O exhibits two dominant peaks at excitation energies of 22.3 and 24.45 MeV. These two peaks are observed in the (γ,n), (p,γ) and (γ,all) reactions and, in the shell model scheme discussed above, they are interpreted as predominantly $d_{5/2}p_{3/2}^{-1}$ and $d_{3/2}p_{3/2}^{-1}$ particle-hole excitations respectively. The other three slightly sharper, small peaks seen in the ^{16}O GDR are supposedly more complicated np-nh excitations[27] but are definitely not well understood.

A feature of the ^{16}O GDR that comes as a surprise is the near constancy of the relative $s_{1/2}$ and $d_{3/2}$ E1 reaction matrix elements on passing from left to right through the two dominant peaks even though the two peaks have entirely different particle-hole configurations. This condition of constancy is also evident in the a_2 and b_2 coefficients. In fact the fluctuations in a_2 arise almost entirely from small

variations in the $(s_{1/2}, d_{3/2})$ relative phase which produces large variations in the interference term in a_2 [see eq. (4)] when the relative
phase is near 90°.

In an attempt to account for the abovementioned constancy calculations have been made using the doorway state ideas of Feshbach, Kerman
and Lemmer[33]. The details of this calculation are discussed in a contribution[34] to this conference. The doorway states are just the single
particle-hole excitations at 22.3 and 24.45 MeV. These are coupled
strongly to the open nucleon channel via the particle hole interaction,
as well as to the γ channel via the El electromagnetic dipole operator[27].
The results of the calculation[34] are shown in fig. 5. The constancy of
the relative $s_{1/2}$ and $d_{3/2}$ reaction matrix elements and the gross
features of the a_2 and b_2 coefficients are quite well reproduced.

Doorway state calculations were first made by Wang and Shakin[27]
in an attempt to account for the three smaller peaks observed in the ^{16}O
GDR. They introduced secondary doorways which had a 3p-3h structure.
There does not seem to be any overpowering physical argument for such
3p-3h structures and the results they obtained for the a_2 and b_2 coefficients were not improved as a result of including the secondary doorways.

The only other calculations that have been made with the aim of
accounting for phenomena observed in the open nucleon channels of the
GDR are based on the coupled channel calculations of Buck and Hill[28].
Although these types of calculations are very sophisticated the results
tend to be a disappointment. Furthermore, the calculations fail to
provide any simple physical picture.

With regard to E2 strength in ^{16}O Bertsch and Tsai[35] and Krewalk
et al.[36] predict a compact isoscalar E2 resonance centered at about
21.5 MeV. This has not been observed in the $(\vec{p}, \gamma)$ experiments although
one must keep in mind that the E2 strength observed in a $(\vec{p}, \gamma)$ reaction
can be either isovector or isoscalar or both.

5. ^{12}C

The GDR of ^{12}C has been studied about as extensively as the GDR of
^{16}O. The first polarization measurements[11] were made on ^{12}C with the
$^{11}B(\vec{p}, \gamma)^{12}C$ reaction but these have since been extended to extract E2
strength.

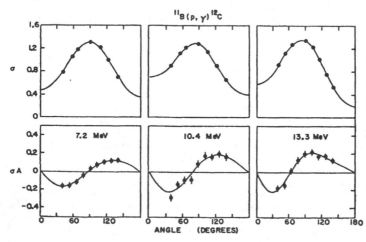

Fig. 7. Angular
distributions σ
and analyzing
powers A (plotted
as σA) for the
reaction
$^{11}B(\vec{p}, \gamma)^{16}O$ at
three proton bom-
barding energies.

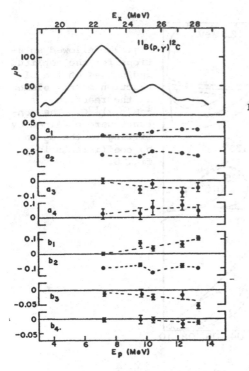

Fig. 8. Values obtained for the coefficients a_k and b_k from the data of fig. 7. The cross section for the reaction $^{11}B(\vec{p},\gamma)^{12}C$ is shown in the upper part of the figure.

Since the spin and parity of ^{11}B is $3/2^-$ there are more proton partial waves contributing compared with the case of ^{16}O.

$$E1 \qquad s_{1/2},\ d_{3/2},\ d_{5/2}$$

$$E2 \qquad p_{1/2},\ p_{3/2},\ f_{5/2},\ f_{7/2}$$

Even all of the a and b coefficients up to $k = 4$ together with A_o will not uniquely determine all of the reaction matrix elements. Nevertheless, the additional constraints provided by the polarization data has proved very valuable for extracting at least a lower limit on the E2 strength.

The same experimental procedure as in the case of ^{16}O has been used in ^{12}C. The data for three energies are shown in fig. 7 and the extracted a and b coefficients are shown in fig. 8. While the GDR shows the usual dominant dipole behaviour it is also evident from fig. 8 that E2 radiation is present at higher energies.

The allowed E1 reaction matrix elements can be determined from a_2, b_2 and A_o. The results are shown in fig. 9 where contours for $s_{1/2}$, $d_{3/2}$ and $\phi_{d_{3/2}}$ are plotted as a function of $|d_{5/2}|^2$ and $\phi_{d_{5/2}}$. Again, relative amplitudes are shown, i.e. the amplitudes are normalized such that $A_o = 1$.

To better define the E1 reaction matrix elements it is necessary to resort to theory. Since the doorway state calculations worked quite well for ^{16}O similar calculations have been performed for ^{12}C. The results are shown in fig. 10. There are two doorway states: a strong one at

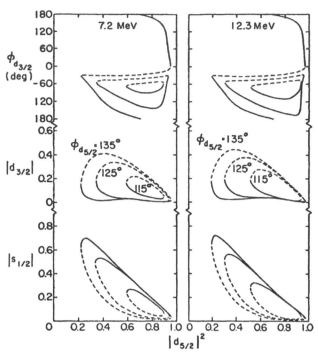

Fig. 9. Allowed solutions for the amplitudes and phases of the El reaction matrix elements in the reaction $^{11}B(\vec{p},\gamma)^{12}C$. The contours were obtained by fitting to the a_2 and b_2 coefficients in fig. 8.

22.5 MeV with a predominantly $d_{5/2}p_{3/2}^{-1}$ structure and a much weaker one at 23.5 MeV with a predominantly $d_{3/2}p_{3/2}^{-1}$ structure. It may be noted that the calculated matrix elements are near allowed solutions of fig. 9 and, moreover, the calculated a_2 and b_2 coefficients agree fairly well with experiment. Of particular interest is the very small calculated value for the $d_{3/2}$ amplitude. The fact that the $d_{3/2}$ phase varies a great deal with energy is of little consequence in view of the near vanishing $d_{3/2}$ amplitude.

In order to determine the E2 amplitudes and hence an E2 cross section a least squares fitting routine was used as for the case of ^{16}O. However the El matrix elements were constrained to agree approximately with the calculated values. The values obtained for the E2 cross section are shown in fig. 11. The E2 strength is centered near 12 MeV and exhausts approximately 40% of the E2 sum rule expressed in eq. (3). There is also evidence for a large amount of E2 strength, at this same energy location, in (α,α') measurements[37].

6. ^{32}S

The GDR of ^{32}S has been studied in much the same way as ^{12}C and ^{16}O, using the reaction $^{31}P(\vec{p},\gamma)^{32}S$. This has been discussed in detail in a contributed paper to this conference[38]. The spin and parities of the particles in the reaction allow a unique determination of the E2 strength as in the case of ^{16}O.

A systematic trend may be observed in the $(\vec{p},\gamma)$ E2 strength as we pass from ^{12}C to ^{32}S. In ^{12}C the E2 strength is noticeably above the GDR, in ^{16}O it is just slightly above, and in ^{32}S it is spread throughout the

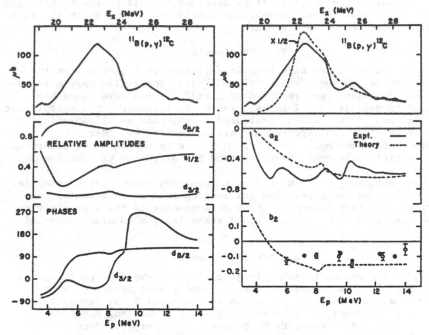

Fig. 10. Calculated values of the E1 reaction matrix elements, and the a_2 and b_2 coefficients for the reaction $^{11}B(p,\gamma)^{12}C$.

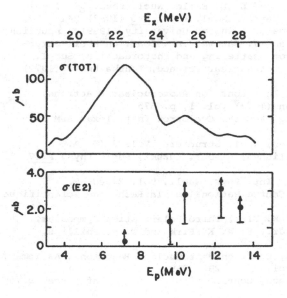

Fig. 11. The total (E1 + E2) and extracted E2 cross section for the reaction $^{11}B(p,\gamma)^{12}C$.

GDR. Of course, the $(\vec{p},\gamma)$ reaction does not see all of the E2 strength.
Indeed, (α,γ) measurements have found substantial E2 strength in light
nuclei at energies below the GDR.

7. Other nuclei

Other nuclei which have been investigated at Stanford with the
$(\vec{p},\gamma)$ reaction are ^{4}He, ^{20}Ne, ^{28}Si and ^{90}Zr. The preliminary measurements on ^{4}He have been reported earlier[39,40]) and have since been
extended to map out the E2 as well as the E1 strength. The measurements
on ^{20}Ne and ^{28}Si have concentrated in the main on E1 radiation[23]).
The case of ^{90}Zr is reported at this conference[41]). Here $T_>$
analogue resonances built on a $T_<$ GDR have been studied to determine
their particle-hole configurations from the extracted E1 reaction matrix
elements.

The isotopes ^{55,57,59}Co of cobalt have been investigated by Weller
et al. using the $(\vec{p},\gamma)$ reaction and are reported at this conference.
A striking result obtained in these measurements is the insensitivity
of the analyzing power to the ground state isospin quantum number of
the cobalt isotopes.

8. Acknowledgements

It is a pleasure to acknowledge colleagues who have participated
at Stanford in this research. Several people were involved in the
experimental work and interpretation of the data: S. S. Hanna, J. R.
Calarco, D. G. Mavis, R. LaCanna, G. King, E. Kuhlmann, R. Avida,
C. C. Chang, G. A. Fisher, E. M. Diener, P. M. Kurjan. Much of the
theoretical work on the doorway state calculations was carried out
by D. G. Mavis.

References

1) N. W. Tanner, G. C. Thomas and E. D. Earle, Nucl. Phys. 52 (1964)
    29; E. D. Earle and N. W. Tanner, Nucl. Phys. A95 (1967) 241
2) W. J. O'Connell, Ph.D. Thesis, Stanford University (1969), unpublished
3) W. C. Barber, E. Hayward and J. Sazawa, Int. Conf. on Nucl. Struc-
    ture Studies using Electron Scattering and Photoreaction, Suppl.
    to Research Report of Lab. of Nuclear Science, Tohuka Univ. (1972)
    Vol. 5, p. 313
4) Y. Torizuka et al., Proc. Int. Conf. on Photonuclear Reactions,
    Asilomar, ed. B. L. Berman (1973) Vol. 1, p. 675
5) D. Youngblood et al., Prog. Report, Cyclotron Inst. Texas A&M Univ.
    (1974) p. 1
6) A. Bohr and B. R. Mottelson, Nucl. Structure, Vol. 2 (W. A.
    Benjamin, N.Y.) to be published; also T. Suzuki, Nucl. Phys. A217
    (1973) 182
7) W. Bertozzi et al., Congr. Int. Phys. Nucl., Vol. 2, ed. P.
    Gugenberger (Editions du Centre National de la Recherche Scientifique,
    Paris, 1964) p. 1026
8) G. W. Cole, Jr. and F. W. K. Firk, Third Polarization Symposium,
    p. 626; also G. W. Cole, Jr., F. W. K. Firk and T. W. Phillips,
    Phys. Lett. 30B (1969) 91
9) R. Nath et al., Proc. Int. Conf. on Photonuclear Reactions, Asilomar,
    ed. B. L. Berman (1973) Vol. 1, p. 2B5, 2B9
10) R. J. Holt and H. E. Jackson, contribution to this conference, p.759

11) H. F. Glavish et al., Phys. Rev. Lett. $\underline{28}$ (1972) 766
12) H. R. Weller et al., Phys. Rev. Lett. $\underline{32}$ (1974) 177
13) H. R. Weller et al., contribution to this conference , p.757
14) W. J. O'Connell, Proc. Int. Conf. on Photonuclear Reactions, Asilomar, ed. B. L. Berman (1973) Vol. 1, p. 71
15) E. Hayward, Nucl. Structure and Electromagnetic Interactions, ed. N. McDonald (Plenum Press, N.Y., 1965) p. 141
16) M. Gell-Mann and V. L. Telegdi, Phys. Rev. $\underline{91}$ (1953) 169
17) S. S. Hanna, Proc. of the Int. Conf. on Nuclear Structure and Spectroscopy, eds. H. P. Blok and A. E. L. Dieperink (Scholar's Press, Amsterdam, 1974) Vol. 2, p. 249
18) M. B. Lewis and F. E. Bertrand, Nucl. Phys. $\underline{A196}$ (1972) 337
19) C. C. Chang, Preprint, 1975
20) S. S. Hanna et al., Phys. Lett. $\underline{40B}$ (1972) 631
21) S. S. Hanna et al., Phys. Rev. Lett. $\underline{32}$ (1974) 114
22) H. F. Glavish, Fifth Symposium on the Structure of Low-Medium Mass Nuclei, eds. J. P. Davidson and B. D. Kern (Univ. Press of Kentucky, 1972) p. 233
23) H. F. Glavish, Proc. Int. Conf. on Photonuclear Reactions, Asilomar, ed. B. L. Berman (1973) Vol. 2, p. 755
24) D. H. Wilkinson, Physica $\underline{22}$ (1956) 1039
25) J. P. Elliot and B. H. Flowers, Proc. Royal Soc. $\underline{242}$ (1957) 57
26) G. E. Brown, Unified Theory of Nuclear Models and Forces (North Holland Publ. Co., Amsterdam, 1967)
27) W. L. Wang and C. M. Shakin, Phys. Rev. $\underline{C5}$ (1972) 1898
28) B. Buck and A. D. Hill, Nucl. Phys. $\underline{A95}$ (1967) 271
29) M. Suffert, W. Feldman, J. Mahieux and S. S. Hanna, Nucl. Instr. and Meth. $\underline{63}$ (1968) 1
30) J. W. Jury, J. S. Hewitt and K. G. McNeill, Can. J. Phys. $\underline{48}$ (1970) 1635
31) T. A. Khan, J. S. Hewitt and K. G. McNeill, Can. J. Phys. $\underline{47}$ (1969) 1037
32) K. A. Snover, E. G. Adelberger and D. R. Brown, Phys. Rev. Lett. $\underline{32}$ (1974) 1061
33) H. Feshbach, A. K. Kerman and R. H. Lemmer, Ann. of Phys. (N.Y.) $\underline{41}$ (1967) 230
34) D. G. Mavis, H. F. Glavish and D. C. Slater, Contribution to this conference, p.749
35) G. F. Bertsch and S. F. Tsai, preprint, 1974
36) S. Krewalk, J. Birkholz, A. Faessler and J. Speth, Proc. Int. Conf. on Nuclear Structure and Spectroscopy (Scholar's Press, Amsterdam, 1974) Vol. 1, p. 31
37) D. Youngblood, private communication
38) E. H. Kuhlmann et al., contribution to this conference, p.753
39) H. F. Glavish et al., Few Particle Problems in the Nuclear Inter-action, eds. I. Slaus, S. A. Moskowski, R. P. Haddock and W. T. H. Van Oers (North Holland, Amsterdam, 1972) p. 632
40) H. F. Glavish et al., Few Body Problems in Nuclear and Particle Physics, eds. R. J. Slobodrian, B. Cujec and K. Ramavataram (Les Presses de L'Université Laval, Quebec, 1975) p. 607
41) J. R. Calarco et al., Contribution to this conference, p.767

DISCUSSION

Von Geramb:
  Is the giant quadrupole strength of ^{16}O that you obtained in the region of 24 MeV of isoscalar or isovector nature ? Can you say something about the isospin property of the E2 peak near 28 MeV in ^{12}C ?

Glavish:
  The $(\vec{p},\gamma)$ experiments by themselves are not capable of distinguishing
  between isoscalar and isovector E2 strength. If one assumes that
  there is similar E2 strength in the neutron channel, to that which
  is found in the proton channel then at least some of the E2 strength
  we observe in ^{16}O and ^{12}C must be isovector in nature, for otherwise
  the isoscalar E2 sume rule would be exceeded.

Slobodrian:
  I am not sure whether I understood your comment on the position of the
  the E2 giant resonance with respect to the E1. Did you imply that the
  former is always above the latter ?

Glavish:
  For nuclei with A<40 $(\vec{p},\gamma)$ reactions show that there is substantial
  E2 strength above the giant E1 resonance in the case of ^{12}C, slightly above in the case of ^{16}O, and in approximately the same position
  as the giant E1 resonance in the case of ^{32}S. For these same nuclei
  (α,γ) reactions indicated there is substantial isoscalar E2 strength
  below the giant E1 resonance. Thus, it appears that there is E2
  strength below, within and above the giant E1 resonance. For nuclei
  with A>40 the situation is not so clear. Same isoscalar E2 strength
  is observed at $63/A^{1/3}$ MeV in electron, proton and alpha scattering,
  which is below the giant E1 resonance, but it is not known at the
  present time where the remainder of the E2 strength (isoscalar and
  isovector) is located.

INELASTIC SCATTERING AND THE COMPOUND NUCLEUS[*]

Charles Glashausser[+]
Rutgers University, New Brunswick, New Jersey

1. Introduction

My subjects today are somewhat old-fashioned: inelastic scattering of light ions, isobaric analog states, the Hauser-Feshbach theory, Ericson fluctuations·, even perhaps intermediate structure. In a sense this is true of most polarization measurements. By the time it is possible to redo the old experiment with a polarized beam the fashion in cross section measurements has changed to a new area where polarized beams cannot yet be used. This tends to make our interests seem somewhat parochial, and it becomes incumbent·upon us to draw important new physics from our measurements, physics which is relevant to the rest of the physics community. I think we have at least partially succeeded in this endeavour and I will tell you about these successes. But I will point out also the problems.

I will talk about two basic types of polarization measurements for inelastic scattering; both of them depend entirely on interference. The first of these, the analyzing power A, is proportional to the left-right asymmetry in the scattering of a polarized beam; A is zero if there is no entrance channel interference. The second is the polarization p of the outgoing particles following the scattering of an unpolarized beam; p is zero if there is no exit channel interference. For inelastic scattering to a particular final state, different channels correspond to different partial waves ℓ, J.

Consider, e.g., an isobaric analog resonance (IAR) of spin and parity $3/2^+$ whose parent state wave function is described by:

$$\psi^{3/2^+} = a|0^+,3/2^+> + b|2^+,1/2^+> + c|2^+,3/2^+> + d|2^+,5/2^+> + e|2^+,7/2^+> + \ldots$$

This resonance is formed in only one entrance channel, a $3/2^+$ proton wave incident on the 0^+ target. It can decay to a final 2^+ state, however, in at least four different exit channels. As far as A and p are concerned, you can distinguish several simple possibilities.

a) If the IAR is formed at a low incident proton energy, below the (p,n) threshold, the off-resonant background will correspond to excitation of many levels in the compound nucleus with many different spins and parities. If these additional entrance and exit channels are described by standard Hauser-Feshbach (HF) theory, A will be zero but p will show a resonance effect due to the interfering exit channels of the IAR. The measured p will be decreased, however, by the effect of the unpolarized HF background beneath the resonance.

b) Two overlapping IAR with different spins and parities

will give rise to a non-zero A which is largest where the interference is greatest, so that A will generally peak between
the two resonances. The polarization p will arise both from
each resonance individually and also from the interference
between the two resonances, so that p will again be quite
different from A.
c) The third case of interest arises at higher energies
above the (p,n) threshold, where the off-resonance background
is described by the DWBA. The analog resonance does interfere with such direct reaction background, so that there are
interfering entrance and exit channels. Both A and p can
then be non-zero and, for example, reach a maximum at the
resonance peak in the cross section. Depending on the wave
function of the resonance, A and p may be different in
magnitude and/or sign.

The analog resonance is used here simply as an example;
most of the above remarks apply equally well to more general
resonances or doorway states. We will apply these simple
pictures to A and p measurements for inelastic scattering in
three specific areas: spectroscopy at IAR, channel-channel
correlations, and Ericson fluctuations and intermediate
structure. This of course will be the heart of the talk but
first we should consider briefly the experimental method for
measuring p.

The defining equations for A and p are as follows:

$$A\sigma = \sigma^{++} + \sigma^{+-} - \sigma^{-+} - \sigma^{--} \tag{1}$$

$$p\sigma = \sigma^{++} + \sigma^{-+} - \sigma^{+-} - \sigma^{--} \tag{2}$$

$$(p-A)\sigma = 2(\sigma^{-+} - \sigma^{+-}) \equiv -2\sigma\Delta S \tag{3}$$

The + and - signs refer to the signs of the spin projections
of the incoming and outgoing particles; σ^{+-} then is the partial cross section for an incoming spin up particle to be scattered and leave with spin down. Up and down are defined with
reference to the normal to the scattering plane.

Since the standard method of measuring p by double scattering is prohibitively time consuming for IAR, almost all
measurements have been performed by the spin-flip asymmetry
method originated at Rutgers[1,2]. Making use of eq.(3) above,
p is determined by measuring A and the spin-flip asymmetry ΔS.
The quantity ΔS is measured directly in a $(\vec{p},p'\gamma)$ experiment
with a polarized incident beam and the gamma ray detector
perpendicular to the reaction plane. This method is limited
to states of spin 1 and 2 which decay to the ground state and
thus it would still be very useful to develop a highly
efficient polarimeter.

2.   Spectroscopy of Isobaric Analog States

At the time of the Madison meeting, inelastic scattering
measurements with polarized beams at IAR had scarcely begun;
the major interest had been elastic scattering. In the intervening time, the experimental methods and the theoretical
analysis have undergone considerable development. Detailed
spectroscopy is possible yet the number of measurements

remains small. I think the reason is related to my introductory comments; sophisticated spectroscopic tools have been developed at a time when interest in spectroscopy is waning.

An inelastic scattering measurement with a polarized beam
at an IAR is an ambitious undertaking: the aim is to determine
the entire wave function of an IAR and its parent state.
Whereas single-particle stripping experiments, for example,
can determine only those components of a wave function in
which the core is not excited, spectroscopy at IAR has always
promised to be the equivalent of stripping on a target in an
excited state. The measurement of polarization parameters
often allows this promise to be fulfilled.

If the off-resonant background is direct, and reasonably
well-described by the DWBA, measurement of either excitation
functions of σ and A at several backward angles or on-resonance angular distributions of these observables is often sufficient to obtain a good description of the wave function.
Both types of measurements have been carried out for many IAR
in ^{138}Ba+p. Angular distributions measured at Rutgers at the
10.00 MeV 7/2$^-$ IAR are shown in Fig.1[3]; an excitation function for E_p=9.8 to 12.0 MeV at 140° is shown in Fig.2. The
latter data were taken as part of a very complete survey of
the N=83 nuclei at Erlangen[4]. The curves shown have been
calculated with a DWBA code in which resonant amplitudes have
been added coherently with the direct reaction background amplitudes. The optical model parameters , the elastic width,
and the total width are fixed by elastic scattering data; the
fits shown were obtained by varying only the magnitudes and

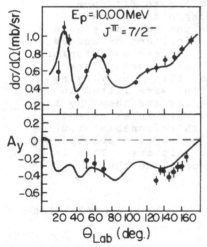

Fig.1. Angular distributions
for ^{138}Ba(p,p')^{138}Ba*(2$^+$)
(Ref.3). The solid curves
are theoretical fits.

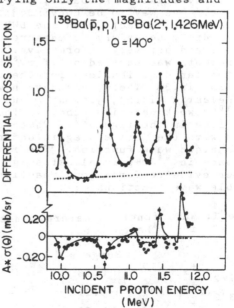

Fig.2. Excitation functions for
^{138}Ba(p,p')^{138}Ba*(2$^+$) (Ref.4).

phases corresponding to the inelastic partial widths. Resonance mixing phases are generally held close to zero. The
quality of the fits shown is typical: good but not perfect.

The results obtained for the 10.00 MeV resonance are [5]
compared with each other and with theoretical calculations
in terms of the unified model in Table 1. It is noteworthy
first that the experimental spectroscopic amplitudes are in
reasonable agreement with each other, although the fits to
the angular distribution and excitation function data are sensitive to resonance and DWBA parameters with quite different
weights. The average discrepancy between the two sets of inelastic amplitudes is comparable to the errors assigned by
the respective authors and gives an estimate of how well the
various components of the wave function can be determined.
On the basis of this evidence, the measurement of σ and A on
resonances like this where the DWBA background is strong does
indeed give spectroscopic information about core-excited components of wave functions which is comparable in quality to
the information provided by elastic scattering or stripping
reactions for the ground state component. The theoretical
wave function is in good agreement with the experimentally
determined one, in both sign and magnitude. By carefully
comparing theory and experiment over the entire range of data
for several nuclei, however, the Erlangen group has noted
that the strength of the core-particle coupling is underestimated in the theory[4]. This is a result of general interest.

Will these wave functions be accepted by unpolarized
nuclear physicists? Probably, but the answer is not yet
definite. One way to be convinced is to try to fit more
observables. In fact, when the DWBA background is small, it
is absolutely necessary to measure other parameters because A
is small and not very informative. The first set of complete
measurements was carried out for ^{88}Sr at Rutgers and Stanford
[2]. Angular distributions for the 7.08 MeV resonance are
shown in Fig.3. These data have now been fitted by varying
the inelastic widths, just as A and σ alone were fitted in
the ^{138}Ba work described above. The results are shown by the
solid line in the figure[6]. These curves are in excellent
accord with the data and again define with reasonable accuracy a unique wave function. The fit at a nearly overlapping
resonance in ^{88}Sr+p is almost as good as this one. This is
further evidence for a good reaction theory, and thus
reliable wave functions.

Table 1: Wave function determined at the 10.00 MeV ($7/2^-$) IAR
         in ^{138}Ba+p

	Rutgers[3]	Erlangen[4]	Theory[5]
$\mid 2f_{7/2},0^+ >$	0.96	0.89	0.94
$\mid 3p_{3/2},2^+ >$	-0.22	-0.17	-0.17
$\mid 2f_{5/2},2^+ >$	-0.25	0.29	-0.04
$\mid 2f_{7/2},2^+ >$	-0.26	-0.15	-0.27

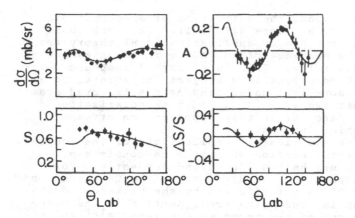

Fig.3. Angular
distributions
for ^{88}Sr(p,p')
^{88}Sr*(2$^+$) meas-
ured on resonan-
ce at 7.08 MeV
(Ref.6).

However, there are several problems. The fit at a higher
isolated resonance in ^{88}Sr+p is not as good, especially for A
which is most sensitive to the DWBA parameters. In general
the influence of an imperfect theory for the direct reaction
background is not well understood; one clearly wants to avoid
changing the values of the inelastic widths to correct for
deficiencies in the DWBA fit to the background. While the
limited study of the effect of changing DWBA parameters on
the values of the inelastic widths has given reassuring
results, the investigation is not yet complete.

Other problems with this method for determining spectroscopic amplitudes are problems inherent in the study of analog states: are the elastic resonance parameters well-known,
especially if the resonances overlap? Are the single-particle widths reasonably determined? The ^{89}Sr wave functions
shown above had to be renormalized by 30 to 50 %. While the
final Rutgers and Erlangen results for the lowest states in
^{139}Ba are in good agreement, two different sets of single-
particle widths were used. The states which are most important for the Erlangen evaluation of the inadequacy of the
unified-model calculations for ^{139}Ba correspond to three
strongly overlapping resonances with other resonances nearby.
Evaluation of errors in individual parameters when so many
parameters are involved is very difficult.

What is needed here then to establish these measurements
of σ, A, S, and ΔS/S as a reliable instrument of spectroscopy
is a new ^{208}Pb for excited-core spectroscopy, a nucleus which
can reliably be described by intermediate-coupling calculations. Such a nucleus, in a region where the background is
large and direct, would provide an ideal test of the
reliability of the wave functions extracted.

3.   Channel-Channel Correlations

In our discussion thus far our emphasis has been on
determining spectroscopic information from the study of IAR
in regions where the background of direct reactions is large.
At energies below the (p,n) threshold where the background

is predominantly compound, the extraction of spectroscopic amplitudes is considerably more difficult. The problem is that the conditions for proper application of HF theory are not fulfilled in the presence of an IAR or in the presence of direct reactions generally. This interplay of compound nucleus (CN) and direct reactions is a subject fundamental to all nuclear physics; theoretical understanding has improved dramatically in the past two years[7,8]. I think polarization experiments in this area definitely represent an attempt to go beyond parochial interests.

The IAR provide the clearest applications of the modified HF theories. The cross section on resonance consists of several parts which may be considered independently. For an IAR of spin and parity J^{π} of $3/2^+$, e.g., the CN background in channels with $J^{\pi} \neq 3/2^+$ is not affected by the presence of the IAR and will be unpolarized. The cross section in the resonant channel is the sum of two terms we can label σ^{IAR} and σ^{CN} with the channel spin and parity index $3/2^+$ assumed. The σ^{IAR} is described by a term in the S matrix, $\langle S \rangle$, which changes relatively slowly with energy. The σ^{CN} is described by a term S^{fl} which fluctuates rapidly with energy; σ^{CN} is well known to be enhanced because of mixing of $T_>$ and $T_<$ states in the vicinity of the resonance. The problem which both experimentalists and theorists have tried to resolve is how to describe σ^{CN}.

The relevance of polarization measurements can be clearly seen if we write the appropriate formulas:

$$S_{CC'} = \langle S_{CC'} \rangle^{IAR} + S_{CC'}^{fl} \ , \ \text{with} \ \langle S_{CC'}^{fl} \rangle \equiv 0 \ . \tag{4}$$

The averaging interval is small compared to the width of the IAR but much larger than the width and spacing of the CN levels, so that the IAR is here a special case of a direct reaction. The CN observables are related only to S^{fl}:

$$\langle \sigma \rangle^{CN} \ \alpha \ \sum_{C'C''} \langle S_{CC'}^{fl}, S_{CC''}^{fl*} \rangle \tag{5}$$

$$\langle \sigma p \rangle^{CN} \ \alpha \ \text{Im} \sum_{C' \neq C''} \langle S_{CC'}^{fl}, S_{CC''}^{fl*} \rangle \tag{6}$$

If $S_{CC'}^{fl}$ and $S_{CC''}^{fl}$ for different exit channels are uncorrelated, then

$$\langle S_{CC'}^{fl}, S_{CC''}^{fl*} \rangle = \langle S_{CC'}^{fl} \rangle \langle S_{CC''}^{fl*} \rangle = 0 \ , \ C' \neq C'' \tag{7}$$

This statistical assumption is at the basis of standard Hauser-Feshbach theory; it is what we shall mean by the statement that there are no channel-channel correlations (CCC). The recent work on HF theory in the <u>absence</u> of direct reactions especially by Weidenmüller and collaborators[7] has shown either analytically or numerically that this is an excellent assumption regardless of the strength of the absorption.

The experimental work has been chiefly directed toward showing that this assumption of no CCC is not valid near an

IAR. Since $<\sigma p>^{CN}$ is proportional to the strength of the channel-channel correlations, a non-zero measured value of $<\sigma p>^{CN}$ is a direct proof of CCC. Unfortunately, the measured value of $<\sigma p>$ on resonance includes contributions also from $<\sigma p>^{IAR}$ which is non-zero but unknown since the wave function of the IAR isn't known. The CN background with $J^{\pi} \neq 3/2^{+}$ is also present but its effect is known. Thus the experiment consists in measuring angular distributions of both σ and p on resonance, and then trying to fit the results with the assumption of no CCC by varying the wave function of the IAR. If the procedure does not give a good fit to all the data, the result is interpreted as evidence that CCC must exist.

A number of examples have now accumulated in which the fits achieved without CCC are definitely poor. Such results for two resonances in ^{90}Zr have been contributed to this Conference by Feist et al.[9]; the dashed line in Fig.4 shows the best fit without CCC at the 6.81 MeV(3/2^{+}) IAR. Note that A is essentially zero because the small DWBA background means there is only one effective entrance channel. The measured value of p here is quite small and it is only by trying to fit

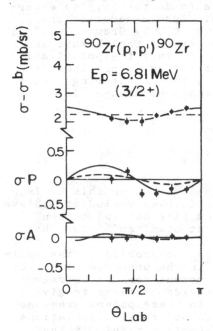

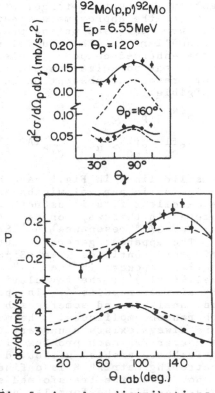

Fig.4. Angular distributions for ^{90}Zr(p,p')^{90}Zr*(2^{+}) measured on resonance at 6.81 MeV (Ref.9). The dashed and solid curves are described in the text.

Fig.5. Angular distributions for ^{92}Mo(p,p')^{92}Mo*(2^{+}) (Ref.10).

σ and p simultaneously that the evidence for CCC becomes
strong. The Rutgers-Heidelberg measurement[10] for ^{92}Mo at
6.55 MeV is shown in Fig.5. The dashed curves represent best
fits without CCC. The upper data are unpolarized (p,p'γ)
cross sections at several proton angles which were fitted
independently of the p and σ data shown below.

Is this result unexpected? The answer seems to depend
on whom you ask. In a simple model, however, with strong
mixing between the $T_>$ IAR and the surrounding sea of $T_<$ CN
states whose widths are assumed negligibly small in the
absence of the IAR, the presence of CCC is clear. Then the
IAR acts as a perfect doorway into the compound nucleus.
Whether the scattering is labeled CN or direct, whether it
is due to <S> or S^{fl}, in this model it all depends exclusively on the wave function of the IAR. If the scattering
experiments were performed with high resolution so that the
fine structure could be resolved, the measured value of p
at each fine structure state would be the same (in the absence of non-resonant CN background). In fact it would be interesting to perform this measurement at an appropriate accelerator where enough time was available to do the $(\vec{p},p'γ)$ experiment on a very thin target of ^{92}Mo. Such a model is the[11]
basis for the approximate treatment of CCC by Graw et al.[11];
their work is based on the previous work of Englebrecht and
Weidenmüller[12]. The crucial result is that the polarization
of the enhanced compound nuclear scattering is precisely the
same as the polarization due to the direct scattering, provided off-resonance absorption in the resonant channel is
negligible:

$$p^{CN} = p^{IAR}$$

$$<S^{fl}_{CC'}, S^{fl*}_{CC''}> \alpha <S_{CC'}>^{IAR}<S^*_{CC''}>^{IAR} \neq 0 \qquad (12)$$

The solid lines in Fig.5 have been calculated on this basis;
the inelastic partial widths have again been varied to achieve
the excellent fits illustrated. Good fits have also been
achieved in this way for data taken at Stanford[13] for the
5.75 MeV 3/2$^+$ resonance in ^{86}Sr+p.

The apparent general solution to the problem of the eval-
uation of quantities like $<S^{fl}_{ab}S^{fl*}_{cd}>$ in the presence of direct
reactions appears in the recent work of Hoffmann
et al. (HRTW)[7]; the formulation of Moldauer seems to give
very similar results[8]. The proofs in these papers are sometimes analytic and sometimes based on numerical calculations
with random amplitudes. The most important result is that
there always exists a unitary transformation U which reduces
the Hauser-Feshbach problem with direct reactions to the Hauser-Feshbach problem without direct reactions. A new
"scattering matrix" $\tilde{S}$ is defined such that $\tilde{S} = U S U^T$; there
are no CCC in the transformed space. HRTW give a recipe for
evaluating all the formulas necessary to actually calculate
σ or p, for example. To someone who does not usually venture
into the kitchen, the recipe looks formidable, but Gerhard
Graw assures me it is quite straightforward. The solid lines
in Fig.4 were calculated according to this formalism[9];

the agreement is excellent here and also for two other resonances in ^{90}Zr+p.

It is useful also to compare the results of the HRTW
formalism with those of the simple model we discussed earlier;
this has been done by Graw and Pröschel[14]. Probably the
main result is that even in the new formalism, the polarization of the enhanced CN contribution remains approximately the
same as the polarization of the direct (IAR) contribution;
this justifies the basic physics of the simple approximation.
However, when you calculate all contributions to <σ> and <σp>
in the two theories, the differences become significant.

To summarize this section, the theorists have given us
an impressive new apparatus which they believe will work in
essentially any situation which involves the interplay of CN
and direct reactions. The theory has been applied thus far
only to the investigation of IAR, and there it has been quite
successful. It is important, however, to test the theory
elsewhere. With the impressive calculational expertise the
theorists have developed, it would certainly be convenient
for the experimentalists if they investigated, for example,
the interaction of DWBA and CN amplitudes at typical energies
for (d,p) or (p,p') reactions. In fact, since so many
channels are important in most reactions even at low energies,
there is reason to believe that standard HF theory is sufficient.

4. Fluctuations and Intermediate Structure

One area where the new HF theories should have considerable impact is Ericson fluctuation theory, both with and
without direct reactions. This theory is well developed
for cross sections. In a region of CN excitation where Γ,
the coherence width of the CN states, is greater than D,
their average separation, the interference between the amplitudes for different CN states gives rise to a fluctuating σ.
The fluctuations are superimposed on a slowly varying background of direct reactions which must be subtracted out.
Standard Ericson theory then states that the cross correlation coefficient Cij, i≠j, between two independent sets of

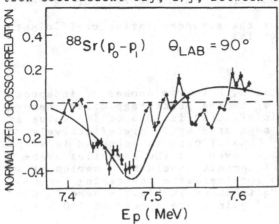

Fig.6. Normalized
cross-corelation bet-
ween elastic and in-
elastic (2$^+$) cross
sections for ^{88}Sr(p,p)
^{88}Sr (Ref.16).

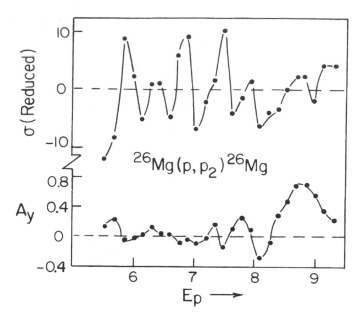

Fig.7. Cross section and analyzing power for inelastic scattering to the 2.94 MeV (2$^+$) state in ^{26}Mg. Each data point
shown represents an average over 150 keV.

data should be zero within finite range of the data (FRD)
errors. The modified HF theories suggest that this is not
necessarily true even in the absence of direct reactions[15].
The group at Bochum has measured the normalized cross correlation between elastic and inelastic cross sections near the
7.5 MeV IAR in ^{88}Sr+p; measurements were made in 1.0 keV
steps. Their results at 90°, shown in Fig.6, are a direct
proof of CCC[16]. The solid curve represents an approximate
calculation based on the HRTW formalism. The good agreement
seems to be another confirmation of this theory, but agreement at 125° is not good.
     The Ericson formula for the autocorrelation coefficient
$C_{ii}(\epsilon)$ for the cross section is:

$$C_{ii}(\epsilon) = \frac{(1-y_D^2)}{N} \quad \frac{\Gamma^2}{\Gamma^2+\epsilon^2}$$

Here ϵ is the energy separation, N is the number of independent reaction channels, and y_D is the percentage direct react-
ion contribution. When ϵ is 3Γ, $C(\epsilon)$ is 10 % of its value at
$\epsilon = 0$, so that data points separated by 3Γ are effectively
independent of each other. Thus if $\sigma(E)$ is measured in steps
of, say, $\Gamma/2$, and then averaged over 3Γ, the resulting averaged data points ($\bar{\sigma}$) at 3Γ intervals should be a series of
random numbers. An example for inelastic scattering to the
second excited state of ^{26}Mg is shown in Fig.7A; here Γ is
about 50 keV and the averaged points are spaced at 150 keV
intervals.

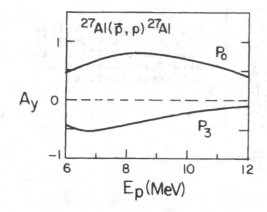

Fig.8. Pure optical model (p_0) and DWBA (p_3) calculations of <A> for scattering from ^{27}Al at 140°.

What, now, should be expected from a measurement of the excitation function of A or p in 25-50 keV steps with a thin target for a nucleus like ^{26}Mg? For pure Ericson fluctuations, each measured value of A may be non-zero and even quite large, but the data should fluctuate between positive and negative values in such a way that the value of A averaged over a sufficient energy range (<A>) is zero. This result is not modified by recent work on HF theory. For pure direct reactions, <A> can be large but it should vary smoothly over an energy region of about 5 MeV. (Optical model and DWBA calculations of A for pure direct elastic and inelastic scattering from ^{27}Al are shown in Fig.8.) In the realistic case in which both CN and direct mechanisms contribute the measured values of A should fluctuate about a smooth background determined by the direct reaction component. Formulas that describe the magnitude of these fluctuations or FRD errors for cross-correlations are not well-established for polarization observables.

If we now average over 3Γ the values of A or σA taken at intervals of Γ/2, then we should again have a series of random numbers. Fig.7B shows the values of Ā plotted at intervals of 3Γ for scattering to the second excited state of ^{26}Mg; compare them with the values of σ̄ in Fig.7A. The bump at the high energy end of the excitation function of Ā does not look like a series of random numbers.

It is just because polarized beam measurements showed promise of revealing intermediate structure that we at Rutgers began to investigate A in the region where statistical fluctuations dominate. Because A is sensitive to interference, it was at least reasonable to expect that intermediate structure might be more visible in an excitation function of A than in an excitation function of σ. To see whether this expectation is confirmed, Haglund et al.[19] have been studying the probability distribution of σA; they are now determining the effect of adding a wide "intermediate structure" state to their random CN amplitudes. We have attacked the problem experimentally, and looked for intermediate structure in A where there was perhaps only a hint in σ. Thus in this region of fluctuating amplitudes, we have been trying to

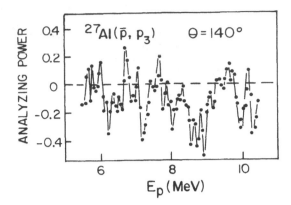

Fig.9. Measured values of A for $^{27}Al(\vec{p},p')^{27}Al$ (2.21 MeV, $7/2^+$) at 140°. Data were taken in 50 keV steps (Ref.18).

extract the physics not out of the fluctuations but out of the possible underlying smooth structures.

We now believe that we have solid evidence that the structure we observe in excitation functions of A for proton inelastic scattering to several states in ^{26}Mg and ^{27}Al at least between 5.5 and 9.5 MeV is non-statistical[18]. This evidence is based on application of the statistical tests for intermediate structure recently described by Baudinet-Robinet and Mahaux[19] and also on the large values of the cross-correlations of A corresponding to several final states.

Data taken in 50 keV steps for $^{27}Al(\vec{p},p_3)^{27}Al^*(7/2^+)$ between 5.5 and 10.5 MeV are shown in Fig.9. These data show the expected Ericson fluctuations. Note there are some very large fluctuations, e.g., at about 7.2 MeV, but these in themselves are not necessarily non-statistical. In fact, in this form it is very difficult to decide whether the fluctuations are non-statistical or not. Note, however, that around 8.5 MeV the values are essentially all more negative than at nearby energies. The presence of bumps is more obvious when the data are smoothed but random data would also show bumps when smoothed.

The quantitative method we use to determine the probability that the fluctuations are random is to average the 50 keV data over an interval of 150 keV and then to decide if the resulting data points constitute a series of random numbers. Data averaged over 150 keV for scattering to the first two 2^+ states in ^{26}Mg for bombarding energies between 5.5 and 17.5 MeV are illustrated in Fig.10. Up to now we have applied statistical tests in detail only for the data between 5.5 and 9.5 MeV which were taken first. When we decided there were real non-statistical effects, we took the higher energy data. The statistical tests are described in Sec. IV.3 and IV.5 of reference 19; the results are listed in Table 2.

The reason why these data are judged non-statistical is that so many large values of A are clustered together. Unless a bump is at least about 12Γ-15Γ wide, it will generally have a fairly high probability. Thus we do not contradict the results of Singh et al.[20] who showed that random compound nucleus amplitudes can lead to fluctuating cross sections which, when smoothed, may appear to the unsophisticated observer like intermediate structure. The sophisticated observer does not trust his eyes and relies on various statistical tests instead. Such an observer finds that Singh's cross sections are indeed random. He also finds that the data shown here are non-random.

Do the results of the statistical tests guarantee that we have found intermediate structure? No - unusual events do occur - but the fact that we observe very low probabilities of randomness in the data for four different states greatly increases our confidence. In addition, the normalized cross-correlation coefficients between the different channels also turn out to be considerably larger than expected for random numbers. This is only true when the coefficients are evaluated over a region of 1.2 MeV, i.e., somewhat larger than the width of the intermediate structure.

Now our task is to relate this anomalous polarization result to more general nuclear physics. There is very little theoretical work on the density of various simple configurations at excitation energies around 20 MeV in these compound nuclei. Of course this is the energy of the giant dipole resonance (GDR) and one possibility worth considering is that the intermediate states are related to the GDR. Evidence that the GDR can be seen as a compound state in elastic scattering excitation functions for A even when it is not visible in σ has been presented to this conference[21]. In fact, the A data for ^{27}Al look qualitatively similar in shape to ^{27}Al(p,γ) cross section data. We were thus stimulated to measure A in scattering from ^{19}F where (p,γ) cross section data reveal especially strong structure which is also correlated in two different exit channels. Our A data, however, appear consistent with the statistical model[22].

Our present effort is directed toward determining whether non-statistical bumps and cross-correlations are systematically present in these light nuclei, and whether some patterns develop. We will also investigate other reaction channels in detail for the compound nucleus ^{27}Al where the most striking structure is observed.

Table 2: Probability that the data are random

Reaction	Angle 120°	140°	160°	Reaction	Angle 145°
^{26}Mg(p,p_0)	>0.05	>0.05	>0.05	^{27}Al(p,p_0)	>0.05
^{26}Mg(p,p_1)	0.01	0.01	>0.05	^{27}Al(p,p_3)	0.01
^{26}Mg(p,p_2)	0.05	0.01	0.02	^{27}Al(p,p_4)	>0.05
				^{27}Al(p,p_{5+6})	>0.01

5.   Summary

The physics content of the polarization measurements I have discussed here should be summarized briefly. In spectroscopy, there is evidence that the core-excited component of nuclear states at low excitation energy can be determined fairly reliably, so that the experimental wave function can be used to test unified-model calculations. For reaction theory, polarization measurements at isobaric analog resonances have shown that channel-channel correlations are important in the presence of direct reactions. These correlations have been included in a new general theory for the interplay of direct reactions and compound nucleus reactions. Polarization measurements are essential in testing this theory. Finally, polarization measurements have indicated the existence of non-statistical structure at high excitation energy in light nuclei; the physics of these structures may be a subject for the next Polarization Symposium.

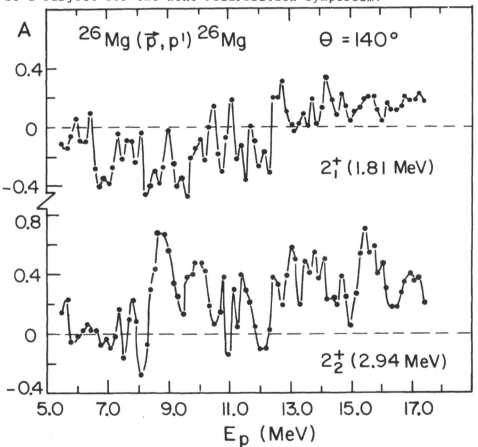

Fig.10. Values of $\overline{A}$ averaged over 150 keV for $^{26}Mg(\vec{p},p')^{26}Mg^{*}$ (Ref.18).

References

* Work supported in part by the National Science Foundation
+ Temporary address: Universität München, Sektion Physik
                  8046 Garching, West Germany

1) R. Boyd et al., Phys. Rev. Lett. 27 (1971) 1590
2) R. Boyd et al., Phys. Rev. Lett. 29 (1972) 955
3) S. Davis, Ph.D. Thesis, Rutgers University (1974) unpubl.
4) H. Clement and G. Graw, Phys. Lett. 57B (1975) 323
5) G. Van den Berghe et al., Phys. Lett. 38B (1972) 467
   and private communication
6) C.F. Haynes, Ph.D. Thesis, Rutgers University (1975)
   unpublished
7) H.M. Hofmann et al., Ann. of Phys. 90 (1975) 403
   and references contained therein
8) P.A. Moldauer, Phys. Rev. C11 (1975) 426 and to be
   published
9) J.H. Feist et al., Contribution to this Symp. p. 769
10) S. Davis et al., Phys. Rev. Lett. 34 (1975) 215
11) G. Graw et al., Phys. Rev. C10 (1974) 2340
12) C.A. Englebrecht and H.A. Weidenmüller, Phys. Rev. C8
    (1973) 859
13) H.T. King and D.C. Slater, Contribution to this Symp. p.763
14) P. Pröschel, Diplomarbeit, University of Erlangen (1975)
    unpublished
15) J.W. Tepel, Z. f. Physik A273 (1975) 59
16) E. Blanke et al., to be published
17) R.F. Haglund et al., Contribution p.785 and private
    communication
18) C. Glashausser et al., Phys. Rev. Lett. 35 (1975) 494
19) Y. Baudinet-Robinet and C. Mahaux, Phys. Rev. C9 (1974)
    723
20) P.P. Singh et al., Phys. Lett. 23 (1967) 255
21) H.R. Weller et al., Contribution to this Symp. p.755
22) J. Eng et al., Contribution to this Symp. p.783

Acknowledgements

The Rutgers experiments were group efforts; I am indebted to my collaborators, particularly S. Davis, C.F. Haynes, A.B. Robbins, and E. Ventura. I am grateful also to Gerhard Graw for many fruitful conversations about channel-channel correlations.

DISCUSSION

Cramer:

I would like to make two comments on the first part of your talk:
a) It would be more impressive to see simultaneous predictions of
excitation functions and angular distributions, rather than one fit
to excitation functions and another fit to angular distributions.
Since you have both kinds of data for barium, why don't you fit them
together ?
b) It has been suggested by others that for cases where spin flip
correlations don't work (i.e. anything but $0^+ \rightarrow 2^+ \rightarrow 0^+$) one could get
equivalent information by studying $(\vec{p}, p'\gamma_0)$ triple correlations. I
would like to suggest that even if DWBA gives satisfactory predictions of spin flip probabilities (which are independent of m-state
phases), it is by no means clear that DWBA is good enough to predict
more general triple correlation functions, where m-state phases are
of great importance.

Glashausser

What I wanted to show by comparing the Rutgers and Erlangen results
was that two independent analyses give essentially the same results.
In fact, the Erlangen results are based on analysing 2 back-angle
excitation functions, which they have shown give good fits also to
excitation functions at 2 forward angles- thus fitting really a
4 point angular distribution.
With respect to your second point, it sounds like it would be very
interesting to try the experiment to see how important such problems
with the DWBA really are.

Donoghue:

I wish to call attention to an additional condition under which
$A_y \equiv 0$. Harney noted that, for spin 1/2 particles on a spin-0 target,
$A_y=0$ unless you have 2 different <u>entrance</u> channels. Recently,
Arnold at Ohio State (see contributed paper p.601) has shown that,
even when this entrance channel condition is satisfied, $A_y=0$ still
holds unless the <u>exit channel-spins of the decaying states are the
same.</u> This new though minor condition can explain some null A_y
measurements in inelastic proton scattering where at least 2 states
are believed to contribute and provides motivation for measuring
spin-flip polarizations.

Hanna:

What can you tell us about the cross correlations in the Mg and Al
experiments ?

Glashausser:

If the correlations are evaluated over a range of about 5 MeV, they
are generally small. If, however, they are evaluated over a sliding
interval of about 1.2 MeV - i.e., an interval only somewhat larger
than the width of the intermediate structure - the correlations turn
out to be quite large. When these values are compared with the cross-
correlations between random numbers over a comparable range, the
cross-correlations in the data, particularly for ^{27}Al, appear non-
statistical.

COMPOUND-NUCLEUS FLUCTUATION EFFECTS [+]

W.J. Thompson

University of North Carolina, Chapel Hill NC 27514 USA,
and Triangle Universities Nuclear Laboratory, Durham NC

1. Introduction

This review concerns the observation and analysis of the effects of fluctuations as a function of bombarding energy of differential cross sections observed with polarized beams in the region of statistical excitation of the compound nucleus (CN). Surveys of fluctuation studies for unpolarized beams have been given recently [1,2]. Although many of the data reported at the Madison symposium probably contained CN effects, no intentional measurements thereof were reported. Probably the first such data, taken 14 years after Ericson's [3] suggestions for cross section studies, are those from Rutgers [4], on $^{27}Al(p,p)^{27}Al$, and from TUNL [5] on $^{28}Si(d,d)^{28}Si$. (By comparison, fluctuation studies in high-energy physics were reported in 1973 [6], on $p(\pi,\pi)p$ at 5 GeV, 9 years after they were suggested [7].)

A motivation for these studies is the expectation that polarized-beam (selected magnetic substate) cross sections will fluctuate more rapidly with energy than will those for unpolarized beams, for which the separate magnetic-substate cross sections (called basic cross sections) are added incoherently. Indeed, a pioneering study by Stephen [8] predicts that for pure CN elastic scattering of spin-1/2 on spin zero the variance of the cross section for an incident polarization P is proportional to $(1+P^2)$. With polarized beams the fluctuating (CN) parts of the scattering amplitudes may be more easily separated from the direct interaction (DI) parts. This is important for: (1) DI model analyses (for example, the optical model and DWBA). (2) CN studies (for example, average level spacings D and widths Γ.

2. Excitation energy regions of interest

The data and analyses considered here are characteristic of strongly-overlapping levels (practically, $\Gamma/D > 2$ is probably sufficient [9]), typically at CN excitation energies above 15 MeV for A < 40 and above 10 MeV for A > 40. Further, to minimize complications from channel-channel correlations: (1) The number of CN decay channels (counting each different particle, energy, J, π, and isospin as a separate channel) should be large (>10, ref. [10]). (2) Dynamic correlations (as in isobaric analog resonances) should be absent.

Two limiting values of the energy resolution in the initial state (determined mainly by the energy resolution of the beam and energy losses in the target) are theoretically tractable: (1) Resolved CN statistical structures, ($\Delta E < \Gamma/3$ gives 10% damping of the autocorrelation function [1]) which are amenable to statistical (Ericson model) analyses. (2) Unresolved CN structures, ($\Delta E = \Delta > 40\Gamma$ gives a standard deviation <20% in average cross sections [9]) which can be treated by variants of the Hauser-Feshbach model. Examples of (1) and (2) are given in contributions by Henneck et al. [11] and by Eastgate et al. [12] respectively. From data of type (1) those of type (2) can be produced; the inverse would require good exit-channel energy resolution and a deconvolution of all energy-loss effects, with reduction of total bombardment time proportional to the target thickness.

3. Resolved fluctuation structures

3.1. TYPICAL EXPERIMENTS

Five contributions to this symposium report well-resolved vector-polarized-beam fluctuation data. Those of Schieck et al. [13] from the Köln-LASL collaboration show strikingly the enhancement of fluctuations with polarized beams. The extracted relative variances of the polarized-beam cross sections are, as expected [14], larger than those with unpolarized beams. Similarly large vector-polarization analyzing power fluctuations in p and d bombardment are reported by the Erlangen [15], Rutgers [16,17], TUNL-Erlangen-München [11], and Madison [18] groups.

Tensor-polarization analyzing power fluctuations in (d,p) reactions, measured in coarse energy steps, have been reported by the Madison group [19].

Fluctuations in total cross sections for polarized neutron elastic scattering from polarized ^{59}Co for energies between 0.3 and 1.0 MeV have been reported [20], although at these energies $\Gamma/D<0.1$.

3.2. METHODS OF FLUCTUATION ANALYSIS

For purposes of CN analyses the formula for the cross section measured with initial polarization components t_{kq} is best written such that $\sigma=\sigma_0(\Sigma t_{kq} T_{kq}^*)=\Sigma t_{kq}^* \sigma_{kq}$, where the sums are over k and q. Each σ_{kq} is a "tensor analyzing cross section", $\sigma_{00}=\sigma_0$ the unpolarized-beam cross section. For energy averaged data $<\sigma>=\Sigma t_{kq}^*<\sigma_{kq}>$. The σ_{kq} are the basic experimental and theoretical quantities.

In statistical studies the finite sample sizes, in the present context called finite-range-of-data (FRD) errors [1,2,9], require that the energy range Δ contains a large enough sample of fluctuations to be statistically significant. This conflicts with the requirement that DI quantities and such parameters as Γ and D be constant in Δ.

Although there are many analysis techniques for unpolarized-beam fluctuation data [1,2], for polarized beams the results of Stephen [8] and of Lambert and Dumazet [21] are applicable only to pure CN reactions, while Kujawski and Krieger [14] treat only the variance of vector-analyzing cross sections. A major difficulty is that Ericson's theory is essentially derived ignoring spins, so that the basic cross sections for each arrangement of projectile and target magnetic substates are assumed equal. For example, in elastic 0(1/2,1/2)0 scattering $var(i\sigma_{11})=c_1(<\sigma_c>)^2$, where $<\sigma_c>$ is the average unpolarized-beam CN cross section with s_1 the fraction of it due to spin flip, and $c_1=2s_1(1-s_1)$. The Ericson assumption is $s_1=1/2$ ($c_1=1/2$, which is the upper limit). Similar restrictions are used to derive the formulas in the contribution from Henneck et al. [11].

Alternative to correlation functions are probability distributions [1] $P(\sigma_{kq}/<\sigma_0>)$. The contribution by Haglund et al. [22] considers $^{26}Mg(p,p_0)^{26}Mg$, with E_p between 5 and 10 MeV, and realistic parameters [23] for the calculation of synthetic excitation functions of σ_0 and $i\sigma_{11}$, and

examination of the resulting probability distributions. Preliminary results suggest that each CN helicity amplitude [24] can be considered as an
independent, normally distributed, random variable. The resulting probability distributions, parameterized by $y_D = \sigma_{DI}/<\sigma_0>$ and s_1, show good
agreement with the data. If these results can be extended to higher spins
and to reactions, then a straightforward method for fluctuation analysis
will have been obtained. For deuterons synthetic excitation functions
will help decide which of the analyzing cross sections σ_{20}, σ_{21}, σ_{22} are
best to measure. The use of an helicity (J_z eigenvalue), rather than a
total angular momentum (J^2 eigenvalue), basis is physics-wise appealing
for $\Gamma >> D$, since then J is not a good quantum number.

Non-statistical (intermediate) structures mixed in with statistical
fluctuations and DI contributions have been suggested in the contribution
by Glashausser et al. [17]. Tests with the above synthetic excitation
functions show that the method of runs tests [25] is applicable to vector-
analyzing cross sections, provided that the data is first smoothed over
an interval $\Delta E_s \approx 3\Gamma$, which removes local correlations.

4. Unresolved fluctuation structures

Data averaged over an energy interval $\Delta >> \Gamma$ are sensitive only to
the energy-averaged CN effects. In the situations considered here $\Gamma >> D$
and many channels are open, so that straightforward extension of Hauser-
Feshbach models for CN cross sections [26,27] may be made for analyzing-
power cross sections. With the usual statistical assumptions about CN
S-matrix elements [26], the following hold:
(1) DI and CN analyzing cross sections combine incoherently
$\quad <\sigma_{kq}> = (\sigma_{kq})_{DI} + <(\sigma_{kq})_{CN}>$.
(2) For unpolarized targets $<(\sigma_{kq})_{CN}> = 0$ if k is odd. (For example,
$\quad$ vector-analyzing CN cross sections vanish.)
(3) $<(\sigma_{kq}(\theta))_{CN}> = (-1)^q <(\sigma_{kq}(\pi-\theta))_{CN}>$.
The quantity $<T_{kq}> = <\sigma_{kq}>/<\sigma_0>$ does not have similar, simple properties.
Since $<\sigma_{kq}>$ data can be taken using thick targets, it can be acquired
much more rapidly than the data discussed in sec. 3. (In principle, but
not in practice, about $\Delta/\Delta E > 100$ times faster.)

The modification of the Hauser-Feshbach analysis for the usual situations in which there is a continuum of final-state energies, has been
described by Eberhard et al. [27]. It requires; transmission coefficients
T_c for each channel contributing to CN formation and decay, the level-
density spin-cutoff factor σ, and the ratio $W_{cc'}D/\Gamma$ as a normalization
factor. Here the width-correlation factor $W_{cc'} \approx 1 + \delta_{cc'}$ if T_c, T_c, >0.5
(refs. 10, 26) is a remnant of channel-channel correlations. Ideally, the
T_c are obtained from analysis of the shape-elastic (SE) scattering from
the target and residual nuclei, Γ can be estimated from cross section
fluctuations, and σ is then limited by the angular distribution of
$<(\sigma_{kq})_{CN}>$ and D is determined from CN normalization.

4.1. ELASTIC SCATTERING

Elastic scattering is attractive to examine CN effects, as compound-
elastic (CE) scattering, because the DI model (the optical model (OM))
has been well parameterized, at least for the central and spin-orbit
potentials. The contribution by Eastgate et al. [12] ($^{28}Si(d,d)^{28}Si$,
$<E_d> = 7.0$ MeV) most nearly satisfies for (d,d) CE studies so far reported
the energy-averaging criterion, since $\Delta/\Gamma = 30$. The CE analysis of the

tensor data is complicated by the presence of tensor potentials in the OM (and vice versa for the extraction of tensor potentials), such that at these energies the CE and tensor-potential effects are of similar magnitude. Complete angular distributions of $<\sigma_{21}>$ arc predicted [11] to separate these effects. The CE contributions maximize beyond 150° and at their maxima have $<(\sigma_{20})_{CE}> > <(\sigma_{22})_{CE}> > <(\sigma_{21})_{CE}>$. The contribution by Bürgi et al.[28] on $^{40}Ar(d,d)^{40}Ar$ near $E_d=10.75$ MeV also shows a strong energy dependence of $\sigma_{20}(175^\circ)$, suggestive of CE effects.

4.2. TRANSFER REACTIONS

The influence of CN effects in (d,p) polarization was felt by Meyer et al. [19] in studying deuteron D-state effects at scattering angles near 0°. They suggest that for a spin-0 target the largest effects will be seen with t_{20} beams, for which $<(\sigma_{20})_{CN}> \approx -\sqrt{2}/4 <(\sigma_0)_{CN}>$ for s-wave neutron transfer, and zero otherwise. Stephenson [29] reports CN effects which overwhelm D-state effects in $^{52}Cr(d,p)^{53}Cr$ at $E_d=6.0$ MeV. In the above examples the transfer-reaction cross sections are at least an order of magnitude smaller than those for elastic scattering; however, one expects $<(\sigma_{kq})_{CN}>$ and $<(\sigma_{kq})_{CE}>$ to be of similar magnitude. Therefore a detailed cross section excitation function and fluctuation analysis to determine y_D for each channel of interest should be made before small polarization effects are investigated.

4.3. DEPOLARIZATION STUDIES

The depolarization parameter D is measured using thick targets. In terms of the spin-flip probability S, $D=1-2S$. Also $<\sigma S> = (\sigma S)_{SE} + <(\sigma S)_{CE}>$. Therefore it is preferable to report σS rather than D. To realize the sensitivity of D to CE effects, consider that usually $0.9<D<1.0$ even at back angles. The value $D=0.9$ could be accounted for by 10% CE cross section half of which was spin-flip. In the contribution of Clegg et al.[30] on $^{14}N(p,p)^{14}N$, $E_p=16.15$ MeV an estimate of $<(\sigma S)_{CE}>$ shows large effects.

Thus CN polarization effects of resolved and energy averaged fluctuations are significant for cross section components less than 0.1 mb/sr at bombarding energies to about 20 MeV and for targets with A up to 90. Analysis by extensions of the Ericson and Hauser-Feshbach models will improve understanding of the interplay of direct-interaction and compound-nucleus effects.

The hospitality of the Universität München during the preparation of this review is gratefully acknowledged.

References

+ Supported in part by the U.S.E.R.D.A.

1) M.G. Braga Marcazzan & L. Milazzo Colli, Prog. Nucl. Phys. 11 (1970) 145

2) A. Richter, in Nucl. Spectroscopy & Reactions, Part B (Academic Press, New York, 1974) p. 343

3) T. Ericson, Phys. Rev. Letters 5 (1960) 430

4) E. Ventura et al., Bull. Am. Phys. Soc. 19 (1974) 498

5) R. Henneck et al., Bull. Am. Phys. Soc. 19 (1974) 478

6) P.J. Carlson, in High-Energy Physics & Nuclear Structure (North-Holland, Amsterdam, 1974) p. 55

7) T. Ericson, CERN TH-406 (1964)
8) R.O. Stephen, Nucl. Phys. 70 (1965) 123
9) P.J. Dallimore & I. Hall, Nucl. Phys. 88 (1966) 193
10) H.M. Hofmann et al., Ann. of Phys. 90 (1975) 403
11) R. Henneck et al., Contribution to this Symposium p. 795
12) R.J. Eastgate et al., Contribution to this Symposium p. 620
    R.J. Eastgate, Ph. D. dissertation, Univ. of. N. C. at Chapel Hill,
    1975 (unpublished), and, to be published
13) H. Paetz gen. Schieck et al., Contribution to this Symposium p. 791
14) E. Kujawski & T.J. Krieger, Phys. Letters 27B (1968) 132
15) K. Imhof et al., Contribution to this Symposium p. 789
16) J. Eng et al., Contribution to this Symposium p. 783
17) C. Glashausser et al., Contribution to this Symposium p. 787
18) E.J. Stephenson and W. Haeberli, Contribution to this Symposium
    p. 663
19) H.O. Meyer & J.A. Thomson, Phys. Rev. C8 (1973) 1215
20) T.R. Fisher et al., Nucl. Phys. A179 (1972) 241
21) M. Lambert & G. Dumazet, Nucl. Phys. 83 (1966) 181
22) R.F. Haglund et al., Contribution to this Symposium p.785
23) O. Häusser et al., Nucl. Phys. A109 (1968) 329
24) M. Jacob & G.C. Wick, Ann. of Phys. 7 (1959) 404
25) Y. Baudinet-Robinet & C. Mahaux, Phys. Rev. C9 (1974) 723;
    H. Levine & J. Wolfowitz, Ann. of Math. Stat. 15 (1944) 58; ibid 163
26) P.A. Moldauer, Phys. Rev. C11 (1975) 426
27) K.A. Eberhard et al., Nucl. Phys. A125 (1969) 673
28) H.R. Bürgi et al., Contribution to this Symposium p. 622
29) E.J. Stephenson & W. Haeberli, to be submitted for publication
30) T.B. Clegg et al., Contribution to this Symposium p. 634

REPORT ON DISCUSSION SESSION:
INTERMEDIATE STRUCTURE, ISOBARIC ANALOG AND GIANT MULTIPOLE

Charles Glashausser
Rutgers University, New Brunswick, New Jersey

Unusual methods for the observation of giant multipoles were discussed at length, in the context of contributions on p.751 and p.755. The former paper was presented by Arvieux; he showed an interpretation of proton inelastic scattering to the 8.88 MeV (2^-) state in ^{16}O. His calculations were based on the method of Geramb (cf.p.751); giant multipole resonances appear as intermediate states in a two-step process. The parameters of the E 1 and E 2 resonances determined in this way were in impressive agreement with photonuclear work; in some sense the agreement was <u>too</u> good to convince some sceptical observers. This was also true for similar results obtained previously for ^{12}C. This method appears so promising that I think it would be very useful to have an independent calculation of the effects of giant resonances in inelastic scattering.

The contribution on p.755, together with the results of additional analysis, was presented by Weller. By doing a phase shift analysis of σ and A_y for proton elastic scattering from ^{56}Fe, he found evidence for excitation of the giant dipole resonance as a compound state in this process. The predominant partial wave was $d_{5/2}$; previous (p,γ) work indicated either a $d_{5/2}$ or $g_{9/2}$ assignment. The evidence for excitation of the GDR comes almost entirely from the A_y data; the GDR is not visible in σ. I find this particularly interesting in the light of our own evidence for intermediate structure in A_y which does not appear in σ for scattering from ^{26}Mg and ^{27}Al.

Several other contributions are also worth mentioning here. Firk noted that there appears to be structure in (γ,n) data which does not appear in (p,γ) data. The interpretation of such a discrepancy remained ambiguous. Firk also asked where the M 1 giant resonance in ^{208}Pb could be found, given the results of the contribution on p.759 which changes the parity assignments for several spin states in ^{208}Pb. Cramer discussed a comparison between spectroscopic factors for (d,p) reactions and proton elastic scattering via IAR. His results for ^{209}Pb, an up-data of the contribution on p.781 , indicated good agreement between the two methods for all but the very highest spin states, when the parameters of both analyses were consistently chosen. The method of Bund and Blair, which is similar to the method of Zaidi and Harney, was used for the IAR analysis. Finally, Kretschmer's contribution (cf. p.765) indicated that the new formulas for width fluctuation corrections in Hauser-Feshbach theory recently proposed by Hoffmann et al. give excellent results for $^{88}Sr(p,n)^{88}Y$ and $^{88}Sr(p,p_0)^{88}Sr$, in contrast to the older formula of Moldauer. This is a useful confirmation of this new theory that I discussed in the invited paper (p.333).

POLARIZED HEAVY IONS

D. Fick

Max-Planck-Institut für Kernphysik, Heidelberg, Germany

1. Introduction

In the past the use of polarized heavy ions has been only a problem of academic interest to the heavy ion physics community. There are theoretical and practical reasons for this: the serious difficulty that in the present state of heavy ion physics, for which our understanding of the phenomena is sometimes crude and incomplete, the use of polarized heavy ions, which often have spin 3/2 and larger, would complicate the interpretation of experiments. This argument is accompanied by the belief that sources for polarized heavy ions which can be used on tandem accelerators are complicated, that the large variety of beams which are necessary for a reasonable investigation of reactions with polarized heavy ions can probably not all be produced in just a few type of sources, that polarization standards for all these ions and because of the large spins, for all the tensor moments must be found which looks like an unsolvable problem. Additionally there was and is the horror that the polarization physicists want more or less stupidly to repeat the experiments which up to now have been done unpolarized, if possible, with a polarized beam.

But in spite of these arguments there has been a successful development of a polarized Li-source[1] which can in principle produce polarized beams of all other alkalides. This development was started years ago[2] at the University of Hamburg and is connected during the whole time with the name of H. Ebinghaus. It is partly to the merit of him and his group that we are dealing with polarization effects in heavy ion interaction during this conference.

Since in Hamburg a suitable accelerator for heavy ions was not available, this source has been installed at the Heidelberg EN-Tandem accelerator. One of the aims of this talk is to report on the first results obtained with this new source. But because of the critical remarks at the beginning and because presently a review talk on the title subject cannot be given a major part will be spent on future aspects of polarization physics with heavy ions.

The interaction of heavy ions is characterized by a rapid onset of strong absorption due to the opening of many reaction channels. This results in a surface localization of the interaction process and therefore in a strong localization of the process in momentum and angular momentum space. In connection with a short wave length this leads either to a classical description of the interaction or to an interpretation in terms of optical language (diffraction scattering). A further characteristic results from the interplay of strong absorption with the Coulomb interaction. In this respect it is interesting that besides the polarized ^{6}Li beam[3], polarized beams for tritons[4], helium-3 [5,6] and GeV-protons[7] came in operation. Reaction with these particles are dominated by strong absorption, too, and can be described in terms of diffraction models. But heavy ions, such as alkalides, possess besides a much larger Coulomb interaction generally a spin larger than 1/2 and therefore a more or less large quadrupole moment. An essential part of the further discussion will deal with this property.

Physics with polarized heavy ions is at the very beginning of investigation. Therefore throughout the following discussion the simplest and most naive pictures which can be found will be used.

2.   Experiments with a vector polarized ^{6}Li-beam

The source for a vector polarized ^{6}Li$^-$-ions has been working for one year at the Heidelberg EN-Tandem. For the last six months it has been used to obtain data mostly on the elastic scattering of ^{6}Li on several isotopes.

The source is of the classical atomic beam type[1,3]. The only special aspect is the ionizer which uses a surface ionization process on heated tungsten[1] to produce the ^{6}Li$^+$-ions from the atomic beam. The great advantage of this design is its simplicity together with an almost 100% ionization efficiency. Polarized ^{6}Li$^+$-beams up to 20 µA are typically achieved with the atomic beam and ionization device of this source. During the operation of the source at the Heidelberg EN-Tandem the beam intensity on target has steadily increased from 2 nA to a maximum of 150 nA ^{6}Li^{3+}. Parallel to the increase of the beam intensity the continuous running time of the source has increased from a few hours up to a maximum of 2.5 days. The breakdown of the source is usually caused by a closure of the aperture of the six pole magnet ($\emptyset$ = 3 mm) by condensed Li.

During the experiments it turned out that a change of the sign of polarization by reversing the direction of the magnetic field over the ionizer or in the Wien-filter is very time consuming and tedious. Therefore the polarization data have been obtained in successive runs with a polarized and unpolarized beam (high frequency transition on and off, respectively). This enhances the danger of systematic errors.

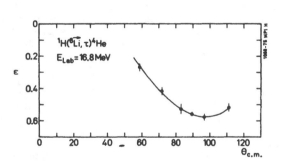

Fig.1

Asymmetry observed with a vector polarized ^{6}Li beam for the reaction ^{1}H(^{6}Li,^{3}He)^{4}He at E_{Li} = 16.8 MeV.

The polarization of the beam had been monitored in the beginning of the experiments with the ^{1}H(^{6}Li,^{3}He)^{4}He reaction which displays at E_{Li} = 16.8 MeV and Θ_{CM} = 90° a fairly high asymmetry of ε = 0.65 $\pm$ 0.04. Fig.1 displays for this reaction an angular distribution of the asymmetry. In the later stage of investigation the elastic scattering of ^{6}Li on ^{12}C at E_{Li} = 22.8 MeV and Θ_{CM} around 45° has been used as a monitor. There a flat plateau is seen in the analyzing power (Fig.2) which occurs at a rather large cross section and a high enough asymmetry.

The polarization of the ^{6}Li beam is not known exactly. Of course, there exist no polarization standards for vector polarized ^{6}Li. But the highest up to now observed asymmetry in the ^{6}Li-^{4}He elastic scattering[8] of ε = 0.79 $\pm$ 0.01 at Θ_{CM} = 75.8° and E_{Li} = 21.3 MeV established a lower limit of the polarization of P $\geq$ 0.52. The finite magnetic fields in the ionization and charge exchange region together with a now increased transition probability of the rf-transition[9] gives an upper limit of the polarization of P $\leq$ 0.64. During the short running time of the source it was not tried to get a better determination of the beam polarization.

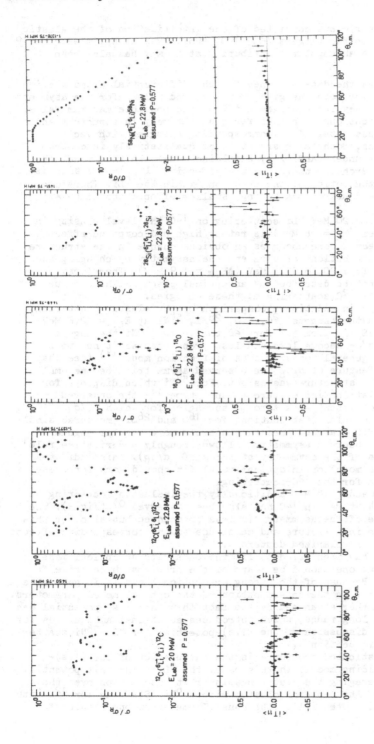

Fig.2 Angular distributions of σ/σ_R and $\langle iT_{11} \rangle$ for the scattering of vector polarized ^6Li on ^{12}C at 20 MeV and on ^{12}C, ^{16}O, ^{28}Si, ^{58}Ni at 22.8 MeV. The polarization of the ^6Li beam is known to be between $0.52 < P < 0.64$. For $\langle iT_{11} \rangle$ essentially the observed asymmetry had been plotted. For $P = 1/\sqrt{3} = 0.577$ which lies in the limits given above the observed asymmetry is just equal to $\langle iT_{11} \rangle$. For scattering on ^{12}C at 22.8 MeV σ/σ_R has been slightly renormalized in respect to the data displayed in a contribution to this Conference10).

The first experiments consisted of the investigation of the elastic
scattering of ^{6}Li on the targets ^{12}C, ^{16}O, ^{28}Si, ^{58}Ni at E_{Li} = 22.8 MeV[10).
For ^{12}C as a target an angular distribution at 20 MeV has also been
measured[10).

Fig.2 displays the data obtained for the differential cross section
and the asymmetry for both energies. The observed values for the asymmetry
are rather large. They reach at least 60% of the possible maximum value.
The angular distributions do not differ qualitatively in structure and
magnitude from those obtained at corresponding energies with vector
polarized deuterons. Both facts are at least qualitatively in contradiction to a naive understanding by a folding model. In this model the
motion of the deuteron and therefore the effective (l.s.)-potential is
smeared out over the motion of the deuteron within the ^{6}Li. Therefore,
as compared to the deuteron scattering, smaller effects should be expected.

Even though for 20 MeV ^{6}Li scattering on ^{12}C the level density in
the compound system (E_x = 26 MeV) is rather high and compound effects
should not be expected, one observes an obvious change in the structure
of the angular distributions of σ/σ_R and the assymetry by changing the
bombarding energy from 20 to 22.8 MeV (ΔE_{CM} = 1.9 MeV). Therefore,
effects which cannot be described by an optical potential have to be
expected in the ^{6}Li-^{12}C scattering at these energies.

Fig.2 displays in addition a comparison of the elastic scattering
data observed with the targets ^{12}C, ^{16}O, ^{28}Si, ^{58}Ni at E_{Li} = 22.8 MeV[10).
With decreasing E/E_{coul} from 3.5 for ^{12}C to 1.4 for ^{58}Ni the angular
distribution of σ/σ_R become less and less pronounced as it has to be
expected from the general features of a diffraction model[11) (see chap.4).
Parallel to this feature of σ/σ_R the observed asymmetries become smaller
and smaller and the structure washes out. For ^{58}Ni which displays for
σ/σ_R an angular distribution of pure Fresnel type [11) the observed
asymmetries are consistent with zero up to angles for which σ/σ_R = 10^{-2}.

The observed angular distributions for ^{16}O and ^{28}Si (measured simultaneously with a SiO_2 target) seem to exhibit a more regular angular
distribution. The observed asymmetry follows roughly a derivative rule[11)
(maxima and minima of the asymmetry at zeros of σ/σ_R). This indicates
that a diffraction model or an optical model fit should work for these
nuclei better than for the ^{12}C-scattering.

For E_{Li} = 20 and 22.8 MeV additionally the inelastic scattering to
the 2^+- state at 4.44 MeV in ^{12}C has also been observed[10) (Fig.3). The
observed structure of the asymmetry is less pronounced but also for this
example comparable in structure and magnitude to the corresponding effects
observed with vector polarized deuterons.

Naturally it has been tried to fit the data with an optical model
code. In doing that one should be aware of the problems which arise from
such a procedure. Because of the strong absorption there is in general a
continuous ambiguity in the determination of the optical model parameters.
The problem is further enhanced by the fact that the (l.s)-potential has
probably a rather long range, which introduces additional ambiguities for
the parameters. A discussion of the (l.s)-potential for ^{6}Li-^{58}Ni scattering can be found in section 4.

A more sophisticated procedure is the calculation of the (l.s)-
potential by a folding model. In this model the ^{6}Li spin-orbit potential
is obtained by averaging the deuteron spin-orbit interaction over the
relative motion of the deuteron cluster within ^{6}Li. Starting for ^{12}C with
deuteron spin-orbit potential of the usual Thomas-form one obtains for

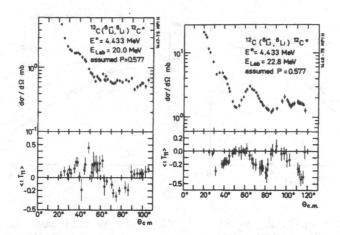

Fig.3 Angular distributions of $d\sigma/d\Omega$ and $\langle iT_{11} \rangle$ for the inelastic scattering of vector polarized ^{6}Li to the 4.44 MeV state in ^{12}C. (For further comments see Fig.2).

the ^{6}Li spin orbit potential more or less a volume potential which extends much further outside[12] (Fig.4). According to the different assumptions on the density distribution ρ of ^{6}Li one gets a different dependence on the target mass A (Fig.5). If the deuteron cluster is concentrated on the surface of ^{6}Li one obtains more or less a $A^{-1/3}$ dependence. For a Fermi-distribution or for the solution of a realistic Woods-Saxon potential[13,15] the A-dependence shows a $A^{-2/3}$ dependence. Independent of these assumptions the strength of the (l.s)-potential is always decreasing with increasing target mass number A which is in qualitative agreement with the present results (Fig.2).

On the basis of this folding model there are two numerical predictions for the ^{6}Li (l.s)-potential for scattering on ^{12}C [12,13]. They differ at least by a factor ten in strength if they are compared at the projectile plus target radius. The smaller one[12] uses for ^{6}Li a 2s-nonexchange wave function[14] whereas the other one uses the solution for a realistic Woods-Saxon potential with the correct separation energy. But it is not clear, if this large discrepancy results from the use of these different ^{6}Li-wave functions. For the second approach there are calculations for the present experimental results on ^{12}C and ^{58}Ni in a contribution to this conference[15]. For ^{12}C the (l.s)-potential calculated by their folding procedure seems to give the correct order of magnitude of the observed asymmetry. It is not surprising that by arguments given before the flat maximum around 50° and the data at backward angles could not be reproduced.

3. Future polarized beams

A large variety of different ion beams available is one of the most important requirements for the investigation of heavy ion reactions. Especially the tandem accelerators offer here the best possibilities.

As was pointed out already in the original papers[1,2] a source which produces polarized ^{6}Li-beams can in principle be used to produce polarized

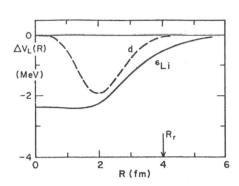
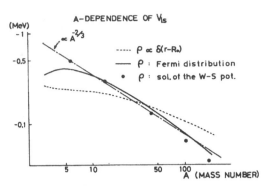

Fig.4 Spin orbit splitting $V_1(R)=$ Fig.5 Projectile mass number
       $(21+1)V_{1S}(R)$ for $1=9$ calcu-   dependence at the pro-
       lated from a folding model   jectile plus target radius
       for ^{6}Li scattering (36 MeV)   of a folded spin-orbit
       on ^{12}C [12]). The dashed curve   potential for different
       displays the spin orbit   density assumptions for
       splitting for deuterons ^{6}Li [13].
       with the same velocity.

beams of the other alkalide isotopes: ^{7}Li, ^{23}Na, ^{39,41}K, ^{85}Rb and ^{133}Cs
(Tab.1) The surface ionization process[1] works for Na, K, Rb and Cs as
well as for Li and no principal reasons can be seen which limit the
production of atomic beams of these elements. Concerning Na some months
ago an atomic beam was produced with the source at the Heidelberg EN-
Tandem which could be ionized without any further troubles. But the high
critical fields of 0.3 T and safety problems may play a role for Cs and
Rb, respectively. (The critical field is defined as of the nuclear magnetic
moment at the electron.)
        Concerning the charge exchange process from positive to negative ions
nothing is known quantitatively for other alkalide beams except for Li,
where K or Cs-vapor is used[16]. In this respect experiments are in progress
at the University of Hamburg[18]. Except for ^{6}Li all other alkalide nuclei
have a spin larger than one (tab.1). As a consequence the situation
concerning the magnification of beam polarization and the selection of
tensor moments by means of certain combinations of high frequemcy transitions become more and more complex. The magnitude of vector polarization
which can be achieved after a weak field transition decreases with increasing spin I. ($P=-2/2I+1$)).   In addition such a device produces also
higher odd rank tensor polarization (For example $I=3/2$: $P_z = -1/2$,
$P_{zz} = 0$, $P_{zzz} = -1/6$. Definition of the P_α for $I = 3/2$ from ref. 17). A
combination of a weak field and a σ-strong field transition produces
$|P_{zz}| = 1/2$ only with an additional amount of odd rank tensor polarization.
But for $I = 3/2$ the inclusion of a weak field transition in combination
with two strong field π-transitions allows for the production of a pure
even rank tensor polarized beam ($P_z = P_{zzz} = 0$, $P_{zz} = + 1$). This example
may be enough to demonstrate that such problems can be solved at least in
principle by the present techniques.

element	abundance (%)	B_{crit} (mT)	I^π	μ/μ_N	r_{rms} (charge)(fm)	Q (mb)	100 Q/Z_eR^2
^{6}Li	7	8.2	1^+	0.8	2.5	-0.8	-0.4
^{7}Li	93	28.8	$3/2^-$	3.3	2.4	-40 (58)	-23
^{23}Na	100	63.3	$3/2^+$	2.2	2.9	+140(100)	+15
^{39}K	93	16.5	$3/2^+$	0.4	3.4	+55 (49)	+2.5
^{41}K	7	9.1	$3/2^+$	0.2	3.4	+67(60)	+3.0
^{85}Rb	100	108	$5/2^-$	1.4	4.2	+260	+3.6
^{133}Cs	100	328	$7/2^+$	2.6	4.8	-3	0.0
^{19}F	100	71.7	$1/2^+$	2.6	2.8	—	—
^{35}Cl	75	21.0	$3/2^+$	0.8	3.3	-79	-3.9
^{37}Cl	25	17.5	$3/2^+$	0.7	3.3	-62	-2.9
^{79}Br	50	102	$3/2^-$	2.1	4.1	+310(330)	+4.8
^{81}Br	50	108	$3/2^-$	2.3	4.1	+260(280)	+4.0
^{127}I	100	124	$5/2^+$	2.8	4.7	-790(700)	-5.9

1466-75 MPIH

Tab.1

Besides the polarized alkalide beams it should also be possible with the same type of source to produce intense beams of polarized negative halogen nuclei (Tab.1) For halogen atomic beams, except fluorine, surface ionization on heated tungsten gives directly negative ions[19]. (It is not clear up to now what kind of material has to be used in the ionizer to obtain negative fluorine beams.) From this point of view sources for polarized negative halogens seems to be very promising, since one can avoid the intensity loss during the charge exchange process into negative ions. But there will be other difficulties. The very active chemical nature of these elements will certainly cause problems in the production of intense atomic beams. Since all halogens have j = 3/2 the separation of one fine structure component will be more complicated than for the alkalides. It is not clear whether the Stern-Gerlach method or the application of the Laser technique will give the better results.

But altogether the technical problems which exist for a polarized alkalide plus halogen heavy ion source seem not to be so severe that they cannot be solved in the (near) future. It is probably more a question of man power than of fundamental difficulty.

Coming back to the present source at the Heidelberg EN-Tandem it is planned to have available a tensor polarized ^{7}Li and ^{23}Na-beam within the next half year.

But what kind of experiments should be done with these new beams ? Indeed, there are experiments, for which one will get no new additional information, except perhaps more complexity if they are done with polarized heavy ions. For example, as I will show in the next section, polarized heavy ions are in general not necessary to determine the j-value for a nucleon transfer reaction initiated with heavy ions. On the other hand there are really new types of heavy ion experiments if one will use polarized heavy ions. Some of these experiments depend on the polarization of the heavy ions in respect to their spin and magnetic moment. But - at

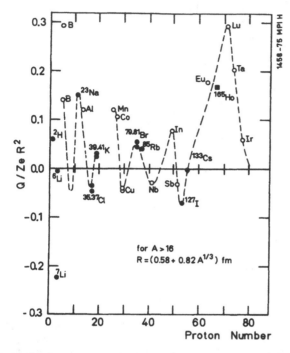

Fig.6

Q/ZeR^2 for different isotopes with an uneven proton number. The nuclei which can be produced by an alkalide-halogen polarized ion source are indicated by black dots.

least it is my personal feeling - the more important and really new feature is that heavy ion beams will have aligned quadrupole moments, if the beam is aligned with respect to the spin. The alignment axes for both are the same.

Because of the strong absorption heavy ion interactions take place predominantly on the interacting surfaces. Therefore, one can expect strong effects from the alignment of the quadrupole moments. In section 5 some typical examples will be discussed.

Conceptionally these ideas were developed by the Madison group in the discussion of subcoulomb stripping experiments initiated by tensor polarized deuterons[20]. This experiment demonstrates in a nice way how to think about such effects and how to link the "conventional" physics of tensor polarized deuterons with the physics of tensor polarized heavy ions. It displays that the main effect of tensor polarized deuterons in this experiment can be ascribed to its deviation from a spherical shape which is in a one to one correspondence to its quadrupole moment.

For a uniformly charged rotational ellipsoid (axis a,b; symmetry axis along b) the quadrupole moment Q is given by:

$$Q = \frac{2}{5} Ze \ (b^2 - a^2) \text{ with } a^2 b = 3R^3$$

R is the radius for a volume equivalent nucleus. For small deformations one finds up to second order

$$a = R(1 - \delta/2 + 3\delta^2/8) \qquad b = R(1 + \delta + 3\delta^2/2)$$

$\delta = \frac{5}{6} Q / ZeR^2$ gives in first order the deviation of b from a spherical form. For a deuteron (Q_d = 2.8 mb) one obtains with the charge radius of r_{ms} = 2.2 fm a δ = 0.05.

Fig.6 displays Q/ZeR^2 for other nuclei with uneven proton number (for R the electric r_{ms}-radius had been chosen). The nuclei which can be produced by an alkalide or halogen polarized heavy ion source are indicated by a black point (Tab.1). ^{6}Li has a very small deformation. But ^{7}Li and ^{23}Na are in this respect very promising objects. ^{7}Li has an extremely large deformation and the deformation of ^{23}Na is comparable to that of ^{165}Ho, which had been used for experiments with aligned quadrupole moments[21] only. The list of Tab.1 shows that an alkalide-halogen source would cover the whole variety of polarized heavy ions concerning spins, magnetic moments and quadrupole-moments. Besides almost spherical nuclei, one with a small (^{6}Li) and one with a large spin (^{133}Cs) there are in each mass region nuclei with prolate (Q>0) and oblate (Q<0) deformation. If the problem concerning the ionization of ^{19}F can be solved the magnetic moments (0.2<μ/μ_N<2.8) and spins (1/2 $\lesssim$ J $\lesssim$ 7/2) cover a large region. Because of the large Coulomb-force for an EN-Tandem (U = 6 MV) the lightest nuclei of this list (up to ^{23}Na) can be used after acceleration in nuclear interaction experiments. For an upgraded MP-Tandem (U = 12 MV) the chlorine and potassium beams as well can be used additionally (Fig.7). In this connection the ideas concerning a postaccelerator device for the upgraded MP-Tandem in Heidelberg should be mentioned[22]. For a 10 MV-version also Br-ions become interesting whereas the use of polarized ^{127}I and ^{133}Cs needs at least the larger version of 20 MV for an application in nuclear reaction studies.

Before going into the discussion of some examples of typical experiments the question of polarization standards for all these beams and all the possible tensor moments have to be discussed. It is obvious that if a source which produces the alkalides and halogen nuclei can be realized one could, of course, not look for individual polarization standards for

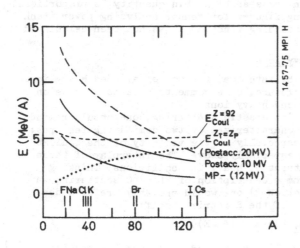

Fig.7

Energy per nucleon which can be reached by different versions of a postaccelerator on a MP-Tandem[22]. On the abscissa these nuclei are indicated which can be produced by an alkalide-halogen polarized ion source.

each of these beams (including all tensor moments). But fortunately
there seem to be an universal polarization standard for all these nuclei:
Coulomb excitation of the polarized projectile nuclei. In first order
perturbation theory the analyzing powers do not depend (for pure multipole transitions λ) on the transition matrix elements. They can be calculated from the Coulomb excitation orbital integrals. With the z-axis along
the normal of the scattering plane (as opposed to the Madison conventions[23])
one obtains, for example, for a projectile transition $I_0 \rightarrow I_f$ the vector
analyzing power:

$$A = (-1)^{I_0+I_f+\lambda} (2\lambda+1) \sqrt{\overline{(2I_0+1)(I_0+1)/I_0}} \begin{Bmatrix} \lambda & \lambda & k \\ I_0 & I_0 & I_f \end{Bmatrix} \begin{pmatrix} \lambda & \lambda & k \\ \lambda & -\lambda & 0 \end{pmatrix} \Pi_\lambda(\vartheta,\xi)$$

($\xi = \eta\Delta E/2E_{CM}$, η Sommerfeldparameter)

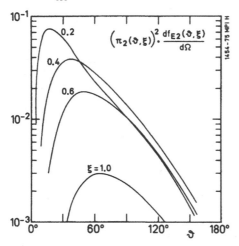

Fig.8

$(\Pi_2(\vartheta,\xi))^2 \, df_{E2}(\vartheta,\xi)/d\Omega$ which
is proportional to $A^2 \; d\sigma/d\Omega$
for E2-Coulomb excitation of
vector polarized heavy ions[24].

The function $\Pi_\lambda(\vartheta,\xi)$ had been calculated by Alder and Winther for $\lambda=1,2$[27].
Fig.10 displays $(\Pi_2(\vartheta,\xi))^2 \, df_{E2}(\vartheta,\xi)/d\Omega$. Since $df/d\Omega$ is proportional
to the differential excitation cross section this quantity is proportional
to $A^2 \cdot d\sigma/d\Omega$. The corresponding figures for tensor analyzing power (rank 2)
are not shown since the corresponding function Π has not been calculated
until now.

4. Vector polarized heavy ions

Even though I personally believe that the use of aligned heavy ions
will be more important in the future, some comments have to be made on
experiments with vector polarized heavy ions.

In a discussion of heavy ion elastic scattering, one should remind
that heavy ion collisions are characterized by two dominating features:
a strong Coulomb interaction especially in the vicinity of the Coulomb
barrier and strong absorption especially above. The interplay of these
two effects determine the features of heavy ion collisions. Thinking
about the problems arising from a description of ^{6}Li-^{12}C scattering in
terms of an optical model (section 2) one would probably prefer a description by a parametrization of the S-matrix itself:

$$S_1(E) = \eta_1(E)\exp(2i\delta_1(E))$$

For heavy ion scattering (large 1) , $\eta_1(E)$ and $\delta_1(E)$ can be approximated[11] by functions $\eta(\lambda,E)$ and $\delta(\lambda,E)$ which depend on the continuous variable

$$\lambda = 1 + 1/2$$

The effect of strong absorption on the elastic scattering function is a more or less smooth increase of $\eta(\lambda,E)$ from zero to unity with increasing 1 . The effect of strong absorption can then be described (at least in first order) by the so-called critical angular momentum $\Lambda = l_{cr} + 1/2$ [11]) for which

$$\eta(\Lambda) = 1/2 .$$

Heavy ion scattering is, therefore, roughly characterized by two features and corresponding parameters: the Sommerfeldparameter η which measures the Coulomb interaction and the critical angular momentum which characterizes the strong absorption. For $\Lambda \gg 1$ the nuclear scattering is diffractive. This condition is fulfilled for the present ^{6}Li experiments (section 2) where $\Lambda = 10$ and 12 for $E_{Li} = 22.8$ MeV and scattering on ^{12}C and ^{58}Ni, respectively. One finds ^{Li}especially simple conditions if the Coulomb interaction satisfies the conditions:

p << 1 Fraunhofer-diffraction scattering   or

p >> 1 Fresnel-diffraction scattering   with

$$p = 2\eta \, (1-(2E/V_c-1)^{-2})$$

For the ^{6}Li experiments at 22.8 MeV p=3 for ^{12}C and p=12 for ^{58}Ni as a target.

The ^{6}Li-^{58}Ni elastic scattering at $E_{Li} = 22.8$ MeV is, therefore, of Fresnel type with a typical structureless dependence of σ/σ_R on the scattering angle Θ. For Fresnel scattering Λ can be determined by [11]:

$$tg(\Theta_{cr}/2) = \eta/\Lambda \quad \text{with} \quad \Theta_{cr} \text{ from } \sigma/\sigma_R(\Theta_{cr}) = 1/4 .$$

For such a type of scattering there exist typical polarization angular distributions (Fig.9) for spin 1/2 particles[25]) (^{19}F on ^{159}Tb, $\eta=32$, $\Lambda=92$).

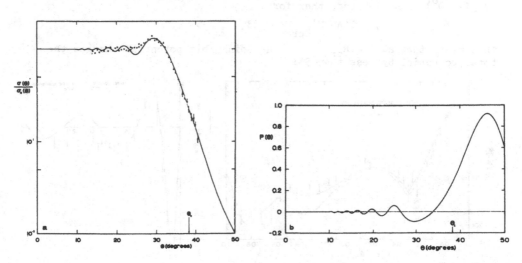

Fig.9   Differential cross section (σ/σ_R) and polarization P calculated for 150 MeV ^{19}F scattering on ^{159}Tb [25]).

The amount of polarization is directly proportional to the splitting of η_1 due to the different spin orientation $1 \pm 1/2$ of the spin 1/2 particles

$$P \propto (\eta_+ - \eta_-)/(\eta_+ + \eta_-)$$

For the calculation of Fig.9 a splitting as used for 180 MeV protons has been taken. This figure displays the unfavourable situation for strongly absorbed heavy ions in respect to vector polarization effects, which results from the interplay of strong absorption and strong Coulomb interaction. Although strong absorption gives rise to strong polarization effects it occurs at large angles where the scattering cross section is several orders of magnitude smaller than σ_R. At small angles the polarization is strongly suppressed by Coulomb damping.

One can say very roughly that the polarization starts at $\theta = \theta_{cr}$ to deviate from zero (Fig.9). In this respect the ^{6}Li-^{58}Ni interaction displays a very small effect (Fig.2). Even at angles where $\sigma/\sigma_R = 10^{-2}$ no clear deviation from zero could be detected.

Using the concept of critical angular momentum and the "quarter point method" one can obtain an estimate of the relative change of the potential due to the (l.s)-potential at $R_{cr} = \Lambda_{cr}/k$.

The (l.s)-potential changes for a given spin orientation (e.g. up) the critical angular momentum Λ and, therefore, the wave number k at $r = R_{cr}$ by $\Delta k \doteq \Delta\Lambda/R_{cr}$. From that one obtains for the change of θ_{cr}:

$$\Delta\theta_{cr} = -2\sin^2(\theta_{cr}/2)(R\Delta k/\eta)$$

For maximum polarization the analyzing power A is given by $A = \Delta\sigma/\sigma$. Using the above equations and $d\ln\sigma/d\theta$ at $\theta=\theta_{cr}$ from the measured unpolarized angular distribution (Fig.2) one obtains for ^{6}Li-^{58}Ni scattering

$$(\Delta U/U)_{R_{cr}} = 2(\Delta k/k_{R_{cr}}) = 0.6 \; A \; (\theta = \theta_{cr})$$

From Fig.2 one estimates for the analyzing power at the "quarter point" ($\theta_{cr} \approx 60°$) $A < 0.04$ and, therefore,

$$(\Delta U/U)_{R_{cr}} < 0.03$$

This means that at $R = R_{cr}$ the spin orbit potential changes the total potential by less than 3%!

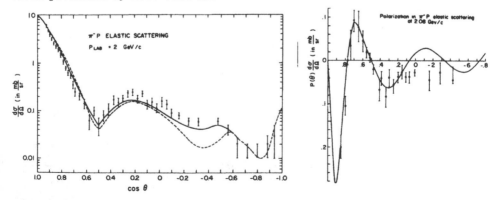

Fig.10 Differential cross section and proton polarization for 2.1 GeV/c π^--scattering on protons[26]. The solid line is a prediction of a simple diffraction model (Fraunhofer scattering)[27]

This example exhibits, furthermore, that a discussion in terms of "phenomenological models" which parametrize the S-matrix seems to be a better way to discuss polarization effects in heavy ion diffraction scattering.

Unfortunately the ^{6}Li-^{12}C scattering (Fig.2) do not satisfy the condition of Fraunhofer scattering (p << 1) even though the unpolarized angular distribution is of oscillating nature. But there is a nice example in the high energy πp-scattering[26] which is essentially dominated by strong absorption but very weak (negligible) Coulomb interaction. Fig.10 displays rather old results for the differential cross section and the polarization observed in π-p scattering at P_{lab} = 2.1 GeV/c[26]. The elastic differential cross section is described very well by a parametrization of the scattering matrix[27]. Since it is Fraunhofer diffraction scattering, it is mainly proportional to the first Bessel function squared

$$d\sigma/d\Omega \propto (J_1(\Lambda\theta))^2$$

with some correction terms depending on three other parameters. The polarization can be understood quite well by a proper splitting $\eta^+ - \eta^-$ of the absorption parameter at the critical angular momentum Λ. Further examples can be found in the recent Argonne ZGS-symposium on high energy physics with polarized beams[28].

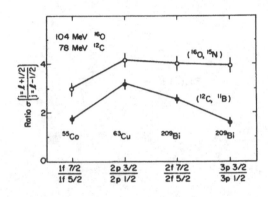

Fig.11

Peak cross section ratios for j=1 ± 1/2 proton single particle states in the residual nucleus indicated[29].

Generally in heavy ion transfer reactions the transferred nucleus is neither in the entrance nor in the exit channel in a relative S-state. Therefore, the heavy ion transfer cross sections are in an implicit way via the dependence on l sensitive to the transferred total angular momentum. As an example Fig.11 displays the ratio of cross section $\sigma(j=1 + 1/2) / \sigma(j=1 - 1/2)$ for (^{16}O, ^{15}N) and (^{12}C, ^{11}B) transitions with different transferred angular momenta[29]. They show a clear j-dependence with a preference of the j=1+ 1/2 transition. This is an example where opposite to classical (d,p) or (d,n)-reactions no polarization measurements are needed to obtain the full spectroscopic information.

Concerning the inelastic scattering of heavy ions nothing special can be expected in general (see for example Fig.3). This changes, at least for the polarization of inelastically scattered heavy ions in the

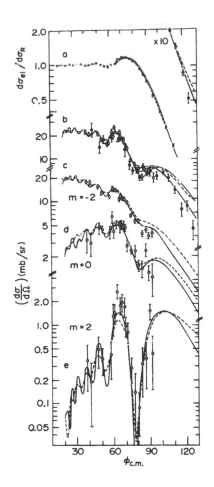

region of the nuclear–Coulomb interference. Fig. 12 displays besides
the unpolarized elastic and inelastic differential cross section
(a,b) the m-substate population
determined from the γ-ray correlation following the excitation of
the 0.86 MeV 2^+-state in ^{56}Fe by
43 MeV ^{16}O ions[30]. (Also here the
z-axis is taken along the normal of
the scattering plane.) Dramatic
polarization effects, not only for
the vector part, had been observed
in the nuclear–Coulomb interference
region ($\theta_{CM} \approx 70^\circ$) in this beautiful
experiment. At forward angles where
the excitation is purely Coulomb
predominantly the m=-2 substate is
populated. This polarization effect
is of the same type as described at
the end of section 3 for Coulomb
excitation of polarized projectiles.

Fig.12

5. Aligned heavy ions

Naturally there exist effects in heavy ion interactions similar to
that in section 4 which arise from the alignment of the spins of the
heavy ions. But in this section effects, which occur from the simultaneous alignment of the quadrupole moments of the heavy ions, will be discussed. For example, Rutherford scattering, Fresnel scattering and the
total reaction cross section will be considered.

For the elastic scattering of aligned quadrupole moments (spin I) in
a pure Coulomb field the deviation from pure Rutherford scattering which
is due to the change of the Coulomb-interaction has been calculated up to
second order for aligned beams (with tensor polarization ρ_{20} [31])

$$\sigma/\sigma_R - 1 = \sqrt{(2I+1)(I+1)(2I+3)} / \sqrt{45I(2I-1)} \; \tilde{f}_{20}(\vartheta) \; (Q/Zea^2) \, \rho_{20}$$

$$+ \text{second order}$$

a is the closest distance between projectile and target, $\tilde{f}_{20}(\vartheta)$ is calculated from the Coulomb excitation orbital integrals[31] $\tilde{f}_{20}$ and reaches its maximum at $\vartheta = 180°$ (Fig.13). Qualitatively the result can be understood easily: except for the statistical factors depending on I the analyzing power is proportional to the quadrupole moment, and because of the necessity of a relative large electric field gradient proportional to $1/a^2$. By geometrical arguments it is obvious that the effect is largest at $\vartheta = 180°$ (f_{20} and $1/a^2$!) The second order contributions[31] and dynamical effects[32] investigated. The latter can be neglected completely, the former are in the worsest case 10% of the first order part.

For an estimate of the effect one has to take into account that pure Coulomb-scattering requires a>R, with R the nuclear interaction radius ($R \approx 1.4 \times A^{1/3}$fm). Therefore, one prefers light targets to minimize a. For aligned ^{7}Li and ^{23}Na beam effects between 5 and 10% with ^{4}He as a target can be expected. This method of determination of quadrupole moments is free of the principal uncertainties (Tab.1) which arise from the classical methods[33].

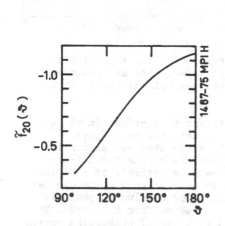

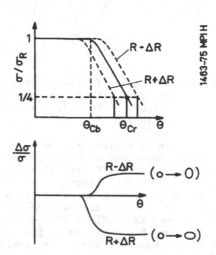

Fig.13 The function $\tilde{f}_{20}(\vartheta)$. It is correlated to f_{20} from ref.31 by $f_{20}( ) = (1/n)\tilde{f}_{20}(\vartheta)$.

Fig.14 Qualitative description of Fresnel scattering with aligned nuclei.

The discussion of elastic scattering of aligned quadrupole moments in the region of Fresnel diffraction scattering ($\Lambda \gg 1$, $p > 1$) can be done similar to that of section 4 . The simplest picture for the effect of aligned quadrupole moments is to take into account the change of the nuclear interaction radius ΔR, which is assumed to be proportional to the quadrupole moment Q. Depending on the sign of ΔR the critical angle θ_{cr} and the angle θ_{Cb} at which the scattering deviates from pure Rutherford scattering changes according to Fig.14. Therefore, one has to expect qualitatively the dependence of $\Delta\sigma/\sigma$ as indicated in the lower part of Fig. 14.

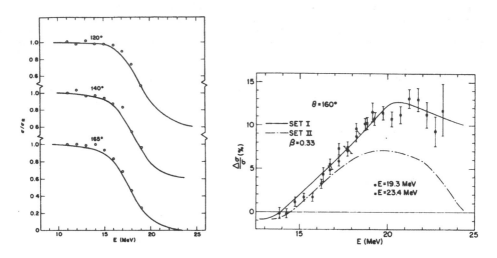

Fig.15   Excitation function for α-^{165}Ho scattering (left). Relative
          change in the cross section if the α-particles are scattered on
          aligned ^{165}Ho [21]. The solid curves are optical model calcu-
          lations.

There is a nice experiment which displays this effect in the excitation functions of backward scattered α-particles on aligned ^{165}Ho-
target [21]. The left part of Fig.15 displays the observed excitation
functions for α-Ho backward scattering which are of the Fresnel-type
(θ and E interchanged, arguments are the same as before). In addition,
the experiment was done with Ho-nuclei which were aligned perpendicular
to the beam axis. Since the quadrupole moment of Holmium is positive
(Tab.1) one has a change of the interaction radius from R to R-ΔR. The
observed change in the cross section as a function of bombarding energy
(Fig.15 right) is in agreement with the qualitative results displayed
in Fig.14.

The above discussion shows that the total reaction cross section is
also changed if the interaction is initiated by heavy ions with aligned
quadrupole moments. First theoretical attempts exist for the interaction
of deformed nuclei [34] naturally for the unpolarized ions only. For low
enough energies the total reaction cross section σ can be parametrized
as [34,35]

$$\sigma = \pi R_c^2 \, (1 - V(R_c)/E)$$

where R_c is the so-called critical radius. $V(R_c)$ can be taken roughly to
be the Coulomb barrier. Fig. 16 displays for the interaction of ^{35}Cl
with Ni-isotopes [35] that such a parametrization is meaningful. In this
experiment the deviation from a $1/E$ dependence for small energies was
taken as an indication of the deformation of the Ni-isotopes. But it
was impossible to decide whether this is an effect of the change in
the effective interaction radius or in the interaction potential or in
both. An aligned quadrupole beam would be a great help, since the interaction radius can be changed in a definite way.

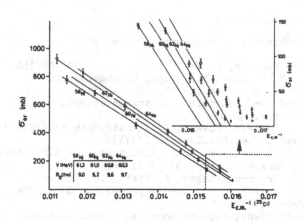

Fig.16

Total reaction cross section for ^{35}Cl on ^{58,60,62,64}Ni as a function of $1/E_{CM}$[35).

^{23}Na	R (fm)	V MeV	$2\frac{\Delta R}{R}$	E = 50 MeV σ^{tot}(mb)	$\frac{\Delta\sigma}{\sigma}$	E = 75 MeV σ^{tot}(mb)	$\frac{\Delta\sigma}{\sigma}$
◯	9.3	41.3	---	473	---	1221	---
⬭	9.8	40.2	0.11	592	0.25	1400	0.15
⬮	9.0	42.3	-0.07	392	-0.17	1110	-0.09

Tab. 2

From the change ΔR of the critical radius, which results from the change in the alignment of the quadrupole moments, one obtains

$$\Delta\sigma/\sigma = 2\Delta R/R_c - \Delta V/(E-V(R_c))$$

Obviously the dependence on E allows one to distinguish between a change of the interaction radius and interaction barrier. For the total reaction cross section of aligned ^{23}Na (along the beam axis) on spherical ^{58}Ni and E=75 MeV one obtains with R = 1.4 A$^{1/3}$fm a $\Delta\sigma/\sigma$ = 0.28. If one takes a more realistic model which uses a refined liquid drop formula for deformed nuclei[36) one can take more properly into account for aligned nuclei the change in R_c and $V(R_c)$. The results for $\Delta\sigma/\sigma$ are collected in Table 2[37).

In all the experiments discussed in this section the change of $\Delta\sigma/\sigma$ can be studied as a function of the orientation of the alignment axis. This will give for the nuclear part of the experiments additional information on the deviation of nuclear interaction, from a spherical shape. Since the charge and nuclear interaction radius are quite different (r_o = 1.0 fm and 1.4 fm, respectively), it is not obvious that the "quadrupole moments" of the charge and the nuclear matter distribution should be the same. Hopefully the use of aligned heavy ion beams can give a contribution to this problem.

References

1) U. Holm, E. Steffens, H. Albrecht, H. Ebinghaus, H. Neuert,
    Z.Physik, 233 (1970) 415
2) H. Ebinghaus, U. Holm, H.V. Klapdor, H. Neuert, Z.Physik, 199(1967)68
3) E. Steffens, H. Ebinghaus, F. Fiedler, K. Bethge, G. Engelhardt,
    R. Schäfer, W. Weiss, D. Fick, Nucl.Instr.Meth. 124 (1975)601
4) R.A. Hardekopf, Operation of the LASL polarized triton source,
    contribution to this Conference, p.865
5) W.E. Burcham, O. Karban, S. Oh, W.B. Powell, Nucl.Instr.Meth.116(1974)1
6) O. Karban, S. Oh, W.B. Powell, Phys.Rev.Lett. 33(1974)1438
7) E. Parker, L.G. Ratner, B.C. Brown, S.W. Gray, A.D. Krisch, H.E.Miettine
    J.B. Roberts, J.R. Fallon, Phys.Rev.Letts. 31(1973) 783
8) P. Egelhof, J. Barrette, P. Braun-Munzinger, D. Fick, C.K. Gelbke,
    D. Kassen, W. Weiss,Elastic scattering of vector polarized ^{6}Li on ^{4}He,
    contribution to this Conference, p.825
9) D. Kassen, P. Egelhof, E. Steffens, Population of an atomic substate
    by the weak field transition and vector polarization; contribution
    to this Conference, p.873
10) W. Weiss, Thesis Heidelberg 1975
11) W.E. Frahn, Fundamentals in Nuclear Theory, IAEA, Vienna 1967, 1 and
    IAEA-SMR-14/13, to be published.
12) W.J. Thompson, Proc. of 2nd Int. Conf. on Clustering Phenomena in
    Nuclei, Maryland 1975, to be published.
13) H. Amakawa, K.I. Kubo; Proc. of Int. Symp. on Cluster Structure and
    Heavy Ion Reactions; Tokyo 1975, to be published.
14) I.V. Kurdymnov, V.G. Neudatchin, Y.E, Smirnov, V.P. Korennoy;
    Phys.Letts. 70B (1972) 607
15) K.I. Kubo, H. Amakawa, Study of polarized ^{6}Li scattering on ^{12}C and
    ^{58}Ni nuclei; contribution to this Conference, p.829
16) H. Ebinghaus, D. Ghoneim, E. Steffens, F. Wittchow; Annual report 1972
    I. Inst. für Experimentalphysik der Universität Hamburg.
17) P.W. Keaton, Proc. 3rd Int. Symp. on Polarization Phenomena in Nuclear
    Reactions, Madison 1970, 422
18) F. Wittchow, private communication
19) J.W. Trischka, D.T.F. Marple, A. White; Phys.Rev. 85 (1952) 137
    M. Kaminsky; Atomic and ionic impact phenomena on metal surfaces;
    Springer Verlag 1965
20) L.D. Knutson, E.J. Stephenson, W. Haeberli, Phys.Rev.Lett.32(1974)690
21) D.R. Parks, S.L. Tabor, B.B. Tripletts, H.T. King, T.R. Fisher,
    B.A. Watson; Phys.Rev.Letts.29 (1972)1264
22) Developments within the postaccelerator project; Annual Report,
    MPI für Kernphysik, Heidelberg 1974, 122
23) The Madison Convention; Proc.3rd Int. Symposium on Polarization
    Phenomena in Nuclear Reactions, Madison 1970
24) K. Alder, A. Winther; Nuclear Electromagnetic Excitation with Heavy
    Ions / Coulomb Excitation, North-Holland-Publishing Company; to be
    published.
    K. Alder, A. Bohr, T. Huus, B. Mottelson, A. Winther, Rev.Mod.Phys.
    28 (1956) 432
25) W.E. Frahn, R.H. Venter; Ann.Physics(N.Y.)27 (1964) 135
26) D.E. Damouth, L.W. Jones, M.L. Perl; Phys.Rev.Letts. 11 (1963) 287
    S. Suwa, A. Yokosawa, N.E. Booth, R.J. Eskerling, R.E. Hill;
    Phys.Rev.Letts. 15 (1965) 560
27) A. Dar, B. Kozlowsky; Phys.Lett. 20 (1960) 314
28) Proc. of the Summer Studies on high energy physics with polarized
    beams, Argonne 1974; ANL/HEP 75-02

29) D.G. Kovar, F.D. Becchetti, B,G. Harvey, F. Pühlhofer, J. Mahoney,
     D.W. Miller, M.S. Zisman; Phys.Rev.Lett. 29 (1972) 1023
30) S.G. Steadman, T.A. Belote, R. Goldstein, L. Grodzins, D. Cline,
     M.J.A. de Voigt, F. Videbaek; Phys.Rev.Letts. 33 (1974) 499
31) K. Alder, F. Roesel, U. Smilansky; Ann.Phys. (N.Y.) 78 (1973) 518
32) R. Beck, M. Kleber; Z.Physik 246 (1971) 383
33) H. Kopfermann, Kernmomente; Akademische Verlagsanstalt Frankfurt 1956
34) J.O. Rasmussen, K. Sugawara-Tanabe; Nucl.Phys. A171 (1971)497
     C.Y. Wong, Phys.Letts. 42B (1972) 186
     J.M. Alexander; L.C. Vaz, S.Y. Liu; Phys.Rev.Letts. 33 (1974) 1487
35) W. Scobel, A. Mignerey, M. Blann, H.H. Gutbrod, Phys.Rev. C11 (1975)1701
36) H.J. Krappe, J.R. Nix in "Physics and Chemistry of Fission 1973",
     Vol.I IAEA Vienna (1974) 159 .
37) J.R. Nix, private communication

DISCUSSION

Conzett:

   Certainly Frahn and his collaborators have developed the parametrized
   S-matrix (or parametrized phase-shift) method of analysis to its
   greatest extent; but, in the interest of historical accuracy, it
   should be noted that it all started with the "sharp-cutoff" model of
   John Blair in the early 1950's, used to describe the elastic scattering of α-particles from nuclei. McIntyre et al. later modified it to
   the smooth cutoff form, again applied to α-particle scattering. The
   first application of this method to heavy-ion elastic scattering was
   then made about 1960 by some unnamed investigators at Berkeley.

Johnson:

   I should like to emphasize that in general there is not expected to
   be as close a correspondence between the quadrupole moment of the
   projectiles involved and the existence of tensor polarization effects
   in transfer reactions as your talk might suggest. The physical processes involved in the large effects seen in the tensor asymmetries
   observed at Birmingham and Madison in the reactions $(\vec{d},p)$ and $(\vec{d},t)$
   are very similar but in the latter case they have little to do
   directly with the quadrupole moment of the triton (which is zero)
   or the deuteron. This can be clearly seen in the calculations of
   Santos and Knutson reported at this Conference. What is involved
   is the cross-term between the S and D state components of the relative motion of the transferred neutron and the proton in the (d,p)
   case, and the deuteron in the (d,t) case, evaluated at the relevant
   momentum transfers. In the (d,p) case, because of the special
   properties of the deuteron, this product is approximately proportional to the deuteron quadrupole moment. There is no analogous
   connection with the triton or deuteron quadrupole moments in the
   (d,t) case.

POLARIZATION STUDIES INVOLVING NEUTRONS

Richard L. Walter
Dept. of Physics, Duke University
and Triangle Universities Nuclear Laboratories, Durham, N.Carolina 27706

Perhaps the first questions about the scientific program of this conference were "Why should there be a separate talk again about neutron polarization phenomena? Can't this work be assimilated into the rest of the presentations?" In fact, to a small degree neutron results have appeared here occasionally in the review papers but typically the poorer accuracy of neutron data places such results into a category of lesser significance than charged particle studies of similar phenomena. Although a similar situation existed prior to the advent of <u>intense</u> polarized-ion sources, the gap between the quality of much of the charged particle data and neutron data has widened since that point in time. However, the understanding of certain aspects of nuclear interactions must be approached through neutron experiments and it is data related to this special role that the present paper is intended to cover. The approach that I will take is to review 1) reasons why neutron experiments are so difficult to perform, 2) new approaches that are being pursued in conventional neutron polarization studies, 3) some recent scattering results, 4) recent reaction experiments, 5) new sources of polarized neutrons, and lastly, 6) polarized-beam polarized-target results of unique interpretation. Because of time limitations only highlights of the last few years will be touched upon in most cases. Furthermore the presentation will be restricted to energies below 50 MeV and the topics of neutron inelastic scattering, spin flip, and n-p and n-d triple scattering studies will not be mentioned here.

Now, consider a typical neutron double-scattering arrangement. Polarized neutrons are produced in a nuclear interaction, either directly in a charged-particle induced reaction or else by elastic scattering of unpolarized neutrons produced previously in some charged particle induced reaction. After such a polarized neutron beam is scattered by a sample, the neutrons must be detected by means of another nuclear process, e.g., n-p scattering in a plastic scintillator. Counting all of the nuclear processes involved, one faces a product of three or four interaction cross sections just to perform what historically has been called a "double-scattering" measurement. An extreme case, shown schematically in Fig. 1, is the setup[1] of Firk's group at Yale. Here an

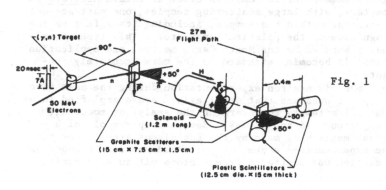

Fig. 1

electron beam from the linac collides with a uranium-lead target to produce bremsstrahlung which in turn photodisintegrates the U-Pb. The (γ,n)
neutrons are polarized by scattering from ^{12}C, then scattered from the
target of interest, and eventually detected in proton-recoil detectors.
So here is a case where one faces a product of _five_ cross-sections in
order to obtain a single analyzing power parameter. Note also the enormous scatterer to scatterer distance of 27 m here. Contrast measurements
of this type to the ease of measuring an analyzing power distribution
for an element by scattering a proton beam which had been produced in a
polarized-ion source from a target into an array of solid state detectors.

     To add to the complexity, most neutron detectors are not effective
energy-spectrometers, so some method of energy discrimination must be
contrived in order to obtain polarization information for all elastic
scattering measurements when the inelastic channel is open, or for the
highest-energy neutron group produced in a charged-particle induced
reaction. This problem is the main reason that neutron scattering experiments have concentrated on scattering from H, D, ^{4}He, ^{12}C and ^{16}O,
and that most reaction studies involve light nuclei where the groups of
neutrons are well separated in energy. Both time of flight techniques,
and coincidence spectrometers, when the scatterer or analyzer can be
prepared in a scintillator form, have been employed to obtain varying
degrees of energy resolution. Lastly, neutrons are also hard to colli-
mate. This makes it difficult to shield the detectors from the direct
flux and it necessitates large separation distances between source and
scatterer.

     On the positive side one has the compensating feature that large
separation distances lend themselves naturally to convenient placement
of spin-precession solenoids (or dipole magnets) which have now become
an integral part of most neutron polarimeters, and to large area, or
correspondingly, large volume detectors. The primary reason for any
success in neutron polarization experiments, however, is that because
neutrons are non-ionizing, massive amounts of material can be effectively employed as analyzers or scatterers. For example, a typical liquid
^{4}He polarimeter is a cylinder 6 cm in diameter and 7 cm in length.
Since liquid ^{4}He corresponds to a gas pressure of about 700 atm, this
amounts to an enviable number of ^{4}He nuclei. In the Yale experiments[1],
for the scattering samples, rectangular slabs 7.5 cm x 15 cm with thicknesses of about 1.5 cm have been employed. This is to be compared with
typical targets for charged particle experiments of 0.4 cm x 0.4 cm in
area and having thicknesses of 10^{-3}cm. In order to obtain _accurate_ polarization parameters, with large scattering samples, one must account
for multiple scattering within the sample, including its effect on the
direction and magnitude of the polarization vector. This type of calculation is usually handled by the Monte Carlo method, and application
of such corrections is becoming standard in the more accurately conducted experiments.

     The progress made in neutron experimentation during the past five
years, and the type of problems that still exist, are exemplified in a
succession of careful attempts to determine the analyzing power $A_y(\theta)$
for n-p scattering around 16 and 21 MeV. Several reports of quite
accurate n-p measurements were made by the Los Alamos group prior to
1972. For these experiments[2] a beam of polarized neutrons produced in
the T(d,n)^{4}He reaction was scattered from protons within an organic

scintillator. The scattered neutrons were detected in coincidence with
the recoil protons in order to eliminate most of the background events
and also to provide some energy discrimination. Although $A_y(\theta)$ was
measured to better than 1%, the results did not conclusively differentiate which of the two contending phase-shift sets were valid. Since then
three other laboratories have produced data at about the same energies
using the identical neutron-source reaction, and likewise, organic
scintillators as the proton scatterer. Jones and Brooks[3] took advantage
of the anistropic properties of the structure of the organic crystal
anthracene and differences in the linear signal and the ratio of the
fast-decay to slow-decay components of the light output to distinguish
right from left scattering from protons constituting the anthracene
scintillator. Ideally, the advantages of this method were that they
could obtain asymmetry values for broad bands of scattering angles simultaneously and also have an effective counting rate that was a factor
of about 100 times greater than that of the earlier measurements because
all scattering events were automatically recorded. However, analysis of
the three-dimensional spectra required very elaborate care. In the experiment of Garrett et al.[4] a setup similar to the Los Alamos experiments was employed but an additional constraint to avoid some of the
neutron background was applied before coincident pulses were accepted.
That is, coincidence was also required between the ^{4}He pulse associated
with the neutron produced in the T+d reaction, the so-called associated-
particle technique. In the most recent publication, Morris et al.[5]
employed a superconducting spin-precession solenoid and incorporated a
neutron-gamma discrimination system for the organic scintillator pulses.
    The summary of all these efforts is shown in Figs. 2 and 3. The
newer data clearly favor the Livermore (LRL X) prediction[5] at 16 MeV although the data set of Morris et al. which has the lowest error bars
suggest a systematically lower curve. At 21 MeV (see Fig. 3), the more
accurate values of Morris et al. still favor the Livermore solution for
angles less than 90° but at the larger angles, they favor the Yale IV
parametrization. This behavior is just the opposite to that of the data
of Jones and Brooks. The earlier LASL data appear to strike a happy
medium. Some of these differences may eventually be shown to be associated with multiple scattering effects inside the organic scintillator,

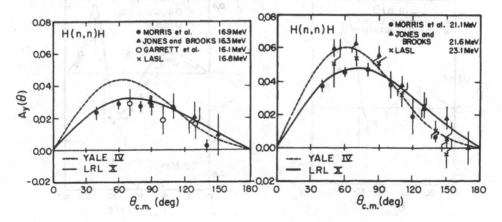

Fig. 2   Fig. 3

but I presently doubt it. Of course, one is free to draw his own conclusions from this display of data. However, recognizing that deviations
shown here represent differences of only 0.015, one must agree that neutron polarization experiments are approaching an accuracy of ±0.01, but
then only when considerable effort is spent in minimizing the background
and carefully scrutinizing the multiparameter information.

The motivation for studying low energy n-p polarization is to aid
in determining the T=0 nucleon-nucleon phase shifts. This information
cannot be obtained from the more accurate p-p scattering data. Furthermore, as Mutchler and Simmons[2] discuss, the n-p polarization is sensitive to the spin-orbit splitting of the T=0 triplet D waves and of the
T=1 triplet P waves and should provide information on any charge-dependent splitting of the T=1 phase parameters. Unfortunately, an interpretation of the summary given in ref.5) suggests that there is still
too much inaccuracy in the data to draw significant conclusions about
this splitting. (See table III in ref. 5.)

Since the Madison Conference, Doleschall[6] and Pieper[7] have provided a series of predictions based on the Faddeev equations of polarization effects in <u>nucleon-deuteron scattering</u> in the 3- to 50-MeV energy
range. In particular, the general shape of $A_y(\theta)$ has been predicted
now but closer agreement will clearly require additional refinements in
the approximation methods and more complete handling of D-waves. In
1974 Doleschall addressed the role that n-d polarization data will have
when compared to the p-d data, and he requested that accurate n-d $A_y(\theta)$
data be obtained in the 20-30 MeV region, particularly at forward angles
where such data are especially sensitive to the tensor force. Recently,
three experiments[8,9] on $D(\vec{n},n)D$ were published for neutron energies of
17, 21, and 35 MeV, but none went to angles forward of 45°(c.m.). (Note
also the seven contributions to this conference related to $D(\vec{n},n)D$.)
These new data have been compared to older $A_y(\theta)$ data at 16 and 23 MeV
and to $D(\vec{p},p)D$ data at 17, 20 and 35 MeV. Except for a couple of data
points, the new n-d and the p-d results are nearly identical. In Fig. 4
the data of Morris <u>et al.</u>[8] are compared to the earlier data and to a
calculation made in 1972 by Pieper[7]. The new results presented at this

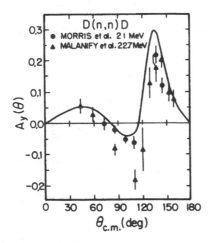

Fig. 4

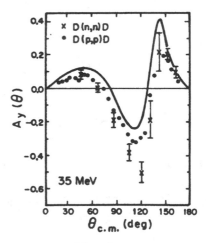

Fig. 5

conference favor the Morris et al. data in the region of 120°, the location of the minimum of the differential cross section. This new data
indicates that $A_y(\theta)$ for D(n,n) and D(p,p) are indeed quite similar
around 22 MeV. The data at the highest energy, i.e., 35 MeV, is compared to D($\vec{p}$,p) results and to another calculation of Pieper in Fig. 5.
Here there is significant difference in $A_y(\theta)$ for the two systems in
the region of the cross-section minimum near 120°. It seems very important to carefully investigate the existence of such differences in
this angular range, probably in the entire range from 20 to 50 MeV.

Results for elastic scattering of polarized neutrons from nuclei
with A ≥ 3 are much sparser than for similar experiments utilizing proton or deuteron beams. In fact, only a very meager amount of data exists
for neutron energies above 5 MeV for nuclei heavier than ^{4}He. The primary reasons for this are the lack of suitable equipment to generate a pro-
lific yield of polarized neutrons at those few laboratories interested
in pursuing fast-neutron studies and the problem alluded to earlier of
discriminating inelastically scattered neutrons from elastically-scat-
tered ones.

Since the Madison Conference, new data on $A_y(\theta)$ for ^{3}He(n,n)^{3}He
have been reported by Hollandsworth et al.[10] at 3 MeV and by Lisowski
et al.[11] at 8 MeV and 12 MeV. The 3-MeV data proved that the ^{4}He level
predictions of Werntz and Meyerhof[12] needed modifications, and the new
8-MeV distribution eliminated the sizeable discrepancy noted earlier between $A_y(\theta)$ for ^{3}He(n,n)^{3}He and the $A_y(\theta)$ for the mirror reaction
^{3}H(p,p)^{3}H. The recent phase shift representation for T + p of Hardekopf
et al.[13] was used by Lisowski et al. as initial parameters in a search
on all the available polarization and cross-section data for the ^{3}He + n
interaction from 0 to 20 MeV. The final nuclear phase shifts and mixing
parameters for both systems now look very similar, giving credence to
the newer polarization data, although neither parametrization claims to
be the unique solution.

The one nucleus for which the most complete set of neutron analyzing
power data exists is ^{4}He and the reason is obvious if one searches for
the best means to measure the polarization of neutron beams with energies
exceeding 0.5 MeV. In the last five years, several polarization experiments have been conducted in order to provide better data for global
^{4}He(n,n)^{4}He phase shift searches and to lead to a better understanding
of the A = 5 level structure. A contribution to this conference from
Bond and Firk gives new $A_y(\theta)$ results in the 2- to 6- MeV range. The
most recent values for $A_y(\theta)$ at higher energies were obtained at 11, 18,
24, 26, 27, and 30 by Broste et al.,[14] at 14 and 17 MeV by Lisowski
et al.,[15] and at 21 MeV by Walter et al.[16]. The measurements of
Broste et al.[14] at 27 and 30 MeV were unique in that they utilized neutrons emitted near 28° from the D(t,n)^{4}He reaction as a source of neutrons. The neutron polarization for this reaction had previously been
calibrated through a measurement on the same reaction but with the role
of the target and projectile reversed, i.e., T(d,n)^{4}He. The calibration
was conducted at an equivalent center of mass reaction angle in which
case the neutron energies were in the 14-MeV range where the analyzing
properties of ^{4}He(n,n)^{4}He are better known. In their ^{4}He + n experiment
Lisowski et al.[15] utilized the high polarization transfer properties of
the D($\vec{d}$,$\vec{n}$)^{3}He reaction at 0° to provide a clean and intense source of
polarized neutrons. The results of this latter experiment are shown in
Fig. 6 along with some phase-shift predictions. The third experiment

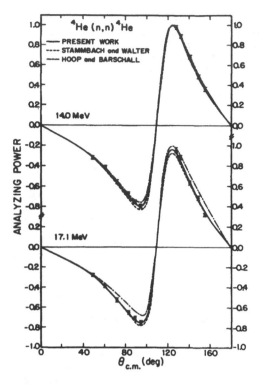

Fig. 6

mentioned above[16] employed the property of the longitudinal polarization transfer coefficient for the T(d,n)^{4}He to give a calibrated source of polarized neutrons as will be reviewed later. The method and the results of measurements of $A_y(\theta)$ for energies above 20 MeV are given in two papers to this conference.[16,17] Calibration data in this energy range was extremely valuable because of the large uncertainties in the analyzing power here. The results suggest that there is little difference between the values of $A_y(\theta)$ for ^{4}He + p and ^{4}He + n in this region if one shifts the energy scales about 1.1 MeV to account for the difference in ^{5}Li and ^{5}He masses. This was originally considered by Hoop and Barschall when they generated their n–^{4}He phase shifts from early p–^{4}He phase shifts. The multi-level R-matrix approach to analyze the mass –5 system by Dodder and Hale at Los Alamos hopefully will assist in determining the analyzing power from 0 to 30 MeV for ^{4}He(n,n)^{4}He, but additional polarization and cross-section data will be required before one can trust that the analyzing power is known to within 2%.

One other new ^{4}He(n,n)^{4}He experiment is particularly noteworthy. Tornow[18] measured the product $p_y(75°, 15\ \text{MeV})A_y(76°, 10\ \text{MeV})$ by scattering neutrons emitted at 0° from the T(d,n)^{4}He reaction successively from two ^{4}He gas scintillators. Somewhat similar experiments had been attempted twice before, but both were of limited usefulness because of poor statistical accuracy. After studying 600 hours of useable data, Tornow obtained a value of 0.571 ± 0.017 for the product after all the corrections were computed. This value is more than three standard deviations from values predicted with the more recent phase-shift sets which causes some concern, but agrees with the older prediction of the Hoop-Barschall phase shifts.

As a side comment, to my knowledge, no one has attempted to pin down the exact location in energy and angle of the absolutely known $A_y = 1.0$ points which Plattner and Bacher[19] have shown must exist for ^{4}He(n,n)^{4}He near 0.5, 4.1 and 11.4 MeV.

Investigations of the analyzing power for most nuclei between ^{6}Li and ^{16}O have been conducted at selected energies in the 0.5 to 5 MeV region. Lane's group at Ohio University has recently reported[20] thorough R-matrix analyses of available n + ^{10}B and n + ^{11}B data between 1.5 and 4.5 MeV which led to the assignments of J, π, and ℓ values for

states in ^{11}B and ^{12}B. The linac group at Yale has doubly-scattered a
continuous distribution of neutrons produced in the (γ,n) reaction on a
Pb - U target in order to obtain[1,21] the analyzing powers of ^{4}He, ^{6}Li,
^{9}Be, ^{12}C and ^{16}O. By scattering twice from ^{12}C they were able to calibrate the polarization properties of ^{12}C(n,n)^{12}C in the 2 to 5 MeV
region. A sample of the counting-rate asymmetry for ^{12}C is shown by the
data at the top half of Fig. 7 for double scattering at 30° and 50°.
From such a product, it is possible to extract p_y as a function of neutron energy. Then , by replacing the second ^{12}C scatterer with other
nuclei, $A_y(\theta)$ was obtained for them, the first ^{12}C scattering in this
case serving as the calibrated neutron polarizer. The ^{6}Li, ^{12}C and ^{16}O
data has been incorporated into R-matrix searches of all the available
$\sigma(\theta)$ and $A_y(\theta)$ data. A sample of the ^{12}C(n,n)^{12}C fits are shown in
Fig. 8 for 4 of many angular distributions and in Fig. 9 for the total
cross section.

   Polarization data for neutron-nucleus optical model analyses have

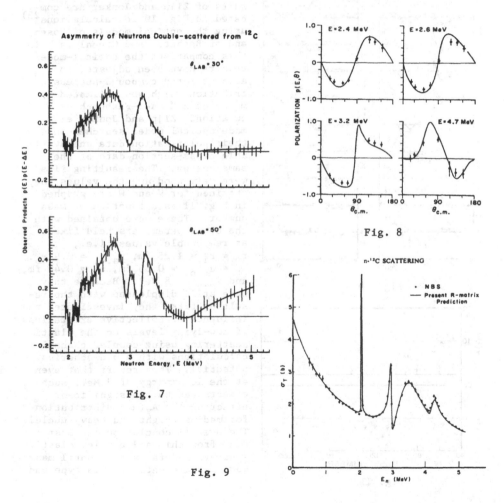

Fig. 7

Fig. 8

Fig. 9

been obtained predominantly below 3 MeV. The most accurate data exists
below 1.5 MeV where neutron detection is easiest. However, the magnitude
of the observed $A_y(\theta)$ is typically less than 0.2 in this energy range
and the $A_y(\theta)$ determination is usually no better than ± 0.02. (See for
example the report by Korzh et al.[22]). Furthermore, the compound-nucleus
contributions always complicate the low-energy optical model interpretation; one really needs higher-energy elastic scattering data. The most
recent data on intermediate and heavy nuclei has been reported by Zijp
and Jonker[23] who used a beam of 3-MeV neutrons produced with a 15% polarization in the D(d,n)^{3}He reaction. Their data agreed fairly well with
earlier 3.2-MeV results reported by Ellgehausen et al.[24] for several
identical targets but there is sufficient discrepancy in some angular

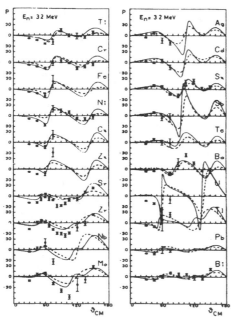

Fig. 10

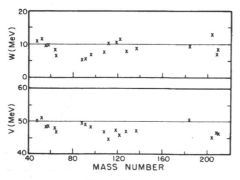

Fig. 11

regions that a third independent
experiment would be in order so
that the corrections applied to the
raw data can be validated. The results of Zijp and Jonker are compared in Fig. 10 to calculations[23]
using the optical models of Rosen
and of Bechetti and Greenlees. In
this comparison the optical-model
results have been adjusted to
account for a compound-nucleus contribution which was estimated by
means of a Hauser-Feshbach calculation. Zijp and Jonker also
made optical model searches on
their polarization data and available cross-section data at the
same energy. The resulting fits
were very good and the values they
obtained for V and W are graphed
in Fig. 11 as a function of mass
number. These were obtained with
the other parameters held fixed,
at reasonable values, i.e.,
$r_o = r_i = 1.25$ fm, $r_{s.o.} = 0.9\ r_o$.
$a_r = a_{s.o.} = 0.65$ fm, $a_i = 0.48$ fm,
and $V_{s.o.} = 8$ MeV. Because they
were still displeased with the resulting fits, they investigated the
influence of collective excitations
of low-lying levels on the elastic
scattering using coupled channel
calculations with a non-spherical
potential. They report that even
at the low energy of 3 MeV, such
effects can have a significant
effect on the $A_y(\theta)$ distribution
for medium weight and heavy nuclei.
The overall conclusion one must
draw from this and earlier elastic-
scattering data is that until many
more measurements of this type and

quality are made at a large number of energies, one will always be
skeptical about the interpretation of such data if the conclusions are
at odds with those to be extrapolated from the beautiful proton data
that already exists.

Elastic scattering of neutrons is noticeably affected at small
angles by the interaction of the magnetic moment of the neutron with the
Coulomb field of the nucleus. This effect, called Mott-Schwinger scattering, causes the polarization to exceed 80% for scattering angles
near 1° for heavy nuclei. Several measurements[25,26] have been reported
in the past few years in addition to two at this conference and most of
the newer data are in agreement with the predicted curves. The discrepancy reported by Drigo et al. at 2.5 MeV remains unexplained.

Most of the early polarization measurements in neutron producing
reactions such as (p,n), (d,n) and (^{3}He,n) were directed at determining
the outgoing neutron polarization when the incident beams were unpolarized. A few years ago, these studies broadened to measuring the analyzing power when the reactions were produced with beams originating in
polarized ion sources. More recently, concentrated efforts on polariza-
tion transfer coefficients in $(\vec{p},\vec{n})$ and $(\vec{d},\vec{n})$ reactions have gotten
underway.

The majority of the $A(a,\vec{n})B$ studies in the past fifteen years involved $(d,\vec{n})$ reactions to investigate DWBA predictions, and secondarily
to discover the best sources of polarized neutrons. (See for example
ref. 27.) Although a large amount of effort was spent in these studies,
most of it was carried out at energies available with single-stage electrostatic accelerators and on targets lighter than Si. From a direct
reaction analysis viewpoint, therefore, the interpretations of the data
was complicated by compound-nuclear effects and the lack of suitable
optical-model parameters for the DWBA calculations. Although the quality of the data continued to gradually improve through better instrumentation, for example, as is illustrated in the recent papers by the
groups of Thornton[28] and Zeitnitz[29], it is safe to say that very
little nuclear physics understanding was generated through these hard
labors. This is not true however, in the very light nuclei cases, that
is for $D(d,\vec{n})$ and the $T(d,\vec{n})$ where some physical insight has been coming
from multi-faceted 4- and 5- body calculations.

It probably is worth mentioning here that the $D(d,\vec{n})^3$He polarization
functions for E_d < 500 keV, which were in a state of confusion at the
time of the Madison Conference, have been fairly well determined now.
The recent papers by Sikkema and Steendam[30], Pospiech et al.[31], and
contributions to this conference seem to have resolved the magnitude
of the polarization and the fact that there is probably no narrow
resonance structure in the D + d reaction near 100 keV. New $D(d,\vec{n})^3$He
polarization data have been obtained at higher energies also, i.e.,
0.9 to 5 MeV, 6 to 14 MeV, and 16 to 22 MeV respectively in refs. 32,
33, and 34 and the polarization in this region appears to be well known
now also.

A few measurements of tensor analyzing powers for $(\vec{d},n)$ reactions
were made around 10 MeV at Wisconsin to check DWBA predictions and to
compare the j-dependence sensitivity of $(\vec{d},n)$ to that of $(\vec{d},p)$.
Hilscher et al.[35] reported on the targets ^{11}B, ^{12}C, ^{14}N and ^{15}N,
Quin et al.[36] for Yt and Cu and Hichwa et al.[37] for ^{58}Ni and ^{60}Ni, the
latter at this conference. The data for ^{89}Y$(\vec{d},n)$ are compared in
Fig. 12 to DWBA results and the j-dependence for ℓ=1 (d,n) and (d,p)

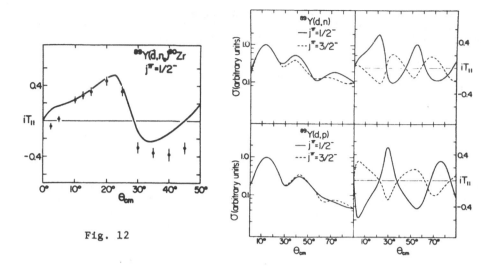

Fig. 12

Fig. 13

transfer reactions as calculated in DWBA are shown in Fig. 13. In this
case the j-dependences at forward angles is opposite for the (d,n) and
(d,p) reactions. For the Ni-Fe region, Hichwa et al. showed that the
ℓ=1 j-dependence are very similar for the ($\vec{d}$,n) and ($\vec{d}$,p) vector analyzing powers. These observations as well as the dependence for ℓ=3 transfer in ^{58}Ni($\vec{d}$,n) have been borne out in DWBA calculations. Vector
analyzing power measurements for ($\vec{d}$,n) reactions have also been made at
Basel[38] at low energies on D, T, ^{6}Li, ^{7}Li and ^{11}B targets to test reaction calculations, and on D and T up to 12 MeV at Los Alamos[39] and
Zurich[40]. Discussion about some of the polarization transfer experiments will follow below.

Some (^{3}He,$\vec{n}$) studies were made for several targets to investigate
the success of DWBA in predicting polarizations produced in diproton
transfer reactions, but again most were done below 6 MeV. (See ref. 41
and references therein.) Large polarizations are exhibited but the
energy dependence of the $p_y(\theta)$ distributions are not reproduced well by
DWBA codes. One case, ^{12}C(^{3}He,n), has been studied[42] in at least
2-MeV steps up to 20 MeV. A sample of the high energy data is shown in
Fig. 14. The curves are only a guide to the eye. Although the variations with energy were quite gradual, and the DWBA calculations could be
made to agree at some energies, the overall agreement was dissatisfying.
See for example the DWBA comparisons at 20 MeV illustrated in Fig. 15.
Because the ^{3}He inelastic scattering cross section is quite significant
for both targets, until coupled-channel calculations are investigated in
these reactions, the conclusions about the quality of the fits will need
to be restrained. Also, the effect on the polarization of two-step
processes for these reactions should be investigated carefully.

No reactions of the (α,$\vec{n}$) type have been performed for alpha energies above 6 MeV, to my knowledge. Below 6 MeV these reactions proceed
primarily through compound-nucleus formation. The recent neutron polarization measurements[43] at Ohio State University on the ^{9}Be(α,n)^{12}C
and ^{13}C(α,n)^{16}O reactions have helped resolve ambiguities in spin and

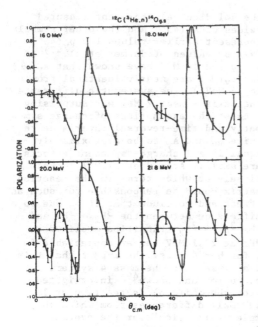

Fig. 14

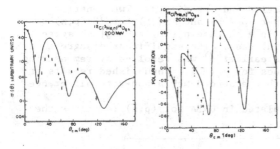

Fig. 15

parity assignments for several states in ^{13}C and ^{17}O respectively. No neutron polarization measurements have been reported where projectiles heavier than ^{4}He have induced the reactions.

Data on $(p,\vec{n})$ reactions have also been reported in the past few years, but most of it was obtained to recheck earlier work. Except for one $^{51}V(p,\vec{n})^{51}Cr$ experiment[44], all the data involved targets of A < 16. The category of $(\vec{p},n)$ experiments induced with polarized proton beams have become feasible at several labs now and it is hoped that this new type of data will add to understanding isospin-dependent terms of the nucleon-nucleus optical model. Although $\vec{t}\cdot\vec{T}$ terms have been used in optical model fits to elastic scattering data, it is felt that the best means to experimentally investigate such terms is through (p,n) studies. Moss et al.[45] studied the quasi-elastic scattering by looking at analyzing powers of analog states for (p,n) reactions on ^{11}B, ^{27}Al, ^{60}Ni, ^{116}Sn, and ^{120}Sn for protons around 25 MeV. For the Sn isotopes, it was necessary to include an isospin-dependent spin-orbit term to the optical model. The sensitivity of the DWBA calculations for one set of optical

model parameters is shown in Fig. 16 where $V_{s\ell}$ = -1 indicates a term of $[V_{s\ell}(r)\ \vec{\ell}\cdot\vec{s}]\ (\vec{t}\cdot\vec{T})/A$ in the optical model potential with $V_{s\ell}$ = 1 MeV and with a sign opposite to the nuclear $\vec{\ell}\cdot\vec{s}$ term. However, Moss et al. feel that the magnitude of the term was not well determined because of existing optical model ambiguities. This study has been extended to other nuclei by Gosset et al. in a report to this conference.

A measurement of Haight et al.[46] of the T(p,n)^{3}He analyzing power stimulated theoretical interest when the values of $A_y(\theta)$ were shown to differ from values for $p_y(\theta)$ for the T(p,n)^{3}He reaction near 2 MeV. Calculations using the parameters of Werntz and Meyerhof showed that A_y and p_y should be nearly equal as shown in Fig. 17 for these parameters at 45° c.m. Here the closed dots (which have been connected by

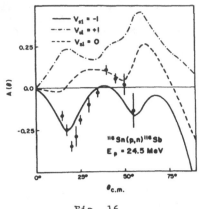

Fig. 16

the solid curve) represent measured values of $A_y(45°)$ and the other symbols represent earlier values for $p_y(45°)$. (New p_y values of Jarmer et al.[47] at 6, 10 and 14 MeV have shown that $A_y(\theta)$ and $p_y(\theta)$ are nearly identical from 6 to 14 MeV.) The systematic difference that exists near 2 MeV was surprising because the assumptions of charge symmetry and time-reversal invariance require p_y and A_y to be approximately equal if all charge-dependent effects are small. Arnold et al.[48] looked closely at which terms in the reaction matrix could be responsible for such an effect and conclude that it is due to a difference between the $^3P_2 \rightarrow {}^3F_2$ and the $^3F_2 \rightarrow {}^3P_2$ transition amplitudes,

and suggest that the 2^- state in ^{4}He at 22.1 MeV has an appreciable f-wave partial width. Arnold et al. further discuss the impact that this conclusion has on other models and analyses for the mass-4 system.

Prior to Arnold's calculation, Rohrer and Brown[49] investigated the differences between p_y and A_y for the ^{7}Li(p,n)^{7}Be reaction at low energies, and they found relatively small differences from which they asserted that there must be a "simple explanation" for the overall agreement. Conzett[50] discussed in general the differences between p_y and A_y for (p,n) reactions to mirror states and points out that differences are expected if differences exist in the bound states of the "extra-core" nucleon in the reaction and its mirror. Two contributions[51] to this conference compare $p_y(\theta)$ and $A_y(\theta)$ for ^{9}Be(p,n)^{9}B and ^{15}N(p,n)^{15}O around 10 MeV. In each case $p_y(\theta)$ and $A_y(\theta)$ agree fairly closely, as in the ^{7}Li(p,n) case. The significant disagreement for ^{9}Be(p,n)^{9}B that was reported earlier[52] was resolved by remeasuring with better accuracy[51] the values for $p_y(\theta)$. Unpublished results of Donoghue et al. however verify that around 3 MeV a non-zero difference does exist for ^{7}Li(p,n).

Madsen et al.[53] have calculated in DWBA for $(\vec{p},\vec{n})$ reactions the

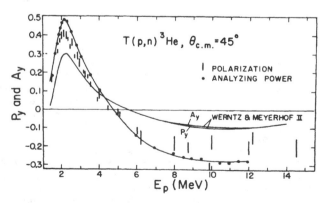

Fig. 17

sensitivity of the (vector) polarization transfer coefficient $K_y^{y'}$ at 0°
to various types of two-body spin-dependent terms, i.e., spin-orbit,
spin-spin and tensor interactions. For those not familiar with this coefficient, $K_y^{y'}$ for (p,n) reactions would be similar to the old
Wolfenstein D or depolarization parameter in nucleon-nucleus elastic
scattering. Madsen et al. point out that for transitions that proceed
entirely by spin-flip, e.g., $^{14}C(p,n)^{14}N_{g.s.}$, $K_y^{y'}(0°)$ should be negative.
For reactions in which both spin-dependent and spin-independent forces
contribute, $K_y^{y'}(0°)$ can be either positive or negative, which can yield
information about the two-body interaction. A sample of their calculations for $K_y^{y'}(0°)$ is drawn in Fig. 18 for $^{11}B(p,n)^{11}C$ for different
mixtures and combinations of the two-body forces. The parameters used

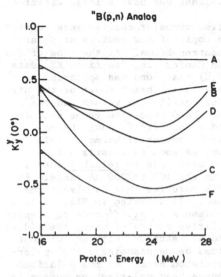

Fig. 18

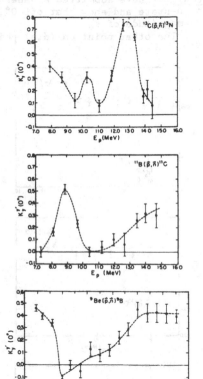

Fig. 19

to generate these curves require too
much description to permit coverage
here but one can refer to the contribution to this conference of
Graves et al. for a terse explanation.
These latter authors report values
near 0.6 for $K_y^{y'}(0°)$ at 22 and 26 MeV.
Earlier a set of $(\vec{p},\vec{n})$ experiments[54]
to observe $K_y^{y'}(0°)$ had been carried
out at TUNL from about 6 to 15 MeV
for targets of ^9Be, ^{11}B, and ^{13}C.
Considering the fact that the ex-
citation functions for these reactions exhibit compound-nucleus resonance formation, it is probably not surprising that appreciable structure in $K_y^{y'}(0°)$ was observed. The data is shown in Fig. 19 where it can
be seen that $K_y^{y'}(0°)$ is usually positive, and usually less than 0.5.
Only for $^9Be(\vec{p},\vec{n})$ has $K_y^{y'}(0°)$ leveled off at the highest energies, and
the value of the plateau is about 0.4. No attempts to fit this set of

data have been made yet.

While I am on the topic of polarization transfer coefficients, several other experiments should be mentioned here. The vector-vector polarization transfer coefficient for $(\vec{d},\vec{n})$ reactions has been measured for targets of D, T, ^{4}He, ^{12}C, ^{14}N, ^{16}O, and ^{28}Si and for $(\vec{t},\vec{n})$ reactions on H and D. The results for some of these studies are presented in reports to this conference. As shown in Fig. 20 from Lisowski et al., for the three heaviest nuclei, $K_y^{y'}(0°)$ was a slight bit under 2/3, the value which one expects from a simple stripping picture in which the neutron (and the neutron spin) is merely a spectator in the interaction. The D-state of the deuteron of course is neglected in this model and the inclusion of it would tend to lower the magnitude of $K_y^{y'}(0°)$. Basak et al.[55] have submitted a paper here on a DWBA calculation including the D-state and show that off 0°, the D-state can have a large effect on the magnitude of $K_y^{y'}$.

The other point on $(\vec{d},\vec{n})$ polarization transfer coefficients relates

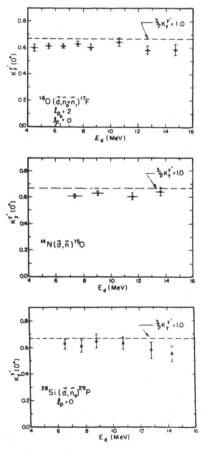

Fig. 20

to the topic of new sources of polarized neutron beams. At the time of the Madison Conference, the first $K_y^{y'}$ data for $(\vec{d},\vec{n})$ reactions was appearing and it was evident that because of the high transfer coefficient values and the size of the 0° reaction cross sections, the $D(\vec{d},\vec{n})$ and $T(\vec{d},\vec{n})$ would become practical sources of beams of polarized neutrons as soon as polarized ion sources were made to reliably output beams in the 100 to 200 nA range. At some laboratories this is almost the case now. Illustrated in Fig. 21 is a graph Simmons et al.[55] made to compare the relative merits of using the T(d,n) and D(d,n) reactions with and without polarized beams assuming a 100% polarized deuteron beam. As the polarized ion source beam polarization enters in as the square when calculating the figure of merit, a typical ion-source polarization of 75% of maximum would lower the plotted values by a factor of 2. Also, the plot makes the comparison for equal intensities for polarized and unpolarized beams. This can be deceiving if one merely glances at the graph. Nevertheless, transfer polarization reactions are valuable sources and whenever a beam of 50 to 100 nA is available on target, the merits of these sources of polarized neutrons probably outweigh the use of other means. Results of the neutron elastic scattering measurements employing the $D(\vec{d},\vec{n})^3$He reaction as a source are beginning to appear now[11,15]. In order to make the $D(d,n)^3$He a more useful source,

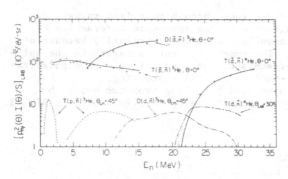

Fig. 21

Fig. 22

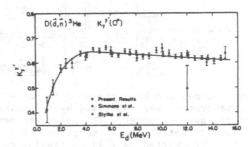

Lisowski *et al.*[57] recently re-measured $\overline{K_y'}(0°)$ for this re-action in closer intervals and at lower energies than had been done before. The results are shown in Fig. 22 to exhibit the accuracy to which K_y' can be measured now for neutrons.

Another source of polar-ized neutrons utilizing the polarization transfer properties of the $T(\vec{d},\vec{n})$ reaction has been investigated at Los Alamos and is reported at this conference. The unique feature of this proposal is that the source was calibrated <u>independently</u> of any neutron polariza-tion analyzer, such as $^4He(\vec{n},n)^4He$. This was possible through the feature that Ohlsen *et al.*[58] pointed out at the Madison Conference. That is, if one knows $A_{zz}(0°)$ for a reaction of the spin structure $1/2 + 1 \to 1/2 + 0$ as occurs for the $T(d,n)^4He$ reaction, one can cal-culate the longitudinal polarization transfer coefficient $K_z^{z'}(0°)$. Then, from the knowledge of the polarization status of the incident <u>longitudi-nally</u> polarized deuteron beam, one can generate a neutron beam of known polarization. Two papers[16,17] at this conference relate to this source of neutrons, one reporting accurate values of $A_{zz}(0°)$ and the other employing the source in a scattering experiment. One should be aware of two aspects of this source of neutrons however. First, for a longitudi-nally polarized beam, the 0° cross section for $T(d,n)$ drops to 1/3 of that for an unpolarized deuteron beam, and secondly, if one generates a beam of longitudinally polarized neutrons, their spins must be pre-cessed transversely in a dipole magnet before a normal double-scattering experiment can be conducted.

This $T(\vec{d},\vec{n})$ data along with other Los Alamos results for polariza-tion transfer coefficients for the same reaction and the $D(\vec{t},\vec{n})$ reaction have been included in Dodder and Hale's mammouth search code for des-cribing the mass-5 system.

Another type of experiment for which new polarization data and analyses are forthcoming is the $X(\gamma,\vec{n})$. Results for $D(\gamma,\vec{n})$ and $^{16}O(\gamma,\vec{n})$ have been reported by the Yale group[59] and for $Pb(\gamma,\vec{n})$ by Holt and Jackson at Argonne.[60] These results were useful in establishing the absence[61] of an (E2) giant quadrapole resonance for ^{16}O and in investi-gating[60] the structure of the (M1) giant resonance in ^{208}Pb.

Transmission of polarized neutrons through polarized targets can
lead to information on the magnitude of the spin-spin interaction but
the interpretation has been inconclusive to date, primarily because of
the compound nucleus effects. The same statement in regard to the spin-
spin term can be made about attempts to interpret all of the depolarization measurements for neutron-nucleus elastic scattering.

The final results to be presented here come from two papers by
Keyworth et al.[62). A schematic of their setup is illustrated in Fig.23.

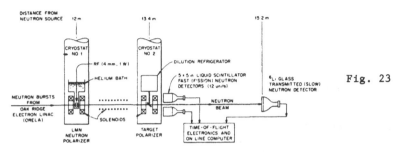

Fig. 23

A pulsed neutron beam produced in the Oak Ridge Electron Linear Accelerator Laboratory was polarized by transmission through a dynamically polarized proton target. The polarized neutrons then struck a second polarized target which was comprised of a fissionable element, ^{237}Np or
^{235}U. Transmitted as well as fission neutrons emitted at 10° and 90°
were detected. The transmission through the second target depends on
the spin of the compound nuclear resonances and on the relative orientation of the neutron spin and the target polarization vector. The top
half of Fig. 24 gives a plot of the arithmetic difference in the transmitted beam intensity for the parallel and anti-parallel configurations.
Note in the top half of Fig. 24 that J = 3 states show a negative going
structure and J = 2 states a positive structure. This result for the
region near 10 eV should be compared to the results in Fig. 25 for the
region near 40 eV. Here is shown a result for fission neutrons emitted
at 90°, in contrast to the previous transmission data. Note here that
all of the resonances in this energy region give a positive going difference, which indicates all are J = 3 resonances. The interpretation of
this concentration of J = 3 states is related to sub-threshold fission

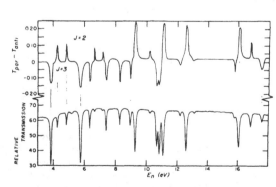

Fig. 24

from the "class II" states
in the second minimum of
the fission barrier. In
this case, the coupling of
the "class II" states in
the second well to the compound nuclear levels in the
first well selects reso-
nances of a single-spin
state, that of the "class
II" state. Additional information obtained in these
experiments proved that the
K quantum number for these
fissioning states is mixed,
with K = 2 and K = 3 con-

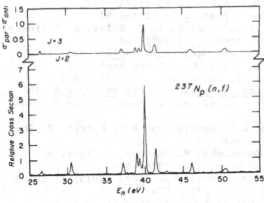

Fig. 25

tributing. Here K is the projection of the total angular momentum J onto the nuclear symmetry axis. This is the first time such K quantum number assignments have been able to be made and the first time that it was shown that such intermediate-structure states do not proceed through a K = 0 transition as had been assumed in fission theories.

References

*Work supported by U.S.E.R.D.A.

1) R.J. Holt, F.W.K. Firk, R. Nath, and H.L. Schultz, Nucl. Phys. A213 (1973) 147
2) G.S. Mutchler and J.E. Simmons, Phys. Rev. C4 (1971) 67
3) D.T.L. Jones and F.D. Brooks, Nucl. Phys. A222 (1974) 79
4) R. Garrett, A. Chisholm, D. Brown, J.C. Duder and H.N. Burgisser, Nucl. Phys. A196 (1972) 421
5) C.L. Morris, T.K. O'Malley, J.W. May, Jr., and S.T. Thornton, Phys. Rev. C9 (1974) 924
6) P. Doleschall, Nucl. Phys. A220 (1974) 491 and refs. therein
7) S.C. Pieper, Nucl. Phys. A193 (1972) 519 and refs. therein
8) C.L. Morris, R. Rotter, W. Dean, and S.T. Thornton, Phys. Rev. C9 (1974) 1687
9) J.Z. Amudio-Cristi, B.E. Bonner, F.P. Brady, J.A. Jungerman, and J. Wang, Phys. Rev. Letters 31 (1973) 1009
10) C.E. Hollandsworth, M. Gilpatrick, and W.P. Bucher, Phys. Rev. C5 (1972) 395
11) P.W. Lisowski, R.L. Walter, C.E. Busch, and T.B. Clegg, (submitted to Nucl. Phys.)
12) C. Werntz and W.E. Meyerhof, Nucl. Phys. A121 (1968) 83
13) R.A. Hardekopf, P.W. Lisowski, T.C. Rhea, R.L. Walter and T.B. Clegg, Nucl. Phys. A191 (1972) 481
14) W.B. Broste, G.S. Mutchler, J.E. Simmons, R.A. Arndt, and L.D. Roper, Phys. Rev. C5 (1972) 761
15) P.W. Lisowski, T.C. Rhea, R.L. Walter, and T.B. Clegg, contribution to this conference, p.534
16) R.L. Walter, P.W. Lisowski, G.G. Ohlsen and R.A. Hardekopf, contribution to this conference, p.536
17) P.W. Lisowski, R.L. Walter, R.A. Hardekopf, and G.G. Ohlsen, contribution to this conference, p.887
18) W. Tornow, Z. Physik 266 (1974) 357
19) G.R. Plattner and A.D. Bacher, Phys. Lett. 36B (1971) 211
20) S.L. Hausladen, C.E. Nelson, and R.O. Lane, Nucl. Phys. A217 (1973) 563 and 546

21) G.T. Hickey, F.W.K. Firk, R.J. Holt, and R. Nath, Nucl. Phys. A225 (1974) 470 and A237 111, and report to this conference, p.532
22) I.A. Korzh, T.A. Kostyuk, V.A. Mishchenko, N.M. Pravidvii, and I.E. Sanzhur, Bull. Acad. Sci. U.S.S.R. 35 (1971) 757
23) E. Zijp and C.C. Jonker, Nucl. Phys. A222 (1974) 93
24) E. Ellgehausen, E. Baumgartner, R. Gleyvod, P. Huber, A. Stricker, and K. Wiedeman, Helv. Phys. Acta. 42 (1969) 269
25) R.B. Galloway and R.M.A. Maayouf, Nucl. Phys. A212 (1973) 182
26) L. Drigo, C. Manduchi, G. Moschini, M.T. Russo-Manduchi, G. Tornielli and G. Zannoni, Il Nuovo Cim. 13A (1973) 867
27) R.L. Walter, in Nuclear Spectroscopy and Reactions, Part B (1974), Acad. Press, ed. by J. Cerny, p. 635 and in Proc. of the Madison Conf. p. 317
28) S.T. Thornton, R.C. Jordan, C.L. Morris, and R.P. Rotter, Z. Phys. 266 (1974) 329 and refs. therein
29) G. Meyer-Kretschmer, P.B. Dunscombe, R. Maschuw, P. Suhr, and B. Zeitnitz, Nucl. Inst. and Meth. 120 (1974) 469
30) C.P. Sikkema and S.P. Steendam, Nucl. Phys. A245 (1975) 1; see also F.A. Guckel and D. Fick, Z. Phys. 271 (1974) 39
31) G. Pospiech, H. Genz, E.H. Marlinghaus, A. Richter, and G. Schrieder, Nucl. Phys. A239 (1975) 125
32) J.R. Smith and S.T. Thornton, Can. Jour. Phys. 50 (1972) 783; R.B. Galloway, A.S. Hall, R.M.A. Maayouf, and D.G. Vass, Nucl. Phys. A242 (1975) 122
33) G. Spalek, R.A. Hardekopf, J. Taylor, Th. Stammbach, and R.L. Walter, Nucl. Phys. A191 (1972) 449
34) R.A. Hardekopf, T.C. Rhea, P.W. Lisowski, R.L. Walter and J. Joyce, Nucl. Phys. A191 (1972) 460
35) D. Hilscher, J.C. Davis, and P.A. Quin, Nucl. Phys. A174 (1971) 417
36) P.A. Quin, J.C. Davis, and D. Hilscher, Nucl. Phys. A183 (1972) 173 see also contribution to this conference from W.H. Wong and P.A. Quin, p.657
37) B.P. Hichwa, L.D. Knutson, J.A. Thomson, W.H. Wong, and P.A. Quin, contribution to this conference, p.655
38) See, for example, U. Von Mollendorff, A. Janett, F. Seiler and H.R. Streibel, Nucl. Phys. A209 (1973) 323 and references therein
39) G.C. Salzman, G.G. Ohlsen, J.C. Martin, J.J. Jarmer, and T.R. Donoghue, Nucl. Phys. A222 (1974) 512 and refs. therein
40) W. Gruebler, V. Konig, P.A. Schmelzbach, R. Risler, R.E. White, and P. Marmier, Nucl. Phys. A193 (1972) 129
41) D.C. Martini and T.R. Donoghue, Phys. Rev. C8 (1973) 621
42) T.C. Rhea, Ph.D. Thesis, Duke Unversity (1973)
43) C.E. Busch, T.R. Donoghue, J.A. Keane, H. Paetz gen. Shieck, and R.G. Seyler, Phys. Rev. C8 (1973) 848 and refs. therein
44) E.H. Sexton, A.J. Elwyn, F.T. Kuchnir, F.P. Mooring, J.F. Lemming, R.W. Finlay, and R.E. Benenson, Bull. Am. Phys. Soc. 16 (1971) 1432
45) J.M. Moss, C. Brassard, R. Vyse, and J. Gosset, Phys. Rev. C6 (1972) 1698
46) R.C. Haight, J.J. Jarmer, J.E. Simmons, J.C. Martin, and T.R. Donoghue, Phys. Rev. Lett. 28 (1972) 1587
47) J.J. Jarmer, R.C. Haight, J.E. Simmons, J.C. Martin, and T.R. Donoghue, Phys. Rev. C9 (1974) 1292
48) L.G. Arnold, R.G. Seyler, T.R. Donoghue, L. Brown, and U. Rohrer, Phys. Rev. Lett. 32 (1974) 310
49) U. Rohrer and L. Brown, Nucl. Phys. A217 (1973) 525

50)   H.E. Conzett, Phys. Lett. 51B (1974) 445; see also L.G. Arnold, contribution to this conference, p.503

51)   R.C. Byrd, P.W. Lisowski, S.E. Skubic, W. Tornow, R.L. Walter, and T.B. Clegg, contribution to this conf.; P.W. Lisowski, G. Mack, R.C. Byrd, W. Tornow, S.E. Skubic, R.L. Walter, and T.B. Clegg, contribution to this conference, p.499

52)   P.W. Lisowski, R.C. Byrd, S.E. Skubic, G. Mack, R.L. Walter, and T.B. Clegg, Bull. Am. Phys. Soc. 20 (1975) 693

53)   V.A. Madsen, J.D. Anderson, and V.R. Brown, Phys. Rev. C9 (1974) 1253

54)   P.W. Lisowski, R.L. Walter, C.E. Busch, and T.B. Clegg, contribution to this conference; see also R.G. Graves et al., Bull. Amer. Phys. Soc. 20 (1975) 694

55)   A.K. Basak, J.A.R. Griffith, O. Karban, J.M. Nelson, S. Roman, and G. Tungate, contribution to this conference, p.661

56)   J.E. Simmons, W.B. Broste, T.R. Donoghue, R.C. Haight, and J.C. Martin, Nucl. Instr. and Meth. 106 (1973) 477

57)   P.W. Lisowski, R.L. Walter, C.E. Busch, and T.B. Clegg, Nucl. Phys. A242 (1975) 298

58)   G.G. Ohlsen, P.W. Keaton, and J.L. Gammel, Proc. Madison Conference, p. 512

59)   R. Nath, F.W.K. Firk, and H.L. Schultz, Nucl. Phys. A194 (1972) 49 and unpublished results

60)   R.J. Holt and H.E. Jackson, submitted to Phys. Rev. Lett.

61)   W.L. Wang and C.M. Shakin, Phys. Rev. C9 (1974) 2144

62)   G.A. Keyworth, J.R. Lemley, C.E. Olsen, F.T. Seibel, J.W.T. Dabbs, and N.W. Hill, Phys. Rev. C8 (1973) 2352 and Phys. Rev. Lett. 31 (1973) 1077

DISCUSSION

Ohlsen:

I would like to remark that the A_{zz} analysing power at $0°$ for the $T(\vec{d},n)^4He$ reaction differs from that for the $^3He(\vec{d},p)^4He$ reaction by as much as 10%. It seems to me that such differences cannot be explained away by Coulomb effects or different thresholds, etc. as was suggested in the discussion session F.
The data to which I refer are contained in the proceedings: $T(\vec{d},n)^4He$ data from the TUNL-Los Alamos collaboration and $^3He(\vec{d},p)^4He$ data from ETH, Zürich.

Santos:

It is worth pointing out that the recent $K_y^{y'}(0°)$ measurements in deuteron stripping reactions show that the spin transfer 3/2 is very small at forward angles, which means that the reaction is appreciably transparent to vector polarization transfer. This transparency in $(\vec{d},n)$ reactions, which is predicted to be particularly pronounced along the x- and y-axis, can therefore be used as a method to obtain beams of polarized neutrons with known polarization. (The quantities mentioned here are defined in Nucl. Phys. A236 (1974) 90.)

Walter:

   Your first statement is quite true but we must be certain of the magnitude of the effect from the spin transfer 3/2, to use your terminology, before we can say these beams have <u>known</u> polarization, unless $K_y^{y'}(0°)$ has been determined experimentally.

Keaton:

   I'm sure that you realize this but I feel it is important to note that there are specific advantages to be gained by using polarization transfer methods to obtain polarized neutron beams which are not taken into account when one considers only the figure of merit P^2I. First, the neutron polarization direction is easily reversed at the source. Second, there are relatively fewer background neutrons produced if one uses the neutrons produced at 0° in the polarization transfer reaction.

Walter:

   Thank you for pointing that out here.

PARTICLE-GAMMA ANGULAR CORRELATION FOR A NUCLEAR
REACTION INITIATED BY A POLARIZED BEAM

R. G. Seyler

Department of Physics, Ohio State University, Columbus, Ohio 43210

Introduction. At the Madison Symposium there was only one contributed paper[1] in this area. Since then the increased availability of polarized-ion sources has led to a sharp upturn in interest. Because of the newness of the subject this presentation will be aimed at the non-expert and will begin with a review of the general theory. Important special geometries will then be considered and some experimental results for these special cases will be mentioned.

General Theory. Several excellent comprehensive treatments of angular correlation theory exist[2]. These treatments are very general and cover any conceivable polarized beam angular correlation situation in the sense that they teach one how to derive the appropriate correlation function although they might not include it explicitly.

More recent expositions on the subject are more specialized and more up to date on recent conventions. Rose and Brink[3] reviewed the theory of angular distributions of gamma rays with strong focus on the important question of phase conventions (e.g. signs of multipolarity mixing ratios). Rybicki, Tamura and Satchler[4] extended the Rose and Brink formalism to the unpolarized particle-gamma angular correlation. Debenham and Satchler[5] hereafter referred to as DS, extended the same formalism to particle-gamma correlations induced by polarized beams. This theory review closely follows DS except for minor notation and normalization alterations.

Consider a reaction of the form $a(s_1,s_2)c(L)d$ where the symbols represent the particles and indicate their spins, e.g., a is the spin of particle a. The symbol L (or L') represents the multipolarity of the gamma ray. The corresponding Greek letter (e.g. α) will refer to the magnetic quantum number.

The polarization of the s_1 beam is described by its density matrix $\rho(s_1)$ here assumed normalized to unit trace, e.g., $\rho(\frac{1}{2}) = \frac{1}{2}(1+\vec{P}\cdot\vec{\sigma})$. Let $A_{\sigma_2\gamma,\sigma_1\alpha}(\theta,\phi)$ be the transition amplitude for the reaction $a(s_1,s_2)c$. This amplitude is the same as the T amplitude of Satchler[6] and is simply related (in ref. 4) to the T amplitude of Tamura[7]. The amplitude A is related to the collision matrix by Lane and Thomas[8].

The $\gamma\gamma'$ element of the density matrix for the state c is given by

$$\rho_{\gamma,\gamma'}(c) = \sum A_{\sigma_2\gamma\ \sigma_1\alpha}(\theta,\phi)\ A^*_{\sigma_2\gamma',\sigma_1'\alpha}(\theta,\phi)\rho_{\sigma_1,\sigma_1'}(s_1)\ , \qquad (1)$$

where θ,ϕ are the center of mass scattering angles of s_2. The statistical tensors[2] corresponding to these density matrices are:

$$\rho_{k\kappa}(c) = \sum_{\gamma\gamma'} (-)^{c-\gamma'}(c\gamma, c-\gamma'|k\kappa)\ \rho_{\gamma,\gamma'}(c)\ , \qquad (2)$$

and

$$\rho_{k_1\kappa_1}(s_1) = \sum_{\sigma_1\sigma_1'} (-)^{s_1-\sigma_1'}(s_1\sigma_1, s_1-\sigma_1'|k_1\kappa_1)\ \rho_{\sigma_1,\sigma_1'}(s_1). \qquad (3)$$

These tensors transform under rotations like the complex conjugate of spherical harmonics of the same rank. Since for spin 1/2 and spin 1 the Madison Convention[9] recommends using tensors which transform like

spherical harmonics, that convention would be satisfied by using the renormalized ($t_{00} = 1$) polarization tensors

$$t_{kK}(c) \equiv \rho_{kK}^*(c)/\rho_{00}^*(c) \quad \text{and} \quad t_{k_1K_1}(s_1) \equiv \rho_{k_1K_1}^*(s_1)/\rho_{00}^*(s_1) \tag{4}$$

From (1) the differential reaction cross section is seen to equal $(\hat{a}\hat{s}_1)^{-2}$
Trace(c) and from (2) Trace $\rho(c) = \hat{c}\rho_{00}(c)$, where $\hat{c} \equiv (2c+1)^{1/2}$. Thus,

$$\frac{d\sigma}{d\Omega} = \hat{c}(\hat{a}\hat{s}_1)^{-2} \rho_{00}(c) , \tag{5}$$

The s_2-gamma direction correlation function is given by[2]

$$W(\theta_\gamma,\phi_\gamma) = \sum_{kK} \rho_{kK}(c) \varepsilon_{kK}^*(c) , \tag{6}$$

where the efficiency tensor describing the c(L)d γ-decay is

$$\varepsilon_{kK}(c) = D_{K0}^k(\phi_\gamma,\theta_\gamma,0) R_k , \tag{7}$$

$\theta_\gamma,\phi_\gamma$ being the polar angles of the γ-ray relative to the arbitrary axes
used for specifying $\rho_{kK}(c)$, and

$$D_{K0}^k(\phi_\gamma,\theta_\gamma,0) = 2\pi^{1/2} \hat{k}^{-1} Y_{kK}^*(\theta_\gamma,\phi_\gamma) \equiv C_{kK}^*(\theta_\gamma,\phi_\gamma) , \tag{8}$$

where C_{kK} is the renormalized ($C_{00} = 1$) spherical harmonic of Brink and
Satchler[10], and R_k, which is real and non-zero only for even k (for
definite parity γ-rays) and is normalized such that $R_0 = 1$, depends only
upon the parameters of the γ-decay,

$$R_k = \sum_{LL'} g_L g_{L'} R_k(LL'cd) . \tag{9}$$

The angular momentum factor $R_k(LL'cd)$ is defined and tabulated in Rose
and Brink[3]. The fraction of the transition resulting in 2^L pole radiation is g_L^2. If as usual, only the L and L+1 multipoles contribute and
if the real mixing ratio g_{L+1}/g_L is renamed δ, (9) becomes

$$R_k = [R_k(LLcd)+2\delta R_k(LL+1cd) + \delta^2 R_k(L+1 \ L+1 \ cd)]/(1 + \delta^2) . \tag{10}$$

Rewriting (6) to exhibit the ϕ_γ dependence,

$$W = \sum_{\substack{k \ even}} R_k \{\rho_{k0}(c) P_k(\cos \theta_\gamma)+2 \sum_{K>0} [\mathrm{Re}\rho_{kK}(c))\cos K\phi_\gamma - (\mathrm{Im}\rho_{kK}(c)\sin K\phi_\gamma]$$

$$\text{times } C_{kK}(\theta_\gamma,0)\} . \tag{11}$$

The density matrices of c and s_1 being related (1) implies that
their corresponding statistical tensors are related. Following DS

$$\rho_{kK}(c) = \sum_{k_1K_1} G_{kK}^{k_1K_1}(\theta,\phi) t_{k_1K_1}^*(s_1) \tag{12}$$

The beam moments $t_{k_1K_1}$ are used rather than $\rho_{k_1K_1}$ to satisfy the Madison
Convention[9]. The reaction angle dependent coefficients $G_{kK}^{k_1K_1}$ are called
the polarization transfer coefficients. Substituting (1), (4) and the
inverse of (3) into (2) and comparing with (12)

$$\tag{13}$$

$$G_{kK}^{k_1K_1} = \sum A_{\sigma_2\gamma\sigma_1\alpha} A_{\sigma_2\gamma'\sigma_1'\alpha}^* (-)^{s_1-\sigma_1'} (s_1\sigma_1,s_1-\sigma_1'|k_1K_1) (-)^{c-\gamma'}(c\gamma,c-\gamma'|kK)/\hat{s}_1 .$$

For an unpolarized beam $t_{k_1\kappa_1}(s_1) = \delta_{k_10}\,\delta_{\kappa_10}$ and therefore

$$G_{k\kappa}^{00} = \rho_{k\kappa}^{unpol}(c) \ . \tag{14}$$

Using (14) and defining $g_{k\kappa}^{k_1\kappa_1} = G_{k\kappa}^{k_1\kappa_1}/G_{k\kappa}^{00}$ one can rewrite (12)

$$\rho_{k\kappa}(c) = \rho_{k\kappa}^{unpol}(c) \sum_{k_1\kappa_1} g_{k\kappa}^{k_1\kappa_1}\, t_{k_1\kappa_1}^{*}(s_1) \ . \tag{15}$$

Evaluating (15) for $k=\kappa=0$, defining $T_{k_1\kappa_1} = g_{00}^{k_1\kappa_1}$ and using (5)

$$\frac{d\sigma}{d\Omega}\bigg) = \frac{d\sigma}{d\Omega}\bigg)^{unpol}\bigg[\sum_{k_1\kappa_1} T_{k_1\kappa_1}\, t_{k_1\kappa_1}^{*}(s_1)\bigg], \tag{16}$$

which identifies $T_{k_1\kappa_1}$ as the analyzing power tensor, following the Madison Convention[9].

<u>Symmetry</u>. The Hermiticity of $\rho_{\sigma\sigma'}$ in (3) implies

$$\rho_{k_1\kappa_1}(s_1) = (-)^{\kappa_1}\,\rho_{k_1-\kappa_1}^{*}(s_1) \ , \tag{17}$$

which is the statement of Hermiticity for the statistical tensors. With the aid of (1), (2) and (3) the corresponding result follows for $\rho_{k\kappa}(c)$,

$$\rho_{k\kappa}(c) = (-)^{\kappa}\,\rho_{k-\kappa}^{*}(c). \tag{18}$$

Obviously (17) and (18) also apply when ρ is replaced by t. The efficiency tensors also have the same symmetry which ensures that the correlation function W is real. The Hermiticity of the tensors in (12) implies a similar symmetry for the transfer coefficients

$$G_{k\kappa}^{k_1\kappa_1} = (-)^{\kappa+\kappa_2}\,G_{k-\kappa}^{k_1-\kappa_1} \tag{19}$$

The reaction amplitude $A_{\sigma_2\gamma,\sigma_1\alpha}$ should be invariant under reflection M through the reaction plane, i.e., $MAM^{-1} = A$. This mirror operation M is equivalent to the parity operation combined with a 180° rotation about the normal to the reaction plane.

Consider first the familiar choice of (right-handed) axes with z along $\vec{k}_1$ and y along $\hat{n} \equiv \vec{k}_1 \times \vec{k}_2$. For this choice the rotation operation reduces to $R_y(180°)|s_1\sigma_1\rangle = (-)^{s_1-\sigma_1}|s_1,-\sigma_1\rangle$ and the parity operation introduces $\Pi_i\Pi_f$ the product of the intrinsic parities of the reaction participants.

$$(-)^{s_2-\sigma_2+c-\gamma-s_1+\sigma_1-a+\alpha}\,A_{-\sigma_2-\gamma,-\sigma_1-\alpha} = \Pi_i\Pi_f\, A_{\sigma_2\gamma,\sigma_1\alpha} \ . \tag{20}$$

Using (20) in (13) one obtains for these axes the DS result $G_{k\kappa}^{k_1\kappa_1} = (-)^{k+k_1}\,G_{k\kappa}^{k_1\kappa_1*}$ (21). Thus the transfer coefficients are real/imaginary as $k+k_1$ is even/odd.

A second useful choice of axes to be designated $\hat{x}\ \hat{y}\ \hat{z}$ and referred to as the "hatted" axes has $\hat{z}$-along $\hat{n}$ and $\hat{y}$ along $\vec{k}_1$. Since $R_{\hat{z}}(180°)|s_1\sigma_1\rangle = (-)^{-\sigma_1}|s_1\sigma_1\rangle$ one finds

$$(-)^{\alpha+\sigma_1-\sigma_2-\gamma}\,\hat{A}_{\sigma_2\gamma\sigma_1\alpha} = \Pi_i\Pi_f\,\hat{A}_{\sigma_2\gamma\sigma_1\alpha} \ . \tag{22}$$

This result is referred to as the Bohr Theorem[11]. Using (22) and (13)

$$\hat{G}^{k_1\kappa_1}_{k\kappa} = (-)^{\kappa-\kappa_1}\,\hat{G}^{k_1\kappa_1}_{k\kappa}\,. \tag{23}$$

Thus the hatted transfer coefficients vanish unless $(\kappa-\kappa_1)$ is even. The tensors $t_{k\kappa}$ transforming like spherical harmonics means

$$t_{k_1\kappa_1'}(x'y'z') = \sum_{\kappa_1} t_{k_1\kappa_1}(xyz)\, D^{k_1}_{\kappa_1\kappa_1'}(xyz \to x'y'z'). \tag{24}$$

Of course (24) would apply with t replaced by ρ^*. For the transfer coefficients using (24) and (12) one finds the DS result

$$G^{k_1\kappa_1'}_{k\kappa'}(x'y'z') = \sum_{\kappa\kappa_1} G^{k_1\kappa_1}_{k\kappa}(xyz) D^{k_1}_{\kappa_1\kappa_1'}\, D^{k*}_{\kappa\kappa'}\,, \tag{25}$$

where the arguments of the D's are the Euler angles taking xyz into x'y'z'.

Applications. Before turning to specific applications it might be appropriate to give a partial answer to the question, "What can an angular correlation measurement do for you?" It can provide information concerning certain combinations of reaction amplitudes which would not be measured by other experiments. Consider the results (1) and (2) summed over $\alpha\sigma_2\gamma\gamma'\sigma_1$ and σ_1'. The unpolarized cross section measures $\rho^{unpol}_{00}(c)$ $\propto \sum |A_{\sigma_2\gamma,\sigma_1\alpha}|^2$. The polarized cross section $\rho_{00}(c)$ involves $\sum A_{\dots\sigma_1\dots} A^*_{\dots\sigma_1'\dots}\rho_{\sigma_1\sigma_1'}$, the other subscripts of A and A* being as in the unpolarized case. The unpolarized correlation measures $\rho^{unpol}_{k\kappa}$ which involves $\sum A_{\dots\gamma\dots}A^*_{\dots\gamma'\dots}(c\gamma,c-\gamma'|k\kappa)$, and the polarized beam correlation permits $\gamma\neq\gamma'$ as well as $\sigma_1\neq\sigma_1'$. These experiments aren't likely to yield unique values for the reaction amplitudes, but they can provide stringent tests of models for these amplitudes. For example, one might calculate A using DWBA and using the multi-step CCBA and compare the results as has been done in two of this session's contributions. Schieb et al. studying ^{24}Mg(d,d'γ) found that the unpolarized correlation favored CCBA whereas the cross section didn't care. Clement et al. studying ^{28}Si($\vec{d}$,pγ) found that only the polarized correlation differentiated, providing evidence for miltistep.

It might be instructive to point out that if the amplitude A were expressed in terms of the coupled collision matrix one would obtain for the correlation function (z along k_1, y along $\hat{n}$)

$$W = \sum_{k_1k_1'k_\ell k_2k} B_{k_\ell k_1'k_2k} \sum_{\kappa\kappa_1\kappa_2} t_{k_1\kappa_1}(s_1)(k_\ell 0,k_1\kappa_1|k_1'\kappa_1')(k_1'\kappa_1',k_2\kappa_2|k\kappa)$$

$$\text{times } C_{k_2\kappa_2}(\theta,0)C_{k\kappa}(\theta_\gamma,\phi_\gamma),$$

where $B_{k_\ell k_1'k_2k}$ is a complicated but angle-independent coefficient which contains angular momentum weighted combinations of all the partial wave matrix elements permitted by angular momentum conservation for the particular set of k's. Here one sees something of the richness of content of the correlation measurement, the possibility of measuring the complete angular dependence so as to extract the large number of B's. The unpolarized correlation involves only B's with $k_\ell = k_1'$ and has a simpler angular dependence ($\kappa_2 = \kappa$), but is still markedly rich in contrast to the unpolarized cross section $\sum_k B_{kkk0}\,P_k$.

Gamma detection along $\vec{k}_1 \times \hat{k}_2$. Choosing the hatted axes and assuming a spin sequence c(L)0 for the gamma decay it is easy to show[4] that the correlation function with gamma detection along $\hat{n}$ is proportional to the sum of the $\gamma = \pm 1$ substate populations of the state c. For inelastic scattering (which is only briefly addressed here since it is the subject of other reviews at this conference) of a spin 1/2 particle on a spin 0 target (of the same parity as c) the Bohr Theorem (22) implies (for $\gamma = \pm 1$) that $\sigma_2 = -\sigma_1$ (i.e., spin-flip). If in addition c $\leq$ 2 the sum of the $\gamma = \pm 1$ substate populations is equal to the spin-flip probability $S(\phi_2)$ and the perpendicular correlation (γ-ray along $\hat{n}$) then becomes proportional to S. Since $C_{\kappa\kappa}(\theta_\gamma, 0)$ in (11) vanishes for $\theta_\gamma = 0$ unless $\kappa = 0$ and from (23) $\hat{G}_{\kappa 0}^{1\mu}$ vanishes unless $\mu = 0$ one can conclude from (12) that only the $\hat{z}$ component of the incident polarization affects the spin-flip correlation.

Boyd et al.[12] point out that for inelastic nucleon scattering on 0^+ targets the spin-flip asymmetry ΔS is related to the analyzing power $A_{\hat{z}}$ and the polarization $p_{\hat{z}}$ of the scattered nucleon. Designating by σ^{+-} the cross section for scattering of a spin up (along z) nucleon into a spin down nucleon (proportional to our $\sum_\gamma |A_{-\frac{1}{2}\gamma}, 0|^2$), they write[12]

$$\frac{d\sigma}{d\Omega} = 1/2(\sigma^{++} + \sigma^{+-} + \sigma^{-+} + \sigma^{--}) = 1/2\sigma(\phi)$$

$$A_{\hat{z}}\sigma(\phi) = \sigma^{++} + \sigma^{+-} - \sigma^{-+} - \sigma^{--},$$

$$p_{\hat{z}}\sigma(\phi) = \sigma^{++} + \sigma^{-+} - \sigma^{+-} - \sigma^{--},$$

$$S\sigma(\phi) = \sigma^{+-} + \sigma^{-+},$$

$$\Delta S\sigma(\phi) = \sigma^{+-} - \sigma^{-+},$$

and conclude that $\Delta S = 1/2(A_{\hat{z}} - p_{\hat{z}})$. $S(\theta)$ and $\sigma(\theta)$ can be measured using an unpolarized beam. $A_{\hat{z}}$ and $\Delta S/S$ require a polarized beam, A reflecting the asymmetry in the counting rate and $\Delta S/S$ the asymmetry in the coincidence counting rate. These quantities have been measured by Boyd and others[13] and used to study isobaric analogue resonances. This highly special polarized correlation merely provides a simpler (than double scattering) method of measuring $p_{\hat{z}}$. The power of the full correlation function with polarized beam appears to be as yet untapped but is perhaps not far away.

Certain transfer reactions where a = s_1 = 1/2 and s_2 = 0 are equally interesting to consider, e.g., (p,$\alpha\gamma$) or (τ,$\alpha\gamma$) on ^{13}C, ^{15}N, ^{29}Si, ^{31}P, etc. If the γ-decay is to a spin zero state and the γ-ray is detected along $\hat{z}$ in coincidence with the α-particle, an argument parallel to that of Boyd et al.[12] leads to the conclusion

$$\Delta S = 1/2(A_{\hat{z}}(s_1) + \pi_i\pi_f A_{\hat{z}}(a),$$

where $\pi_i\pi_f$ is the product of the intrinsic parities and $A(s_1)$ and $A(a)$ are the analyzing powers of the incident beam and target respectively. Here S and ΔS are measured as before but S changes its interpretation from the spin-flip probability to the probability of the spins s_1 and a being antiparallel ($\pi_i\pi_f$ negative) or parallel ($\pi_i\pi_f$ positive). Measuring $\Delta S/S$ is thus an alternative method of measuring target analyzing power for such reactions. The measurement of the unpolarized perpendicular correlation for the ^{13}C(τ,α) reaction is the subject of a contributed paper by Fujisawa et al.

Particle detection along $\hat{k}_1$. Correlation experiments in which s_2 is detected at 0° or 180° (z-axis along $\hat{k}_1$) are very special. They offer the possibility of obtaining spectroscopic information (SI) by a reaction mechanism independent (RMI) procedure. Litherland and Ferguson[14] first proposed such measurements and McCullen and Seyler[15] extended the technique to include polarized beams.

For the unpolarized case the cylindrical symmetry causes the $\rho_{\kappa\kappa}(c)$ to vanish for $\kappa \neq 0$, leading to the correlation function (11)

$$W^{unpol}(\theta_\gamma) = \sum_{k \text{ even}} \rho_{k0}^{unpol} R_k P_k (\cos\theta_\gamma). \qquad (26)$$

Using (2) ρ_{k0} can be expressed in terms of the diagonal elements of the density matrix $\rho_{\gamma\gamma}$ which are proportional to the population parameters $P(\gamma)$ of the state c. The symmetry of the situation requires $P(\gamma) = P(-\gamma)$. Angular momentum conservation along z requires

$$\alpha + \sigma_1 = \sigma_2 + \gamma . \qquad (27)$$

Defining $M = a + s_1 + s_2,$ (28)
leads to $\gamma_{max} \leq M.$ (29)

If $M = 0$ or $1/2$ there are no unknown substate populations and the method is at its best. If $M = 1$ or $3/2$ there is one unknown population ratio but if the correlation can provide two coefficient ratios one retains the possibility of determining one unknown piece of spectroscopic data, say a mixing ratio. On the other hand if $c < 2$ or if the reaction mechanism happens to cause ρ_{40} to nearly vanish, the correlation would yield only one coefficient ratio and the method would fail. The polarized beam methods would then be needed.

To introduce these methods suppose the beam has a polarization symmetry axis which is perpendicular to the beam. It can be chosen as the y-axis since $\vec{k}_1 \times \vec{k}_2$ vanishes. Parity conservation requires the vector polarization to be transverse to be effective.

To evaluate (12) note that from (13) and (27) the $G_{\kappa\kappa}^{k_1\kappa_1}$ vanish unless $\kappa_1 = \kappa$. The $t_{k_1\kappa_1}$ in (12), specified with respect to z along $\vec{k}_1$, can be expressed in terms of $\hat{t}_{10}$ and $\hat{t}_{20}$ the beam moments referred to the polarization symmetry axis using (24): $t_{10} = t_{21} = 0$, $t_{11} = -i\,\hat{t}_{10}/\sqrt{2}$, $t_{20} = -1/2\hat{t}_{20}$, and $t_{22} = -1/2(3/2)^{\frac{1}{2}}\,\hat{t}_{20}$. Inserting these results into (12) one finds

$$\rho_{k0} = \rho_{k0}^{unpol} - 1/2\, G_{k0}^{20}\, \hat{t}_{20}$$

$$\rho_{k1} = i\, G_{k1}^{11}\, \hat{t}_{10}/\sqrt{2} ,$$

$$\rho_{k2} = -1/2(3/2)^{\frac{1}{2}}\, G_{k2}^{22}\, \hat{t}_{20} .$$

Since (even k) G_{k1}^{11} is imaginary for these axes the $\rho_{\kappa\kappa}$ are real and the correlation (11) contains no $\sin\kappa\phi_\gamma$ terms.

$$W(\theta_\gamma\phi_\gamma) + \overline{W}(\theta_\gamma) + W_1(\theta_\gamma\phi_\gamma) + W_2(\theta_\gamma\phi_\gamma), \qquad (30)$$

where W_1 is proportional to the beam vector polarization and has a $\cos\phi_\gamma$ dependence,

$$W_1(\theta_\gamma,\phi_\gamma) = \sqrt{2}\,\hat{t}_{10}\cos\phi_\gamma \sum_k iG_{k1}^{11}R_k\, C_{k1}(\theta_\gamma,0), \qquad (31)$$

W_2 depends on the beam tensor polarization and has a $\cos 2\phi_\gamma$ dependence,

$$W_2(\theta_\gamma \phi_\gamma) = -\sqrt{3/2}\,\hat{t}_{20}\,\cos 2\phi_\gamma \sum_k G^{22}_{k2}\,R_k C_{k2}(\theta_\gamma,0), \qquad (32)$$

and $\overline{W}(\theta_\gamma)$ is the correlation function averaged over ϕ_γ,

$$\overline{W}(\theta_\gamma) = W^{unpol}(\theta_\gamma) - 1/2\,\hat{t}_{20} \sum_k G^{20}_{k0}\,R_k P_k(\cos\theta_\gamma),$$

which is more convenient if written

$$\overline{W}(\theta_\gamma) = W^{unpol}(\theta_\gamma)\,[1-ft_{20}] + t_{20}\sum_k (G^{20}_{k0}+fG^{00}_{k0})\,R_k P_k(\cos\theta_\gamma). \qquad (33)$$

Each of these three equations can serve for the RMI extraction of SI, but to see this one needs the k-dependence of the transfer coefficients. Using channel-spin coupling and the amplitude formalism[15]

$$G^{k_1\kappa}_{k\kappa} = \hat{s}_1 \sum (-)^{c-\gamma'}(c\gamma,c-\gamma'|k\kappa)(-)^{s_1-\sigma_1'}(s_1\sigma_1,s_1-\sigma_1'|k_1\kappa)(c\gamma,s_2\sigma_2|S_2\textstyle\sum)$$

$$(c\gamma',s_2\sigma_2|S_2'\textstyle\sum')(S_2\textstyle\sum,\ell_2 0|b\textstyle\sum)(S_2'\textstyle\sum',\ell_2'0|b'\textstyle\sum')$$

$$(S_1\textstyle\sum,\ell_1 0|b\textstyle\sum)(S_1'\textstyle\sum',\ell_1'0|b'\textstyle\sum')(a\alpha,s_1\sigma_1|S_1\textstyle\sum)$$

$$(a\alpha,s_1\sigma_1'|S_1'\textstyle\sum')\,\hat{\ell}_1\hat{\ell}_1'\hat{\ell}_2\hat{\ell}_2'\,R^b_{S_2\ell_2 S_1\ell_1}\,R^{b'}_{S_2'\ell_2' S_1'\ell_1'}. \qquad (34)$$

For M ($= a+s_1+s_2$) $= 1$ or $3/2$, (34) leads to

$$iG^{11}_{k1} = (cM,\,c1-M|k1)r_1, \qquad (35)$$

and

$$G^{22}_{k2} = (cM,c2-M\ k2)r_2, \qquad (36)$$

where r_1 and r_2 are real RM dependent factors independent of k. Similarly for $s_1 = 1$ or $3/2$ and for certain values of f and γ_f,

$$G^{20}_{k0} + fG^{00}_{k0} = (c\gamma_f,c-\gamma_f|k0)r_0. \qquad (37)$$

the permissable values of f and γ_f are listed below

s_1	$a+s_2$	f	γ_f	
1	$0,\tfrac{1}{2}$	$-2^{-\tfrac{1}{2}}$	$a+s_2$	
1	0	$2^{\tfrac{1}{2}}$	1	(38)
3/2	0	± 1	$1+\tfrac{1}{2}f$	

Using (35) and noting that $\hat{t}_{20}\cos\phi\gamma = [3s_1/(s_1+1)]^{\tfrac{1}{2}}\vec{p}\cdot\hat{n}$ where here $\hat{n}$ is along $\vec{k}_1 \times \vec{k}_\gamma$, (31) can be written

$$W_1 = \vec{p}\cdot\hat{n}\,r_1 \sum_{k\geq 2} (cM,c1-M|k1)\,R_k C_{k1}(\theta_\gamma,0), \quad M = 1 \text{ or } 3/2. \quad (39)$$

Since W_2 is left-right symmetric a left-right correlation asymmetry measurement (assumping $\vec{p}\neq 0$) will be proportional to the sum in (39), which when multiplied by r_1 is the asymmetry function $Y(\theta)$ of Glavish et al.[16]. This group[16] evaluated $Y(\theta)$ by subtracting the correlations measured with $\vec{p}$ parallel and antiparallel to $\hat{n}$. Particle s_2 was detected at 180°. They successfully tested the method using the reaction $^{12}C(p,p'\gamma)^{12}C$, (2^+ 4.43 MeV $\rightarrow 0^+$ g.s.) where the pure E2 transition completely determines the angular dependence of $Y(\theta)$. They then studied the reaction $^{24}Mg(p,p'\gamma)^{24}Mg$, $3^+ \rightarrow 2^+$ where they sought to determine $1/\delta$, the M1/E2 mixing ratio, already known to be small. They obseved that the predicted values of $Y(\theta)$ (which is antisymmetric about $\theta = 90°$) were particularly sensitive near 50° (or 130°) to the (small) values of $1/\delta$. An excellent fit to the data was obtained for $1/\delta = 0.06$. This method requires only that c and L_{max} each be ≥ 2, so that $Y(\theta)$ has at least two terms. The incident particle s_1 may have spin 1/2, 1 or 3/2 so long as M = 1 or 3/2.

A second method for the RMI extraction of SI is based on W_2, (32) which with the aid of (36) becomes for M = 1 or 3/2

$$W_2(\theta_\gamma, \phi_\gamma) = \hat{t}_{20} \cos 2\phi_\gamma \; r_2 \sum_{k \geq 2} (cM, c2-M|k2) R_k C_{k2}(\theta_\gamma, 0) \qquad (40)$$

The value of this sum can be obtained (for a particular θ_γ) from a left+right-up-down asymmetry (a right-up asymmetry would suffice for equal times, for up and down vector polarization) using a spin 1 or 3/2 tensor polarized beam. Again c and $L_{max} \geq 2$ to get two terms. No experiment using this method has been reported.

A third method for the RMI extraction of SI is based on $\overline{W}(\theta_\gamma)$, (33) which after substituting (37) becomes

$$\overline{W}(\theta_\gamma) - W(\theta_\gamma)^{unpol}[1-ft_{20}] = t_{20}r_0 \sum_k (c\gamma_f, c-\gamma_f|k0) R_k P_k (\cos \theta_\gamma), \qquad (41)$$

where the possible values of f and γ_f are given by (38). Measurements of W^{unpol} and $\overline{W}$ (for known t_{20}) when combined according to (41) permit one to determine the R_k. In contrast to the above methods here the sum includes k = 0 and thus will yield R_2 if $c \geq 1$. For c = 1 or 3/2 this is then the only method available.

C.R. Gould and co-workers[17,18] have successfully employed this method for c = 3/2. They use a polarized deuteron beam whose polarization symmetry axis coincides with the beam direction (z-axis) at the target. [Any vector polarization present is then along the beam direction and contributes nothing]. This choice is very advantageous since by symmetry it eliminates the ϕ_γ dependence and thus eliminates the need to average over ϕ_γ. Eq. (41) still applies, but $\overline{W}$ can be replaced by W for any value of ϕ_γ. Instead of measuring $_\alpha W(t_{20})$ and W^{unpol} they elected to measured W for two values t_{20}^α and t_{20}^β of beam polarization. Calling the corresponding correlations W_α and W_β one finds from (41)

$$U(\theta_\gamma) \equiv W_\alpha(\theta_\gamma) - W_\beta(\theta_\gamma) F = gr_0 \sum_k (c\gamma_f, c-\gamma_f|k0) R_k P_k(\cos\theta_\gamma), \qquad (42)$$

where

$$g = t_{20}^\alpha - t_{20}^\beta F \quad \text{and} \quad F = (1-ft_{20}^\alpha)/(1-ft_{20}^\beta) \qquad (43)$$

From (38) for a (d,p) reaction on a spin zero target or a (d,α) reaction on a spin 1/2 target $f = -(2)^{-\frac{1}{2}}$ and $\gamma_f = 1/2$ and the factor

$-ft_{20}$ becomes $\frac{1}{2}p_{zz}$. For (d,α) on spin 0 two values of f and γ_f are possible. The linear combinations controlled by f are such as to eliminate the contribution from one of the two magnetic substate magnitudes. For example, for (d,α) on spin 0 letting $p(0)$ be the population of the zero substate of the deuteron beam (z along $\vec{k}_1$) and using (1) and (2)

$$\rho_{k0}(c) = (c0,c0|k0)(-)^c |A_{0000}|^2 p(0)+(c1,c-1|k0)(-)^{c-1}|A_{0110}|^2(1-p(0))\ (44)$$

Attaching superscripts α and β corresponding to two different beam polarizations, it is obvious that one linear combination will eliminate the $1-p(0) = p(1)+p(-1)$ term and another the $p(0)$ term. Inserting the $\rho_{k0}(c)$ into (11) one has

$$[1-p^\beta(0)]W^\alpha-[1-p^\alpha(0)]W^\beta = [p^\alpha(0)-p^\beta(0)]|A_{0000}|^2\sum_k (c0,c0|k0)R_kP_k \quad (45)$$

and

$$p^\beta(0)W^\alpha-p^\alpha(0)W^\beta = [p^\alpha(0)-p^\beta(0)]|A_{0110}|^2\sum_k (c1,c-1|k0)R_kP_k \quad (46)$$

These are just the equations one finds using the two f and γ_f values of (38) and $p(0) = (1-\sqrt{2}\,t_{20})/3$ in (42).

For (d,p) on spin 0, [or with a slight change in the A's for (d,α) on spin 1/2]

$$\rho_{k0}(c) = (c\tfrac{1}{2},c-\tfrac{1}{2}|k0)(-)^{c-\frac{1}{2}}[(2|A_{-\frac{1}{2}\frac{1}{2}00}|^2-|A_{\frac{1}{2}\frac{1}{2}00}|^2)p(0)+|A_{\frac{1}{2}\frac{1}{2}10}|^2]$$

$$+(c3/2,c-3/2|k0)(-)^{c-3/2}|A_{-\frac{1}{2}\frac{3}{2}10}|^2(1-p(0)\ . \quad (47)$$

Since the A and p parts of the first term don't factor only the second term can be eliminated leading to

$$[1-p^\beta(0)]W^\alpha-[1-p^\alpha(0)]W^\beta=[p^\alpha(0)-p^\beta(0)]2|A_{-\frac{1}{2}\frac{1}{2}10}|^2\sum_k (-)^{-\frac{1}{2}}(c\tfrac{1}{2},c-\tfrac{1}{2}|k0)R_kP_k\ (48)$$

which agrees with (42). Both (45) and (48) would correspond to the result of an ideal experiment with a deuteron beam in which every particle had $\sigma_1 = 0$ i.e., $p(0) = 1$. In the [(d,α) on a 0 target] case, (45), such an experiment would, from (34) require $(-)^c = (-)^{\ell_1+\ell_2+1} = -\pi_i\pi_f$, the familiar non-natural parity result for the residual state. There is no similar parity restriction on c in the case of (48).

Gould et al.[17] have measured $U(\theta_\gamma)$ for several $(d,p\gamma)$ reactions on 0^+ targets leading to spin 3/2 states c. They detect the proton at 180° and use 4 or 5 gamma detectors spaced between 30° and 90° to sample $W(\theta_\gamma)$. They tested the method on ^{12}C studying the $(3/2^- \ 3.68 \to 1/2^-$ g.s.) transition in ^{13}C for which they obtained a δ value in agreement with earlier γ-γ correlation work. They then studied the $(3/2^+ \ 2.32 \to 3/2^+$ g.s.) and $(3/2^+ \ 3.54 \to 1/2^+ \ 0.76)$ transitions[17] in ^{31}Si and the $(3/2^- \ 2.46 \to 3/2^- \ 1.94)$ and $(3/2^- \ 1.94 \to 7/2^-$ g.s.) transitions[18] in ^{41}Ca and most recently[19] several low-lying levels (not just c = 3/2) in ^{33}S and ^{35}S. The results for the various transitions are very similar in that usually two quite different values of δ are obtained consistent with the data. Sometimes one of the two values can be ruled out on the basis of lifetime or transition strength systematics. In cases where one value cannot be rejected an independent measurement, e.g. linear polarization of the γ-ray, could be called upon.

The highly successful experiments of Glavish et al.[16] and Gould et al.[17-19] demonstrate conclusively the utility of two of these polarized beam methods for extending the RMI extraction of SI to cases where $a+s_1+s_2 = 1$ or $3/2$. Hopefully the other methods and/or other spin combinations will also be attempted and found useful.

References

1) G. Bergdolt, Polarization Phenomena in Nuclear Reactions (Univ. of Wisconsin Press, Madison, Wisconsin, 1971) p. 408.
2) S. Devons and L.J.B. Goldfarb, Handbuch der Physik, Vol. 42 (Springer, Berlin, 1957) A.J. Ferguson, Angular Correlation Methods in Gamma-Ray Spectroscopy (North-Holland, Amsterdam, 1965).
3) H.J. Rose and D.M. Brink, Rev. Mod. Phys. 39 (1967) 306.
4) F. Rybicki, T. Tamura, and G.R. Satchler, Nucl. Phys. A146 (1970) 659.
5) A.A. Debenham and G.R. Satchler, Particles and Nuclei 3 (1972) 117.
6) G.R. Satchler, Nucl. Phys. 55 (1964) 1.
7) T. Tamura, Rev. Mod. Phys. 37 (1965) 679.
8) A.M. Lane and R.G. Thomas, Rev. Mod. Phys. 30 (1958) 257.
9) Polarization Phenomena in Nuclear Reactions (ed. Barschall and Haeberli, Univ. of Wisconsin Press, Madison, Wisconsin, 1971).
10) D.M. Brink and G.R. Satchler, Angular Momentum, 2nd Ed. (Oxford Univ. Press, Oxford, 1968).
11) A. Bohr, Nucl. Phys. 10 (1959) 486.
12) R. Boyd, S. Davis, C. Glashausser, and C.F. Haynes, Phys. Rev. Lett. 29 (1972) 1590.
13) R.N. Boyd, D. Slater, R. Avida, H.F. Glavish, C. Glashausser, G. Bissinger, S. Davis, C.F. Haynes and A.B. Robbins, Phys. Rev. Lett. 29 (1972) 955, H.T. King, Ph.D. Thesis, Stanford University (1974) M. Thum, G. Mertens, H. Lesiecki, G. Mack and K. Schmidt, contributed paper to this Symposium, p.799
14) A.E. Litherland and A.J. Ferguson, Can. J. Phys. 39 (1961) 788.
15) J.D. McCullen and R.G. Seyler, Nucl. Phys. A139 (1969) 203.
16) H.F. Glavish, R. Avida, S.S. Hanna, E. Diener, R.N. Boyd, and C.C. Chang, Phys. Rev. Lett. 30 (1973).
17) C.R. Gould, R.O. Nelson, J.R. Williams, D.R. Tilley, J.D. Hutton, N.R. Roberson, C.E. Busch, and T.B. Clegg, Phys. Rev. Lett. 30 (1973) 298.
18) C.R. Gould, D.R. Tilley, C. Camerson, R.D. Ledford, N.R. Roberson and T. B. Clegg, contributed paper to this Symposium, p.807
19) C.R. Gould, private communication. (to be published)
20) G.D. Jones, P.W. Green, J.A. Kuehner and D.T. Petty (this interesting preprint on $(\vec{d},\alpha\gamma)$ was received after completion of the manuscript).

SOME RECENT AND POSSIBLE NEW APPLICATIONS OF POLARIZED BEAMS[†]

Stanley S. Hanna
Department of Physics, Stanford University
Stanford, California 94305 U.S.A.

1. Introduction

In this talk I would like to discuss some fairly recent uses of polarized beams and to speculate a little on some future uses. In the first part of the talk I will describe the production of polarized nuclei by means of polarized beams and the use of these nuclei in hyperfine studies to measure nuclear moments and solid state phenomena. I would then like to suggest some future applications of polarized nuclei produced in this manner. In the second part of the talk I will discuss the present and future work on polarized targets which make possible the study of reactions involving both polarized beams and polarized targets. This discussion will be divided into a description of the use of solid (cryogenic) targets and a look at the possibility of using polarized beams as targets.

2. Polarized nuclei

2.1. METHODS OF PRODUCING POLARIZED NUCLEI

Several methods have been used over the years to produce polarized nuclei which can then be used to study a wide variety of phenomena:

(1) Atomic polarization at low temperature makes use of the atomic hyperfine fields to produce nuclear polarization. This is at present the most widely used method and it has been applied to many problems in nuclear and solid state physics[1].

(2) Circularly polarized resonance radiation in optical pumping has been used to polarize atoms which then polarize the nuclei through the hyperfine interactions[2].

(3) Nuclear reactions have been used to produce nuclear orientation. If the recoil nuclei are restricted to a given angle, vector polarization is produced[3].

(4) Unpolarized nuclei produced in nuclear reactions have been passed through an atomic beam apparatus to produce beams of polarized nuclei[4].

(5) If polarized beams are used to initiate nuclear reactions it has been found that appreciable polarization is transferred to the residual nuclei. It is this method which will be discussed in the next section.

2.2. PRODUCTION OF POLARIZED NUCLEI BY POLARIZED BEAMS

As examples of this method the β-emitting nuclei ^{8}Li, ^{12}B, ^{29}P, and ^{39}Ca have been produced and polarized by $(\vec{d},p)$, $(\vec{d},n)$ and $(\vec{p},n)$ reactions with the beam polarization perpendicular to the beam direction. The targets were Li metal, ZrB_2, Si crystals, and ionic K compounds, respectively, all thick enough (150 mg/cm^2) to stop both the beam and all the recoil nuclei. The net polarization transferred to the recoil nuclei was measured by detecting the resulting β asymmetries. Since all recoil nuclei were used, high β counting rates of $10^3 \rightarrow 10^4$ cps

[†] Supported in part by the U. S. National Science Foundation.

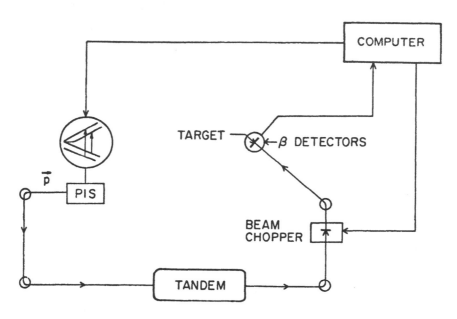

Fig. 1. Schematic arrangement for detecting polarization of nuclei produced in polarized reactions by observing the resultant β-decay asymmetry.

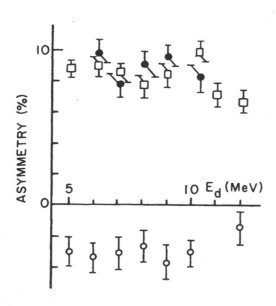

were obtained in two telescopes placed at 0 and 180° to the polarization direction. The incident beam was pulsed and β particles were counted only during the period between pulses. To eliminate possible instrumental asymmetries,counts were accumulated with the beam polarization alternately on and off for deuteron reactions, and alternately up and down for proton reactions.

Fig. 1 illustrates this technique of detecting polarized nuclei. The polarized beam produced by an atomic-beam source is accelerated by a tandem, passes through a beam chopper and then impinges on the target. Counts from the β detectors are stored in a computer. Timing signals from the computer system regulate the beam-pulsing pattern, reverse the direction of polarization in the source (at the appropriate times) and sort the

Fig. 2. Measured β-asymmetries from
^{7}Li($\vec{d}$,p)^{8}Li (solid circles), ^{11}B($\vec{d}$,p)^{12}B
(squares), and ^{28}Si($\vec{d}$,n)^{29}P (open circles).

TABLE 1

Polarization transfer in nuclear reactions

Product(I^{π},$T_{1/2}$)	Target	Reaction[a]	Asymmetry(%)	A[b]	P[c]
^{8}Li(2^{+},0.84s)	Li	$(\vec{d},p)$	+7.5	-0.67	-
^{12}B(1^{+},20ms)	ZrB$_2$	$(\vec{d},p)$	+7.5	-1	-
^{29}P($1/2^{+}$,4.2s)	Si	$(\vec{d},n)$	-2.9	+0.61	-
^{39}Ca($3/2^{+}$,0.87s)	KCaBr$_3$	$(\vec{p},n)$	-2.3	+0.95	-

a) Deuteron polarization = 0.6; proton polarization = 0.65.
b) Angular correlation coefficient in β decay.
c) Sign of polarization transfer.

β counts into designated sections of the computer according to the polarization direction.

Some of the results of these measurements are shown in fig. 2 and summarized in table 1. It is seen that β-asymmetries as high as 8% were observed for a beam polarization of about $P_z = 0.6$ (table 1) and found to be fairly insensitive to beam energy, E = 5 → 10 MeV (fig. 2). We note also that the polarization transfer is antiparallel (-) to the initial beam polarization in the cases studied so far. A program is now underway at Stanford to study the nuclear mechanisms that are responsible for this polarization transfer.

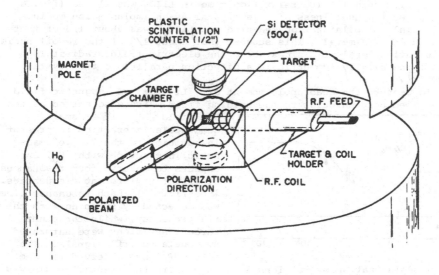

Fig. 3. Apparatus used to detect β-asymmetries and measure NMR of polarized nuclei produced in nuclear reactions with polarized beams.

2.3 NUCLEAR MOMENT MEASUREMENTS

To determine the applicability of this method to the measurement of nuclear moments by means of NMR detection, the known magnetic moment of ^{12}B was measured. The experimental arrangement is shown in fig. 3. The applied magnetic field serves the dual purpose of preserving the nuclear polarization and providing the field H_0 for the NMR measurement. The field H_1 is applied perpendicular to H_0 by means of the coils surrounding the target. The two β-detector telescopes at 0° and 180° are shown schematically.

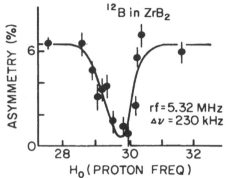

The NMR measurement on ^{12}B was obtained, at a modulated ($\Delta\nu$ = 230 kHz) frequency set at $\overline{rf}$ = 5.32 MHz, by scanning with the H_0 field. The result of the measurement is shown in fig. 4. A pronounced resonance was observed (in which the nuclear polarization was almost completely destroyed) for values of H_0 and $\nu(H_1)$ in good agreement with the known moment of ^{12}B.

Fig. 4. NMR spectrum of polarized ^{12}B nuclei implanted in $ZrBr_2$ following the reaction $^{11}B(\vec{d},p)^{12}B$.

2.4 THE QUADRUPOLE MOMENT OF ^8Li

To illustrate the application of this method to the measurement of a quadrupole moment, a hexagonal single crystal of $LiIO_3$ was used as a target to investigate the electric quadrupole interaction of $^8Li(I = 2, T_{1/2} = 0.84$ sec) by means of NMR detection. A holding field of about 1 kG was necessary to preserve the ^8Li polarization in the crystal, see fig. 5. The spin-lattice relaxation time in $LiIO_3$ was 11 sec (fig. 6) which shows the usefulness of this crystal as an implantation medium. The initial ^8Li polarization measured in $LiIO_3$ was about 1/3 of that measured in Li metal. This suggest that about 1/3 of the recoil nuclei find substitutional Li sites where the electric field gradient is known to be axially symmetric in a direction parallel to the crystal c-axis [5].

To measure the quadrupole moment of ^8Li the rf magnetic field was applied perpendicular to the holding field which also served as the

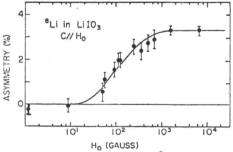

NMR field H_0. If a "small" quadrupole interaction is present in addition to the "large" magnetic interaction with H_0, one expects to observe four resonances of slightly different frequencies. (See fig. 7.) Each resonance line was detected by scanning with the rf frequency, while the other three transitions were saturated by modulated rf signals. The four NMR lines observed for the case where the crystal c-axis was parallel to H_0 are shown in fig. 7. The solid curves in fig. 7

Fig. 5. Polarization of ^8Li ions in $LiIO_3$, as a function of the holding field.

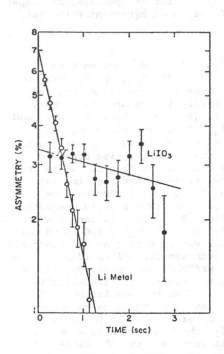

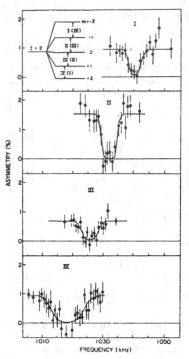

Fig. 6. Polarization of ^{8}Li ions in Li metal and LiIO$_3$, as measured by the β-decay asymmetry, as a function of time.

Fig. 7. The four NMR lines observed for polarized ^{8}Li implanted in a LiIO$_3$ crystal.

are the best fits which take into account the intrinsic Gaussian line shape and the level of the rf intensity. At the center of each resonance it is seen that the polarization is completely destroyed when the other transitions are saturated. This result shows that the nuclear spin is I = 2 (i.e., all magnetic sublevels are accounted for) that the field gradient is unique, and that the ^{8}Li nuclei which keep their polarization sit in equivalent sites in the LiIO$_3$ crystal. The intrinsic line broadening obtained from the line-shape analysis was consistent with the dipolar broadening estimated for ^{8}Li in this crystal. Thus, broadening produced by radiation damage was small, although a holding field of about 1 kG was necessary to decouple the time-dependent interactions due presumably to radiation damage.

The electric quadrupole coupling constant of ^{8}Li in LiIO$_3$, as determined from the resonant frequencies in fig. 6, is $|eqQ| = 29.2 \pm 0.8$ kHz. Using the known[5] coupling constant of ^{7}Li, we obtain the ratio of ground state quadrupole moments

$$|Q(^8\text{Li})/Q(^7\text{Li})| = 0.66 \pm 0.05$$

in fair agreement with the result obtained using polarized neutron capture by Ackerman et al.[6]. If we use the value[5] $Q(^7\text{Li}) = -3.66 \pm 0.03$ fm^2 we obtain

$$|Q(^8\text{Li})| = 2.4 \pm 0.2 \text{ fm}^2 .$$

This small moment indicates that ^{8}Li is rather spherical in shape;
the shell-model theory gives values in fairly good agreement with this
result[8]).

2.5 THE MAGNETIC MOMENT OF ^{39}Ca

Using the same method we have carried out NMR measurements on

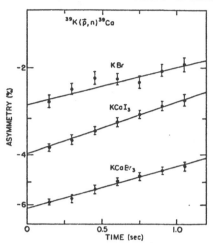

^{39}Ca($I^\pi = 3/2^+$,$T_{1/2} = 0.87$ sec) to
determine its magnetic moment and
ultimately its quadrupole moment.
The targets were KBr, KCaBr$_3$, and
KCaI$_3$. A holding field H$_0$ of
6.5 kG was employed. It was found
that the β-asymmetries depended
very much on the samples, while the
spin-lattice relaxation time was
about 3 sec in all samples (see
fig. 8). The NMR results are shown
in figs. 9 and 10. In the case of
KCaI$_3$ and KCaBr$_3$ a modulated rf
signal was used to scan the resonance.
For both samples a linewidth of about
0.3 MHz(FWHM) was observed. In the
case of KCaBr$_3$ the linewidth is
broader than the modulation. This
broadening we attribute to quadrupole interaction. In the case of
KCaI$_3$ the modulation is wide enough
to saturate all the NMR lines and
the asymmetry is destroyed at the
center of the resonance. For KCaBr$_3$

Fig. 8. Polarization of ^{39}Ca ions
in KBr, KCaI$_3$, and KCaBr$_3$, as
measured by the β-decay asymmetry, as a function of time.

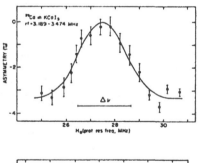

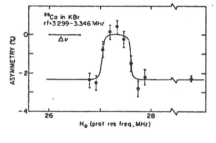

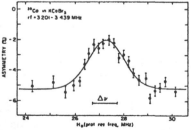

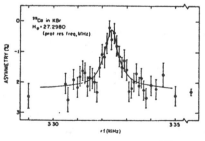

Fig. 9. NMR spectra of ^{39}Ca in
target samples of KCaI$_3$ and KCaBr$_3$.

Fig. 10. NMR spectra of ^{39}Ca in a
target sample of KBr.

TABLE 2

Results of NMR measurements on ^{39}Ca

	Target/Implantation Medium		
	KCaBr$_3$(Fused)	KCaI$_3$(Fused)	KBr(Powder)
β asymmetry[a] (%)	5.5	3.3	2.3
H_0(P.R.F., MHz)	27.286(41)	27.466(69)	27.2980(5)
rf modulation (MHz)	3.201-3.439	3.189-3.474	
rf centroid (MHz)	3.320(10)	3.332(10)	3.3239(6)
Linewidth (kHz)	$\approx 3 \times 10^2$	$\approx 3 \times 10^2$	3.0(7)
μ(nm)[b], uncorrected	1.019(4)[c]	1.016(5)[c]	1.02015(12)
Diamagnetic corr.(nm)			1.5×10^{-3}
μ(nm), corrected			1.0217(2)

a) Proton polarization = 0.65.
b) Proton moment μ_p = 2.792743 used in calculation.
c) Error includes a possible shift due to eqQ interaction.

the modulation is not sufficient to saturate the lines and complete destruction is not achieved.

For KBr, when a modulated signal was used, a resonance was obtained which clearly indicated that the natural line width was considerably smaller than the modulation width (see top of fig. 10). This was confirmed by scanning the resonance with an unmodulated signal (bottom of fig. 10) which produced an extremely sharp resonance (more than 6 times sharper than for the modulated resonance). We interpret the resonance in KBr as showing that many ^{39}Ca ions are implanted in K substitutional sites in the cubic crystal and these ions are doubly ionized in these sites. The fact that the NMR linewidth is very narrow, FWHM = 6 kHz, and the asymmetry is completely destroyed at the center of the resonance supports the assumption of equivalent stable cubic sites for the ^{39}Ca ions.

The values of the magnetic moment determined in KCaI$_3$ and KCaBr$_3$ are less precise but agree with the value determined in KBr (see table 2). The uncorrected value from the resonance in KBr is μ = 1.02015(12) nm. After applying the diamagnetic correction σ = 1.45(10)[9] the corrected magnetic-moment value is

$$\mu(^{39}\text{Ca}) = 1.0216(2) \text{ nm}.$$

The sum of the magnetic moments of mirror states is known to be independent of mesonic effects and also of configuration mixing for any central interaction[10]. Furthermore, the observed sum is usually close to the sum of the Schmidt values. This is illustrated in table 3 where sums for a number of mirror pairs are compared with the Schmidt-value sums. For ^{39}Ca and ^{39}K the observed sum is 1.41 nm, whereas

TABLE 3

Comparison of magnetic moments of mirror nuclei

Nucleus	Spin	μ(Exp)	μ(Schmidt)	Difference	
				Non-Rel.	Rel.
^{15}O	1/2	0.7189	0.638		
^{15}N	1/2	-0.2832	-0.264		
Sum		0.4357	0.374	0.062	0.075
^{17}O	5/2	-1.894	-1.913		
^{17}F	5/2	4.722	4.793		
Sum		2.829	2.880	-0.051	0.033
^{39}K	3/2	0.392	0.124		
^{39}Ca	3/2	1.022	1.148		
Sum		1.413	1.272	0.141	0.178
^{41}Ca	7/2	-1.595	-1.913		
^{41}Sc	7/2	5.43	5.793		
Sum		3.84	3.88	-0.045	0.075

the sum of the Schmidt values is 1.27 nm. The quite large deviation of +0.14 nm increases to 0.18 nm when the relativistic effect[11] on the nuclear moment is taken into account. A further interesting comparison can be made with the moment of ^{38}K. For shell model states we have:

$$\mu(^{39}Ca) + \mu(^{39}K) = \mu(^{38}K).$$

The moment of ^{38}K is 1.3735 nm. The deviation of this value from the sum 1.413 can probably be attributed[12] to a small amount of configuration mixing in ^{38}K.

2.6   OTHER APPLICATIONS OF POLARIZED NUCLEI

In this section we list some other possible applications of this method of producing polarized nuclei.

(1) Nuclear physics: (i) We have already indicated that the polarization transfer observed in these reactions can be studied in order to gain insight on the reaction mechanism. For this purpose thin targets on suitable thick backings (which do not contain the target nuclei) can be used to study the polarization transfer as a function of bombarding energy. (ii) When polarized beams of heavy ions become available these studies can be carried out on heavy-ion reactions. Also, exotic nuclei can be produced and their moments can be measured. (iii) The β-decay process itself can be studied. For example, the β-decay asymmetry can be measured as a function of β energy in order to obtain information on the existence of second class currents. (iv) The moments of gamma-emitting spin-1/2 states could be measured by detecting the circular polarization of the gamma radiation. (v) The moments of very short lived states can be measured by using detection methods other

than NMR, such as perturbed angular correlation (PAC), stroboscopic, recoil-into-vacuum, recoil-into-gas, implantation into ferromagnetic media, etc. (vi) As we have already heard at this conference, polarized nuclei can be used to study symmetry problems such as parity conservation in strong interactions.

(2) Solid state physics: (i) We have already seen that the measurements made on these polarized nuclei reveal many properties of the implantation mechanism, such as the number of substitutional sites obtained compared to interstitial sites or damaged substitutional sites. Thus, the radiation damage produced by the implanted ions can be studied. (ii) Other solid-state properties can be measured, such as the internal magnetic and electric fields produced in impurity ions in a host lattice and the spin-lattice (T_1) and spin-spin (T_2) relaxation times.

3. Polarized targets

3.1. SOLID CRYOGENIC TARGETS

As an extension of the use of polarized beams to study nuclear reactions and properties, we now consider the use of polarized targets. With the availability of both polarized beams and targets one has the freedom to perform the following types of experiments with a projectile "a" and target "A": $a + A$, $\vec{a} + A$, $a + \vec{A}$, and $\vec{a} + \vec{A}$, where the arrows designate polarization of either vector or tensor character. The following experiments with polarized targets have already been performed[13-18]:

$$n + \overrightarrow{^{59}Co}; \quad \vec{n} + \overrightarrow{^{59}Co}$$

$$n + \overrightarrow{^{165}Ho}; \quad \vec{n} + \overrightarrow{^{165}Ho}$$

$$p + \overrightarrow{^{165}Ho}; \quad \vec{p} + \overrightarrow{^{165}Ho}$$

$$\alpha + \overrightarrow{^{165}Ho}$$

These experiments were all carried out with solid targets in which the atoms are polarized at low temperatures and large magnetic hyperfine fields in the atom in turn polarize the nuclei. In the next section we give a brief description of the experiments carried out on $p + {}^{165}\overrightarrow{Ho}$ and $\alpha + {}^{165}\overrightarrow{Ho}$.

3.2. ALPHA AND PROTON SCATTERING FROM $^{165}\overrightarrow{Ho}$

The scattering of alpha particles and protons from oriented and unoriented ^{165}Ho nuclei was measured from 14 MeV to 23 MeV alpha particle energy and from 5.5 MeV to 14.5 MeV proton energy. The target was a thick single crystal of ^{165}Ho attached to the mixing chamber of a ^3He-^4He dilution refrigerator. A temperature of about 0.2°K was maintained in the target crystal; at this temperature the statistical tensor[19] describing the tensor polarization is $B_{20} = -0.35 B_{20(max)}$. Scattered alpha particles were detected at an angle of 160°; scattered protons were detected at 160° and 135°. The particle detectors were ultrapure-germanium diodes operated at a temperature of 10°K.

Differences as large as 13% between the cross sections of the oriented and the unoriented holmium were measured with alpha particles (see fig. 11). For protons, differences as large as 8% were measured (see fig. 12). The alpha-particle data were analyzed in terms of the deformed optical model with a coupled-channel formalism. Two optical model parameter sets, which had provided equally good fits to data on elastic and inelastic scattering

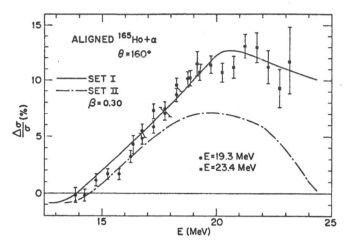

Fig. 11. Effect of alignment of ^{165}Ho on the cross section for α scattering as a function of α energy.

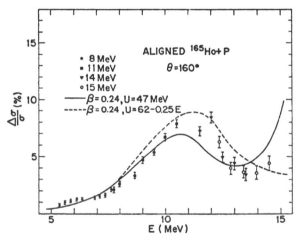

Fig. 12. Effect of alignment of ^{165}Ho on the cross section for p scattering as a function of p energy.

from rare earth nuclei, were used in the analysis of the oriented-target data. The parameter set with a real potential well depth U = 50.0 MeV and imaginary well depth W = 11.2 MeV gave a good fit to the data, while the other set gave a very poor fit. This result resolves the ambiguity in optical model parameters for scattering of low energy alpha particles from rare earth nuclei. The alpha particle data gave an estimate of the nuclear deformation parameter for ^{165}Ho of $\beta_2 = 0.288 \pm 0.012$.

The data taken at proton energies below 8 MeV were analyzed to give an estimate of the intrinsic electric quadrupole moment for ^{165}Ho of $Q_O = 8.80 \pm 0.52$ barn. Optical model parameters, derived from those used previously to fit ^{165}Ho + p elastic and inelastic scattering data, were employed in coupled-channel calculations. The best fit to the proton data was obtained with $\beta_2 = 0.203 \pm 0.008$.

3.3. POLARIZED PROTON SCATTERING FROM ^{141}Pr

More recently, thin polarized targets of ^{141}Pr have been produced and the following sequence of experiments is being carried out at Stanford:

$$p + {}^{141}\text{Pr}; \; \vec{p} + {}^{141}\text{Pr}; \; p + \overrightarrow{{}^{141}\text{Pr}}; \; \vec{p} + \overrightarrow{{}^{141}\text{Pr}}$$

In this case the goal of the experiments is to study the analog states[20,21]) in ^{142}Nd which cannot be analyzed completely unless the spin of the target is specified. Figures 13 and 14 show results obtained on $p + {}^{141}$Pr and $\vec{p} + {}^{141}$Pr. It is seen that large analyzing powers are measured on the analog resonances. Very recently, a successful measurement of $\vec{p} + {}^{141}$Pr has been carried out and a pronounced effect from the target polarization has been observed.

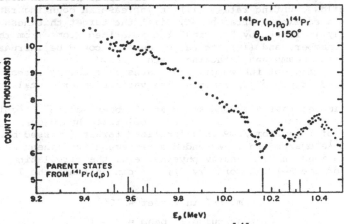

Fig. 13. Elastic scattering of protons from ^{141}Pr showing analog states in ^{142}Nd. Parent states in ^{142}Pr are indicated at the bottom.

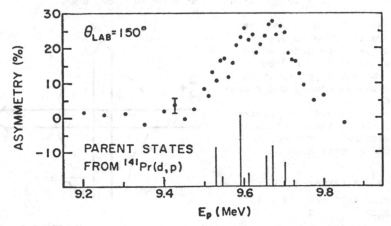

Fig. 14. Analyzing powers in the scattering of polarized protons from ^{141}Pr over the analog states in ^{142}Nd. Parent states in ^{142}Pr are indicated at the bottom.

3.4. POLARIZED BEAM TARGETS

It has been proposed[22,23]) that polarized beams, produced by existing polarized ion sources, could be used as targets in nuclear physics
experiments. With present technology polarized targets of protons, deuterons and even tritons could be produced in this way. Not only are the
hydrogen isotopes difficult to polarize by cryogenic means, but they are
involved in some of the most basic reactions of nuclear physics. Such a
gas jet of polarized atoms would have the advantage of providing a very
thin windowless target. With cryogenic targets the need to avoid excessive heating of the target restricts the beam current to very low values.
With polarized beam targets no such restriction would exist and the "thin-
ness" of the targets could be largely compensated by the use of the maximum beam currents available from existing facilities.

Several methods have been proposed for utilizing gas jet targets:
(i) a simple crossed-beam arrangement would be the simplest configuration
but would give low counting rates, (ii) an increase in counting rate could
be achieved in a co-linear arrangement, (iii) the target thickness could
be substantially increased by "storing" the polarized atoms from the jet
in a suitable chamber, and (iv) the target density could be increased by
use of circulating beams and "stacking" techniques.

In the following outline we give estimates of counting rates which
could be achieved in $\vec{p} + \vec{p}$ experiments with various experimental configurations.

It is estimated that a ground-state atomic-beam source[24]) equipped
with a compressor magnet[25]) would produce a polarized hydrogen target
with a density of 10^{12} atoms/cm^3 in a localized target (crossed beam) and
perhaps 5×10^{11}atoms/cm^3 in an extended arrangement (co-linear beam).

In applications in high-energy physics, e.g. the circulating synchrotron beam of the Fermi laboratory[26]), one can estimate the luminosity
L as follows:

$$\text{atoms/cm}^3 \text{ in target} = 10^{12}$$
$$\text{protons/pulse in beam} = 2 \times 10^{13}$$
$$\text{traversals/sec from circulation} = 5 \times 10^4$$
$$L = 10^{12} \times 2 \cdot 10^{13} \times 5 \cdot 10^4 = 10^{29} \text{ cm}^{-2}\text{sec}^{-1}$$

This luminosity should be adequate
for many investigations and, compared to cryogenic targets, would
provide background-free hydrogen
targets in which the polarization
direction could be easily and rapidly switched.

In low-energy nuclear physics
experiments the luminosity in a
simple crossed-beam arrangement
(fig. 15) can be estimated as
follows:

$$\text{atoms/cm}^3 \text{ in target} = 10^{12}$$
$$\text{beam intensity (500 nA)} = 3 \times 10^{12}$$
$$L = 10^{12} \times 3 \cdot 10^{12}$$
$$= 3 \times 10^{24} \text{ cm}^{-2}\text{sec}^{-1}.$$

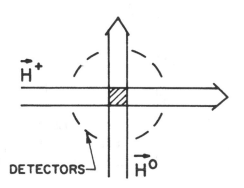

Fig. 15. Schematic diagram of a
crossed-beam arrangement for studying the $\vec{p} + \vec{p}$ reaction.

This luminosity would give

10 counts/hr/ster per mb/ster cross section.

A coaxial-beam arrangement[23] is illustrated in fig. 16. In this set-up the interacting beams could not be concentrated as much as in the crossed-beam configuration, but the extended interaction region would make possible an increase in total counting rate by a factor of 10 to 100. In order to utilize the extended region collimated detectors could be placed along the interaction region, but there would be considerably less freedom to obtain data at different angles in this arrangement.

Figure 17 shows a schematic arrangement of a target in which the polarized atoms from the gas jet are stored in a "bottle". This set-up should produce a significant increase in the target density, but there are several important depolarizing mechanisms inherent in this storage procedure which must be overcome: (i) Collisions of the polarized atom with the container walls can result in serious depolarization. This problem has been studied extensively in maser research and wall coatings have been developed which greatly reduce this mechanism of depolarization. (ii) Atomic collisions in the storage vessel which involve spin exchange can produce appreciable depolarization. This effect, as well as the wall depolarization, can be greatly reduced by using strong magnetic decoupling fields (such as those discussed in chapter 2 above).

Storage of polarized nuclei from a polarized beam can also be achieved by deposition on a solid surface. This has been demonstrated by the Hamburg group[27] in the case of ^{6}Li nuclei produced by a polarized ^{6}Li beam.

3.5. APPLICATIONS OF POLARIZED BEAM TARGETS

We list here only a few of the uses of polarized beam targets that come to mind. The successful use of polarized hydrogen targets will make possible studies of the basic reactions: $\vec{p} + \vec{p}$, $\vec{p} + \vec{d}$, $\vec{p} + \vec{t}$, $\vec{d} + \vec{d}$, $\vec{d} + \vec{t}$, $\vec{t} + \vec{t}$, $\vec{n} + \vec{p}$, $\vec{n} + \vec{d}$ and $\vec{n} + \vec{t}$. The production of beams of polarized nuclei with Z > 1 will extend these studies to all of nuclear physics. We note that already polarized beams of ^{3}He[28] and ^{6}Li[29] have been produced and used and that polarized gas targets of ^{3}He[30] and the ^{6}Li targets mentioned above[27] are available. It is only a matter of time until polarized beams and

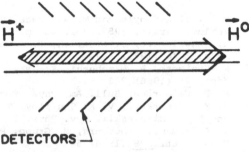

Fig. 16. Schematic diagram of a colinear-beam arrangement for studying the $\vec{p} + \vec{p}$ reaction.

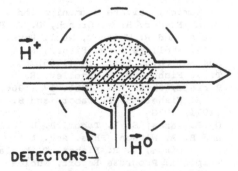

Fig. 17. Schematic diagram for storing polarized protons in a bottle for studying the $\vec{p} + \vec{p}$ reaction.

polarized-beam targets of many nuclei become available. Together with
the cryogenic targets discussed above, these developments will make it
possible to study nuclear reactions quite generally with the spins of
both projectile and target specified. Thus, many of the ambiguities
that have plagued the analysis of nuclear reactions will be removed.
We can, for example, look forward to the time when the interaction between
two heavy deformed nuclei can be studied as a function of the relative
direction of the deformations of the two nuclei.

4. Acknowledgments

It is a pleasure to acknowledge my colleagues at Stanford who have'
taken part in the research discussed in this talk. The experiments on
polarized nuclei were carried out by T. Minamisono, J. W. Hugg, D. G.
Mavis, T. K. Saylor, and H. F. Glavish. The recent experiments on
cryogenic targets are being performed by B. B. Triplett and J. Stanley.
Much of the discussion of polarized beam targets is due to H. F. Glavish
and the development of such a target is being undertaken at Stanford
by him in collaboration with D. G. Mavis and J. Dunham.

References

1) R. J. Blin-Stoyle and M. A. Grace, Handbuch der Physik (Springer
    Verlag, Berlin, 1957) Vol. 42, p. 555
2) H. Schweickert, J. Dietrich, R. Neugart, and E. W. Otten, Nucl.
    Phys. A246 (1975) 187
3) K. Sugimoto, A. Mizobuchi, K. Nakai, and K. Matuda, J. Phys. Soc.
    Japan 21 (1966) 213
4) F. P. Calaprice, Bull. Am. Phys. Soc. 19 (1974) 58
5) V. M. Sarnatskii, V. A. Shutilov, T. D. Levitskaya, B. I. Kidyarov,
    and P. L. Mitnitskii, Sov. Phys. Solid State 13 (1972) 2021
6) H. Ackerman, D. Dubbers, M. Grupp, P. Hertjans and H. -J. Stockman,
    Phys. Lett. 52B (1974) 54
7) S. Green, Phys. Rev. A4 (1971) 251
8) D. Kurath, private communication
9) F. D. Feiock and W. R. Johnson, Phys. Rev. Lett. 21 (1968) 785
10) H. A. Mavromatis and L. Zamick, Phys. Lett. 20 (1966) 171
11) H. Ohtsubo, M. Sano, and M. Morita, J. Phys. Soc. Japan 34 Suppl.
    (1973) 509
12) L. Zamick, private communication
13) T. R. Fisher, H. A. Grench, D. C. Healey, J. S. McCarthy, D. R.
    Parks, and R. Whitney, Nucl. Phys. A179 (1972) 241
14) E. G. Shelley, T. R. Fisher, R. S. Safrata, J. McCarthy, and S. M.
    Austin, Phys. Lett. 19 (1966) 684
15) T. R. Fisher, E. G. Shelley, R. S. Safrata, J. McCarthy, and R. C.
    Barrett, Phys. Rev. Lett. 17 (1966) 36
16) T. R. Fisher, S. L. Tabor, and B. A. Watson, Phys. Rev. Lett. 27
    (1971) 1078
17) D. R. Parks, S. L. Tabor, B. B. Triplett, H. T. King, T. R. Fisher,
    and B. A. Watson, Phys. Rev. Lett. 29 (1972) 1264
18) N. S. Dixon, T. R. Fisher, T. K. Saylor, J. H. Stanley, and B. B.
    Triplett, Progress Report, Nuclear Physics Laboratory, Stanford
    University, 1973, p. 126
19) D. M. Brink and G. R. Satchler, Angular Momentum, second edition
    (Clarendon Press, Oxford, 1968)

20) M. Hasinoff, G. A. Fisher, P. Kurjan, and S. S. Hanna, Nucl. Phys. A195 (1972) 78

21) J. H. Stanley, B. B. Triplett, D. C. Slater, H. F. Glavish, T. K. Saylor, and S. S. Hanna, Progress Report, Nuclear Physics Laboratory, Stanford University, 1974, p. 68

22) H. F. Glavish, private communication

23) J. H. Barker, R. M. Delaney, and J. L. Gammel, Proposal for a Low Energy Physics Facility Employing Colliding Polarized Beams and Targets, St. Louis University, St. Louis, Missouri, 1974

24) H. F. Glavish, Proc. Symp. on ion sources and formation of ion beams, Brookhaven National Laboratory, 1971, ed. Th. J. M. Sluyters (Brookhaven National Laboratory, Upton, N.Y.,1971) p. 207

25) H. F. Glavish, Fourth Int. Symp. on polarization phenomena in nuclear reactions, Zürich, Switzerland, 1975, p.844

26) D. Javonovic, private communication

27) J. Ulbricht, F. Wittchow, U. Holm, K. D. Stahl, and H. Ebinghaus, Fourth Symp. on polarization phenomena in nuclear reactions, Zürich, Switzerland, 1975, p.875

28) W. E. Burcham, J. B. A. England, R. G. Harris, O. Karban, and S. Roman, Nucl. Phys. A246 (1975) 269

29) E. Steffens, H. Ebinghaus, F. Fiedler, K. Bethge, G. Engelhardt, R. Schäfer, W. Weiss, and D. Fick, Nucl. Instr. Meth. 95 (1971) 39

30) W. R. Boydin, D. M. Hardy, and S. D. Baker, Proc. Third Int. Symp. on polarization phenomena in nuclear reactions, Madison, 1971, eds. H. H. Barschall and W. Haeberli (University of Wisconsin Press, Madison, 1971) p. 592

DISCUSSION

Schatz:

In order to explain quantitatively your measured net polarization transfers you have to know what fraction of excited nuclei is sitting at an unperturbed lattice site. Recently we have shown in Stony Brook and Erlangen that at room temperature often only a small fraction of the excited nuclei is unperturbed. This, of course, makes a quantitative interpretation very dependent from the radiation damage problems in your target.

Hanna:

That is correct of course, but we think that in special cases we might be able to make such quantitative interpretations. We think the implantation of ^{8}Li in LiIO$_3$ might be such a case. We hope there will be other cases.

Klein:

1. You showed a decoupling field of 1 kGauss From α–γ correlation attenuation experiments I remember decoupling fields of about 10 kGauss. What is more typical?
2. How many polarized reaction products would one need to generate, say, a nice NMR Signal ?

Hanna:

1. The smallness of the decoupling field is due to the nice properties of the crystal we used. I hope we can find other such nice crystals so that small decoupling fields will become typical.

2. For NMR detection by observing the destruction of an angular correlation, the production of polarized nuclei in the usual nuclear
reaction is adequate. To detect NMR in the conventional way would not
be possible with these methods.

Slobodrian:

It seemed to me that in your nuclear moments determination the diamagnetic correction was rather large. What was the accuracy of the
correction ?

Hanna:

Actually, compared to the numerical result, the diamagnetic correction is quite small and can oe calculated to sufficient accuracy so
as to contribute very little to the error.

Cramer:

There is another kind of polarized gas target which has been used for
a number of years. That is the optically pumped ^{3}He target. With the
advent of high-intensity tuned lasers, it would seem that such a
technique could be extended to a wider class of targets. You have obviously devoted some effort to the consideration of atomic beam targets. Have you also considered the alternative of optically pumped
targets ?

Hanna:

I am certainly aware of the ^{3}He targets and the possibility of using
optical pumping to produce polarized targets, but I haven't really
made a careful study of these targets, for example how general the
method would be. From the work of Otten we know that nuclear polarization by optical pumping has been achieved in the case of alkali-
like atomes. The work is very elegant but the production of polarized
targets seems difficult but perhaps not more so than the targets I
discussed.

Poppema:

I like to announce, as I also briefly communicated in the discussion
session on polarized ion sources, that an experiment along one of the
lines of future applications as indicated by S.S. Hanna is already
in progress in our laboratory of the Eindhoven University of Technology. The idea, put forward by G.J. Witteveen who also is performing
the experiment, is to cross an intense unpolarized proton beam having
good emittance with an electron-polarized atomic sodium beam in order to produce an electron-polarized atomic hydrogen beam of some
keV.
This novel scheme offers promising prospects for the production of
intense polarized hydrogen ion beams. The same method is also applicable to the production of polarized beams of heavy ions.

Hanna:

Thank you. I am very interested to hear about this clever experiment.
Similar experiments have been discussed at Stanford .

EXTENSION TO MADISON CONVENTION

B.A. Robson

The Australian National University, Canberra

During the Symposium a meeting was held to discuss the generalization of the Madison Convention to include polarization-transfer and spin-correlation coefficients. The following recommendations were made by the group for the reaction b(a,c)d.

A. Reference frames

1. The polarization of a target in the lab. frame should be referred to a right-handed coordinate system in which the positive z-axis is either (i) along or (ii) in the opposite direction to the momentum $\vec{k}_a$ of the incident particles and the positive y-axis is along $\vec{k}_a \times \vec{k}_c$.

2. For many reactions it is convenient to define <u>transversity</u> frames in which the z-axes are perpendicular to the reaction plane. In such cases it is recommended that for each beam the positive z-axis be along $\vec{k}_a \times \vec{k}_c$, the positive y-axis along the direction of momentum of the particles and the x-axis chosen to form a right-handed orthogonal frame. For the target in the lab. system, the y-axis should be chosen either (i) along or (ii) in the opposite direction to the momentum $\vec{k}_a$.

B. Polarization of initial and final systems

1. The state of spin orientation of the initial (final) assembly of particles should be denoted by the symbols

$$t_{k_a q_a k_b q_b}(t_{k_c q_c k_d q_d}) \quad \text{or} \quad p_{i_a, i_b}(p_{i_c, i_d}) \tag{1}$$

for the spherical and Cartesian representations respectively. These quantities are the expectation values of the corresponding direct products of spherical tensor operators

$$\left[\tau_{k_a q_a} \otimes \tau_{k_b q_b}\right] \quad \left(\left[\tau_{k_c q_c} \otimes \tau_{k_d q_d}\right]\right) \tag{2}$$

or Cartesian tensor operators

$$\left[\mathcal{P}_{i_a} \otimes \mathcal{P}_{i_b}\right] \quad \left(\left[\mathcal{P}_{i_c} \otimes \mathcal{P}_{i_d}\right]\right) \tag{3}$$

as defined in the Madison Convention for particles with spin $s \leq 1$. Here i_a, etc stand for $0, x_a, y_a, z_a, x_a x_a, x_a y_a, \ldots$, etc and a zero index implies a $(2s + 1) \times (2s + 1)$ unit matrix operator $\mathcal{P}_0$. The comma in the symbol p_{i_a, i_b} is used to differentiate between quantities such as

$$p_{x_a x_a, o} = \langle \mathcal{P}_{x_a x_a} \otimes \mathcal{P}_o \rangle \tag{4}$$

and

$$p_{x_a, x_a} = \langle \mathcal{P}_{x_a} \otimes \mathcal{P}_{x_a} \rangle \tag{5}$$

2. For particles of arbitrary spin s, the spherical tensor operators should be defined to have matrix elements

$$\{\tau_{kq}\}_{\alpha\beta} = (2k + 1)^{1/2} C(sks, \beta q \alpha) \tag{6}$$

where the Clebsch-Gordan coefficients are as defined by Rose (Elementary theory of angular momentum, Wiley, New York 1957). The operators then satisfy the normalization condition

Table 1. Recommended symbols for various one and two arrow experiments

Quantity	Cartesian	Spherical
analyzing power $b(\vec{a},c)d$	$Z_{i_a,o}^{o,o} = A_{i_a}$	$Z_{k_a q_a oo}^{oooo}{}^{*} = T_{k_a q_a}$
polarization $b(a,\vec{c})d$	$Z_{o,o}^{i_c,o} = P^{i_c}$	$Z_{oooo}^{k_c q_c oo}{}^{*} = T^{k_c q_c}$
polarization-transfer coefficient $b(\vec{a},\vec{c})d$	$Z_{i_a,o}^{i_c,o} = K_{i_a}^{i_c}$	$Z_{k_a q_a oo}^{k_c q_c oo}{}^{*} = T_{k_a q_a}^{k_c q_c}$
spin-correlation coefficient initial channel $\vec{b}(\vec{a},c)d$	$Z_{i_a,i_b}^{o,o} = C_{i_a,i_b}$	$Z_{k_a q_a k_b q_b}^{oooo}{}^{*} = T_{k_a q_a k_b q_b}$
spin-correlation coefficient final channel $b(a,\vec{c})\vec{d}$	$Z_{o,o}^{i_c,i_d} = C^{i_c,i_d}$	$Z_{oooo}^{k_c q_c k_d q_d}{}^{*} = T^{k_c q_c k_d q_d}$

$$\mathrm{Tr}\{\tau_{kq}\tau_{k'q'}^{\dagger}\} = (2s+1)\,\delta_{kk'}\delta_{qq'} \ . \tag{7}$$

C. Polarization observables

The initial and final polarization states are related by linear transformations

$$\sigma p'_{i_c,i_d} = \sigma_o \sum_{i_a i_b} Z_{i_a,i_b}^{i_c,i_d} \, p_{i_a,i_b} \qquad \text{(Cartesian)} \tag{8}$$

and

$$\sigma t'_{k_c q_c k_d q_d} = \sigma_o \sum_{k_a q_a k_b q_b} Z_{k_a q_a k_b q_b}^{k_c q_c k_d q_d} \, t_{k_a q_a k_b q_b} \qquad \text{(spherical)}$$

$$= \sigma_o \sum_{k_a q_a k_b q_b} T_{k_a q_a k_b q_b}^{k_c q_c k_d q_d}{}^{*} \, t_{k_a q_a k_b q_b} \tag{9}$$

where σ_o is the cross section for an unpolarized initial system. In terms of the transition matrix M, the transformation coefficients are given by

$$Z_{i_a,i_b}^{i_c,i_d} = \mathrm{Tr}\left\{M\left[\varphi_{i_a} \otimes \varphi_{i_b}\right]M^{\dagger}\left[\varphi_{i_c} \otimes \varphi_{i_d}\right]\right\}/\mathrm{Tr}\{MM^{\dagger}\} \tag{10}$$

and

$$Z{}^{k_c q_c k_d q_d}_{k_a q_a k_b q_b}{}^* = T{}^{k_c q_c k_d q_d}_{k_a q_a k_b q_b}$$

$$= \text{Tr}\left\{ M \left[\tau_{k_a q_a} \otimes \tau_{k_b q_b} \right] M^\dagger \left[\tau_{k_c q_c} \otimes \tau_{k_d q_d} \right]^\dagger \right\} / \text{Tr}\left\{ MM^\dagger \right\}$$

$$(11)$$

Special symbols which are recommended for the various one and two arrow experiments are given in table 1.

EDITORIAL:

The extensions to the Madison Convention was discussed by the interested participants in one particular session set up especially for this purpose. The discussion was based on a proposal by B.A. Robson, which was distributed to all participants at the beginning of the symposium. The above paper was presented as a report on the discussion session in a plenary session. It should be emphasized that the above convention has not been submitted to the I.U.P.A.P. in order to become an offical addition to other polarization conventions. It should also be pointed out that the notation in table 1 is partially in disagreement with the Madison Convention. Furthermore, certain reservations have been formulated during the discussion session against the above extension.

SUMMARY

H.H. Barschall
University of Wisconsin, Madison, Wisconsin, USA

When the Organizing Committee of this Symposium asked me to summarize
the conference, I was reluctant to accept the invitation, because I have
not been active in the field since I helped to prepare the Proceedings of
the last Symposium. I was told that this was an asset because I could take
a more detached view of the progress made in the interim.

This is the fourth in the series of Symposia which have been held
with five year intervals. A summary of the present Symposium is bound to
consider not only the developments since the last Symposium, but also
trends throughout the series.

In his concluding remarks at the third Symposium Professor Fleischmann characterized as the principal subject of the first Symposium the
development of sources of polarized ions, the principal subject of the
second Symposium the many new experimental results, and that of the
third Symposium the increase in the theoretical effort. He expressed the
hope that the next Symposium would connect the pieces of information to
gain better insight into nuclear structure. I will try to look at the
four Symposia and particularly at the present one, in the light of
Professor Fleischmann's remarks. I will try not only to discuss areas of
progress, but also areas in which more work needs to be done.

It is obviously impossible to summarize the content of over two
hundred contributions and of over thirty invited papers in half an hour.
My own interests will clearly influence the selection of subjects which
I will mention. I apologize to the ninety percent of contributors whose
work I will not have time to discuss. To be impartial in the omissions, I
will not mention the names of any of the contributors or speakers at the
present meeting.

It may be of some interest to compare the four Symposia statistically. Table 1 lists attendance and numbers of papers in the areas of
instrumentation, measurements, and theory. I classified all papers as experimental unless they reported no measurements. I noticed, however,
that a larger fraction of the experimental papers at the present Symposium contained extensive analysis than in the previous Symposia and that
a larger fraction of the theoretical papers were written by people who
also do experiments. It is good that experimentalists take a greater
interest in the analysis of their data, but it would not be so good if
this were to result in a decrease in the interest shown by full-time
theorists.

Table 1

SYMPOSIUM	ATTENDANCE	CONTRIBUTED PAPERS			
		EXPERIMENTAL	THEORETICAL	SOURCES	TARGETS
1	175	27	9	14	1
2	> 200	73	13	17	4
3	239	101	38	20	9
4	247	153	51	22	2

Table 1 shows that the attendance at the Symposia has remained fairly constant, but that the number of contributed papers has steadily increased, i.e., that more and more activity in polarization has occurred as more sources of polarized particles have become available. The distribution of the papers does not show any drastic trend.

There was some discussion of the need for new conventions. I do not think that there is at present any confusion comparable to that preceding the adoption of the Basel and Madison conventions to justify a new convention. On the other hand, there is a great need for a comprehensive article which explains the conventions and notations currently used by low energy nuclear physicists and the relation of these to older notations, such as those in the papers by Wolfenstein and by Welton, as well as their relation to the practices of the high energy community. I would like to urge Dr. Simonius to write such an article and to publish it in a regular journal.

Progress in physics, particularly in nuclear physics, is stimulated by the development of new instruments and new experimental methods. At the third Symposium we heard that at several laboratories 0.1 µA beams of polarized protons and deuterons were available on target. This is ample for most charged particle studies, but it is still encouraging to hear at the present Symposium of polarized beams of close to 1 µA. The new development in sources presented at this Symposium is the production of beams of polarized tritons, ^{3}He, and ^{6}Li, and we heard of the results of experiments performed with these sources.

The large currents of polarized deuterons reported at the third Symposium led to the hope that by the use of polarization transfer much improved sources of polarized neutrons would become available. As a neutron physicist I was disappointed by the small number of new results obtained with this technique although a contribution which used this method reported a neutron polarization as high as 60 %. Neutron experiments are more difficult to do than charged particle experiments, and they are therefore generally less popular both when unpolarized and when polarized neutrons are used.

Not only has there been much progress since the third Symposium in the development of new sources of polarized particles, but there has also been substantial progress in the absolute determination of the polarization of particle beams. This has come about through the recognition that under certain conditions of energy and angle the polarization of the scattered particles or reaction products reaches values which can be calculated rigorously.

Now let me turn to some of the new results which were reported. First a negative result which I was glad to learn: for two decades there have been recurring reports of measurements which are not consistent with calculations of the electromagnetic interaction of neutrons with nuclei. The latest measurements agree well with Schwinger's original predictions. Although most of us expected this, it is nevertheless a comforting finding.

Polarization phenomena in nuclear reactions are not likely to have any impact on practical applications, but I was interested to learn of at least one practical result. It is in an area in which I am especially interested, i.e., fusion technology. Polarization studies in the ^{7}Li system have led to a better knowledge of the ^{6}Li(n,α) cross section at or below the 240 keV resonance, information of importance to fusion reactors.

For the remainder of my summary I like to talk about basic nuclear physics. The most basic information pertains to symmetry properties, and one of the first things any worker in the polarization field learns, is the effect of parity and time reversal invariance on the relations between measurable polarization parameters, such as the relation between polarization and asymmetry. Since deviations from invariance are surely small in strong interactions, very high precision in the polarization measurements is needed to show any effect of such deviations. p-p scattering experiments both at low and high energies are being carried out with sufficiently high precision that one might have expected to have found some evidence of parity violation, but so far the results have been negative. A continuation and further refinement of this type of experiment will help to clarify this important area, particularly the inconsistency with Lobashov's results on circular polarization of γ-rays.

The observation of the very small effects caused by parity violation in weak interactions on nuclear reaction analyzing powers is another example of the impact of polarization experiments on the study of symmetry properties.

Isospin invariance does not have as direct a bearing on polarization phenomena as parity and time reversal, nor does it hold rigorously because of Coulomb effects and the neutron-proton mass difference. Nevertheless there are some consequences which can be observed in polarization phenomena. A disturbing result which was reported at the second Symposium was a substantial difference in the polarization in the reactions ^{2}H(d,$\vec{n}$) and ^{2}H(d,$\vec{p}$), and also in ^{3}H(d,$\vec{n}$) and ^{3}He(d,$\vec{p}$). These differences were probably experimental errors but they had not been eliminated at the third Symposium. There were no contributions on the subject at this Symposium, but there have been reports elsewhere which indicate that the remaining difference is not larger than expected from Coulomb effects. Charge symmetry also imposes symmetries on the angular distributions of analyzing powers. One has to be careful, however, in applying these arguments to cases where the incident beam is polarized. In that case the symmetry is expected to apply only in direct reactions.

Beyond the studies of symmetry properties the next most basic nuclear physics studies involve nucleon-nucleon scattering. Polarization measurements are essential for a complete description of the nucleon-nucleon interaction. For the first time good enough polarization data have been obtained for p-p scattering to permit a phase shift analysis at an energy below 25 MeV with errors in the phases of 0.4° or less. This permits the elimination of some of the model-dependent suggested phase shifts. In spite of the improvement in available sources of polarized neutrons no comparable progress was reported for n-p scattering. Not only are there more phase shifts to be determined for the n-p system than for the p-p system, but the measurements are far less accurate. For p-p scattering at ·0 MeV the maximum polarization is about 0.2 %. The only n-p measurements of polarization reported at this conference were at 14 MeV, but the data are controversial and may be interpreted as yielding a polarization of either 0 or 2 %. The only other n-p scattering experiment that was reported was a spin correlation experiment at 50 MeV. More work in n-p polarization is clearly needed.

There are methods, other than n-p scattering for studying the n-p interaction, particularly methods for finding the fraction of D-state in the deuteron. Some of these involve polarization measurements, but there are not many experimental results from which quantitative conclusions can be drawn.

Polarization measurements are important for the understanding of the three-body problem. At the time of the third Symposium there had been no calculations of these polarizations. At this Symposium we saw that calculations based on the nucleon-nucleon potential and using the Faddeev equation can reproduce the experiments qualitatively, but that there are still quantitative differences, probably caused by an inadequate knowledge of the nucleon-nucleon interaction. The calculations can readily be carried out only for neutrons. The experiments on n-d polarization have smaller relative errors than those for n-p polarization, because the n-d polarization is much larger, but the n-d measurements are much less accurate than the p-d measurements and more measurements are needed. The calculations on the n-d system are therefore compared with p-d experiments. Fortunately or unfortunately the difference between neutron and proton measurements is much smaller than the difference between experiment and calculation. There is also a need for a measurement of the polarization of the recoiling deuterons from neutron scattering; there are such measurements for proton scattering, but not for neutrons.

Although there is poor agreement between calculation and experiment for the three-body system, calculations on more complex light nuclei show surprisingly good agreement with experiments. These calculations are based on an effective nucleon-nucleon interaction in a refined cluster model.

Polarized beams are most useful for studying either the spin-orbit or the spin-spin interaction. Since the spin-orbit interaction is larger and more significant, most of the measurements and most of the theory are concerned with the spin-orbit interaction. I will discuss the spin-spin interaction first because it will take less time. At the third Symposium Dr. Postma showed a table summarizing the available data regarding a spin-spin term in the optical potential. None of the results showed definitely what the sign or the magnitude of the spin-spin term is. The situation has not improved much in the meantime. Perhaps it has become worse because it has become clear that neither the polarized-neutron polarized-target transmission experiments nor the measurements of the depolarization can easily be interpreted as due to the spin-spin interaction. Compound nucleus effects and quadrupole spin-flip effects may be responsible for most of the results which had been interpreted as due to the spin-spin interactions. Perhaps measurements on nuclei with spin 1/2 at higher bombarding energy will permit a determination of the spin-spin potential.

Now let me turn to the main subject of the conference, polarization phenomena associated with spin-orbit coupling. In describing the elastic scattering of particles the observed angular distribution is usually fitted with an optical potential which must contain a spin-orbit term if it is to yield polarizations. Either optical model parameters are deduced from the measurements or the data are compared with a potential which contains a dependence of the parameters on energy and mass number. The first approach is really only a convenient parametrization of the data, but does not give much insight into the process. The second approach has the problem that no optical potential fits all the data even if one considers only nucleon elastic scattering up to, say, 15 MeV. The popular Becchetti-Greenlees potential fits proton data better than neutron data, while some of the older potentials appear to fit the neutron data better. There is some evidence in contributions to this meeting that the Becchetti-Greenlees potential yields polarizations which differ from measurements for small angle elastic proton scattering where one might have expected a good fit. There does not appear to be much effort in trying to obtain a better optical model potential for nucleon elastic scattering in spite of the abundance of new data, especially polarization data. Nor does there appear to be much effort to get more information about the spin-orbit term in the optical potential.

There are efforts, however, to find an optical potential for describing the interaction of heavier particles. Attempts to obtain an optical potential which describes the interaction of deuterons with a wide range of nuclides over a wide range of energies have not been very successful, especially if one requires that the potential fit both vector and tensor polarization data. For this purpose a tensor term is added to the usual form of the potential.

The situation is worse for ^{3}H and ^{3}He projectiles even though one does not have to worry about tensor polarization in this case. The observed polarizations in the elastic scattering of ^{3}H and ^{3}He are unexpectedly larger, and this causes difficulties in finding optical model parameters which fit more than one nuclide at one energy. One can fit the ^{3}He polarizations at one energy for one nuclide by using a spin-orbit interaction which is sharply localized at the nuclear surface, but the physical reason for this radial dependence is not clear.

Large polarizations have also been observed in the scattering of polarized ^{6}Li by light nuclei. In this case there has been some success in fitting the data with an optical potential in which the ^{6}Li spin-orbit interaction is ascribed to the deuteron in the d-α cluster.

Polarization studies serve as a powerful tool for determining quantum numbers of nuclear energy levels. The measurement of the angular distribution of the analyzing power in transfer reactions permits a unique assignment of the total angular momentum of states of the final nucleus. This technique was discussed at the third Symposium for $(\vec{d},p)$ and $(\vec{p},d)$ reactions and has since been extended to other one-nucleon transfer reactions.

Polarization studies on inelastic scattering at isobaric analog resonances permit the determination of the major configurations in the wave functions of excited nuclear states. Polarization studies on inelastic scattering also permit the determination of collective properties of nuclear excited states.

Polarization studies have shown that nuclear reactions proceed in a more complicated way than had been hoped. In particular for transfer reactions, but also for scattering processes, there is increasing evidence that a two-step mechanism is important.

Topics which were hardly mentioned at the third Symposium and on which there were a number of contributions at this Symposium, are fluctuation effects and investigations of the giant resonance by using incident polarized particles and observing capture γ-rays.

Fluctuation effects using unpolarized particles were a popular subject for studies about ten years ago. The interpretation of the data in terms of intermediate structure led to many arguments and few useful results. I am not yet convinced that the study of fluctuations in polarization will be a more fruitful subject, except that fluctuations are undoubtedly associated with compound nucleus formation, and therefore their presence should be a warning that direct reaction theory has to be applied with caution.

Actually the use of the energy dependence of polarization for the purpose of identifying intermediate structure is not a new idea, but such measurements and their interpretation were already published ten years ago.

The study of giant resonances with polarized charged particle beams has greatly contributed to clarifying the contribution of various types of transitions to the observed structure of these resonances.

As the experiments have become more complete, such as by including a variety of polarization measurements, and as the data have become more precise, the theories which try to describe the observations have become more and more complicated. The simple-minded models no longer suffice to account for all the observations. As more and more effects are included in the description of the observed phenomena, I find it more and more difficult to have a physical understanding of the process which is being described. With the increasing complexity of the calculations I become worried about the question whether there might be errors in the codes or calculations, since few people like to spend the time and money to repeat somebody else's calculations.

In looking over the contributions I noticed many very precise data on particular energy levels or particular reactions and extensive calculations pertaining to such energy levels or reactions. I have much trouble in trying to summarize these results. Since I am no longer active in this area, I find it often difficult to grasp why the experimental or theoretical effort has been made, which is described in these contributions. Physicists working in other fields occasionally compare nuclear physicists to atomic spectroscopists who try to disentangle the line spectra of the rare earths. I would urge nuclear spectroscopists to explain more clearly the significance of their measurements or calculations for the understanding of nuclei so that physicists in other specialities can appreciate their efforts better.

I have so far carefully avoided mentioning names of contributors. I will now deviate from this practice in order to emphasize the contributions to this meeting made by Dr. Grüebler and the other members of the local committee. This has been one of the best planned and executed conferences that I have attended. Not only has the scientific program been excellent, but even the details of the participants' comfort and convenience have been provided for with exceptional care, ranging from the transportation arrangements to the delicious pastries in the dining room. The organizers of the fifth Symposium will have a hard time if they want to duplicate the quality of the program and of the arrangements. I should like to express the gratitude of all the participants to Dr. Grüebler and his colleagues.

I wish to thank Professor W. Haeberli and Dr. R. Plattner for their help in preparing this summary.

Contributed Papers

Nucleon – Nucleon Interaction

THE ANALYZING POWER IN $\vec{n}$-p SCATTERING

AT E_{lab} = 14.2 MeV AND Θ_{lab} = 45°

B.Th. Leemann, R. Casparis, M. Preiswerk,
H. Rudin, R. Wagner, P.E. Żuprański*
Institut für Physik der Universität Basel

Polarized 14.2 MeV neutrons were obtained from the ^{3}H($\vec{d}$,$\vec{n}$)^{4}He reaction induced by vector-polarized deuterons provided by an atomic beam ion source and accelerated to 140 keV[1]). Utilizing an associated particle-time-of-flight method, the coincident neutron beam defined by the solid angle of the detected associated α-particles emerged at a mean angle of 82° relative to the d⁺-beam. The use of scintillators as scatterers provides additional information which allows to reduce the background substantially. A single neutron detector placed at a fixed location has been used. The analyzing power then has been extracted from the yields obtained for opposite incident neutron polarizations. This has been performed by reversing the magnetic field direction of the deuteron source ionizer. The polarization of the incident neutrons has been determined by scattering of ^{4}He using the phaseshifts of Stammbach and Walter[2]) in order to calculate the analyzing power in n-^{4}He scattering. The average incident neutron polarization was 48 %.

During the n-p experiment, the incident polarization has been checked in short intervals by measuring the deuteron tensor polarization[1]). Five independent n-p scattering measurements were performed using scatterers of different sizes. The results are listed in the following table :

Run No.	Scatterer	$A(\Theta_{cm} = 90°)$ {%}
1,2,4	φ = 3.81 cm L = 7.62 cm	2.12 ± 0.22
3,5	φ = 2.0 cm L = 4.0 cm	0.22 ± 0.36

The results obtained using a large scatterer are in good agreement with predictions from various phaseshift sets[3,4]) as well as with other measurements performed at nearby energies[5,6]). In order to get an estimate of the effects of multiple scattering, particularly on carbon, the experiment has been repeated using a smaller scatterer. The result is spectacular in so far as it differs by more than five standard deviations from the previous one. The quoted errors are purely statistical, the systematic errors are estimated not to exceed the statistical ones. The dependence on the size of the scatterer stresses the importance of multiple scattering. The analyzing power of n-C scattering is a strong function of the scattering angle and the energy, but is not known over the whole energy region which allows kinematically an additional n-C scattering process. Therefore a multiple scattering correction calculation, based on experimental data, is not possible.

438

References

* On leave from The Instytut Badań Jądrowych, Warsaw.

1) B.Th. Leemann, R. Casparis, M. Preiswerk, H. Rudin, R. Wagner, P.E. Żuprański, to be published.
2) Th. Stammbach, R.L. Walter, Nucl.Phys. A180 (1972) 225.
3) M.H. MacGregor, R.A. Arndt, R.M. Wright, Phys.Rev. 182 (1969) 1714.
4) R.E. Seamon, R.A. Friedman, G. Breit. R.D. Haracz, J.M. Holt, A. Prakash, Phys.Rev. 165 (1968) 1579.
5) G.S. Mutchler, J.E. Simmons, Phys.Rev. C4 (1971) 67.
6) R. Garrett, A. Chisholm, D. Brown, J.C. Duder, H.N. Bürgisser, Proc. Symp. Polarization Phenomena, Madison 1970.

NEUTRON-PROTON POLARIZATION MEASUREMENTS AT 14.2 MeV*

W. Tornow, P.W. Lisowski, R.C. Byrd, S.E. Skubic and
R.L. Walter
Duke University and Triangle Universities Nuclear Laboratory
and
T.B. Clegg, University of North Carolina and TUNL

The magnitude of the polarization observed in neutron-proton scattering for E_n < 20 MeV is known to be less than a few percent and has been difficult to measure accurately. The usual experimental arrangement used to obtain n-p polarization data involves organic scintillators as scatterers; consequently, the systematic accuracy depends on the amount of multiple scattering from carbon and hydrogen in the scatterer. Although n-^{12}C scattering yields large polarization in the relevant energy region[1], the background polarization has been either assumed to be zero or, in one instance at 23.1 MeV[2], demonstrated to be small by measurements with two scatterers of different size.

Recently, Leemann et al.[3] measured n-p polarization to an extremely high accuracy. In fig. 1 the open and closed squares represent their data, obtained with 3.8 cm and 2 cm diameter scintillators, respectively. The authors concluded that the polarization measured with the large diameter scintillator was falsified by multiple scattering effects on ^{12}C, implying that the polarization in n-p scattering is very small. In order to investigate the influence of multiple scattering from ^{12}C in n-p polarization data, a new approach to understanding the ^{12}C problem was taken. In this brief report we give the results of a test for measuring n-p asymmetries using a plastic scintillator that could be surrounded by a cylindrical graphite cover. The measurement was carried out in alternate runs with and without the graphite shell.

A 270 keV thick D_2 gas cell was bombarded by an 11.5 MeV polarized deuteron beam ($p_{yy} = p_y = 0.7$). Polarized neutrons ($p_n = 0.6$) of 14.2 MeV were produced at $\theta_1 = 0°$[4]. The separation between the gas cell and scatterer was 66 cm. The plastic scintillator was 2.5 cm in diameter and 2.7 cm in height. The graphite cover, which had wall and lid thicknesses 0.22 and 0.11 cm, respectively, was designed to surround the organic scintillator with an amount of ^{12}C equal to that which it already contained. Neutrons scattered left and right through $\theta_2 = \pm45°$ (lab) were detected by two plastic scintillators which subtended 7.5°. The proton recoil spectrum in the scatterer was gated using neutron time-of-flight between the scatterer and each side detector. To minimize instrumental asymmetries, successive runs were taken with the deuteron quantization axis oriented to produce neutron beams with the spin either up or down.

The results of the present measurement are shown in fig. 1 by open and closed circles for scatterer with and without ^{12}C. The errors shown are only statistical. Also plotted are the Basel results[3] and a prediction based on the LRL X phase shifts[5]. No background corrections have yet been made, but their influence on the interpretation of the results is small. Other uncertainties and corrections are also expected to have a negligible effect. Realizing that the amount of ^{12}C present is in the ratio of 2:1 for the two arrangements, we find that the effect of the ^{12}C in the original scatterer on the extracted n-p polarization

440

value is probably much less than that observed in the earlier study of ref.[3]. Our results therefore imply that the true n-p polarization value is likely to be consistent with the LRL X prediction.

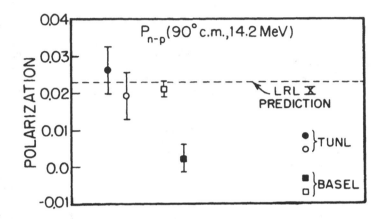

Fig. 1. Polarization in n-p scattering at 14.2 MeV, $\theta = 90°$ (c.m.) for different scattering samples.

References
*Work supported in part by U.S. Energy Research and Development Admini-
stration.

1) G. Mack, Z. Physik 212 (1968) 365
2) R.B. Perkins and J.E. Simmons, Phys. Rev. 130 (1963) 272
3) B. Leemann, R. Casparis, M. Preiswerk, H. Rudin, R. Wagner and P. Zupranski, Helv. Phys. Acta 47 (1974) 479
4) P.W. Lisowski, R.L. Walter, C.E. Busch and T.B. Clegg, Nucl. Phys. A242 (1975) 298
5) M.H. MacGregor, R.A. Arndt and R.M. Wright, Phys. Rev. 182 (1969) 1714

ANALYZING POWER OF n-p AND n-d ELASTIC SCATTERING AT E_n = 14.5 MeV

R. Berendt, H. Dobiasch, R. Fischer, V. Gerhardt, B. Haesner,
F. Kienle, H.O. Klages, R. Maschuw, R. Schrader, P. Suhr and
B. Zeitnitz
II. Institut für Experimentalphysik, Universität Hamburg

Using a polarized neutron beam arrangement [10] the analyzing power of elastic neutron-proton- and neutron-deuteron scattering was determined. Neutrons with energy E_n=14.5 MeV and polarization Q=(42.9 - 1.0)% from the reaction ^{2}H (d,n) ^{3}He were scattered by a C_6H_6- or C_6D_6-scintillator respectively. By use of a spin precessing magnetic field the asymmetry of the scattering was measured for five different angles simultaneousely. In this experiment the energies of the incoming and scattered neutrons (both determined by time-of-flight techniques) and the energy of the recoil particle was detected in coincidence. Fig.1 shows our experimental data of the n-p-analyzing power together with some recent results from other laboratories.

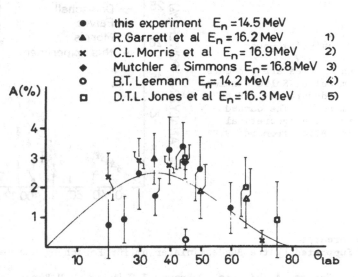

Fig. 1
n-p analyzing power
at 14.5 MeV.
The solid line is
calculated from a
single energy phase
shift analysis.

Precise data of experimentally determined observables of neutron-proton elastic scattering are necessary to perform good phase shift analyses. In the low energy region there are still some ambiguities in the $^1P_1$ nucleon-nucleon phase shift as well as for the $^3S_1$ - $^3D_1$ - mixing parameter ε_1 [6]. Together with the already known n-p-observables the polarization data were used to perform a new single energy phase shift analysis. The solid curve in fig.1 shows the calculated n-p-analyzing power at 14.5 MeV from this analysis. It was not necessary to put any constraints on the $^1P_1$-phase shift nor on the mixing parameter ε_1 which comes out positive.

In another experiment the analyzing power of the elastic neutron-deuteron scattering at $E_n=14.5$ MeV was measured using the same technique as described in the n-p-case. The results are shown in fig.2 together with the n-d-data from Morris et al.[7] at 16.8 MeV and the polarization of p-d-scattering at 14.5 MeV from Faivre et al.[8]. The solid line is a theoretical calculation from Doleschall [9] at 14.1 MeV solving the three particle integral equations.

Such calculations are now able to take into account nucleon-nucleon interaction in higher partical wave states and also tensor forces but have difficulties to include the coulomb force. Therefore neutron-deuteron polarization experiments are necessary. The comparison with p-d polarization data has to confirm whether or not there are differences between n-d- and p-d polarization observables. Compared to the data of Faivre et al. there seem to be small deviations in the minimum region about 80° c.m. which may be more prominent at higher energies where further experiments are planned. Also the maximum value of the analyzing power near 128° c.m. must be determined more accurately.

Fig. 2
n-d analyzing power
at 14.5 MeV.
Open circles are p-d-data
from ref.8. Solid circles
are n-d-data at 16.8 MeV
from ref.7. The dashed
curve is a theoretical
calculation from ref.9.

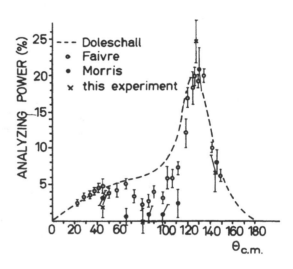

References:
* Supported by the Bundesministerium für Forschung und Technologie

1) R.Garret, A.Chisholm, D.Brown, J.C.Duder, H.N.Bürgisser
   Nucl.Phys. A 196 (1972) 421
2) C.L.Morris, T.K.O'Malley, J.W.May, Jr. and S.T.Thornton
   Phys.Rev. C 9 (1974) 924
3) G.S.Mutchler, J.E.Simmons - Phys.Rev. C 4 (1971) 67
4) B.T.Leemann Thesis 1974 Universität Basel
5) D.T.L.Jones, F.D.Brooks - Nucl.Phys. A 222 (1974) 79
6) J.Binstock, R.Bryan - Phys.Rev. D 9 (1974) 9
7) G.L.Morris, R.Rotter, W.Dean, S.T.Thornton - Phys.Rev. C9 (1974)1687
8) J.C.Faivre, D.Garetta, J.Jungermann, A.Papineau, J.Sura, A.Terrats
   Nucl.Phys. A 127 (1969) 169
9) P.Doleschall - Nucl.Phys. A 201 (1973) 264
10) R.Fischer et al. - this conference, p.835

MEASUREMENT OF THE n-p SPIN CORRELATION PARAMETER A_{yy} AT 50 MeV[*]

S.W. Johnsen , F.P. Brady, N.S.P. King, and M.W. McNaughton
Crocker Nuclear Laboratory and Department of Physics
University of California, Davis 95616
and
Peter Signell
Department of Physics
Michigan State University, East Lansing, Michigan 48824

Recent phase-shift fits to nucleon-nucleon scattering data have revealed that the phase parameter ϵ_1 is poorly determined for neutron-proton scattering near 50 MeV. Values of χ^2, for the whole data set, show a nearly flat minimum when plotted against ϵ_1, apparently corresponding to solutions near $\epsilon_1 = -8°$ and $0°$ [1]). A measurement of the observable A_{yy}, would be valuable in determining the ϵ_1 parameter[1]). The only previous measurement of A_{yy} for neutron-proton scattering has been at 23 MeV[2]).

Using the polarized neutron beam and polarized proton target at CNL, the parameter A_{yy} has been measured at 50 MeV. Fig. 1 shows the experimental layout for performing these measurements. The y axis is perpendicular to the scattering plane (the paper in fig. 1). The polarized neutron beam is produced via the reaction $T(d,n)^4He$ using 38.0 MeV deuterons incident on a cooled high pressure gas tritium target. The spin-precession magnet can reverse the beam polarization. The absolute polarization has been determined in a double reaction scattering experiment[3] to be 0.45±.015. The polarized proton target contains a hydrated lanthanum magnesium nitrate (LMN) crystal of ≈3mm thickness. The free hydrogen in the LMN is polarized dynamically to average polarizations of 40%. Recoil protons from the LMN are detected in four dE-E telescopes. Beam peak neutrons are selected by TOF.

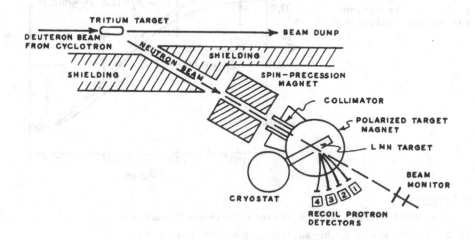

Fig. 1. Set up of scattering experiment to measure A_{yy}.

If the target and beam polarizations, p_t and p_b respectively, are measured along the positive y-axis, the intensity for scattering to the left is given by:

$$I(\theta) = I_o(\theta) * [1 + (p_t+p_b)P(\theta) + p_t p_b A_{yy}(\theta)] .$$

$P(\theta)$ is the neutron-proton polarization function and I_o is the (un-polarized) differential cross section. Values of A_{yy} were determined from the asymmetry

$$e = \frac{(\uparrow\uparrow) - (\uparrow\downarrow) + (\downarrow\downarrow) - (\downarrow\uparrow)}{(\uparrow\uparrow) + (\uparrow\downarrow) + (\downarrow\downarrow) - (\downarrow\uparrow)} = p_1 p_2 A_{yy}$$

which not only eliminates the requirement of knowing P, but also reduces systematic errors.

The measured values of A_{yy} are shown in fig. 2. The uncertainties shown were calculated from run to run dispersions. There is an additional normalization uncertainty of $\approx 25\%$ due to the uncertainty in target polarization. The curves plotted are the phase shift predictions for the cases of $\varepsilon_1 \cong -8°$, $+0.2°$ and $+ 4°$. The value $\varepsilon_1 = 0.2 \pm 1.7°$ corresponds to the minimum χ^2 for all 50 MeV n-p data. The solution for $\varepsilon_1 = -8°$ appears to be ruled out. Investigation revealed that if one datum point, $\sigma(90°)$ at 60.9 MeV (see ref. 1), is removed ε_1 for the minimum χ^2 goes to $+2.1 \pm 1.8°$ in better agreement with nucleon-nucleon interaction models.

Fig. 2. Measurements of the parameter A_{yy} compared to phase parameter predictions for several values of ε_1.

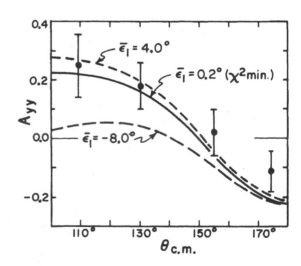

References

* Supported in part by the National Science Foundation

1) J. Binstock and R. Bryan, Phys. Rev. D9 (1974) 2528.
2) J.J. Malanify et al, Phys. Rev. Letters 17 (1966) 481.
3) A. Sagle et al, "International Conference on Few Body Problems in Nuclear and Particle Physics", Laval 1974 (to be published).

TENSOR POLARIZATION OF DEUTERONS IN p(n,$\vec{d}$)γ
AND THE D- AND D*- PROBABILITIES OF THE DEUTERON+

J.P. Svenne and S.F.J. Wilk
University of Manitoba, Winnipeg, Canada, R3T 2N2

The wavefunction of the deuteron has been known[1] since 1941 to contain a significant $3D_1$-state component. The size of this component, however, is a matter of considerable uncertainty: On the one hand, static electromagnetic moments (magnetic dipole and electric quadrupole) require a D-state probability $P_d=4\%$. On the other hand, the best phenomenological nucleon-nucleon potentials[2] predict $P_d=6.5\%$. It would be most useful in sorting out this discrepancy, if a direct determination of this quantity could be made in an independent experiment.

The tensor polarization $\langle T_{20} \rangle$ of deuterons produced by radiative capture of neutrons by protons, p(n,$\vec{d}$)γ has been shown by Czyz and Sawicki[3] to be sensitive to the D-state probability of the deuteron wavefunction. In a series of experiments proposed[4] for the University of Manitoba cyclotron, it is intended to measure this quantity, as well as other observables in the p(n,d)γ reaction. In support of this experimental programme, theoretical calculations are in progress to improve upon the old estimate of ref. (3). In particular, the existing theory must be extended to the higher energies to be used here (neutron lab. energy=20-45 MeV). The additional features to be included are higher partial waves in the NN interaction, higher multipoles in the electromagnetic field. and the possible presence of nucleon isobars[5] in the deuteron. This latter effect can be expected to have particular importance in attempting to resolve the discrepancy in P_d. According to Kisslinger and co-workers[5,6] as much as 1% of the deuteron wavefunction may contain one or two of the excited states of the nucleon. It is intended to include these components using the coupled channels deuteron wavefunctions of Jena and Kisslinger[6]. The calculations are being done in the helicity representation in accordance with the Madison convention.

+ Work supported in part by the National Research Council and A.E.C.B. of Canada.

1) W. Rarita and J. Schwinger, Phys. Rev. 59 (1941) 436.
2) e.g.: R.V. Reid, Jr., Ann. Phys. (N.Y.) 50 (1968) 411.
3) W. Czyz and J. Sawicki, Nucl. Phys. 8 (1958) 621.
4) J.S.C. McKee, University of Manitoba Internal Report #759, (1974) and J.S.C. McKee and C.O. Blyth, AEC Report NTIS/NP-18861 (1969).
5) A.K. Kerman and L.S. Kisslinger, Phys. Rev. 180 (1969) 1483.
6) S. Jena and L.S. Kisslinger, Ann. Phys. (N.Y.) 85 (1974) 251.

ON THE d→p+n VERTEX CONSTANT

I. Borbély[+] and F. Nichitiu[*]
Joint Institute for Nuclear Research
Laboratory for Theoretical Physics, 141980 Dubna

It was as long ago as 1958 that Chew proposed to continue the differential cross section as a function of the reaction angle to the nonphysical region and in this way to extract spectroscopic information[1]. The analyticity of the differential cross section follows from that of the amplitude. It is well known that the singularities of the amplitude in the $z=\cos\theta$ plane represent various direct reaction mechanisms, the simplest and well-studied one is the transfer pole[2,3]. The residue at the pole contains information on the asymptotical normalization of the wave function of the transferred particle in both nuclei.

A possible way of extracting the residue is to fit a polynomial to $(z-z_p)^2 d\sigma/d\Omega$ and simply insert $z=z_p$ into it. To increase the effectiveness of the method all these are carried out in a variable $x=x(z)$ received by the so called "optimal conformal mapping"[4,5], which takes into account some information on the location of other singularities. This method was applied for studying the structure of the lightest nuclei as well as for analysing reactions on light nuclei[6,7,8,9].

The G(d→p+n) vertex constant could be determined by several ways. First of all one can calculate it theoretically with the aid of the "realistic" nucleon-nucleon potentials, as it is proportional to the asymtotical normalization of the deuteron wave-function[3]. The Hamada-Johnston potential[10] gives $G_d^2=0.44$ fermi, while both the hard and soft core potentials of Reid[11] give 0.43 f. The analyticity of the nucleon-nucleon scattering amplitude in the energy plane provides the second possibility. The simple extrapolation by the effective range formula and the dispersion relation calculation[12] give 0.43 f. The later has an error of 3 %. And finally we have the analyticity in the $\cos\theta$ plane. The analysis of the n-d elastic scattering data was attempted by Kisslinger[7] and Dubnicka-Dumbrajs[6]. But due to their poor accuracy these data are unsuitable for extracting the correct residue: Kisslinger ambiguously included nonsignificant terms, while Dubnicka and Dumbrajs claimed to have got a correct result, but they used a wrong normalization.

We used the very accurate p-d scattering data, where the neutron exchange gives a pole. The handling of the Coulomb singularity on the edge of the physical region at forward angles presents some difficulty[8]. We used the method proposed in ref.5); in addition by a factor of $(1-\cos\theta)^2$ we removed the Rutherford pole in the differential cross section and strongly reduced the effect of the interference cut. Our results are presented in table 1.

E_p	ref.	N	$G, 10^{-3}f^2$	χ^2
3.0	13	11	155±5	1.17
6.78	14	11	172±7	0.41
16.24	15	11	156±4	1.00
19.92		11	153±7	0.76
22.0	16	11	157±6	1.37
35.0		12	183±8	1.11
46.3		11	150±7	1.20
Average			159±4	

Table 1.
Polynomial approximation results. E_p is the proton lab. energy, N is the order of the fitted polynomial, while $G=G_d^4$. The normalization errors of the exp. data are included into the errors.

As our result significantly differs from the results of other methods, we used rational function approximation too. Rational functions provide a better tool for continuation than polynomials[17]. We used the conformal mapping, but in this case it was not necessary to remove the Rutherford pole. The results, presented in table 2, agree with the polynomial approximation results.

E_p	N	M	$G, 10^{-3} f^2$	χ^2
6.78	3	3	161±2	0.51
	5	2	161±3	0.50
	5	1	155±2	0.57
Average			159±5	
46.3	4	4	139±13	1.5
	5	3	164±15	1.5
	5	4	147±19	1.5
Average			158±10	

Table 2.
Results for rational function approximations. N and M are the orders of the nominator and denominator polynomials. The errors of the averages contain the normalization errors.

So we have got a different deuteron vertex constant from that provided by the realistic nucleon-nucleon potentials. We think that the whole problem should be studied carefully, with other methods too, before jumping into conclusions. But if our result proves to be correct, then it has probably far-reaching consequences for nucleon-nucleon potentials.

References

+   On leave from Central Research Inst. for Physics, Budapest.
*   Present address: Inst. for Atomic Physics, Bucharest.

1) G.F. Chew, Phys. Rev. 112 (1958) 1380
2) I.S. Shapiro, Proc. Int. School Phys. Enrico Fermi, course 38, (Academic Press, New York) 1966, p.210
3) E.I. Dolinsky et al., Nucl. Phys. A202 (1973) 97
4) S. Ciulli, Nuovo Cimento A61 (1969) 787; A62 (1969) 301
5) R.E. Cutkosky, B.B. Deo, Phys. Rev. 174 (1968) 1859
6) S. Dubnicka, O.V. Dumbrajs, Nucl. Phys. A235 (1974) 417
7) L.S. Kisslinger, Phys. Rev. Lett. 29 (1972) 505
8) L.S. Kisslinger, Phys.Lett. 47B (1973) 93
9) I. Borbély, Nuovo Cim. Lett.; Dubna preprint E4-8445(1974)
10) T. Hamada, I.D. Johnston Nucl. Phys. 34 (1962) 382
11) R.V. Reid, Ann. of Phys. 50 (1968) 411
12) M.P. Locher, Nucl. Phys. B23 (1970) 116
13) D.C. Kocher, T.B. Clegg, Nucl. Phys. A132 (1969) 455
14) R. Grotzschel et al., Nucl. Phys. A174 (1971) 301
15) T.A. Cahill, J. Greenwood, Phys. Rev. C4 (1971) 1499
16) S.N. Bunker et al., Nucl. Phys. A113 (1968) 461
17) I. Borbély, F. Nichitiu, Dubna preprint

POLARIZATION IN PROTON-PROTON SCATTERING AT 10 MeV*

J.D.Hutton, W. Haeberli and L.D.Knutson
University of Wisconsin, Madison, Wisconsin 53706
P. Signell
Michigan State University, East Lansing, Michigan 48824

The polarization in proton-proton scattering at an incident energy of 10.0±0.05 MeV has been measured at seven angles with an accuracy of ±0.02%. A gaseous hydrogen target (1-2 atm) was bombarded with polarized protons. The sign of the beam polarization was reversed with a spin-precession solenoid between the ion source and the accelerator. The beam polarization (75-80%) was monitored continuously with a p-α polarimeter which was calibrated in terms of a precision p-α measurement of Ohlsen, et.al.[1].

The left-right asymmetry of scattered protons was observed with two surface barrier detectors placed symmetrically to the right and the left of the incident beam. The pulse height spectra showed a flat unstructured background on the low-energy side of the p-p peak and small peaks corresponding to elastic proton scattering by deuterium and contaminants (C,N,O). The peak to background ratio was 10^3 and the asymmetry of the background was <0.5%. The level of impurities (<0.01%) was monitored by observing protons scattered at 60°. At θ_{lab}=10°, where the elastic scattering from the contaminants was not resolved from p-p scattering, a correction of (0.005±0.002)% was applied, based on the values of the analyzing powers of the contaminants.

The beam was collimated by a 1 to 3 mm wide slit 2.6 m from the target and a 0.6 mm wide slit 0.15 m from the center of the target. The beam was kept centered on both sets of slits by a feedback system. To investigate effects of beam shifts, the source polarization was set to zero and the asymmetry from reversal of the precession solenoid was measured. The mean of 11 asymmetry measurements made mostly at forward angles was (0.002±0.009)%.

For each scattering angle, the polarization measurements were divided into a number of separate runs consisting of about 10^6 counts in each detector for each spin orientation. The scatter of the measurements about their mean for each angle was consistent with the statistical error. However, an analysis of the scatter for the entire data set showed that an additional random error of 0.025% was present in the individual measurements. The error bars in fig.1 were calculated by including this error in quadrature with the statistical error.

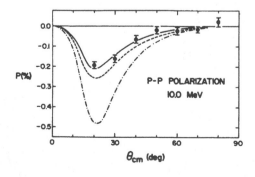

Fig.1. Proton-proton polarization versus c.m. scattering angle at E_p=10.0 MeV. The curves are explained in the text.

The p-wave phase shifts for low-energy p-p scattering have traditionally been obtained by extrapolation from higher energies using phenomenological representations. The 10 MeV polarizations predicted from
two "energy dependent" analyses are compared to the present measurements
in fig.1. The dashed line is based on the phase shifts of MacGregor,et.
al.[2]). The prediction from the Yale phase shifts[3]) is essentially identical. In the recent 1-27.6 MeV analysis of Arndt,et.al.[4]) the phase
shifts depend critically on the decision whether the absolute cross section scale of the measurements should be "floated"(i.e. treated as an
adjustable parameter) or not. The prediction from the unfloated 1-27.6
MeV analysis (dot-dash curve in fig.1) is entirely inconsistent with the
measurements, while the results of the floated analyses (not shown) are
in reasonable agreement with the data.

The present measurements permit for the first time a model independent determination of the phase shifts at an energy below 25 MeV. The
analysis made use of the 9.918 MeV cross section data of Jarmie,et.al.[5])
and employed the method described in ref.[6]). The resulting phase shifts
and uncertainties are:

$$^1S_0: 55.38°\pm0.15° \qquad\qquad ^3P_0: 2.62°\pm0.40°$$

$$^3P_1: -1.94°\pm0.10° \qquad\qquad ^3P_2: 0.64°\pm0.09°$$

The calculated polarization is shown as the solid curve in fig.1. The
tensor p-wave phase shift combination Δ_T[6]) is found to be $-0.812°\pm0.055°$
which is consistent with $\Delta_T = -0.91°\pm0.28°$ at 9.69 MeV which Noyes and
Lipinski[7]) deduced on the basis of a single A_{yy}/A_{xx} spin correlation
measurement.

When the analysis was repeated with the cross section normalization
floated, the quality of the fit was unchanged (normalized χ^2 changed
from 0.93 to 0.94). Thus a single energy analysis of the new 10 MeV
data set shows no preference for floating the 9.918 MeV cross section
normalization[5]).

References

*Work supported in part by the U.S.Atomic Energy Commission
1) G.G. Ohlsen, J.L. McKibben, G.P. Lawrence, P.W. Keaton and D.D.
   Armstrong, Phys. Rev. Lett. 27 (1971) 599
2) M.H. MacGregor, R.A. Arndt and R.M. Wright, Phys. Rev. 182 (1969)
   1714
3) R.E. Seamon, K.A. Friedman, G. Breit, R.D. Haracz, J.M. Holt and
   A. Prakash, Phys. Rev. 165 (1968) 1579
4) R.A. Arndt, R.H. Hackman and L.D. Roper, Phys. Rev. C9 (1974) 555
   The caption for fig.3 is incorrect. The floated curve is the upper
   one for fig.3a and the lower one for figs.3b and 3c. The numbers on
   the vertical scale in figs.2b and 3b should be negative and the values in fig.3b should be integer multiples of 0.2 (R.A. Arndt, private
   communication).
5) N. Jarmie, J.L. Jett, J.L. Detch and R.L. Hutson, Phys. Rev. Lett.
   25 (1970) 34
6) M.S. Sher, P. Signell and L. Heller, Ann. Phys. (N.Y.) 58 (1970) 1
7) H.P. Noyes and H.M. Lipinski, Phys. Rev. 162 (1967) 884

PROTON-PROTON ANALYZING POWER MEASUREMENTS AT 16 MeV.[*]

P.A. Lovoi, G.G. Ohlsen, Nelson Jarmie, C.E. Moss, and D.M. Stupin.

Los Alamos Scientific Laboratory, Los Alamos, New Mexico, 87545

Proton-proton elastic scattering analyzing powers are very small at low energies because of the nature of the P-wave forces. Only recently has this observable been measured with significant precision: by Hutton, Haeberli and Knutson at Wisconsin[1] at 10 MeV; and by our group at 16 MeV[2,3]. Both groups attained accuracies of ±0.0002.

Our measurements were carried out in a precision scattering chamber, the "Supercube"[3]. A left-right detector configuration (not in coincidence) was used together with spin-up and spin-down runs to reduce systematic errors. The target was a 2.5 μm Havar-foil gas cell 9.7 cm in diameter, filled with 350 Torr of hydrogen gas. Both normal spin reversal and a rapid reversal up to 1000 Hz were used. The results of both methods agreed.

Various tests to detect or eliminate the possibility of systematic errors were done. Two auxiliary measurements were the most important. Many runs were taken (at 15° lab) with the beam polarization in the plane of the scattering giving an asymmetry of -0.00016 ±0.00020. Since the result is consistent with a null measurement, no correction for false asymmetry was made of the data. The second auxiliary measurement used a target of helium to study p-^{4}He scattering to verify the beam polarization and the ability of the system to measure an asymmetry. At 112° (lab) and 16 MeV, an R-Matrix analysis[4] predicts an analyzing power of 0.9911. The measured value was 0.9936 ±0.0040.

Our results are given in table I and figure 1. The errors given are dominated by the statistical error. The beam energy was 16 MeV ±15 keV. The solid line on the figure is the LRL-X phase shift prediction[5] which is in relatively good agreement. The Wisconsin results at 10 MeV were somewhat smaller than the LRL prediction[5]. The significance of the analyzing power measurements, in particular their effect on the tensor and spin-orbit terms of the nuclear force, await detailed analysis.

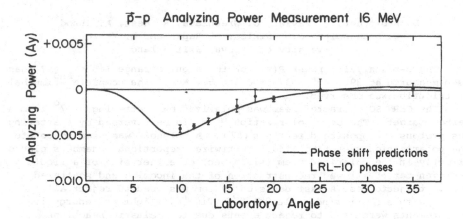

Figure 1. p-p analyzing power measurement at 16 MeV.

Table 1

θ(lab) deg.	θ(cm) deg.	analyzing power
10.00	20.08	-0.0043 ±0.0004
11.50	23.09	-0.0041 ±0.0002
13.00	26.10	-0.0035 ±0.0003
15.00	30.11	-0.0027 ±0.0002
16.00	32.12	-0.0018 ±0.0007
18.00	36.14	-0.0007 ±0.0008
20.00	40.15	-0.0010 ±0.0003
25.00	50.18	-0.0000 ±0.0012
35.00	70.22	+0.0001 ±0.0007

References

* Work done under the auspices of the U.S. ERDA

1) J.D. Hutton, W. Haeberli, and L.D. Knutson, Bull. Am Phys. Soc. 20, 576, (1975)
2) P.A. Lovoi, N. Jarmie, G.G. Ohlsen, C.E. Moss, Bull. Am. Phys. Soc. 20, 85 (1975)
3) P.A. Lovoi, "proton-proton analyzing power measurements at 16 MeV (unpublished) Ph.D. dissertation, U. of New Mexico (1975)
4) G. Hale, Los Alamos, private communication.
5) M.H. MacGregor, R.A. Arndt, and R.M. Wright Phys. Rev. 182, 1714 (1969)

POLARIZATION MEASUREMENTS IN P-P ELASTIC SCATTERING FROM 400 TO 580 MeV[*]

D. Aebischer, B. Favier, G. Greeniaus, R. Hess, A. Junod[+]
C. Lechanoine, J.C. Niklès, D. Rapin and D. Werren
University of Geneva, Switzerland

The p-p analyzing power $P(\theta)$ for the angular range $1.5^\circ < \theta_{LAB} < 7^\circ$ has been measured at 398, 455, 497, 530 and 572 MeV. The complete azimuthal distribution was observed.

The CERN SC extracted beam was polarized by scattering at 7° from a carbon target. The beam polarization P_o could be reversed by scattering the protons in opposite directions $(\pm 7^\circ)$. A solenoid was used to rotate the polarization vector by $\sim \pm 40^\circ$. Multiwire proportional chambers placed directly in the polarized beam $(< 10^5 p/sec)$ on either side of a liquid hydrogen target allowed reconstruction of the incoming and scattered proton trajectories. A fast decision system was used to reject all events with a scattering angle $\theta < 1.5^\circ$. Time of flight and energy loss measurements were used to reject events due to inelastic reactions. A complete description of the beam and experimental apparatus can be found in Ref.1 where data for the analyzing power of carbon are given.

Scattering of a polarized beam on an unpolarized target produces an azimuthal distribution of the form

$$N(\theta,\phi) = \left[1 + |\vec{P}_o| \, P(\theta) \, \cos\phi\right] N(\theta)/2\,\pi$$

where $N(\theta)$ is the number of particles scattered at angle θ, and ϕ the azimuthal angle between $\vec{P}_o$ and the normal to the scattering plane.

The procedure used to determine the beam polarization is described in Ref.1. The beam polarizations were $P_o = .376 \pm .015$ at 398, 455 and 572 MeV, $P_o = .432 \pm .025$ at 497 MeV and $P_o = .314 \pm .030$ at 530 MeV. These three values are not completely independent.

The p-p analyzing power was determined using the statistics

$$S(\theta) = \Sigma \sin \phi_i = 0 \qquad C(\theta) = \Sigma \cos \phi_i = N(\theta) \cdot P_o \cdot P(\theta)/2$$

This also provided a check for systematic errors in the shape of the azimuthal angle distribution. Non-overlapping bins in θ (1° wide) were used. The data are shown in fig.1. The errors are purely statistical.

There are three main sources of systematic errors: a) Uncertainties in P_o can be considered as a scale error common to all data points at a given energy. b) Our data show a small positive asymmetry for the contamination from 3 body inelastic reactions. The maximum effect on $P(\theta)$ is estimated by assuming they have 0 asymmetry. The relative errors $\Delta P/P$ are $\leqslant 2\%$ at 400 MeV and $\leqslant 6\%$ at 580 MeV. c) Errors in the reconstruction of the scattering angle are less than $.01^\circ$ and have a negligible effect on $P(\theta)$ after averaging over the 6 different spin directions.

These results represent the first polarization measurements at very small angles in this energy range. In the region where coulomb scattering is dominant, the analyzing power decreases rapidly.

References

* Supported by Swiss National Funds, CERN MSC Division, CICP and SIN
+ Swiss Federal Institute of Technology, Zurich
1) D. Aebischer et al., Nucl. Instr. and Meth. 124 (1975) 49.

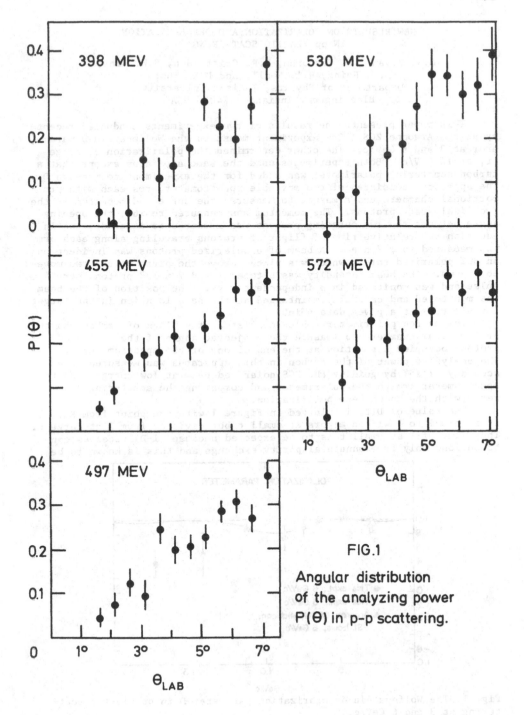

FIG.1

Angular distribution
of the analyzing power
P(Θ) in p-p scattering.

NEW RESULTS ON POLARIZATION AND DEPOLARIZATION
IN pp ELASTIC SCATTERING*

G.W. Bryant, M. Corcoran, R.R. Crittenden, S.W. Gray,
R.M. Heinz, H.A. Neal[+], and D.R. Rust
Department of Physics, Indiana University
Bloomington, Indiana, 47401 USA

This paper presents the results of two experiments conducted recently at the Argonne ZGS. One experiment measured the Wolfenstein D parameter at 3 and 6 GeV/c. The other determined the polarization at large
|t| at 12 GeV/c. Both experiments used the same apparatus except that a
carbon scattering polarimeter was added for the experiment to measure D.
The apparatus consisted of two moveable spectrometer arms each with proportional chambers and a magnet to measured the angle and momentum of the
two final state protons. The momentum was measured to ±6%, the opening
angle between the two outgoing protons was measured to ±4 mrad., and in
addition the relative time of flight of protons traveling along each arm
was measured to ±.7 nsec. A beam of unpolarized protons was incident on
an 80% polarized target which was placed above the pivot of the two moveable arms. The beam intensity was between 1 and 7×10^9 protons per
pulse and was monitored in 3 independent ways. The position of the beam
was monitored and carefully maintained at the same location in the target
for all runs at a given data point.

The carbon polarimeter, which consisted of a block of carbon and two
proportional chambers to measure the projected angle of the scattered
proton, occupied a position at the end of one of the spectrometer arms.
The analyzing power of the carbon in this apparatus was measured to an
accuracy of ±5% by guiding the ZGS polarized beam at low energy down the
spectrometer arm to the polarimeter and comparing the scattering asymmetry with the known beam polarization.

The value of D(t) is plotted in Figure 1 with one point from Ref. 2.
It is equal to 1 within errors at small t but deviates from 1 at large t.
The value of 1 at small t is to be expected because (1-D) receives con-
tributions only from unnatural parity exchange and this is known to be

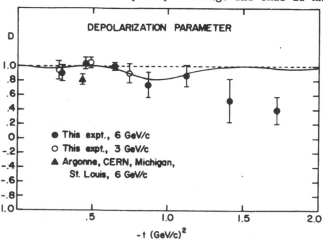

Fig. 1. The Wolfenstein depolarization parameter D in pp elastic scattering at 3 and 6 GeV/c.

small in PP scattering.

The polarization is shown in Figure 2 with 10 GeV/c data from Ref. 3. There appears to be good evidence for a third maximum in polarization around t = -2.8 (GeV/c)^2.

The Chu-Hendry optical model[1] is used to obtain the curves shown in the figures. The differential cross-section, polarization, and D are fit simultaneously to obtain these curves.

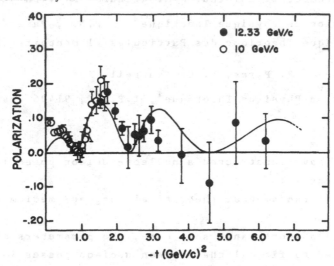

Fig. 2. The polarization in pp elastic scattering at large |t|. Solid circles from this experiment at 12.33 GeV/c. Open circles at 10 GeV/c from Ref. 3.

References

* Work supported by the U.S. Energy Research and Development Administration under Contract No. AT(11-1)-2009, Task A, and by the Alfred P. Sloan Foundation.
+ Alfred P. Sloan Foundation Fellow

1) S.-Y. Chu and A.W. Hendry, Phys. Rev. D6, 190 (1972); T.Y. Cheng, S.-Y. Chu and A.W. Hendry, ibid. 7, 86 (1973).
2) R.C. Fernow, S.W. Gray, A.D. Krisch, H.E. Miettinen, J.B. Roberts, K.M. Terwilliger, W. DeBoer, E.F. Parker, L.G. Ratner and J.R. O'Fallon, Phys. Lett. 52B, 243 (1974).
3) M. Borghini, L. Dick, J.C. Olivier, H. Aoi, D. Cronenberger, G. Gregoire, Z. Janout, K. Kuroda, A. Michalowicz, M. Poulet, D. Sillou, G. Bellettini, P.L. Braccini, T. Del Prete, L. Foa, P. Laurelli, G. Sanguinetti and M. Valdata, Phys. Lett. 36B, 501 (1971).

NEW SEMIPHENOMENOLOGICAL SOFT CORE AND VELOCITY DEPENDENT
NUCLEON-NUCLEON POTENTIAL

M. Lacombe, B. Loiseau, J-M. Richard, R. Vinh Mau

Division de Physique Théorique[*], I.P.N. , Paris[**]
and Physique Théorique des Particules Elémentaires,Paris[**]

P. Pires, R. de Tourreil

Division de Physique Théorique[*], I.P.N. , 91406 Orsay

We have constructed a nucleon-nucleon potential[1] which possesses :

    i) a fundamental theoretical long and medium range part.

    ii) a short range soft core, the parameters of which are adjusted to fit all the nucleon nucleon phases ($J \leqslant 6$) up to 350 MeV as well as the deuteron parameters.

    iii) an energy dependence which, to a good approximation, is linear. This simple energy dependence can be, in turn, easily transformed into a velocity dependence.

    Although the number of free parameters is small (six in each isospin state) the quality of the fit is very good (X^2/data = 2.5 for proton-proton scattering and X^2/data = 3.7 for neutron-proton scattering). The fits are shown in figure I.

---

* Laboratoire associé au C.N.R.S.
** Postal address : Tour 16, 1er étage - 4, Place Jussieu
          75230 Paris Cedex 05   -   France

Calculations of nuclear structure parameters (namely binding and saturation properties of nuclear matter) are in progress.

References

1) This report is a summary of the following works :
   W.N. Cottingham, M. Lacombe, B. Loiseau, J-M. Richard and R. Vinh Mau, Phys. Rev. D8 (1973) 800.
   M. Lacombe, B. Loiseau, J-M. Richard, R. Vinh Mau, P. Pires, and R. de Tourreil, preprint IPNO/TH 75-09, to be published in Phys. Rev. D.

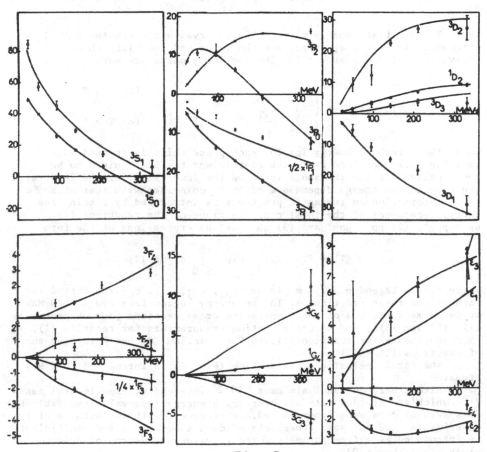

Fig. I

LOCALLY ENERGY DEPENDENT PHASE SHIFT ANALYSIS OF PROTON-PROTON

SCATTERINC IN THE 20 - 750 MeV REGION

J. Bystricky and F. Lehar

Département de Physique des Particules Elémentaires

CEN - Saclay, France

Our approach to phase shift analysis is characterized as being intermediate between a conventional analysis at a fixed energy, and an analysis with energy dependence of the phase shifts through the entire interval [1,2]. We use the usual formalism [3] together with corrections due to the electromagnetic effects [4]. Our method consists in approximating the real part of each phase shift, within a given interval of incident particle energies T, by a cubic function

$$\text{Re } \delta_\ell(T) = \sum_{i=0}^{3} \frac{a_{\ell i}}{i!} (T - T_o)^i$$

where T_o is a fixed energy within this interval and where the fitted parameters a_{ℓ_o}, ..., $a_{\ell 3}$ represent the value and the first three derivatives of $\delta_\ell(T)$ at $T = T_o$. The imaginary parts are parametrized by

$$\text{Im } \delta_\ell(T) = \begin{cases} 0 & \text{for } T \leqslant T_o \\ \sum_{i=2}^{3} \frac{b_{\ell i}}{i!} (T - T_{o\ell})^i & \text{for } T > T_o \end{cases},$$

where the threshold energy $T_{o\ell}$ for each phase shift is treated as a variable parameter, and where the coefficient $b_{\ell 2}$ is required to be non-negative. If the threshold is below the lower limit of the interval, the form of the energy dependence of Im δ_ℓ coincides with that of Re δ_ℓ.

Information on inelastic processes is introduced by fitting the energy dependence of the total cross sections of the reactions (1) pp → ppπ°, (2) pp → pnπ⁺ and (3) pp → dπ⁺ by expressions of the form

$$\sigma_j(T) = (T - T_{oj})^2 \exp \sum_{i=0}^{n} c_{ij} P_i(T)$$

where P_i are Legendre polynomials and T_{oj}, c_{oj}, ..., c_{nj} are fitted for each of the three reactions j. In the energy region from 290 to 970 MeV we use 8 measurements of total inelastic cross sections [sum of processes (1), (2) and (3)], and 44 cross section measurements for reaction (1), 19 measurements for reaction (2), 60 for reaction (3) and 6 for the sum of reactions (2) + (3).

The total inelastic cross section (Fig. 1) is introduced in 5 MeV steps.

Altogether, our analysis uses 2852 independent experimental data [5,6] which are divided into six overlaping energy intervals (see Table 1). Some sets of data were renormalized, in agreement with the authors of the experiments, and for some other sets of data a normalization coefficient is introduced as a free parameter. The χ^2 value per degree of freedom is always less than 1.03.

Fig. 2 shows the three contributions σ_0, σ_1 and σ_2 to the σ_T calculated from the imaginary parts of the three amplitudes which do not vanish in the forward direction.

The cross section σ_0 is measured in an unpolarized transmission experiment, σ_1 can be measured with beam and target polarized perpendicular to the incident momentum but paralel to each other, and the sum of σ_1 and σ_2 can be obtained from measurements with a longitudinally polarized beam and target.

The overall total cross section σ_T is given by

$$\sigma_T = \sigma_0 + \sigma_1 \ (\vec{P}_B \cdot \vec{P}_T) + \sigma_2 \ (\vec{P}_B \cdot \vec{k})(\vec{P}_T \cdot \vec{k}) \quad ,$$

where $\vec{P}_B$ and $\vec{P}_T$ are the polarizations of the beam and the target respectively and where $\vec{k}$ is the direction of the beam [7]. Neither σ_1 nor σ_2 have so far been measured in our energy range.

The ratio of the real to the imaginary part of the spin independent forward amplitude strongly depends on the precision of differential cross section measurements at small angles in particular the results of paper [6]. On Fig. 3 the ratio obtained in our analysis is compared with dispersion relation predictions [8], showing excellent agreement up to 600 MeV.

Figs. 4 to 6 represent the energy dependence of some of the phase shifts. In the interval 220 to 300 MeV there exist only data on differential cross sections and polarizations and very few measurements of total cross sections σ_0. Consequently the phases are not well determined in this region. Almost all phase shifts show a structure between 350 and 400 MeV and between 650 and 700 MeV independently of the position and the width of the energy intervals used to introduce local energy dependence.

In view of the number and the quality of the data in the region from 350 to 400 MeV the structure observed in this region can be considered as being well established. The structure around 650 to 700 MeV had been observed previously in an energy independent analysis [9]. We confirm this result which may however be effected by the limited number of measurements above 680 MeV.

Energy Interval	Measurements	Measurements in Overlaping Regions		Parameters	χ^2
19 – 156	816	–	499	67	766.83
95 – 270	698	499	144	61	583.97
170 – 350	423	144	241	56	355.18
270 – 460	716	241	462	78	636.50
380 – 610	959	462	264	90	835.65
560 – 750	850	264	–	92	719.26

- Table I -

References

1) M.M. Mac Gregor, R.A. Arndt, R.M. Wright, Phys. Rev. $\underline{169}$ (1967) 1128
   Phys. Rev. $\underline{182}$ (1969) 1714

2) R.A. Arndt, R.H. Hackman, L.D. Roper, Phys. Rev. $\underline{C9}$ (1974) 555

3) P. Cziffra, M.M. Mac Gregor, M.J. Moravcsik, H.P. Stapp,
   Phys. Rev. $\underline{114}$ (1959) 880

4) G. Breit, R.D. Haracz, High Energy Physics (E.H.S. Burlop, Academic
   Press Inc., New York, 1967) Vol. 1, p. 21

5) J. Bystricky, F. Lehar, Z. Janout, Note CEA-N-1547(E) (1972),
   An updated edition will be published

6) D. Aebischer, B. Favier, G. Greeniaus, R. Hess, A. Junod, C. Lechanoine,
   J.C. Nikles, D. Rapin, C. Serre, D. Werren, To be presented on this
   Conference, p.452

7) S.M. Bilenkii, L.I. Lapidus, R.M. Ryndin, Usp. Fiz. Nauk $\underline{84}$ (1964) 243,
   Soviet Physics Uspekhi $\underline{7}$ (1965) 721

8) V.S. Barashenkov, V.D. Toneev, Preprint JINR P2-3850, Dubna 1968

9) G. Cozzika, Thesis, Note CEA-N-1720 (1974)
   J. Bystricky and F. Lehar, International Conference on Few Body Problems
   in Nuclear and Particle Physics, Quebec, August 27-31, 1974.

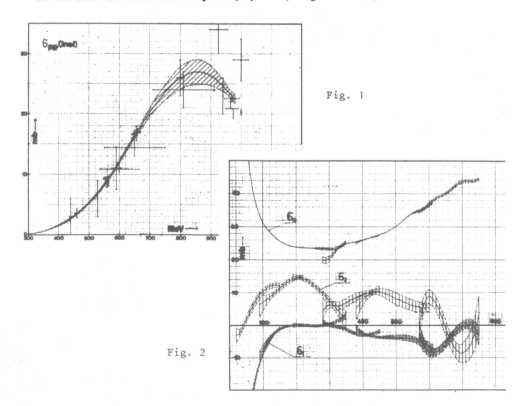

Fig. 1

Fig. 2

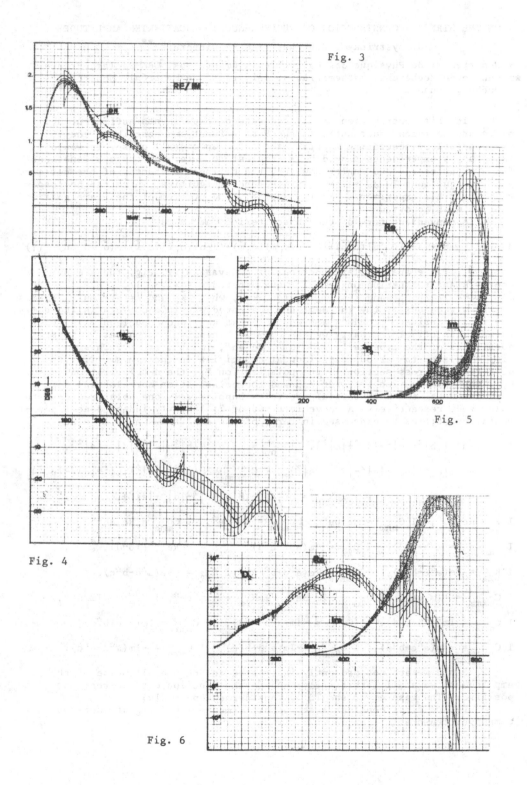

Fig. 3

Fig. 5

Fig. 4

Fig. 6

ON THE DIRECT RECONSTRUCTION OF NUCLEON-NUCLEON SCATTERING AMPLITUDES

J. Bystricky[*], F. Lehar[*] and P. Winternitz[**]

[*] Département de Physique des Particules Elémentaires, CEN-Saclay, France.
[**] Centre de Recherches Mathématiques, Université de Montréal, Montréal, Québec, Canada.

In this contribution we discuss sets of experiments sufficient and necessary to reconstruct uniquely the 9 real functions describing the nucleon-nucleon scattering matrix, except for an overall phase. It would seem that 9 experiments should be sufficient to determine these 9 functions for a given energy and angle. However, the bilinear character of the corresponding equations leads to certain discrete ambiguities, which can be resolved by further experiments. It is precisely the problem of these ambiguities that we wish to discuss.

A convenient form of the NN scattering matrix is

$$M = \tfrac{1}{2} \{(a+b)+(a-b)(\vec{\sigma}_1 \vec{n})(\vec{\sigma}_2 \vec{n})+(c+d)(\vec{\sigma}_1 \vec{m})(\vec{\sigma}_2 \vec{m})+(c-d)(\vec{\sigma}_1 \vec{\ell})(\vec{\sigma}_2 \vec{\ell})+e(\vec{\sigma}_1+\vec{\sigma}_2,\vec{n})\} \quad (1)$$

where $a, \ldots, e$ are complex functions of two variables, e.g. the c.m.s. energy k and scattering angle θ. We have denoted $\vec{n} = (\vec{k}_i \times \vec{k}_f)/|\vec{k}_i \ \vec{k}_f|$, $\vec{m} = (\vec{k}_f-\vec{k}_i)/|\vec{k}_f-\vec{k}_i|$ and $\vec{\ell} = (\vec{k}_i+\vec{k}_f)/|\vec{k}_i+\vec{k}_f|$, where $\vec{k}_i$ and $\vec{k}_f$ are unit vectors in the directions of the incident and scattered particle momenta in the c.m.s. The Pauli matrices $\vec{\sigma}_1$ and $\vec{\sigma}_2$ act on the first and second nucleon wave functions, respectively. Clearly we can form 25 linearly independent bilinear quantities of the type $|a|^2, \ldots |e|^2$, Re ab^*, Im ab^*, etc ... These are simply related to various experimental quantities in the c.m.s., i.e. the differential cross-section I, the polarization P_i, the depolarization tensor D_{ik}, the polarization transfer K_{ik} and the polarization correlations C_{ik}, $C_{ik\ell}$ and $C_{ik\ell m}$ for both initial particles unpolarized, one polarized or both polarized, respectively. A convenient set of 25 linearly independent experimental quantities is expressed in terms of the amplitudes $a, \ldots, e$ as

$$I = \tfrac{1}{2}\{|a|^2+|b|^2+|c|^2+|d|^2+|e|^2\}, \quad I\,C_{nn} = \tfrac{1}{2}\{|a|^2-|b|^2-|c|^2+|d|^2+|e|^2\},$$

$$I\,D_{nn} = \tfrac{1}{2}\{|a|^2+|b|^2-|c|^2-|d|^2+|e|^2\}, \quad I\,C_{mmmm} = \tfrac{1}{2}\{|a|^2+|b|^2+|c|^2+|d|^2-|e|^2\},$$

$$I\,K_{nn} = \tfrac{1}{2}\{|a|^2-|b|^2+|c|^2-|d|^2+|e|^2\},$$

$$I\,P = \text{Re } ae^*, \quad I\,C_{mnm} = \text{Re } be^*, \quad I\,C_{nmm} = \text{Re } ce^*, \quad I\,C_{mmn} = \text{Re } de^*,$$

$$I\,D_{mm} = \text{Re}(a^*b+c^*d), \quad I\,D_{\ell\ell} = \text{Re}(a^*b-c^*d), \quad I\,K_{mm} = \text{Re}(a^*c+b^*d),$$

$$I\,K_{\ell\ell} = \text{Re}(a^*c-b^*d), \quad I\,C_{mm} = \text{Re}(a^*d+b^*c), \quad I\,C_{\ell\ell} = -\text{Re}(a^*d-b^*c),$$

$$I\,C_{m\ell mm} = -\text{Im } ae^*, \quad I\,D_{\ell m} = -\text{Im } be^*, \quad I\,K_{\ell m} = -\text{Im } ce^*, \quad I\,C_{\ell m} = -\text{Im } de^*,$$

$$I\,C_{mn\ell} = \text{Im}(a^*b+c^*d), \quad I\,C_{\ell nm} = -\text{Im}(a^*b-c^*d), \quad I\,C_{nm\ell} = \text{Im}(a^*c+b^*d),$$

$$I\,C_{n\ell m} = -\text{Im}(a^*c-b^*d), \quad I\,C_{\ell mn} = -\text{Im}(a^*d+b^*c), \quad I\,C_{m\ell n} = -\text{Im}(a^*d-b^*c).$$

Since we are not presently interested in the overall phase of the amplitudes, we shall (arbitrarily) choose the function e to be real and positive. (The case $e = 0$ should be considered separately).

Let us first consider a set of experiments involving at most two component tensors.

We shall demonstrate that 9 experiments, properly chosen, are sufficient to reconstruct the amplitudes with at most one two-fold ambiguity. In
terms of the quantities I, IP, $ID_{\ell m}$, $IK_{\ell m}$, $IC_{\ell m}$, $ID^{\pm} \equiv \frac{1}{2} I(D_{mm} \pm D_{\ell\ell})$ and
$IK^{\pm} \equiv \frac{1}{2} I(K_{mm} \pm K_{\ell\ell})$ we obtain :

$$\text{Re } a = IP/e, \quad \text{Re } b = X(e^2 K^- - I\, C_{\ell m}D_{\ell m})/Pe, \quad \text{Re } c = X(e^2 D^- - I\, C_{\ell m}K_{\ell m})/Pe,$$

$$\text{Re } d = IP/Xe, \quad \text{Im } a = (e^2 Y - I\, C_{\ell m}X)/e, \quad \text{Im } b = -ID_{\ell m}/e, \quad \text{Im } c = -IK_{\ell m}/e,$$

$$\text{Im } d = -IC_{\ell m}/e \tag{3}$$

where $\quad X = (D_{\ell m}K^+ - K_{\ell m}D^+)/(D_{\ell m}D^- - K_{\ell m}K^-), \quad Y = (K^+K^- - D^+D^-)/(D_{\ell m}D^- - K_{\ell m}K^-) \tag{4}$

Formulas (3) determine a, b, c and d uniquely in terms of experimental
quantities and e. Adding up the squares of all expressions in (3). we obtain
an expression for $|a|^2 + |b|^2 + |c|^2 + |d|^2$. Combining this with the differential
cross-section I we obtain a quadratic equation for e^2 which has two real and
positive roots $e^2_{1,2}$. To remove the remaining two-fold ambiguity, a convenient
quantity would be $IC^- = \frac{1}{2} I(C_{mm} - C_{\ell\ell}) = \text{Re } a^* d$ yielding uniquely

$$e^2 = I(P^2 + X^2 C^2_{\ell m})/X(C^- + C_{\ell m}Y) \tag{5}$$

Alternatively, a measurement of IC_{mm} alone would provide a new quadratic equation for e^2, which together with the one mentioned above would
also yield e^2 uniquely. Note that we have used 10 c.m.s. experimental quantities corresponding to 11 experiments in the relativistic case (at least
if we are interested in experiments for more than one angle). If we had
replaced e.g. $D_{\ell m}$, $K_{\ell m}$ or $C_{\ell m}$ by a quantity like D_{nn}, we would have introduced an additional sign ambiguity, thus necessitating one more experiment.
In particular if we choose the four simplest experiments I, D_{nn}, K_{nn} and C_{nn}
plus any five others we obtain an eightfold ambiguity which can be removed
by measuring three more quantities bringing the total to twelve in the c.m.s.
(and 14 in the lab. frame in the relativistic case).

It can be seen from the formulas (2) that the three and four component
tensors are not really needed and that they cannot even be used to reduce the
number of necessary c.m.s. experiments. However we can obtain reconstruction
formulas that are considerably simpler than formulas (3)-(5), thus minimizing
the influence of experimental errors on the accuracy with which we reconstruct the amplitudes.

From this point of view the most advantageous set of experiments
would be one allowing a "linear reconstruction" of the scattering matrix.
This would involve 10 specifically chosen experiments, yielding :

$$e^2 = I(1 - C_{mmmm}), \quad \text{Re } a = IP/e, \quad \text{Re } b = I\, C_{mnm}/e, \quad \text{Re } c = I\, C_{nmm}/e,$$

$$\text{Re } d = I\, C_{mmn}/e, \quad \text{Im } a = -I\, C_{m\ell mm}/e, \quad \text{Im } b = -I\, D_{\ell m}/e, \quad \text{Im } c = -I\, K_{\ell m}/e,$$

$$\text{Im } d = -I\, C_{\ell m}/e \tag{6}$$

(the 10 c.m.s. experiments figuring in (6) of course represent a much
larger number of lab. frame experiments).

DIRECT RECONSTRUCTION OF THE N-N
SCATTERING MATRIX IN THE ISOSPIN STATE T = 1 PLUS T = 0
AT 90° c.m., BETWEEN 310 MeV AND 670 MeV

J.M. Fontaine[*]
DPh-N/ME, C.E.N. Saclay, France

For an isospin state T = 0 or T = 1, if one takes into account the invariance with respect to space rotation, space reflection and time reversal, the N-N scattering matrix can be expressed in the following form[1]

$$M = B.S. + C(\sigma_1 + \sigma_2)\vec{N} + N(\sigma_1 \vec{N} \sigma_2 \vec{N})T + \frac{1}{2} G(\sigma_1 \vec{K} \sigma_2 \vec{K} + \sigma_1 \vec{P} \sigma_2 \vec{P})T + \frac{1}{2} H(\sigma_1 \vec{K} \sigma_2 \vec{K} - \sigma_1 \vec{P} \sigma_2 \vec{P})$$

where $\vec{N}$, $\vec{K}$, $\vec{P}$ are unit vectors in the directions $(p \wedge p')$, $(p'-p)$ and $(p'+p)$, respectively, p' and p are outgoing and incident momenta in the c.m. system, S and T are singlet and triplet projection operators, and σ_1 and σ_2 are Pauli matrices of the incident and target nucleons. This form of the scattering matrix has the advantage to separate at 90° c.m. the amplitudes of the isospin T = 0 and T = 1.

For each isospin state T = 0 or T = 1, B, C, N, G and H are five complex functions of the nucleon energy and scattering angle. The exchange of two nucleons must leave the scattering matrix unchanged, so at 90° c.m. some amplitudes are equivalent to zero and the number of real quantities to be determined is smaller. Thus we have for the isospin state T = 1, N(90° c.m.) = G(90° c.m.) = 0, and for the isospin state T = 0, B(90° c.m.) = C(90° c.m.) = H(90° c.m.) = 0. At this angle, it is enough to determine five real quantities in the isospin state T = 1 (pp system) and nine real quantities in the isospin state T = 1 and T = 0 (np system) with an arbitrary phase.

To have a minimum of nine independent measurements in each case[2], we occasionally had to do some extrapolations for one or two experimental quantities. Thus, at 310 MeV, the angular distributions of parameters D and R, measured up to 80° c.m., have been continued to 90° c.m. as well as the measurements of the R parameter at 670 MeV have been interpolated to 90° c.m. The measurement of the A parameter have been used at 600 MeV and 635 MeV.

The experimental data have been taken from ref. 3. As the overall phase cannot be determined, the amplitude C is set to be real and positive. For the calculation, the least square method, χ^2 criterion and relativist formulae were used.

The results presented here are only at 90° c.m.,for the isospin state T = 1 and T = 0 at 310, 430, 520 and 600 MeV, and for the isospin state T = 1 at 635 and 670 MeV. In figs. 1 to 5 a comparison with the up-to-date phase shift analysis[4] is given. To compare the results, the phase C is set equal to the phase C of the phase shift analysis. The different solutions are in good agreement except at 520 MeV for the phase of the amplitude B, but some ambiguities still remain due to the lack of measured quantities.

References

* Centre National de la Recherche Scientifique.
1) L. Wolfenstein, Phys. Rev. 96 (1954) 1654.
2) C.R. Schumacher and H.A. Bethe, Phys. Rev. 124 (1961) 1934.
3) J. Bystricky et al., Note CEA N 1547(E).
4) J. Bystricky and F. Lehar, Communication to this Conference, p.458

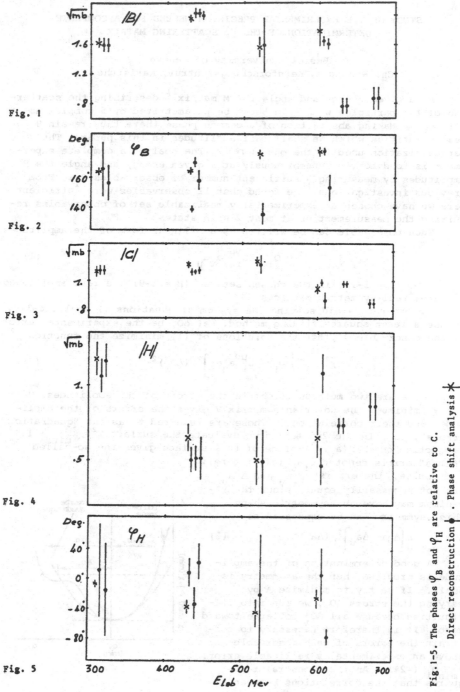

Fig. 1

Fig. 2

Fig. 3

Fig. 4

Fig. 5

Fig. 1-5. The phases φ_B and φ_H are relative to C.

Direct reconstruction ● . Phase shift analysis ✳

STUDY OF THE EXPERIMENTAL PRECISION NEEDED FOR A COMPLETE
DETERMINATION OF THE pp SCATTERING MATRIX

D. Besset, University of Geneva
Ch. Weddigen, Kernforschungszentrum, Karlsruhe

At a given energy and angle the M matrix[1] describing the scattering of two particles with spin $\frac{1}{2}$ can be parametrized by 5 complex amplitudes. Making abstraction of a common phase there thus remain 9
real quantities which we shall call amplitudes in this paper. The
parametrization used is the one of ref.2. The goal of a complete experiment is to determine unambiguously at a given energy and angle the 9
amplitudes by measuring a sufficient number of observables O_i. From
previous investigations[3] we found that 11 observables were sufficient.
Here we have chosen an experimentally realizable set of observables requiring the measurement of at most 2 spin states.

Each observable can be written as a bilinear form of the amplitudes

$$O_i = L_i^{\nu\mu} a_\nu a_\mu \qquad (1)$$

where O_i (i = 1-11) is the chosen set, a_μ (μ = 1-9) are real amplitudes
$L_i^{\nu\mu}$ are real symmetric matrices.

In order to avoid solving the system of equations(1) analytically,
we use a least squares fitting method. Let ΔO_i be the experimental error
of the observable O_i, then the solutions of (1) minimize the function χ^2

$$\chi^2 = \sum_i \left[O_i - L_i^{\mu\nu} a_\mu a_\nu \right]^2 / (\Delta O_i)^2 \qquad (2)$$

There are two methods to obtain the errors of the amplitudes. Near
the χ^2 minimum, the covariance matrix V gives the errors of the amplitudes and their correlations. Those are referred to as the "quadratic
errors", Δa_ν. In the 2nd method we evaluate the surface $\chi^2 = \chi^2_{min} + 1$.
Projection on the (a_ν, χ^2)-plane of this surface gives the so-called
"Minos" errors denoted $\Delta_\pm a_\nu$ (cf. figure).

Of course, the errors $\Delta_+ a_\nu$ and $\Delta_- a_\nu$
are not necessarily equal, since the χ^2-
minimum may have an asymmetric shape.
This asymmetry can be expressed as

$$A = \sum_{M=\pm} \sum_\nu \left[\Delta_M a_\nu - \Delta a_\nu \right]^2 / (\Delta a_\nu)^2 \qquad (3)$$

A good determination of the amplitude is provided when the asymmetry is
small. If we try to minimize A by
varying the errors ΔO_i, we run into difficulties because all ΔO_i collapse toward
zero. It is therefore necessary to con-
strain the errors of each observable
above an experimentally realizable error,
$\Delta_{min} O_i$ ($\approx$2%). Another constraint requires that the correlations between am-

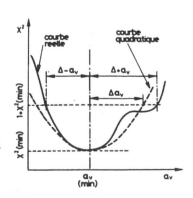

plitudes be a minimum. Under these conditions the function to minimize
is:

$$F = A + \sum_{\nu} \dot{\rho}_{\nu}^2 + \sum_{i} H (\Delta_{min} O_i - \Delta O_i) \qquad (4)$$

where ρ_{ν} is the global correlation coefficient for a_{ν}, $H(z) = 10^9 z^2$ if
$z > 0$, $H(z) = 0$ elsewhere.

The search was performed for 9 different energy-angle cases. To be
as close as possible to reality, we calculated values for O_i from phase
shifts[4]. The minimizing process was stopped when A was below 40%. The
absolute errors on the amplitudes were $\simeq .25$. The corresponding ΔO_i are
presented in table I.

We conclude that for this set of observables the required precision
for all observables but I_o is approximately the same. The precision of
I_o is not involved in the search, since multiplication of the amplitudes
by a common factor leaves all observables except I_o unchanged; however,
the precision of the amplitudes depends on the precision of I_o.

Observables	Absolute errors of observables for 9 cases									average error $\pm\sigma$	
I_o	.966	.861	.809	.756	.843	.805	.578	.669	.736	.086	.015*
P	.085	.059	.072	.046	.058	.102	.064	.117	.109	.079	.025
DNN	.094	.068	.164	.102	.103	.099	.060	.107	.127	.103	.031
KNN	.088	.063	.088	.098	.081	.131	.047	.046	.059	.078	.027
CNN	.080	.067	.101	.071	.111	.154	.070	.077	.091	.091	.028
KPP	.131	.048	.050	.097	.077	.059	.129	.056	.047	.077	.034
CPP	.110	.053	.056	.119	.105	.062	.058	.058	.057	.075	.027
DQQ	.069	.075	.112	.075	.066	.092	.097	.095	.049	.081	.019
DPQ	.156	.095	.082	.081	.098	.079	.068	.067	.130	.095	.030
KPQ	.071	.058	.054	.050	.058	.061	.047	.062	.046	.056	.008
CPQ	.108	.068	.063	.050	.060	.062	.049	.084	.048	.066	.019
Energy (MeV)	290	350	410	450	450	450	450	450	450	* For I_o mean is on relative error	
Angle (CM Deg.)	45	45	45	25	35	45	55	65	70		

Table I

References
1) H.P. Stapp, T.J. Ypsilantis and N. Metropolis, Phys.Rev.105(1957)302
2) C.R. Schumacher and H.A. Bethe, Phys. Rev., 121 (1961) 1534
3) Ch. Weddigen, SIN spring school, Zuoz, 1975
4) R.A. Arndt, R.H. Hackman & L.D. Roper, Phys. Rev., 9c (1974) 555.

Three Body Problem

ANALYZING POWER OF THE $\vec{n}$-d SCATTERING AT 14.2 MeV

M. Preiswerk, R. Casparis, B.Th. Leemann, H. Rudin, R.
Wagner, P.E. Żuprański*
Institut für Physik der Universität Basel

We have measured the analyzing power of deuterium for 14.2 MeV neutrons at ten angles between 50° and 152° (CM).

The experiment has been done with the same source and set up described in the contribution of B.Th. Leemann et al. to this conference. A cylindrical deuterated benzene scintillator (NE 213, diameter 3.8 cm, length 7.8 cm) served as scatterer. Time-of-flight spectra of the scattered neutrons and energy spectra of the deuteron recoils were recorded simultaneously. The polarization of the incoming neutrons was determined with the ^{4}He($\vec{n}$,n)^{4}He scattering at a backward angle using a high pressure scintillation chamber.

The results of several calculations[1,2] suggest that p-wave interaction and tensor forces must be taken into account simultaneously in order to obtain correct vector and tensor polarization in the $\vec{n}$-d scattering. However, up to 15 MeV the neutron polarization is mainly sensitive to p-wave interaction. It is not surprising, therefore, that our data agree better with the calculations of Pieper[3] than with those of Doleschall[4] (fig.1). In the framework of a perturbation technique

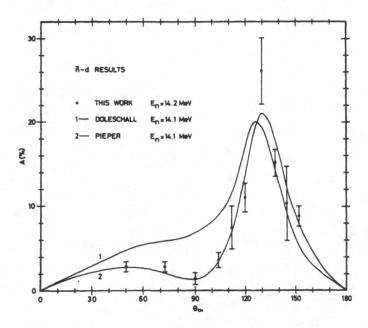

Fig. 1. Analyzing power A(Θ) of the $\vec{n}$-d scattering. Comparison of our experimental results with two theoretical calculations.

Pieper's calculations take s-, p- and d-wave nucleon-nucleon interaction into account but no tensor force, whereas Doleschall's method is based on a tensor force and on a simplified p-wave interaction.

The agreement of our results with $\vec{p}$-d data at 14.5 MeV of Faivre et al.[5] is good.

References

* On leave from the Instytut Badan Jądrowych, Warsaw

1) J.C. Aarons and I.H. Sloan, Nucl. Phys. A182 (1972) 369
2) P. Doleschall, Phys. Lett. 40B (1972) 443
3) S.C. Pieper, Nucl. Phys. A193 (1972) 529
4) P. Doleschall, Nucl. Phys. A201 (1973) 264
5) J.C. Faivre et al., Nucl. Phys. A127 (1969) 169

ANALYZING POWER OF 14 MEV NEUTRONS ON LIGHT NUCLEI

J.E. Brock, A. Chisholm, R. Garrett and J.C. Duder
Physics Department, University of Auckland, Private Bag, Auckland.

We are at present measuring the analyzing powers for the scattering of 14 MeV polarized neutrons for the following nuclei:
  (a) protons
  (b) deuterons
  (c) carbon (ground state and 4.4 MeV level)
Our main interest is in the 2- and 3-body systems. We hope later to make measurements on n-p radiative capture and on the n-d break-up.

Polarized neutrons are generated by the $T(d,n)^4He$ reaction, with 150 keV vector polarized deuterons on a tritiated titanium target. The deuteron polarization is produced by a conventional "atomic beam" type of polarized ion source. The deuteron vector polarization, and hence the neutron polarization, is reversed by the switching of three R.F. transition units. The neutron polarization is about 55% and remains within a few percent of this over several weeks. At present we infer the neutron polarization from regular measurements of the tensor polarization component, P_{zz}, of the deuterons. We also monitor continuously the neutron polarization by recording the asymmetry in the neutron scattering from carbon.

A neutron beam is defined by detecting the recoiling alpha-particles with a thin NE102A scintillator and photomultiplier.

The scattering targets are various organic scintillators in cylindrical form, typically 5 cm diameter and 8 cm high. The scattering both from protons and from the 4.4 MeV level of carbon can be seen in the NE213 target by the use of pulse-shape discrimination; the de-excitation gammas signal the carbon scattering. Scattering from deuterons is observed in NE230 and from carbon (elastic) in NE102A. The neutrons scattered to left and right are detected in NE102A scintillators of 5 cm × 10 cm × 10 cm. We use three pairs of these to increase the counting rate. All scintillators are coupled to 56 AVP photomultipliers. The side detectors are at distances of 40 cm to 60 cm from the scatterer, which is at 120 cm from the neutron source. Use of the associated alpha-particle technique enables us to measure the scattering angles to high accuracy ($\sim\frac{1}{4}°$).

After a triple coincidence has been formed we record on magnetic tape six parameters for each event specifying: scatterer recoil energy, scatterer pulse shape, scatterer to side detector time of flight, side detector recoil energy, scattering angle and neutron polarization state. Since these data are recorded event by event, various "windows" can be placed on each parameter off-line from the experiment. After this treatment of the two dimensional matrices, the backgrounds are reduced to a very low level; for the n-p and n-d elastic scatterings, the events outside of the elastic peaks can be accounted for in terms of multiply-scattered neutrons.

The results obtained so far for the analyzing powers (in percentages) for scatterings from protons and carbon are given in the following table, and from deuterons are given in Fig. 1, where p-d analyzing powers[1] at a similar energy are also plotted.

θ_{lab} (degrees)	30	40	50	60	70	80	90
Proton	3.9 ±1.1		1.0 ±0.8		1.9 ±1.4		
Carbon Q = 0		-40 ±8	-62 ±8	-57 ±11	11 ±13	9 ±13	11 ±15
Carbon Q = -4.4 MeV	-15.1 ±6.6		-10.4 ±3.1		-19.2 ±8.3		

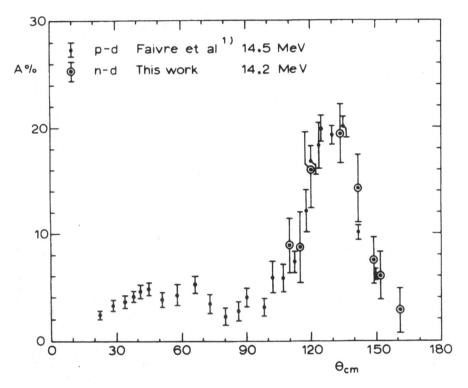

Fig. 1. Analyzing powers in n-d scattering (open circles, present work) and in p-d scattering (closed circles, Faivre et al[1]).

Reference

1. J.C. Faivre, D. Garreta, J. Jungerman, A. Papineau, J. Sura and A. Tarrats, Nucl. Phys. A127 (1969) 169

POLARIZATION IN n-d SCATTERING AT 16.4 AND 21.6 MeV

M. Steinbock, F.D. Brooks and I.J. van Heerden
Department of Physics, University of Cape Town, Rondebosch,Cape,
South Africa.

A comparison of the polarization asymmetries observed in n-d and p-d elastic scattering provides a test of the charge symmetry of nuclear forces. Discrepancies have been observed between the p-d and n-d polarizations measured at incident energies near to 22 MeV [1-3]) and 35 MeV [1,4]) respectively. However Morris et $al.$[5]) have recently measured P_{nd} at incident neutron energies of 16.8 and 21.1 MeV and have obtained results in good agreement with the p-d polarization data. The present work has been motivated by the need for further n-d data in this region.

The measurements were made using a ^{3}H(d,n) source with pulsed 5.0 MeV deuterons from the Van de Graaff accelerator of SUNI, Faure, C.P. South Africa. Neutrons emitted at angles 20° and 80° respectively, with energies 21.6 and 16.4 MeV and polarizations +0.21 and -0.42 respectively were used. Time-of-flight gating was used to select the primary component in the neutron spectrum.

The polarization analyser was a deuterated anthracene scintillation crystal. Recoil deuterons from n-d elastic scattering within the crystal were studied and pulse shape discrimination (PSD) was used to determine the left-right asymmetry of the recoils. The experimental method is similar to that used previously [6,7]) to study the polarization in n-p scattering and depends on the fact that the PSD response of the crystal is sensitive to the direction of the recoil deuteron relative to the crystal axes. The scintillation detector provides a fast output for timing, a pulse height output L, from which the deuteron energy may be deduced, and a PSD output S. Fig.1 shows an LS spectrum from dual-parameter analysis of data obtained using 21.6 MeV neutrons entering the crystal in a direction parallel to its artificial c'-axis. The forward recoil deuteron peak F which lies at a relatively low S value in fig.1 would move to the position F' in the LS plane for neutrons entering parallel to the b - axis. Thus for neutrons entering in a direction lying within the bc'-plane and making an angle of 50° with the c'-axis recoils to the right (towards the b-axis)and to the left (towards the

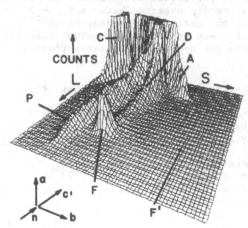

Fig. 1 LS spectrum for 21.6 MeV neutrons directed parallel to the c'-axis, showing ridges due to: Compton electrons (C);breakup protons (P); recoil deuterons (D); and α-particles (A) from reactions on carbon. The peak F corresponds to forward recoiling deuterons. The point F' shows the position of this peak in the spectrum obtained for neutrons directed parallel to the b-axis.

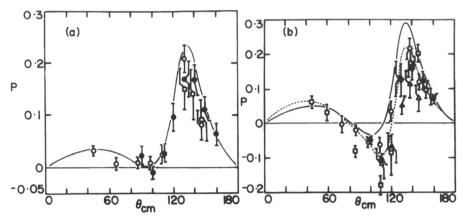

Fig.2 Polarization data at energies near to (a) 16.4 and (b) 21.6 MeV.
Solid circles show present work;squares ref.[2]);triangles ref.[3]);
open circles, ref.[5]). See text for curves.

c'-axis) produce higher and lower S outputs respectively. The dispersion
of the deuteron component over the LS plane at this orientation of the
crystal provides a basis for determining the asymmetry.

The recoil deuteron energy deduced from L is correlated with the
associated neutron centre-of-mass scattering angle θ_{cm} hence the asymmetry may be determined as a function of θ_{cm} . Methods based on those
developed for the natural anthracene analyser [6,7]) have been used, with
modifications to correct for the partial overlap of the proton and
deuteron components in the LS spectra.

The results of the measurements are shown in fig.2 together with
other data [2,3,5]) at nearby energies. The solid curves are based on the
work of Pieper [8]) and were adjusted to 16.4 and 21.6 MeV by interpolating
his results. The dashed curve in fig. 2b shows the trend of the p-d
polarization data [1]) at 20.2 MeV. The trend of the p-d data [1]) at 17.5
MeV lies close to the solid curve in fig.2a. The data obtained at 16.4
MeV (fig.2a) are consistent with the other data at nearby energies,
except possibly in the region near $\theta_{cm} = 130^\circ$. The data obtained at
21.6 MeV (fig.2b) are consistent with the recent measurements of Morris
et al.[5]) and confirm that the negative dip observed between $\theta_{cm} = 80^\circ$ –
120° is shallower than indicated by the earlier n-d measurements [2,3])
and consistent with the p-d data at these angles. We conclude therefore
that there is no discrepancy between the n-d and p-d polarizations at
these energies.

References
1) J.C. Faivre,Garreta,Jungerman,Papineau,Sura and Tarrata,Nucl.Phys.
A127 (1969) 169
2) J.J.Malanify,Simmons,Perkins and Walter,Phys.Rev. 146 (1966) 632
3) R.L.Walter and Kelsey, Nucl. Phys. 46 (1963) 66
4) J. Zamudio-Cristi,Bonner, Brady,Jungerman and Wang,Phys.Rev.Letts. 31
(1973) 1009
5) C.L. Morris,Rotter,Dean and Thornton,Phys.Rev.C,9(1974) 1687
6) F.D.Brooks and Jones,Nucl.Instr. & Meth.,121 (1974) 69 & 77
7) D.T.L. Jones and Brooks,Nucl. Phys. A222 (1974) 79
8) S.C. Pieper, Nucl. Phys. A193 (1972) 529

ANGULAR VARIATION OF THE DEPOLARIZATION FACTOR D(θ) IN THE
D($\vec{n}$,$\vec{n}$)D REACTION AT 2.45 MeV*

M. Ahmed[+], D. Bovet, P. Chatelain and J. Weber
Institut de Physique, Université de Neuchâtel,
2000 Neuchâtel, Switzerland

From effective range approximation (ERA) and phase shifts analyses of the published data for the n,d elastic scattering, one obtains different sets of phase shifts. The predictions for D(θ) obtained from some of these sets are sufficiently different so that one may hope to be able to discriminate between them[1].

The present triple scattering experiment has been started with this aim. D(θ) has been measured at the following laboratory angles : $30°$; $40.5°$; $55°$ and $70°$. The experimental set-up is shown in Fig. 1. The relevant data are : E_d= 2.88 MeV; C-target thickness : 400 µg/cm^2; $\theta_1 = 25°$; E_n = 2.45 MeV; polarization of the neutron incident beam : $P_1 = -0.45\pm0.03$ [2]. The liquid-He polarimeter has been described elsewhere [3] as well as the usefullness of the spin precession method by a magnetic field.

The electronics were of the classical time-of-flight, fast-slow spectrometer type.

Each event, recorded on a PDP 15/20 computer, was defined by the occurence of a triple coincidence between events in the D, He and H scintillators and by a set of five parameters : the two time-of-flight T1 and T2 (measured on the D-He and He-H distances resp.) and the energies of the recoil in the three scintillators.

The scheme which has been followed to extract D from the experimental data is :

1) From the bi-dimensional representation of T1 and T2, one can obtain a good estimate of the background, especially of the contribution of T2 uncorrelated to T1. By integration of the "left" and "right" peaks (after background subtraction) the experimental left-right asymmetries ε($\mathcal{H}$ = 0) and ε($\mathcal{H}$ ≠ 0) are calculated. Their arithmetic mean gives then a value from which the false asymmetries are eliminated[3].

2) A sophisticated Monte-Carlo code is then used to simulate the experiment. The parameter D is adjusted until the experimental time-of-flight spectra and the experimental asymmetry are correctly reproduced. In doing this, D is assumed to be constant.

The results are listed in Table 1.

θ_{Lab}	$30°$	$40.5°$	$55°$	$70°$
D(θ)	0.31±0.15	0.34±0.17	0.54±0.17	0.66±0.15

Table 1. Depolarization factor D(θ) in the D($\vec{n}$,$\vec{n}$)D reaction at E_n = 2.45 MeV. The S.D. are essentially statistic. The result of our test measurement of D in the (n,p) elastic scattering at 2.45 MeV is D = 0.67±0.16. The expected value calculated from published ERA parameters [4] is 0.64.

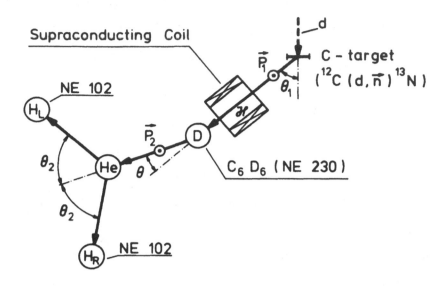

Fig. 1. Experimental set-up as described in the text.

ERA and phase shifts analyses which include these results in the set of data are in progress and will be published shortly.

The authors are thankful to Dr Nussbaum for writing the data acquisition program. The Monte-Carlo code has been run on the CDC-Cyber 7326 computer of the EPFL-computer center.

References

* Supported by the Fonds National Suisse de la Recherche Scientifique.
+ Present address : Building 8, AERE Harwell, Oxfordshire, OX11 ORA.

1) S. Jaccard and R. Viennet, Nucl. Phys. A182 (1972) 541
2) R.L. Walter, Proceedings of the 3rd Int. Symp. on Polarization Phenomena in Nuclear Reactions, p. 317, Madison 1970
3) J. Piffaretti, Helv. Phys. Acta 40 (1967) 805
4) H.P. Noyes, ARNS (1972) 476

ERA ANALYSES OF THE n-d ELASTIC SCATTERING : COMPARISON OF PREDICTED AND MEASURED VALUES OF THE DEPOLARIZATION FACTOR D(θ)*

D. Bovet, S. Jaccard[+] and R. Viennet
Institut de Physique, Université de Neuchâtel,
2000 Neuchâtel, Switzerland

The situation for the n-d scattering below the inelastic threshold is not yet completely clarified, especially as far as the doublet states are concerned [1][2]. Our first ERA analysis[1] did not lead to standard results. In particular, the pole of Van Oers and Seagrave in the effective range function of the ^2S state could not be reproduced. Therefore, two new analyses were performed with the following parametrisations :

$$k^{2L+1} \, ctg \, \delta_{L,S} = A_{L,S} + \tfrac{1}{2} R_{L,S} \, k^2 + P_{L,S} \, k^4$$

except for the ^2S state, which was parametrized by :

set I : $k \, ctg \, \delta_{0,\frac{1}{2}} = A_{0,\frac{1}{2}} + \tfrac{1}{2} R_{0,\frac{1}{2}} \, k^2 + P_{0,\frac{1}{2}} \, k^4$

set II : $k \, ctg \, \delta_{0,\frac{1}{2}} = A'_{0,\frac{1}{2}} + \tfrac{1}{2} R'_{0,\frac{1}{2}} \, k^2 + P'_{0,\frac{1}{2}}/(1+k^2/W^2)$

The experimental values used for the fit were the same as in ref.[1] except that :
i) The triton residue was not taken into account.
ii) The most recent values at zero energy were used :

total cross sections : $\sigma_0 = 3.390 \pm 0.012$ b [3]
scattering lengths : $a_4 = 6.47 \pm 0.14$ fm, $a_2 = 0.57 \pm 0.14$ fm [4]
coherent scattering length : $a_4 + \tfrac{1}{2}a_2 = 6.672 \pm 0.007$ fm [3]

The parameters obtained are listed in Table I. Set I presents the same caracteristics as our previous result. Set II is quasi-standard, except for the ^2D phase, which is negative in the present case.

Fig. 1 shows the calculated depolarization factor $D(\theta_{cm})$ for $E_{Lab} = 2.45$ MeV, together with the recently measured values [5]. Although the experimental absolute values do not agree with any set of parameters, the general trend seems to be in favour of set II.

Table I.

Set I $\chi^2 = 98$; reduced $\chi^2 = 1.46$

	^2S	^4S	^2P	^4P	^2D	^4D
$A_{L,S}$	-1.533	$-1.576 \cdot 10^{-1}$	1.720	$3.835 \cdot 10^{-3}$	$9.169 \cdot 10^{-4}$	$-4.405 \cdot 10^{-4}$
$R_{L,S}$	$-5.243 \cdot 10^{1}$	2.969	$-1.273 \cdot 10^{2}$	$9.355 \cdot 10^{-1}$	$0.$	$0.$
$P_{L,S}$	$-1.034 \cdot 10^{2}$		$6.137 \cdot 10^{2}$		1.113	-2.550

Set II $\chi^2 = 110$; reduced $\chi^2 = 1.72$

	^2S	^4S	^2P	^4P	^2D	^4D
$A_{L,S}$	$-2.265 \cdot 10^{-1}$	$-1.576 \cdot 10^{-1}$	-1.478	$1.988 \cdot 10^{-3}$	$-2.727 \cdot 10^{-2}$	$-3.134 \cdot 10^{-3}$
$R_{L,S}$	2.727	2.885	$1.053 \cdot 10^{2}$	1.254	$6.329 \cdot 10^{-1}$	$3.787 \cdot 10^{-1}$
$P_{L,S}$	-1.315		$-4.890 \cdot 10^{2}$		$-8.544 \cdot 10^{-1}$	-4.813
W	$8.780 \cdot 10^{-2}$					

480

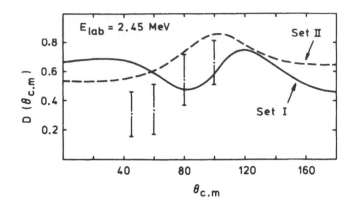

Fig. 1

The programs were run on the CDC–Cyber 7326 of the EPFL–Computer center.

References

* Supported by the Fonds National Suisse de la Recherche Scientifique.
+ Present address : British Columbia University, Department of Physics,
   Vancouver, Canada.

1) R. Viennet, Nucl. Phys. A189 (1972) 424
2) D. Bovet, S. Jaccard and J. Weber, Helv. Phys. Acta 46 (1973) 659
3) W. Dilg, L. Koester and W. Nistler, Phys. Letters 36B (1971) 208
4) S.J. Nikitin, W.T. Smolyankin, W.Z. Kolganow, A.W. Lebedev and
   G.S. Lomkazy, The First International Conference on Peaceful Uses
   of Atomic Energy (Geneva, 1955) (The United Nations, New York, 1956)
   Vol. 2, p. 81
5) M. Ahmed, D. Bovet, P. Chatelain and J. Weber, Fourth Polarization
   Symp. (1975), p.477

POLARIZATIONS IN ELASTIC n-d SCATTERING WITH THE REID POTENTIAL

C. Stolk and J.A. Tjon
Instituut voor Theoretische Fysica der Rijksuniversiteit Utrecht
Princetonplein 5, Utrecht, The Netherlands

We present here results of a n-d scattering calculation using the full Reid soft core potential [1]). The method of the calculation is analogous to that of ref 2; the S-wave contributions are calculated exactly by solving the Faddeev equations for this case using the Padé technique [3]), and the non-S-wave parts of the two-nucleon T-matrix and the deuteron D-wave are treated perturbationally. We calculated at various energies from 5.5 MeV to 46.3 MeV the polarization of the neutron and the Wolfenstein parameters [4]) for the neutron spin transfer, and the vector and tensor polarizations of the deuteron; furthermore we expressed our three-particle S-matrix in terms of split phase shifts and mixing parameters.

In figures 1 and 2 we show some results for the neutron polarization at resp. 14.1 and 46.3 MeV neutron lab energy. With varying energy the contribution from the P-wave forces behaves on a reduced scale very similar to Pieper's results for his separable potential sets C and D [2]). There are very significant contributions from the tensor force, in the backward hemisphere mainly through the deuteron D-wave. When all forces are included, the calculated curve approximately fits the data at 46.3 MeV [5-9]), although the minimum at intermediate angles should be deeper and broader. However, the calculated height of the backward maximum is energy-dependent whereas experimentally it is found to be nearly constant between 15 and 50 MeV [10]). We note that the various contributions seem to be additive.

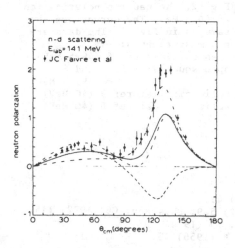

For the deuteron vector polarization we find a reasonable agreement to the data at 5.5 and 10.85 MeV, but at 14.1 and 22.7 MeV the dip at intermediate angles is not deep enough and the backward maximum is too low.

Fig. 1. The neutron polarization at 14.1 MeV. The data are from ref 10. Potentials included are:

dash: $^1S_0, ^3S_1 - ^3D_1$.

dash-dot: $^1S_0, ^3S_1, ^1P_1, ^3P_0, ^3P_1, ^3P_2$, no deuteron D-wave.

dash-dot-dot: $^1S_0, ^3S_1 - ^3D_1, ^1P_1, ^3P_0, ^3P_1, ^3P_2$.

full: $^1S_0, ^3S_1 - ^3D_1, ^3D_2, ^1D_2, ^1P_1, ^3P_0, ^3P_1, ^3P_2$.

The gross behaviour of the
deuteron tensor polarizations is
set by the deuteron D-wave, with
significant corrections from the
P-wave and D-wave forces. They do
not differ profoundly from the
results obtained by Doleschall[11]);
the agreement with the data is
better in the backward hemisphere
than in the forward directions.
The Wolfenstein parameters vary
strongly with the different
potential sets, although the trend
is set by the S-wave forces. The
agreement with the few available
data is reasonable.

A comparison of our calculation to the phase shift analysis
by Arvieux[12]) gives discrepancies
that are of the same order as the
splittings in the phase shifts.
Although there is only a small
overlap in the energy ranges
considered, our results are in
reasonable accord with the split
phase analysis by Schmelzbach et
al[13]), with the exception of the
spin-and-orbit mixing parameters.

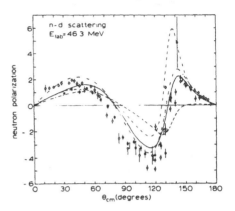

Fig. 2. The neutron polarization
at 46.3 MeV. The curves are the
same as in fig. 1. The data are
p-d measurements:
open circles: ref 6 (49.5 MeV)
open squares: refs 7 and 8
                       (49.23 MeV)
solid circles: ref 9 (40 MeV)
solid squares: ref 5 (40 MeV)

References.

1. R.V.Reid Jr., Ann.Phys. 50 (1968) 411.
2. S.C.Pieper, Nucl.Phys. A193 (1972) 529; Phys.Rev. C6 (1972) 1157.
3. W.M.Kloet and J.A.Tjon, Ann.Phys. 79 (1973) 407.
4. L.Wolfenstein, Ann.Rev.Nucl.Sci. 6 (1956) 43.
5. H.E.Conzett et al, Phys.Lett. 11 (1964) 68.
6. S.J.Hall et al, Comptes Rend.du Congr.Intern.du Phys.Nucl.,
              Paris 1964, II.219.
7. A.R.Johnston et al, Phys.Lett. 21 (1966) 309.
8. W.R.Gibson et al, Proc.Intern.Nucl.Phys.Conf.(Acad.Press,
              New York 1967) 1016.
9. S.N.Bunker et al, Nucl.Phys. A113 (1968) 461.
10. J.C.Faivre et al, Nucl.Phys. A127 (1969) 169.
11. P.Doleschall, Nucl.Phys. A220 (1974) 491.
12. J.Arvieux, Nucl.Phys. A221 (1974) 253.
13. P.A.Schmelzbach et al, Nucl.Phys. A197 (1972) 273.

LOCAL POTENTIAL CALCULATION OF d-n OBSERVABLES

H. Stöwe, H.H. Hackenbroich and P. Heiss
Institut für Theoretische Physik
Universität zu Köln, 5 Köln 41, Germany

In recent years many efforts have been made to investigate the three nucleon system. Parallel to the progress in
experimental physics, powerful theoretical methods have been
developed, especially for solving the Fadeev equations for
separable potentials. For local potentials there are only
few results because of mathematical difficulties. Nevertheless one has to test realistic local potential models by
calculating three nucleon data.

We started from the soft core spin-dependent Eikemeier-
Hackenbroich potential and calculated d-n scattering data by
variational methods[1] tested on several systems of heavier
nuclei. In this framework it is not possible to treat the
three-body break-up process explicitly. Instead, we included channels with dineutron and singlet deuteron substructures described by square-integrable functions.

For given J and π our wave function is an antisymmetrized superposition of square integrable functions Ψ_{si}
describing distortion effects and of channel functions of
the form

$$\{\{\psi_I\psi_{II}\}^S R^L Y_L(\Omega_{\vec{R}})\}^{J\pi} \chi(R).$$

Here ψ_I and ψ_{II} denote the wave functions of the fragments
(deuteron-neutron, singlet deuteron-neutron, dineutron-
proton). Our calculations showed that the d-p phase shifts
are very sensitive to the description of the substructures.
We therefore used superpositions of eight Gaussians for each
radial part of the substructures giving energy expectation
values of $E_d = -2.222$ MeV and $E_{nn} = 0.754$ MeV.

A triton binding energy of -7.302 MeV is obtained using
the functions Ψ_{si}. The calculation for the S-matrix was restricted to orbital angular momenta $L \leq 3$. The ^{2}S phase
shift (fig.1) decreases more rapidly than the experimental
phase shift, corresponding to the difference in the triton
binding energy. The structure at about 3 MeV is due to our
simplified description of the break-up channels. All other
phases of fig.1 show the same quantitative behavior and the
nontrivial order of the phase splitting as the p-d analysis
of Schmelzbach et al.[2] Our results differ from the analysis
mainly in the magnitude of the splitting and in some non-
diagonal S-matrix elements not shown in fig.1.

The calculated differential cross section and vector
polarisations (fig.2) show fair agreement with experimental
data. The vector polarisations are in general too small. For
the tensor polarisations - which are small and very sensitive to the nondiagonal S-matrix elements - calculation and
experiment do not agree. We think that these polarisations
can be obtained reliably only by introducing correct wave
functions for the three body break-up.

484

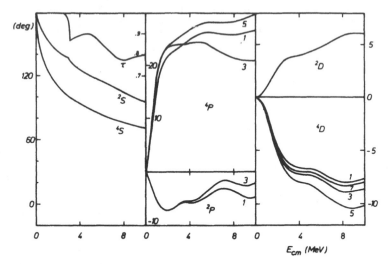

Fig.1 d-n diagonal phase shifts and magnitude τ of the ^{2}S S-matrix element. The curves are labeled by 2J, the symbols by 2S + 1.

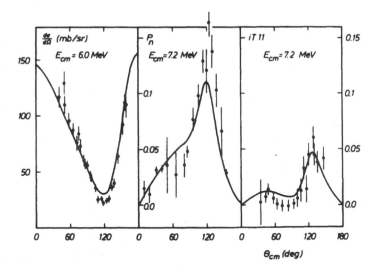

Fig.2 Elastic d-n cross section and vector polarisations

References

1) P. Heiss and H.H. Hackenbroich, Z. Phys. <u>235</u> (1970) 422
2) P.A. Schmelzbach, W. Grüebler, R.G. White, V. König,
   R. Risler and P. Marmier, Nucl. Phys. <u>A197</u> (1972) 273

PHASE-SHIFT ANALYSIS OF p-d ELASTIC SCATTERING

P.A. Schmelzbach, W. Grüebler, V. König and R. Risler
Laboratorium für Kernphysik, Eidg. Techn. Hochschule
8049 Zürich, Switzerland

Recently, a new phase-shift analysis of p-d elastic scattering has been completed between 3.0 and 5.75 MeV incident proton energy[1]. Good fits to the cross section, the proton polarization and the vector and tensor analysing powers of the H($\vec{d}$,d)H scattering have been obtained by splitting the (complex) phase shifts, coupling the states of different orbital momentum, and allowing for channel spin non-conservation. The full freedom was restricted to phase shifts for $L \leqslant 2$. The F and G waves were also taken into account, but as unsplit phase shifts without mixing.

The strongest mixing is observed between the $^2S_{1/2}$ and $^4D_{1/2}$ states, in which case both orbital angular momentum and channel spin are not conserved. Channel-spin mixing is present between the quartet and doublet P-states, and, with smaller values of the mixing parameters, between the quartet and doublet D-states.

The study of the sensitivity of the analysed observables to uncorrelated changes of 2° in the phase shifts and the mixing parameters and to alteration of the sequences of the multiplets shows that the uncertainty in the determination of the mixing parameters is fairly large and that there are still ambiguities in the sequences of the split quartet phase shifts. Hence, more experimental information is needed to refine the analysis.

The Wolfenstein parameters for proton-to-proton polarization transfer have only a very small sensitivity to changes in the mixing parameters and in the sequences of the multiplets. The proton-to-deuteron vector polarization transfer coefficients have fairly large values, but their sensitivity to the changes considered in the phase shifts and mixing parameters is relatively small.

The predicted vector-to-tensor polarization transfer coefficients $K^{i'j'}_{\frac{1}{4}}$ are generally smaller than 0.1 in the energy range of this analysis. They are comparable with the measurements of Michell and Ohlsen[2] but in most cases, since the errors of the experimental data are as large as the predicted effects, little information on the quality of the predictions can be gained.

After completion of the phase-shift analysis of ref. 1) measurements of angular distributions of the analysing powers of the d-p elastic scattering became available up to 16 MeV [3]. An attempt has then been made to extend the analysis of the p-d elastic scattering up to $E_p = 8$ MeV. However, some structure of the analysing power T_{22} around 90° and the behaviour at larger angles could not be reproduced in a satisfying way.

In the fig.1 one can see preliminary fits to the new measurements of the analysing powers of the d-p elastic scattering. The curves cover the range of the analysed data. At 14 MeV only the analysing power data have been fitted, while at 16 MeV cross section and analysing power data for the protons are available at the corresponding energy and have also been fitted.

The phase shifts obtained are compatible with the results at lower

486

energy but because of the limited number of data points and limited angu
lar range covered by the measurements they are still very inaccurate.
More experimental and computational work will be necessary to clear up
the situation. The introduction of a more powerful search routine in the
program will hopefully make the calculations faster and allow one to
introduce the new measurements between 8 and 12 MeV [4] into the analysis
also.

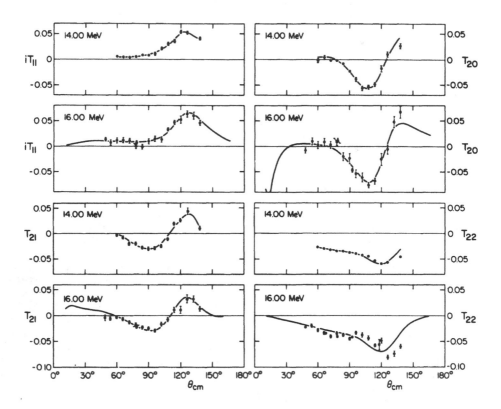

Fig. 2. The analysing powers of d-p elastic scattering at 14 and 16
MeV. The dots are the new measurements of ref.3. When the errors are not
indicated they are smaller than the symbol used. The curves are preliminary fits to the data.

References

1) P.A. Schmelzbach, W. Grüebler, R.E. White, V. König, R. Risler and
   P. Marmier, Nucl. Phys.A197 (1972) 273
2) C.K. Mitchell and G.G. Ohlsen, Proc. Int. Conf. on Few Particles
   Problems in the Nuclear Interaction (North-Holland, 1972) 460
3) G.G. Ohlsen and W. Grüebler, unpublished measurements performed
   during a visit of W.G. at LASL
4) C.K. Mitchell, Thesis, LASL (1973)

MEASUREMENT OF THE PROTON ANALYZING POWER IN THE
p-d BREAKUP REACTION*

F. N. Rad, H. E. Conzett, R. Roy[+], and F. Seiler[‡]
Lawrence Berkeley Laboratory, University of California
Berkeley, California 94720

We report in this paper on the measurement of the proton analyzing power A_y in the $^2H(\vec{p},p)d^*$ reaction at E_p = 22.7 MeV, which corresponds to the same center of mass energy as that in the $^1H(\vec{d},p)d^*$ reaction reported by Conzett[1]. Here, d* denotes final-state np pairs with relative energy $E_{np} \leqslant$ 1 MeV, in both the singlet and triplet states. Previously, there had been measurements of the vector analyzing power in the $^2H(\vec{p},2p)n$ (ref.2) and $^1H(\vec{d},2p)n$ (ref.3) transitions to the np final-state interaction (FSI) region at E_p = 10.5 MeV and E_d = 12.2 MeV, respectively. In both instances the analyzing powers were very small or consistent with zero. The higher energy results at E_d = 45.4 MeV[1] showed values of A_y reaching 0.15±0.02. The exact three-body calculations[4,5], which have been successful in fitting N-d breakup cross sections, have so far been limited to S-wave N-N input interactions and, thus, provide no polarization.

Our experimental results for A_y in the reaction $^2H(\vec{p},p)d^*$ are shown in Fig. 1, where the errors are purely statistical. For comparison the elastic scattering analyzing power is shown as the smooth curve. As can be seen, A_y in the breakup reaction reaches substantial values at angles greater than 70° and its angular distribution is quite similar to that of the elastic $A_y(\theta)$. A comparable behavior of the deuteron vector analyzing power was observed in the $^1H(\vec{d},p)d^*$ reaction[1]. This similarity between the elastic and inelastic analyzing power is rather unexpected in view of the results that were reported by Brückmann et al.[6]. In the analysis of their $^1H(d,d^*)p$ data at 52.3 MeV, they identified the separate contributions to the cross section from the production of singlet and triplet np pairs,

$$(d\sigma/d\Omega) = (d\sigma/d\Omega)_s + (d\sigma/d\Omega)_t. \tag{1}$$

They found that between Θ_{cm} = 85° and 150° the production of singlet pairs exceeded that of triplet pairs ($E_{np} \leqslant$ 1 MeV), with the ratio $(d\sigma/d\Omega)_s/(d\sigma/d\Omega)_t$ reaching a value of about 10 near 120°. Thus, even though the angular dependence of $(d\sigma/d\Omega)_t$ was similar in shape to that of the elastic cross section, the d* production cross-section (given in eq.(1)) showed little resemblance to the elastic angular distribution. Furthermore, the recent exact three-body calculation of Kluge et al.[5] very nicely reproduces the ratio of the singlet to triplet d* production cross sections for E_{np} = 0.

We can, in the same way, express our analyzing power results as the sum of the singlet and triplet d* production contributions,

$$A_y = \frac{A_y^s (d\sigma/d\Omega)_s + A_y^t (d\sigma/d\Omega)_t}{(d\sigma/d\Omega)_s + (d\sigma/d\Omega)_t}. \tag{2}$$

We then see, from the similar angular distributions of A_{y,d^*} and $A_{y,elastic}$ shown in Fig. 1 that a tenable conclusion is that,

(a) $A_{y,d*}^{t}(\Theta)$ is similar to that of $A_{y,elastic}(\Theta)$ and, (b) $A_{y,d*}^{s}(\Theta)$ is similar to $A_{y,d*}^{t}(\Theta)$. Condition (a) is certainly reasonable. However condition (b) is, a priori, quite unexpected, particularly in view of the marked dissimilarities of $(d\sigma/d\Omega)_{s}$ and $(d\sigma/d\Omega)_{t}$ at the nearby equivalent energy $E_{p} = 26.1$ MeV. Thus, there is an obvious need to include in the exact three-body calculations the more realistic N-N tensor and P-wave interactions in an effort to explain these polarization results.

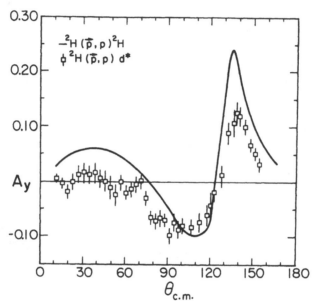

Fig. 1

References

* Work performed under the auspices of the U.S. Energy Research Development Administration.
+ National Research Council of Canada, Postdoctoral Fellow.
‡ On leave of absence from the University of Basel, Switzerland.

1) H. E. Conzett, in Proc. Int'l. Conf. on Few Body Problems in Nuclear and Particle Physics, Quebec, Canada, 27-31 August 1974 (to be published); F. N. Rad, et al., Bull. Amer. Phys. Soc. 20 (1975) 578.
2) J. Arvieux, et al., Nucl. Phys. A150 (1970) 75.
3) C. O. Blyth, et al., Proc. Int'l. Conf. on Few Body Problems in Nuclear and Particle Physics, Quebec, Canada, 27-31 August 1974.
4) R. T. Cahill and I. H. Sloan, Nucl. Phys. A165 (1971) 161; W. Ebenhöh, Nucl. Phys. A191 (1972) 97.
5) W. Kluge, et al., Nucl. Phys. A228 (1974) 29.
6) H. Brückmann, et al., Nucl. Phys. A157 (1970) 209.

POLARIZATION TRANSFER IN THE D($\vec{p}$,$\vec{n}$)2p REACTION FROM 10 TO 15 MeV[*]

R.L. Walter, R.C. Byrd, and P.W. Lisowski,
Duke University and Triangle Universities Nuclear Laboratory
and
T.B. Clegg, University of North Carolina and TUNL

Recent advances in the field of few-nucleon theory have made it possible to calculate accurate representations of cross section and polarization data for the mass-three system. Although most of the theoretical work has dealt with two-particle final-state reactions, detailed calculations of cross sections for the N-D break-up channel are beginning to appear[1]. As a benchmark for future theoretical calculations which include spin-dependent forces, a determination of the polarization transfer coefficient $K_y^{y'}$ for the D($\vec{p}$,$\vec{n}$)2p reaction at θ = 0° has been made at five proton energies from 10.5 to 15.0 MeV.

About 50 nA of proton beam with 80% polarization was incident on a 700 keV thick deuterium gas target. A technique similar to that given in ref.[2] was used to measure the polarization of the neutrons, which in this reaction are emitted with a continuous energy distribution. Briefly, the pulse-height spectrum of recoil energies in a ^{4}He scintillator was gated by pulses from two side detectors positioned at θ = 120° (lab). The resulting left- and right-gated spectra were then used to determine the polarization and energy distribution of neutrons emitted at 0° in the D($\vec{p}$,$\vec{n}$)2p reaction. Such a spectrum generated with a 15 MeV proton beam incident on deuterium is shown in fig. 1a. To avoid computational difficulty in obtaining a preliminary analysis, a gaussian function which gave a good representation of the polarimeter resolution was used to separate the high-energy p-p final-state interaction neutron contribution from the neutron contribution arising from other break-up processes. In fig. 1b is shown the observed asymmetry calculated for each channel in the energy spectrum. Figure 1c presents the values of $K_y^{y'}$(0°) calculated from a five-channel average of the asymmetry data.

Neutrons associated with the p-p final state interaction give rise to the highest energy pulses in the spectrum. From the values for $K_y^{y'}$(0°), it is clear that these neutrons have undergone a spin flip, while neutrons from the other break-up processes have not. A negative value for the p-p final state region is predicted by a qualitative argument based on the Pauli exclusion principle: p-p pairs of low relative energy must be in a singlet state and must therefore have been formed from an incident polarized proton and a proton of opposite spin from the target deuteron. To show the dependence on proton bombarding energy, $K_y^{y'}$(0°) was extracted from the regions of the maxima in the yield at each energy using the gaussian representation discussed above. The results are shown in fig. 2. It is interesting to note that the opposite polarizations seen in fig. 2 occurred at all incident proton energies; in fact, the resulting polarization transfer coefficients show no energy dependence over the 10.5 to 15.0 MeV range.

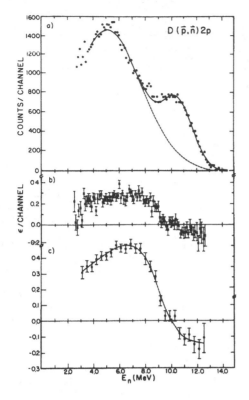

Fig. 1. Experimental results for the $D(\vec{p},\vec{n})pp$ reaction at E_p = 15 MeV. The curve in 1c is a guide to the eye.

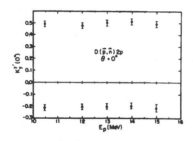

Fig. 2. Polarization transfer coefficients extracted from the two regions of maximum neutron yield of the spectra. The triangles represent data from the p-p interaction region.

References
*Work supported in part by U.S. Energy Research and Development Administration.

1) J.M. Wallace, Phys. Rev. 9C (1974) 897
2) P.W. Lisowski, R.L. Walter, C.E. Busch and T.B. Clegg, Nucl. Phys. A242 (1975) 298

POLARIZATION TRANSFER IN p-d BREAKUP REACTIONS*

R. G. Graves, M. Jain,[†] H. D. Knox,[††] J. C. Hiebert, E. P. Chamberlin,
R. York and L. C. Northcliffe
Cyclotron Institute and Physics Department
Texas A&M University, College Station, Texas 77843, U.S.A.

As part of a comprehensive study of the neutrons produced in the p-d breakup reaction at $E_{cm} \approx 14$ MeV and $\theta_n(lab) = 0°$ and $18°$, we have previously reported measurements of the neutron differential cross section[1] and the neutron polarization[2] for the D(p,n)2p and H(d,n)2p reactions. To complement these results we have now used the polarized proton and deuteron beams from the cyclotron to obtain data on the polarization transfer in both reactions at $E_{cm} \approx 14$ MeV and $\theta_n(lab) = 0°$ and $18°$. While the primary objective is to determine the transverse polarization transfer coefficient $K_y^{y'}(E_n)$, the data are analyzed in a way which also provides a simultaneous determination of the analyzing power $A_y(E_n)$ and the polarization parameter $P(E_n)$ for the reaction. The analysis of the multiple detector multi-parameter data has been completed for the $D(\vec{p},\vec{n})2p$ measurement at $\theta_n(lab) = 18°$ and is in progress for the other three experiments.

The experimental values of $K_y^{y'}(E_n)$ for the $D(\vec{p},\vec{n})$ reaction at $E_p = 20.4$ MeV and $\theta_n = 18°$ are shown in fig. 1(a), and the values of $A_y(E_n)$ for the corresponding $D(\vec{p},n)$ reaction are shown in fig. 1(b). The values obtained for $P(E_n)$ for the $D(p,\vec{n})$ reaction are not shown but are consistent with our previous measurements of P using an unpolarized beam[2]. Note that both $K_y^{y'}$ and A_y are negative for the neutrons of highest energy (the region associated with both quasi-free scattering and the p-p final state interaction), while they are positive for the lower energy neutrons (including those in the region around 7.5 MeV associated with the n-p final state interaction).

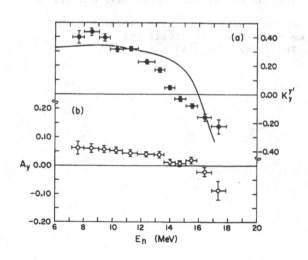

Fig. 1. Spin observables for the D(p,n)pp reaction at $\theta_n = 18°$ for $E_p = 20.4$ MeV. (a) Measured transverse polarization transfer coefficient $K_y^{y'}$ (closed circles) and the prediction (solid line) given by a three-body separable-potential model; (b) Measured analyzing power A_y (open circles).

The measured $K_y^{y'}$ values are in agreement with existing impulse approximation (IA) calculations[3,4] for the highest energy neutrons but differ from the IA predictions in the region of low neutron energy where the IA is expected to be unreliable[4]. Better qualitative agreement with the trend of the data is shown by a calculation of $K_y^{y'}$ based on an exact solution of the three-particle Faddeev equations for separable spin-dependent S-wave nucleon-nucleon potentials[5], which is shown by the solid line in fig. 1(a). The agreement is only qualitative, however, and there are significant discrepancies between the calculated and measured values in the region of intermediate neutron energy where $K_y^{y'}$ changes sign. The inadequacy of the theory is due in part to the fact that the M-matrix used includes only S-waves. One consequence of this restriction to the S-wave interaction is that the theory cannot reproduce the A_y values shown in fig. 1(b) since it can only give zero for P and A_y. As a result, terms in the expression for $K_y^{y'}$ involving P and A_y vanish, and the prediction for $K_y^{y'}$ assumes a zero-order value: the ratio of the outgoing neutron polarization to the incoming proton polarization. The discrepancy between the theoretical and experimental $K_y^{y'}$ values must involve more than the effect of non-zero terms containing P and A_y, however, because the discrepancy is greatest at $E_n \approx 14$ MeV where A_y and P are very small. Higher partial waves must be responsible for this discrepancy as well as for the nonzero A_y and P values. The present data clearly indicate the need for a three-body theory with more realistic NN forces.

References

*   Supported in part by the U. S. National Science Foundation
†   Present address: Los Alamos Scientific Lab., Los Alamos, NM
†† Present address: Accelerator Lab., Ohio University, Athens, OH

1) R. G. Graves, M. Jain, L. C. Northcliffe and F. N. Rad, Proceedings of the International Conference on Few-Body Problems in Nuclear and Particle Physics, Quebec, Canada (1974)
2) F. N. Rad, L. C. Northcliffe, J. G. Rogers and D. P. Saylor, Phys. Rev. Lett. 31 (1973) 57; F. N. Rad et al., Phys. Rev. C 8 (1973) 1248
3) K. Ramavataram and Q. Ho-Kim, Nucl. Phys. A156 (1970) 395
4) G. V. Dass and N. M. Queen, J. Phys. A1 (1968) 259
5) M. Jain and G. Doolen, Phys. Rev. C 8 (1973) 124

"DECAY" OF A TWO-NUCLEON SYSTEM IN FINAL STATE INTERACTION

R. van Dantzig, J.R. Balder, G.J.F. Blommestijn,
W.C. Hermans, B.J. Wielinga
IKO, Amsterdam, the Netherlands.

Polarized nuclear systems with non-zero total orbital angular momentum decay non-isotropically in their own CM-system. Conversely, an observed anisotropy in the decay obviously can only arise from polarization and thus from higher than $l=0$ partial wave components. In this sense, anisotropy may be a measure for the relative importance of $l>0$ amplitudes.

Here we report on an application of this simple idea to the three-nucleon reaction $p+d \rightarrow p+(pn)$ at $E_d = 26.5$ MeV, where (pn) is a proton-neutron subsystem in final state interaction (FSI), with given relative energy T_{pn}. As is well known, the cross section for this process is sharply peaked near $T_{pn} = 0$ due to the $^1S_0$ FSI. A p-n pair in a state corresponding to the $^1S_0$ FSI peak is often denoted as a nearly bound singlet "deuteron" (d^*). For the d^* concept to be a valid one, the system should be in a pure $l=0$ state; consequently, an isotropic decay of the (pn) system should be expected. The degree of anisotropy as a function of T_{pn} could thus yield information on the reaction mechanism.

The present investigation aims in particular at experimental evidence for interference of different components of the doublet break-up amplitude, corresponding to permutations of different nucleon pairs.

Aaron and Amado[1] were the first to show that in separable potential Faddeev calculations the FSI peak contains - contrary to expectation based on the Watson-Migdal model [2] - essential contributions from the other FSI-diagrams, which through interference actually determine the width of the FSI peak. This increased the already existing conceptual difficulties connected with the FSI process. In particular, one should note that the FSI is not truly a two-step process. There is no resonant intermediate state (with definite quantum numbers) but rather a peculiar threshold behaviour, that was previously considered to be simply and almost exclusively linked to the low-energy nucleon-nucleon scattering phase shifts. If the p-n $^1S_0$ FSI peak in the present experiment would contain significant contributions from interactions with the third (spectator) nucleon, then $l>0$ components might be manifest in the directional correlation of the "d^*"-decay.

The measurements[4] presented here, were obtained with the BOL 4π-type multidetector system[3] and the synchrocyclotron at our institute..The cross section was deter-

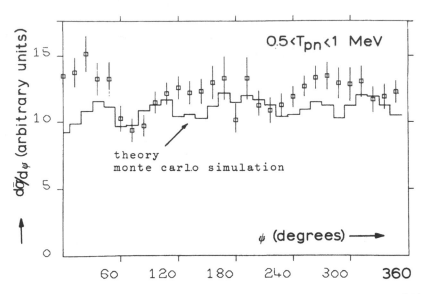

$0.5 < T_{pn} < 1$ MeV

$\frac{d\sigma}{d\psi}$ (arbitrary units)

theory
monte carlo simulation

ψ (degrees) ⟶

mined as a function of the polar and azimuthal angle (ψ) in the (pn) CMS, the z-axis being taken in the recoil direction. The data studied were integrated over all recoil directions and sliced in T_{pn}-bins of 0.5 MeV. All measurements are compared with a Faddeev calculation[5] using a local S-wave potential. The data with $T_{pn} > 1$ MeV are quite anisotropic as one would expect since this is outside the FSI region. The data relevant for the present investigation are below $T_{pn} = 1$ MeV. In the lower energy region, $T_{pn} < 0.5$ MeV, the decay was found to be isotropic and in quantitative agreement with the theory. In the interval $0.5 < T_{pn} < 1$ MeV a discrepancy develops (see figure). Here the theory still predicts isotropy but the data are anisotropic ($\sim 20\%$) and fall about the same percentage above the theoretical prediction. Apparently, a "l>0 source" not included in the theory (therefore another one than the interaction with the spectator nucleon) is responsible for the observed anisotropy.

References
1) R. Aaron and R.D. Amado, Phys. Rev. 150 (1966) 857
2) K.M. Watson, Phys. Rev. 88 (1952) 1163
3) L.A.Ch. Koerts et al., Nucl. Instr. and Meth. 92 (1971) 157
4) B.J. Wielinga, thesis 1972, Amsterdam
   B.J. Wielinga et al., submitted to Phys. Rev.
   Collaboration with prof. Ivo Slaus is greatfully acknowledged.
5) W.M. Kloet and J.A. Tjon, Nucl. Phys. A210 (1973)380
   Annals of Physics 79 (1973) 407.
We are indebted to these authors for their theoretical contributions to this work.

Symmetry Properties

DEVIATIONS FROM CHARGE-SYMMETRY IN THE REACTION ^{4}He($\vec{d}$,t)^{3}He[+]

W. Dahme and P.J.A. Buttle[+]
Sektion Physik der Universität München,
8046 Garching, Germany

H.E. Conzett, J. Arvieux[§], J. Birchall[■],and R.M. Larimer
Lawrence Berkeley Laboratory, University of California,
Berkeley, California 94720, U.S.A.

For nuclear reactions in which the final nuclei belong to the same isospin multiplet, Simonius[1] showed that the concept of charge symmetry of the nuclear forces requires the angular distributions of the cross section to be symmetric and the vector analyzing power to be antisymmetric with respect to 90° in the c.m. system. Experimental evidence of deviations from this theorem was reported e.g. by Gross et al.[2] for the cross section of the reaction ^{2}H(α,t)^{3}He. In order to provide additional information for a more detailed study of the symmetry violating mechanism we measured the vector analyzing powers iT_{11} of the mirror reactions ^{4}He($\vec{d}$, t)^{3}He and ^{4}He($\vec{d}$,^{3}He)^{3}H at $\bar{E}_d$ lab = 32.0 and 41.0 MeV[3] The angular distributions of iT_{11} for the ^{3}He and triton channel differ by as much as 0.08 (fig.1).

The simplest method of taking into account Coulomb effects is a coherent superposition of slightly different proton and neutron transfer amplitudes in a DWBA calculation, as also considered by Gross et al.: $T(\vec{k}_f) = T_p(\vec{k}_f)+T_n(-\vec{k}_f)$. Choosing the coordinate system according to the Madison Convention we get, using Satchler's[4] notation:

$$T_{M_B m_b}^{M_A m_a}(\Theta,0) = T_{pM_B m_b}^{M_A m_a}(\Theta,0) + (-1)^m\, T_{nm_b M_B}^{M_A m_a}(\pi-\Theta,0)$$

The reaction in question also is of a spin type not usually treated in DWBA formalisms: $1 + 0 \rightarrow 1/2 + 1/2$. Because of the similarity of ^{3}He and ^{3}H we assume their LS interactions to be identical and use channel spin $\vec{S}$ in the final channel. The zero-range DWBA code DWUCK was modified to sum over S = 0,1 and to add the two transfer amplitudes coherently.

The optical model parameters used in these DWBA calculations were obtained by fitting to the 15 to 45 MeV elastic $\vec{d}$+^{4}He vector analyzing power data of Leemann et al.[5] along with the cross sections of Willmes[6] and the 6 to 32 MeV elastic ^{3}He+^{3}H cross section data of Bacher et al.[7] and Batten et al.[8]. No ^{3}He+^{3}H analyzing power data are available at present. The derived parameters are listed in table 1 for

Table 1. Optical model parameters (r_c = 1.3 fm)

	E_d	V_o	r_o	a_o	W_V	W_S	r_I	a_I	V_{LS}	r_{LS}	a_{LS}
d+^{4}He	32.0	68.7	1.18	0.56	0.0	2.55	2.56	0.5	4.6	1.18	0.5
	41.0	62.2	1.18	0.56	0.0	3.27	2.56	0.5	4.6	1.18	0.5
^{3}H+^{3}He	32.0	192.5	1.30	0.60	0.74	0.0	1.60	0.5	8.0	1.30	0.5
	41.0	205.0	1.30	0.60	17.5	0.0	1.60	0.5	8.0	1.30	0.5

498

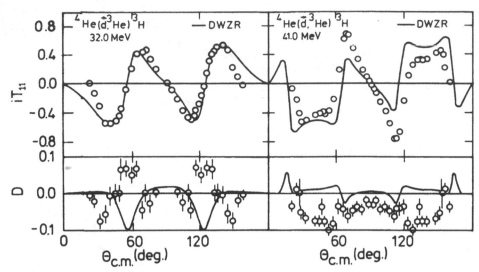

Fig.1. Vector analyzing powers $iT_{11}(^3He)$ (top) and differe-rence $D=iT_{11}(^3He)-iT_{11}(^3H)$ (bottom) with DWBA calculations.

the two energies of interest here. Good fits are achieved over the entire energy region of the $\vec{d}+^4He$ data. Fits of less quality were obtained for the $^3He+^3H$ data. The bound state form factors were calculated as by Gross et al.

Fig.1 shows a comparison of our preliminary calcula-tions with the experimental data. The structure is reprodu-ced but the quantitative agreement is not satisfactory, es-pecially when expressed in terms of the Barshay-Temmer quan-tity D. The fit to the cross section is of similar quality. At present, the discrepancies seem to be due to the uncer-tainties of the $^3He+^3H$ optical potential, but it is still questionable whether such a few nucleon problem can be ade-quately described by a simple DWBA calculation.

We are grateful to J. Raynal for providing the optical model code MAGALI and to P.D. Kunz for the DWBA code DWUCK4.

References

† Work performed in part under the auspices of the U.S. AEC
  and supported in part by the BMFT
+ Permanent address: University of Manchester, England
§ Present address: I.S.N., University of Grenoble, France
⌧ Present address: University of Basel, Switzerland

1) M.Simonius, Phys. Letters 37B(1971)446
2) E.E. Gross et al., Phys.Rev. C5(1972)602
3) W. Dahme et al., Proc.Int.Conf.on Nucl.Phys., Munich 1973
   (North Holland, Amsterdam 1973) p.444
4) G.R. Satchler, Nucl.Phys. 55(1964)1
5) Ch. Leemann et al., Bull.Am.Phys.Soc. 17(1972)562
6) H. Willmes, private communication
7) A.D. Bacher et al., Nucl.Phys. A119(1968)481
8) R.J. Batten et al., Nucl.Phys. A151(1970)56

COMPARISON OF P_y AND A_y FOR ^{9}Be(p,n)^{9}B AT 8.1 AND 9.1 MeV[*]

P.W. Lisowski, G. Mack, R.C. Byrd, W. Tornow,
S.E. Skubic and R.L. Walter
Duke University and Triangle Universities Nuclear Laboratory
and
T.B. Clegg, University of North Carolina and TUNL

Differences between P_y and A_y for mirror reactions may be capable of yielding information about isospin violation terms in nuclear interactions [see ref.[1]]. Thus far such studies have been reported only for T(p,n)^{3}He and ^{7}Li(p,n)^{7}Be. In the first case, large differences were observed; in the second, considerable similarities were reported, particularly at 3 MeV, the highest energy for which A_y was measured. In order to extend the comparison of these two observables to heavier targets, the ^{9}Be(p,n)^{9}B and ^{15}N(p,n)^{15}O reactions have been investigated at our laboratory. This report gives the preliminary results of the ^{9}Be(p,n)^{9}B experiment, which obtained A_y data in 10° steps from 20° to 100° at 8.1 and 9.1 MeV and P_y data at 8.1 MeV for the same angles.

Two different experimental arrangements were used in the A_y measurements. In the first, asymmetries were determined from measurements of the relative cross section for spin-up and spin-down protons incident on a single organic scintillator. The A_y values thus acquired were significantly different from the P_y values reported earlier by Walker et al.[2]. To verify the accuracy of this method, the measurement was repeated using a left-right pair of high-pressure ^{4}He recoil scintillators. Such detectors have a low sensitivity to γ-rays and a well-known response to monoenergetic neutrons; these features should allow a better estimate of the contribution from background neutrons, that is, those not from the ^{9}Be(p,n$_o$)^{9}B reaction. For the present reaction this aspect is particularly important since 3- and 4-body break-up produces a continuous distribution of neutron background which can extend into the neutron energy region of interest. A preliminary analysis of the second set of data showed good agreement to our earlier results and correspondingly poor agreement with the P_y data of Walker et al. Therefore, in order to obtain measurements of P_y and A_y under as nearly identical circumstances as possible, an angular distribution of P_y at 8.1 MeV was measured using the experimental arrangement outlined in the ^{15}N(p,n)^{15}O paper presented at this conference[1]. In fig. 1 our 8.1 MeV A_y data is compared to this P_y measurement and to that of Walker et al. Also shown are the A_y values obtained in the present work and the 9.1 MeV P_y data from ref.[2].

The A_y data is still preliminary. The contribution from the break-up neutrons has not yet been suitably subtracted, but is expected to increase the magnitude of the final values by less than one standard deviation. Such a shift clearly would tend to bring the A_y values into even closer agreement with those for P_y at 8.1 MeV.

In view of the 8.1 MeV results, it is difficult to assess the significance of the sizeable differences between A_y and P_y in the 9.1 MeV data. A check on the P_y values reported in ref.[2] will therefore be conducted in the near future at our laboratory.

500

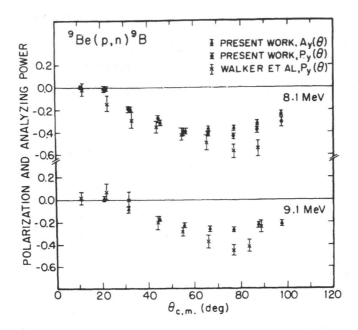

Fig. 1. Comparison of polarization and analyzing power angular distributions for ^{9}Be(p,n)^{9}B at 8.1 and 9.1 MeV.

References
*Work supported in part by U.S. Energy Research and Development Adminis-
tration.

1) R.C. Byrd, P.W. Lisowski, S.E. Skubic, W. Tornow, R.L. Walter and
    T.B. Clegg, these proceedings, p.501
2) B.D. Walker, C. Wong, J.D. Anderson and J.W. McClure, Phys. Rev.
    137 (1965) B1504

NEUTRON POLARIZATION AND ANALYZING POWER IN THE ^{15}N(p,n$_o$)^{15}O REACTION*

R.C. Byrd, P.W. Lisowski, S.E. Skubic, W. Tornow, and R.L. Walter
Duke University and Triangle Universities Nuclear Laboratory
and
T.B. Clegg, University of North Carolina and TUNL

Although the (p,n) reaction between mirror nuclei has long been used to explore the effective neutron-proton interaction[1,2], special attention has recently been given to differences between the polarization P and analyzing power A[3,4]. Following Conzett[4], we consider a measurement of P in the reaction B(p,$\vec{n}$)B', for which invariance under time reversal I gives

$$B(p,\vec{n})B' = B'(\vec{n},p)B \qquad\qquad (P = A_I).$$

To compare P and A, we must somehow transform A_I into A, i.e.,

$$B'(\vec{n},p)B \text{ into } B(\vec{p},n)B'$$

Noting that an isospin rotation R of T_z into $-T_z$ will interchange neutrons and protons, we apply R to A_I to obtain

$$B'(\vec{n},p)B = B_R'(\vec{p},n)B_R \qquad\qquad (A_I = A_{IR}).$$

Finally, comparing A_{IR} to A, i.e.,

$$B_R'(\vec{p},n)B_R \text{ to } B(\vec{p},n)B' \qquad\qquad (A_{IR} \rightarrow A),$$

we see that equality of P and A requires $B_R' = B$ and $B_R = B'$. For (p,n) reactions on mirror nuclei, however, these conditions follow from the charge symmetry of (n-n) and (p-p) forces, a hypothesis violated only by the electromagnetic part of the interaction between the core and the outer nucleon. Therefore, any difference between P and A in these reactions should be due to deviations from exact charge symmetry between the extra-core bound-state wave functions for the initial neutron and final proton. To investigate these deviations, we have measured values of P and A in the ^{15}N(p,n$_o$)^{15}O reaction to complement those for ^{9}Be(p,n$_o$)^{9}B presented in another paper contributed to this conference. The preliminary ^{15}N results are reported here.

Analyzing powers A_y in ^{15}N($\vec{p}$,n$_o$)^{15}O were measured at 10° intervals from 20° to 100° at energies of 10.3 and 11.3 MeV, and at 60° in 100 keV steps for the energies in between. Sixty-five nA of 75% polarized proton beam was incident on a 150 keV thick ^{15}N$_2$ gas target, which was viewed by a pair of high-pressure helium recoil scintillators subtending less than 5°. False asymmetries from detection efficiency and beam current integration were reduced by reversing the proton spin direction at the ion source.

Data reduction for the A_y values included combination of left and right detector spectra for opposite spin directions and gas-out background subtraction. Asymmetries were calculated for the same regions about the recoil edge in each spectrum; dividing these values by the averaged incident beam polarization for each angle produced the analyzing powers A_y shown in fig. 1. Further corrections to these values are expected to be slight.

The polarization P_y of neutrons in ^{15}N(p,$\vec{n}_o$)^{15}O was measured in 10° steps from 10° to 100° at 10.3 and 11.3 MeV. A polarimeter with angular resolution better than 5° was set up with a spin precession solenoid, a high-pressure helium scintillator, and a pair of plastic side detectors. This system allowed the separation of neutrons from gammas by time-of-

flight methods. The cancellation
of instrumental asymmetries due to
both charge integration and detector
efficiencies was achieved by neutron
spin precession.

To obtain the polarization results, left and right spectra were
combined, backgrounds were subtracted,
and asymmetries were calculated.
The resulting values, shown in fig. 1,
include correction factors for geometry, multiple scattering, and ^{4}He
analyzing power in the polarimeter;
assigned errors reflect statistical
and background uncertainties. The
final analysis should yield nearly
identical results.

These measurements are expected to support the conclusion
that any breaking of charge symmetry between ^{15}N and ^{15}O in this
reaction causes at most slight
differences between P and A.
Further discussion of these results
is presented in a related paper by
L.G. Arnold.

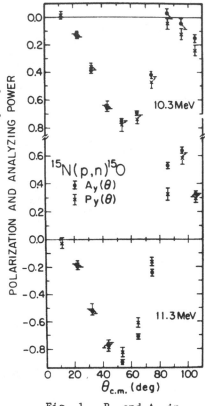

Fig. 1. P_y and A_y in
^{15}N(p,n$_o$)^{15}O.

References
*Work supported in part by U.S. Energy Research and Development Admini-
stration.

1) S.D. Bloom, N.K. Glendenning and S.A. Moszkowski, Phys. Rev. Letters
   3 (1959) 98
2) B.D. Walker, C. Wong, J.D. Anderson, J.W. McClure and R.W. Bauer,
   Phys. Rev. 137 (1965) B347
3) L.G. Arnold, R.G. Seyler, T.R. Donoghue, L. Brown and U. Rohrer,
   Phys. Rev. Letters 32 (1974) 310
4) H.E. Conzett, Phys. Rev. Letters 51E (1974) 445

POLARIZATION PHENOMENA AND ISOSPIN CONSERVATION
IN QUASI-ELASTIC REACTIONS

L. G. Arnold
Department of Physics
The Ohio State University, Columbus, Ohio USA

An isospin multiplet is a set of states with the same spin S, iso-spin T, and nucleon number $A = N + Z$. The $2T + 1$ members of the multiplet differ by their isospin projection $T_3 = (N - Z)/2$. Transitions between adjacent members of a multiplet are a class of reactions in which the quantum numbers (S,T,A) of the target nucleus and (s,t,a) of the incident projectile are not changed by the reaction process; the particles emerging from the reaction have quantum numbers (S,T,A) for the residual nucleus and (s,t,a) for the detected particle. This class of reactions is often called quasi-elastic. It differs from elastic scattering in the following way: the isospin projections, T_3 for the target and t_3 for the projectile, are not changed by an elastic process; T_3 is increased (decreased) one unit and t_3 is decreased (increased) one unit by a quasi-elastic process.

A general description of elastic and quasi-elastic processes is obtained by constructing an M matrix that incorporates both spin and isospin degrees of freedom. Conditions placed on M are charge conservation and invariance under space rotation, space reflection, and time reversal. Linearly independent operators in spin space are combined with functions of the center of mass momenta in the usual way[1] to yield forms invariant under space rotation and reflection; forms invariant under time reversal are denoted by M_+, forms which change sign by M_-. These forms are combined with charge conserving linearly independent operators in isospin space $O_\pm$ to yield product forms, M_+O_+ and M_-O_-, that are invariant under time reversal. The operators $O_\pm$ may be constructed from irreducible spherical isospin tensors. Tensors T_{kq} with $q = o$ conserve charge, tensors T_{ko} with k even conserve charge parity, and tensors T_{oo} conserve isospin.

Expressions for the observables are obtained from the standard density matrix formalism by replacing the isospin operators $O_\pm$ with matrix elements appropriate to the process being considered. The validity of this procedure is a consequence of charge conservation.

The elastic scattering M matrix and the polarization — analyzing power equality for elastic scattering are recovered by noting that the diagonal matrix elements of the operators O_- are identically zero.

If product forms M_-O_- are not allowed, the polarization $P_{y'}$ in a quasi-elastic reaction initiated by unpolarized projectiles is identically equal to the analyzing power A_y of the reaction initiated by polarized projectiles[†]. Product forms M_-O_- are not allowed under the following conditions:

1. <u>Isospin Conservation</u> Tensors T_{oo} are time reversal invariant; hence, operators O_- are not allowed if isospin is conserved.
2. <u>Charge Parity Conservation for Mirror Transitions ($t = \frac{1}{2}$ on $T = \frac{1}{2}$)</u> Tensors T_{ko} with k even are time reversal invariant in this special case; hence, operators O_- are not allowed.
3. <u>Spin $s = \frac{1}{2}$ on spin S = 0</u> Space reflection invariance does not allow forms M_- for this spin system.

[†]The notation for observables is taken from a review by Ohlsen[2].

Conditions 1 and 2 are due to Conzett[3] and are referred to as Conzett's Theorem[4]. Condition 3 is a further illustration of the limitations of this spin system for tests of fundamental symmetries.

The elastic scattering polarization transfer coefficient identity, $K_x^{z'} = -K_z^x$, also holds in a quasi-elastic reaction if any of conditions 1-3 are satisfied.

The simplest product forms M_-O_- which violate isospin conservation are linear in the spins and isospins of the projectile and target. They are

$$(\vec{s} \cdot \hat{x}\, \vec{S} \cdot \hat{z} \pm \vec{s} \cdot \hat{z}\, \vec{S} \cdot \hat{x})(\vec{t} \times \vec{T})_3$$

where $\hat{x}$ is in the direction of the momentum transfer, $\hat{y}$ is perpendicular to the reaction plane, and $\hat{z} = \hat{x} \times \hat{y}$. These are the only forms allowed for spin $\frac{1}{2}$ on spin $\frac{1}{2}$ mirror transitions. The operator $(\vec{t} \times \vec{T})_3$ is of the same order in the isospins of the projectile and target as the isospin conserving $\vec{t} \cdot \vec{T}$ operator, and would be present in Lane's model[5] of quasi-elastic reactions if that model were complete to terms linear in the isospins of the projectile and target.

Within the past three years comparisons of $P_{y'}$ and A_y have been made for the quasi-elastic reactions ^{3}H(p,n)^{3}He, ^{7}Li(p,n)^{7}Be, ^{9}Be(p,n)^{9}B, and ^{15}N(p,n)^{15}O. To the extent that isospin impurities of the target nucleus ground state can be ignored, these comparisons involve mirror transitions and are therefore tests of charge parity conservation. The comparisons have been made for energies corresponding to the 10-20 MeV range of excitation energies. According to Wilkinson's hypothesis[6], isospin mixing of compound states and isospin violating reaction mechanisms are most likely to be observed in this energy range for light nuclei. Differences between $P_{y'}$ and A_y have been found in the vicinity of the 21.9 MeV 2^- state of ^{4}He, the 19.4 MeV 1^- state of ^{8}Be, and between 14 and 17 MeV in ^{10}B where several broad states have been tentatively identified. It should also be noted that $P_{y'}$ and A_y were found to be equal in the vicinity of other states in each of these nuclei, the most interesting case being the isospin mixed 19.02 and 19.22 MeV 3^+ states of ^{8}Be. General trends have not as yet emerged from these studies; however, an analysis[7] of the difference between $P_{y'}$ and A_y for the ^{3}H(p,n)^{3}He reaction indicates that comparisons of these observables can lead to useful information about the structure of light nuclei.

Additional comparisons of $P_{y'}$ and A_y for quasi-elastic reactions would be desirable.

References

1) L. Wolfenstein and J. Ashkin, Phys. Rev. 85 (1956) 947
2) G. G. Ohlsen, Rep. Prog. Phys. 35 (1972) 717
3) H. E. Conzett, Phys. Letters 51B (1974) 445
4) L. Brown, U. Rohrer, L. G. Arnold, and R. G. Seyler, Carnegie Inst. Wash. Yearbook 73 (1974) 194
5) A. M. Lane, Nucl. Phys. 35 (1962) 676
6) D. H. Wilkinson, Phil. Mag. 1 (1956) 379
7) L. G. Arnold, R. G. Seyler, T. R. Donoghue, L. Brown, and U. Rohrer, Phys. Rev. Letters 32 (1974) 310

Light Nuclei

POLARIZATION IN p-t AND t-t SCATTERING

N.N. Putcherov, S.V. Romanovsky, A.E. Borzakovsky
and T.D. Chesnokova
Institute of Nuclear Research,
Ukranian Academy of Science, Kiev, URSS

The purpose of the present double scattering experiment was to measure the polarization in p-t and t-t scattering. The protons scattered by carbon at 45° hat the energy 6,02±0,16 MeV and polarization -0,56±0,05. After that polarized protons were scattered by the second (Ti-T) target and left-right asymmetry was measured. The values of the proton polarization are represented in the table 1.

Table 1. Polarization of p-t scattering

θ_L^o	P
40	-0,18 ± 0,07
45	-0,14 ± 0,07
50	0,04 ± 0,05
65	0,05 ± 0,07

These experimental results agree well with experimental data at the nearest energies.

The second part of the experiment concerned obtaining polarized tritons using 27 MeV α-particles. After scattering the recoil tritons at the angle 45° relative to the initial α-particle direction have energy 13,04±0,4 MeV. They were collimated and then they were scattered at the second (Ti-T) target. Scattered tritons and products of other nuclear reactions were detected with the telescope of semi-conductor counters ΔE+E and then they were analysed according to their mass and energy. The telescopes of the counters arranged on the left and on the right from the target made it possible to determine asymmetry in the t-t scattering. The table 2 gives the values of asymmetry at the different angles of scattering.

Table 2. Asymmetry A and Polarization P in t-t scattering

θ_{CM}^o	A	P
42	0,04 ± 0,03	0,05 ± 0,03
60	-0,11 ± 0,03	-0,13 ± 0,03
90	-0,01 ± 0,04	-0,01 ± 0,05

Evaluating the possible recoil tritons polarization the differential cross section of the α-t scattering was measured at energy 27,2 MeV. The phase shift analyses allowed to estimate the recoil tritons polarization. The values of polarization in t-t scattering if P_t= 0,86 are represented in the table 2.

ANALYZING POWERS OF p-^{3}H AND p-^{3}He ELASTIC SCATTERINGS FOR E_p < 4 MeV[†]

J.C. Volkers, L. Arnold, T.R. Donoghue, Sr. Mary A. Doyle, L.J. Dries,
W.S. McEver, and J.L. Regner
Department of Physics, The Ohio State University, Columbus, Ohio 43210

Analyzing powers for p-^{3}H and p-^{3}He elastic scattering processes have been measured for E_p < 4 MeV. The motivation for these measurements was the continuing uncertain nature of the level structure of the A = 4 nuclei first delineated in the analysis of Werntz and Meyerhof[1] who reported two orderings of the A = 4 levels were possible. More recently, it was noted that neither of these sets of levels can describe the analyzing power[2] or polarization transfer coefficients[3] in the T(p,n)^{3}He reaction and hence some revision in the A = 4 level structure is clearly required. It was suggested that the comprehensive multi-channel, charge-independent R-matrix calculations at Los Alamos[4] could benefit from accurate polarization data on all reaction channels, particularly at the lower energies where the number of open reaction channels is low and the number of contributing states is few. As this makes the analysis simpler, a firm base for analyses of higher energy data can be established.

The measurements were made using the atomic beam polarized ion source[5] that operates in the high voltage terminal of our CN Van de Graaff accelerator. The polarized protons of ~ 30 nA with p_y = 0.60 had spins oriented normal to the reaction plane in a scattering chamber that could be rotated about the beam axis. Solid state detectors arranged symmetrically in pairs about the beam axis were used, with four such pairs located at fixed 10^0 intervals. Chamber rotation techniques interchanged the detectors in the measurements. For the p-^{3}He measurements, a 3.8 cm dia. gas cell with 2.5 μm havar foils was pressurized to 2/3 atm. For the p-^{3}H measurements, two titanium-tritium targets were used: a 1.5 μg/cm^2 Ti layer on a 1 μm Ni backing or a 0.5 μg/cm^2 Ti layer on a 0.5 μm Al backing. Both targets suffered from the wrinkling characteristic of these targets and this prevented reliable measurements in the 70^0-110^0 range from being obtained at the lower energies. A very flat thin target has now been obtained to complete this data set.

Angular distributions of the p-^{3}He analyzing powers were measured at nine energies between 2.07 and 3.75 MeV for the 45^0 to 120^0 (lab) angular interval. Data at 2.28, 3.54 and 3.74 MeV are shown in the figure as solid dots where the uncertainties in A_y are typically ±0.01. Also shown here (as open circles) are the earlier double scattering polarization data of Drigo et al.[6]. Substantial deviation between the two data sets is evident at forward angles at 2.28 and 3.74 MeV, and some deviations at back angles at 3.54 MeV. We believe that their technique of measuring the particle spectrum with a large detector subtending the entire angular range of measurement and relying on spectrum stripping techniques to deduce the polarizations as a function of angle is the source of this disagreement. Also shown here are R-matrix calculations[4] made using the T = 0 levels for A = 4 determined in a re-analysis of all available data, but not including our data set. The agreement between our data and the calculations is reasonably good, although the calculations underestimate A_y at higher energies and vice-versa at lower energies. Some changes in the fits are expected when our data set is included, particularly at the lower energies where there has been little accurate data to date to guide the calculations.

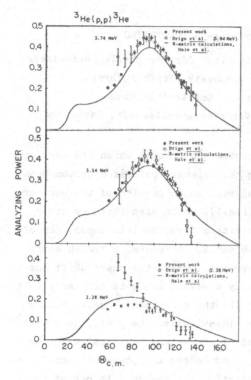

$^3He(p,p)^3He$

ANALYZING POWER

$\Theta_{c.m.}$

3.74 MeV

• Present work
○ Drigo et al.   (3.84 MeV)
— R-matrix calculations,
   Hale et al.

3.54 MeV

• Present work
○ Drigo et al.
— R-matrix calculations,
   Hale et al.

2.28 MeV

• Present work
○ Drigo et al.   (2.38 MeV)
— R-matrix calculations,
   Hale et al.

Our p-^{3}H data have been measured as excitation curves at nine angles from 1.9 to 3.9 MeV with some gaps at angles around 90°, angles where A_y is expected to maximize. At the higher energies, comparison with earlier double scattering polarization data of Drigo et al.[7] shows the two data sets are in agreement, although some deviations at back angles are developing. Our data at 3.27 MeV are in good agreement with the double scattering ^{1}H(t,t) data of Veeser et al.[8] at E_t = 9.65 MeV. Additional polarization data are required, particularly at low projectile energies to determine the properties of the low lying 0^- and 0^+ states and work on this problem is continuing.

References

†   Supported in part by U.S. National Science Foundation.

1) C. Werntz and W.E. Meyerhof, Nucl. Phys. A121, (1968) 38.
2) R.C. Haight, J.J. Jarmer, J.E. Simmons, J.C. Martin, and T.R. Donoghue, Phys. Rev. Lett. 28, (1972) 1587; also, Phys. Rev. C 9 (1974) 1292.
3) T.R. Donoghue, R.C. Haight, G.P. Lawrence, J.E. Simmons, D.C. Dodder, and G.M. Hale, Phys. Rev. Lett. 27, (1971) 974: also, Phys. Rev. C 5 (1972) 1826.
4) G.M. Hale, private communication.
5) T.R. Donoghue, et al., this conference, p.840.
6) L. Drigo, et al., Nucl. Phys. 89 (1966) 632.
7) L. Drigo, et al., Nuovo Cimento 51B, (1967) 43.
8) L.R. Veeser, D.D. Armstrong, and P.W. Keaton, Nucl. Phys. A140, (1970), 177.

PHASE SHIFTS FOR ELASTIC p-^{3}H SCATTERING

D. Fick and K. Kilian, Max-Planck-Institut für Kernphysik, Heidelberg

J.C. Fritz, Institut für Biophysik, Freiburg/Brsg.

A. Neufert, Strahlenzentrum, Universität Gießen

R. Kankowsky, Berthold-Frieseke Vertriebsgesellschaft, Karlsruhe

The elastic scattering data of protons on tritons which had been obtained between 4 and 12 MeV [1] using the Erlangen polarized proton beam [2] has been extensively phase shift analysed. An inspection of the contour plot of the analyzing power A(θ,E) (fig.1), which also includes other data [3,4] and ranges up to 20 MeV exhibits a structureless dependence of the analyzing power with respect to θ and E. Therefore, a smooth energy behaviour of the phases should be expected. During the phase shift analysis, which included S,P and partially D-waves, two sets of phases (set I and II) have been found which describe the experimental data equally well (fig.2). The phases of set I are identical to the preliminary ones [5]. At higher energies (E$_p$>8MeV) both set of phases are identical with the two sets found by Hardekopf et al. [3]. And indeed the phases of set II of the present analysis were found by starting the search with one of these phase sets. In the present analysis it was also tried to get an estimate of the sensitivity of the fit with respect to the determination of the various phases. For this purpose χ^2-contour plots were calculated for each ($\delta\eta$)- combination separately. (An element of the S-matrix is related to δ and η by S=η exp (2iδ)). The dashed areas in fig.2 correspond to a 10% change of χ^2 with respect to its minimum value. Generally, the absorption parameters are less accurately determined by the analysis than the real phases. For the different partial waves the fit is less sensitive to the determination of the S-waves compared with the determination of the P-waves. The D-waves are rather small and not displayed in fig.2 . Additionally fig.2 displays a comparison with theoretical phases [6] (solid lines) which were obtained by a resonating group calculation. They exhibit an overall good agreement with the experimental ones. Within these calculations the coupling of the dd-channel to the p^{3}H-channel has been neglected. The experimental results seem to justify this approximation. But there is a problem concerning the absolute value of the P-phases. The theoretical calculations [6] predict no crossing of the δ^0_{11}-phase through 90°. Therefore, one gets a level sequence 2$^-$,0$^-$,1$^-$ because δ^2_{11}>δ^0_{11}>δ^1_{11} .

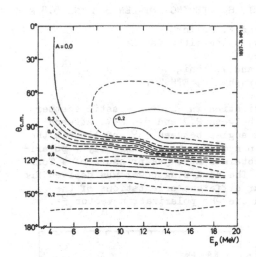

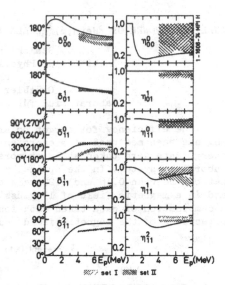

Fig.1 Experimentally determined contour plot $A(\theta,E)$ for elastic p-^{3}H
scattering[1,3,4)

Fig.2 Phases for the elastic p-^{3}H
scattering together with theoretical predictions[6]) (solid lines)

But the level sequence $2^-,0^-$ is in complete disagreement with all up to
now known features of the excited states in ^{4}He which give uniquely 0^-,
2^- [7]). Therefore, one may suppose that in the calculations a pass of δ^0_{11}
through 90° was missed because of a too wide net of basis points in the
calculations.

1) J.C.Fritz, R.Kankowsky, K.Kilian, A.Neufert, D.Fick, Proc. of the 3rd
   Int. Symposium on Polarization Phenomena in Nuclear Reactions, Madison
   1970, 482

2) G. Clausnitzer et al. Nucl. Instr. Meth. 80 (1970 245

3) R.A. Hardekopf, P.L.Lisowski, T.C.Thea, R.L. Walter, Nucl.Phys. A191
   (1972) 481

4) R. Darves-Blanc, Nguyen van Sen, J.Arvieux, A.Fiore, J.C.Gondrand,
   G. Perrin, Letts. al Nuovo Cimento 4 (1972) 16

5) K.Kilian, J.C.Fritz, F.A.Gückel, R.Kankowsky, A.Neufert, J.Niewisch,
   D.Fick, Proc. of the Int.Conf. on Few Body Problems in the Nuclear
   Interaction, Los Angeles 1972, 667

6) P.Heiss, H.H.Hackenbroich, Z.Physik 251 (1972) 168

7) S. Fiarman, W.E. Meyerhof, Nucl.Phys. A206 (1972) 1

ANALYZING POWER OF $\overrightarrow{^3\text{He}}$(p,p)^{3}He ELASTIC SCATTERING BETWEEN 2.3 AND 8.8 MeV

G. Szaloky and F. Seiler
Institut für Physik, Universität Basel
and
W. Grüebler and V. König
Laboratorium für Kernphysik, ETH Zürich

The efficiency for target polarization in ^{3}He + p elastic scattering has been measured at 5 energies between 2.3 and 8.8 MeV. At each energy data were taken at 12 or more angles between 35° and 145° in the laboratory system. In the optically pumped ^{3}He target assembly[1] polarizations between 0.16 and 0.22 were obtained at a gas pressure of 4 Torr and in a magnetic field of 15 gauss. The analyzing power was derived from a combination of data taken with four different orientations of the directions of the magnetic field and the ^{3}He polarization vector [2].

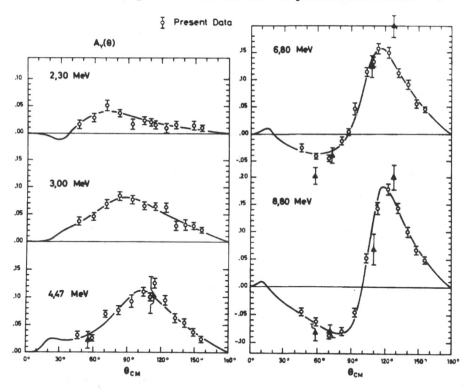

Fig. 1. The angular distribution of the efficiencies $A_y(\theta)$ at 5 proton energies (given in the laboratory system). The triangles represent the values of Baker et al., θ_{cm} are given in the center-of-mass system. The lines connecting the measured points are produced by the best phase-shift sets.

The present data are shown as circles in fig. 1, together with previous measurements by Baker et al.[2,3], marked by triangles. The latter are not corrected by the scale factor between 0.77 and 1.04 mentioned by the authors. Thus the agreement is satisfactory. At 2.3 MeV the efficiency A_y is considerably smaller than predicted by existing phase shifts. Up to 4.5 MeV the analyzing power increases without basic changes in the angular distribution. Above that energy, however, a different distribution is found, while the peak values continue to increase up to $A_y \simeq 0.18$.

Since these measurements represent a substantial increase of the first order polarization data available, a phase shift analysis yields considerably smaller bands for each phase shift[4]. Although the multiplicity of solutions is not reduced, a more solid basis for an analysis of second order polarization data, such as efficiency correlation coefficients, is now available. A set of such measurements should allow a unique solution. The present phase shift analysis leads to several second order observables which are substantially different for different solutions[4], thus indicating the direction of the most profitable effort.

References

1) U. Rohrer, P. Huber, Ch. Leemann, H. Meiner and F. Seiler,
   Helv. Phys. Acta 44 (1971) 846
2) S.D. Baker, D.H. McSherry and D.O. Findley,
   Phys. Rev. 178 (1969) 1616
3) D.H. McSherry and S.D. Baker,
   Phys. Rev. C1 (1970) 888
4) G. Szaloky and F. Seiler, to be published

POLARIZATION TRANSFER COEFFICIENTS FOR ^{3}He($\vec{p},\vec{p}$)^{3}He SCATTERING
BELOW 11 MeV

W.G. Weitkamp[*], W. Grüebler, V. König, P.A. Schmelzbach, R. Risler and
B. Jenny
Laboratorium für Kernphysik, ETH, Zürich, Switzerland

The polarization transfer coefficients [1,2] $K_x{}^x$, $K_z{}^x$ and $K_y{}^y$ for the
elastic scattering of protons from ^{3}He have been measured at incident
energies of 6.82, 8.82 and 10.77 MeV at up to 4 angles. A double scattering technique similar to that described elsewhere in these proceedings[3]
has been used.

The measurements were performed with a typically 15-20 nA proton
beam with a polarization of 80 %. The first scattering cell contained up
to 20 atm of ^{3}He cooled by liquid nitrogen. The scattered particles were
focused by a magnetic quadrupole triplet into a vane-type polarimeter 2 m
away, to reduce background. This polarimeter utilized scattering from ^{4}He
at either 60° or 113°, and contained 20 atm of ^{4}He. The angular acceptance of the magnetic triplet was 4.6° .

The component of beam polarization perpendicular to the incident momentum was monitored continuously with a polarimeter employing elastic
scattering from ^{4}He at 113°. The analyzing power of this polarimeter could
be calculated precisely from available measurements [4]. This monitor polarimeter was also used to calibrate the vane-type polarimeter, with the
incident polarized beam passing directly through the monitor polarimeter
into the vane-type polarimeter.

The instrumental asymetry of the monitor polarimeter was measured
with unpolarized beam. It remained constant through the measurements. The
instrumental asymetry of the vane-type polarimeter was cancelled by taking
half the data at each point with the polarimeter rotated 180°.

During $K_x{}^x$ and $K_z{}^x$ measurements, the beam polarization direction must
be perpendicular or parallel, respectively, to the incident momentum direction, and must be precisely aligned since errors enter into the resulting value in first order. The source Wien filter, which controls this direction, was calibrated daily. In the $K_x{}^x$ measurement, the direction could
be held to ± 2°, while in the $K_z{}^x$ case, the beam polarization monitor
measured any misalignment directly so the error could be held to ± 0.3°.

The magnitude of the beam polarization remained constant to ± 0.4%
during the measurement of a given point.

For most measurements, backgrounds in the vane-type polarimeter were
negligibly small. In those cases where neutron-produced background was
evident, the background was measured directly by switching off the magnetic quadrupole, reducing the solid angle for charged particles by a factor
of 1600.

The presently available data are shown in fig. 1. In all cases, the
angles are center-of-mass angles, but the coefficients are refered to the
outgoing particle's laboratory helicity frame. Error bars in the figure
indicate total errors, except that no correction has been made for the
(presumably small) effects of finite geometry in the first scattering.

Also shown in fig. 1 are predictions of polarization transfer coefficients from a recent R matrix analysis of p-^{3}He scattering[5]. The agree-

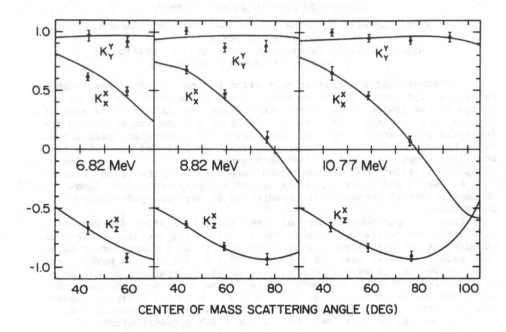

Fig. 1. Polarization transfer coefficients for proton elastic scattering from ^{3}He.

ment is generally good. This is somewhat unexpected since the combinations of amplitudes forming these coefficient are mathematically independent of the combinations forming those observables, cross section, p and ^{3}He analysing power, and spin correlation coefficient $C_{x,z}$, for which data was previously available below 11 MeV.

We wish to express our appreciation to G.M. Hale for making his R matrix calculations available prior to publication.

<u>References</u>

*   On leave from the University of Washington, Seattle, Wash.

1) G.G. Ohlsen, Rep. Prog. Phys. <u>35</u> (1972) 717
2) R.A. Hardekopf and D.D. Armstrong, to be published in Phys. Rev. C
3) W.G. Weitkamp <u>et al</u>, Fourth Polarization Symposium , p.595
4) P. Schwandt, T.B. Clegg and W. Haeberli, Nucl. Phys. <u>A163</u> (1971) 432
5) G.M. Hale, J.J. Devaney, D.C. Dodder and K. Witte, Bull. Am. Phys. Soc. <u>19</u> (1974) 506

THE VECTOR ANALYZING POWER IN ELASTIC DEUTERON-DEUTERON
SCATTERING BETWEEN 20 AND 40 MeV*

H. E. Conzett, W. Dahme[+], R. M. Larimer, Ch. Leemann,
and J. S. C. McKee[‡]
Lawrence Berkeley Laboratory, University of California
Berkeley, California 94720

The subject of d-d elastic scattering has not received very much attention in the past. The cross sections are quite smooth functions of
energy in the region up to 20 MeV where there are the most data available.
However, the complexity of the spin structure and the low threshold for
inelastic processes has made any meaningful phase-shift analysis impossible because of the large number of parameters involved even with the
restrictive assumption of channel-spin conservation[1]. A resonating-
group calculation[2] has obtained good agreement with the cross-section
data between 5 and 20 MeV, but the use of a purely central nucleon-
nucleon potential precludes the prediction of any spin-polarization
observables.

From an experimental point of view, the polarization experiments
in elastic d-d scattering have raised a qualitative question. Previous
measurements of the vector analyzing power iT_{11} in $\vec{d}$-d scattering have
been made at several energies below 12 MeV.[1,3] and at 21.4 MeV[4]. Non-
zero but very small values of iT_{11} were obtained, reaching a maximum
value of about 0.04 at 21.4 MeV. These values are almost an order of
magnitude smaller than the nucleon and deuteron vector analyzing powers
found in other elastic processes involving few nucleon systems, e.g.
p + ^{2}H, ^{3}He, ^{4}He and $\vec{d}$ + ^{3}He, ^{4}He. Since sizable contributions of
S, P, and D-waves were required to fit the d-d data[1,2], the rather
insignificant polarization effects could not be explained as a consequence of a predominance of S-wave scattering. Thus, its reason
remained unexplained.

We have extended the measurements of vector analyzing powers in $\vec{d}$-d
scattering to 40 MeV to examine whether or not its anomalously small
value persists at these higher energies. Also, another determination
near 20 MeV was desired, since the older measurement at 21.4 MeV was
rather uncertain because of lack of knowledge of the beam polarization.
We used the axially injected vector-polarized deuteron beam from the
Berkeley 88-Inch Cyclotron. Left-right asymmetry data were taken
simultaneously at two angles separated by 20°, using pairs of ΔE-E silicon detector telescopes. A polarimeter, consisting of a gas target and
a pair of ΔE-E counter telescopes, was placed downstream of the main
scattering chamber and provided continuous monitoring of the beam polarization. The analyzer used was ^{4}He, whose vector analyzing power in
d-^{4}He elastic scattering has been measured in detail[5]. The differential
cross section for vector-polarized deuterons is given by

$$\sigma(\theta) = \sigma_o(\theta) \left[1 + 2 \ (it_{11}) \ (iT_{11}) \right], \tag{1}$$

where $\sigma_o(\theta)$ is the differential cross-section for unpolarized deuterons
and it_{11} is the beam polarization. A left-right asymmetry measurement
gives

$$\epsilon(\theta) = 2 \ (it_{11}) \ (iT_{11}), \tag{2}$$

and the simultaneous determination of the beam polarization yields the
vector analyzing powers iT_{11}. Figure 1 shows our data at E_d = 20, 30,

and 40 MeV; the particle symmetry requires that $iT_{11}(\theta) = - iT_{11}(\pi-\theta)$. Our 20-MeV values are a factor of two larger than the previous results at 21.4 MeV, and clearly the vector analyzing powers increase rapidly with increasing energy. These values, when compared with the analyzing powers in $\vec{d}$-p elastic-scattering measured at comparable center-of-mass energies[6], can no longer be considered anomalously small.

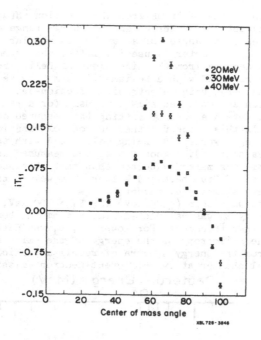

Center of mass angle

XBL 729-3846

References

* Work performed under the auspices of the U. S. Energy Research and Development Administration.
+ Sektion Physik der Universität München, 8046 Garching, Germany.
‡ University of Manitoba, Winnipeg, Manitoba, Canada.

1) G. R. Plattner and L. G. Keller, Phys. Letters 30B, (1969) 327; P. D. Lien, Nucl. Phys. A178 (1972) 375; H. O. Meyer and P. Schiemenz, Nucl. Phys. A197 (1972) 259.
2) F. S. Chwieroth, Y. C. Tang, and D. R. Thompson, Nucl. Phys. A189 (1972) 1.
3) W. Grüebler, V. König, R. Risler, P. A. Schmelzbach, R. E. White, and P. Marmier, Nucl. Phys. A193 (1972) 149.
4) J. Arvieux, J. Goudergues, B. Mayer, and A. Papineau, Phys. Letters 22 (1966) 610.
5) H. E. Conzett, W. Dahme, Ch. Leemann, J. A. Macdonald and J. P. Meulders, Fourth Polarization Symp., p.566.
6) J. S. C. McKee, H. E. Conzett, R. M. Larimer and Ch. Leemann, Phys. Rev. Lett. 29 (1972) 1613.

THE POLARIZATION OF NEUTRONS FROM THE REACTION ^{2}H(d,n)^{3}He,
MEASURED WITH A NEW TYPE OF HELIUM RECOIL POLARIMETER[*])

C.P. Sikkema and S.P. Steendam
Laboratorium voor Algemene Natuurkunde der Rijksuniversiteit
Groningen, Groningen, the Netherlands

The polarization of the neutrons from the reaction ^{2}H(d,n)^{3}He has
been measured for unpolarized deuterons in the energy range E_d = 50 –
700 keV and neutron emission angles of about 50° in the lab. system (52°
c.m.s). A new type of polarimeter is used[1]. Left-right asymmetries in
n-^{4}He scattering are deduced from the directions of helium recoil
particles, which are recorded with a helium-filled multiwire proportional chamber (drift chamber), using electronic direction-sensitive methods.
The asymmetries, which are obtained simultaneously for a range of
scattering angles θ, were analyzed by fitting the measured data with
functions $P_n P_{He}(\theta)$. In this procedure the neutron polarization P_n was
adjusted, the analyzing power $P_{He}(\theta)$ being calculated with the phase
shifts of Hoop and Barschall[2]. A good fit to the measured asymmetries
was obtained. This fit was markedly better than that obtained when
using the phase shifts of Stammbach and Walter[3]. However, the resulting
values of P_n were only slightly different.

The Ti-D targets were thin ($\leq$ 80 keV) for E_d > 200 keV, and the
average reaction energy $\bar{E}_d$ was calculated simply from the bombarding
energy E_d and the target thickness. For lower E_d, $\bar{E}_d$ was determined experimentally from the anisotropy of the energy of the emitted neutrons,
which was deduced from the energy spectra of recoil α-particles recorded
with the proportional chamber at two different neutron emission angles.

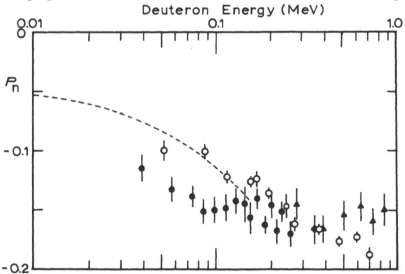

Fig. 1. Polarization of neutrons from the ^{2}H(d,n)^{3}He reaction. Analysis
by n-^{4}He scattering. ● Behof et al.[4], θ_n = 47°; ◆ Mulder[5], θ_n = 46.5°;
▲ Davie and Galloway[6], θ_n = 46°; O present work, θ_n = 46-50°, θ_n(c.m.s.)
= 52°. Dashed curve: Theoretical according to Boersma[9].

The measured values of P_n are plotted as a function of $\overline{E}_d$ in fig. 1, together with various earlier measurements. For $\overline{E}_d < 250$ keV, our results are in fig. 1 only compared with those of Behof et al.[4]. The latter are also plotted versus $\overline{E}_d$ for the thick D_2O ice targets employed instead of the bombarding energy E_d given in the original paper. In a review at the Madison Symposium in 1970 by Walter[7], the polarizations of Behof et al. were considered as the most reliable for low E_d, rather than other results indicating a resonance-like behaviour of P_n around $E_d = 100$ keV (cf. ref. 7). These results also mutually disagree. Our measured polarizations do not indicate the presence of a resonance, but they definitely disagree with those of Behof et al., especially for $E_d < 200$ keV. Polarizations measured most recently by Galloway et al.[8] (not shown in the fig. 1) agree quite well with our values for $E_d = 50 - 200$ keV. The curve in fig. 1 shows the energy dependence of P_n expected according to the theoretical treatment of Boersma[9] which should hold for $E_d = 0-150$ keV. An unknown constant has been adjusted to fit our values of P_n in this region.

References

* Supported in part by the Netherlands Foundation for Fundamental Research of Matter (FOM)

1) C.P. Sikkema, Nucl. Inst. <u>122</u> (1974) 415
   C.P. Sikkema, these Proceedings, p.885
2) B. Hoop and H.H. Barschall, Nucl. Phys. <u>83</u> (1966) 65
3) Th. Stammbach and R.L. Walter, Nucl. Phys. <u>A180</u> (1972) 225
4) A.F. Behof, T.H. May and W.I.Mc. Garry, Nucl. Phys. <u>A108</u> (1968) 250
5) J.P.F. Mulder, Thesis, University of Groningen, 1968
6) H. Davie and R.B. Galloway, Nucl. Inst. <u>108</u> (1973) 581
7) R.L. Walter, Polarization Phenomena in Nuclear Reactions, University of Wisconsin Press, Madison 1971, p. 327
8) R.B. Galloway, private communication
   A. Alsoraya, R.B. Galloway and A.S. Hall, these Proceedings, p.520
9) H.J. Boersma, Nucl. Phys. <u>A135</u> (1969) 609

THE POLARIZATION OF NEUTRONS FROM THE ^{2}H(d,n)^{3}He REACTION

A. Alsoraya, R.B. Galloway and A.S. Hall
Department of Physics, University of Edinburgh, Scotland

Measurements are in progress of the polarization of the neutrons from the ^{2}H(d,n)^{3}He reaction for deuteron energies below 500 keV because of the need for thin target data for comparison with low energy reaction theory[1] and because of the discrepancies between such data as is available[2,3]. The neutron polarimeter consists of a ^{4}He gas scintillator operating at a pressure of 70 atmospheres which scatters neutrons into a pair of liquid scintillators with pulse-shape discrimination against gamma-rays. Pulse height spectra due to ^{4}He recoil nuclei detected in coincidence with scattered neutrons are recorded and analysed as in earlier work at higher deuteron energies[4,5], using the phase shifts of Stambach and Walter[6]. The energy dependence of the polarization of neutrons emitted at 45° is shown in fig. 1 along with higher energy measurements made with the same equipment[4,5] and other thin target measurements [7-11]. Measurements made with ^{12}C as analyser have been omitted because of the difficulty caused by resonance in the n-^{12}C system[12]. An acceptable fit to the measurements below 200 keV is provided by the theory of Boersma[1] as shown by the curve in fig. 1 which is normalised to the measurement at 135 keV. Similar agreement has been found recently by Sikkema and Steendam[13].

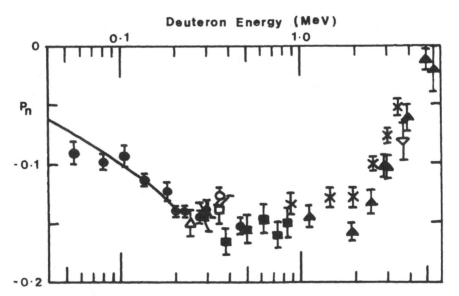

Fig. 1. Energy dependence of the polarization of neutrons emitted at 45° -47° from the ^{2}H(d,n)^{3}He reaction. ● present measurements, ■ ref.[4], ▲ ref.[5], △ ref.[7], ○ ref.[8], □ ref.[9], ▽ ref.[10], × ref.[11], ——— calculation[1].

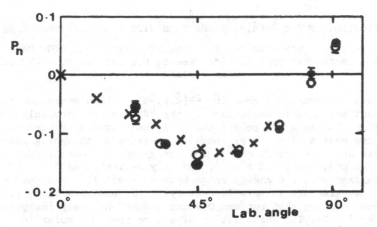

Fig. 2. Angular dependence of polarization. ● 460 keV, ○ 290 keV,
× 350 keV[8].

The angular dependence of polarization below 500 keV deuteron
energy is important in relation to the approach cross-section description of the reaction[14,5]. The new measurements and the only previous
thin target one[8] are compared in fig. 2.

References

1)   H.J. Boersma, Nucl. Phys. A135 (1969) 609
2)   R.L. Walter, Polarization Phenomena in Nuclear Reactions (Ed. H.H.
     Barschall and W. Haeberli, Univ. Wisconsin Press Madison 1970)p.317
3)   R.B. Galloway, Nucl. Instr. and Meth. 92 (1971) 537, errata 95
     (1971) 393
4)   H. Davie and R.B. Galloway, Nucl. Instr. and Meth. 108 (1973) 581
5)   R.B. Galloway, A.S. Hall, R.M.A. Maayouf and D.G. Vass, Nucl. Phys.
     A242 (1975) 122
6)   Th. Stammbach and R.L. Walter, Nucl. Phys. A180 (1972) 225
7)   R.B. Galloway and R. Martinez Lugo, these proceedings, p.881
8)   H.J. Boersma, C.C. Jonker, J.G. Nijenhuis and P.J. Van Hall, Nucl.
     Phys. 46, (1963) 660
9)   J.P.F. Mulder, Phys. Lett. 23 (1966) 589
10) F.O. Purser, J.R. Sawers and R.L. Walter, Phys. Rev. 140 (1965)
     B870
11) J.R. Smith and S.T. Thornton, Can. J. Phys. 50 (1972) 783
12) H. Hansgen and M. Nitzsche, Nucl. Phys. A165 (1971) 401
13) C.P. Sikkema and S.P. Steendam, private communication
14) F.O. Purser, G.L. Morgan and R.L. Walter, Proc. 2nd Int. Symp.
     Polarization Phenomena of Nucleons (Birkhauser Verlag Basel 1966)
     p.514

INVESTIGATION OF THE ^{2}H($\vec{\text{d}}$,p)^{3}H REACTION WITH POLARIZED DEUTERONS

W. Grüebler, P.A. Schmelzbach, V. König, R. Risler and D.O. Boerma
Laboratorium für Kernphysik, Eidg. Techn. Hochschule, 8049 Zürich,
Switzerland

For several years the reaction ^{2}H($\vec{\text{d}}$,p)^{3}H has been among the favourite reactions used in investigating the ^{4}He nucleus. Not only the cross section, but also the polarization of the outgoing protons have been measured over a wide energy range[1]. A series of analysing power measurements have been performed at low energy up to about 0.5 MeV by the Basel group[2], Franz and Fick[3], and Ad'yasevich et al.[4].

Measurements in the energy range between 3 and 11.5 MeV have been performed some time ago in this laboratory[5]. The analysis in refs. 5) suffers from the fact that between 0.5 and 3.0 MeV no analysing power date were available. The installation of a more powerful polarized ion source at the 12 MeV EN-tandem accelerator made it possible to measure the necessary data in the energy range between 1.0 and 3.0 MeV. Measurements of the differential corss section σ_0, the vector analysing power iT_{11}, and the three tensor analysing powers T_{20}, T_{21}, and T_{22} have been carried out in energy steps of 0.5 MeV.

For T_{20} and T_{21} the measurements cover the angular range between 7.5° and 157.5° in the Laboratory system in step of 5°.

For iT_{11} and T_{22} the smallest angle is 22.5°. The statistical error is generally smaller than 0.015.

The angular distribution of the cross section and of the quantities $\sigma_0 \cdot T_{kq}$ has been fit by Legendre polynomial expansions.

The normalized Legendre polynomial coefficients[5]

$$d_{kq}(L) = \frac{4\pi \lambda^2 \, a_{kq}(L)}{\sigma_{tot}} = \frac{a_{kq}(L)}{a_{oo}(0)}$$

are presented as a function of the incident deuteron energy in fig. 1. The dots are the results of the calculations; the solid lines are drawn by hand for the sake of clarity. The size of the dots are in all cases larger than or equal to the error found in the fit. Legendre coefficients which are too small to appear different from zero at all energies on the scale used in fig. 1 are not shown. At low energy the results of the Basel group[2] and at energies higher than 2.5 MeV the results of[5] are indicated. The most prominent variation in the value of the coefficients occurs in the energy range up to about 3 MeV. The present experimental data were needed to determine the behaviour in this energy range quantitatively. A search for one or more resonances in ^{4}He in this energy region, as well as the assignment of spin and parity can only be done in a careful, detailed analysis, which is in preparation in this laboratory.

References

1) G. Spalek, R.A. Hardekopf, J. Taylor, Th. Stammbach and R.L. Walter, Nucl. Phys. A191 (1972) 449
R.A. Hardekopf, T.C. Rhea, P.W. Lisowski, R.L. Walter and J.M. Joyce, Nucl. Phys. A191 (1972) 460

R.A. Hardekopf, P.W. Lisowski, T.C. Rhea and R.L. Walter,
Nucl. Phys. A191 (1972) 468

2) K. Jeltsch, A. Janett, P. Huber and H.R. Striebel,
Helv. Phys. Acta 43 (1970) 279

3) H.W. Franz and D. Fick, Nucl. Phys. A122 (1968) 591

4) B.P. Ad'yasevich, V.G. Afonenko and D.E. Fomenko, Sov. Journ.
Nucl. Phys. 11 (1970) 411

5) W. Grüebler, V. König, P.H. Schmelzbach, R. Risler, E.E. White
and P. Marmier, Nucl. Phys. A193 (1972) 129

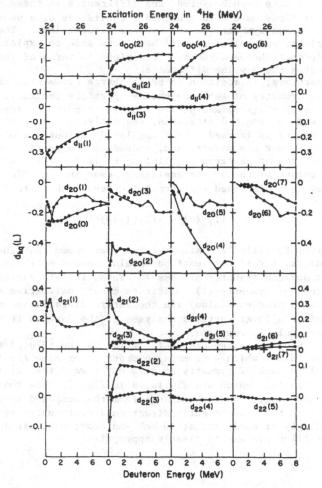

Fig.1. Normalized Legendre polynomial expansion coefficients $d_{kq}(L)$ for the differential cross section, and the vector and tensor analysing powers. The dots are larger than the uncertainties of the fits. At energies below 0.5 MeV the results are from ref. 2).

THE VECTOR ANALYZING POWER IN THE ^{2}H($\vec{\text{d}}$,p)^{3}H REACTION BETWEEN 15 and 25 MeV*

H. E. Conzett, R. M. Larimer, F. N. Rad, R. Roy$^+$ and F. Seiler$^+$
Lawrence Berkeley Laboratory, University of California
Berkeley, California 94720

The charge-symmetric reactions ^{2}H(d,p)^{3}H and ^{2}H(d,n)^{3}He have been studied in considerable detail at energies up to 15 MeV[1,2]. Differential cross sections, nucleon polarizations, and deuteron vector and tensor analyzing powers have been measured, and differences in these observables for the two reactions have been examined for evidence of a possible deviation from the charge symmetry of the nuclear interactions. The more recent comparisons and calculations[3] have been able to explain the observed differences between the two reactions in terms of the Coulomb effect, including the Q-value difference.

These reactions, in addition to being charge symmetric, each possess the additional symmetry of entrance-channel particle identity. This requires that $\sigma(\theta) = \sigma(\pi-\theta)$ and $p(\theta) = -p(\pi-\theta)$ for the differential cross sections and the nucleon polarizations, respectively. In general, no comparable symmetry is imposed on the angular distributions of the deuteron analyzing-power components, and, indeed, no suggestion of symmetry is seen in the data at deuteron energies up to 11.5 MeV[2]. However, recent measurements of the vector analyzing power in the ^{2}H($\vec{\text{d}}$,p)^{3}H reaction at 30 MeV have disclosed the surprising result that, there, the symmetry

$$iT_{11}(\theta) = -iT_{11}(\pi-\theta) \tag{1}$$

is approximately fulfilled[4]. Also, it has been shown that the condition (1) holds exactly if the reaction should proceed entirely by way of the direct nucleon-transfer process[5]. Thus, in this particular case deviations from the symmetry (1) constitute clear qualitative evidence that the reaction proceeds (also) via the intermediate (compound) nucleus ^{4}He. Thus, at the lower energies analyses of the data in terms of states in ^{4}He are certainly appropriate.

We report here on measurements of $A_y(\theta)$ in the ^{2}H($\vec{\text{d}}$,p)^{3}H reaction at 15, 20, and 25 MeV which were made in order to examine the transition from the complete lack of symmetry at 11.5 MeV toward that of (1) at 30 MeV. Our results, which are displayed in fig. 1, show that the transition is a gradual one. Thus, the change from the compound-nucleus reaction mechanism to the predominantly direct nucleon-transfer reaction mode is correspondingly gradual, and at 30 MeV and above analysis in terms of the direct-reaction process is clearly appropriate.

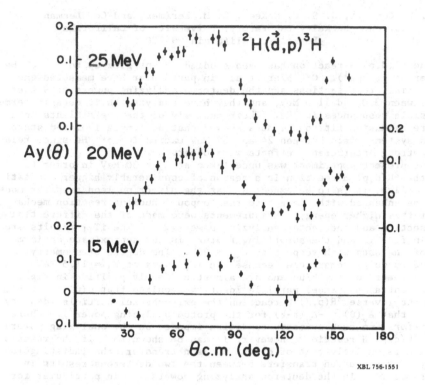

^{2}H$(\vec{\text{d}},\text{p})^3$H

25 MeV

20 MeV

15 MeV

Ay(θ)

θ c.m. (deg.)

XBL 756-1551

References

*   Work performed under the auspices of the U.S. Energy Research and
Development Administration.
+   National Research Council of Canada Postdoctoral Fellow.
†   On leave of absence from the University of Basel, Switzerland.

1)   R. A. Hardepopf, R. L. Walter and T. B. Clegg, Phys. Rev. Lett. <u>28</u>
(1972) 760 and references therein.
2)   W. Grüebler et al., Nucl. Phys. <u>A193</u> (1972) 129.
3)   Reference 41 (unpublished) of ref. 2.
4)   H. E. Conzett, J. S. C. McKee, R. M. Larimer and Ch. Leemann,
Fourth Polarization Symp., p.526
5)   H. E. Conzett, J. S. C. McKee, R. M. Larimer and Ch. Leemann, to
be published.

THE VECTOR ANALYZING POWER IN THE ^{2}H($\vec{\text{d}}$,p)^{3}H REACTION AT 30 MeV.*

H. E. Conzett, J. S. C. McKee[+], R. M. Larimer, and Ch. Leemann
Lawrence Berkeley Laboratory, University of California
Berkeley, California 94720

The ^{2}H(d,p)^{3}H reaction has been studied at energies up to 12 MeV by a number of groups[1]). Grüebler et al. in particular have measured the differential cross sections and the deuteron analyzing powers at 9 energies between 3.0 and 11.5 MeV, and they have analyzed their data in terms of possible resonances in ^{4}He. Their analysis of the coefficients in a Legendre expansion fit to the data showed that no simple isolated state of this system exists between 24 and 30 MeV excitation of ^{4}He since relatively strong interference effects were observed.

The present experiment was undertaken at E_d = 30 MeV in order to study the ^{2}H(d,p)^{3}H reaction in a region of considerably higher excitation of ^{4}He. Also, it is to be expected that the direct nucleon transfer mode should be enhanced with respect to the compound-nucleus reaction mechanism at this higher energy. Measurements were made of the differential cross-section and the vector analyzing power iT_{11}. The iT_{11} results are shown in fig. 1, and the surprising feature is the approximate antisymmetry of the data with respect to θ_{cm} = 90°. The degree of symmetry observed is quite remarkable because the iT_{11} data at E_d = 11.5 MeV show little symmetry of any kind and are almost uniformly positive in sign.

The entrance channel particle identity requires that $\sigma(\theta) = \sigma(\pi-\theta)$, and in the inverse ^{3}H($\vec{\text{p}}$,d)^{2}H reaction the exit-channel particle identity requires that $A_y(\theta) = -A_y(\pi-\theta)$ for the proton analyzing power[2]). There is, *a priori*, no such symmetry condition imposed on the analyzing powers in the ^{2}H(d,p)^{3}H reaction. However, it can be shown that if the reaction mechanism is entirely that of direct nucleon transfer, the indistinguishability of the neutron transfers between the two deuterons results in exact symmetries in the deuteron analyzing powers[3]). In particular for our purposes

$$iT_{11}(\theta) = - iT_{11}(\pi-\theta), \qquad (1)$$

so the near antisymmetry of our data is clear evidence of a predominantly direct nucleon-transfer reaction mode at this higher energy. Of course, the *certain* conclusion is that deviations from eq. (1) show that other than the direct nucleon-transfer process is contributing to the reaction. It is possible, in principle, for the compound-nucleus reaction mechanism to give the result (1) if the reaction should proceed entirely through a single J^π state of ^{4}He or through states of the same parity so that only the even-L terms would contribute in the Legendre expansion

$$\sigma(\theta)iT_{11}(\theta) = \sum a_L P_L^1 (\cos \theta).$$

In view of the data and analysis of Grüebler et al.[1]), this circumstance is most unlikely in this reaction.

In summary, we have found in the ^{2}H($\vec{\text{d}}$,p)^{3}H reaction that the entrance-channel particle identity imposes definite symmetries on the polarization observables that are clear signatures of the direct-reaction process. We know of no other example of a condition by which this process can be identified so clearly.

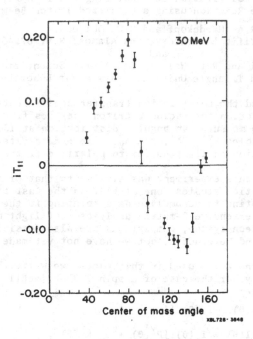

References

* Work performed under the auspices of the U.S. Energy Research and Development Administration.
+ University of Manitoba, Winnipeg, Manitoba, Canada

1) W. Grüebler et al., Nucl. Phys. A193 (1972) 129 and refs. therein.
2) We follow the Madison convention, Polarization Phenomena in Nuclear Reactions, eds. H. H. Barschall and W. Haeberli (Univ. of Wisc. Press, Madison, 1971) p. xxv.
3) H. E. Conzett, J. S. C. McKee, R. M. Larimer, and Ch. Leemann, to be published.

Neutron Polarization and Analyzing Power in the
H($\vec{t}$,n)³He Reaction using a Polarized Triton Beam*

R.A. Hardekopf and G.G. Ohlsen
Los Alamos Scientific Laboratory, Los Alamos, N.M. 87545, USA
and
P.W. Lisowski and R.L. Walter, Los Alamos Scientific
Laboratory and Triangle Universities Nuclear Laboratory

We have measured the polarization transfer coefficient $K_y^{y'}$ at $\theta = 0°$ in the H($\vec{t}$,n)³He reaction for incident triton energies from 6.5 to 16 MeV. In addition, we have measured an angular distribution at 13.6 MeV of the polarization observables $K_y^{y'}$, $P^{y'}$, and A_y. These data extend the study of neutron polarization in light nuclei to polarized triton beam observables.

Our purpose in this experiment was similar to that of many previous few-nucleon polarization studies done at LASL in the last five years. In short, we are attempting to accumulate data relating to the spin-dependent amplitudes for extensive R-matrix analyses[1] of light nuclear systems. In the 4-nucleon system, comparisons are also possible to the analysis of Werntz and Meyerhof[2], but we have not yet made these calculations.

To clarify the notation used in this paper, we write the following equations which apply for the case of a spin 1/2 projectile and a spin 1/2 outgoing particle:

$$I(\theta) = I_o(\theta) \, [1 + p_y \, A_y(\theta) \,]$$

$$p_{y'} \, I(\theta) = I_o(\theta) \, [P^{y'}(\theta) + p_y \, K_y^{y'}(\theta) \,].$$

$I(\theta)$ and $I_o(\theta)$ are the polarized and unpolarized cross sections, respectively. An unprimed subscript refers to the incident particle helicity reference frame while the primed axes refer to the outgoing laboratory frame.[3] These distinctions are not important for the present case since we measured only y-axis polarization components and $\hat{y}'$ is along $\hat{y}$, but we retain the more general notation for consistency. The lower case letters describe the polarization component of the incident beam (p_y) or outgoing neutrons ($p_{y'}$) while the upper case letters $A_y(\theta)$, $P^y(\theta)$, and $K_y^y(\theta)$ are properties of the reaction. $P^y(\theta)$ is the polarization function, as would be measured with an unpolarized beam incident, and $K_y^y(\theta)$ is the polarization transfer coefficient, analogous to the Wolfenstein D parameter in nucleon-nucleon scattering.

At $\theta = 0°$, A_y and $P^{y'}$ are zero since they are odd functions of θ. The outgoing neutron polarization is therefore related to the incoming triton polarization by

$$p_{y'} = p_y \, K_y^{y'}(0°).$$

For these measurements the neutron polarization was reversed by reversing the triton polarization at the ion source. The neutron asymmetry was measured in a conventional manner by scattering from liquid helium at an analyzing angle of 115°(lab).

At non-zero angles, a four-fold measurement sequence was used in which the neutron polarization was precessed ± 90° by a solenoid, with triton spin-up and spin-down runs for each solenoid polarity. This sequence allows extraction of $A_y(\theta)$, $P^y(\theta)$, and $K_y^y(\theta)$ from the coincidence counts accumulated and depends on reproducible current integration for the four runs. $A_y(\theta)$ was also obtained, independent of solenoid polarity, by summing part of the helium-scintillator recoil spectrum for each triton spin-up, spin-down sequence. These results agreed well with the values calculated from the coincidence counts.

The values for $K_y^{y'}$ at $\theta = 0^{\circ}$ are plotted in Fig. 1. The triton energy range shown on the abscissa corresponds to neutron energies from 2.5 to 11.5 MeV. Although $K_y^{y'}$ reaches a maximum of over 0.3 at the lowest energy measured, in general the polarization transfer is much smaller than that for $(\vec{p},\vec{n})$ and $(\vec{d},\vec{n})$ reactions[4]. This might be expected since the spins of the two neutrons in the triton are anti-aligned. It must be mentioned, however, that even with the low polarization transfer this reaction might be a useful source of polarized neutrons at low energies because of the very large laboratory differential cross section (about 500 mb/sr) at 0°.

Figure 2 shows the 3 polarization observables as a function of θ_{cm} at 13.65 ± 0.25 MeV. Since the polarization function $P^y(\theta)$ does not depend on beam polarization, the data are directly comparable to that for the $T(\vec{p},\vec{n})^3$He reaction allowing for kinematic considerations. A comparison to 4.4 MeV $T(\vec{p},\vec{n})^3$He measurements[5] shows good agreement in the region of overlap.

In the $T(\vec{p},n)^3$He reaction $A_y(\theta)$ and $P^{y'}(\theta)$ are nearly equal because of charge symmetry considerations. There is probably a simple geometric explanation as to why $A_y(\theta)$ and $P^y(\theta)$ have opposite signs in the $H(t,n)^3$He reaction, but the reason that their magnitudes are nearly the same is not obvious to us at the present time.

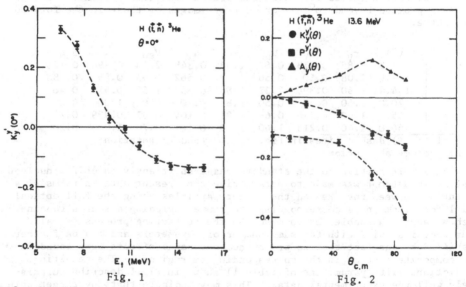

Fig. 1   Fig. 2

References
* Supported by the U.S. Energy Research and Development Adminsitration
1) D.C. Dodder, these proceedings, p.167
2) Carl Werntz and Walter E. Meyerhof, Nucl. Phys. A121 (1968) 38
3) G.G. Ohlsen, Rep. Prog. Phys. 35 (1972) 717
4) J.E. Simmons, et al., Nucl. Instr. Meth. 106 (1973) 477, and contributions to this conference from TUNL, p.499 and p.501
5) J.R. Smith and S.T. Thornton, Nucl. Phys. A186 (1972) 161

SCATTERING OF POLARIZED ^{3}He BY ^{1}H, ^{2}H, ^{3}He AND ^{4}He AT 32 MeV

O. Karban, N.T. Okumusoglu, A.K. Basak, C.O. Blyth, J.B.A. England,
J.M. Nelson, S. Roman and G.G. Shute
Department of Physics, University of Birmingham, Birmingham B15 2TT,
England

A large amount of experimental information on the polarization of protons and deuterons in interactions involving small number of nucleons contributed significantly to the present understanding of few-nucleon systems. However, similar data on the polarization of ^{3}He particles are practically limited to only few experiments with polarized ^{3}He targets. With the availability of the 33 MeV polarized ^{3}He beam from the University of Birmingham Radial Ridge Cyclotron it was of prime interest to obtain such data for targets with A=1 to 4.

The results[1] for ^{1}H were obtained with a 20 mg/cm^2 polyethylene foil while a 6 mg/cm^2 deuterated polyethylene foil was used as a deuteron target. The data for both helium isotopes were obtained using a 2 atm. gas cell[2] in the main scattering chamber. The mid-target scattering energy varied from 31.5 MeV for protons to 32.5 MeV for deuterons. The experimental details were described elsewhere[3]).

The ^{3}He polarization angular distributions together with the differential cross section data at the corresponding cm energies (ref.4) for protons, deuterons, ^{3}He and ^{4}He, are shown in fig.1. The measured maximum polarization varies from 0.20 for protons to -0.67 for ^{4}He. The ^{3}He data plotted beyond 90°(cm) are mirror images of those in the forward hemisphere.

Table 1

	$V^{a)}$	r_V	a_V	W_D	r_W	a_W	V_{so}	r_{so}	a_{so}
p	51.3	1.85	0.878	0.67	1.98	0.348	2.43	0.994	0.123
*)	164.0	2.08	0.50	0.56	1.09	0.562	0.93	0.86	0.188
d	194.4	1.60	0.396	2.37	1.83	0.933	1.54	0.91	0.40
^{3}He	70.0	1.50	0.50	0.50	1.80	0.70	1.0	1.30	0.20
^{4}He	65.3	1.40	0.248	0.6$^{b)}$	4.91	0.104	0.97	0.939	0.746
*)	66.0	1.40	0.213	1.60	1.50	0.162	0.70	1.0	0.20

r_c=1.4; *) dashed lines in fig.1; a) standard notation;
b) volume absorption

As an alternative to the standard phase-shift analysis of few-nucleon systems an attempt was made to parametrise the present data in terms of a potential between the ^{3}He and the target particles using the full optical model formalism and a corresponding programme including a search routine. Such an approach enables us to describe the scattering process for all four nuclei studied with the same number of parameters and can be further justified by the fact that no pronounced resonance effects were found in the composite systems in the corresponding cm regions. The resulting predictions (with parameters of table 1) shown in fig.1 describe surprisingly well the experimental data. This may indicate that any target spin effects represent only corrections to the basic set of phase shifts which can be deduced from the formal optical potential. These sets can be then used as a starting point in more refined phase-shift analyses.

References
1) C.O. Blyth et al., Nucl.Phys., to be published.
2) R.C. Brown et al., Nucl.Phys., A207 (1973) 456
3) W.E. Burcham et al., Nucl.Phys. A245 (1975) 170

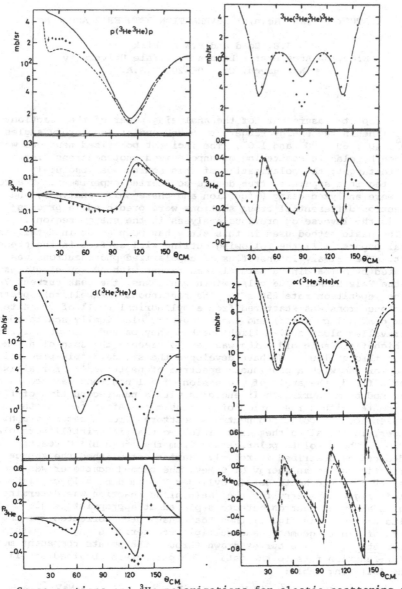

Fig.1. Cross sections and ^{3}He polarizations for elastic scattering of ^{3}He particles by ^{1}H, ^{2}H, ^{3}He and ^{4}He.

4) D.G. MacDonald, W. Haeberli and L.W. Morrow, Phys.Rev. 133 (1964) B1778; T.R. King and R. Smythe, Nucl.Phys. A183 (1972) 657; J.G. Jenkin, W.D. Harrison and R. Brown, Phys.Rev.C1 (1970) 1622; C.G. Jacobs Jr., and R.E. Brown, Phys.Rev. C1 (1970) 1615.

A STUDY OF THE ^{4}He$(\vec{n},n)$^{4}He REACTION BETWEEN 2 AND 6 MeV*

J.E. Bond and F.W.K. Firk

Electron Accelerator Laboratory, Yale University
New Haven, Ct. 06520, U.S.A.

We report measurements of the analyzing power of the reaction ^{4}He$(\vec{n},n)$^{4}He in the energy range 2 to 6 MeV and at laboratory angles of 20°, 40°, 60°, 80° and 110°. The incident polarized neutrons were obtained by elastic scattering of unpolarized photoneutrons from a graphite target; the polarization of this source was absolutely calibrated in a separate, true double-scattering experiment. The results were analyzed using R-function and phase-shift analyses, and the neutron differential cross sections were predicted (a process which is the inverse of previous analyses in the energy region).

The basic method used in this study has been given in detail in several papers [1,2]; the following outline will suffice in the present report. The primary intense flux of unpolarized photoneutrons was generated by bombarding a natural lead target with 40-MeV electrons from the Yale LINAC. The pulse width was 25 ns, the peak current 7A and the repetition rate 250 s^{-1}. The continuous (Maxwellian) spectrum of photoneutrons was scattered from a cylindrical shell of graphite (10 cm dia. x 1 cm thick) and those neutrons elastically scattered at 50° travelled along a 27-m flight path. They passed through a 1.2 m long high-field solenoid which was set to precess the spin of a 3-MeV neutron through 180°. We have developed the standard spin-precession method for use with a continuous spectrum of neutrons [3]: for a fixed magnetic field, the angle of precession of all neutrons was accurately determined by measuring their energies with a nanosecond time-of-flight spectrometer with a resolution of 0.75 ns.m^{-1}. After passing through the solenoid, the polarized neutrons scattered from a second graphite cylinder, identical to the first, into two plastic scintillators placed at $\pm$ 50°. The absolute polarization from the ^{12}C$(n,\vec{n})$^{12}C reaction was deduced as described in ref. 1. Having established the source polarization up to an energy of 6 MeV, the second scatterer was replaced by a liquid He target (a cylinder 7.5 cm dia. x 15 cm high) and the left-right asymmetry for the ^{4}He$(\vec{n},n)$^{4}He reaction was determined using an array of neutron detectors placed at appropriate angles. The results are shown in fig. 1; these data have been corrected for the effects of finite geometry and multiple scattering using the method of Stinson et al. [4]. The curves drawn through the points represent multi-level R-function fits to the data. The parameters obtained are listed in the following table:

ℓ_j	$E_{\lambda\ell j}$ (MeV)	$\gamma^2_{\lambda\ell j}$ (MeV)	$R_{0\ell j}$	$R_{1\ell j}$ (MeV)$^{-1}$	$B_{\ell j}$
$s_{\frac{1}{2}}$	-----	-----	0.18	0.01±0.01	0.00
$p_{\frac{1}{2}}$	7.01±0.24	15.66±1.26	0.37±0.26	0.00±0.12	-0.29
$p_{3/2}$	0.99±0.03	7.61±0.45	0.28±0.25	0.01±0.07	-0.74

Interaction radius =3 fm.

Here, all the symbols have their usual meanings. We have used a mildly energy dependent form for the effects of distant levels: $R^{\infty}_{\ell j} \approx R_{0\ell j} + R_{1\ell j}E$.

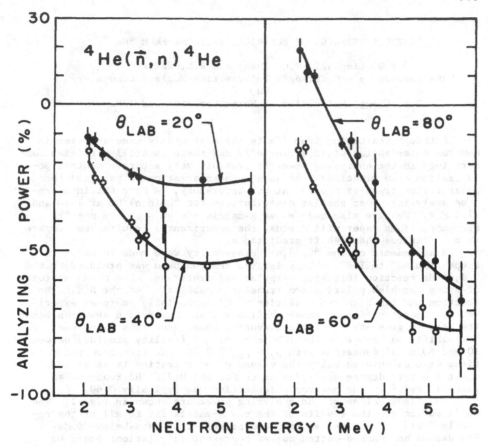

Fig. 1. The measured point analyzing power for the reaction $^4\mathrm{He}(\vec{n},n)^4\mathrm{He}$ at four angles. The curves represent R-function fits to the data using the listed parameters.

The uncertainties given in the table were obtained using the correlated error method in ref. 5). The fit was found to be insensitive to d-wave contributions from states at higher energies. The overall agreement between the present parameters and those given in ref. 6) is good; however, we do obtain a slighly larger reduced width for the $p_{\frac{1}{2}}$ state at 7 MeV.

References

*Work supported by E.R.D.A.

1) R.J. Holt, F.W.K. Firk, R. Nath and H.L. Schultz, Phys. Rev. Lett. 28 (1972) 114 and Nucl. Phys. A 213 (1973) 147
2) G.T. Hickey, F.W.K. Firk, R.J. Holt and R. Nath, Nucl. Phys. A 225 (1974) 470
3) R. Nath, F.W.K. Firk, R.J. Holt and H.L. Schultz, Nucl. Instr. 98 (1972) 385
4) G.M. Stinson, S.M. Tang and J.T. Sample, Nucl. Instr. 62 (1968) 13
5) D. Cline and P.M.S. Lesser, Nucl. Instr. 82 (1970) 291
6) Th. Stammbach and R.L. Walter, Nucl. Phys. A180 (1972) 225

ELASTIC SCATTERING OF POLARIZED NEUTRONS FROM ^{4}He *

P.W. Lisowski, T.C. Rhea, and R.L. Walter
Duke University and Triangle Universities Nuclear Laboratory
and
T.B. Clegg, University of North Carolina and TUNL

Although scattering from ^{4}He is the most widely used analyzer in neutron polarization studies, above 12 MeV there is still sufficient uncertainty in the analyzing power to prohibit very accurate neutron polarization determinations. Because such information is required for polarization transfer studies at our laboratory, we have obtained precise analyzing power angular distributions for ^{4}He($\vec{n}$,n)^{4}He at 14.0 and 17.1 MeV. We have also made a new R-matrix analysis of the n + ^{4}He system[1]. This paper will discuss the experimental results and compare them to various phase shift predictions.

Measurements of the ^{4}He($\vec{n}$,n)^{4}He asymmetry were made in about 10° steps from 40° to 148°. The polarized neutron beam was produced by the D($\vec{d}$,$\vec{n}$)^{3}He reaction initiated by polarized deuterons. The novel approach of using the high polarization transfer capabilities of the D($\vec{d}$,$\vec{n}$)^{3}He reaction and its high cross section at 0° appreciably improves experimental conditions[2] over those available when unpolarized deuteron beams are used to generate a polarized neutron beam. For this experiment, the Lamb-shift ion source of the TUNL accelerator facility provided between 10 and 60 nA of deuterons with $p_z = p_{zz} \cong 0.70$. Neutron beam polarizations were calculated using the values of polarization transfer coefficient and zero-degree analyzing power for the D($\vec{d}$,$\vec{n}$)^{3}He reaction as given in ref.[3]. The neutron polarization was typically 0.60.

The final values of the analyzing power are shown in fig. 1. The solid curves are the results of the new R-matrix fit to all of the reliable ^{4}He(n,n)^{4}He cross section and polarization data below 20 MeV[1]. The dashed and dashed-dotted curves represent calculations based on Stammbach-Walter[4] and Hoop-Barschall[5] phase shift predictions, respectively. The new R-matrix calculation gives the best representation at both energies. The Stammbach-Walter calculation agrees with the data better at 14.0 MeV than at 17.1 MeV, and the older Hoop-Barschall predictions, which are useful at energies above 20 MeV, do not describe the present results well at 17.1 MeV.

Although the new R-matrix analysis provided an improved representation of ^{4}He(n,n)^{4}He observables in the energy region from 10 to 18 MeV, it also indicated that additional accurate differential cross section measurements were needed to better determine the values of ^{4}He(n,n)^{4}He phase shifts. As a result, such cross section measurements have been undertaken at our laboratory[6]. Hopefully, when values of the cross section data are included in the R-matrix analysis, the phase shift solution for the n + ^{4}He system will be determined well enough to permit a very accurate calculation of the analyzing power from 10 to 18 MeV.

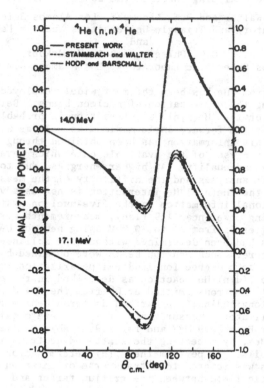

Fig. 1. Comparison of data with phase shift predictions at 14.0 and
17.1 MeV.

References
*Work supported in part by U.S. Energy Research and Development Admini-
stration.

1) P.W. Lisowski, Ph.D. thesis, Duke University, 1973 (unpublished)
2) J.E. Simmons, W.B. Broste, T.R. Donoghue, R.C. Haight and
   J.C. Martin, Nucl. Instr. and Meth. 106 (1973) 477
3) P.W. Lisowski, R.L. Walter, C.E. Busch and T.B. Clegg, Nucl. Phys.
   A242 (1975) 298
4) Th. Stammbach and R.L. Walter, Nucl. Phys. A180 (1972) 225
5) B. Hoop and H.H. Barschall, Nucl. Phys. 83 (1966) 63
6) G. Mack, P.W. Lisowski, and R.L. Walter (to be published)

^{4}He(n,n)^{4}He Analyzing Power in the 20 to 30 MeV Energy Range*

R.L. Walter and P.W. Lisowski, Los Alamos Scientific
Laboratory and Triangle Universities Nuclear Laboratory
and
G.G. Ohlsen and R.A. Hardekopf
Los Alamos Scientific Laboratory, Los Alamos, N.M. 87545, USA

Scattering from ^{4}He has been the most widely employed method for accurately measuring the polarization of nucleon beams. Below 18 MeV the
analyzing power for the ^{4}He(n,n)^{4}He interaction is probably known to within about ± 0.03 at the forward angle minimum as well as at the backward
angle maximum. This information has been obtained through energy-dependent phase-shift analyses of the available data which have gradually
grown in quantity and quality. At higher energies, due to the complexity
of the phase-shift analyses and the difficulty of the measurements, the
understanding of the ^{4}He(n,n)^{4}He interaction is not as advanced. In order
to provide additional information on the five-nucleon system and to calibrate the analyzing power near 119°(lab), measurements were made of $A_y(\theta)$
for neutron energies E_n from 20 to 30 MeV using neutron beams whose polarization values had been determined without any reliance on neutron polarization analyzers. Such neutron beams were obtained by utilizing the
properties of the zero degree longitudinal polarization transfer coefficient K_z^z of the T(d,n)^{3}He reaction as described in the contribution of
Lisowski et al.[1] Neutrons emitted at 0° from the T + d reaction initiated by a 75% longitudinally polarized deuteron beam were incident on a
liquid ^{4}He scintillator. Measurements of $A_y(\theta)$ were obtained for θ = 60°
and 70° (lab) for E_n = 20.9 MeV and for 119° (lab) at five energies between 20 and 30 MeV, by detecting the scattered neutrons in coincidence
with the ^{4}He recoils. To perform the $A_y(\theta)$ determination the neutron-
polarization axis was rotated to a transverse orientation by means of a
transverse magnetic field between the tritium target and the ^{4}He scatterer. The deuteron beam polarization was determined to within ± 0.01 by
the quench-ratio method and A_{zz} values were obtained from previous measurements.[1]

The results of preliminary analyses of the data are illustrated in
Figs. 1 and 2 where comparisons are made with calculations based on the

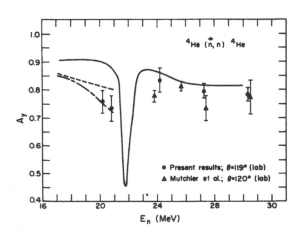

Fig. 1

A_y (119°) for ^{4}He(n,n)^{4}He
elastic scattering. The
curves represent calcula-
tions using the phase
shifts of ref. 2) (———),
ref. 3) (-----), and
ref. 4) (——— — ———).

Hoop-Barschall phase-shifts[2] and with calculations based on several phase-shift sets that have appeared recently.[3-5] Data of Mutchler et al.[6] are also included in Fig. 1 for comparison. Some of these latter data were obtained using neutrons from the T(d,n) reaction emitted at θ≈30° whose polarization had been calibrated by measuring the polarization in the D(t,n) reaction at like center of mass energies and angles. This calibration, however, relies on accurate knowledge of $A_y(\theta)$ for E_n ≈ 11 MeV as well as on accurate techniques to measure the neutron polarization for the D(t,n) reaction.

The present data verify that A_y (119°) is near 0.80 above the $d_{3/2}$ resonance and about 0.75 immediately below it. These data will be combined with other data obtained previously at LASL to obtain better values of the phase-shifts for ^{4}He(n,n)^{4}He in the 20 to 30 MeV region.

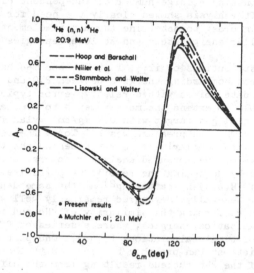

Fig. 2 $A_y(\theta)$ for ^{4}He(n,n)^{4}He elastic scattering. The curves represent calculations based on the phase shift of ref. 2) (—— — ——), ref. 3) (—— —— ——), ref. 4) (------) and ref. 5) (————).

References

*   Supported by the U.S. Energy Research and Development Administration
1) P.W. Lisowski, R.L. Walter, R.A. Hardekopf, and G.G. Ohlsen, these proceedings, p.558.
2) B. Hoop, Jr. and H.H. Barschall, Nucl. Phys. 83 (1966) 65.
3) P.W. Lisowski and R. L. Walter (to be publihsed); P. W. Lisowski, Ph.D. thesis, Duke University (1973).
4) Th. Stammbach and R.L. Walter, Nucl. Phys. A18 (1972) 225.
5) A. Niiler, M. Drosg, J. C. Hopkins, J.D. Seagrave and E.C. Kerr, Phys. Rev. C4 (1971) 36.
6) G. S. Mutchler, W. B. Broste, and J. E. Simmons, Phys. Rev. C3 (1971) 1031

MEASUREMENT OF THE ANALYSING POWERS IN ^{3}He($\vec{\text{d}}$,d)^{3}He SCATTERING

B. Jenny, W. Grüebler, V. König, P.A. Schmelzbach, R. Risler,
D.O. Boerma and W.G. Weitkamp
Laboratorium für Kernphysik, Eidg. Techn. Hochschule, 8049 Zürich
Switzerland

Some time ago, a phase shift analysis based on the cross section and all analysing power components of d-^{3}He elastic scattering has been tried. Measurements of iT_{11}, T_{20}, T_{21} and T_{22} exist in the energy range between 4.0 and 11.5 MeV[1]. Due to the relatively complicated spin structure of the scattering problem and the existence of open reaction channels, the scattering matrix contains a large number of independent elements. In fact, the phase shift analysis showed clearly the need for extendet and more accurate data in order to determine the matrix elements accurately. Measurements in smaller energy steps and at lower energies were also desirable.

New angular distributions of iT_{11}, T_{20}, T_{21} and T_{22} have been measured for ten deuteron bombarding energies between 2.0 and 11.5 MeV, each containing up to 50 data points with a relative error typically of 0.005. It was possible to extend the measurements to c.m. angles of 12.5° and 165° using a special gas target with 0.9 mg/cm^2 mylar foil entrance and exit windows. The ^{3}He gas pressure was 200 Torr. The beam polarization was determined with a polarimeter mounted behind the target.

Fig. 1 shows two of the measured analysing powers. Circles indicate data points obtained by detecting the recoil ^{3}He particles. A comparison with measurements of ^{3}H($\vec{\text{d}}$,d)^{3}H scattering[2] at the same deuteron energies is interesting. Generally they agree remarkably well with ^{3}He($\vec{\text{d}}$,d) ^{3}He scattering, thus confirming the similarity of ^{5}He and its mirror nucleus at higher excitation energies, where a series of T = 1/2 levels is predicted by theoretical calculations (3,4,5). One part of the occuring significant deviations can probably be ascribed to the 0.24 MeV higher excitation of the ^{5}He nucleus resulting from the different threshold energies. The discrepances of T_{22} at 5.0 MeV between 30° and 60° and those at 8.0 MeV seem to be caused by other reasons. A phase shift analysis of the new data is now in progress at this laboratory.

References

1) V. König, W. Grüebler, R.E. White, P.A. Schmelzbach and P. Marmier, Nucl. Phys. A185 (1972) 263
2) A.A. Debenham, V. König, W. Grüebler, P.A. Schmelzbach, R. Risler and D.O. Boerma, Nucl. Phys. A216 (1973) 42
3) P. Heiss and H.H. Hackenbroich, Nucl. Phys. A162 (1971) 530
4) R.F. Wagner and C. Werntz, Phys. Rev. C4 (1971) 1
5) F.S. Chwieroth, R.E. Brown, Y.C. Tang and D.R. Thompson, Phys. Rev. C8 (1973) 938

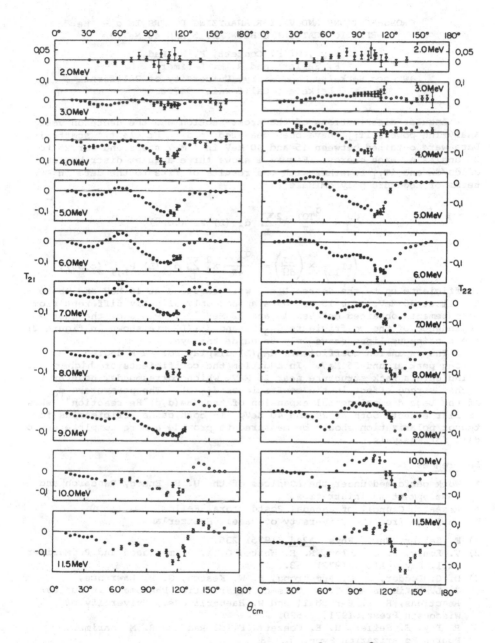

Fig. 1. Tensor analysing powers T_{21} and T_{22} in $^3\text{He}(\vec{d},d)^3\text{He}$ scattering.

CROSS-SECTIONS AND VECTOR ANALYZING POWERS IN $\vec{d}$ - ³He
ELASTIC SCATTERING BETWEEN 15 AND 40 MeV.*

R. Roy[+], H. E. Conzett, F. N. Rad,
F. Seiler[‡] and R. M. Larimer
Lawrence Berkeley Laboratory, University of California
Berkeley, California 94720

Angular distributions of the cross-section and the deuteron vector
analyzing power, iT_{11}, have been measured in $\vec{d}$ - ³He elastic scattering.
Data were obtained between 15 and 40 MeV in 5 MeV steps and for 25 to
35 angles at each energy. Figure 1 shows three angular distributions
of $d\sigma/d\Omega$ and iT_{11} together with the results of fits to the data in
terms of Legendre polynominals[1],

$$\frac{d\sigma}{d\Omega} = \frac{\sigma_{TOT}}{4\pi} \lambda^2 \sum_L d_{00}(L) \, P_L (\cos\theta)$$

$$\left(iT_{11}\right) \times \left(\frac{d\sigma}{d\Omega}\right) = \frac{\sigma_{TOT}}{4\pi} \lambda^2 \sum_L d_{11}(L) \, P_{L,1} (\cos\theta).$$

Coefficients up to the degree L = 9 were required at 15 MeV and up to
L = 11 at 30 MeV. The 15 MeV data appear only slightly different from
measurements obtained between 10 and 12 MeV[2,3]. Plots of the energy
dependence of the coefficients $d_{00}(L)$ and $d_{11}(L)$ are shown in figure 2;
the continuous lines serve only to guide the eye.

The expansion coefficients $d_{00}(L)$ display for each degree L a
strong peak around 35 MeV. In addition the coefficients of higher degree
L indicate a weak structure near 20 MeV, which corresponds, however, to a
rather strong effect in the coefficients $d_{11}(L)$. The parameters $d_{00}(L)$
of the Legendre polynomial expansion of the ³He(d,p)⁴He reaction[4] show
similar effects both at 20 and 40 MeV. In both cases efficiencies for
tensor polarization should be measured to provide a more complete set of
data for analysis.

References

* Work performed under the auspices of the U. S. Energy Research and
Development Administration.
+ Research Council of Canada, Postdoctoral Fellow.
‡ On leave from the University of Basel, Switzerland.

1) F. Seiler, Nucl. Phys. A244 (1975) 236.
2) V. König, W. Grüebler, R. E. White, D. A. Schmelzbach and P. Marimer,
   Nucl. Phys. A185 (1972) 263.
3) D. C. Dodder, D. D. Armstrong, P. W. Keaton, G. P. Lawrence,
   J. L. McKibben and G. G. Ohlsen, Polarization Phenomena in Nuclear
   Reactions, H. H. Barschall and W. Haeberli eds., University of
   Wisconsin Press, 1971, p 520.
4) R. Roy, F. Seiler, H. E. Conzett, F. N. Rad and R. M. Larimer,
   Fourth Polarization Symp., p.548

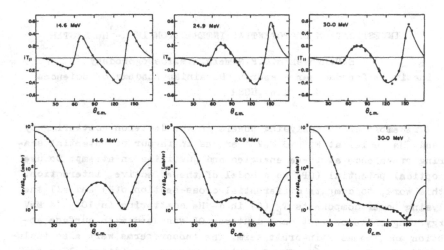

Fig. 1

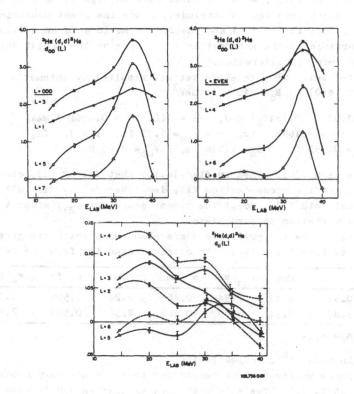

Fig. 2

INVESTIGATION OF POTENTIAL INTERACTION IN d-^{3}He SYSTEM

E.B.Lyovshin,O.F.Nemets,A.M.Yasnogorodsky
Institute for Nuclear Research, Ukrainian Academy of Sciences
Kiev,USSR

A resemblance of analysing powers in the deuteron scattering by ^{3}He and ^{4}He nuclei at $E_d^{lab}>8$ MeV[1] argues in favour of potential scattering prevalence at these energies and justifies an attempt to use the optical potential (OP) as a model of the effective interaction. In this work, to compute differential cross-section $\sigma(\theta)$ and all the analysing power components $T_{kq}(\theta)$ in d-^{3}He scattering in 10 - 14 MeV energy range, the conventional deuteron OP was used with surface absorbtion and Thomas spin-orbit term. The tensor terms were also included of T_r and T_L forms[2]. The T_r matrix elements, offdiagonal in orbital angular momentum, were accounted for. An explicit dependence of the OP on target spin was not included, since the great resemblance of analysing powers in ^{3}He(d,d) and ^{4}He(d,d) reactions is an evidence for small importance of the interaction dependence on the target spin for quantities under consideration.

The following OP parameters set was resulted by automatic search routine for $\sigma(\theta)$ at E_d = 10 - 14 MeV[3]

$$V = -(41.625 + 0.124E_d) \text{ MeV}, \quad W = -(0.610 + 0.056E_d) \text{ MeV},$$
$$r_w = (4.458 - 0.165E_d) \text{ fm}, \quad r_v = r_{so} = 1.75 \text{ fm}, \quad r_c = 1.3 \text{ fm}, \qquad (1)$$
$$a_v = a_{so} = 0.593 \text{ fm}, \quad a_w = 0.526 \text{ fm}, \quad V_{so} = 2.15 \text{ MeV}.$$

It is seen (full lines in figs.1a,1b) that the OP (1), obtained as a result of the cross-section fit, describes fairly well all the experimental data[1,4], except the tensor component T_{21}, without any additional variation of parameters.

In fig.1a the the predictions are also shown which are given by A and B sets (see the table) both obtained starting from the sets

	V Mev	W Mev	V_{so} MeV	a_v fm	r_v fm	a_v fm	r_w fm
A	90.37	2.43	7.54	0.456	2.28	0.526	2.15
B	163.80	3.54	5.00	0.482	2.36	0.501	2.71

$r_v = r_{so}, \; a_v = a_{so}.$

proposed in refs.[5,6], respectively.

The phase shifts (PS) corresponding to set A are very close to those of set B. All three sets used give similar values of weakly splitted odd-L PS. However, the even-L PS given by set (1) are greatly different from those of A and B sets. This may contradict the opinion[7] that the smallness of the spin orbit interaction in the states of odd-L is the only reason for $iT_{11}(90°) = 0$.

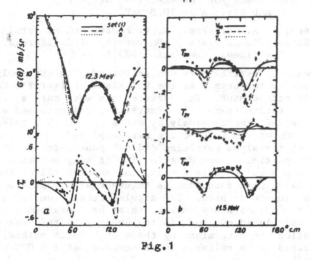

Fig.1

The PS energy dependence, given by set (1), shows the overall similarity to that of the "refined cluster model"[8], but optical D-wave PS are more splitted, and there is a strong D-wave absorbtion.

Thus, the optical model describes successfully the observable quantities and the OP parameters change smoothly over 10 - 14 MeV energy range. This may be an additional indication of the potential scattering prevalence in the $d-{}^3He$ system at the energies under consideration and the lack of the pronounced isolated resonances in ${}^5Li$ at the 22 - 25 MeV excitation energies.

Within the above computational procedure the effect of tensor terms is nearly the same as in the case of scattering by haevier nuclei: they give perhaps some improvements to the T_{20} and T_{22} fit and an almost correct behavior of T_{21}, while cross-section and vector analysing power are practically unaffected[3], see fig.1b.

References
1) V.König at al., Nucl.Phys. A185(1972)263.
2) G.R.Satchler, Nucl.Phys. 21(1960)116.
3) E.B.Lyovshin,O.F.Nemets,A.M.Yasnogorodsky, Phys.Lett.52B(1974)392.
4) J.C.Allred et al.,Phys Rev. 88(1952)425;
   T.A.Tombrello,R.J.Spiger,A.D.Bacher, Phys.Rev. 154(1967)935;
   J.E.Brolley,T.M.Putnam,L.Rosen,L.Stewart, Phys.Rev.117(1960)1307.
5) E.M.Henley,E.C.Richards,D.U.L.Yu, Nucl.Phys. A103(1967)361.
6) W.Tobockman, Proc.Rutherford Jub. Conf.(Manchester,1961), p.465.
7) M.Tanifuji,K.Yazaki, Few Particle Problems, North Holland PC, 1972, p.739.
8) P.Heiss,H.H.Hackenbroich, Nucl.Phys. A162(1971)530.

ANALYZING POWER FOR THE REACTIONS D($\vec{t}$,t)D AND D($\vec{t}$,α)n AT 10.5 MeV[*]

G. G. Ohlsen, R. A. Hardekopf, R. V. Poore, and N. Jarmie
Los Alamos Scientific Laboratory, University of California
Los Alamos, New Mexico 87545

The experiment we describe here was performed with the newly available polarized triton beam[1] at the Los Alamos Scientific Laboratory Van de Graaff accelerator. The measurements were carried out in the "supercube" scattering chamber[2]. Deuterons, tritons, and alpha particles were detected simultaneously by means of two ΔE-E telescopes. 48 micron ΔE and 1000 micron E detectors were employed. Over the angular range for which the alpha particles did not penetrate the ΔE detectors, the alpha particles were detected in "ΔE singles" mode. Over the remainder of the angular range for alphas, and for all of the deuteron and triton data, ΔE-E coincidences together with software mass identification routines were used. The angular resolution for the measurements was ± 1° (laboratory full width at half maximum). The asymmetries were calculated on-line via the usual geometric mean method. The beam polarization was determined by the quench ratio method, which has been demonstrated to be reliable to an accuracy of ± 0.005 or better for triton beams[3].

D. C. Dodder and G. M. Hale[4] have kindly provided predictions for these two observables based on their large scale R-matrix analysis of the five nucleon system. The choice of a bombarding energy of 10.5 MeV for the present data, which corresponds to 7 MeV in the T(d,n)^{4}He reaction, was chosen because of the large amount of experimental information already available at that energy. In fig. 1 the experimental data is plotted along with the predictions at 10.5 MeV (solid curve), 9 MeV (short-dashed curve) and 12 MeV (long-dashed curve). There appears to be only a slight energy dependence in this region. The data are only qualitatively fit by the predictions, in spite of the fact that the predictions are based on a large amount of data, including polarization transfer data in the T($\vec{d}$,n)^{4}He reaction[5]. We therefore conclude that the availability of triton data will make a significant contribution to the determination of the correct phenomenological parameters for a description of the five-nucleon system.

We have compared the present measurements with the available measurements in the mirror $^3\vec{He}$(d,d)^{3}He and $^3\vec{He}$(d,p)^{4}He reactions[6-9]. Although no data is available at the precise energy of the present experiment, interpolated values agree with the present data. An accurate comparison of $^3\vec{He}$(d,p)^{4}He and $\vec{T}$(d,n)^{4}He analyzing powers would possibly afford an interesting test of charge symmetry, since the large Q-value would presumably make the Q-value difference between the two reactions insignificant and thus would remove one of the major complications in the interpretation of such a comparison[10].

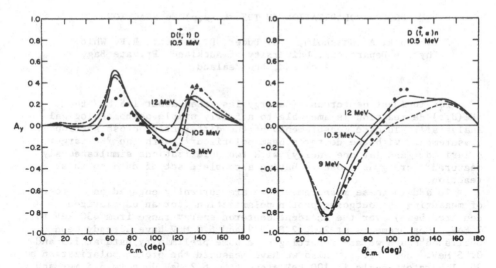

Fig. 1. Observed analyzing powers for the reactions D($\vec{t}$,$\vec{t}$)D and
D($\vec{t}$,α)n. The solid curves represent R-matrix predictions at the
same and adjacent energies. For the D($\vec{t}$,t)D graph, dots refer
to the detection of tritons and triangles to the detection of recoil
deuterons.

References

*
  Work performed under the auspices of the U.S. ERDA

1) R. A. Hardekopf, these proceedings, p.865
2) G. G. Ohlsen and P. A. Lovoi, these proceedings, p.907
3) R. A. Hardekopf, G. G. Ohlsen, R. V. Poore, and N. Jarmie,
   these proceedings, p.903
4) D. C. Dodder and G. M. Hale, private communication
5) J. W. Sunier, R. V. Poore, R. A. Hardekopf, L. Morrison,
   G. C. Salzman, and G. G. Ohlsen, to be published
6) D. M. Hardy, S. D. Baker, D. H. McSherry, and R. J. Spiger,
   Nucl. Phys. A160 (1970) 154
7) B. E. Watt and W. T. Leland, Phys. Rev. C2 (1970) 1677
8) B. E. Watt and W. T. Leland, Phys. Rev. C2 (1970) 1680
9) G. G. Ohlsen, R. A. Hardekopf, D. P. May, S. D. Baker, and
   W. T. Armstrong, Nucl. Phys. A233 (1974) 1
10) R. A. Hardekopf, R. L. Walter, and T. B. Clegg, Phys. Rev. Lett. 28
    (1972) 760

PROTON POLARIZATION IN THE ^{3}He(d,$\vec{p}$)^{4}He REACTION

J.E. Brock, A. Chisholm, J.C. Duder, R. Garrett, R.E. White
Physics Department, University of Auckland, Private Bag,
Auckland, New Zealand.

For incident deuterons of energy less than about 1.5 MeV the ^{3}He(d,$\vec{p}$)^{4}He reaction is amenable to a fairly simple phenomenological analysis[1]. To date measurements of the differential cross section[2], measurements with the deuteron beam polarized[3], with the ^{3}He target polarized[4] and (at one energy) with two polarizations simultaneously measured[5] bring us close to having a complete set of data on this reaction.

To add to these measurements we are currently engaged on a program of measuring the outgoing proton polarization (for an unpolarized deuteron beam) over the incident deuteron energy range from 430 keV to 7 MeV. Measurements at 2.0, 2.8, 3.9 and 6.0 MeV have already been published[6] and we are reporting here some preliminary data at 1.5 and 0.75 MeV. See fig. 1. Also we have measured the proton polarization at 30° laboratory angle in 100 keV steps from 6.2 MeV down to 3.5 MeV and at four energies below this. This measurement was carried out to see if the states thought to exist at 18 and 20 MeV excitation might produce a discontinuity in P(30°) with energy. These results are shown in fig. 2. Also plotted on the same graph are neutron polarizations from the ^{3}H(d,$\vec{n}$)^{4}He reaction[7] compared at the same entrance channel energy.

No measurements of proton polarization at angles greater than 150° (lab.) have been made. This we are planning to do in the near future by accelerating a ^{3}He beam on to a deuterium target.

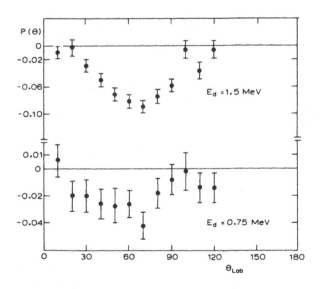

Fig. 1. Measured proton polarization for the ^{3}He(d,$\vec{p}$)^{4}He reaction.

Below 1.5 MeV the reaction is dominated by the 430 keV resonance. Since rank one polarizations require interference between two different matrix elements, the existence of proton polarization indicates a deviation from the simple reaction mechanism involving only the $^4S_{\frac{3}{2}}$ entrance channel with its large 430 keV resonance. Quantitative analysis has not yet been attempted as the experimental results are incomplete. A qualitative analysis suggests that below 1.5 MeV the proton polarization is produced by interference between the $\frac{3}{2}^+$ state responsible for the 430 keV resonance and $\frac{1}{2}^-$ and $\frac{3}{2}^-$ states. The latter two are produced with entrance channel spin $\frac{3}{2}$ and orbital angular momentum $\ell = 1$.

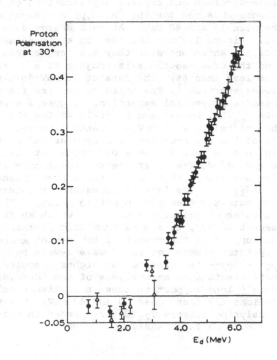

Fig. 2. Proton polarization at 30° as a function of energy for the ^{3}He(d,$\vec{p}$)^{4}He reaction. The open triangles are neutron polarizations from the ^{3}H(d,$\vec{n}$)^{4}He reaction at the same entrance channel energy.

References

1) F. Seiler and E. Baumgartner, Nucl. Phys. A153 (1970) 193
2) J.P. Conner, T.W. Bonner and J.R. Smith, Phys. Rev. 88 (1952) 468
3) R. Garrett and W.W. Lindstrom, Nucl. Phys. A224 (1974) 186
4) V. Rohrer, P. Huber, Ch. Leemann, H. Meiner and F. Seiler, Helv. Phys. Acta 44 (1971) 846
5) Ch. Leemann, H. Bürgisser, P. Huber, V. Rohrer, H. Paetz gen. Schieck and F. Seiler, Helv. Phys. Acta 44 (1971) 141
6) J.F. Clare, Nucl. Phys. A217 (1973) 342
7) G.S. Mutchler, W.B. Broste and J.E. Simmons, Phys. Rev. C3 (1971) 1031

548

CROSS-SECTIONS AND VECTOR ANALYZING POWERS IN THE
^{3}He($\vec{d}$,p)^{4}He REACTION BETWEEN 15 AND 40 MeV.*

R. Roy[+], F. Seiler[‡], H. E. Conzett,
F. N. Rad and R. M. Larimer
Lawrence Berkeley Laboratory, University of California
Berkeley, California 94720

Differential cross-sections and angular distributions of the vector
analyzing power iT_{11} were obtained for the ^{3}He($\vec{d}$,p)^{4}He reaction in
intervals of 5 MeV between 15 and 40 MeV. At each energy data were
taken at 25-35 angles. Figure 1 shows the results at three energies.
The statistical errors are shown wherever they are larger than the
symbols. The scale of the cross-sections is subject to a systematic
error, estimated to be less than 6%. The data at 15 MeV join smoothly
to measurements of lower energies[1]. The solid curves are the results
of fitting with a Legendre polynomial expansion. Figure 2 shows the
expansion coefficients $d_{kq}(L)$, normalized to yield 4π for the total
cross-section[2]. The points below 12 MeV are taken from ref. 1).
The coefficients of the cross-section for unpolarized particles
$d_{00}(L)$ for even degree L show some evidence of a broad structure near
20 and 40 MeV, while the odd-degree coefficients repeat only the 20 MeV
structure. This coincides with a shift away from a predominance of the
coefficients $d_{11}(2)$ to $d_{11}(1)$. The latter indicates large interference
terms between reaction matrix elements of opposite parity. This is
also visible in the angular distributions of iT_{11}, which shift from
antisymmetry with respect to 90°, to a more symmetric distribution.
This observation adds support to the result of two recent analyses[2,3]
which postulate mostly interference between d-wave levels below 11.5
MeV and a strong $d_{7/2}^{+}$ - $f_{7/2}$- interference at higher energies.
The Legendre coefficients from an analysis of the ^{3}He($\vec{d}$,d)^{3}He
elastic scattering data[4] lend support to these tentative conclusions
because similar variations are found near 20 and 40 MeV. Clearly,
measurements of the analyzing tensors $T_{2q}(\theta)$ are needed in order to
provide the data for a more definite analysis.

References

* Work performed under the auspices of the U. S. Energy Research and
Development Administration.
+ Research Council of Canada, Postdoctoral Fellow.
‡ On leave from the University of Basel, Switzerland.

1) W. Grüebler, V. König, A. Ruh, P. A. Schmelzbach, R. E. White and
P. Marmier, Nucl. Phys. A176 (1971) 631.
2) F. Seiler, Nucl. Phys. A244 (1975) 236.
3) D. D. Dodder and G. M. Hale, Fourth Polarization Symp., p.167
4) R. Roy , H. E. Conzett, F. N. Rad, R. Seiler and R. M. Larimer,
Fourth Polarization Symp., p.540

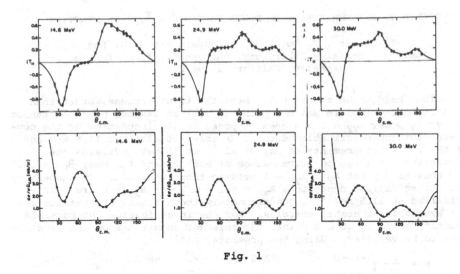

Fig. 1

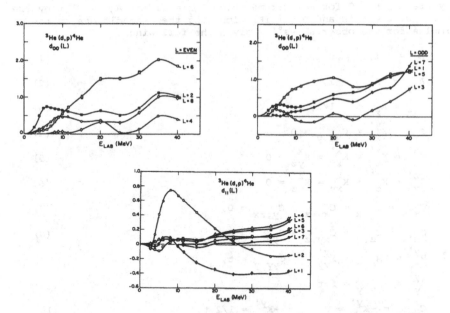

Fig. 2 Energy dependence of the coefficients $d_{00}(L)$ and $d_{11}(L)$.
The smooth curves are drawn to guide the eye.

POINTS OF MAXIMUM ANALYZING POWER IN THE ^{3}He$(\vec{d},p)^4$He REACTION*

F. Seiler$^+$, R. Roy‡, H. E. Conzett and F. N. Rad
Lawrence Berkeley Laboratory, University of California
Berkeley, California 94720

The ^{3}He$(\vec{d},p)^4$He reaction has been the first process, other than elastic scattering, in which a point (E_o,Θ_o) of maximum possible tensor analyzing power $A_{yy} = 1$ has been proposed[1]. An inspection of the complete deuteron polarization data of Grüebler et al.[2] at 11.5 MeV and of the i T_{11} measurements of Roy et al.[3] at 15 MeV indicates the possibility of large positive values of both A_y and A_{yy} near $\Theta_{cm} = 120°$ and thus of a point $A_y = A_{yy} = 1$ between these energies. The necessary but not sufficient conditions $A_{xx} = A_{zz} = -1/2$ and $A_{xz} = 0$ are nearly fulfilled at 11.5 MeV[1,4]. Unfortunately these tensor observables are not available at higher energies. For the investigation and possible identification of such a point, the relevant conditions on the M-matrix have to be verified. Using the presentation[5,6,7]

$$M = \frac{1}{\sqrt{2}} \begin{pmatrix} -iA-D & \sqrt{2}\,F & -iA+D & -B-C & \sqrt{2}\,E & -B+C \\ B-C & \sqrt{2}\,E & B+C & -iA+D & \sqrt{2}\,F & -iA-D \end{pmatrix}.$$

they are $A = B = 0$ for an extreme value $A_{yy} = 1$; for $A_y = \pm 1$ they are $A = B = 0$, $C = \mp iE$ and $D = \mp iF$. Imposing these conditions on the formulae for the observables[6,7] gives the following

$$A_y = \pm 1, \tag{1}$$

$$A_{yy} = -K^{y'}_{o,y} = 1, \tag{2}$$

$$A_{xx} = A_{zz} = -1/2, \tag{3}$$

$$A_{xz} = K^{y'}_{xz} = C_{xz,y} = 0 \tag{4}$$

$$K^{x'}_{x} = K^{x'}_{z} = K^{x'}_{xy} = K^{x'}_{yz} = 0 \tag{5}$$

$$K^{z'}_{x} = K^{z'}_{z} = K^{z'}_{xy} = K^{z'}_{yz} = 0 \tag{6}$$

$$C_{x,x} = C_{z,x} = C_{xy,x} = C_{y,zx} = 0, \tag{7}$$

$$C_{x'z} = C_{z,z} = C_{xy,z} = C_{yz,z} = 0, \tag{8}$$

$$P^{y'} = -A_{o,y} = K^{y'}_{yy} = -C_{yy,y} = t, \tag{9}$$

$$K^{y'}_{y} = -C_{y,y} = \pm t, \tag{10}$$

$$C_{xx,y} = -K^{y'}_{xx} = C_{zz,y} = -K^{y'}_{zz} = 1/2\,t, \tag{11}$$

$$K^{x'}_{o,x} = -K^{z'}_{o,z} = u, \tag{12}$$

$$K^{x'}_{o,z} = K^{z'}_{o,x} = v. \tag{13}$$

Here σ_o, A, P, K and C denote the unpolarized cross-section, analyzing power, particle polarizations, polarization transfer coefficients and efficiency correlation coefficients, respectively. The first sub-scripted index stands for the beam, the second for the target polariza-tion. Thus 24 polarization observables involving two or less particle polarizations are numerically determined, while the other 14 are given by the 3 parameters t, u and v. With the cross-section for unpolarized particles there are thus 4 parameters that can be determined experimen-tally. By a careful selection of the experiments, through an inspec-tion of the general formulae[5,6], a verification of an extreme point of the components A_y and A_{yy} should be feasible. The establishment of such a point would be very important in an analysis of the process, due to the restrictions imposed on some elements of the M-matrix.

References

* Work performed under the auspices of the U. S. Energy Research and Development Administration.
+ On leave of absence from the University of Basel, Switzerland.
‡ Research Council of Canada, Post-doctoral fellow
1) F. Seiler, LBL-report, LBL-3474, and to be published
2) W. Grüebler, V. König, A. Ruh, P. A. Schmelzbach, R. E. White and Schmelzbach, R. E. White and P. Marmier, Nucl. Phys. A176 (1971) 631.
3) R. Roy, F. Seiler, H. E. Conzett, F. N. Rad and R. M. Larimer, Fourth Polarization Symp., p.548
4) F. Seiler, F. N. Rad and H. E. Conzett, LBL-report, LBL-3496.
5) P. L. Csonka, M. J. Moravcsik and M. D. Scadron, Phys. Rev. 143 (1966) 775.
6) G. G. Ohlsen, Rep. Prog. Phys. 35 (1972) 717.
7) J. L. Gammel, P. W. Keaton and G. G. Ohlsen, Los Alamos Scientific Laboratory, Report No. LA-4492-MS, July 1970.

MAXIMUM TENSOR ANALYZING POWER A_{yy} = 1
IN THE ^{3}He($\vec{d}$,p)^{4}He REACTION*

F. Seiler[+]
Lawrence Berkeley Laboratory, University of California
Berkeley, California 94720

Data points for the cartesian analyzing power $A_{yy}(\theta)$, derived from measurements of Grüebler et al.[1], reach and even exceed unity near E_0 = 9 MeV and θ_0 = 27° in the center of mass system (figure). This suggests the possibility of an extreme value of the analyzing power A_{yy} = 1 in this region[2]. For this case the conditions on the transition matrix M are

$$M_{1,\ 1/2;\ 1/2} = -M_{-1,\ 1/2;\ 1/2}$$
$$M_{1,\ -1/2;\ 1/2} = -M_{-1,\ -1/2;\ 1/2'} \quad (1)$$

where the indices denote the magnetic quantum numbers of the deuteron, ^{3}He and proton spins, respectively. The data show that these four real equations are nearly fulfilled at 8 and 10 MeV. Actual compliance at some critical point (E_0, θ_0) cannot be demonstrated directly, since a reliable analysis of the reaction is not yet available. An alternate method[2] consists of inserting eqs. (1) into the formulae for the observables[3,4] and testing the resulting conditions experimentally.

$$P^{y'} = -A_{o,y} = K^{y'}_{yy} = -C_{yy,y} \quad (2)$$
$$K^{y'}_{xx} = -C_{xx,y} \quad (3)$$
$$K^{y'}_{zz} = -C_{zz,y} \quad (4)$$
$$K^{x'}_{x} = K^{x'}_{z} = K^{z'}_{x} = K^{z'}_{z} = 0 \quad (5)$$
$$K^{x'}_{xy} = K^{x'}_{yz} = K^{z'}_{xy} = K^{z'}_{yz} = 0 \quad (6)$$
$$C_{x,x} = C_{z,x} = C_{x,z} = C_{z,z} = 0 \quad (7)$$
$$C_{xy,x} = C_{yz,x} = C_{xy,z} = C_{yz,z} = 0 \quad (8)$$

Some values for the proton polarization $P^{y'}$, the $^3\vec{\text{He}}$ analyzing power $A_{o,y}$ and the polarization transfer coefficients K^j_i are available near the critical point (E_0, θ_0) and lend support to the assignment of an extreme value A_{yy} = 1 [2]. Better statistical accuracy can be obtained for efficiency correlation coefficients $C_{i,j}$, however no such data are yet available near 9 MeV.

By a proper selection of observables, measured at and around the critical point, eqs.(1) can be tested thoroughly. In this respect conditions (5) to (8) are particularly important, since they are independent of the calibration of any polarization. Once the existence of an extreme value A_{yy} = 1 is thus established, eqs. (1) can be used to restrict the solution space of an analysis. By analyzing angular distributions at the critical energy E_0, the <u>linear</u> conditions (1) can be imposed directly on the usual system of bilinear equations. In view of the problems encountered in the analysis of reactions, this potential use of extreme points of the analyzing power may be more important than the more obvious application as a calibration point for deuteron tensor polarization.

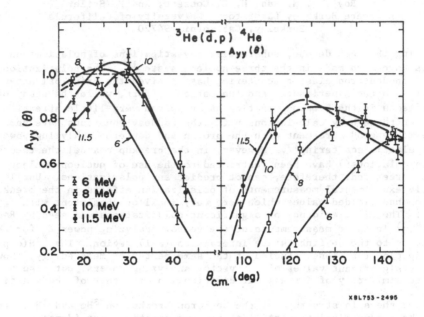

^{3}He$(\vec{d},p)$ ^{4}He

$A_{yy}(\theta)$

6 MeV
8 MeV
10 MeV
11.5 MeV

$\theta_{c.m.}$ (deg)

XBL753-2495

References

* Work performed under the auspices of the U. S. Energy Research and Development Administration.
+ On leave from the University of Basel, Switzerland.

1) W. Grüebler, V. König, A. Ruh, P. A. Schmelzbach, R. E. White and P. Marmier, Nucl. Phys. A176 (1971) 631.
2) F. Seiler, Lawrence Berkeley Laboratory report LBL-3474, March 1975, and to be published.
3) P. W. Keaton, J. L. Gammel and G. G. Ohlsen, Ann. of Phys. 85 (1974) 152.
4) G. G. Ohlsen, Rep. Progr. Phys. 35 (1972) 717.

VECTOR ANALYZING POWER IN THE ^{3}He($\vec{d}$,^{3}He)np REACTION*

R. Roy$^+$, F. N. Rad, H. E. Conzett, and F. Seiler†
Lawrence Berkeley Laboratory, University of California
Berkeley, California 94720

During the past decade, substantial investigations of polarization effects have been made in the three-nucleon system[1]. The polarization in nucleon-deuteron elastic scattering has received the principal attention of both the experimental and theoretical effort, while the study of such effects in the breakup reaction has received very little attention. The exact three-body calculations using the Faddeev equations have shown a remarkably good agreement with the proton and deuteron analyzing powers in p-d elastic scattering[1]. However, in the breakup channel the theoretical calculations[2] have been restricted to the use of nucleon-nucleon S-wave forces and, therefore, cannot predict any polarizations. Until recently experimental measurements of polarization effects in the breakup reaction had yielded values which were very small or consistent with zero[3]. The first evidence of significant polarizations was seen by Rad et al.[4,5] in their measurements of the vector analyzing power A_y for the transition to the np final state interaction (FSI) region. The ^{1}H($\vec{d}$,p)np and ^{2}H(p,p)np reactions, studied at the same center-of-mass energy, showed not only significant values of the vector analyzing powers, but also a definite similarity of its angular distribution with that of the elastic channel.

Since the spin structure of the deuteron breakup on ^{3}He and ^{1}H is the same, the nondynamical properties of the two reactions are identical. Thus, it is of interest to look for similar polarization effects in the ^{3}He($\vec{d}$,^{3}He)d* reaction, where d* denotes final-state np pairs with low relative energy E_{np}, in both singlet and triplet states. We report here measurements of the vector analyzing power A_y in this reaction at E_d = 30, 35, and 40 MeV for E_{np} <2.0 MeV. Our results are shown in fig. 1. The statistical errors are smaller than the symbols. For comparison, the analyzing powers in $\vec{d}$-^{3}He elastic scattering at the same energies are shown as the smooth curves. It is seen that A_y in the ^{3}He($\vec{d}$,^{3}He)d* reaction reaches substantial values and follows that of the elastic channel at d* production angles beyond 90° c.m. The peak values near θ_{cm} = 135° and 155° are quite constant in magnitude and position over the 10 MeV energy interval studied. In a comparison of our results with the previous measurements[4], near θ_{cm} = 135° the ratio A_y (elastic)/A_y (d*) is $\simeq$ 3 in the ^{3}He($\vec{d}$, d*) ^{3}He and ^{1}H($\vec{d}$, d*) ^{1}H reactions respectively, and the ratio is $\simeq$ 1.4 and $\simeq$ 1.9, respectively, near θ_{cm} = 155°. We are presently involved in a comparison of these data with DWBA calculations of $A_y(\theta)$ in transitions which produce both the $^1S_0$ and $^3S_1$ final state d*.

555

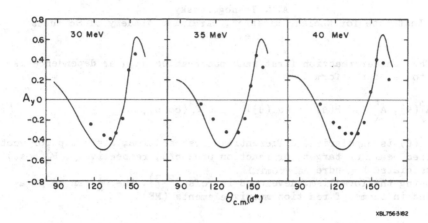

XBL7563182

References

*   Work performed under the auspices of the U.S. Energy Research and
    Development Administration.
+   National Research Council of Canada, Postdoctoral Fellow.
†   On leave of absence from the University of Basel, Switzerland.

1) H. E. Conzett, in Proc. Int'l Conf. on Few Body Problems in Nuclear
    and Particle Physics, Quebec, Canada, 27–31 August 1974 (to be
    published).
2) R. T. Cahill and I. H. Sloan, Nucl. Phys. A165 (1971 161; W. Ebenhöh,
    Nucl. Phys. A191 (1972) 97.
3) F. N. Rad, L. C. Northcliffe, J. G. Rogers, and D. P. Saylor, Phys.
    Rev. Lett. 31 (1973) 57; Phys. Rev. C8 (1973) 1248; J. Arvieux, et
    al., Nucl. Phys. A150 (1970) 75; C. O. Blyth, et al., Proc. Int'l
    Conf. on Few Body Problems in Nuclear and Particle Physics, Quebec,
    Canada, 27–31 August 1974.
4) F. N. Rad, R. Roy, H. E. Conzett, J. Birchall, F. Seiler and R. M.
    Larimer, Bull. Amer. Phys. Soc. 20 (1975) 578.
5) F. N. Rad, H. E. Conzett, R. Roy and F. Seiler, Fourth Polarization
    Symposium, p. 487.

MODEL-FREE ANALYSIS OF ^{3}He(d,p)^{4}He REACTION

A.M. Yasnogorodsky

Institute for Nuclear Research, Ukrainian Academy of Sciences
Kiev, USSR

The nuclear reaction first rank observables angular dependence is known to be of the form

$$A_y^b(\theta), \; A_y^t(\theta), \; P(\theta) \; = [\sigma_o(\theta)]^{-1} \sum_k C_k^i P_k^1(\cos\theta), \tag{1}$$

here $\sigma_o(\theta)$ is unpolarized differential cross-section; $i = b, t, p$ for vector polarized beam, or target, or reaction products, respectively; $P_k^1(\cos\theta)$ are associated Legendre polynomials.

Using the formalism, developed in refs. [2,3], the C_k^i can be parametrized in terms of reaction matrix elements (ME)

$$C_k^{b,t} = \frac{\mathchar'26\mkern-10mu\lambda^2}{4\hat{a}\hat{s}_1} \left[\frac{2(j+1)}{3j}\right]^{1/2} \sum_{\substack{mn}} (-)^{l_n + J_m + S'_m + 1} \; \delta_{\substack{S'_m S'_n}} \; A_{mnk} \begin{Bmatrix} s_1 & a & S_m \\ \lambda & \Lambda & 1 \\ 1 & kk \end{Bmatrix} \begin{Bmatrix} S_m l_m J_m \\ S_n l_n J_n \\ 1 & k & k \end{Bmatrix} \text{Im}(R_m R_n^*), \tag{2}$$

here $R_m = \langle S'_m l'_m J_m | R | S_m l_m J_m \rangle$ is the matrix element corresponding to the transition from a state of channel spin S_m and orbital angular momentum l_m to a state of channel spin S'_m and orbital angular momentum l'_m; $\hat{x} = (2x+1)^{1/2}$; $\mathchar'26\mkern-10mu\lambda$ is the incident particle cm wavelength; $\lambda = 1$, $\Lambda = 0$, $j = s_1$ ($\lambda = 0$, $\Lambda = 1$, $j = a$) for vector polarized beam (target); C_k^p are equal to C_k^b of corresponding inverse reaction; s_1, a are spins of projectile and target, respectively.

One may write for ^{3}He(d,p)^{4}He reaction

$$C_k^b / C_k^t = (\; f_k + \tfrac{1}{2} e_k + d_k \;)/(\; f_k - e_k - \tfrac{1}{2} d_k \;), \tag{3}$$

here f_k is the sum of terms with $S_m = S_n = 3/2$; e_k is that with $S_m = 3/2$, $S_n = 1/2$; d_k is that with $S_m = S_n = 1/2$.

As it has been concluded [4] on the basis of some nondynamical arguments, the experimental equation $A_y^b(\theta) = A_y^t(\theta)$ gives an indication of the possible channel spin selection rule $\Delta S = \pm 1$. However, it can be seen from eq. (3) that the existence of type e and d terms, i.e. transitions from $S_- = 1/2$ states, is not excluded, but $e_k = -d_k$ is required.

At deuteron energies near 10 MeV A_y^b and P have opposite sign at almost all angles [5,6,7]. There are no type e terms in C_k^p, and the f_k^p, d_k^p sums contain the same ME combinations $R_m R_n^*$, as f_k^b, d_k^b. Since $d_k^p / d_k^b = \text{const} > 0$ for any k, and accounting for eq.(3) and above consideration of $A_y^b = A_y^t$, one has to look for f_k terms as the only source of A_y^b and P sign difference. The numerical factors a_{mn} of the same $R_m R_n^*$ in f_k^p, f_k^b are related as follows

$$\frac{a_{mn}^p}{a_{mn}^b} = \kappa \; \frac{(l'_m - l'_n)(l'_m + l'_n + 1) - (J_m - J_n)(J_m + J_n + 1)}{(l_m - l_n)(l_m + l_n + 1) - (J_m - J_n)(J_m + J_n + 1)}, \; \kappa > 0 \;. \tag{4}$$

We evaluated eq.(4), as an example, for $C_2^{b,p}$ coefficients since just

Table 1

J^π	Matrix element
$1/2^+$	$\alpha_{10} = \langle 1/2\ 0\ 1/2 \vert R \vert 1/2\ 0\ 1/2 \rangle$
$1/2^+$	$\gamma_{10} = \langle 1/2\ 0\ 1/2 \vert R \vert 3/2\ 2\ 1/2 \rangle$
$1/2^-$	$\alpha_{11} = \langle 1/2\ 1\ 1/2 \vert R \vert 1/2\ 1\ 1/2 \rangle$
$1/2^-$	$\beta_{11} = \langle 1/2\ 1\ 1/2 \vert R \vert 3/2\ 1\ 1/2 \rangle$
$3/2^+$	$\alpha_{32} = \langle 1/2\ 2\ 3/2 \vert R \vert 1/2\ 2\ 3/2 \rangle$
$3/2^+$	$\beta_{32} = \langle 1/2\ 2\ 3/2 \vert R \vert 3/2\ 2\ 3/2 \rangle$
$3/2^+$	$\gamma_{32} = \langle 1/2\ 2\ 3/2 \vert R \vert 3/2\ 0\ 3/2 \rangle$
$3/2^-$	$\alpha_{31} = \langle 1/2\ 1\ 3/2 \vert R \vert 1/2\ 1\ 3/2 \rangle$
$3/2^-$	$\beta_{31} = \langle 1/2\ 1\ 3/2 \vert R \vert 3/2\ 1\ 3/2 \rangle$
$5/2^+$	$\alpha_{52} = \langle 1/2\ 2\ 5/2 \vert R \vert 1/2\ 2\ 5/2 \rangle$
$5/2^+$	$\beta_{52} = \langle 1/2\ 2\ 5/2 \vert R \vert 3/2\ 2\ 5/2 \rangle$
$5/2^-$	$\gamma_{53} = \langle 1/2\ 3\ 5/2 \vert R \vert 3/2\ 1\ 5/2 \rangle$
$7/2^+$	$\gamma_{74} = \langle 1/2\ 4\ 7/2 \vert R \vert 3/2\ 2\ 7/2 \rangle$

Table 2

$R_m R_n^*$	g
$\beta_{11}\ \beta_{32}^*$	+1
$\beta_{32}\ \beta_{52}^*$	+1
$\beta_{32}\ \gamma_{10}^*$	−1
$\beta_{52}\ \gamma_{10}^*$	+1/4
$\beta_{32}\ \gamma_{32}^*$	0
$\beta_{52}\ \gamma_{32}^*$	−5
$\beta_{32}\ \gamma_{74}^*$	−1/6
$\beta_{11}\ \gamma_{53}^*$	−1/4
$\beta_{31}\ \gamma_{53}^*$	−1/5
$\gamma_{10}\ \gamma_{32}^*$	−1/3
$\gamma_{32}\ \gamma_{74}^*$	−1/3

those fix angular dependence of A_y^b and P at deuteron energies around 10 MeV. Table 2 contains the ME combinations, allowed in $f_{\frac{1}{2}}$ by eq.(2), together with the corresponding values of $(1/\kappa)a_{mn}^p/a_{mn}^b = g$. It is seen the A_y^b and P sign difference takes place only if the ME of γ-type (see table 1) are accounted for *). Since all γ's correspond to the transitions between states of different 1, it becomes clear that interchannel tensor interaction is to be present. (The approximate relation $A_y^t \approx -P$ alone is not a sufficient basis for previously made conclusions [8] of the significance and specific form of the tensor force).

References

*) The Greek symbols, used for ME in table 1, are indexed as follows:
$\alpha_{2J,1}$; $\beta_{2J,1}$; $\gamma_{2J,1'}$

1) Proc. 3d Intern. Symp. on Pol. Phenomena, Press of Wisc. Univ.,1971
2) L.J.B. Goldfarb, Nucl. Reactions, v. 1, Amsterdam, 1959
3) T. Welton, Fast Neutron Physics, v. 2, ch. 5F (Interscience N.Y.),1959
4) P.W. Keaton et al., see ref. [1], p. 528
5) N.I. Zaika et al., Izv. AN SSSR, ser.fiz. 32 (1968) 257
6) G.R. Plattner, L.S. Keller, Phys. Letters 29B (1969) 301
7) R.J. Brown, W. Haeberli, Phys. Rev. 130 (1963) 1163
8) M. Tanifuji, Phys. Rev. Letters 15 (1965) 113

POLARIZATION TRANSFER IN THE $D(\vec{t},\vec{n})$ ⁴He REACTION

G. G. Ohlsen and R. A. Hardekopf
Los Alamos Scientific Laboratory, Los Alamos, N.M. 87545, USA
and
R. L. Walter and P. W. Lisowski, Los Alamos Scientific
Laboratory and Triangle Universities Nuclear Laboratory

Measurements have been made of the neutron polarization from the $D(\vec{t},\vec{n})$ ⁴He reaction with polarized tritons[1] incident. The neutron polarization was measured by means of a liquid helium neutron polarimeter[2].

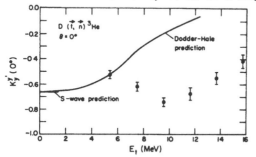

Fig. 1. Measured values of $K_y^{y'}(0°)$ for the D(t,n)⁴He reaction. The smooth curve is the prediction of Dodder and Hale. At very low energies, where the system is dominated by an S-wave, $J = 3/2^+$ resonance, $K_y^{y'}$ must approach $-2/3$.

The bulk of the data ware taken at a reaction angle of 0°. These results are shown in fig. 1, along with the predictions furnished by D. C. Dodder and G. M. Hale from their R-matrix analysis of the 5-nucleon system. The lack of agreement is surprising in view of the past successes of their parameterization. In fig. 2 an angular distribution of $K_y^{y'}(\theta)$ at 10.5 MeV is presented, again together with the Dodder-Hale prediction. Here there is at least qualitative agreement with the shape of the curve. For the $1/2 + 1 \rightarrow 1/2 + 0$ spin structure of the present reaction, one can show[3] that

$$2K_y^{y'}(0°)+K_z^{z'}(0°) = -1 \qquad (1)$$

from which it follows that $-1 \leqslant K_y^{y'}(0°) \leqslant 0$. From a simple "pick-up" argument, one predicts that $K_y^{y'} = -1/3$, as follows: the deuteron target can be considered to be an equal mixture of $m_I = 1$, $m_I = 0$, and $m_I = -1$ states. Neglecting the D-state of the deuteron and possible spin-exchange forces, a spin-up triton must combine with a spin-down proton to form ⁴He. The $m_I = +1$ deuteron therefore cannot react, the $m_I = -1$ deuteron results in a spin-down neutron, and the $m_I = 0$ deuteron results in a spin-up neutron. Based on the fraction of the $m_I = 0$ spin amplitude which corresponds to a proton with spin-down, one would expect the $m_I = 0$ cross sectionsto be 1/2 of the $m_I = -1$ cross section, and thus the outgoing neutrons would have a polarization of $-1/3$. For very low energies, where the $J = 3/2^+$ resonance at ~ 100 keV dominates the reaction, $K_y^{y'}(0°)$ must approach $-2/3$.

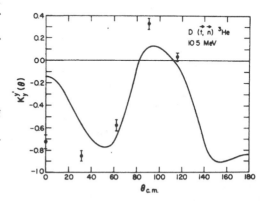

Fig. 2. Measured values of $K_y^{y'}(\theta)$ at a triton energy of 10.5 MeV. The smooth curve is the prediction of Dodder and Hale.

One can show from the M-matrix
formalism for this spin system
(specialized to 0° or 180°) that
the 0° $D(\vec{t},n)^4He$ polarization transfer coefficient is related to the
deuteron tensor analyzing power
in $T(\vec{d},n)^4He$ by:

$$K_y^{y'}(0°,\text{tritons}) = -(2A_{yy}+1)/3. \quad (2)$$

In principle, A_{yy} can be accurately
determined using a polarized deuteron beam at the same center of
mass energy and angle. To compare
to the $D(t,n)$ reaction at 0°,
then, one must detect either the
0° alpha particles in the $T(d,\alpha)n$
reaction or the 180° neutrons in
the $T(d,n)^4He$ reaction. The
former experiment can be done
more easily and offers the possibility of providing a neutron
beam with known polarization.

In fig. 3, the figure of
merit for the $D(t,n)$ reaction as
a source of polarized neutrons is

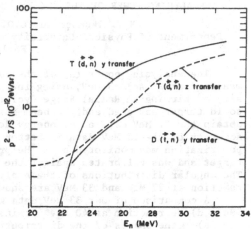

Fig. 3. Figure of merit for several
reactions as sources of polarized
neutrons. For the triton beam case,
a polarization of 0.8 is assumed.
For the deuteron beam cases, the
vector and tensor polarization are
each taken to be 0.8.

compared to another reaction that can be calibrated via an analyzing
power measurement[4], namely, longitudinal polarization transfer in the
$T(\vec{d},n)^4He$ reaction. It is seen that the figure of merit for the
$D(t,n)^4He$ reaction is not quite as good as that for the longitudinal
spin transfer in the $T(\vec{d},n)^4He$ reaction. Nonetheless, it provides
transversely polarized neutrons and thus may actually be superior for
applications, since a spin precession magnet is not needed. For comparison, we also show in fig. 3 the figure of merit for transverse spin
transfer in the $T(\vec{d},n)^4He$ reaction[5].

*Work performed under the auspices of the U.S. Energy Research and
Development Administration.

References

1) R. A. Hardekopf, these proceedings
3) G. S. Mutchler, W. B. Broste, and J. E. Simmons, Phys. Rev. C3
   (1971) 1031
3) G. G. Ohlsen, Reports on Prog. in Physics 35 (1972) 717
4) P. W. Lisowski, R. L. Walter, R. A. Hardekopf, and G. G. Ohlsen,
   these proceedings, p.887
5) W. B. Broste, G. P. Lawrence, J. L. McKibben, G. G. Ohlsen, and
   J. E. Simmons, Phys. Rev. Letters 25 (1970) 1040

ANALYSING POWER OF THE ^{2}H($^3\vec{\text{He}}$,^{4}He)^{1}H REACTION AT 27 AND 33 MEV

N.T. Okumuşoğlu, C.O. Blyth and W. Dahme*
Department of Physics, University of Birmingham, Birmingham B15 2TT,
England

The analysing power (A) of the reaction ^{2}H($^3\vec{\text{He}}$,^{4}He)^{1}H has been measured at 27 and 33 MeV, using the polarized ^{3}He beam from the University of Birmingham Radial Ridge Cyclotron.   In the experiment a (CD$_2$)$_n$ solid target was used and the beam intensity on the target was 50 pA. To obtain the 27 MeV data an Al energy degrader was used.   The statistical accuracy of the 33 MeV data is better than that of 27 MeV.   The beam polarization was monitored by a ^{3}He-proton polarimeter located after the target and was calibrated against the ^{3}He + ^{4}He double scattering results. The angular distributions of the analyzing power of the ^{2}H($^3\vec{\text{He}}$,^{4}He)^{1}H reaction at 27 MeV and 33 MeV are shown in figs. 1 and 2 respectively.

A comparison of our 33 MeV data with the analysing power of the ^{4}He($\vec{\text{p}}$,d)^{3}He reaction at 40 MeV[1] brings out the following points:
a)   the shapes of the distributions are similar but the result $-P=A$, obtained by Brückmann and Schmidt[2] and Beckmann et al[3], does not seem to hold for the present results, which give: $- P \simeq 2A$. This point could be cleared up by measuring the proton polarization in the ^{2}H($^3\vec{\text{He}}$,$\vec{\text{p}}$)^{4}He reaction to our energy.
b)   A bump around Θ_{cm} = 90° in the 33 MeV data is also present in the 40 MeV data[1].   It is however, not observed in 27 MeV.

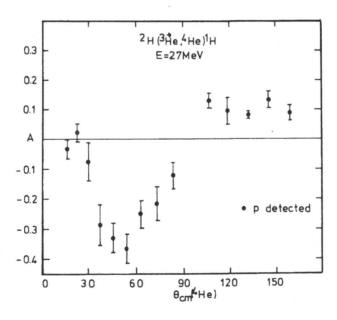

Fig. 1

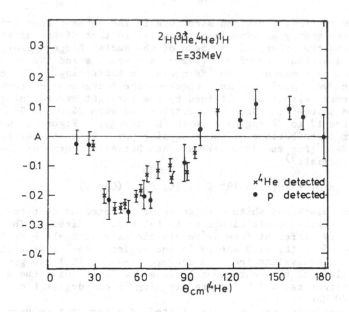

Fig. 2

* On leave from the University of Munich

References

1) A.L. Sagle, B.E. Bonner, W.B. Broste, N.S.P. King, H.E. Conzett, W. Dahme and Ch. Leemann, Private communication.
2) H. Brückmann and F.K. Schmidt, Nucl. Phys. A136 (1969) 81.
3) R. Beckman, H. Holm and K. Lorenzen, Z. Physik 271 (1974) 89.

VECTOR ANALYSING POWERS IN $^3\vec{\text{He}}$-d ELASTIC SCATTERING BETWEEN 24 AND 33 MeV

N.T. Okumusoglu, C.O. Blyth, O. Karban, W. Dahme† and G.G. Shute*
Department of Physics, University of Birmingham, Birmingham B15 2TT, England

In order to study the broad structure in the d+^{3}He system around 20 MeV deuteron energy an experiment parallel to that of Roy et al[1], has been done using the polarized ^{3}He beam of the Radial Ridge Cyclotron.

Angular distributions of relative cross-sections and ^{3}He analysing power (A) have been measured, in $^3\vec{\text{He}}$-d elastic scattering. The cross-sections which were monitored as a check on the background correction are in good agreement with those obtained by the straightforward interpolation from King and Smythe[2]. Data were obtained between 24.0 and 33.0 MeV in 3 MeV steps for 12 to 27 angles at each ^{3}He energy. Figure 1 shows the three angular distributions of the analysing power with the results of fits to the data (for the 33.0 MeV data see Ref.4) in terms of the Legendre Polynomials[3]

$$\frac{d\sigma}{d\Omega} (\Theta) \times A(\Theta) = (\sigma_{tot}/4\pi)\lambda^2 \sum_L d_{11}(L) \, P_{L,1} (\text{Cos } \Theta)$$

In this figure black and white circles refer to detected deuterons and ^{3}He respectively. Coefficients up to L=6 were required. The 24 MeV data is slightly different from d+^{3}He elastic data at E_d=14.6 MeV[1]. The variation of $d_{11}(L)$ with energy in the region under consideration is shown in fig.2; equivalent incoming deuteron energy, (E_d), is given on the upper scale, and the continuous lines are only to guide the eye. They show a structure, which is more striking in odd degree L coefficients around E_d = 20 MeV.

The cross-section data was also fitted to a Legendre polynomial series:

$$\frac{d\sigma}{d\Omega} (\Theta) = (\sigma_{tot}/4\pi) \, \lambda^2 \sum_L d_{oo} (L) \, P_L (\text{Cos } \Theta)$$

The coefficients, $d_{oo}(L)$, are tabulated in Table 1.

The phase shift analysis of the ^{3}He+d elastic scattering observables[5] was successfully tested at 11.5 MeV incident deuteron energy and a similar analysis of the present data is in progress.

†   On leave from University of Munich, Germany
*   On leave from University of Melbourne, Australia

References

1) R. Roy, H.E. Conzett, F.N. Rad, F. Seiler and R.M. Larimer, Fourth Polarization Symp., p.540
2) T.R. King and R. Smythe, Nucl.Phys. A183 (1972) 657
3) F. Seiler, Nucl. Phys. A244 (1975) 236
4) O. Karban, A.K. Basak, C.O. Blyth, J.B.A. England, J.M. Nelson, S. Roman and G.G. Shute. Fourth Polarization Symp., p.530
5) C.O. Blyth, N.T. Okumuşoğlu, G.G. Shute, University of Birmingham, Nucl. Structure Gp. Ann.Prog.Report, (1975) 68

TABLE 1

E(^{3}He) d_{oo}(L)	24 MeV	27 MeV	30 MeV	33 MeV
d_{oo}(1)	1.68 ± 0.05	1.775 ± 0.02	1.90 ± 0.07	2.00 ± 0.11
d_{oo}(2)	2.32 ± 0.07	2.45 ± 0.06	2.46 ± 0.10	2.69 ± 0.17
d_{oo}(3)	2.22 ± 0.07	2.38 ± 0.07	2.54 ± 0.12	2.72 ± 0.18
d_{oo}(4)	2.06 ± 0.08	2.20 ± 0.07	2.12 ± 0.12	2.41 ± 0.18
d_{oo}(5)	0.64 ± 0.06	0.87 ± 0.08	1.02 ± 0.11	1.31 ± 0.15
d_{oo}(6)	0.52 ± 0.05	0.70 ± 0.05	0.64 ± 0.11	0.93 ± 0.11
d_{oo}(7)	0.09 ± 0.03	0.16 ± 0.03	0.19 ± 0.06	0.34 ± 0.06
d_{oo}(8)	0.11 ± 0.02	0.19 ± 0.02	0.11 ± 0.03	0.105±0.04

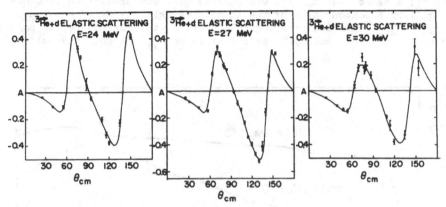

FIGURE 1

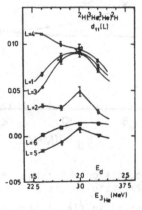

FIGURE 2

ANALYSING POWER MAXIMA A_y = -1 AND A_{yy} = 1 IN d-α ELASTIC SCATTERING

W. Grüebler, P.A. Schmelzbach, V. König, R. Risler, B. Jenny,
D.O. Boerma and W.G. Weitkamp
Laboratorium für Kernphysik, Eidg. Techn. Hochschule,
8049 Zürich, Switzerland

Investigation of the conditions for which a component of analysing power reaches its theoretical maximum value is not only interesting for practical applications but gives a deeper insight into the structure of the compound nucleus. In a recent paper[1] the conditions which the M-matrix elements have to fulfil in order to obtain such maxima has been tabulated for all analysing power components. The simplest relation was found for A_{yy} = 1:

$$M_{11} + M_{1-1} = 0 \qquad\qquad (1)$$

Vector analysing power maxima A_y = ± 1 require two additional relations besides the condition (1). Therefore the search in this case is more complicated than for A_{yy} = 1.

Based on a phase shift analysis in the energy range between 3 and 17 MeV[2], a search for A_{yy} = 1 was carried out. The prediction is shown in fig. 1 by a contour plot of the quantities $\text{Re}(M_{11} + M_{1-1})$ = 0 and $\text{Im}(M_{11} + M_{1-1})$ = 0. The crossing points of the two curves are the points in question. It can readily be seen that three A_{yy} = 1 points should exist.

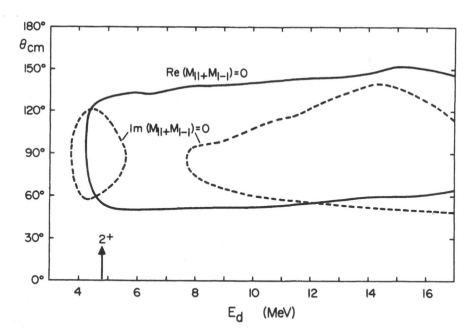

Fig.1. Contour plot of the condition $M_{11} + M_{1-1}$ = 0. The arrow indicates the deuteron energy of the 2+ state in ^{6}Li.

The three regions predicted in fig. 1 have been investigated experimentally[1]. The resulting angles and energies are listed in table 1.

Table 1. Energies and angles found for $A_{yy} = 1$.

Ed_{lab} (MeV)	4.30	4.57	11.88
θ_{cm}	120.7	58.0	55.1
θ_{lab}	90.5	39.4	37.3

The investigation of the other two relations for $A_y = 1$ or -1 shows that they are fulfilled near the 11.88 MeV point of $A_{yy} = 1$. Since all three relations show in the corresponding contour plots a very small angle between real and imaginary trajectories at the crossing points, there is a large uncertainty in the precise angle and energy of the crossing. It has been suspected that a possibly point with $A_y = -1$ coincides with $A_{yy} = 1$ at 11.88 MeV. In this case one would have $A_{xz} = o$ and $A_{xx} = A_{zz} = -1/2$. In order to clear up the situation, an excitation function of A_y was measured at $\theta_{cm} = 55.3^o$. The result is shown in fig.2. Although the A_y reaches a value fairly near to -1, this maximum is shifted to higher energy. From this excitation curve it is obvious that no point exists where A_y is equally -1. The present results are measured with a high absolute precision, since the $A_{yy} = 1$ point nearby could be used as a calibration of the beam polarization.

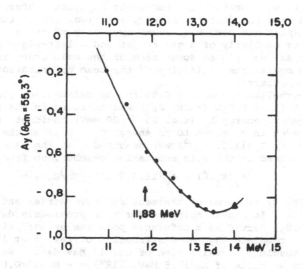

Fig.2. Excitation curve of A_y at $\theta_{cm} = 55.3^o$. The triangle is from ref.3.

References

1) W. Grüebler, P.A. Schmelzbach, V. König, R. Risler, B. Jenny and D.O. Boerma, Nucl. Phys. A242 (1975) 285
2) W. Grüebler, P.A. Schmelzbach, V. König, R. Risler and D.O. Boerma, Nucl. Phys. A242 (1975) 265
3) G.G. Ohlsen, P.A. Lovoi, G.C. Salzman, U. Meyer-Berkhout, C.K. Mitchell and W. Grüebler, Phys. Rev. C8 (1973) 1262

THE VECTOR ANALYZING POWER IN $\vec{d}$ - ^{4}He ELASTIC SCATTERING BETWEEN 15 AND 45 MeV*

H. E. Conzett, W. Dahme[+], Ch. Leemann,
J. A. Macdonald[‡], and J. P. Meulders[‡]
Lawrence Berkeley Laboratory, University of California
Berkeley, California 94720

Phase shift analyses of d - ^{4}He elastic scattering data below 17 MeV have led progressively to improved agreement with the data and to the determination of level parameters for those states of ^{6}Li that appear as resonances in the d + ^{4}He channel[1,2]. These data have shown substantial values of the vector and tensor analyzing powers, so we selected ^{4}He as a potentially good analyzer for the higher energy polarized deuteron beams. Also, two independent determinations[2,3,4] of the absolute vector analyzing power $A_y(E_d,\theta_{cm})$ have been made which are in very good agreement. In the scattering of a purely vector-polarized deuteron beam the measured left-right asymmetry is given by

$$\varepsilon(\theta) = (3/2) \, p_y \, A_y(\theta), \tag{1}$$

where p_y is the beam polarization. Asymmetry data were taken at 5 MeV intervals from 15 to 45 MeV at center-of-mass angles from 30° to 165°. At all angles the elastically scattered deuterons were detected with ΔE-E counter telescopes placed at equal angles to the left and right of the beam axis. Particle identification served to separate deuterons from protons which came from the deuteron breakup reaction. A gas target of ^{4}He at $\simeq$ 1 atm. pressure was used with beams typically of 50 nA. A polarimeter, consisting similarly of a gas target and a left-right pair of counter telescopes, was placed downstream of the main scattering chamber, and it provided continuous monitoring of the beam polarization during the course of the experiment.

Absolute normalization of our data to the calibration point A_y(11.5 MeV, 118°) = - 0.410 ± 0.010 (refs. 2,3) was achieved in the following manner. The beam of energy E_1 (e.g. 15 or 20 MeV) incident at the first target was degraded in aluminum to an energy E_2 = 11.5 at the polarimeter where the asymmetry ε_2(11.5, 118°) was measured. At the same time, $\varepsilon_1(E_1,\theta)$ was measured in the main scattering chamber, so from (2)

$$A_y(E_1,\theta) = A_y(11.5, \, 118°) \, \varepsilon_1/\varepsilon_2 \, , \tag{2}$$

which gives $A_y(E_1,\theta)$ in terms of the measured asymmetries and the A_y calibration value. At higher energies for E_1 a previously determined value for $A_y(E_2,\theta_2)$ served as a reference polarimeter analyzing power. We also used the complete angular distribution of ref. 5 at 15 MeV for an independent absolute normalization of our 15 MeV data. We found this to correspond to a value of A_y(11.5 MeV, 118°) = - 0.415±0.010, in very good agreement with the other calibration value.

Our 20-45 MeV results are shown in fig. 1, and our 15 - MeV data are in excellent agreement with those of ref. 5. It is seen that $\vec{d}$ - ^{4}He elastic scattering serves as a very good vector-polarization analyzer over a wide range of deuteron energies.

Of particular interest is the fact that A_y(155°) reaches values near 1.0 at 24.9 and 29.9 MeV. It has been shown recently that at an energy and angle for which $A_y(E,\theta)$ = ± 1.0 in $\vec{d}$ - ^{4}He scattering, all the analyzing-power components are fixed and determined[6]. This result has been extended to apply to all two-body reactions with incident polarized deuterons, and it is independent of the spins of the target and reaction products[7]. In view of our results, a point near $\theta_{cm} = 155°$ between 25 and 30 MeV in d - ^{4}He scattering is the best candidate for reaching the extreme value $A_y = 1$ (ref. 8). The possibility of an $A_y = -1$ value at $\theta_{cm} = 55°$ at 11.88 MeV[6] is not supported by the $A_y(E,\theta)$ data in that region[4].

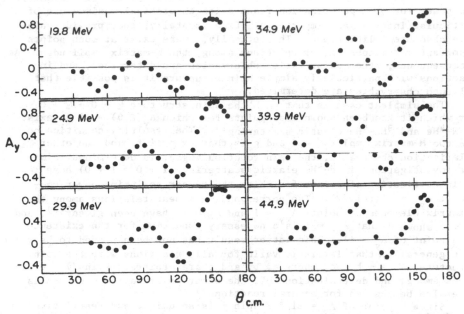

Fig. 1

XBL756-3155

References

* Work performed under the auspices of the U. S. Energy Research and Development Administration.
+ Sektion Physik der Universität München, 8046 Garching, Germany
‡ Chalk River Nuclear Laboratories, Chalk River, Ontario, Canada
⧧ Centre de Physique Nucleaire, Université de Louvain, Belgium

1) P. A. Schmelzbach et al., Nucl. Phys. A184 (1972) 193, and refs. therein.
2) W. Grüebler et al., Nucl. Phys. A242 (1975) 265.
3) W. Grüebler et al., Nucl. Phys. A134 (1969) 686; V. König et al., Nucl. Phys. A166 (1971) 393.
4) G. G. Ohlsen et al., Phys. Rev. C8 (1973) 1262.
5) R. A. Hardekopf et al., Los Alamos Report LA-5051 (1972).
6) W. Grüebler et al., Nucl. Phys. A242 (1975) 285.
7) F. Seiler, F. N. Rad and H. E. Conzett, Fourth Polarization Symp., p.897.
8) H. E. Conzett, F. Seiler, F. N. Rad, R. Roy, and R. M. Larimer, Fourth Polarization Symp., p.568

MEASUREMENT OF THE VECTOR ANALYZING POWER NEAR THE VALUE $A_y \approx 1.0$ IN d-⁴He ELASTIC SCATTERING[*]

H. E. Conzett, F. Seiler[+], F. N. Rad, R. Roy[+] and R. M. Larimer
Lawrence Berkeley Laboratory, University of California
Berkeley, California 94720

Points in energy and angle (E,θ) at which a component of the analyzing power for spin-polarization reaches its theoretical maximum are of particular interest and importance. Experimentally, they provide valuable absolute calibrations. Theoretically, there exist at such points important conditions on, or relations among, the M-matrix amplitudes. As a consequence, other polarization observables are determined, and in reactions with particularly simple spin structure it is possible that all such observables are determined.

The simplest case is that with the spin structure $\frac{1}{2} + 0 \rightarrow \frac{1}{2} + 0$, for which it has been shown that points for which $A_y(E,\theta) = \pm 1$ do exist in N-⁴He and ³He-⁴He elastic scattering[1]. The resulting condition on the two M-matrix amplitudes f and g is that $g = \pm i \, f$, and the other polarization observable, the spin rotation angle, is determined. Similar investigations in d-⁴He elastic scattering $(1 + 0 \rightarrow 1 + 0)$ have led to identification of points for which the tensor analyzing-power component $A_{yy} = 1$ (ref. 2). The corresponding linear relations among the M-matrix elements at points $A_{yy} = 1$ and $A_y = \pm 1$ have been given[2,3], and it was shown[2] that $A_{yy} = 1$ is a necessary condition for the existence of a point of $A_y = \pm 1$. This latter result has now been proved to be valid more generally; that is, it is valid for all reactions with the spin structure $1 + a \rightarrow b + c$, where a, b, and c are arbitrary spins[4]. Relations among, and determinations of, the other polarization observables have also been given for several reactions[5].

Since a point of $A_y = \pm 1$ for spin 1 is so unique and restrictive, and a priori perhaps so unlikely, the location and identification of such a point would be of considerable value. The best candidate for such a point was suggested in the measurements of $A_y(\theta)$ in d-⁴He elastic scattering between 15 and 45 MeV[6]. There, values of $A_y(\theta_L = 135°) \geqslant 0.97$ were found near 25 and 30 MeV. Therefore, we have examined this energy region to determine the maximum value of $A_y(E,\theta)$. The experimental details and the calibration procedures are described elsewhere[6]. We find that A_y reaches its maximum value near 28.6 MeV, and we show in fig. 1 our preliminary results for $A_y(28.6 \text{ MeV}, \theta_L)$. Although finite geometry corrections and polynomial fits to all our data in this energy and angular region must be made before the maximum value of A_y is determined, it is clear from fig. 1 that A_y does indeed reach a value very close to unity near $\theta_L = 135°$ at 28.6 MeV.

We have also measured $A_{yy}(E,\theta)$ over the same regions, using the ⁴He(d,d)⁴He A_{yy} data of ref.[3] at 17 MeV for the determination of the tensor polarization of our beam. Again, our preliminary result is that A_{yy} reaches a maximum value approaching unity near 26.8 MeV and $\theta_L = 135°$, as would be required to fulfill the necessary condition that $A_{yy} = 1$ at a point for which $A_y = 1$.

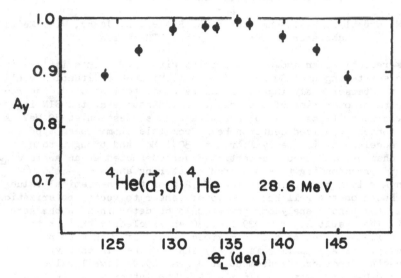

XBL 756-1531

References

*   Work performed under the auspices of the U.S. Energy Research and
     Development Administration.
+   On leave of absence from the University of Basel, Switzerland.
†   National Research Council of Canada Postdoctoral Fellow.

1)   G. R. Plattner and A. D. Bacher, Phys. Lett. 36B (1971) 211.
2)   W. Grüebler et al., Nucl. Phys. A242 (1975) 285.
3)   G. G. Ohlsen et al., Phys. Rev. C 8 (1973) 1262.
4)   F. Seiler, F. N. Rad and H. E. Conzett, Fourth Polarization Symp.,p.897
5)   F. Seiler, Fourth Polarization Symp.; F. Seiler, H. E. Conzett,
     F. N. Rad and R. Roy, ibid.; F. Seiler, R. Roy, H. E. Conzett and
     F. N. Rad, ibid.; F. Seiler, F. N. Rad, H. E. Conzett and R. Roy,
     ibid., p.552, p.571, p.550, p.587.
6)   H. E. Conzett et al., Fourth Polarization Symp., p. 566

570

MEASUREMENT OF THE TENSOR ANALYSING POWER A_{yy} NEAR THE VALUE $A_y=1$ IN d-α ELASTIC SCATTERING

G. Heidenreich, J. Birchall
Institut für Physik, Universität Basel
and
V. König, W. Grüebler, P.A. Schmelzbach, R. Risler, B. Jenny, W.G. Weitkamp
Laboratorium für Kernphysik, ETH Zürich

It has recently been shown that a point with $A_y=1$ may exist[1,2] in d-α elastic scattering near 30 MeV and $\theta_{cm}= 155°$. Such a situation would imply that the tensor analysing power A_{yy} is equal to 1 also [3]. The recent beginning of operation of the polarized ion source at the SIN injector cyclotron made it possible to investigate this question at 30 MeV. A vector and tensor polarized deuteron beam from this atomic beam type source was accelerated in the cyclotron to 30.1 MeV and brought to a scattering chamber. The recoil α-particles were detected at an angle θ_{lab} equal to 12° corresponding to a deuteron scattering angle $\theta_{cm}= 156.0°$. Assuming an $A_y= 1$, a beam polarization of 72 % of the theoretical value was found. Based on the well known ratio of tensor to vector polarization in the beam, the tensor analysing power could be determined in the same measurement. The result $A_{yy} = 0.829 \pm 0.030$ shows clearly that at the energy and angle measured no double maximum exist. If such a point is present nearby, the A_{yy} maximum is fairly sharp peaked in energy as well as in angle since another measurement at $\theta_{cm} = 125.9°$ reveal values $A_{yy}= 0.392 \pm 0.041$ and $A_y = - 0.316 \pm 0.042$. The latter value is in agreement with measurements of the Berkeley group[1].

An experimental search in an extended angle and energy range will be done in the near future.

References

1) H.E. Conzett, W. Dahme, Ch. Leemann, J.A. McDonald and J.P. Meulders, Contribution to this Symposium, p.566
2) H.E. Conzett, F. Seiler, F.N. Rad, R. Ray and R.M. Larimer, Contribution to this Symposium, p.568
3) W. Grüebler, P.A. Schmelzbach, V. König, R. Risler, B. Jenny and D.O. Boerma, Nucl. Phys. A242 (1975) 285

POINTS OF MAXIMUM POLARIZATION EFFICIENCY IN ^{4}He$(\vec{d},d)^4$He*

F. Seiler$^+$, H. E. Conzett, F. N. Rad and R. Roy‡
Lawrence Berkeley Laboratory, University of California
Berkeley, California 94720

Several points (E_i, Θ_i) of maximum possible analyzing power have recently been identified in d-^{4}He elastic scattering. From an analysis of data between 3 and 17 MeV, three points with $A_{yy} = 1$ were found below 12 MeV and another with $A_y = -1$ may exist above that energy[1]. An experimental indication that extreme values of the analyzing power A_y may occur comes from measurements near $E_d = 28$ MeV and $\Theta_{cm} \simeq 154°$, where $A_y \simeq A_{yy} \simeq 1$ [2].

Reaching an extreme value of some component of the analyzing power imposes certain conditions on the elements of the transition matrix M. In the parametrization of Ohlsen et al.[3]

$$M = \frac{1}{2}\begin{pmatrix} A-B & \sqrt{2}\,D & -A-B \\ -\sqrt{2}\,D & 2\,C & \sqrt{2}\,D \\ -A-B & -\sqrt{2}\,D & A-B \end{pmatrix}$$

an extreme value of $A_{yy} = 1$ requires $B = 0$. For an extreme value $A_y = \pm 1$, however, $A = C = \pm iD$ and $B = 0$ is necessary. Thus $A_{yy} = 1$ is a prerequisite for $A_y = \pm 1$ [1,4], and both components must peak at the same critical point (E_i, Θ_i). A definite identification of points with extreme values of the analyzing power can be carried out by using a phase-shift analysis to compute the amplitudes A through D and to verify the validity of the relevant conditions[1]. In an alternate approach[5] these conditions are introduced into the formulae for the observables[3] and can then be checked experimentally. The requirement $B = 0$ for $A_{yy} = 1$ yields

$$A_{yy} = P^{y'y'} = K_{yy}^{y'y'} = 1 \tag{1}$$

$$A_{xx} = P^{x'x'} = K_{yy}^{x'x'} = K_{xx}^{y'y'} \tag{2}$$

$$A_{zz} = P^{z'z'} = K_{zz}^{y'y'} = K_{yy}^{z'z'} \tag{3}$$

$$K_x^{x'} = K_z^{x'} = K_{xy}^{x'} = K_{yz}^{x'} = 0 \tag{4}$$

$$K_x^{z'} = K_z^{z'} = K_{xy}^{z'} = K_{yz}^{z'} = 0 \tag{5}$$

$$K_x^{x'y'} = K_z^{x'y'} = K_{xy}^{x'y'} = K_{yz}^{x'y'} = 0 \tag{6}$$

$$K_x^{y'z'} = K_z^{y'z'} = K_{xy}^{y'z'} = K_{yz}^{y'z'} = 0 . \tag{7}$$

Eqs. (4) to (7) are particularly important, since a value of zero at (E_i, Θ_i) does not depend on the calibration of any polarizations. For an extreme value $A = \pm 1$, the relative values of <u>all</u> M-matrix elements are determined and the common constant is given by the cross section σ_o at (E_i, Θ_i). Thus all 51 polarization observables are numerically determined, and in addition to eqs. (4) to (7) the following conditions hold

$$A_y = P^{y'} = K_y^{y'y'} = K_{yy}^{y'} = \pm 1 \tag{8}$$

$$A_{yy} = P^{y'y'} = K^{y'}_y = K^{y'y'}_{yy} = 1 \tag{9}$$

$$A_{xz} = P^{x'z'} = K^{x'z'}_{xz} = 0 \tag{10}$$

$$K^{y'}_{xz} = K^{x'x'}_{xz} = K^{y'y'}_{xz} = K^{z'z'}_{xz} = 0 \tag{11}$$

$$K^{x'z'}_y = K^{x'z'}_{xx} = K^{x'z'}_{yy} = K^{x'z'}_{zz} = 0 \tag{12}$$

$$K^{z'z'}_y = K^{y'}_{zz} = K^{x'x'}_y = K^{y'}_{xx} = \mp\, 1/2 \tag{13}$$

$$K^{z'z'}_{zz} = K^{z'z'}_{xx} = K^{x'x'}_{zz} = K^{x'x'}_{xx} = 1/4 \tag{14}$$

$$A_{xx} = A_{zz} = K^{y'y'}_{xx} = K^{y'y'}_{zz} = -\,1/2 \tag{15}$$

$$P^{x'x'} = P^{z'z'} = K^{x'x'}_{yy} = K^{z'z'}_{yy} = -\,1/2 \tag{16}$$

The quantities P,A and K denote the polarization of the scattered deuterons, the analyzing power and the polarization transfer coefficients respectively. With these conditions a straighforward experimental identification of a point $A_y = \pm 1$ or $A_{yy} = 1$ is possible. Care should be taken to verify each condition in several ways, choosing the observables by an inspection of the general formulae and taking into account the relations between observables required by other symmetries[3].

References

* Work performed under the auspices of the U.S. Energy Research Development Administration
\+ On leave of absence from the University of Basel, Switzerland
‡ National Research Council of Canada, Postdoctoral fellow

1) W. Grüebler, P. A. Schmelzbach, V. König, R. Risler, B. Jenny, and D. Boerma, Nucl. Phys. A242 (1975) 285.
2) H. E. Conzett, F. Seiler, F. N. Rad, R. Roy and R. M. Larimer, Fourth Polarization Symp., p.568
3) G. G. Ohlsen, J. L. Gammel and P. W. Keaton, Phys. Rev. C5 (1972) 1205
4) F. Seiler, F. N. Rad and H. E. Conzett, Fourth Polarization Symp. p.897
5) F. Seiler, LBL-report, LBL-3474, and to be published.

DETERMINATION OF THE POLARIZATION TRANSFER COEFFICIENT K_y^{y} (0°)
FOR THE BREAK-UP REACTIONS ^{4}He($\vec{d},\vec{n}$)^{4}He,p AND D($\vec{d},\vec{n}$)D,p *

P.W. Lisowski, R.C. Byrd, and R.L. Walter
Duke University and Triangle Universities Nuclear Laboratory
and
T.B. Clegg, University of North Carolina and TUNL

The polarization of the continuous neutron distribution produced at
θ = 0° by the break-up of vector-polarized deuterons on D and ^{4}He has
been measured at 8.7, 11.4, and 14.4 MeV. Our interest in this study
was generated by an earlier D($\vec{d},\vec{n}$)^{3}He measurement[1] of the polarization
transfer coefficient K_y^{y} (0°). In that experiment large asymmetries in
the continuous break-up neutron background were observed. However,
values of $K_y^{y'}$ (0°) for only the ground-state neutron group could be extracted because the deuteron beam was both vector and tensor polarized.
Those conditions required knowledge of the tensor analyzing power
A_{yy}(0°) for the neutron continuum, which has never been measured. In
order to avoid this problem, the present measurements were made using
a pure vector polarized beam.

The general procedure used in this experiment was nearly identical
to that described in both ref.[1] and in our contribution[2] to this conference on the D($\vec{p},\vec{n}$)2p reaction. This method determines the neutron
polarization by scattering from ^{4}He into plastic side detectors at 120°
(lab) and recording the ^{4}He recoil energy in coincidence with the side
detector pulses. One such spectrum for neutrons from the D($\vec{d},\vec{n}$)^{3}He reaction at a deuteron energy of 11.4 MeV is shown in fig. 1. (The
D(d,n_0)^{3}He monoenergetic peak at about channel 95 shows the experimental
resolution, which was about 25% due largely to the polarimeter geometry
and target thickness chosen for this experiment.) Since these spectra
are equivalent to those for the neutron energy E_n, analysis of the observed asymmetry permits a determination of the energy-dependent polarization of the break-up continuum.

Plotted under the spectrum in fig. 1a are the extracted values of
the polarization transfer coefficient $K_y^{y'}$ (0°) as a function of channel
number, which is linearly related to E_n. Over much of the break-up region $K_y^{y'}$ (0°) has a value slightly larger than 0.5. In fig. 1b the same
information is shown for ^{4}He($\vec{d},\vec{n}$)^{4}He,p; for this reaction the value of
$K_y^{y'}$ (0°) is near 0.63 throughout the highest energy region of the spectrum. Finally, in fig. 2 we show $K_y^{y'}$ (0°) at each bombarding energy for
both reactions; these values were calculated from data in the region of
the maximum in each break-up spectrum. The closeness of $K_y^{y'}$ (0°) to the
value of 2/3 predicted for a simple stripping mechanism (shown by the
dashed line) suggests that the outgoing neutron has nearly the same
polarization as the "neutron in the incident deuteron." For both reactions it appears that the neutron emitted at 0° during the three-
particle break-up process suffers little spin perturbation, just as was
observed by Simmons et al.[3] for the two-particle final state D($\vec{d},\vec{n}$)^{3}He
reaction.

574

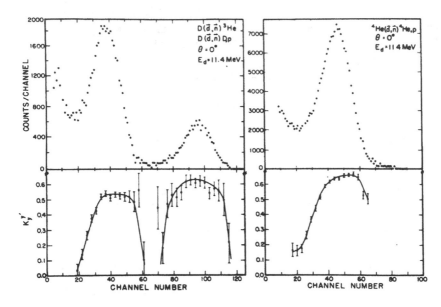

Fig. 1a. Experimental results for
the D($\vec{d}$,$\vec{n}$)D,p and D($\vec{d}$,$\vec{n}$)^{3}He reactions.

Fig. 1b. Experimental results for
the ^{4}He($\vec{d}$,$\vec{n}$)^{4}He,p reaction.

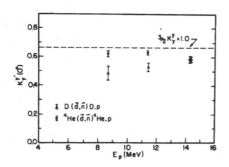

Fig. 2. Polarization transfer
coefficients extracted from the
regions of the maximum break-up
yield.

References

*Work supported in part by U.S. Energy Research and Development Adminis-
tration.

1) P.W. Lisowski, R.L. Walter, C.E. Busch and T.B. Clegg, Nucl. Phys.
A242 (1975) 298
2) R.L. Walter, R.C. Byrd, P.W. Lisowski, G. Mack and T.B. Clegg,
these proceedings, p.489
3) J.E. Simmons, W.B. Broste, G.P. Lawrence, J.L. McKibben and
G.G. Ohlsen, Phys. Rev. Letters 27 (1971) 113

ELASTIC SCATTERING OF ^{3}He FROM POLARIZED ^{3}He*

A. D. Bacher
Indiana University, Bloomington, Indiana 47401

E. K. Biegert, S. D. Baker, and D. P. May
Rice University, Houston, Texas 77001

K. Nagatani
Texas A & M University, College Station, Texas 77843

Experimental evidence now exists for highly excited levels in ^{6}Be. Measurements of the differential cross section for ^{3}He-^{3}He elastic scattering[1,2,3] show a broad resonance at about 24 MeV excitation energy in ^{6}Be which has been identified with the $\ell=3$ partial wave. Resonant structure has also been observed in the ^{3}He-^{3}He capture reaction[4] corresponding to a $J^\pi = 3^-$ level at 23 MeV excitation. Resonating group calculations[5] employing a phenomenological imaginary potential predict an unsplit ^{3}F level at 25 MeV excitation.

To study further this resonance region, we have measured the polarization analyzing power A_y^T for the elastic scattering of ^{3}He from a polarized ^{3}He target at laboratory angles 20°, 35°, and 50° and for laboratory energies between 8.9 and 34.8 MeV, corresponding to excitation energies in ^{6}Be between 16 and 29 MeV. The ^{3}He beam from the Texas A & M cyclotron was incident on a ^{3}He target constructed at Rice. Detectors were placed on either side of the beam at each angle to cancel certain instrumental asymmetries. Details of the construction of the target and of the determination of the ^{3}He polarization (typically 0.13 in this experiment) are given in refs.[6,7]. Repeated measurements at each energy reproduced satisfactorily.

The excitation function of A_y^T at $\theta_{lab} = 20°$ is shown in fig. 1. The error bars represent the combined statistical uncertainties and do not include a systematic uncertainty in the normalization which is estimated to be less than ±15%. The negative maximum of -0.37 ± 0.02 occurs at $E_{cm} = 11.3$ MeV. Data of similar quality have been obtained at 35° and 50° although values of A_y^T are smaller at these angles (typically +0.13 and -0.28, respectively).

Fig. 1 shows the result of a recent resonating group calculation by Reichstein[8] which uses the imaginary potential of Thompson et al.[5] and includes a phenomenological nucleon-nucleon spin-orbit interaction[9], the strength of which provides a good description of the nucleon-^{4}He phase shifts below the inelastic threshold. This calculation predicts splitting of the p-wave phase shifts and of the f-wave phase shifts; the latter exhibit resonant behavior near 25 MeV excitation. Since the calculation was done before these data were taken, no adjustment of the spin orbit strength was made to fit our data. There is qualitative agreement between this calculation and our data at all three angles. This suggests that the resonating group calculations provide a reasonable starting point for an analysis of the splitting in the $\ell=3$ phase shifts in this region of excitation in ^{6}Be.

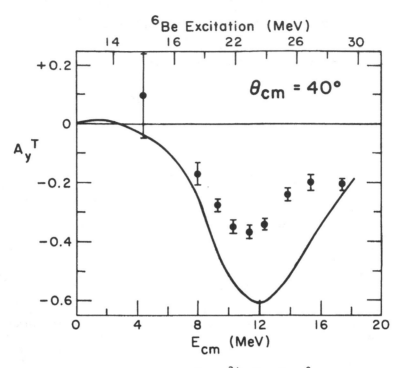

Fig. 1. Excitation function of $A_y{}^T$ in $\overrightarrow{^3\text{He}}(^3\text{He},^3\text{He})^3\text{He}$ at $\theta_{cm} = 40°$ versus the center-of-mass energy (bottom) and the ^6Be excitation energy (top). The solid curve represents the resonating group prediction of Reichstein described in the text.

References

*   Supported in part by the NSF and by the US AEC.

1)   A. D. Bacher, R. J. Spiger and T. A. Tombrello; Nucl. Phys. A119 (1968) 481.

2)   A. D. Bacher, T. A. Tombrello, E. A. McClatchie and F. Resmini, Lawrence Berkeley Laboratory Report UCRL - 18667 (1969) 104.

3)   J. G. Jenkin, W. D. Harrison and R. E. Brown, Phys. Rev. C1 (1970) 1622.

4)   E. Ventura, J. Calarco, C. C. Chang, E. M. Diener, E. Kuhlmann and W. E. Meyerhof, Nucl. Phys. A219 (1974) 157.

5)   D. R. Thompson, Y. C. Tang, J. A. Koepke and R. E. Brown, Nucl. Phys. A201 (1973) 301.

6)   D. M. Hardy, S. D. Baker and W. R. Boykin, Nucl. Instr. Meth. 98 (1972) 141.

7)   S. D. Baker, D. H. McSherry and D. O. Findley, Phys. Rev. 178 (1969) 1616.

8)   I. Reichstein, private communication.

9)   I. Reichstein and Y. C. Tang, Nucl. Phys. A158 (1970) 529.

INVESTIGATION OF THE POLE-GRAPH DOMINANCE IN THE ^{6}Li($\vec{p}$,pα)^{2}H REACTION

J.L. Durand, J. Arvieux, J. Chauvin, C. Perrin, G. Perrin
Institut des Sciences Nucléaires, BP 257, 38044 Grenoble Cédex, France

Direct knock-out reactions are often used to calculate nuclear properties such as the reduced widths of clusters in nuclei. This is usually done through the quasi-free scattering (QFS) approximation. But the kinematics favoring such simple mechanisms allows also more complicated rescatterings to take place which will make the extracted nuclear information incorrect. One type of experiment which indicates if the one-pole approximation is correct is the so-called Treiman-Yang (TY) test [1,2]. But this method has serious drawbacks : its validity is a necessary but not sufficient condition and, moreover, the corresponding experiments are rather difficult to make. Two such experiments have been done on the ^{6}Li(p,pd)^{4}He reaction, one between 40 and 50 MeV [3] for which the TY test was shown to be valid and one at 19 MeV [4] where it was also found valid except for high momentum transfer ($|\vec{q}| > 40$ MeV/c).

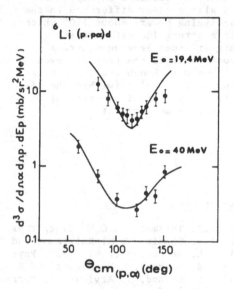

Fig. 1. Triple differential cross section at the QFS peak.

Another way of testing the validity of the QFS is through the use of incident polarized projectiles. If the reaction is described by the one-pole diagram the analyzing power at the QFS must be identical to the one of the free scattering. If any rescattering takes place the polarization of the incident particle will change and the final analyzing power will not be any more identical to the free scattering one [5]. We have applied this method to the ^{6}Li($\vec{p}$,pα)^{2}H reaction at 19 and 40 MeV since p-α scattering exhibits a well-known high analyzing power [6]. The experiment has been done at the Grenoble isochronous cyclotron with a $\sim$ 80 % polarized proton beam of 10-20 nA scattered by self-supporting ^{6}Li enriched targets. Experimental details can be found in ref. 5). The results are given in Fig. 1 for the triple differential cross-section taken at the QFS peak (transferred momentum $\vec{q}$ = 0). The curves are calculated using

$$\frac{d^3\sigma}{d\omega_p d\omega_\alpha dE_p} = K |\phi_0|^2 \left(\frac{d\sigma}{d\omega}\right)_{p-\alpha}$$

where K is a kinematic factor, ϕ_0 the momentum wave function of the α-cluster in the ^{6}Li at $\vec{q}$ = 0 and $(d\sigma/d\omega)_{p-\alpha}$ the free p-α cross section. The value of $|\phi_0|^2$ is 4.2 Sr^{-1} fm^3 at 19.4 MeV and 0.65 Sr^{-1} fm^3 at

578

40 MeV. It can be seen at first
sight that at 40 MeV the shape of
the cross-section is very similar
to the free p-α cross-section. At
40 MeV the quasi-free and free p-α
(curve) analyzing powers shown in
fig. 2 are also almost identical.
This is a confirmation that the
⁶Li(p,pα) reaction is well described by a pole mechanism at 40 MeV.
At 19 MeV the shape of the cross-
section is not well fitted using
the free p-α cross section especially at backward angles. There
is also a clear difference in the
analyzing powers above 110° which
is a strong indication that
higher-order Feymann diagrams
contribute to the reaction even
at $\vec{q}$ = 0 although the TY test was
shown to be valid (within the limited statistical accuracy of
this last measurement).

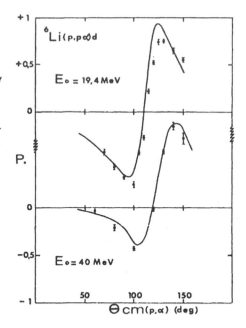

Fig. 2. Analyzing power of the
QFS reaction.

References

1) S.B. Treiman and C.N. Yang, Phys. Rev. Lett. 8 (1962) 140
2) I.S. Shapiro, V.M. Kolybasov and G.R. Augst, Nucl. Phys. 61 (1965) 353
3) J.L. Beveridge et al. Helv. Phys. Acta 47 (1974) 211
4) R.B. Liebert, K.H. Purser and R.L. Burman, Nucl. Phys. A216 (1973) 335
5) J.L. Durand, J. Arvieux, C. Perrin and G. Perrin, Phys. Lett. 53B
   (1974) 57
6) A.D. Bacher et al. Phys. Rev. C5 (1972) 1147

ANALYZING POWER FOR ^{4}He(t,t)^{4}He ELASTIC SCATTERING*

R. A. Hardekopf, N. Jarmie, G. G. Ohlsen, and R. V. Poore
Los Alamos Scientific Laboratory, University of California
Los Alamos, New Mexico 87545

We have obtained analyzing power data for elastic triton-alpha scattering at θ_{cm} = 49.6° and θ_{cm} = 123° over the energy range 7-14 MeV (see fig. 1). In addition, we have obtained angular distributions of the analyzing power at 12.25 MeV, 10.82 MeV, and 8.79 MeV (see fig. 2). In both figures, dots represent data obtained by detecting tritons and triangles represent the data obtained by detecting alpha particles.

The resonance near 8.8 MeV is of particular interest, since it corresponds to a resonance at ~ 240 keV in the ^{6}Li(n,α)T reaction. The absolute cross section for the ^{6}Li(n,α)T reaction near this resonance is of particular interest because of its usefulness as a neutron flux monitoring standard, and because of the potential use of ^{6}Li as a tritium-breeding "blanket" in a controlled thermonuclear reactor. Hale[1] has shown that recently obtained absolute triton-alpha elastic scattering cross section data[2] probably suffice to determine the ^{6}Li(n,α)T total cross section in this region to an accuracy of about 3%. The curves on the figures are from Hale's best fit to all available cross section data. This fit corresponds to a resonant cross section value of 3.4 ± 0.1 barn. The analyzing power data appears to be sufficiently well predicted in this region that they are not expected to have a large impact on this value. The analysis did not include information at energies higher than about 10 MeV, so the poor agreement at higher energies is not significant.

A Wigner cusp effect is evident in the region of the threshold for the reaction ^{4}He(t,n)^{6}Li. It is most apparent for θ_{cm} = 123°. For the cross section measurements[3], a cusp effect was most apparent near 49.6°. Limited measurements were also made at θ_{cm} = 110° and θ_{cm} = 60.6°, in the hope of locating a more pronounced effect, but none was seen.

In fig. 2, the square points at 12.25 MeV are from the double scattering measurements by Keaton et al.[3]. It is seen that the agreement with the present data is highly satisfactory.

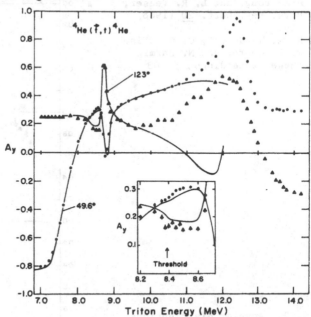

Fig. 1. Excitation function for the ^{4}He($\vec{t}$,t)^{4}He analyzing power.

Additional data on triton-
alpha elastic scattering near
11 MeV and $\theta_{cm} = 95°$, where an
$A_y = -1$ point exists, are presented elsewhere in these
proceedings[4].

References

* Work performed under the
auspices of the U.S. ERDA

1) G. M. Hale, Bull. Am. Phys.
Soc. 20 (1975) 148 and Proceedings of the Conference
on Nuclear Cross Sections
and Technology, 3-7 March
1975, to be published

2) N. Jarmie, G. G. Ohlsen,
P. A. Lovoi, D. M. Stupin,
R. A. Hardekopf, A. N.
Anderson, J. W. Sunier,
R. V. Poore, R. J. Barrett,
G. M. Hale, and D. C.
Dodder, Bull. Am. Phys.
Soc. 20 (1975) 596

3) P. W. Keaton, Jr., D. D.
Armstrong, and L. R. Veeser,
Phys. Rev. Lett. 20 (1968)
1392

4) R. A. Hardekopf, G. G. Ohlsen,
R. V. Poore, and N. Jarmie,
these proceedings, p.903

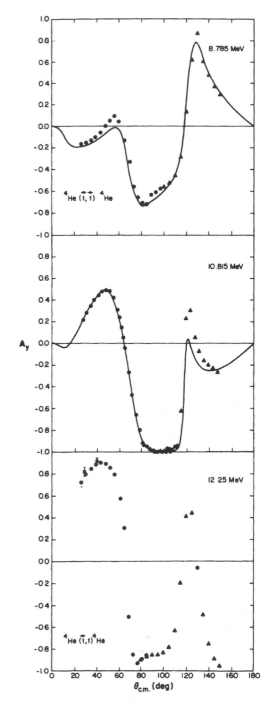

Fig. 2. Angular distributions
of the analyzing power for
^{4}He($\vec{t}$,t) ^{4}He elastic scattering.

POLARIZATION FOR REACTIONS IN THE ^{7}Li SYSTEM*

G. M. Hale and D. C. Dodder
Los Alamos Scientific Laboratory, University of California
Los Alamos, New Mexico 87545

Recent measurements[1,2] indicate large polarization effects are present in the ^{4}He($\vec{t}$,t)^{4}He and ^{4}He(t,$\vec{n}$)^{6}Li reactions at energies corresponding to E_t between 10 and 11 MeV. We show that these large polarization effects are predicted by a multichannel R-matrix analysis[3], the parameters of which are consistent with the current ^{7}Li level scheme. Other polarization predictions from the analysis are presented and discussed.

Large values of triton polarization for the ^{4}He(t,t) reaction were evident previously in the phase shift analysis of Spiger and Tombrello[4] and in the measurements of Keaton, et al.[5] The new measurements[1], made with a polarized triton beam, are shown in Fig. 1 at E_t = 8.79 and 10.82 MeV, along with curves calculated from the R-matrix analysis. The double minimum in the analyzing power at 10.82 MeV is substantiated by predictions of $A_y \sim -1$ points at E_t = 11.1, θ_{cm} = 93°, and E_t = 11.2 MeV, θ_{cm} = 111°. In addition, an $A_y \sim +1$ point is predicted at E_t = 6.25 MeV, θ_{cm} = 156°. The magnitude of the calculated analyzing power becomes large, but does not closely approach unity, at several other points in the vicinity of the resonances between 3 and 11 MeV.

Large values of neutron analyzing power for the ^{6}Li($\vec{n}$,t)^{4}He reaction are also evident in recent measurements of Karim and Overley[2]. Fig. 2 shows preliminary data[2] at $\theta_{cm} \simeq$ 90° (approximately where the maximum polarization occurs) along with predictions from the R-matrix analysis. The peak in the calculated analyzing power at E_n = 1.1 MeV appears to come mostly from interference between a 3/2$^-$ level having most of its width in the ^{4}P(n-^{6}Li) channel and the 1/2$^+$ s-wave transition.

Although the negative parity levels at low energies in ^{7}Li are well established, the nature of the s-wave transitions which presumably form the underlying background is not well understood, particularly in the vicinity of the 5/2$^-$ resonance at 240 keV neutron energy. Polarization measurements in this region should be sensitive to the interference between the resonant 5/2$^-$ and underlying (s-wave) transitions. Unfortunately, the interference near the resonance does not appear to produce large neutron analyzing powers; however, calculated triton polarizations, shown as analyzing powers for the ^{4}He($\vec{t}$,n)^{6}Li reaction on the contour plot of Fig. 3, attain moderate size (~ 30% at 110°) near the resonance.

Finally, we mention that the R-matrix calculations predict sizeable triton-to-neutron polarization transfer effects in the reaction near the resonance at E_n = 240 keV (E_t = 8.75 MeV). For instance, K_y^y becomes -.98 at zero degrees for E_t = 9.0 MeV. Because of the small reaction cross section (~10mb cm at zero degrees) however, it is doubtful that the large polarization transfer could be exploited as a source of low-energy polarized neutrons.

In summary, we find that the large polarizations measured recently for ^{4}He(t,t)^{4}He and ^{6}Li(n,t)^{4}He are reproduced by calculations based on the known level structure of ^{7}Li. Different interference terms appear to be responsible for the large polarizations observed in the two reactions, the first coming mainly from 5/2$^-$, 1/2$^+$ interference and the second from 3/2$^-$, 1/2$^+$ interference. Measurements of other polarization effects that are predicted to be moderately large could provide detailed information above the ^{7}Li level structure at low energies.

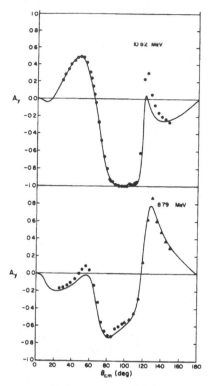

Fig. 1. Triton analyzing power
for ^{4}He$(\vec{t},t)^4$He.

References

* Work performed under the auspices of
the United States Energy Research and
Development Administration.

1) R. A. Hardekopf et al., these proceedings, p.579
2) M. Karim and J. C. Overley, Bull. Am.
Phys. Soc. 20 (1975) 167, and private
communication.
3) G. M. Hale, Bull. Am. Phys. Soc. 20
(1975) 148.
4) R. J. Spiger and T. A. Tombrello,
Phys. Rev. 163 (1967) 964.
5) P. W. Keaton et al., Phys. Rev. Lett.
20 (1968) 1392.

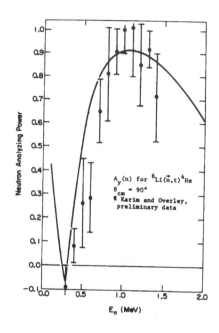

Fig. 2. Neutron analyzing power
for ^{6}Li$(\vec{n},t)^4$He.

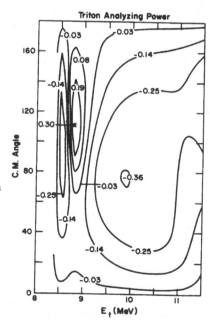

Fig. 3. Triton analyzing power
for ^{4}He$(\vec{t},n)^6$Li.

ELASTIC SCATTERING OF ^{4}He FROM POLARIZED ^{3}He*

A. D. Bacher.
Indiana University, Bloomington, Indiana 47401

S. D. Baker, E. K. Biegert, and D. P. May
Rice University, Houston, Texas 77001

E. P. Chamberlin
Texas A & M University, College Station, Texas 77843

Below an excitation energy of 12 MeV in ^{7}Be, the ^{3}He-^{4}He elastic scattering differential cross sections[1] provide clear evidence for the expected shell model states with p-shell configurations. Only minor changes in the phase shifts determined from the cross section data are required to fit polarized ^{3}He measurements[2] up to 9 MeV excitation. In higher energy differential cross section measurements[3,4] up to 42 MeV excitation some broad structure is apparent, but attempts to extend the phase shifts to excitation energies above 12 MeV have not been successful. Resonating group calculations[5] employing an ℓ-dependent phenomenological imaginary potential reproduce the cross section data over a wide energy range and predict a broad $\ell = 2$ level at 11.6 MeV and both $\ell = 4$ and $\ell = 5$ levels near 25 MeV excitation.

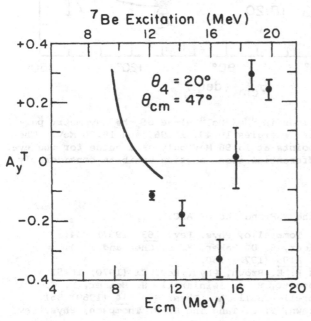

Fig. 1. Excitation function in $^3\vec{\text{He}}(^4\text{He},^4\text{He})$ ^{3}He for A_y^T at $\theta_{cm} = 47°$ versus the center-of-mass energy (bottom) and the ^{7}Be excitation energy (top). The solid curve shows the behavior predicted by ref.[1].

We have measured the polarization analyzing power A_y^T for the elastic scattering of ^{4}He from a polarized ^{3}He target at bombarding energies from 24.5 to 42.0 MeV, corresponding to excitation energies in ^{7}Be between 12 and 20 MeV. The experimental configuration is similar to that of ref[6]. Detection of ^{3}He and ^{4}He at laboratory angles 20° and 35° and of ^{3}He at 50° provides measurements at five center-of-mass angles.

The excitation function of A_y^T at $\theta_{cm}=47°$ is shown in fig. 1 along with a solid curve which indicates values predicted by the phase shift analysis[1] at lower energies. The error bars represent only statistical uncertainties[6]. The range of variation of A_y^T at the other angles is indicated by three representative angular distributions in fig. 2. The variations in the values of A_y^T at the forward and backward angles (see also fig. 1) are presumably related to structure that appears in the differential cross sections in this same energy range.

Preliminary results of resonating group calculations[7] which include a phenomenological nucleon-nucleon spin-orbit interaction also predict structure for A_y^T in these regions and are expected to provide a useful guide in extending the phase-shift analysis to higher energies.

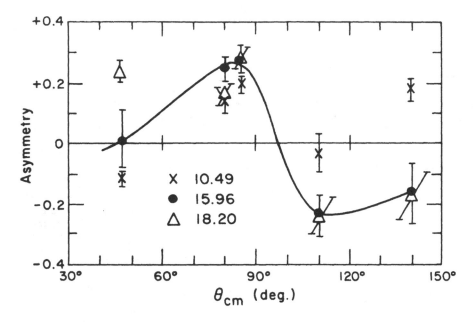

Fig. 2. Angular distributions in $^3\mathrm{He}(^4\mathrm{He},^4\mathrm{He})^3\mathrm{He}$ of the asymmetry parameter A_y^T at center-of-mass energies 10.49, 15.96, and 18.20 MeV. The solid curve connects the points at 15.96 MeV only as a guide for the eye. (The derivative of the differential cross section exhibits considerably sharper oscillations.)

References

*   Supported in part by the NSF and the US AEC.

1) R. J. Spiger and T. A. Tombrello, Phys. Rev. <u>163</u> (1967) 964.
2) D. M. Hardy, R. J. Spiger, S. D. Baker, Y. S. Chen and T. A. Tombrello, Nucl. Phys. <u>A195</u> (1972) 250.
3) C. G. Jacobs, Jr., and R. E. Brown, Phys. Rev. <u>C1</u> (1970) 1615.
4) A. D. Bacher, H. E. Conzett, R. de Swiniarski, H. Meiner, F. G. Resmini and T. A. Tombrello, Bull. Am. Phys. Soc. <u>14</u> (1969) 851.
5) J. A. Koepke, R. E. Brown, Y. C. Tang and D. R. Thompson, Phys. Rev. <u>C9</u> (1974) 823.
6) A. D. Bacher, E. K. Biegert, S. D. Baker, D. P. May, and K. Nagatani, contribution to Fourth Polarization Symp., p.575
7) R. D. Furber, R. E. Brown, Y. C. Tang and D. R. Thompson, University of Minnesota Annual Report (1974) 14.

MEASUREMENT OF THE DIFFERENTIAL CROSS SECTION AND THE ANALYZING POWERS iT_{11}, T_{20}, T_{21} AND T_{22} OF THE REACTION ^{6}Li$(d,\alpha)^4$He

R. Risler, W. Grüebler, A.A. Debenham, D.O. Boerma, V. König
and P.A. Schmelzbach
Laboratorium für Kernphysik, Eidg. Techn. Hochschule
8049 Zürich, Switzerland

Measurements of the ^{6}Li$(d,\alpha)^4$He reaction contribute valuable information on the ^{8}Be nucleus above 22 MeV excitation energy. Furthermore polarization data for this reaction can be used for calibrating polarized ^{6}Li beams.

At about 12 energies in the region between 1.5 and 11.5 MeV, angular distributions of the differential cross section and the four analyzing powers were measured. The results show large effects with strong variations over the energy range considered.

Legendre polynomials were fitted to the data points. Because of the two identical particles in the outgoing channel, only even coefficients had to be included. The experimental data in fig. 1 show that some of the coefficients for the tensor analyzing powers are linearly related to each other, i.e. have the same energy dependence, over a considerable energy range. Based on this fact a preliminary evaluation of the data using the methods pointed out by Seiler[1] was carried out. If only a few R-matrix elements are necessary to describe the main properties of a reaction, such linear relations give information on the matrix elements involved.

The results show three levels with spin and parity 2^+, 2^+ and 4^+ between 22 and 26 MeV excitation energy in the compound nucleus ^{8}Be. This agrees with the $\alpha-\alpha$ phase-shift analysis[2].

References

1) F. Seiler, Nucl. Phys. A187 (1972) 379
2) A.D. Bacher, F.G. Resmini, H.E. Conzett, R. de Swiniarski,
   H. Meiner and J. Ernst, Phys. Rev. Letters 29 (1972) 1331
3) R. Neff, P. Huber, H.P. Nägele, H. Rudin und F. Seiler,
   Helv. Phys. Acta 44 (1971) 679

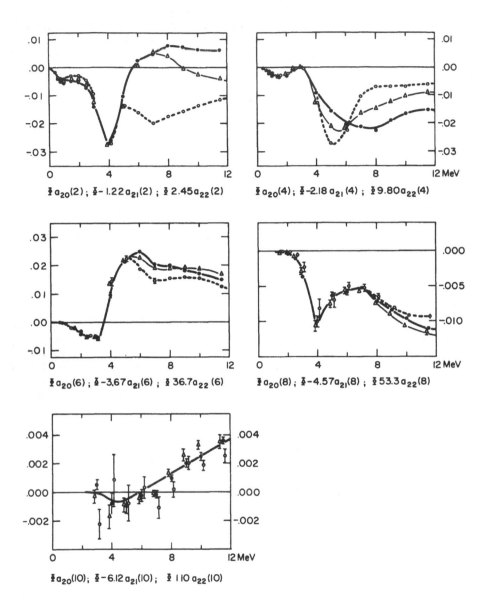

Fig. 1. Linear relations between the Legendre polynomial coefficients $a_{20}(L)$, $a_{21}(L)$ and $a_{22}(L)$ for the reaction $^6Li(d,\alpha)^4He$. Below 1 MeV data from ref.3) are used. Where the curves coincide the points of $a_{21}(L)$ and $a_{22}(L)$ are drawn slightly shifted. The curves only link corresponding points and are not the result of a calculation.

EXTREME VALUES OF THE ANALYZING POWER
IN THE ^{6}Li$(\vec{d},\alpha)^4$He REACTION*

F. Seiler[+], F. N. Rad, H. E. Conzett, and R. Roy[+]
Lawrence Berkeley Laboratory, University of California
Berkeley, California 94720

Extreme values of the analyzing power have been identified in $\vec{p}$ - ^{4}He
and $\vec{d}$-^{4}He elastic scattering[1,2]. Recently it has been proposed that
similar points exist also in reactions[3]. A calculation of the cartesian
tensor A_{yy} from the ^{6}Li$(\vec{d},\alpha)^4$He data of Grüebler et al.[4] shows experimental values with $A_{yy} \simeq 1$ at two points (fig.). Although more data are
clearly needed, extreme values $A_{yy} = 1$ are possible. A definite identification requires the elements of the M-matrix to fulfill certain conditions. For $A_{yy} = 1$ the M-matrix[5]

$$M = -\frac{1}{2} (a-e, \sqrt{2}\, f,\ a+e;\ \sqrt{2}\, h,\ -2k,\ -\sqrt{2}\, h;\ a+e,\ -\sqrt{2}\, f,\ a-e) \quad (1)$$

must satisfy $a = 0$. If an analysis is not available, the influence of
this condition on certain observables can be tested. Beside the efficiencies A^1_{ij} for deuteron polarization and A^2_{kl} for ^{6}Li polarization, only the
efficiency correlation coefficients $C_{ij,kl}$ are available. The 25 linearly independent observables[5] yield

$$A^1_{yy} = A^2_{yy} = 1 \quad (2)$$

$$C_{x,z} = C_{z,x} = C_{x,x} = C_{z,z} = 0 \quad (3)$$

$$C_{xy,x} = C_{xy,z} = C_{yz,x} = C_{yz,z} = 0. \quad (4)$$

Near 6 MeV all efficiencies undergo rapid changes and near 30° A_y is
large and positive. The possibility $A_y = 1$ [ref.[6]] should therefore be
investigated with closely spaced efficiency measurements. If $A_y \simeq 1$ is
found, the conditions

$$a = 0, \qquad h = \pm\, ie, \qquad k = \pm\, if \quad (5)$$

should be tested. Experimentally this yields

$$A^1_y = \pm 1$$

$$A^1_{yy} = A^2_{yy} = 1$$

$$A^1_{xx} = A^1_{zz} = -1/2$$

$$A^1_{xz} = C_{xz,y} = C_{xz,xz} = C_{xz,zz} = 0$$

$$2C_{zz,y} = \pm\, C_{y,y}$$

$$2C_{zz,xz} = \pm\, C_{y,xz}$$

$$2C_{zz,zz} = \pm\, C_{y,zz}$$

in addition to eqs. (3) and (4). Since efficiency correlation experiments
are difficult, it is likely that a verification of the conditions will be
done using the results of an analysis. The conditions above may help in
the selection of the data base necessary, since measurements near a suspected extreme point are a powerful tool in an analysis[3]. It is also
very likely that similarly large values of A_{yy} and A_y will be found for

polarized ^{6}Li-beams, since these maxima seem correlated with certain excited states in the intermediate system.

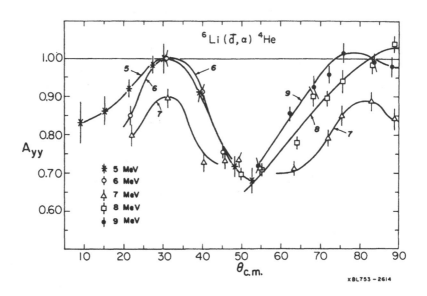

References

* Work performed under the auspices of the U. S. Energy Research and Development Administration.
+ On leave of absence from the University of Basel, Switzerland.
† Research Council of Canada, Postdoctoral Fellow.

1) G. R. Plattner and A. D. Bacher, Phys. Lett. 36B (1971) 211.
2) W. Grüebler et al., Nucl. Phys. A242 (1975) 285.
3) F. Seiler, Fourth Pol. Symp. and LBL-report, LBL-3474, March 1975.
4) W. Grüebler et al., Results of Measurements and Analyses, Eidgenössische Technische Hochschule, Zürich, May 1973.
5) W. E. Köhler and D. Fick, Z. für Phys. 215 (1968) 408.
6) F. Seiler, F. N. Rad and H. E. Conzett, LBL-report, LBL-3496, and to be published.

THE $^{15}N(\vec{p},\alpha)^{12}C$ REACTION BELOW 1.25 MeV

George H. Pepper and Louis Brown
Department of Terrestrial Magnetism
Carnegie Institution of Washington, Washington, D. C.

and

Louis G. Arnold
Department of Physics
The Ohio State University, Columbus, Ohio

Reactions initiated by protons on ^{15}N have been the object of intensive study for many years. The continuing interest in these reactions is indicative of their importance in several areas. First, at low energies, the reactions are an important hydrogen burning process in stellar nucleosynthesis. The $^{15}N(p,\alpha)^{12}C$ reaction is part of the carbon-nitrogen cycle, while the $^{15}N(p,\gamma)^{16}O$ reaction allows a loss of catalyst from the cycle. Second, the ^{16}O compound system is one of the more important testing grounds for theories of nuclear structure. Several states of ^{16}O near the ^{15}N + p threshold are members of isospin multiplets that contain the four lowest levels of ^{16}N and ^{16}F. These states are nearly pure one particle - one hole configurations and are isospin mixed with nearby T = O states from the same configurations. Finally, the abundance of experimental data on these reactions and the relatively simple spin structures of the initial and final states affords the opportunity for reasonably simple tests of nuclear reaction theories.

In this experiment angular distributions of the $^{15}N(p,\alpha)^{12}C$ reaction initiated by polarized protons have been measured at a proton energy of 0.34 MeV and at five energies between 0.90 and 1.25 MeV where differential cross section data were available. Excitation functions have been measured at 90° c.m. between 0.34 and 1.21 MeV, and at 63.4° and 116.6° c.m. between 1.04 and 1.20 MeV. Measurements were also taken near 45° and 135° lab at 0.60 and 0.80 MeV. The asymmetries obtained in this experiment are in agreement with the measurements by Ad'yasevich et al.[1] over the 0.3 to 0.5 MeV range of energies covered by their experiment.

The experiment used the Carnegie Van de Graaff with a Basel polarized proton source. Except for the ^{15}N target cell the equipment is described elsewhere.[2] The analyzing power data have been fitted with associated Legendre functions. Below 1.02 MeV only P_1^1 and P_2^1 are required for good fits; higher energies need the addition of P_3^1. This result is expected from known level structure. The 1^- state at 0.34 MeV is predominantly s-wave; other partial waves are also seen at this energy through interference. The values of the Legendre coefficients at 0.34 MeV are $a_1 = 0.018 \pm 0.006$ and $a_2 = -0.021 \pm 0.004$.

These two coefficients change sign near 0.4 MeV, show a gradual variation with energy up to about 0.8 MeV, then oscillate with increasing amplitudes in the interval from 0.8 to 1.25 MeV. The coefficient a_3 is negative and attains its maximum value near 1.16 MeV. The rapid variations of the coefficients with energy reflects the number of states of the compound system. A multi-level analysis of the data is in progress.

References

1) B. P. Ad'yasevich, V. G. Antonenko, D. A. Kuznetsov, Yu. P. Polunin and D. E. Fomenko, J. Nucl. Phys. (USSR) 3 (1966) 290
2) L. Brown and C. Petitjean, Nucl. Phys. A117 (1968) 343

POLARIZATION IN ^{12}C($\vec{n}$,n)^{12}C AND ANALYZING POWER IN
^{12}C($\vec{n}$,n')^{12}C*(4.44 MeV) AT E_n =15.85 MeV

H. Lesiecki, G. Mack, G. Mertens, K. Schmidt and M. Thumm
Physikalisches Institut der Universität Tübingen,
D-7400 Tübingen, Morgenstelle, Germany

According to Boreli et al.[1] the cross section for elastic
n-^{12}C scattering shows in the energy region of 14 to 21 MeV resonances at 15.8 and 19.5 MeV. For the 15.8 MeV resonance only a
slight indication is present in the inelastic scattering leading to
the first excited state at 4.44 MeV. Therefore the polarization p(Θ)
in the elastic channel and the analyzing power A(Θ) in the inelastic channel were measured at a neutron energy of 15.85 MeV.

The neutrons produced in the ^{3}H(d,$\vec{n}$)^{4}He reaction by 1.9 MeV
deuterons under 70° had a primary polarization of - 0.135[2,3] and
were scattered by the ^{12}C nuclei of a plastic scintillator. The
scattering asymmetry was measured in a time-of-flight polarimeter.
The carbon recoil method was used for timing of the scattering in
the scintillator. Elastic and inelastic scattering was well separated. At a given backward scattering angle the beam leg allowed only
measurements with one side detector. Therefore the measurements
were done at two different angles simultaneously and the primary
polarization of the neutrons was inverted in every second run by a
superconducting spin-precession solenoid. The measurements were
normalized to a monitor using time-of-flight technique for n-p
scattering in a plastic scintillator.

From the measured asymmetries and the known primary polarization the polarization p(Θ) and the analyzing power A(Θ) were deduced. In fig. 1 and fig. 2 the results are shown together with
data which have been already published[4,5].

Polarization in elastic scattering. The polarization data in
fig. 1 are shown together with calculations of Zombeck[6] who used
the optical model parameters of Frasca et al.[7] (solid curve) and a
calculation with optical model plus a resonance term at 15.8 MeV.
The best reproduction of our data at forward angles was achieved
with a $f_{5/2}$ resonance of 0.5 MeV width (dashed curve). However at
medium and backward angles the agreement with both calculations is
poor. The situation is similar as for protons scattered by ^{12}C. A
number of measurements between 15 and 20 MeV show angular distributions of the polarization comparable to our neutron data. Optical
model calculations for these data in general also show stronger
angular variations than the measurements. The same is true for 14
MeV neutron data.

Analyzing power in inelastic scattering. In fig. 2 the results
are shown together with earlier published data[4,5] and with data
of Casparis et al.[8] at 14.2 MeV. The dashed curve represents a
coupled-channel analysis of Hodge and Tamura for the 14.2 MeV data.
In the forward angle region the measurements at 15.85 MeV seem to
differ remarkably from those at 14.2 MeV which are definitely out
of the resonance.

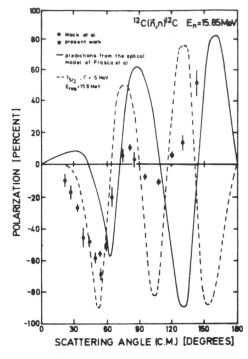

Fig. 1. Elastic scattering
polarization p(θ).

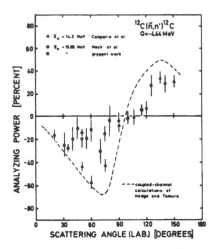

Fig. 2. Inelastic scatte-
ring analyzing power A(θ).

References

1) F. Boreli, B.B. Kinsey and P.N. Shrivastava, Phys. Rev. 174 (1968) 1147
2) G. Hentschel, G. Mack and G. Mertens, Z. Naturf. 23a (1968) 1401
3) W. Tornow, G. Mack, G. Mertens and H. Spiegelhauer, Phys. Lett. 44B (1973) 53
4) G. Mack, Z. Phys. 212 (1968) 365
5) G. Mack, G. Hentschel, C. Klein, H. Lesiecki, G. Mertens, W. Tornow and H. Spiegelhauer, Proc. of the 3rd Int. Symposium on Polarization Phenomena in Nuclear Reactions, Eds. H.H. Barschall and W. Haeberli, Madison 1970, p. 615
6) M.V. Zombeck, Thesis, MIT (1969)
7) A.J. Frasca, R.W. Finlay, R.D. Koshel and R.L. Cassola, Phys. Rev. 144 (1966) 854
8) R. Casparis, B. Leemann, M. Preiswerk, H. Rudin, R. Wagner and P. Zupranski, private communication , to be published

SCATTERING OF POLARIZED PROTONS FROM CARBON FROM 11.5 TO 18 MeV *

W.G. Weitkamp**
Laboratorium für Kernphysik, ETH, Zürich, Switzerland

H.O. Meyer
University of Basel, Basel, Switzerland

J.S. Dunham, T.A. Trainor and M.P. Baker[†]
University of Washington, Seattle, Washington

As an aid in the study of the structure of ^{13}N, and to determine the suitability of ^{12}C as a proton polarization analyser above 11 MeV, we have measured angular distributions of analyzing power and differential cross section for the scattering of protons from carbon in the energy region from 11.5 to 18 MeV. We have analyzed these data in terms of phase shifts and resonance parameters.

The measurements, which extend similar data taken at lower energies[1], were carried out at the University of Washington tandem Van de Graaff accelerator. Natural carbon targets of 1-2 mg/cm^2 were used. The beam polarization was monitored continuously with a ^{4}He polarimeter, the analyzing power of which was taken from the measurements of ref. 2. Measurements were made at 100 keV intervals at up to 19 angles; a contour plot of the analyzing power data is shown in fig. 1.

The analyzing power and cross section data were analyzed using complex phase shifts with $\ell \leq 3$ below 13.7 MeV and $\ell \leq 4$ above. Because of

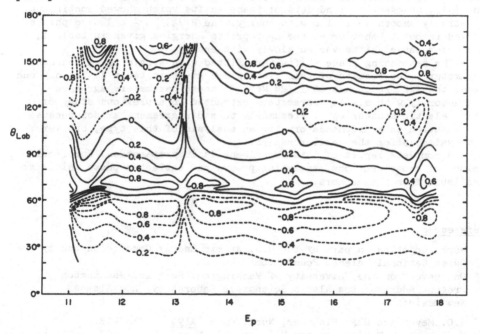

Fig. 1. Contour plot of the analyzing power of proton elastic scattering from carbon. Scattering angle and proton energy are in the laboratory system. Data below 11.5 MeV are taken from ref. 1.

TABLE 1

^{13}N level parameters from the ^{12}C(p,p)^{12}C analysis

$E_{p,lab}$ (MeV)	$E_x(^{13}N)$ (MeV)	J^π	Γ_{cm} (keV)	Γ_p/Γ	γ^2 (keV)	α
13.13 ± 0.02	14.06	3/2$^+$	180±35	0.29 ± 0.07	12±2	73°
15.24 ± 0.04	16.00	7/2$^+$	135±90	0.05 ± 0.04	5±2	-12°
17.58 ± 0.03	18.16	3/2$^+$	322±75	0.08 ± 0.02	4.8±0.7	212°
17.60 ± 0.02	18.18	1/2$^-$	225±50	0.24 ± 0.06	8.1±1.4	-70°

the large number of parameters involved, there were often several sets of phase shifts at a given energy which fit the data equally well. It was possible, however, to find sets of phase shifts which showed continuous, relatively smooth variation with energy. The $P_{1/2}$, $D_{3/2}$ and $G_{1/2}$ phases showed resonant behavior at the appropriate energies given in table 1; the other phase shifts varied slowly with energy.

The resonating phase shifts were fitted with single level resonance parameters. The results of this fitting are given in table 1. In only one case, the 7/2$^+$ resonance at 15.24 MeV, do these parameters agree satisfactorily with similar parameters extracted in a previous study of p+^{12}C elastic scattering[3]. Presumably this disagreement arises because polarization data are mandatory for an analysis of this type, but were not available for the earlier analysis.

A complete report of this experiment is in preparation. Numerical values of cross sections, analyzing powers and extracted phase shifts are available from the authors.

References

* Work supported in part by the U.S. Atomic Energy Commission and by the Swiss National Science Foundation
** On leave from the University of Washington, Seattle, Washington
† Present address: Los Alamos Scientific Laboratory, Los Alamos, New Mexico

1) H.O. Meyer and G.R. Plattner, Nucl. Phys. A199 (1973) 413
2) P. Schwandt, T.B. Clegg and W. Haeberli, Nucl. Phys. A163 (1971) 432
3) M.J. LeVine and P.D. Parker, Phys. Rev. 186 (1969) 1021

THE SPIN ROTATION ANGLE IN ^{12}C(p,p)^{12}C SCATTERING

W.G. Weitkamp[*], W. Grüebler, V. König, P.A. Schmelzbach, R. Risler
and B. Jenny
Laboratorium für Kernphysik, Eidg. Techn. Hochschule
8049 Zürich, Switzerland

In describing proton elastic scattering from carbon at incident
energies above 11 MeV, it is necessary to employ complex phase shifts
with $\ell \leqslant 4$. The resulting large number of parameters (18) makes it difficult to extract unambiguous phase shifts from polarization and cross
section data alone. We have investigated the possibility of using measurements of the spin rotation angle β to improve the quality of phase
shifts extracted from ^{12}C(p,p)^{12}C scattering data.

That the spin rotation angle is sensitive to small changes in the
phase shifts may be seen from fig. 1. The two curves shown are values of
β predicted by two different sets of phase shifts. The dotted curve is
calculated from a set of published phase shifts[1], the solid curve from a
set fitted to the same data using a different fitting procedure. Both
sets fit the available data at 11.3 MeV equally well. In fact, the phase
shifts in the two sets are quite similar; the largest difference is in
the $S_{1/2}$ phase shift, where the absolute value of the difference in $\eta e^{2i\delta}$
is 0.16. It is clear that an easily measurable difference exists in the
predicted values of β.

The technique used in measuring β is illustrated schematically in
fig. 2. The incident beam, with polarization $\vec{P}_i$ (typically 0.8 with an
average intensity of 20 nA) was scattered from a natural carbon target

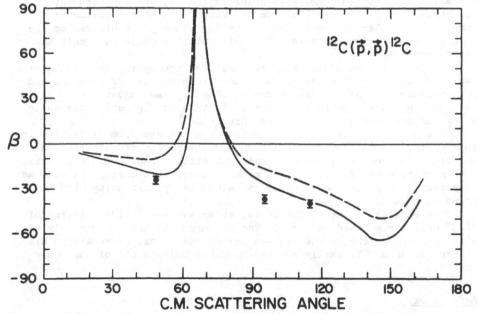

Fig. 1. Spin rotation angle β as a function of proton center-of-mass
scattering angle. Both angles are in degrees.

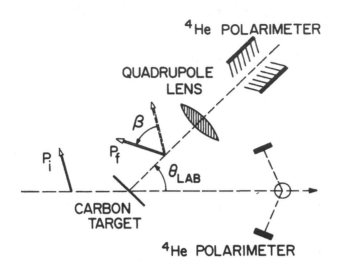

Fig. 2. Experimental arrangement for the spin rotation angle measurement.
The polarization vectors are in the scattering plane. For clarity, the
polarimeters are shown rotated 90° out of position.

0.5 MeV thick. The component of polarization in the scattering plane perpendicular to the beam momentum was monitored by a ^{4}He polarimeter downstream. The scattered beam, with polarization $\vec{P}_f$, was focussed into a
vane-type ^{4}He polarimeter 2 m from the target. In this polarimeter the
component of polarization in the scattering plane perpendicular to the
outgoing momentum was measured. The spin rotation angle is simply the angle between $\vec{P}_i$ and $\vec{P}_f$.

To measure β as efficiently as possible, the approximate value of β
was first determined by making short measurements with $\vec{P}_i$ parallel and
perpendicular to the incident momentum. Then $\vec{P}_i$ was adjusted so that $\vec{P}_f$
was parallel to the outgoing momentum. In this configuration the angle
of $\vec{P}_f$ was measured with the largest precision.

The major uncertainty in the measurement arises from difficulty in
controlling the incident polarization direction. The ETH atomic-beam
polarized ion source produces a beam with stable, reproducible polarization magnitude and direction, but minor changes in steering the beam as
it passes through the accelerator can alter the polarization direction
by as much as ±2°.

The results of the measurements, at an average incident energy of
11.37 MeV, are plotted in fig. 1. The measurements agree moderately well
with the predictions of the revised set of phase shifts; remaining discrepancies should be easily removed by minor adjustments of the phase
shifts.

References

*   On leave from the University of Washington, Seattle, Washington
1)   H.O. Meyer and G.R. Plattner, Nucl. Phys. A199 (1973) 413

INVESTIGATION OF THE GROUP OF LEVELS OF THE NUCLEI ^{21}Na

V.I. Soroka, A.I. Mal'ko, M.V. Arzimovich, N.N. Putcherov
and N.V. Romantchuk
Institute of Nuclear Research
Ukrani n Academy of Science, Kiev, URSS

This work contains results of the investigation of excited states of nuclei ^{21}Na near 5 MeV in reaction p-Ne^{20} with polarized protons.

The polarized beam of protons was obtained by elastic scattering protons on the carbon at 60° near the large resonance at E_p=2,88 MeV[1]. The polarized protons striked the second target. It was natural neon about 35 keV thick. The polarization excitation functions were measured at 45°, 60° and 90° (see fig.1). The uncertainties of the results included the statistical errors, the uncertainties of the analysing power and accuracy of alignment (less then 1%).

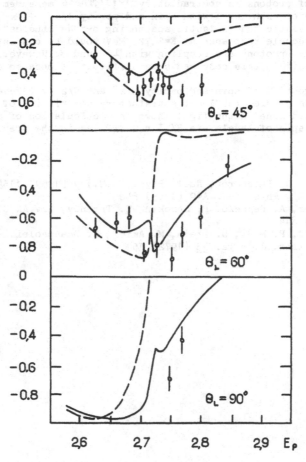

Fig. 1. Polarization of elastically scattered protons by ^{20}Ne.

The solid curves in fig. 1 show the calculated polarization. The parameters are given in table 1 and represent essentially the best-fit values.

Table 1. Best-fit parameters of ^{21}Na levels

E_p(MeV)	$E_r(^{21}$Na)	l	J^{π}	Γ(keV)	Γ_p/Γ
2,720	5,02	1	$1/2^-$	200	1
2,725	5,03	2	$5/2^+$	15	0,21
2,725	5,03	3	$5/2^-$	6	0,35

The widths of the levels of nuclei ^{21}Na-5,03 MeV have been derived from these data only.

The previous information about the existence of the large resonance at 2,72 MeV of protons is contradictory[2,3,4].These measurements confirm the existence of such level with spin and parity 1/2.

Recent elastic and inelastic scattering p-^{20}Ne studies[4] have revealed that the double resonance at E=2,725 MeV would be consistent with the assumption that protons are coupling with l=2 and 3. However, a more precise analysis in this case had some difficulties because of a lot of parameters.

The suggestion of spin and parity $5/2^+$ and $5/2^-$ for levels 5,03 MeV of nuclei ^{21}Na provide the least satisfactory description of our data.

The dotted line in the fig.1 shows that calculation of analyzing power if the spin of levels are $3/2^+$ and $7/2^-$ as in the previous analysis[4].

References

1) V. Soroka, N. Putcherov, Nucl. Phys. (S.U.) 9 (1969) 1159
2) W. Haeberli, Phys. Rev. A99 (1955) 640
3) A. Val'ter, A. Dejneko, P. Sorokin, T. Taranov, Izv.AN USSR (phys.) 24 (1960) 844
4) M. Lambert, P. Midi, D. Drain, M. Amiel, H. Beanmeoleille, A. Dancky, C. Meynaolier, J. Phys. 33 (1972) 155

NUCLEAR REACTION BY POLARIZED PROTONS AND DEUTERONS
AT LOW ENERGY

V.I. Borovlev, I.D. Lopatko, R.P. Slabospitsky, M.A. Chegoryan
Kharkov Physical-Technical Institute, Kharkov, U.S.S.R.

Polarization and vector analyzing power (VAP) were studied in the proton and deuteron interaction with atomic nuclei to determine quantum characteristics of nuclei and the nuclear reaction mechanism. In measurements we used polarized proton and deuteron beams accelerated by the tandem generator. Beam intensity was $\sim$ 150 pA and monochromacy was better than 0.1%. The experimental data [1-3] were analyzed by the formulae given in ref. [4].

To determine quantum characteristics of excited states of the nucleus ^{27}Al and to investigate the peculiarities in the polarization behaviour in the case of multilevel interference we have studied the polarization of protons elastically scattered by the nuclei ^{26}Mg. It is shown that at the energy range $E_p=1.8\div2.2$ MeV one has observed complex $p(E_p)$ dependence due to the interference of two wide overlapping resonances with a number of narrow-resonances in their background. The best fits obtained to the polarization with resonance parameters are presented in table 1. It should be noted that the data could be obtained only due to using a polarized beam with a high monochromacy.

The analysis of the VAP of the reaction $^{24}Mg(p,p')^{24}Mg^*$ has shown that our experimental data are well described under the assumption of three excited states, the parameters of which are presented in table 1. The experimental VAP-values in the region of the isolated resonance at $E_p=2.415$ MeV turned out to be zero within the experimental errors. This suggests a conclusion that in this case the reaction proceeds via the compound nucleus.

The comparison of the polarization tensor component values obtained from the phase analysis of the elastic deuteron scattering by ^{12}C with the experimentally observed tensor polarization values enabled to determine the total angular moments and level parities of the level ^{14}N in the investigated energy range (table 1). The analysis of the (d,p)-reaction VAP by the R-matrix theory allowed to determine the remaining quantum characteristics of ^{14}N nuclear levels. Similar results were obtained from the cross section and VAP analysis of the reaction $^{16}O(d,p1)^{17}O(0.87MeV)$.

The VAP in the interaction of polarized deuterons with the nuclei ^{28}Si was studied with an aim of discovering and investigating intermediate resonances in the nucleus ^{30}P. Taking into account the fact that the mean distance between the levels of ^{30}P $D\sim1$ keV, the mean level widths ~30 keV, and the experimental energy resolution >100 keV(for (d,p)-reactions), it should be expected that the VAP value would fluctuate within several percent. The observed considerable value of VAP and its energy dependence may be explained only by assuming the occurrence of excited number (~6) resonances of the intermediate structure with J<4 and $\Gamma\leq 200$ keV in the nucleus ^{30}P.

Target nucleus	E_p, E_d(keV)	E_x(MeV)	J^π	Γ(keV)	Γ_p(keV)	$\Gamma_p{}'$(keV)
^{24}Mg	2925±3	5.10	$1/2^-$	15	3	10
	2950±3	5.12	$3/2^+$	20	1	16
	2970±5	5.14	$(9/2^+)$	50	1	45
^{26}Mg	1890±2	10.09	$3/2^+$	2±0,5		
	2021±8	10.22	$3/2^-$	25±6		
	2055±13	10.25	$1/2^+$	63±1		
	2058±3	10.26	$3/2^-$	0.5		
	2100±3	10.29	$3/2^+$	0.5		
	2135±4	1033	$1/2^-$	2±0.5		
	2145±4	10.34	$5/2^+$	2±0.5		
	2185±4	10.37	$3/2^+$	0.8±0.4		
	2190	10.38	$3/2^+$	0.2		
^{12}C	1624	11.65	2^-			
	1682	11.71	1^+			
	1781	11.80	1^+			
	1950	11.94	(1^+)			
	2200	12.24	(3^+)			
	2498	12.417	4^-			
	2622	12.54	2^+			
	2726	12.606	3^+			
	2817	12.68	3^+			

References

1) I.D. Lopatko, V.I. Borovlev, G.B. Andreev, R.P. Slabospitsky, Izv. AN SSSR, ser.fis. 35 (1971) 1666
2) V.I. Borovlev, A.P. Klyucharev, R.P. Slabospitsky, G.B. Andreev, I.D. Lopatko, Izv. AN SSSR, ser.fis. 35 (1971) 1669
3) V.I. Borovlev, A.S. Dejneko, A.P. Klyucharev, Yu. G. Mashkarov, R.P. Slabospitsky, Izv. AN SSSR, ser.fis. 36 (1972) 2228
4) G.L. Vysotskij, M.A. Chegoryan, R.P. Slabospitsky, Preprint KFTI 72-9 (1972)

ANALYZING POWERS FOR p + ^{28}Si SCATTERING BETWEEN 2.0 AND 3.8 MeV[†]

W.S. McEver, L.G. Arnold and T.R. Donoghue
Department of Physics, The Ohio State University, Columbus, Ohio 43210

Excitation curves of A_y for p + ^{28}Si elastic scattering have been measured between 2 and 3.8 MeV to resolve conflicting J^π assignments for states in ^{29}Si near 6 MeV excitation energy. These conflicts arose partly in various analyses of cross section data[1,2], but most recently in the analysis[3] of (p,p'γ) angular correlation data at our laboratory.

Our A_y measurements utilized the polarized ion source[4] that operates in the high voltage terminal of our pressurized Van de Graaff accelerator. Experimental procedures similar to those of Volkers et al.[5] were used. A 50 nA beam with p_y = 0.58 was incident on a 3.5 keV thick natural silicon target. Excitation curves of A_y at several angles are shown in Fig. 1, where ΔA_y is typically ±0.01. The sharp energy dependence of A_y reflects the presence of known resonances, indicated by arrows. The solid curve is a single level approximation calculation that represents a best visual fit to the data made using on an on-line computer with display capabilities. Published resonance parameters[1] were used as trial inputs, where possible, and were varied to obtain this fit. The parameters in the table were used to calculate the curves in Fig. 1. The A_y analysis confirms some earlier assignments, but more importantly demands that some previously "firm" assignments be changed. For instance, the resonance at 3.1 MeV was previously assigned[1,2] J^π = 5/2$^-$ whereas the A_y data clearly dictate a 7/2$^-$ assignment. The 3.575 and 3.711 MeV resonances are 3/2$^-$ and 3/2$^+$ in agreement with the recent angular correlation analysis[3] and others[2] and conflict with some earlier analyses[1,6] one of which included some double scattering P(Θ) data.[6] In summary, we have measured a minimum of new data and carried out a simple analysis to unambiguously determine spectroscopic information and this re-emphasizes the value of polarization data for this purpose.

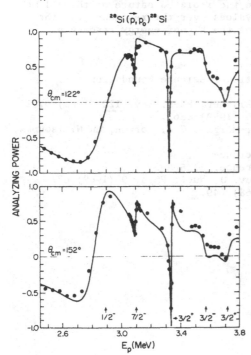

^{28}Si($\vec{p},p_0$)^{28}Si

Fig. 1

RESONANCE PARAMETERS FOR p + ^{28}Si

E_p (MeV)	$E_x(^{29}P)$ (MeV)	J^π	Γ (keV)	Γ_p (keV)
2.895	5.53	1/2$^-$	450	450
3.102	5.74	7/2$^-$	14	3.5
3.335	5.97	3/2$^+$	9.5	7
3.575	6.20	3/2$^-$	98	20
3.715	6.33	3/2$^+$	74	9
3.950	6.56	1/2$^+$	225	225

602

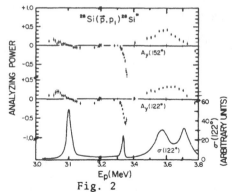

Fig. 2

Below 2.8 MeV, A_y is large and varies only gradually with energy, making this elastic scattering process useful as a polarization analyzer for say polarization transfer measurements when a thin solid state detector is used as the scatterer. The sharp variation at higher energy make it less useful for this purpose.

We also measured A_y for the protons scattered inelastically from the first 2^+ state of ^{28}Si for E_p = 3.05 to 3.8 MeV. The data plotted in Fig. 2 reveal some unusually large values of A_y for an inelastic process, reaching -0.76 at 3.355 MeV. Harney[7] noted that this can occur only if states of different J^π interfere in the entrance channel. One can show further that only product terms of the elements of the scattering matrix that have $s'_1 = s'_2$ contribute to A_y, and hence $A_y \neq 0$ serves as a measure of the degree of overlap of the components of states of different J^π that decay by the same channel spin. Here s'_1 (or s'_2) is the exit channel spin for $J = J_1$ (or J_2), where $J_1^\pi \neq J_2^\pi$. An analysis of the inelastic data is complicated by the double ℓ-valued decays possible and by the additional partial level widths and potential phase shifts that must be determined and was not attempted at this time in light of the firm J^π assignments made in the elastic data analysis. One does note the isolated nature of the 3.1 MeV resonance evidenced by the small A_y values over the resonance vs. the overlapping and interfering effects of the higher energy resonances.

References

†   Supported in part by the U.S. National Science Foundation.

1) T.A. Belote, E. Kashy, and J.R. Rissher, Phys. Rev. 122 (1961) 920.
2) J. Vorona, et al., Phys. Rev. 116 (1959) 1563.
3) N.L. Gearhart, H.J. Hausman, J.F. Morgan, G.A. Norton, and N. Tsoupas, Phys. Rev. C 10 (1974) 1739.
4) T.R. Donoghue, et al., this conference, p. 840
5) J.C. Volkers, et al., this conference, p. 508
6) V.I. Soroka and N.N. Pucherov, Sov. J. Nucl. Phys. 9 (1969) 677.
7) H.L. Harney, Phys. Lett. 28B (1968) 249.

Optical Potentials and Elastic Scattering

THE ELASTIC SCATTERING OF 2.9 AND 16.1 MeV NEUTRONS[*]

R.B. Galloway and A. Waheed[+]

Department of Physics, University of Edinburgh, Scotland

Measurements of the angular dependence of the polarization due to elastic scattering can serve as a sensitive test of optical model parameters. However data for neutron elastic scattering are sparce[1]. The first results are presented of a programme of simultaneous measurement of differential cross-section and polarization for elastic scattering by medium and heavy nuclei in which a wide range of neutron energy is to be covered. The measurements to date concern the elastic scattering of 2.9 MeV neutrons by Fe, Cu, I, Hg and Pb and of 16.1 MeV neutrons by Cu and Pb. They have as yet been compared only with published optical model calculations.

The angular dependence of polarization due to scattering of 2.9 MeV neutrons by Fe (fig. 1) agrees well with previous measurements at the nearby energies of 3.25 Mev[2] and 3.2 MeV[3] and confirms the positive swing around 160° of the Rosen[4] optical model curve as corrected for compound elastic scattering by Ellgehausen et al.[2]. These calculations are much less successful in fitting the differential cross-section data.

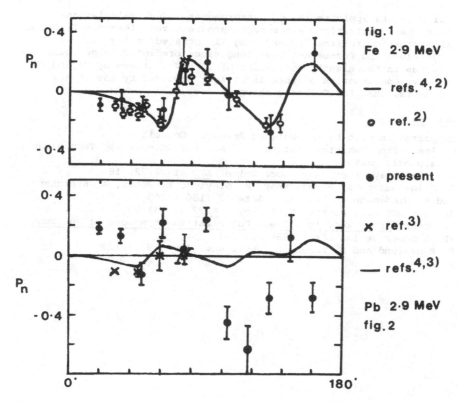

fig.1
Fe 2·9 MeV

— refs.[4,2)]

o ref.[2)]

● present

x ref.[3)]

— refs.[4,3)]

Pb 2·9 MeV
fig.2

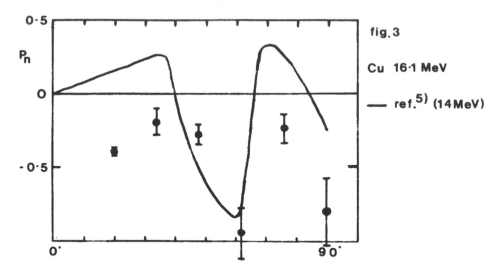

fig. 3

Cu 16·1 MeV

—— ref.[5] (14 MeV)

Similar comments apply to the Cu measurements. Such calculations resemble the polarization measurements progressively less for I, Hg, and Pb, the situation for the latter being illustrated in fig. 2.

Bjorklund and Fernbach[5] have long since predicted large polarization values in the elastic scattering of 14 MeV neutrons by Cu and Pb. Qualitative confirmation of this is at last provided by the 16 MeV measurements, that for Cu being shown in fig. 3.

References

* Supported in part by the Science Research Council
+ On leave from Pakistan Institute of Nuclear Science and Technology, Rawalpindi, Pakistan
1) R.B. Galloway, Proc. Roy. Soc. Edinb. A70 (1971/72) 181
2) D. Ellgehausen, E. Baumgartner, R. Gleyvod, P. Huber, A. Stricker and K. Wiedemann, Helv. Phys. Acta 42 (1969) 269
3) E. Zijp and C.C. Jonker, Nucl. Phys. A222 (1974) 93
4) L. Rosen, Proc. 2nd. Int. Symp. Polarization Phenomena of Nucleons (Birkhauser Verlag Basel 1966) p. 253
5) F. Bjorklund and S. Fernbach, Phys. Rev. 109 (1958) 1295

607

POLARIZATION OF 10 MeV NEUTRONS SCATTERED
FROM BISMUTH AND LEAD AT SMALL ANGLES†

J.M. Cameron, A.H. Hussein, S.T. Lam, G.C. Neilson,
J.T. Sample, H.S. Sherif, and J. Soukup

Nuclear Research Centre, University of Alberta, Edmonton, Canada

Polarization in the scattering of neutrons from nuclei at small
angles is expected to be dominated by effects due to the Mott-Schwinger
(M-S) interaction[1]. Furthermore, the maximum value of the polarization
approaches 1.0 for heavy nuclei. This makes the use of small angle scattering from such nuclei attractive as a neutron polarimeter as the analyzing power may then be determined in a model independent manner[2].
Recent measurements on the small angle scattering from bismuth (Bi) and
lead (Pb) have, however, yielded results in which the Bi data agree
satisfactorily with calculations which include the M-S interaction but
those for Pb do not[3]. To check the above discrepancy we have made measurements of the polarization in the elastic scattering from Pb and Bi
using 10 MeV neutrons.

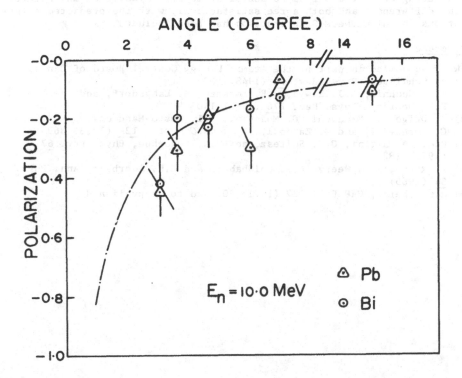

Polarization in the small angle scattering of neutrons at 10 MeV. The
theoretical curve is the result of a calculation in which the M-S
interaction is included.

Polarized neutrons were produced by the $^9Be(\alpha,n)^{12}C$ reaction at 30°
Lab., using a pulsed alpha beam. This resulted in neutrons of polarization P = 0.44 $\pm$ 0.03 [4]). The experiment was carried out in a shielded
source geometry and a superconducting solenoid used to rotate the neutron
spin by 180°. Scattered neutrons were detected in an array of three
NE-213 scintillation detectors and pulse shape discrimination was used to
separate events due to neutrons from those due to gammas hitting the
detectors. The angular resolution of the system for the scattering
measurements was $\pm$ 0.6 deg.

The results of the present small angle polarization measurements for
Pb and Bi are shown in the figure. The dashed curve is the result of a
calculation in which the nuclear part of the scattering amplitude is
calculated using the optical model parameters of ref. 5), while the
effect of the M-S interaction is calculated in the distorted wave Born
approximation[6]).

An average of the difference between the polarization for Pb and Bi
determined from these measurements for all the experimental points is
ΔP = .02 $\pm$ .037, whereas the same quantity calculated for the previous
measurements[3]) yields ΔP = .09 $\pm$ .04.

In conclusion, measurements of polarization in the small angle
scattering of neutrons from Pb and Bi at 10 MeV do not show any significant differences and both agree satisfactorily with the predicted polarizations in which the M-S interaction has been included.

References

†Work supported in part by the Atomic Energy Control Board of Canada.
1) J. Schwinger, Phys. Rev. 73 (1948) 407
2) F.T. Kuchnir, A.J. Elwyn, J.E. Monahan, A. Langsdorf, and
   F.P. Mooring, Phys. Rev. 176 (1968) 1405
3) L. Drigo, C. Manduchi, G. Moschini, J.T. Russo-Manduchi,
   G. Tornielli, and G. Zannoni, *Il Nuovo Cimento*, 13A (1973) 867
4) D.C. De Martini, C.R. Soltesz, and T.R. Donoghue, Phys. Rev. 67
   (1973) 1824
5) L. Rosen, J.G. Beery, A.S. Goldhaber, and E.H. Auerbach, Ann. Phys.
   34 (1965) 96
6) H.S. Sherif, CAP Bull. 27 (1971) 70, and to be published

THE COMPOUND ELASTIC SCATTERING OF NEUTRONS
AND THE OPTICAL MODEL POLARIZATION

V.M. Morozov, N.S. Lebedeva, Yu.G. Zubov
Kurchatov Institute of Atomic Energy, Moscow, USSR

In accordance with the optical model concepts the neutron-nucleus elastic cross section experimentally observed in the region of non-overlapping resonances of the nucleus is the sum of two items

$$\sigma_e = \sigma_{se} + \sigma_{ce} \; . \tag{1}$$

The first describes the shape elastic cross section calculated with the optical potential, the second is the compound elastic cross section. Since only the shape elastic scattering is responsible for neutron polarization (at least for even-even nuclei) then experimentally observed polarization is represented by the following expression

$$P_e(\theta) = \frac{P_{se}(\theta) \, \sigma_{se}(\theta)}{\sigma_e(\theta)} \tag{2}$$

According to the widespread interpretation the presence of two items in (1) is connected with the existence of two physically distinct and therefore incoherent mechanisms of elastic scattering. The shape elastic scattering is the scattering by the nucleus taken as a whole. It should have diffraction structure and weak energy dependence. The angular distribution $\sigma_{ce}(\theta)$ should be symmetric with respect to the plane $\theta = \pi/2$, which is inherent to the decay of the free longliving system. The probability of the compound nucleus formation is assumed to be in direct dependence on the number and strength of resonances in the total cross section. The neutrons emitted to the elastic channel after the compound nucleus formation should come later to those undergone the shape elastic scattering.

There is an opinion that though the direct division of these processes is at present impossible there is an indirect evidence confirming the validity of such a physical picture: the fraction of σ_{ce} in σ_e increases for energy regions and nuclei for which the contribution from resonances is large, and the number of open inelastic channels is small.

The differences in angular and energy dependences of σ_{se} and σ_{ce} permit to verify the validity of the idea of the existence of two elastic scattering mechanisms. The neutron beam after passing the filter, that is the plate of the same substance as the sample under investigation, contains a comparatively small number of neutrons which have energies corresponding to the resonances in the region of the beam energy spread. Therefore in accordance with the idea of two mechanisms of elastic scattering the cross section and polarization observed with the filtered beam should have the following view:

$$\sigma_e^f(\theta) = \sigma_{se}(\theta) + \alpha\sigma_{ce}(\theta) \tag{3}$$

and

$$P^f_e(\theta) = \frac{P_{se}(\theta)\ \sigma_{se}(\theta)}{\sigma^f_e(\theta)} \qquad (4)$$

where $\alpha<1$ describes the degree of suppression of the compound nucleus formation. If the hypothesis of the mechanisms of elastic scattering is valid then the difference $\sigma_e(\theta) - \sigma^f_e(\theta)$ should be symmetric with respect to $\theta = \pi/2$ and $P^f_e(\theta)$ should exceed $P_e(\theta)$.

The experiments were performed in which the differential elastic cross sections were investigated for some nuclei with ordinary and filtered neutron beams [1,2]. The experimental results for ^{208}Pb and the neutron beam with the energy (1.8 ± 0.2) MeV (the beam polarization $\sim 25\%$) are listed in fig. 1. The filter transmission was ~ 0.1. The difference $\sigma_e(\theta) - \sigma^f_e(\theta)$ (curve III of fig. 1) has sharp diffraction structure and cannot be interpreted as the compound elastic cross section. The value of $P^f_e(\theta)$ is rather lower than $P_e(\theta)$ and in any case is not higher.

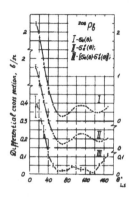 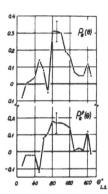

Fig. 1

The above experiment has shown that the supposition of the existence of two mechanisms of the neutron elastic scattering differing experimentally and comparable in their intensities is not valid. The experiment has provided an evidence of the coherence between shape elastic and compound elastic scattering. This is equivalent to the statement that such a division of the elastic cross section is formal. Therefore the lack of agreement between the model prediction and the experimental polarization data is not surprising.

References

1) V.M. Morozov, Yu.G. Zubov, N.S. Lebedeva, Yad. Fiz. <u>17</u> (1973) 734
2) V.M. Morozov, Yu.G. Zubov, N.S. Lebedeva, Yad. Fiz. <u>21</u> (1975) 945

DETERMINATION OF THE REACTION CROSS SECTION FOR P+^{90}ZR AROUND THE (P,N) THRESHOLD USING POLARIZED PROTONS[+]

W. Kretschmer, G. Böhner, E. Finckh

Physik. Inst. der Universität Erlangen-Nürnberg; Erlangen West-Germany

Usually the reaction cross section σ_R is measured at energies which are some MeV above the (p,n) threshold using the fact that the compound nucleus (CN) decays mainly via the neutron channels, so that $\sigma_R \approx \sigma_{pn}$. In the energy range around the (p,n) threshold σ_R is dominated by the compound elastic cross section σ^{CE} which is a part of the experimental cross section. The use of polarized protons enables a rather model independent separation of the shape elastic (or direct) and the compoundelastic cross section[1] and hence an inclusion of σ^{CE} to the reaction cross section. This method is based on a careful determination of the energy-averaged proton elastic scattering matrix $\langle S_{cc} \rangle$ obtained from an analysis of the polarization dependent cross section $\langle \sigma A \rangle$.

This concept was applied to the proton induced CN-reaction on the target nucleus ^{90}Zr(Q_{pn}= -6.97 MeV) in an energy range (4.5 - 9 MeV) where many isobaric analog resonances (IAR) are excited. Figure 1 shows an angular distribution of $\langle \sigma A \rangle$ for the reaction ^{90}Zr($\vec{p},p_0$) at an off resonant energy E_p= 6.5 MeV with an optical model fit curve. The optical model was obtained in an iterative way insofar as first $\langle \sigma A \rangle$ was fitted and than after the extraction of σ^{CE} both $\langle \sigma A \rangle$ and $\sigma_R = \sigma^{CE} + \sum_{p \neq p_0} \sigma_{pp'}$ were fitted simultaneonsly. In this way the ambiguities of the optical model were reduced substantially. The striking feature of this potential at 6.5 MeV (V_R= 60, r_R= 1.13, a_R= 0.75; W_D= 9.8, r_I= 1.2, a_I= 0.40; V_{SO}= 6.2, r_{SO}= 1.01, a_{SO}= 0.87; r_C= 1.21 in the usual units) is the drastic reduction of the diffuseness a_I of the imaginary potential.

Figure 2 shows the experimental magnitudes $\langle \sigma A \rangle$, $\sigma^{CE}(\theta)$ and σ_R as a sum of σ_{pn}, $\sigma_{pp'}$ and σ^{CE} together with calculated curves. The IAR's designed in the figure by arrows, are described in the usual way by a resonance term (resonance energy E_R , partial width Γ_p, total width Γ and resonance mixing phase ϕ_R) in $\langle S_{CC} \rangle$. The parameters of the most pronounced resonances are given in table 1, the weaker IAR's also contributing to the structure of $\langle \sigma A \rangle$ are investigated in more detail in the future.

Table 1: Resonance paraters of the IAR's observed in ^{90}Zr($\vec{p},p_0$)

E_R^{lab} (MeV)	J^{π}	Γ_p (keV)	Γ (keV)	ϕ_R (rad)	
4.707	$5/2^+$	3.8	24	.09	4)
5.92	$1/2^+$	44	88	.04	
6.81	$3/2^+$	17	53	.02	
7.027	$7/2^+$	1.0	20	-.04	
7.66	$3/2^+$	2.4	36	.04	
7.86	$3/2^+$	6.4	48	.02	
8.09	$3/2^+$	5.0	45	0	*
8.46	$3/2^+$	2.5	30	0	*

* preliminary results

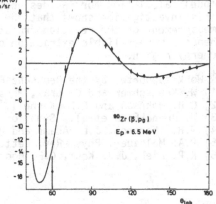

Fig.1: angular distribution of $\langle \sigma A \rangle$ for ^{90}Zr($\vec{p},p_0$) at E_p=6.5MeV

612

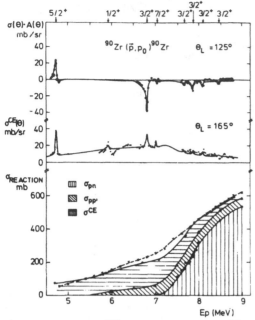

Fig. 2: $\langle\sigma A\rangle$, $\sigma^{CE}(\theta)$ and σ_R for p+^{90}Zr

For simplicity the optical potential was taken to be energy independent besides the imaginary diffuseness a_I (0.4 for $5 < E_p < 7$MeV, 0.5 - 0.55 for $7 < E_p < 8$MeV 0.6 for $8 < E_p < 9$MeV). A similar energy dependence of a_I has been observed for the tin isotopes by Johnson and Kernell[2] and by Drenckhahn et al.[3]. The compound elastic cross section for θ=165° is shown in the middle of figure 2. In the region of the $s_{1/2}$ IAR at 5.92MeV σ^{CE} was obtained by subtracting the calculated shape-elastic cross section from the experimental cross section given in ref. 4. The compound elastic cross section shows the expected energy behaviour: it increases slowly with increasing energy due to the better transmission through the Coulomb barrier and it decreases for $E_p > Q_{pn}$ (effective) $\approx$7.2MeV. The fluctuations in $\sigma^{CE}(\theta)$ for $E_p > 7.5$MeV may

be caused by additional IAR's which are not yet taken into account. Due to the isospin mixing σ^{CE} is enhanced in the region of the IAR's. The solid curve is a Hauser Feshbach calculation, where the proton transmission coefficients are determined by the above mentioned optical potential and the IAR-parameters, and the neutron transmission coefficents by a neutron potential given by Moldauer[5]. The lower part of fig. 2 shows the experimental reaction cross section (taken at off-resonant energies) as a sum of σ_{pn}(measured with 3 longcounters, $\Delta\sigma \approx$ 15-20%), $\sigma_{pp'}$ given by Lieb et al.[5]($\Delta\sigma \approx$ 10%) and σ^{CE} ($\Delta\sigma \approx$10-15%), compared with an optical model calculation for σ_R (designed by x--x--x in fig. 2).
This investigation shows that the use of polarized protons enables a measurement of the reaction cross section below the (p,n) threshold and hence a more reliable extraction of optical model parameters in the low energy region.

+ Work supported by the Deutsche Forschungsgemeinschaft
1) W. Kretschmer and G. Graw, Phys. Rev. Lett. 27 (1971) 1294
2) C.H. Johnson and R.L. Kernell, Phys. Rev. C2 (1970) 639
3) W. Drenckhahn et al., Symp. on Polar. Phenomena, Zürich 1975, p.613
4) W.H. Thompson, J.L. Adams and D. Robson, Phys. Rev. 173 (1968) 975
5) P.A. Moldauer, Phys. Rev. Lett. 9(1962) 17
6) K.P. Lieb, J.J. Kent, C.F. Moore, Phys. Rev. 175(1968) 1482

OPTICAL MODEL PARAMETERS FOR ENERGIES BELOW THE COULOMB BARRIER*

W. Drenckhahn, A. Feigel, E. Finckh, R. Kempf, M. Koenner,
P. Krämmer and K.-H. Uebel

Physikalisches Institut der Universität Erlangen-Nürnberg
D 8520 Erlangen, West Germany

The parameters of the optical model are usually determined by measurements of the elastic cross section and the polarization at energies around and above the Coulomb barrier. For the analysis of compound reactions, these parameters are extrapolated to lower energies. In this energy range, the nuclear effects are seen much stronger in the absorption than in the scattering and polarization. About 1 MeV above the neutron threshold, the absorption cross section can be well approximated by the (p,n) cross section because the emission of charged particles is strongly hindered. For the comparison of measured and calculated absorption cross sections, both are divided by the transmission through the coulomb and angular momentum barrier. This reduced cross section corresponds to the neutron strength function, but all partial waves are added.

We have measured the (p,n) cross section of the even tin isotopes for E_p = 4 to 10 MeV, the differential cross section and the analyzing power of the elastic proton scattering at E_p = 8.8 MeV. Our (p,n) data agree with the results of Johnson and Kernell (1) who measured only from E_p = 3 to 5.5 MeV, and the reduced cross section is well reproduced in the energy dependence and absolute value by calculations using their parameter set C or D, see fig. 1a. These sets have surface absorption with a rather small diffuseness since the absorption cross section could only be fitted by this change. Fig. 1b shows that the global set of Becchetti and Greenlees (2) can not reproduce the absorption data, neither by reducing the imaginary potential.

The differential cross section of the elastic scattering at E_p = 8.8 MeV is in agreement with the calculations using the global parameters of Becchetti and Greenlees and set D of Johnson and Kernell. Set C with a diffuseness of the imaginary potential of a_w = 0.24 fm can be excluded since the diffraction pattern oscillates too strongly.

In fig. 2, the analyzing power is compared with calculations. The best over all agreement with the data for the three isotopes and the different angles is obtained using set D. Below the Coulomb barrier, it has an energy dependent diffuseness, a_w = 0.093 + 0.052·E, the normal value is reached around 12 MeV. For energies around the Coulomb barrier, a reduction in the value of a_w has also been found by Rathmell and Haeberli (3). In the tin isotopes, only an energy dependent diffuseness of the imaginary potential can reproduce the absorption cross section as well as the scattering and polarization data.

References:
(1) C.H. Johnson and R.L. Kernell, Phys. Rev. C2 (1970) 639

(2) F.D. Becchetti, Jr., and G.W. Greenlees, Phys. Rev. 182 (1969) 1190
    G.W. Greenlees et al. Phys. Rev. C3 (1971) 1231

(3) R.D. Rathmell and W. Haeberli, Nucl. Phys. A178 (1972) 458

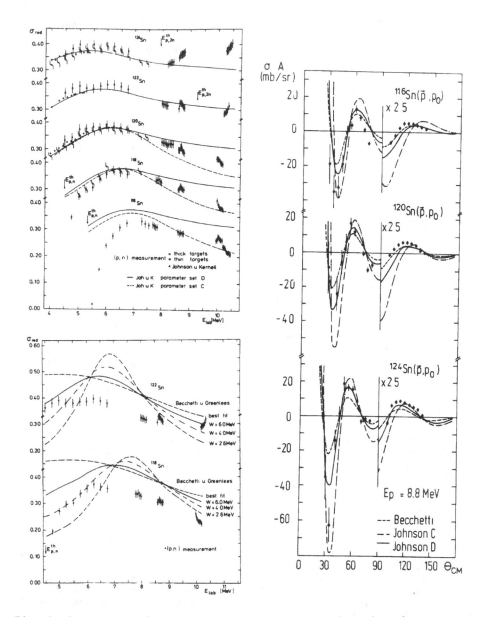

Fig. 1: Comparison of measured and calculated reduced cross sections, optical model parameters of Johnson and Kernell (a) and of Becchetti and Greenlees (b) with different values of the imaginary potential.

Fig. 2: Polarized (p,p₀) cross sections for three tin isotopes compared with different optical model calculations (see fig 1).

* Work supported by Deutsche Forschungsgemeinschaft

ANGULAR DISTRIBUTIONS OF THE POLARIZATION
ELASTICALLY SCATTERED PROTONS

N.N. Putcherov, S.V. Romanovsky, T.D. Chesnokova
Institute of Nuclear Research, Kiev, URSS

As it has been shown by Spencer[1] polarization of protons elastically scattered from a spin-zero target in the theory of complex angular
momentum could be written

$$P(\theta) = -2\Gamma e^{\beta\theta}[\text{Sin}(K\theta+F+\phi)+D\text{Sin } \phi]/A(\theta) \qquad (1)$$

where

$$A(\theta) = [1+\Gamma^2 e^{\beta\theta}]D+\text{Cos}(K\theta+F)-\Gamma^2 e^{\beta\theta}\cdot\text{Cos}(K\theta+F+2\phi)$$

Γ,β,F,K,D,ϕ - the parameters of calculations.

The parameters of equation (1) are slightly different from original. A
similar expression has been obtained by Gubkin[2]. We have made an attempt
to fit eq.(1) to experimental data in the widely range of mass numbers
(27-209) and energy (5-50 MeV). About 150 angular distributions of polarization have been investigated.

The first time we changed the parameters Γ,β,K,F,D. The calculations
have shown a good agreement of eq.(1) with experimental data of the
elastic scattering of protons. Particularly the regularity of the polarization patterns, i.e. maximum or a minimum and their position are well
fitted. The best mean values of the parameters $\Gamma=0,2$ $\beta=0,6$ are in all
cases.

In the second part of the calculations the parameters Γ and β were
fixed and other were changed. It has been shown that

$$K = 2kA^{1/3} \qquad (2)$$

where k is the wave number and A is the mass number. The parameter K de-
fines the periodicity of the patterns of the angular dependence of the
polarization.

Let the spin-orbit interaction be small. Adding it to the central
potential leads to the turn of the angular distribution. Zeros of the
polarization will be set where there were maximum or minimum differential
cross section. The periodicity of the oscillation is clearly observed.
The parameter F is the energy and mass number function. It has been shown
empirically that

$$F = -(6+22,25 \text{ } k/A^{1/3}) \text{ } . \qquad (3)$$

For protons the parameter D approaches the value

$$D = \exp [1,5/(E/B)^2] \qquad (4)$$

Where E is energy of protons and B is energy of Coulomb barrier. Thus,
only one parameter ϕ remains in the eq. (1). It means a lot in the
energy region higher 20 MeV.

References

1) M.B. Spencer, Nucl. Phys. A102 (1967) 545
2) I.A. Gubkin, Nucl. Phys. A134 (1969) 209

PRECISION T_{20} AND T_{22} DATA IN FORWARD HEMISPHERE FOR POLARIZED DEUTERON ELASTIC SCATTERING ON ^{50}Cr AND ^{66,67}Zn

O. Karban, A.K. Basak, J.A.R. Griffith, S. Roman and G. Tungate

Department of Physics, University of Birmingham, Birmingham B15 2TT, England

As a by-product of our studies of the (d,p) reaction tensor analysing powers on ^{50}Cr, ^{67}Zn and recently ^{66}Zn with 12.3 MeV polarized deuteron beam, large number of counts was accumulated in the deuteron elastic channel, particularly at angles forward of 60°. Thus, precision data on the T_{20} and T_{22} tensor analysing powers (TAP) were obtained for medium weight nuclei in an angular region where the polarization effects at deuteron energies about 12 MeV are usually small and considered of no particular interest due to the dominant Coulomb scattering.

A standard experimental technique described, e.g. in ref.1, was used to detect deuterons elastically scattered from ≈2 mg/cm² targets at four azimuthal angles φ=0, 90, 180 and 270°. The experimental results for ^{67}Zn are shown in fig.1 together with the vector analysing power (VAP) measured separately with a pure vector polarized beam. While the experimental iT_{11} and T_{20} follow a regular oscillatory pattern, the T_{22} data show an unexpected behaviour in the region forward of Θ_{cm}=60°, which was first observed on ^{50}Cr and later confirmed on both ^{66}Zn and ^{67}Zn. The appearance of an additional minimum in T_{22} at Θ_{cm}~30° is well outside statistical errors of ±0.001 and might be of a physical importance since it certainly cannot be reproduced by a standard optical model calculation. A comparison of all ^{66}Zn and ^{67}Zn data shows that they are within statistical errors identical, which excludes any target-spin effects.

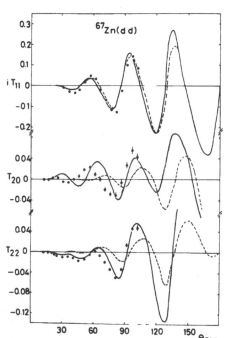

In order to investigate a possible source generating the 30° minimum in T_{22}, an optical model programme[2]) with tensor terms[3]) T_R and T_L was used to analyse the experimental data. First, a standard optical model programme with an automatic search was employed to fit the cross section and VAP data. The results are shown as dashed lines in fig.1. It is apparent that there is not even a remote fit to the TAP's with a potential resulting from the cross section and VAP analysis. Difficulties similar to those described in ref.3 were encountered when searching on T_{20} and T_{22} separately, since the experimental VAP and TAP's oscillate in phase, while the standard optical model predictions for TAP's are always shifted towards larger angles.

The optical model calculation with the tensor term T_R requires coupling of partial waves differing by $\Delta L=2$ and it is therefore substan-

Fig.1. Optical model predictions without (dashed lines) and with the tensor potential (solid lines).

ially longer than a standard calculation, making any automatic search
rather unpractical.   Thus, a grid search on six tensor parameters
(strengths and geometries of both T_R and T_L terms) was undertaken to find
a combination, which would affect T_{20} and T_{22} while maintaining a good
fit to the cross section and VAP.   This requirement practically excludes
the T_L potential and results in the following parameters for T_R corresponding to predictions in fig. 1 (solid lines): V_R=0.30, r_T=1.05 and
a_T=0.636.

There is a distinct improvement to the TAP's fits, in particular to
the T_{22} angular distribution, when the tensor term T_R is included in the
calculation.   Moreover, the 30° minimum is sensitive to the geometrical
parameters of the T_R term, with a strong preference for the central real
geometry.   The form of the T_R potential used and its radial dependence
is that suggested in ref.3, i.e.

$$T_R = - V_R \, 4 \, a_T^2 \, r^2 d_g^2(r)/dr^2 \left[(\underline{s} \cdot \underline{r})^2/r^2 - 2/3 \right]$$

where g(r) is the Saxon-Woods form factor.   No need for introducing an
imaginary tensor potential[4]) was found.

The present results can be interpreted as a direct experimental confirmation for an existence of the tensor term T_R in a deuteron-nucleus
potential.   Moreover, the requirement for a central-real potential geometry in T_R supports the validity of the folding model for a deuteron
potential[5]).   It was also shown in ref.5 that such a potential is a
direct consequence of the deuteron D-state since without the latter the
T_R vanishes during the folding procedure.   The fact that the elastic
data are affected particularly at small scattering angles could reflect
a long-range character of the tensor potential.

References

1) A.K. Basak, et al., Nucl.Phys. A229 (1974) 219
2) B.A. Robson, private communication
3) M. Irshad and B.A. Robson, Nucl.Phys. A218 (1974) 504
4) L.D. Knutson and W. Haeberli, Phys.Rev.Lett. 30 (1973) 986
5) P.W. Keaton and D.D. Armstrong, Phys.Rev. C8 (1973) 1692.

MASS-NUMBER SYSTEMATICS OF DEUTERON FOLDING-MODEL POTENTIALS*

J. A. Ramirez and W. J. Thompson
University of North Carolina, Chapel Hill, N. C. 27514 USA &
Triangle Universities Nuclear Laboratory, Durham, N. C.

The real parts of the optical model (OM) potential for deuterons, as calculated in the folding model (FM), [1,2] have been fitted to analytic forms convenient for use in standard OM codes. The target mass number range $6 \leq A \leq 216$ was used in a linear least-squares fit of each of the FM potential parameters as a function of $A^{1/3}$ for 15-MeV deuterons. Differences between observables calculated with FM potentials and with fitted potentials were <2%. These potentials can be used as starting values for OM searches.

The real part of the FM potential includes central, spin-orbit, and spin-radial tensor terms

$$V_d(\vec{R}) = V_d^c(R) + V_d^{so}(R)\,\vec{L}\cdot\vec{S} + V_d^{T_R}(R)\left[(\vec{S}\cdot\vec{R})^2 - \tfrac{2}{3}\right]. \quad (1)$$

The folding integrals were performed numerically using a modified version of code FMODEL.[3] Two parametrizations of the nucleon-nucleus OM potentials were used; that of Watson et al.[4] for $6 \leq A \leq 16$, and that of Becchetti and Greenlees[5] for $40 \leq A \leq 216$. The nucleon potentials were evaluated at an energy of 7.5 MeV.

Three internal deuteron wave functions with distinctly different short-range properties were used; Reid's soft-core, Reid's hard-core[6] and Erkelenz et al.'s OBEP model.[7] However, all give the same deuteron scattering potentials to within 1%.

The central potential V_d^c (R) was fitted to a Woods-Saxon potential

$$V_d^c(R) = \frac{-V_0}{1 + \exp(x)}\,, \quad x = \frac{R - R_0}{a_0}\,, \quad R_0 = r_0 A^{1/3}\,, \quad (2)$$

the spin-orbit potential V_d^{so} (R) was fitted to

$$V_d^{so}(R) = \frac{-2 V_{so}\exp(x)}{a_{so} R (1 + \exp(x))^2}\,, \quad x = \frac{R - R_{so}}{a_{so}}\,, \quad R_{so} = r_{so} A^{1/3}\,, \quad (3)$$

and the radial-spin tensor potential to

$$V_d^{T_R}(R) = -2 V_{T_R} R \frac{d}{dR}\left[\frac{1}{R}\frac{d}{dR}(1 + \exp(x))^{-1}\right], \quad x = \frac{R - R_{T_R}}{a_{T_R}}$$

$$= \frac{-2 V_{T_R}\exp(x)}{a_{T_R}^2(1 + \exp(x))^2}\left[\frac{a_{T_R}}{R} - \frac{1 - \exp(x)}{1 + \exp(x)}\right]. \quad R_{T_R} = r_{T_R} A^{1/3} \quad (4)$$

Volume integrals per nucleon, whose significance for central potentials is discussed in ref.[8], are given in table 1. These integrals are approximately 20% higher for the central potential and agree very well for the spin-orbit potential, in comparison with phenomenological potentials.[9] The present integrals vary smoothly with A and can be used to restrict the parameters for 16<A<40.

Table 1. Parameters for deuteron FM potential ($c=A^{1/3}$).

Parameter		$6\leq A\leq16$	$40\leq A\leq216$
V_0	(MeV)	$52\pm3+(20\pm1)c$	$89\pm2+(3.4\pm.3)c$
a_0	(fm)	$.705\pm.005+(.042\pm.003)c$	$.96\pm.01+(.004\pm.002)c$
r_0	(fm)	$1.11\pm.001$	$1.14\pm.005$
J_0	(MeV.fm^3)	$593c^{-2}+299c^{-1}+178+58c$	$1933c^{-2}+90c^{-1}+277+11c$
V_{so}	(MeV.fm^2)	$1.7\pm.1+(1.10\pm.06)c$	$4.7\pm.2+(.21\pm.03)c$
a_{so}	(fm)	$.55\pm.01+(.101\pm.005)c$	$.88\pm.02+(.024\pm.004)c$
r_{so}	(fm)	$1.13\pm.02$	$.98\pm.0002$
J_{so}	(MeV.fm^3)	$24c^{-2}+16c^{-1}$	$57c^{-2}+2.5c^{-1}$
V_{TR}	(MeV.fm^2)	$4.1\pm.1+(1.54\pm.04)c$	$7.5\pm.2+(.17\pm.04)c$
a_{TR}	(fm)	$.68\pm.02+(.047\pm.008)c$	$.94\pm.02+(.011\pm.003)c$
r_{TR}	(fm)	$1.11\pm.01$	$1.12\pm.01$
J_{TR}	(MeV.fm^3)	$172c^{-2}+65c^{-1}$	$317c^{-2}+7.2c^{-1}$

In table 1 the error bars are the standard deviations from the linear-squares fits of the fitted analytic forms to $A^{1/3}$. The radius parameters r_0, r_{so}, r_{TR} are independent of A to within 1%.

The above values are for 15-MeV deuterons, but can be approximately scaled to other deuteron energies by using the appropriate energy dependences[4,5] of the nucleon-nucleus central-potential depths.

We thank Dr. L. D. Knutson for the folding-model code.

References

*Supported in part by the U.S.E.R.D.A. and by the UNC Research Council.

1) S. Watanabe, Nucl. Phys. 8 (1958) 484
2) P. W. Keaton and D. D. Armstrong, Phys. Rev.C8 (1973) 1692;
   P. W. Keaton, et al., LASL Report No. LA-4379-MS, 1970, unpublished
3) L. D. Knutson, Univ. of Wisconsin 1974, private communication
4) B. A. Watson, P. P. Singh and R. E. Segel, Phys. Rev. 182 (1969) 977
5) F. D. Becchetti and G. W. Greenlees, Phys. Rev. 182 (1969) 1190
6) R. V. Reid, Ann. Phys. (N. Y.) 50 (1968) 411
7) K. Erkelenz, K. Holinde and K. Bleuler, Nucl. Phys. A139 (1969) 308
8) G. W. Greenlees, G. J. Pyle and Y. C. Tang, Phys. Rev. C1 (1970) 1145
9) J. M. Lohr and W. Haeberli, Nucl. Phys. A232 (1974) 381; J. Childs
   and W. W. Daehnick, Bull. Am. Phys. Soc. 20 (1975) 626

COMPOUND–NUCLEAR POLARIZATION EFFECTS IN
DEUTERON ELASTIC SCATTERING*

R. J. Eastgate, T. B. Clegg, R. F. Haglund & W. J. Thompson
University of North Carolina, Chapel Hill, N. C. 27514 USA &
Triangle Universities Nuclear Laboratory, Durham, N. C.

Compound-elastic (CE) effects in low-energy, energy-averaged
deuteron elastic scattering are shown to be as important as those of the
spin-radial-tensor (T_R) coupling term of the optical-model (OM)
potential.

Data on ^{28}Si$(d,d)^{28}$Si, obtained at TUNL, are shown in fig. 1.
Measurements were made at 6.75, 7.25, and 7.75 MeV with a 0.5-MeV-thick
natural-Si wafer and combined to produce an effective energy-averaging
interval of 1.5 MeV, centered on 7.0 MeV.

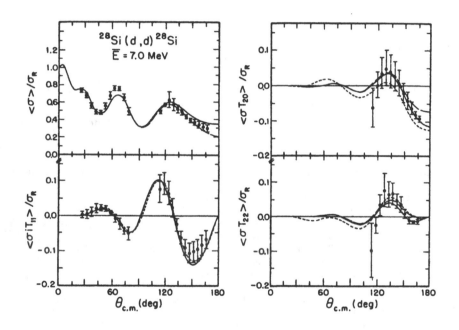

Fig. 1 Energy-averaged data are plotted ratio-to-Rutherford, as are
fits with OM (— —), with OM plus CE (—), and with OM including T_R
potential (— — —). Error bars are standard deviations among the 3
combined energies; statistical errors are negligible.

For energy-averaged data, the random-phase approximation (RPA)
implies that averages of products, either of CE amplitudes with
differing angular momenta and parity, or of CE and OM amplitudes,
vanish.

Thus, in RPA,

$$\langle \sigma T_{kq} \rangle = (\sigma T_{kq})_{OM} + \langle \sigma T_{kq} \rangle_{CE}, \quad k = 0, 2, \tag{1}$$

$$\langle \sigma i T_{11} \rangle = (\sigma i T_{11})_{OM},$$

with $T_{00} = 1$. For spin-s particles scattered at c.m. angle θ the CE analyzing cross sections in RPA are[1]), in the Madison convention,

$$\langle \sigma T_{kq} \rangle_{CE} = (-)^{k+q} \sum_{|q'| \le k} \langle \sigma t_{kq'} \rangle_{CE}\, d^{(k)}_{q'q}(\theta), \tag{2}$$

where

$$\langle \sigma t_{kq'} \rangle_{CE} = \frac{\lambda^2 (2k+1)^{1/2}}{16\pi^2(2s+1)^{1/2}} \sum_{J\ell\ell'} (2J+1)^2 (2\ell+1)(2\ell'+1)\, T^J_{\ell'\ell} \times$$

$$\times \sum_{m m_1} (-)^{m_1-s} \begin{pmatrix} \ell & s & J \\ 0 & m & -m \end{pmatrix}^2 \begin{pmatrix} s & s & k \\ m_1 & q-m_1 & -q \end{pmatrix} \begin{pmatrix} \ell' & s & J \\ m-m_1 & m_1 & -m \end{pmatrix} \times$$

$$\times \begin{pmatrix} \ell' & s & J \\ m-m_1+q & m_1-q & -m \end{pmatrix} C^*_{\ell'm-m_1}(\theta)\, C_{\ell'm-m_1+q}(\theta). \tag{3}$$

Here the $d^{(k)}_{q'q}(\theta)$ are reduced rotation matrices. λ is the entrance-channel wavelength, and 3-j symbols and modified spherical harmonics $C_{\ell m}$ appear. Compound-nucleus (CN) information is contained in the $T^J_{\ell'\ell}$, for which we use a modified Hauser-Feshbach formula[1,2,3]) involving OM transmission coefficients and a Fermi-gas, level-density formula[4]).

Previous estimates of OM parameters were refined by optimizing the fit to $< \sigma i T_{11} >$. CE parameters, in good agreement with level-density estimates[4]), were obtained by requiring a good fit to $<\sigma T_{20} >/\sigma_R$, and a fair fit to $< \sigma >/\sigma_R$, at backward angles. The predicted angular distribution of $<\sigma T_{22}>/\sigma_R$ agrees closely with the data.

A spin-radial-tensor potential V_{TR} was calculated in a folding-model approximation, then fitted to an analytic form factor. Using nucleon-scattering potentials from Watson et al.[5])

$$V_{TR}(r) = -V_R\, r \frac{d}{dr} \left\{ \frac{1}{r} \frac{d}{dr} \left[1 + \exp\left(\frac{r-R_R}{a_R}\right) \right]^{-1} \right\}, \tag{4}$$

with $V_R = 8.5$ MeV.fm^2, $R_R = 3.25$ fm, $a_R = 0.81$ fm. The changes in the OM cross sections and vector analyzing powers are negligible, but the effects on T_{20} at backward angles are similar to CE effects. However, V_{TR} did not reproduce the negative values of T_{22} at backward angles. Thus, V_{TR} and CE effects are of comparable importance for the system studied. Similar effects are expected for the 4n target nuclei with $A \le 40$ at deuteron energies below 10 MeV.

References
*Supported in part by the U.S.E.R.D.A. and by the UNC Research Council.

1) R. J. Eastgate, Ph.D. dissertation, UNC 1975, unpublished
2) K. A. Eberhard et al., Nucl. Phys. A125 (1969) 673
3) P. A. Moldauer, Phys. Rev. C11 (1975) 426
4) A. Gilbert & A. G. W. Cameron, Can. J. Phys. 43 (1965) 1248, 1446
5) B. A. Watson et al., Phys. Rev. 182 (1969) 977

THE OPTICAL MODEL AND BACKANGLE SCATTERING OF DEUTERONS FROM ^{40}A

H.R. Bürgi, W.Grüebler, P.A. Schmelzbach, V. König, B. Jenny,
R. Risler and W.G. Weitkamp

Laboratorium für Kernphysik, Eidg. Techn. Hochschule, 8049 Zürich
Switzerland

Polarization phenomena in the elastic scattering of deuterons by complex nuclei have been investigated in the past years in many laboratories. Analyses in terms of optical-model potentials have been carried out in order to find the important tensor terms in deuteron optical-model. While an optical potential including a vector spin-orbit term has been successful in describing the differential cross section σ_0 as well as the vector analysing power iT_{11} data for many nuclei and a large range of deuteron bombarding energies, the experimental results for the three tensor analysing powers T_{20}, T_{21}, T_{22} have not been reproduced satisfactorily. In particular, it seems to be difficult to fit the data at backward angles. A satisfactory fit in this angular range seems the more important, because the values of the polarization observables are in general larger than at forward angles and the analyses more sensitive to the possible tensor interactions, since they are not diluted by Coulomb scattering. On the other hand the compound nucleus reaction may also contribute more strongly at backward angles.

A recent analysis studied the elastic scattering of deuterons on ^{40}A at an energy of 10.75 MeV[1]. The data between 10 and 11 MeV were found to be insensitive to energy variation in a angular range up to θ_{lab} = 120°; therefore the compound nucleus contribution was expected to be small. The five observables σ_0, iT_{11}, T_{20}, T_{21} and T_{22} were measured at this laboratory in a large angular range. An extended investigation of the influence of the optical potential tensor terms was performed. The tensor terms T_R and T_L were included in the calculation. Particular care was given to investigating of the sensitivity of the different observables to the optical potential tensor terms and to finding the sensitive angular regions in order to be able to measure the corresponding quantities more accurately in a future experiment.

The result of this investigation for T_{20} is shown in fig. 1. The dotted curve shows the best fit using an optical potential without any tensor term but including the spin-orbit interaction. The dashed line is the best fit found by adding a T_L term, and the curve with long dashes represents the best result with an additional T_R term alone. The solid line finally is the best fit with both tensor terms. Substantial sensitivity of the optical potential tensor terms to the tensor analysing powers T_{20} at 180° and T_{21} between 150° and 175° are observed.

In order to obtain from these predictions information about the nature of the tensor interaction, measurements at θ_{lab} = 175° between 8.8 and 12.4 MeV have been carried out. The results are shown in fig. 2. The measurement at 10.75 MeV agrees fairly well with the prediction of the solid curve (T_R and T_L term) of fig. 1. The curves shown in fig. 2 are calculated with the potential parameters found at 10.75 MeV. The same signatures are used as in fig 1. Although one cannot expect the optical-model to reproduce the fluctuations, it is remarkable that the general behavior of the T_{20}, the decrease from + 0.1 to - 0.9 is in accordance with the calculations. Experimentally a similar decrease of the cross section in the same energy range from about 3.0 to 0.2 mb/sr was observed, which also is in general agreement with the calculations. In spite of this result, one has to be very careful in extracting tensor interaction information from data at only one energy. An analysis requires complete data sets at several energies and explanation of the variation of the observables as a function of energy.

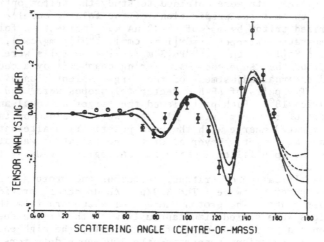

Fig.1. Tensor analysing power T_{20} for elastic deuteron scattering by ^{40}A. The curves are explained in the text.

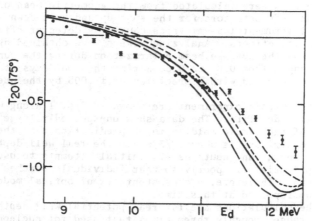

Fig.2. Excitation function of the analysing power T_{20} at θ_{lab} = 175°. The curves are explained in the text.

References

1) H.R. Bürgi, W. Grüebler, P.A. Schmelzbach, V. König and R. Risler, Nucl. Phys. A247 (1975) 322

ELASTIC SCATTERING OF POLARIZED TRITONS FROM
^{52}Cr, ^{60}Ni, ^{90}Zr, ^{116}Sn, and ^{208}Pb*

R.A. Hardekopf, L.R. Veeser, and P.W. Keaton, Jr.
Los Alamos Scientific Laboratory, Los Alamos, New Mexico 87545, USA

Angular distributions of the triton analyzing power at 15.0 MeV for five target nuclei of atomic masses 52, 60, 90, 116, and 208 are presented. These initial data were obtained to study the triton optical-model potential and to test the predications of the folding model[1] for tritons.

A polarized triton beam[2] of 20-40 nA was incident on foils of the following separated isotopes: ^{52}Cr[1 mg/cm^2], ^{60}Ni[1 mg/cm^2], ^{90}Zr-[5 mg/cm^2], ^{116}Sn[10 mg/cm^2], ^{208}Pb[3 mg/cm^2]. The foils were suspended in the center of the "supercube" scattering chamber[3] by a mechanism that allowed external adjustment of the target orientation with respect to the beam. Two pairs of ΔE-E detector telescopes were used with angular separation 18°. Each pair viewed the target at equal angles right and left of the beam axis. All data were obtained by taking runs of equal integrated charge with the beam polarization alternately "up" and "down" at the target. Reversal of the polarization was accomplished by reversing the spin filter and ionization region fields at the polarized triton source.

Mass identification of tritons, deuterons and protons was accomplished by computer software. The 300/μm ΔE detectors provided well-separated mass peaks. The proton and deuteron spectra from the (t,p) and (t,d) reactions were stored simultaneously with the triton spectra but have not been analyzed. For most of the targets, the high energy protons from (t,p), reactions penetrated the 1000 μm E detectors and these spectra are not useful. Inelastic triton peaks were also observed in the spectra but have not been analyzed.

The asymmetries were calculated from the geometric mean of counts in the right and left detectors for the spin up and spin down runs. This method cancels instrumental asymmetries caused by detector efficiency and current integration effects. Analyzing powers were obtained by dividing the asymmetries by the average beam polarization during the runs. Beam polarizations ranged from 0.80 to 0.86 during the four days of measurements and were determined with a precision of $\pm$ .005 by the quench-ratio method[4]

The results of this experiment are shown in Fig. 1 along with preliminary optical model fits. The data show unexpectedly large analyzing powers in view of the simple folding model predication that the spin-orbit potential for tritons is about 1/3 while the real well depth is three times that for protons and neutrons[5]. Initial attempts to use the folding model potential fit poorly so that individual searches on each element are shown. Therefore, a consistent set of optical model parameters cannot be reported at this time.

Most of the calculations are for real potential well depths in the 150 to 170 MeV range, roughly three times that used for nucleon scattering. For ^{90}Zr, good fits were found using either a volume or a surface imaginary potential and spin orbit potentials of 8.1 and 5.1 MeV respectively. For ^{52}Cr and ^{116}Sn, volume absorption fit the data better than surface absorption. For ^{60}Ni the fits were poorer. For ^{208}Pb, the triton energy is so far below the Coulomb barrier that the polarization and the structure in the cross section angular distribution are too small to determine optical model parameters. The spin-orbit well depth, VSO, is in the range 5-16 MeV for the fits shown.

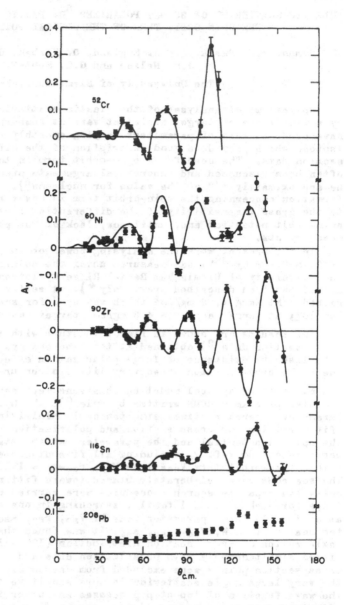

Fig. 1 Elastic
scattering triton
analyzing powers
are shown. The
lines are optical
model fits to
four of the five
nuclei studied.

References

* Supported by the U.S. Energy Research and Development Administration
1) P.W. Keaton, Jr., E. Aufdembrink, and L.R. Veeser, Los Alamos
   Scientific Laboratory Report LA-4379-MS (1970)
2) R.A. Hardekopf, these proceedings, p.865
3) G.G. Ohlsen, these proceedings, p. 907
4) R.A. Hardekopf, G.G. Ohlsen, R.V. Poore, and N. Jarmie, these
   proceedings, p. 903
5) P.W. Keaton Jr., these proceedings, p. 173

ELASTIC SCATTERING OF 33 MeV POLARIZED ^{3}He PARTICLES BY ^{58}Ni AND THE SPIN-ORBIT TERM OF THE OPTICAL POTENTIAL

S. Roman, A.K. Basak, J.B.A. England, O. Karban, G.C. Morrison, J.M. Nelson and G.G. Shute[+]

The University of Birmingham, England

Optical model analyses of the elastic scattering of ^{3}He particles by a wide range of target nuclei at various bombarding energies [1-3] have produced sets of parameters varying smoothly from isotope to isotope which provide a good description of the differential cross section data. The need for a spin-orbit term in the potential has often been discussed and theoretical arguments predict its depth to be approximately 1/3 of the value for nucleons[5]. The empirical information concerning the spin-orbit term is however inconclusive, due to the gross insensitivity of the differential cross sections to the spin-orbit potential and, until now, lack of ^{3}He polarization measurements.

In the present work the analysing power of the elastic scattering of 33 MeV ^{3}He by ^{58}Ni was measured using the polarized ^{3}He beam from the University of Birmingham Radial Ridge Cyclotron. The experimental method has been described previously[4]. A selfsupporting, 99 % enriched ^{58}Ni target, 2 mg/cm^2 thick was used for scattering angles below 30°; at larger angles a 4.5 mg/cm^2 target was used.

The results are shown in fig. 1 together with the differential cross section data [2] obtained at the same energy. The measurements establish the existence of large polarization effects in a region where the cross section shows very little structure.

Exploratory optical model calculations were performed using the computer program RAROMP written by Pyle [6], which includes automatic parameter search routines simultaneously minimising the goodness-of--fit parameter for cross section and polarization data. The form of the optical potential and the parameters at the start of the searches were taken from ref. 2, including all five discrete potential families with a spin-orbit term V_{so}= 4.25 MeV, r_{so}=r_R = 1.110, a_{so}= 0.45 fm. The searches were deliberately biased toward fitting the polarization data. Two separate search procedures were carried out in multiple loops for each potential family, searching in one or two parameters at a time: 1) Three parameter search: V_{so}, r_{so}, a_{so}; 2) Five parameter search: V_R, W, V_{so}, r_{so}, a_{so}. It was found that the very large angle region of cross section distribution (130 - 170°) could not be fitted simultaneously with polarization data and subsequently these cross section points were excluded from searches. It is likely that the very large angle scattering is more sensitive to contributions to the wave function of two-step processes and other interference effects.

The calculations showed that the optical model can successfully describe the main features of ^{3}He scattering by Nickel including polarization data. Some details of the polarization distribution could not be reproduced with the limited search procedures adopted: notably the region around 50°. As expected [1] the depth of the spin-orbit potential could not be reliably extracted from an analysis of data for just one nucleus. Most of the searches produced V_{so} values between 3 and 4 MeV but some initial parameter sets resulted in V_{so} around 5.5 MeV. No strong preference for one potential family could be established but the sets with J_R=340 and 440 MeV·fm³ were slightly favoured.

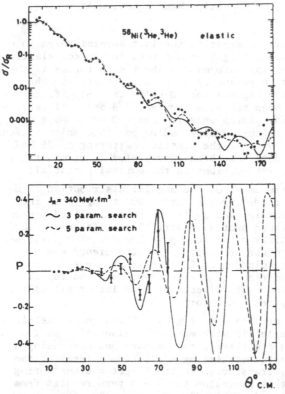

Fig.1. Analysing power of the elastic scattering of ³He by ⁵⁸Ni at 33 MeV. The differential cross section data are those of Cage et al. [2]. The theoretical curves are optical model predictions with two potentials derived from the J_R= 340 MeV·fm³ set of ref. 2 as follows: 3 param. search: V_{so}=3.39 MeV, r_{so}=1.052, a_{so}=0.272 fm; 5 param. search: V_R=128.7, W_D=21.41 MeV, V_{so}=5.27 MeV, r_{so}=1.210, a_{so}=0.333 fm. r_R was fixed at 1.110 fm throughout.

In agreement with the analysis of the scattering of polarized ³He by light nuclei [4], one unambiguous finding of the present work is the shape of the spin-orbit potential for ³He, which appears to be sharply localized at the nuclear surface. Irrespective of the method of search or the discrete family chosen, the spin-orbit radius converged to a value equal to the central potential radius with a small spin-orbit difuseness parameter, typically $0.2 \leqslant a_{so} \leqslant 0.3$ fm. However, it is clear that much more data are needed before the spin-orbit potential depth can be deduced with confidence.

References
+ On leave from the University of Melbourne, Victoria
1) P.E. Hodgson, Adv. in Phys. 17 (1968) 563
2) M.E. Cage et al., Nucl. Phys. A183 (1972) 449
3) C.B. Fulmer and J.C. Hafele, Phys. Rev. C7 (1973) 631
4) W.E. Burcham, J.B.A. England, R.G. Harris, O.Karban and S. Roman, Nucl. Phys. A245 (1975) 170
5) A.Y. Abul-Magd and M. El-Nadi, Prog.Theor.Phys. 35 (1966) 798
6) G.J. Pyle, Univ. of Minnesota report COO-1265-64 (1964) 1.

DEPOLARIZATION IN THE ELASTIC SCATTERING OF 17-MeV PROTONS FROM ^{9}Be

M. P. Baker,[†] J. S. Blair, J. G. Cramer, T. Trainor, and W. Weitkamp
University of Washington, Seattle, Washington, USA*

There has been considerable interest in the last several years in determining the role of nucleon spin-target spin forces in nucleon elastic scattering. One method of deriving information about such forces is the measurement of the depolarization parameter D which is related to the spin-flip probability (S) by the expression: D = 1 - 2S. Significant deviations from unity in D have been reported for ^{9}Be, ^{10}B and ^{27}Al targets[1] at a single scattering angle for proton energies near 20 MeV. More recently, Birchall et al.[2] have measured values of D less than unity at four angles in the range $50 < \theta_{cm} < 90°$ for the elastic scattering of 25-MeV protons from ^{9}Be. The latter data have been interpreted in terms of spherical and tensor spin-spin interactions in the optical potential.

Here we report on the measurement of the angular distribution of D for the elastic scattering of 17-MeV protons from ^{9}Be at angles in the range $55 < \theta_{cm} < 165°$. The reaction was initiated by a polarized proton beam and the polarization of those protons elastically scattered by the ^{9}Be analyzed using a high-resolution, silicon polarimeter. The polarimeter has a measured analyzing power of ~+0.4 and an efficiency equal to $1-2 \times 10^{-5}$ with an intrinsic resolution of 170 keV. Data were taken at scattering angles both left and right of the incident beam and for alternate orientations of the incident spin direction. The latter allows the cancellation of first-order systematic errors.

The experimental results are shown in fig. 1. The angular distribution is dominated by a back-angle dip with measured values of D as low as 0.64 ± 0.06. The curves shown are theoretical comparisons made using a multipole expansion of the elastic-scattering amplitude in terms of the amount of angular momentum (J) transferred to the target nucleus during the scattering process.[3] In this expansion the J = 0 term results from the standard optical potential and can produce no spin flip, i.e., D = 1. Only the terms in the expansion with J > 0 can produce spin flip, and to the extent that such terms form an important part of the elastic scattering amplitude, substantial spin flip is expected.

In the prediction for the J = 1 term,[4] the largest reasonable estimates[5] for the strengths of the spherical and tensor spin-spin interactions have been assumed. However, the data generally indicate much larger departures from unity in D than those predicted for the J = 1 contribution. The predictions for the J = 2 contribution to the spin-flip probability, on the other hand, are more sizable. A ratio technique[3] has been developed for determining the role of the J = 2 term in the spin-flip process. The technique requires an estimate of the intrinsic spin-flip probability for the J = 2 component of the elastic scattering amplitude alone. This estimate has been made using experimental measurements of inelastic spin-flip probability for the 4.43-MeV state of ^{12}C.[6]

At the more forward angles, the predictions for the J = 2 contribution to the spin-flip probability more closely approximate the magnitude observed. However, there is still a large discrepancy between experiment and theory at the back angles. One possible explanation is that there

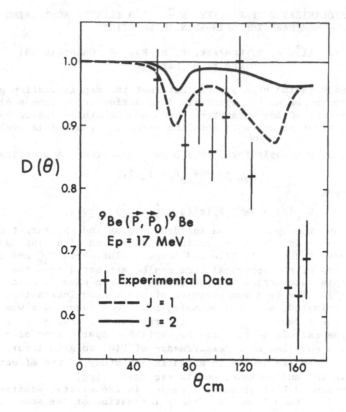

Fig. 1. Depolarization angular distribution for
the reaction $^9\text{Be}(\vec{p},\vec{p}_0)^9\text{Be}$ at 17 MeV.

exists a nonnegligible, compound-nuclear contribution to the elastic scattering amplitude which has a substantial spin-flip component. However,
measurements of the differential cross-section excitation functions reveal no evidence of such a compound nuclear contribution.[7]

REFERENCES

†   Present address: Los Alamos Scientific Laboratory, University of
    California, Los Alamos, NM 87545.
*   Supported in part by ERDA.
1) R. Beurtey, P. Catillon, and P. Schnabel, J. de Phys. 31 Supp. C2
    (1970), 96.
2) J. Birchall, H. E. Conzett, J. Arvieux, W. Dahme, and R. M. Larimer,
    Phys. Lett. 53B (1974), 165.
3) J. S. Blair, M. P. Baker, and H. S. Sherif, to be published.
4) A. H. Hussein and H. S. Sherif, Phys. Rev. C8 (1973), 518.
5) G. R. Satchler, Particles and Nuclei 1 (1971), 397.
6) M. A. D. Wilson and L. Schecter, Phys. Rev. C4 (1971), 1103.
7) M. P. Baker, Ph.D. Thesis, University of Washington, unpublished
    (1975).

THE DEPOLARIZATION PARAMETER IN $\vec{p}$ - ^{10}B ELASTIC SCATTERING AND THE SPIN-SPIN INTERACTION*

J. Birchall[+], H. E. Conzett, F. N. Rad, S. Chintalapudi[+],
and R. M. Larimer

Sherif and Hussein[1] have pointed out that the depolarization parameter D is a sensitive probe of the spin-spin interaction in nucleon-nucleus elastic scattering. Other parameters, such as the cross-section and asymmetry, polarization and spin rotation parameters, are relatively insensitive.

Two types of spin-spin force have been considered. A spherical term:

$$U_{ss}(r) = V_{ss}F_0(r)\underline{\sigma}.\underline{I},$$

and a tensor term:

$$U_{st}(r) = -\tfrac{1}{2}V_{st}F_t(r)\{3(\underline{\sigma}.\hat{\underline{r}})(\underline{I}.\hat{\underline{r}}) - \underline{\sigma}.\underline{I}\},$$

where $\underline{\sigma}$ and $\underline{I}$ are the spins of the incident proton and the target nucleus, respectively, and $\underline{r}$ is a unit vector in the direction of a line connecting the centers of the projectile and target. The depth V_{ss} and the form factor $F_0(r)$ of the spherical term can be estimated from the nucleon-nucleon spin-spin interaction and the single nucleon wave function in the target nucleus[2]). The form and strength of the tensor interaction have not yet been estimated, so a phenomenological Woods-Saxon form was taken for $F_t(r)$ and the strength V_{st} of the interaction was treated as a free parameter in the calculated fits to the available sparse depolarization data[3]). It is clear that more measurements of $D(\theta)$ to good accuracy over wider angular ranges are needed in a continuing study of the effects of the target spin in nucleon-nucleus elastic scattering.

We have measured $D(\theta)$ at several angles in the elastic scattering of 26-MeV polarized protons from ^{10}B. The polarization of the scattered protons is given by

$$p(\theta) = \{A(\theta) + D(\theta)\ p_o\}/\{1 + p_oA(\theta)\},$$

where p_o is the beam polarization and $A(\theta)$ is the target analyzing power. The polarization of the beam was continuously monitored by scattering from a ^{4}He gas target downstream from the ^{10}B. The polarization of the elastically scattered protons was measured by a polarimeter with high figure of merit and good energy resolution[4]). The polarimeter used a 1-mm thick silicon solid state detector as polarization analyzer and two side detectors at $\pm 27^o$ to the polarimeter axis. Protons which passed unscattered through the analyzer detector were stopped in a "zero degree" detector. The zero degree collimation had the same angular width as the analyzer, with respect to the target center, but much reduced angular height. The analyzing power of the target was deduced from the spin up--spin down count ratio in the zero-degree detector.

Geometrical errors in the determination of D were minimized by careful monitoring and adjustment of beam alignment during the runs, by deducing D from spin-up/spin-down ratios in each side detector and by obtaining results with the silicon polarimeter placed on each side of the beam. As a check on these procedures the D-parameter of ^{12}C was measured at a number of angles (D for elastic scattering from a spin zero nucleus should be identically 1.0). Values of D consistent with 1.0 were found in each case.

Results of our D-parameter measurements are shown in fig. 1. The curves are not fits to our data. They are calculations from ref. 1, where the values of V_{st} were chosen to reproduce a data point from Saclay[5] at 65° c.m. and 19.8 MeV. It was pointed out by Sherif and Hussein that the tensor strengths V_{st} extracted in their fits to the data were rather large. As a result, very recent theoretical effort has disclosed another contribution to deviations from unity of $D(\theta)$, which has been termed the quadrupole spin-flip effect[6]. This effect can be present for nuclei that have ground-state quadrupole deformations, and, as such, $I \geqslant 1$. Hence, further investigations are required to determine the separate effects from the explicit spin-spin interaction and from the quadrupole deformations.

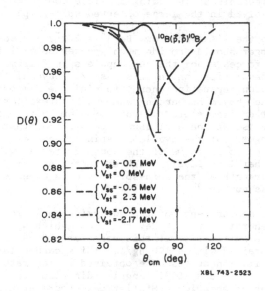

XBL 743-2523

References

* Work performed under the auspices of the U.S. Energy Research and Development Administration.
+ University of Basel, Switzerland.
† Bhabha Atomic Research Centre, Trombay, India.

1) H. S. Sherif and A. H. Hussein, Phys. Lett. 41B, (1972) 465; Phys. Rev. C8 (1973) 518.
2) G. R. Satchler, Particles and Nuclei 1, (1971) 397.
3) J. Birchall et al., Phys. Lett. 53B (1974) 165, and references therein.
4) J. Birchall et al., Nucl. Instr. and Meth. 123 (1975) 105.
5) R. Beurtey, P. Cattillon and P. Schnabel, J. de Phys. 31, Supp. C2 (1970) 96; P. Schnabel, thesis, University of Paris, 1971 (unpublished).
6) J. S. Blair and H. S. Sherif, Bull. Amer. Phys. Soc. 19, (1974) 1010; J. S. Blair, M. P. Baker, and H. S. Sherif, to be published.

ANALYZING POWER FOR ELASTIC SCATTERING OF PROTONS FROM ^{13}C*

T.A. Trainor, N.L. Back, J.E. Dussoletti, and L.D. Knutson
Nuclear Physics Laboratory
University of Washington, Seattle, Washington, USA

Recent measurements of the Wolfenstein D parameter for ^{9}Be($\vec{p},\vec{p}_0$)^{9}Be[1] have shown large deviations from unity at backward angles, corresponding to a considerable spin-flip probability. Such deviations cannot be accounted for by spin-spin forces with realistic strengths and are substantially larger than those predicted by the recently developed "quadrupole" spin-flip mechanism[2]. It is therefore possible that several different mechanisms, including possible compound-nuclear contributions, are involved in the large observed spin-flip probability.

In contrast to the situation for ^{9}Be($J^\pi = 3/2^-$) + p scattering from a spin-1/2 target should provide much greater sensitivity to strictly spin-spin forces since the quadrupole spin-flip mechanism does not contribute to depolarization in this case[2]. Several factors, including a large proton separation energy to minimize compound nuclear contributions at the chosen bombarding energy, high excitation energy of the target first excited state to avoid problems with energy straggling in thick targets, and natural abundance are important in selection of the spin-1/2 target. Of the available spin-1/2 nuclei ^{13}C and ^{15}N represent the best cases in terms of the above criteria, with ^{29}Si, ^{31}P and ^{89}Y being somewhat less ideal. Lighter target nuclei are to be preferred since the strength of the spin-spin force is expected to decrease with increasing mass number[3].

We report here measurements of the analyzing power for ^{13}C + p for proton bombarding energies from 10 MeV to 18 MeV which were made in order to examine the suitability of this case for a D-parameter measurement. Data were obtained with a 100 μg/cm^2 thick carbon target enriched to 97% in ^{13}C. Excitation curves were obtained at laboratory angles of 30°, 50°, 70°, 110°, 130° and 150°. Angular distributions were also measured at a number of energies. Statistical errors are less than 0.01 for all cases.

Representative data are shown in fig. 1. The broad structures in the backward-angle excitation curves indicate significant compound nuclear contributions over the entire energy range. These results indicate that ^{13}C is not suitable for a depolarization measurement at presently attainable tandem energies.

In contrast to the structured backward-angle excitation curves that at θ_{lab} = 50° is relatively structure-free. The magnitude of the analyzing power at this angle is large (-0.6 to -0.8) over a broad energy range from 11.5 MeV to 18 MeV. We suggest, therefore, that ^{13}C is a good alternative polarization analyzer in cases where a gas-target analyzer is impractical. Buildup of ^{12}C contamination is, in practice, not a serious problem at typical polarized beam intensities, and inelastic scattering to the 4.43 MeV state in ^{12}C serves as a convenient monitor of ^{12}C content in the target.

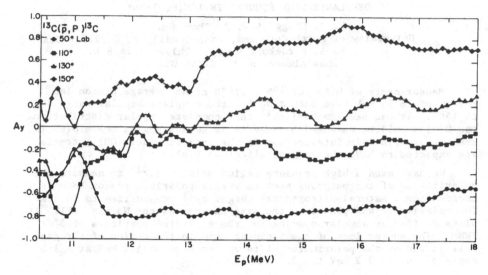

Fig. 1. Analyzing power excitation curves for $^{13}C(\vec{p},p_0)^{13}C$.

References

*   Work supported in part by USERDA
1)   M.P. Baker et al., Nucl. Phys. Lab. Annual Report, Univ. of Washington (1975) 93
2)   J.S. Blair and H.S. Sherif, Bull. Am. Phys. Soc. 19 (1974) 1010
3)   G.R. Satchler, Particles and Nuclei 1 (1971) 397

DEPOLARIZATION STUDIES IN ^{14}N(P,P)^{14}N*

T. B. Clegg[+] & W. J. Thompson
University of North Carolina, Chapel Hill, N. C. 27514 USA
R. A. Hardekopf & G. G. Ohlsen, L.A.S.L.
Los Alamos, N. M. 87544 USA

Measurements of $D(\theta)$ for ^{14}N(p,p)^{14}N at an average proton lab energy of 16.15 MeV have been made for the complete angular range 30° to 150°. It has been emphasized[1] that complete angular distributions for $D(\theta)$ should be measured because it is at back angles where $\sigma(\theta)$ is small that spin-spin interactions in the optical model (OM) potential are expected to have the largest effect on $D(\theta)$.

We have used a high pressure helium polarimeter[2] to measure the polarization of the outgoing protons when a polarized proton beam is incident on a natural-nitrogen gas target cell pressurized to 10 atm. The polarimeter was placed in the same chamber as the N_2 gas cell. Its slits defined an angular opening for the scattered particles of 5° FWHM. The energy loss of the incident beam in the region of the gas cell viewed by the polarimeter slits varied between 1.1 MeV at a lab angle of 90° to 2.2 MeV for 30° and 150°.

Data were collected at each angle for a spin-up-spin-down sequence of the incident beam polarization with the polarimeter first on the left and then on the right of the incident beam. In a separate experiment a complete angular distribution for the vector analyzing power A_y was measured at the same energy. All these results were combined to calculate a value for $D(\theta)$ at each angle by a quadratic method which eliminates first order instrumental asymmetries[3]. Our experimental results are shown in fig. 1 together with previous cross section results[4] at the same energy. Two angular regions of values of $D(\theta)$ smaller than unity appear, indicating the probable occurrence of spin-flip for an incident proton.

The present results might be explained by a spin-spin interaction in the OM potential, but there are other contributions to D : (1) A quadrupole spin-flip effect, in which the proton is spin-flipped by spin-orbit plus quadrupole (LSQ) coupling, has been proposed[5]. Preliminary estimates of this contribution for ^{14}N(p,p)^{14}N (fig. 1) do not produce effects on $D(\theta)$ as large as those which we observe. The LSQ mechanism is effective only for target spin I>1/2. (2) Compound-elastic (CE) scattering may contribute to $D(\theta)$. For example, in the excitation energy region of strongly-overlapping levels Hauser-Feshbach type calculations may be appropriate. In fig. 1 are shown the results of such a calculation. The overall CE normalization is such that the CE cross section is 1% of the experimental cross section at 90° and 10% at 160°, which is consistent with available data (fig. 3 of ref. 6). However, CE spin flip is seen to have large effects on $D(\theta)$.

Therefore, it is important in estimating spin-spin interactions from $D(\theta)$ that the target spin I=1/2 and that the bombarding energy be sufficiently high that CE scattering is <u>completely</u> negligible.

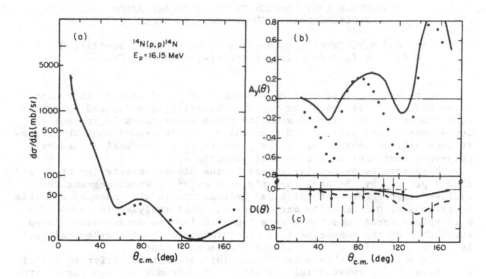

Fig. 1. (a) Cross section, (b) analyzing power, and (c) depolarization for ^{14}N(p,p)^{14}N at 16.15 MeV. Error bars include statistical errors plus uncertainties in beam polarization (1%) and in analyzing powers of the ^{14}N and ^{4}He (1% each). Coupled-channels calculations use the OM parameters of ref.[6] to produce the solid curves. Addition of CE scattering produces the dashed curve in (c).

References

* Research supported in part by the U.S.E.R.D.A.
+ Visiting scientist at L.A.S.L. summer 1974.

1) A. H. Hussein & H. S. Sherif, Phys. Rev C8 (1973) 518
2) R. A. Hardekopf, D. D. Armstrong & P. W. Keaton, Jr., Nucl. Instr. Meth. 114 (1974) 17
3) R. A. Hardekopf & D. D. Armstrong, L.A.S.L. preprint 1975
4) N. Jarmie & J. H. Jett, Phys. Rev. C10 (1974) 54
5) J. S. Blair & H. S. Sherif, Bull. Am. Phys. Soc. 19 (1974) 1010
6) H. F. Lutz et al., Nucl. Phys. A198 (1972) 257

QUADRUPOLE CONTRIBUTION TO THE DEPOLARIZATION
OF NUCLEONS SCATTERED BY DEFORMED NUCLEI

J.S. Blair and M.P. Baker,[+] University of Washington, Seattle, U.S.A.[*];
H.S. Sherif, University of Alberta, Edmonton, Canada[§]

In recent years several measurements of the depolarization parameter $D(\theta)$ have been made for nucleons elastically scattered by nuclei
with non-zero spin. The idea behind these experiments is to look for
deviations of the parameter D from unity. It is argued that these deviations will be indicative of the presence of a spherical or tensor
spin-spin component in the optical potential[1].

Measurements on depolarization in the elastic scattering of 17 and
25 MeV protons by ^{9}Be have recently been made[2,3] which augment the
earlier measurements with 21.4 MeV protons.[4] In all these, appreciable
deviations of D from unity are observed. When analyzed in terms of
spin-spin interactions these were found to lead to unreasonably large
strengths for the tensor part of the interaction if the spherical part
is calculated using an approximate microscopic model.[5]

The purpose of this note is to point out that for deformed nuclei
with spin > 1/2, substantial deviations of D from unity can result from
quadrupole effects in the scattering process. For such deformed targets, large J=2 contributions to the elastic scattering process are
possible, where J is the momentum transferred to the nucleus. Unlike
the monopole (J=0) contributions for which spin-flip along the normal
to the scattering plane is forbidden, these quadrupole parts of the
scattering amplitude can contain considerable spin-flip and hence will
contribute to the deviation of D from unity. The interaction responsible for this process is the same as that for the familiar strong J=2
inelastic excitation of low-lying collective levels. It has been a
familiar observation that excitations of the first 2^+ state in even
nuclei are accompanied by large spin-flip probabilities.[6]

The intrinsic quadrupole spin-flip probability in elastic scattering may be calculated using DWBA or coupled channel calculations.
Alternatively, this may be approximated by the experimentally measured
spin-flip probabilities for the lowest 2^+ levels of the neighboring
even nuclei. Both procedures have been carried out for ^{9}Be. In this
nucleus, the ground state is considered as the lowest member of a
K = 3/2 band, with the easily excited $5/2^-$ level at 2.43 MeV as the
first excited member of the band.

The results are shown in Fig. 1 for incident proton energy of 25
MeV. Details of the calculations will be published elsewhere.[7] We
note that the quadrupole contributions to the deviations of D from
unity are substantial, and are in approximate accord with the data. At
the least, their magnitude makes it rather difficult to extract information about the spin-spin interaction. We, therefore, urge caution
in analyzing depolarization data from deformed targets with spin > 1/2.
On the other hand, spin - 1/2 targets are good candidates for study
since in this case quadrupole effects are not present. Also the quadrupole spin-flip effect decreases with increasing energy so it may be
possible to look for the contribution to depolarization from spin-spin
interactions alone using intermediate energy beams.

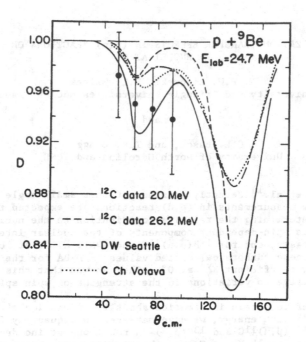

Fig. 1. Depolarization of protons on ⁹Be at 24.7 MeV. The curves show
the predicted values of D due to the quadrupole spin-flip effect. The
spin-flip probabilities are based on measurements for ^{12}C (solid and
dashed curves), on DW calculations (dash-dot curve), and on coupled
channel calculations (dotted curve).

References

†Present address: Los Alamos Scientific Laboratory
*Supported in part by the U.S. Atomic Energy Commission
§Supported in part by the Atomic Energy Control Board of Canada.

1) A.P. Stamp, Phys. 153 (1967) 1052
   H.S. Sherif and A.H. Hussein, Phys. Lett. 41B (1972) 465
   A.H. Hussein and H.S. Sherif, Phys. Rev. C8 (1973) 518
2) M.P. Baker, Ph.D. Thesis, University of Washington (1975),
   unpublished
3) J. Birchall et al., Phys. Lett. 53B (1974) 165
4) R. Beurtey, P. Catillon, and P. Schnabel, J. Phys. (Paris) 31,
   Supp. 5-6, C2 (1970) 96 and P. Catillon, Polarization Phenomena in
   Nuclear Reactions, H.H. Barschall and W. Haeberli, eds. (Univ.
   of Wisconsin Press, Madison (1971) p.657
5) G.R. Satchler, Particles and Nuclei 1 (1971) 397
6) W.A. Kolasinski, J. Eenmaa, F.H. Schmidt, H.S. Sherif, and
   J.R. Tesmer, Phys. Rev. 180 (1969) 1006
7) J.S. Blair, M.P. Baker, and H.S. Sherif, to be published.

POLARIZATION TRANSFER EFFECTS IN $(\vec{p},\vec{n})$ REACTIONS ON LIGHT NUCLEI AT 0° *

P.W. Lisowski, R.L. Walter
Duke University and Triangle Universities Nuclear Laboratory

and

C.E. Busch, and T.B. Clegg
University of North Carolina and TUNL

Madsen et al.[1] recently pointed out that small angle polarization transfer measurements in $(\vec{p},\vec{n})$ reactions are expected to provide a means of determining the relative contribution to the nuclear force of the various spin-dependent components of the nuclear interaction. For several cases, and for $^{11}B(\vec{p},\vec{n})^{11}C$ and $^{15}N(\vec{p},\vec{n})^{15}O$ at 16 MeV in particular, these authors calculated values in DWBA for the polarization transfer coefficient $K_y^{y'}$ at 0°. They showed that this observable is very sensitive to variations in the strengths of spin-spin and tensor terms.

In order to obtain information about the variation of the magnitude of $K_y^{y'}(0°)$ with energy, we have measured this quantity for the $^9Be(\vec{p},\vec{n})^9B$, $^{11}B(\vec{p},\vec{n})^{11}C$ and $^{13}C(\vec{p},\vec{n})^{13}N$ reactions at incident proton energies from 7 to 15 MeV. Target thicknesses were chosen to be approximately 800 KeV in order to average over any rapid changes in the transfer polarization as well as to provide a reasonable counting rate. For these experiments the proton beam polarization averaged about 0.8. The neutron polarizations were analyzed by scattering through 120° (lab) from ^4He gas contained in a high pressure scintillator.

The results of the three sets of measurements are shown in figs. 1, 2 and 3. The errors shown represent a combination of statistical and background subtraction uncertainties. All the transfer coefficients show rather large changes with energy. At the highest energies, $K_y^{y'}(0°)$ for $^9Be(\vec{p},\vec{n})^9B$ appears to have levelled off at about +0.4. The $K_y^{y'}(0°)$ for $^{11}B(\vec{p},\vec{n})^{11}C$ is rising gradually and seems to extrapolate well to the value recently reported by Graves et al.[2] at 21 MeV. The data clearly show that resonances in these (p,n) reactions [on light nuclei] have a large effect on $K_y^{y'}(0°)$ at energies

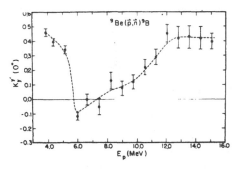

Fig. 1. Experimental results of $K_y^{y'}(0°)$ for $^9Be(\vec{p},\vec{n})^9B$. The line is a guide to the eye.

below 15 MeV. This behavior indicates that direct reaction calculations, such as DWBA, would be of limited value unless compound nuclear effects were correctly included.

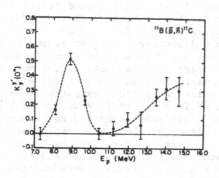

Fig. 2. Experimental results of $K_y^{y'}(0°)$ for $^{11}B(\vec{p},\vec{n})^{11}C$. The line is a guide to the eye.

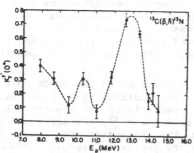

Fig. 3. Experimental results of $K_y^{y'}(0°)$ for $^{13}C(\vec{p},\vec{n})^{13}N$. The line is a guide to the eye.

References

* Work supported in part by U.S. Energy Research and Development Administration.

1) V.A. Madsen. J.D. Anderson and V.R. Brown, Phys. Rev. C9 (1974) 1253
2) R.G. Graves, L.C. Northcliffe, J.C. Hiebert, E.P. Chamberlin, R. York and J.M. Moss, Bul. Am. Phys. Soc. 20 (1975) 694

QUASI-ELASTIC (p,n) REACTIONS INDUCED BY POLARIZED PROTONS

J. Gosset, B. Mayer and J.L. Escudié
DPhN/ME, C.E.N. Saclay,
BP.2, 91190, Gif-sur-Yvette, France

Quasi-elastic (p,n) reactions[1], i.e. (p,n) reactions leaving the residual nucleus in the analogue state of the target ground state, are very useful to study the isospin dependent part U_1 $\vec{t}.\vec{T}$. of the optical potential [2], within the framework of the Lane model [3]. One expects to get informations about a possible spin dependence of U_1 by measuring the analysing power (AP) in such reactions. A previous experiment [4] favoured a spin-orbit component $V_{so}^{(1)}$ in U_1. In order to draw quantitative conclusions on $V_{so}^{(1)}$, we measured the AP angular distribution in quasi-elastic (p,n) reactions induced by 22.8 MeV polarized protons extracted from the Saclay variable energy cyclotron, on ^{49}Ti, ^{56}Fe, ^{64}Ni, ^{70}Zn, ^{90}Zr, ^{96}Zr, ^{117}Sn, ^{165}Ho and ^{208}Pb. To facilitate these measurements a new time-of-flight neutron spectrometer has been set up at the Saclay cyclotron. We used eight NE 213 liquid scintillator detectors, 1.6 liter each, coupled to eight 58 DVP photomultipliers. Data were stored and analysed by an on-line computer. The beam time structure and polarization were continuously monitored. We did not measure the efficiency of our detectors so that only relative cross sections were obtained, in good agreement with the absolute differential cross sections measured at Boulder [5] at the same energy. For comparison with the Distorted Wave Born Approximation calculations, we used the Boulder cross section data together with our AP measurements.

Preliminary macroscopic DWBA calculations using the DWUCK code [6], showed that the AP is very sensitive to the addition of a spin-orbit component $V^{(1)}$ in the form factor U_1, especially to its magnitude but not so much to the geometrical parameters used for $V_{so}^{(1)}$. The effect of this component is an overall displacement of the AP towards more positive or negative values, according to the sign (respectively negative or positive) of $V_{so}^{(1)}$. The data favour a negative $V_{so}^{(1)}$ meaning that the spin-orbit potential must be greater for neutrons than for protons, in agreement with a simple model [4]. Unfortunately the AP was also found to be sensitive to the geometrical parameters used for the spin-orbit distorting potentials in the entrance and exit channels.

Therefore we fixed this geometry and tried to determine quantitatively the $V_{so}^{(1)}$ component by adjusting it in order to fit our AP data. The starting values for the optical parameters, with $V_{so}^{(1)}$ equal to zero, were taken from the energy-dependent, Lane-model, nucleon-nucleus optical potential determined by Patterson et al.[7] from their quasi-elastic (p,n) reaction cross section measurements. The results of this adjustment are quite satisfactory since $V^{(1)}$ is concentrated, for all the nuclei, around a mean value of 4.3 MeV, with a radius of 1.01 fm and a diffuseness of 0.75 fm. Remembering the possible ambiguity with respect to the geometrical parameters used for the spin-orbit distorting potential, we tried another geometry (radius 1.17 fm and diffuseness 0.60 fm) which was also found to give a good agreement to the proton elastic scattering data [8]. With this geometry we obtained a mean value of 6 MeV for $V_{so}^{(1)}$ instead of 4.3 MeV with the first geometry. This ambiguity corresponds approximately to a constant product of the depth and the diffuseness of this $V_{so}^{(1)}$ term, which seems reasonable for a surface-peaked potential.

The results obtained with the second geometry are compared on fig.1 with
the experimental AP data for ^{49}Ti and ^{90}Zr. This figure shows, and it
is the same for the other nuclei, that the spin-orbit component $V_{so}^{(1)}$ of
the isospin dependent part of the optical potential improves very much
the global agreement between the experimental and calculated AP but does
not yet give a perfect account for the data, which present less structure
than the calculations.

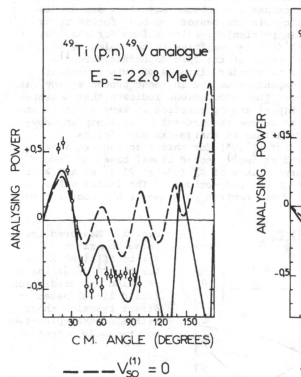
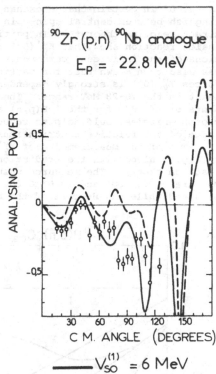

^{49}Ti (p,n) ^{49}V analogue
$E_P = 22.8$ MeV

^{90}Zr (p,n) ^{90}Nb analogue
$E_P = 22.8$ MeV

ANALYSING POWER

C.M. ANGLE (DEGREES)

$------ V_{so}^{(1)} = 0$

$\underline{\hspace{2cm}} V_{so}^{(1)} = 6$ MeV

Références

1) J.D. Anderson and C. Wong, Phys. Rev. Lett. **7** (1961) 250.
2) G.R. Satchler, Isospin in Nuclear Physics (North-Holland Publ. Comp.
   Amsterdam (1969) p. 391.
3) A.M. Lane, Phys. Rev. Lett. **8** (1962) 171 and Nucl. Phys. **35** (1962) 676
4) J.M. Moss, C. Brassard, R. Vyse and J. Gosset, Phys. Rev. **C6** (1972)
   1698.
5) R.F. Bentley, J.D. Carlson, D.A. Lind, R.B. Perkins and C.D. Zafiratos
   Phys. Rev. Lett. **27** (1971) 1081.
6) P.D. Kunz, DWUCK code, private communication.
7) D.M. Patterson, R.R. Doering and Aaron Galonsky, Michigan State
   University Cyclotron Laboratory, preprint (1974).
8) F.D. Becchetti and G.W. Greenless, Phys. Rev. **182** (1969) 1190.

POLARIZATION TRANSFER IN THE ^{11}B($\vec{p},\vec{n}$) ANALOG REACTION*

R. G. Graves, J. C. Hiebert, J. M. Moss, E. P. Chamberlin, R. York
and L. C. Northcliffe
Cyclotron Institute and Physics Department
Texas A&M University, College Station, Texas 77843, U.S.A.

Measurements of the transverse polarization-transfer coefficient $K_y^{y'}$ at 0° in certain charge-exchange reactions may allow one to distinguish between central spin-spin and tensor two-body forces using direct reaction theory[1]. In particular, calculations for the ^{11}B(p,n) analog reaction show that $K_y^{y'}$(0°) is sensitive to the relative contributions of the individual components of the spin-dependent force[1]. It has also been shown that for particular mixtures of spin-dependent forces $K_y^{y'}$(0°) is strongly dependent on the incident proton energy, at least in the 16-28 MeV range. The calculations indicate that measurements of $K_y^{y'}$(0°) for the ^{11}B(p,n) analog reaction at several incident proton energies would help to reduce the present uncertainty in knowledge of the relative contributions of charge-exchange forces.

Two of our measurements of $K_y^{y'}$(0°) for this reaction are shown in fig. 1 along with the predictions[1] (solid lines) based on direct reaction theory. The measured values of $K_y^{y'}$(0°) at 21.3 MeV and 26.5 MeV are +0.59±0.04 and +0.67±0.06, respectively. The latter value was obtained while the He-cell of the neutron polarimeter had poor resolu-

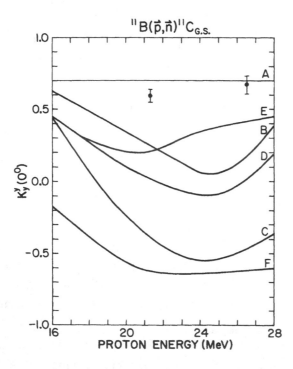

Fig. 1. Measured transverse polarization transfer coefficient $K_y^{y'}$(0°) for the ^{11}B($\vec{p},\vec{n}$) analog reaction and predictions (solid lines) based on direct reaction theory with a central-plus-tensor two-body potential.

tion and should be regarded as preliminary[2]. These results, along with
that for a recent run at 16.4 MeV currently being analyzed, will provide
a determination of the energy dependence of $K_y^{y'}$ in the 16.4-26.5 MeV
range.

The predictions shown in fig. 1 correspond to various combinations
of potentials used in the effective central-plus-tensor two-body interaction. The charge-exchange part of this interaction is given by

$$V_{eff} = \vec{\tau}_o \cdot \vec{\tau}_i [f_c(\vec{r}_{oi})(V_\tau + V_{\sigma\tau}\vec{\sigma}_o \cdot \vec{\sigma}_i) + V_T S_{12} f_T(\alpha',\beta,\vec{r}_{oi})]$$

where, respectively, τ, σ, and S_{12} are the isospin, spin and two-body
tensor operators, f_c and f_T are the central and tensor radial form factors, and V_τ, $V_{\sigma\tau}$, and V_T are the isospin, spin-isospin and tensor-iso-
spin potentials. Curve A is the prediction when only L = 0 orbital angu-
lar momentum transfer is considered. Curves B through F correspond to
various values of the potentials V_τ, $V_{\sigma\tau}$, and V_T. For example, curves
E, B, and D are predictions for purely central forces ($V_T = 0$). Curves
C and F correspond to central-plus-tensor interactions, the difference
between the two curves being due to slightly different values of the
potentials. Comparison of curves A and B shows the effect of higher
L transfers. Comparison of curve C with B shows the effect of including
the tensor force. Our measurements are in best agreement with the simple L = 0 central force prediction, curve A. These results are surprising in view of existing evidence favoring a substantial tensor
force[3] in the effective two-nucleon interaction. It is premature
to conclude that the tensor force is substantially smaller than current estimates of its strength. The effect of an energy dependence
in the two-body force itself has not been included in the calculation
of ref. 1. The present results do however show that knowledge of K_y^y
places a substantial constraint on the allowed strength and form of the
spin-dependent parts of the effective interaction.

References

*Supported in part by the U. S. National Science Foundation

1) V. A. Madsen, J. D. Anderson and V. R. Brown, Phys. Rev. C **9** (1974)
1253
2) Although it is not certain that the poor resolution decreased the
accuracy of the neutron polarization determination, it is planned
to repeat the 26.5-MeV measurement
3) G. M. Crawley et al., Phys. Lett. **32B** (1970) 92

Transfer Reactions

POLARIZATION TRANSFER AT 0° IN
THE $^{12}C(\vec{d},\vec{n}_0)^{13}N$ AND $^{12}C(\vec{d},\vec{n}_1)^{13}N^*$ REACTIONS[†]

R.K. TenHaken and P.A. Quin

University of Wisconsin, Madison, Wisconsin 53706, USA

We have measured polarization transfer coefficients, $K_y^{y'}$, at 0° for the $^{12}C(\vec{d},\vec{n})$ reaction for proton transfers to the ground and first excited states of ^{13}N. A purely vector polarized deuteron beam was accelerated to seven different incident energies in the energy range of 5.7 to 9.8 MeV. These beams were used to bombard a 5 mg/cm² elemental carbon target. The beam polarization was continuously monitored by measuring the neutron asymmetry from the $^{12}C(\vec{d},n_0)^{13}N$ reaction at 15° in the lab. The $^{12}C(\vec{d},n_0)^{13}N$ analyzing powers were measured in a previous experiment[1] at the same energies using the same target. Using these analyzing powers and noting the neutron asymmetry the incident beam polarization, $P_y(d)$, is easily obtained in the standard way.

The neutrons produced at zero degrees impinge upon a high pressure ⁴He gas scintillator[2] located 30 cm in front of the carbon target. Neutrons scattered by the helium nuclei were detected in two symmetrically placed liquid scintillators. These scintillators were located at a distance of one meter from, and at a lab angle of ±118° with respect to, the high pressure cell. At this lab angle and in our energy range the ⁴He(n,n) analyzing power is known to be both large and constant[3]. The neutron groups were separated by measuring the neutron time of flight over the one meter flight path from the analyzer to the neutron detectors. The average analyzing powers of the polarimeter were determined by a Monte Carlo computer code, MOCCASINS[4]. This program also corrects for finite geometry and multiple scattering. The transferred neutron polarization, $P_y(n)$, was then readily obtained by noting the left-right asymmetry of the scattered neutrons.

For a purely vector polarized incident deuteron beam, the polarization transfer coefficients, $K_y^{y'}$, at zero degrees may now be calculated through a simple relationship shown in fig. 1 along with our results. We note that the values we have obtained are quite large and near their maximum value of 2/3, similar to results reported by Lisowski, et al.[5]. This is consistent with the simple model of the neutron as a spectator to the reaction.

The measured values of $K_y^{y'}$ fluctuate as a function of energy, particularly for the transition to the ground state. Although these fluctuations may arise from a direct reaction mechanism, the 0° $^{12}C(d,n_0)^{13}N$ excitation function measured by Davis and Din[6] exhibits considerable energy dependence. This suggests that the compound nucleus contribution to the cross section is not negligible. The data of ref.[6] is made compatible with our measurements by averaging the 0° cross section over our target thickness. These values are plotted in fig. 2 as closed circles. Note that the cross section fluctuations are opposite in phase to the fluctuations in $K_y^{y'}$.

To make the contribution of the direct reaction more transparent we form the product of our measured values of $K_y^{y'}$ (0°) multiplied by the average cross section at that energy. For a sufficient energy average only the direct part should remain, and would be constant in the simple spectator model. These values are plotted in the bottom half of fig. 2 as open boxes and are indeed more constant with energy than either the 0° cross section or polarization transfer measurements. Therefore it would appear that the fluctuations in the ground state values of $K_y^{y'}$ (0°)

can be explained to a great extent as being due to compound nucleus contributions.

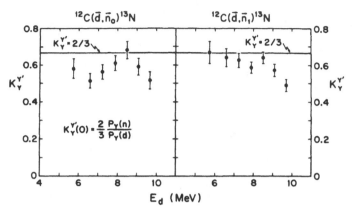

Fig. 1.

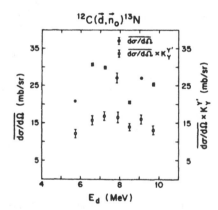

Fig. 2.

References

†   Work supported in part by the U.S. Atomic Energy Commission
[1] R.K. TenHaken and P.A. Quin, Bull. Am. Phys. Soc. 20 (1975) 694
[2] R.E. Shamu, Nucl. Inst. Meth. 14 (1961) 297
[3] B. Hoop and H.H. Barschall, Nucl. Phys. 83 (1966) 65
[4] P.W. Lisowski, private communication; J.R. Sawyers, Ph.D. dissertation, Duke University, 1966; Th. Stammbach, unpublished TUNL report
[5] P.W. Lisowski, et al., Bull. Am. Phys. Soc. 19 (1974) 1076
[6] J.R. Davis and G.U. Din, Nucl. Phys. A179 (1972) 101

MEASUREMENTS OF $K^{y'}_{y}(0°)$ OVER A RESONANCE REGION IN $^{12}C(d,n)^{13}N$ *

R.L. Walter, R.C. Byrd, P.W. Lisowski, and G. Mack
Duke University and Triangle Universities Nuclear Laboratory
and
T.B. Clegg, University of North Carolina and TUNL

In the energy range from threshold to 12 MeV, the $^{12}C(d,n)^{13}N$ reaction and the mirror reaction $^{12}C(d,p)^{13}C$ exhibit considerable structure in the forward angle cross section excitation function. This structure must be due to compound nucleus effects in the interaction, so it is not surprising that DWBA calculations have never been able to describe suitably the angular dependence of the differential cross section, the polarization, and the asymmetry. In addition, the $^{12}C(d,n)^{13}N$ reaction has the interesting feature that although the magnitude of the differential cross section changes appreciably with energy in the region of the peaks in the excitation function, its angular dependence, or pattern, is not grossly altered. This observation, along with the fact that the reaction cross section exhibits an $\ell_p = 1$ stripping shape, suggests that the effect of the resonances is primarily to enhance the probability that the stripping reaction should take place.

Originally, a measurement of $K^{y'}_{y}(0°)$ was conducted for $^{12}C(\vec{d},\vec{n}_o)^{13}N$ because it was felt that such a measurement would aid in understanding the relative strengths of the reaction processes in the d + ^{12}C system. Because the $^{12}C(d,n_1)^{13}N*$ reaction is an $\ell_p = 0$ transfer which has a large yield at 0° it was convenient to determine $K^{y'}_{y}(0°)$ values for this reaction also. Targets for the studies were generally 100 to 200 keV in thickness. A measure of the 0° yield for the (d,n_o) reaction was also obtained with targets of about these thicknesses, again to perform energy averaging similar to that in the $K^{y'}_{y}(0°)$ measurements.

The results are shown in figs. 1, 2 and 3. Although the cross sections vary by about a factor of nearly 50% in some regions, sizeable changes in $K^{y'}_{y}(0°)$ are not observed for the $(\vec{d},\vec{n}_o)$ case. Tests were made of some of the high and low $K^{y'}_{y}(0°)$ values by repeating chosen points with different target thicknesses. This did not seem to affect the general results since the data reproduced well. Some of the data presented here have recently been verified by TenHaken and Quin[1], who are also reporting their findings in a contribution to this conference. The conclusion drawn from the $(\vec{d},\vec{n}_o)$ data, and to a lesser extent from the $(\vec{d},\vec{n}_1)$ data, is that the resonance contribution does not shift the $K^{y'}_{y}(0°)$ values very far from the average of about 0.6 which we observed for other $(\vec{d},\vec{n})$ reactions on ^{14}N, ^{16}O, and ^{28}Si reported in ref.[2]. Since the other reactions were generally $\ell_p = 0$ processes, it appears that either the same direct reaction effects occur in the $^{12}C(\vec{d},\vec{n})^{13}N$ reaction or else the compound-nuclear processes occur in such a way that little vector depolarization results.[3]

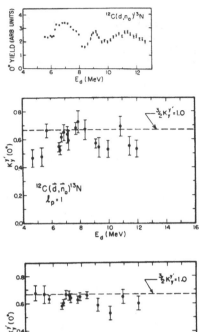

Fig. 1. Zero-degree yield for $^{12}C(d,n_o)^{13}N$.

Fig. 2. Experimental results for the ground-state neutron group for $^{12}C(\vec{d},\vec{n})^{13}N$.

Fig. 3. Experimental results for the first excited state neutron group for $^{12}C(\vec{d},\vec{n})^{13}N$.

References
*Work supported in part by U.S. Energy Research and Development Administration.

1) R.K. TenHaken and P.A. Quin, Bul. Am. Phys. Soc 20 (1975) 694
2) P.W. Lisowski, R.C. Byrd, R.L. Walter and T.B. Clegg, these proceedings, p. 653
3) Following discussions about these data, W.J. Thompson privately communicated results of a simple calculation which show that, for the ideal case of (d,n) reactions which proceed through a single compound nucleus level, values of $K_y'(0°)$ may be near the maximum value and of either sign, depending on the spin and parity of both the compound level and the final state in the residual nucleus.

VECTOR ANALYZING POWER OF THE ^{12}C(D,N)^{13}N AND ^{28}SI(D,N)^{29}P REACTIONS*

P. Krämmer, W. Drenckhahn, E. Finckh and J. Niewisch[+]

Physikalisches Institut der Universität Erlangen-Nürnberg
D 8520 Erlangen, West-Germany

The vector analyzing power (VAP) of ($\vec{d}$,n) reactions has been measured at deuteron energies around 10 MeV for the nuclei ^{11}B,^{12}C,^{14,15}N (1) and ^{89}Y (2) only. The angular distribution corresponds to the VAP of the ($\vec{d}$,p) reaction and shows the same j-dependence for the light elements; for Y, the sign is changed due to Coulomb effects.

We have measured the ($\vec{d}$,n) reaction for ^{12}C and ^{28}Si at E_d = 11 MeV with the Erlangen Lamb shift source. The neutrons were detected by the proton recoils in six 1.5" x 1.5" NE-213 scintillators by appling n-γ pulse shape discrimination. The pulse height distribution was differentiated and the lines were fitted by Gaussian curves in a display program.

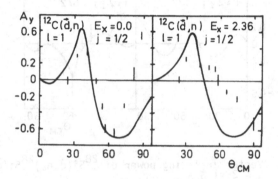

Fig. 1 Vector analyzing power of ^{12}C($\vec{d}$,n$_o$) and ^{12}C($\vec{d}$,n$_1$)

Fig. 1 shows the VAP of the ^{12}C($\vec{d}$,n$_o$ and n$_1$) reactions, which are ℓ =1, j = 1/2 transitions. DWBA calculations with the same parameters as in ref. 1 are given as solid lines. The calculation for the ground state transition is in quantitative agreement with the data, while for the first excited state only qualitative agreement is obtained.

In fig. 2, the VAP for the ^{28}Si($\vec{d}$,n$_o$) reaction which is a ℓ = 0 transition is compared with DWBA calculations. The deuteron parameters are the same as those used by Debenham et al. (3) in their investigation of the ^{28}Si($\vec{d}$,p) reaction. The neutron parameters are taken from Becchetti and Greenlees (4). The angular distribution of the analyzing power in the ($\vec{d}$,n) reaction has a similar structure as in the ($\vec{d}$,p) reaction, both distributions disagree with the calculations.

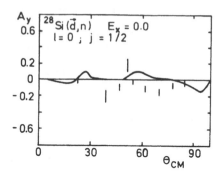

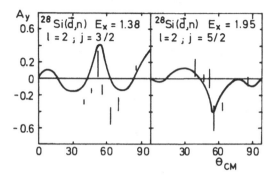

Fig. 2 and 3. Vector analyzing power of $^{28}Si(\vec{d},n_o)$ $^{28}Si(\vec{d},n_1)$ and $^{28}Si(\vec{d},n_2)$

VAP for the to the $\ell = 2$, $j = 3/2^+$ and $5/2^+$ transitions to the first and second excited states are shown in fig. 3. The $(\vec{d},n)$ angular distributions correspond to the $(\vec{d},p)$ data and are well reproduced by the DWBA calculations. The j-dependence is clearly seen in these $\ell = 2$ transitions.

References:

(1) D. Hilscher et al., Nucl. Phys. A174 (1971) 417

(2) P.A. Quinn et al., Nucl. Phys. A183 (1972) 173

(3) A.A. Debenham et al., Nucl. Phys. A167 (1971) 289

(4) F.D. Becchetti, Jr. and G.W. Greenlees, Phys. Rev. 182 (1969) 1190

* Work supported by Deutsche Forschungsgemeinschaft

+ CERN, Genf

MEASUREMENTS OF VECTOR POLARIZATION TRANSFER COEFFICIENTS FOR $(\vec{d},\vec{n})$ REACTIONS ON ^{14}N, ^{16}O, AND ^{28}Si *

P.W. Lisowski, R.C. Byrd, W. Tornow, and R.L. Walter
Duke University and Triangle Universities Nuclear Laboratory
and
T.B. Clegg, University of North Carolina and TUNL

In a $(\vec{d},\vec{n})$ reaction, the polarization transfer coefficient $K_y^{y'}$ measures the fraction of vector polarization of the incident deuteron transferred to the outgoing neutron. For the $D(\vec{d},\vec{n})^3$He reaction at $\theta = 0°$, it is known that this parameter remains at nearly 90% of a simple stripping model value of 2/3 over the energy range extending from 4 to 15 MeV[1,2]. In order to investigate the systematics of the same polarization transfer coefficient for heavier nuclei, we have made similar measurements for targets of ^{12}C, ^{14}N, ^{16}O, and ^{28}Si. In this paper we report our results for the reactions ^{16}O$(\vec{d},\vec{n}_0+\vec{n}_1)^{17}$F, ^{28}Si$(\vec{d},\vec{n}_0)^{29}$P, and ^{14}N$(\vec{d},\vec{n})^{15}$O (leading to states around 5 MeV in excitation of ^{15}O.) In each of these reactions, it is believed that the dominant contribution to the measured neutron groups is $\ell_p = 0$.

The TUNL Lamb-shift ion source was used to supply a pure vector polarized deuteron beam with $p_y \cong 0.47$. As discussed in ref.[1], the use of a pure vector polarized deuteron beam avoids measurement of the tensor analyzing power $A_{yy}(0°)$, which otherwise enters into the $K_y^{y'}(0°)$ determination. Relatively thick targets of approximately 500 keV energy loss were employed not only to generate a suitable counting rate, but to produce some averaging over any resonance structure that may exist in these reactions. The neutron polarization was analyzed by scattering from ^{4}He contained in a high pressure scintillator. The experimental arrangement and analysis methods used with the neutron polarimeter are similar to those indicated in ref.[1].

The results are shown in figs. 1, 2, and 3. All of the reactions exhibit $K_y^{y'}(0°)$ values which have a structureless energy dependence and are near 2/3. Other than that discussed in ref. 3, the only previously reported data in this mass region was a single point at 12.1 MeV[4] for ^{28}Si$(\vec{d},p)^{29}$Si. In that case $K_y^{y'}(0°)$ was also seen to be large (0.67 ± 0.07). The results for each of these $(\vec{d},\vec{n})$ reactions at 0° seem to indicate that the interaction is approximately transparent to the incident vector polarization. For $\ell = 0$ direct reactions, this might be a necessary condition [see ref. 5]. On the other hand, no explanation has been given for the significance of the fact that for all the reactions the $K_y^{y'}(0°)$ values systematically lie about 10% below the stripping model limit of 2/3. If this deviation does not arise from other reaction complexities, perhaps it is a manifestation of the deuteron D-state, which Gammel suggested[2] might be responsible for the similar effect observed in the $D(d,n)^3$He reaction.

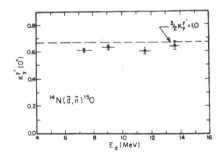

Fig. 1. Values of the polarization transfer coefficient $K_y^{y'}(0°)$ for the $^{14}N(\vec{d},\vec{n})^{15}O$ reaction.

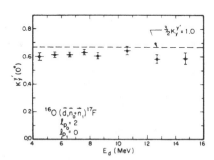

Fig. 2. Values of the polarization transfer coefficient $K_y^{y'}(0°)$ for the $^{16}O(\vec{d},\vec{n}_0+\vec{n}_1)^{17}F$ reaction.

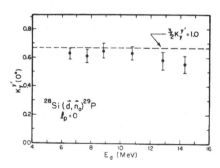

Fig. 3. Values of the polarization transfer coefficient $K_y^{y'}(0°)$ for the $^{28}Si(\vec{d},\vec{n})^{29}P$ reaction.

References

*Work supported in part by U.S. Energy Research and Development Admini-
stration.

1) P.W. Lisowski et al., Nucl. Phys. A242 (1975) 298
2) J.E. Simmons et al., Phys. Rev. Letters 27 (1971) 113
3) R.L. Walter, R.C. Byrd, P.W. Lisowski, G. Mack and T.B. Clegg, these
   proceedings, p. 649
4) R.C. Brown et al., Polarization Phenomena in Nuclear Reactions, ed.
   by H.H. Barschall and W. Haeberli, (1971), p. 782
5) P.D. Santos, Phys. Letters, 51B (1974) 14

THE J-DEPENDENCE OF THE VECTOR ANALYZING POWER
OF $(\vec{d},n)$ REACTIONS ON ^{58}Ni AND ^{60}Ni[+]

B.P. Hichwa, L.D. Knutson[++], J.A. Thomson[+++], W.H. Wong and P.A. Quin
University of Wisconsin, Madison, Wisconsin 53706

It has been shown[1]) that the vector analyzing power for $(\vec{d},p)$ strip-ping reactions on f-p shell nuclei is a reliable indicator of the total angular momentum of the transferred neutron, j_n. For a given value of ℓ_n, iT_{11} for $j_n = \ell_n + 1/2$ and $j_n = \ell_n - 1/2$ is opposite in sign. This j-dependence has been used to make spin assignments for nuclear levels populated by the $(\vec{d},p)$ reaction. In this paper we report measurements of angular distributions of cross section and vector analyzing power iT_{11} for the reactions ^{58}Ni$(\vec{d},n)^{59}$Cu and ^{60}Ni$(\vec{d},n)^{61}$Cu. The purpose of these measurements was to investigate the j-dependence of iT_{11} for transitions with $\ell_p = 1$ and 3.

The analyzing power of transitions to the low-lying states in ^{59}Cu and ^{61}Cu was measured by observing the left-right asymmetry of neutrons produced by an 8 MeV polarized deuteron beam. The energy of the out-going neutrons was measured by the time-of-flight technique[2]). The meas-urements covered an angular range from 5° to 45° in 5° steps.

The results for $\ell_p = 1$ transitions are shown on the left in fig. 1. For the $j_p^\pi = 3/2^-$ transition to the ground states of ^{59}Cu and ^{61}Cu, the observed vector analyzing powers show an oscillatory pattern with a clear minimum near 30°. For the $j_p^\pi = 1/2^-$ transitions, which populate the 497 keV state in ^{59}Cu and the 477 keV state in ^{61}Cu, the vector an-alyzing powers are opposite in sign and tend to be somewhat larger in magnitude. The curves show the results of DWBA calculations for the ^{58}Ni$(\vec{d},n)^{59}$Cu reaction. The agreement is good for the $j_p^\pi = 3/2^-$ tran-sitions, but less accurate for the $j_p^\pi = 1/2^-$ transitions. The calcu-lation overestimates the magnitude of the j = 1/2 peak near 30° by a factor of at least two.

It is interesting to compare iT_{11} for $(\vec{d},n)$ reactions with similar measurements for $(\vec{d},p)$ reactions. The ^{58}Ni$(\vec{d},n)^{59}$Cu and ^{60}Ni$(\vec{d},n)^{61}$Cu measurements are compared in fig. 1 with measurements[3]) for the ^{54}Fe$(\vec{d},p)^{55}$Fe reaction at 8 MeV. The shape and magnitude of the meas-ured vector analyzing powers are similar for the two reactions. In ad-dition, the discrepancy between the measurements and the DWBA predic-tions for the $j^\pi = 1/2^-$ transitions near 30° is present in both cases.

We also measured the analyzing power to the $j_p^\pi = 5/2^-$ states at 921 keV in ^{59}Cu and 972 keV in ^{61}Cu. Forward of 50° iT_{11} is negative. In fig. 2 these measurements are compared to measurements[4]) for a $j_p^\pi = 7/2^-$ transition in ^{51}V$(\vec{d},n)^{52}$Cr. These transitions clearly show j-dependence in this angular region.

References

+   Work supported in part by the U.S. Atomic Energy Commission
++   Present address: Dept. of Physics, University of Washington,
     Seattle, Washington 98195
+++ Present address: Office of Director of Defense Program Analysis and
     Evaluation, The Pentagon, Washington, D.C. 20301
1)   T.J. Yule and W. Haeberli, Phys. Rev. Lett. 19 (1967) 756 and Nucl.
     Phys. A117 (1968) 1
2)   D. Hilscher, J.C. Davis and P.A. Quin, Nucl. Phys. A174 (1971) 417
3)   P.J. Bjorkholm, Ph.D. thesis, University of Wisconsin (1969)
4)   P.A. Quin and J.A. Thomson, to be published

Fig. 1.

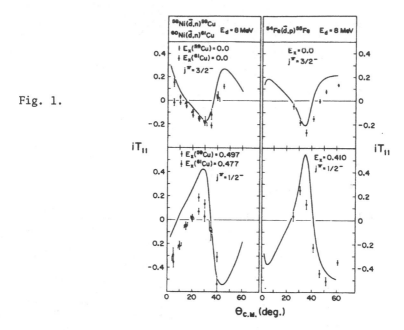

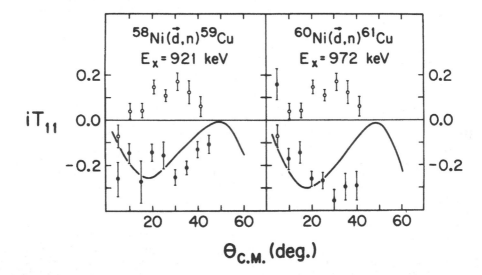

Fig. 2. The solid points and curves are measurements and DWBA calculations for the $j_p^\pi = 5/2^-$ transitions in $^{58,60}Ni(\vec{d},n)^{59,61}Cu$. The open points are measurements for the $j_p^\pi = 7/2^-$ transition to the ground state in $^{51}V(\vec{d},n)^{52}Cr$.

STUDY OF THE ^{63}Cu($\vec{d}$,n)^{64}Zn and ^{65}Cu($\vec{d}$,n)^{66}Zn REACTIONS[†]

W.H. Wong and P.A. Quin

University of Wisconsin, Madison, Wisconsin 53706, U.S.A.

Measurement of the vector analyzing power for ($\vec{d}$,p) reactions on spin zero target nuclei has been shown to be a useful technique for obtaining the total angular momentum of the transferred neutron[1]). When target ground states have non-zero spin, in general more than one value of j-transfer is allowed. In 1969 Kocher and Haeberli[2]) used the vector analyzing power measurement to deduce the spectroscopic factor for two j-values which contribute to a transition in the ^{53}Cr($\vec{d}$,p)^{54}Cr reaction. The present ($\vec{d}$,n) experiment on ^{63}Cu and ^{65}Cu was done to compare with the above mentioned ($\vec{d}$,p) reaction since in all three cases the transferred nucleon is the second nucleon outside Z or N = 28 shell.

Differential cross section and vector analyzing power of transitions to the ground and first excited states of ^{64,66}Zn were studied in the ($\vec{d}$,n) reaction on ^{63,65}Cu with an 8 MeV polarized deuteron beam. The energy of the outgoing neutrons was measured by the time-of-flight technique. Distorted-wave calculations are compared with the data.

The j_p = 3/2 transitions to the ground states of ^{64}Zn and ^{66}Zn are shown in fig. 1. The cross-section show a typical ℓ = 1 transfer and in good agreement with DWBA calculation. The vector analyzing powers iT_{11} show a qualitative agreement with DWBA calculation for j_p = 3/2 transfer. A study[3]) of the ^{58,60}Ni($\vec{d}$,n)^{59,61}Cu reactions showed that for j_p = 3/2 the DWBA calculation of iT_{11} was in good agreement but for the j_p = 1/2 transfer the calculation was too large by a factor of 2.

In the transition to the 2^+ first excited states of ^{64}Zn and ^{66}Zn, the transferred proton coupling to the 3/2 spin of the target nucleus can have ℓ_p = 1 or 3. The cross section measured here and by others[4]) is dominated by ℓ_p = 1. As a result only j_p = 1/2, 3/2 transfers are allowed. The vector analyzing power results for the 2^+ transition in ^{63,65}($\vec{d}$,n)^{64,66}Zn are shown in fig. 2. In both cases, the results are consistent with a pure j = 3/2 transfer. A 20-30 percent admixture of j_p = 1/2 would give a zero vector analyzing power according to the DWBA calculation.

The results of Kocher and Haeberli[2]) for ^{53}Cr($\vec{d}$,p) indicated that the ratio of j_p = 3/2 to j_p = 1/2 is 1.6 to 1. Our results for a proton transfer to the same configuration indicate that the transition is consistent with a pure j_p = 3/2 and limits the j_p = 1/2 to less than approximately 15%.

References

†   Work supported in part by the U.S. Atomic Energy Commission

1)   W. Haeberli, in Polarization Phenomena in Nuclear Reactions, eds. H.H. Barschall and W. Haeberli (University of Wisconsin Press, 1971) p. 235

2)   D.C. Kocher and W. Haeberli, Phys. Rev. Lett. 23 (1969) 315

3)   B.P. Hichwa, L.D. Knutson, P.A. Quin, J.A. Thomson, W.H. Wong, The Fourth International Symposium on Polarization Phenomena in Nuclear Reactions, Zürich, Switzerland, to be published, p.655

4)   V.V. Okorokov, V.M. Serezhin, V.A. Smotryaev, D.L. Tolchenkov, I.S. Trostin, Yu. N. Cheblukov, Yad. Fiz. 8 (1968) 869

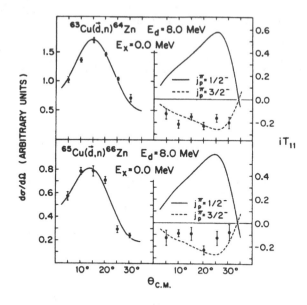

Fig. 1. Cross section and vector analyzing power for transitions to the ground states of ^{64}Zn and ^{66}Zn. These transitions are pure $J_p^{\pi} = 3/2^-$ proton transfers.

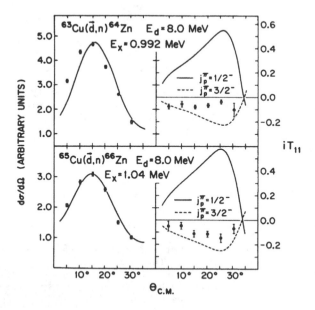

Fig. 2. Cross section and vector analyzing power for transitions to the $J^{\pi} = 2^+$ first excited states in ^{64}Zn and ^{66}Zn. These transitions are consistent with $\ell_p = 1$, $j_p = 3/2$. A limit of 15% admixture of $j_p = 1/2$ is not excluded.

STUDY OF TWO-STEP REACTION MECHANISM AND COMPOUND NUCLEUS
CONTRIBUTIONS IN (d,p) REACTIONS

G. Berg, S. Datta, E.J. Stephenson and P.A. Quin*
University of Wisconsin, Madison, Wisconsin 53706

The (d,p) reaction mechanism involved in populating low-lying non-single particle 7/2+ states in s-d shell nuclei has been discussed in the literature.[1-2] For s-d shell nuclei which possess low-lying collective states, a two-step process involving the excitation of the 2+ state in the target nucleus followed by an ℓ = 2 neutron transfer has been suggested as the reaction mechanism. The compound nuclear reaction mechanism is a possible alternative.

With a view to gaining more insight into the reaction mechanism the differential cross section and the vector analyzing power in the $^{20}Ne(d,p_2)^{21}Ne^*$ and $^{24}Mg(d,p_4)^{25}Mg^*$ reactions were studied. The angular distributions of cross section and analyzing power were measured in the angular range 15 to 165 degrees. The energies of the incident beam were varied from 10.0 to 12.0 MeV, in steps of 330 keV for ^{20}Ne and 250 keV for ^{24}Mg. In addition, the (d,p) cross section excitation functions were measured in the energy range 8 to 13 MeV at four backward angles.

Fluctuation analysis of the excitation functions was carried out[3] by calculating the autocorrelation function with a modulated background. The mean width Γ was found to be 126±12 keV for neon and 95±7 keV for magnesium. The amount of direct and compound nuclear (CN) contributions to the cross section was computed for the first three states in ^{21}Ne and the first five states in ^{25}Mg. The Hauser-Feshbach (HF) calculations were normalized using a least-squares criterion to the computed values of the compound cross section. The uncertainty in the normalization factor was 15%.

A couple-channels (CC) code[4] was used to calculate the cross section and vector analyzing power assuming a two-step reaction mechanism. The spectroscopic amplitudes for single particle transfer were obtained from a shell model[5] and Nilsson model calculations. Amplitudes for the inelastic transitions were taken from ref. 6. Optical model parameters in the deuteron channel were determined by a coupled-channels fit of the elastic data. The proton optical parameters were obtained from literature.[6,7] The deuteron absorptive potential was then varied to improve the quality of the coupled-channels fit to the (d,p) data.

Calculations and the energy averaged data of the cross section and the product of the cross section and the analyzing power ($\sigma \cdot iT_{11}$) are shown in Fig. 1. The uncertainty due to fluctuations in these energy-averaged data were estimated to be less than 10% in the cross section and less than 0.003 in $\sigma \cdot iT_{11}$. The ^{24}Mg cross section is mainly compound while for ^{20}Ne the direct contribution is dominant. We conclude that in the presence of compound and two-step processes it is necessary to determine the CN contribution precisely. Then energy averaged cross sections are in agreement with the sum of the CC and HF calculation. Since the compound nuclear mechanism does not contribute[8] to $\sigma \cdot iT_{11}$, the right hand side of the figure shows only the CC calculation. While the period of the energy averaged data is approximately reproduced the magnitude and phase are in poor agreement. We have found that the vector analyzing power is quite sensitive to the details of the CC calculation whereas the cross section is not. As a result, measurements of the vector analyzing power or $\sigma \cdot iT_{11}$ provide a more stringent test of the CC theory than cross section measurements only.

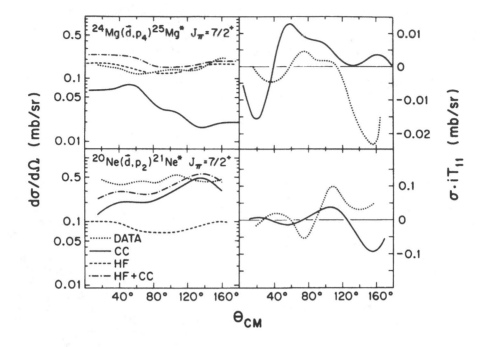

Fig. 1. The dotted curves are drawn through the energy averaged data
taken in 10° steps. Full lines show the CC calculations and
the dash-dot curves are the sum of CC and HF calculations.

References

*Work supported in part by the U.S. Atomic Energy Commission
1) D. Braunschweig, T. Tamura and T. Udagawa, Phys. Lett. 35B(1971)273
2) G. Brown, A. Denning and J. G. B. Haigh, Nucl. Phys. A225(1974)267
3) G. Pappalardo, Phys. Lett. 13(1964)320
4) P. D. Kunz, CHUCK2, private communication.
5) B. H. Wildenthal, private communications.
6) R. de Swiniarski et. al. Phys. Rev. Lett. 23(1969)317
7) F. D. Becchetti, Jr. and G. W. Greenlees, Phys. Rev. 182(1969)1190
8) W. J. Thompson, Phys. Lett. 25B(1967)454

POLARIZATION TRANSFER IN $^{28}\text{Si}(\vec{d},\vec{p})^{29}\text{Si}$ REACTION AND DEUTERON D — STATE EFFECTS

A.K.Basak, J.A.R.Griffith, O.Karban, J.M.Nelson, S.Roman
and G.Tungate
The University of Birmingham, England

The vector polarization transfer coefficient[1] $(z^{11}_{11} + z^{11}_{1-1})$ for the $^{28}\text{Si}(d,p)^{29}\text{Si}$ g.s. reaction was measured at laboratory angles of $(10.0 \pm 0.7)^{\circ}$ and $(20.0 \pm 0.8)^{\circ}$ using Si — polarimeters with a geometry similar to that described in ref.2.

The experiment was made possible after the deuteron source on the Birmingham University Radial Ridge Cyclotron was improved to provide a beam upto 70 nA on target with it_{11}= 0.51. The 12.4 MeV vector-polarized deuteron beam was transported onto a natural Si — target, 12.2 mg/cm^2 thick. A melinex foil of 161 mg/cm^2 was used to stop deuterons and degrade the energy of the protons on the polarimeter target to 12 MeV. Two polarimeters were used at $\pm\theta$ with respect to the beam direction. Five detectors were used in each polarimeter and in addition to the target–detector and two side detectors there were trigger and anti–coincidence detectors for effective elimination of background. The polarimeter target thickness was 116 mg/cm^2.

The analysing powers of the polarimeters were first calculated from the data of Martin et al.[3]. The calculated value at 9.7 MeV agreed quite well with a separate calibration measurement using $^1\text{H}(d,d)^1\text{H}$ at 0° for which the transfer coefficient is known[4]. DWBA calculations were carried out with an extended version of DWCODE[5].

The polarization transfer coefficients at 10° and 20° were found to be (0.857 ± 0.074) and (0.129 ± 0.141) respectively, the uncertain -ties being due to statistics only. The DWBA calculations presented in fig.1 show that the prediction including the D–state contribution

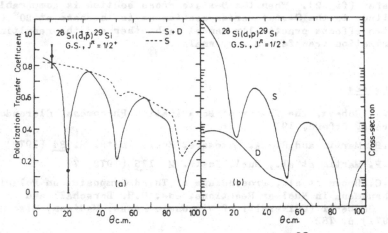

Fig.1. (a) Polarization transfer coefficient for $^{28}\text{Si}(d,p)^{29}\text{Si}$ g.s. reaction and DWBA predictions; (b) DWBA predictions of S– and D– state cross sections for the reaction.

(S + D) is in very good agreement with the measurements. The measurements at 10° could be equally well described by calculations with or without the D-state whereas the 20° point definitely requires the D-state contribution.

This behaviour can be visualised in terms of a simple spectator model of the reaction[4]). In the deuteron S-state, the proton after the stripping process carries the incident beam polarization (i.e. positive polarization transfer) while in the D-state the

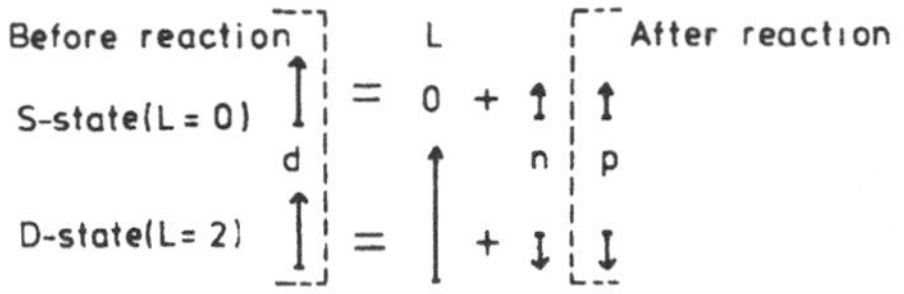

Fig.2. Polarization transfer in the spectator model of (d,p) reaction.

proton spin has to be anti-parallel to the deuteron spin and being a spectator to the reaction gives rise to a negative polarization transfer (fig.2). When the D-state cross section is comparable in magnitude to the S-state cross section as is the case at 20° (fig.1b) the two effects practically cancel each other and the resulting polarization transfer is very small.

References

1) B.A. Robson, The Theory of Polarization Phenomena, Clarendon Press, Oxford, 1974

2) J.P. Martin and R.J.A. Levesque, Nucl. Instr. M. 84 (1970) 211

3) J.P. Martin et al., Nucl. Instr. M. 113 (1973) 77

4) R.C. Brown et al., Proceedings of Third Symposium on Polarization Phenomena in Nuclear Reactions, eds. H.H. Barschall and W. Haeberli (University of Wisconsin Press, Madison, Wisconsin, 1971) p. 782

5) J.D. Harvey and F.D. Santos, DWCODE, University of Surrey, 1970, unpublished.

TENSOR ANALYZING POWERS IN $(\vec{d},p)$ REACTIONS ON ^{46}Ti AND ^{52}Cr

E.J. Stephenson and W. Haeberli
University of Wisconsin, Madison, Wisconsin 53706[†]

Measurements were made of the cross section and the vector and tensor analyzing powers on ^{52}Cr$(\vec{d},p)^{53}$Cr at 6 MeV and ^{46}Ti$(\vec{d},p)^{47}$Ti at 6 and 10 MeV. The experimental technique is discussed in ref. 1. The measurements were compared with DWBA calculations which included the deuteron D-state using the method of ref. 2.

The following conclusions can be drawn from this work:

(1) As was found at higher energies[1-6], the deuteron D-state is essential in understanding the measured tensor analyzing powers. In particular, the large values of $T_{21}(\theta)$ are almost entirely an effect of the deuteron D-state. The D-state effects on the cross section and the vector analyzing power are small and increase with increasing ℓ-transfer. The tensor analyzing powers are similar at 6 and 10 MeV (see ref.1 for 10 MeV measurements on ^{52}Cr), and the quality of agreement between the calculations and the measurements shows no significant change with energy.

(2) The j-dependence observed previously[3-5] at forward angles in $T_{22}(\theta)$ is also present at 6 MeV. Sample ℓ = 1 and 3 transitions are shown in fig. 1, along with examples for ℓ = 2 and 4 from ref. 1. By choosing the abscissa to be the ratio of the scattering angle $\theta_{C.M.}$ to the angle of the stripping peak $\theta_{S.P.}$, the relationship between the stripping peak angle and the j-dependent features of $T_{22}(\theta)$ becomes clear. The measurements for the E_x = 0.159 state on ^{47}Ti represent the first observation of the j-dependent features for a j^π = 7/2$^-$ transition.

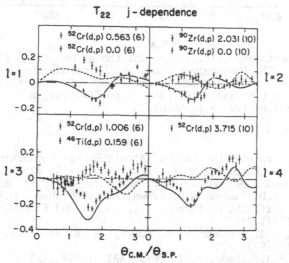

Fig. 1. Measurements and calculations including the D-state for $T_{22}(\theta)$. The solid (open) points and curves are for j = ℓ+1/2 (j = ℓ-1/2) transitions. Each transition is identified by the target, final state excitation energy (MeV), and deuteron beam energy (MeV). A calculation for an imaginary j^π= 7/2$^+$ transition in ^{52}Cr$(\vec{d},p)^{53}$Cr at E_x = 3.715 MeV has been included.

(3) The present results show that the j-dependence rule as proposed in ref. 3-5 does not change with deuteron bombarding energy, and a comparison of our measurements to those on lighter targets[3]) reveals that the rule is not influenced by mass number. In addition, the calculations of $T_{22}(\theta)$ are less sensitive to the choice of optical potential parameters than $iT_{11}(\theta)$ (see fig. 2). Since measurements of $iT_{11}(\theta)$ do not always yield definitive spin assignments (e.g. $\ell = 4$ state in ref. 7), these observations make the use of the j-dependence in $T_{22}(\theta)$ to extract spin information very promising.

(4) The compound nucleus contributions to the reaction were investigated. The magnitude of the compound cross section was obtained from the size of the variance of the fluctuations in excitation functions of cross section and vector analyzing power measured from 5 to 7 MeV on each target. In contrast to ref. 9, the compound cross section proved to be too small in all cases to significantly affect the calculations.

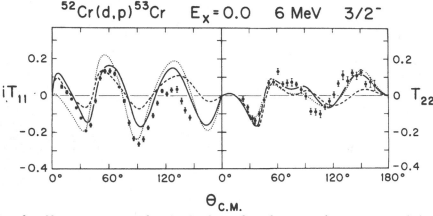

$$^{52}Cr(d,p)^{53}Cr \quad E_x = 0.0 \quad 6 \text{ MeV} \quad 3/2^-$$

Fig. 2. Measurements and calculations for the ground state transition in $^{52}Cr(d,p)^{53}Cr$. The DWBA calculations which include the deuteron D-state were made for the deuteron potential of ref. 4 (dotted), the same potential with a radial cutoff at 4.4 fm. (solid), and the adiabatic potential of ref. 8 (dashed).

References

†)   Work supported in part by the U.S. Atomic Energy Commission
1)   N. Rohrig and W. Haeberli, Nucl. Phys. A206 (1973) 225
2)   R.C. Johnson and F.D. Santos, Part. and Nucl. 2 (1971) 285
3)   R.C. Johnson, F.D. Santos, R.C. Brown, A.A. Debenham, G.W. Greenlees, J.A.R. Griffith, O. Karban, D.C. Kocher and S. Roman, Nucl. Phys. A208 (1973) 221
4)   L.D. Knutson, E.J. Stephenson, N. Rohrig and W. Haeberli, Phys. Rev. Lett. 31 (1973) 392
5)   G. Delic and B.A. Robson, Nucl. Phys. A232 (1974) 493
6)   L.D. Knutson, J.A. Thomson and H.O. Meyer, Nucl. Phys. A241 (1975) 36
7)   J.A. Aymar, H.R. Hiddleston, S.E. Darden and A.A. Rollefson, Nucl. Phys. A207 (1973) 596
8)   R.C. Johnson and P.J.R. Soper, Phys. Rev. C1 (1970) 976
9)   H.O. Meyer and W.A. Friedman, Phys. Lett. 45B (1973) 441

VECTOR POLARIZATION OF DEUTERONS, J-DEPENDENCE OF DIFFERENTIAL CROSS SECTIONS FOR DEUTERON INDUCED REACTIONS AND SPIN-ORBIT TERM OF INTERACTION POTENTIAL

N.I. Zaika, Yu.V. Kibkalo, A.V. Mokhnach, O.F. Nemets, V.S. Semenov,
V.P. Tokarev and P.L. Shmarin
Institute for Nuclear Research, Ukrainian Academy of Sciences, Kiev, USSR

The cyclotron of the Institute for Nuclear Research in Kiev has been to measure the vector analysing power (VAP) of elastically scattered deuterons at E_d=12.6 MeV from ^{9}Be, ^{10}B, ^{12}C, ^{13}C, Ti, Ni, Ta nuclei[1]) and to investigate j-dependence of the differential cross sections for (dp) stripping reactions from ^{10}B, ^{12}C, ^{13}C, ^{14}N, ^{26}Mg, ^{90}Zr, ^{94}Zr nuclei and (dt) pick-up reactions from ^{9}Be, ^{10}B, ^{13}C nuclei at E_d=13.6 MeV. The experimental data were analysed by optical and diffractional models of scattering and distorted wave Born approximation for potential with spin-orbit term.

It was found that a fairly good agreement of calculations with the vector analysing power for medium nuclei can be obtained when using optical model if one can add in the potential determined from the fits to the elastic scattering angular distributions with a V≈100 MeV depth of real part the spin-orbit term of the Thomas type V_{so}~5-7 MeV. For the light nuclei (A=9-13) angular dependence of vector analysing power is quite well predicted by the calculations but agreement is only qualitative. From this fact as well as from remarkable magnitude changes of vector analysing power for the light nuclei region and from no monotonous behaviour of excitation function in elastic scattering (see, for example, for ^{12}C(dd)^{12}C [2]) the conclusion can be drawn that compound-processes play a definite role though the direct mechanism of scattering prevails.

The calculations of vector analysing power when using diffractional model developed for deuteron scattering in [3] with spin-orbit interaction in the first approximation give qualitative agreement for differential cross sections and vector analysing power. The typical theory to the experimental data are shown on fig. 1.

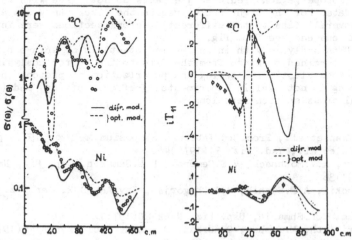

Fig. 1. The differential cross sections of elastically scattered deuterons (a) and VAP (b). The optical model calculation on Ni uses parameters from [5] (- - - -) and [6] (.......).

From the investigations of j-dependence of the differerential cross sections the following conclusions can be done:
a) For the light nuclei region j-effects become apparent (for (dp) as well as for (dt) reactions) in spite of possible contribution of indirect reaction mechanism. J-effects show more expressed diffractional structure of angular distributions for $j=l_n-1/2$ as compared with $j=l_n+1/2$ transitions.
b) J-effects for $l_n=2$ transitions for nuclei 1d(^{26}Mg) and 2d(^{90}Zr, ^{92}Zr, ^{94}Zr) shell differ as it can be seen from fig. 2.

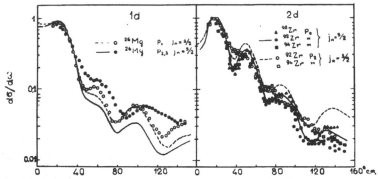

Fig. 2. Angular distributions of protons from (dp) reactions for the 1d shell and the 2d shell. Solid line is DWBA calculation for j=5/2, dotted line is for j=3/2. Proton parameters were taken from [7], deuteron parameters from [5].

DWBA analysis was carried out with spin-orbit potential in both incoming ($V_{so}^d \sim 5-7$ MeV) and exit ($V_{so}^p = 2.5-5.5$ MeV) reaction channels[4]. The optical model parameters were taken from the fits to the elastic scattering data at corresponding energies. It was found that for Zr isotope region the theory did not reflect quantitatively experimental j-effects. For the Mg isotope region predicted j-effects do not agree with the experimental data. For the light nuclei region the discrepancy was also found. The overall fits to the differential cross sections were in good agreement as one can see from fig. 2.

Thus, DWBA analysis with inclusion in potentials of spin-orbit terms which were determined obtained from the fits to the vector analysing power of deuterons elastic scattering and polarization of protons in elastic scattering do not explain experimental j-effects of (dp) and (dt) differential cross section reactions.

References
1) Yu.V.Gofman et al., Proc.3rd Internat.Symposium Madison,1970, p.673
   N.I.Zaika et al., Yad. Fiz. 21(1975)460
2) N.I.Zaika, A.V.Mokhnach, V.S.Semenov, P.L.Shmarin, Usp. Fiz. Nauk
   19(1974)1136
3) G.L.Vysockij, A.P.Soznik, M.A.Chegorjan, Izv.An.SSSR, Ser.Fiz.37(1973)
   1742
4) N.I.Zaika, P.L.Shmarin, Usp. Fiz. Nauk 20(1975)
5) J.A.R.Griffith, M.Irshad, O.Karban, S.Roman, Nucl.Phys.A146(1970)193
6) V.V.Tokarevsky, V.N.Shcherbin, 23th Conf.on Nuclear Spectroscopy and
   Nuclear Structure, Leningrad (1973), p.268
7) F.G.Perey, Phys.Rev.131(1963)745

MEASUREMENT OF THE VECTOR ANALYZING POWER
IN REACTIONS WITH DEUTERONS ON ^{14}N, ^{28}Si, ^{32}S, ^{34}S AND ^{40}Ca

M.V. Pasechnik, V.G. Egoshin, L.S. Saltykov and V.I. Chirko
Institute for Nuclear Research, the Ukrainian Academy
of Sciences, Kiev, USSR

We measured the vector analyzing power A_y of (d,p) reactions on ^{14}N, ^{28}Si, ^{32}S, ^{34}S targets and asymmetry in reactions ^{40}Ca(d,p)^{41}Ca for transitions to the ground and the first excited states ^{41}Ca.

The vector polarized deuterons at 11.6 MeV were produced under elastic scattering of the first deuteron beam of cyclotron on carbon at 25° lab. [1,2]. The vector polarization deuteron beam value is $p_y = -0.22 \pm 0.02$ and tensor components are negligible [3]. The asymmetry of the protons angular distribution in the reaction ^{40}Ca(d,p)^{41}Ca was measured with the polarized beam received under elastic scattering deuteron on carbon at 30° lab. The beam polarization characteristics were measured with the help of the reaction ^{3}He(d,p)^{4}He and elastic scattering on carbon. The beam had the notable value of the vector and tensor polarization [4].

The received experimental data are shown in fig. 1. The measured angular dependence form of the vector analyzing power for light nucleus ^{14}N (the orbital and total angular momenta of the captured neutron are $l = 1$ and $j = 1/2^-$) considerably differs from that one which is observed for the reaction with $l = 1$ and $j = 1/2^-$ in region of the intermediate atomic weight nuclei. Perhaps, it is connected with the fact that nitrogen is the light nucleus and the resonance effect dependence connected with the process of compound nucleus formation is notable in this reaction. A_y value has negative meaning in the region of the main stripping maximum for reactions on the intermediate atomic weight nuclei ^{28}Si(d,p)^{29}Si ($E_x = 1.28$ MeV) and ^{32}S(d,p)^{33}S (g.s.) (the orbital and total angular momenta of the captured neutrons are $l = 2$ and $j = 3/2^-$ respectively). This corresponds to the vector analyzing power sign for reactions with the total angular momentum of the captured neutron $j = 1 - 1/2$ [5]. Asymmetries in the proton angular distributions in the reaction ^{40}Ca(d,p)^{41}Ca for transitions leading to the ground state (black circles) and to the excited states of nuclei ^{41}Ca($E_x = 1.95$ MeV and $E_x = 2.47$ MeV), which are unresolved in our experiment (light circles), are shown too. Angular dependences of the vector analyzing power for reactions ^{28}Si(d,p)^{29}Si (g.s.) and ^{32}S(d,p)^{33}S ($E_x = 0.84$ MeV) are also shown in the figure. In these reactions the neutron is captured with the orbital momentum $l = 0$ and $j = 1/2^+$. The measured analyzing powers in the given angle range have negative values. The analyzing power in the reaction ^{34}S(d,p)^{35}S refers to the total transition to the excited states of the nucleus ^{35}S ($E_x = 1.99$ MeV and $E_x = 2.34$ MeV). Transitions to the states with $E_x = 2.34$ MeV and $E_x = 1.99$ MeV are characterized by different orbital momenta of the captured neutrons $l = 1$ and $l = 3$ respectively. The total angular momentum is $j = 3/2^-$ for the state with $E_x = 2.34$ MeV and for the state with $E_x = 1.99$ MeV is $j = (5/2, 7/2)^-$ [5]. Taking into account the value of the differential cross-section at deuteron short energy 10 MeV in the main stripping maximum range for $l = 1$ [6], the received negative value of the analyzing power A_y in the angles place of 20°-50° is characteristic for the case of the parallel spin orientation and the orbital momentum of the captured neutron $j = 1 + 1/2$.

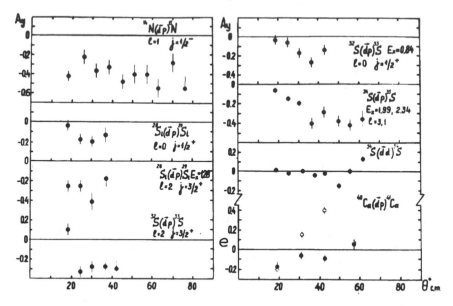

Fig. 1. Measured vector analyzing powers (d,p) reactions on ^{14}N, ^{28}Si, ^{32}S, ^{34}S and asymmetry in reaction ^{40}Ca(d,p)^{41}Ca.

References

1) L.S. Saltykov, Izv. Akad. Nauk SSSR, Ser. Fiz. 31 (1967) 260
2) V.G. Egoshin et al. Ukr. Fiz. Zh. 20 (1975) 509
3) N.I. Zaika et al. Yad. Fiz. 7 (1968) 754
4) A.S. Klimenko et al. Ukr. Fiz. Zh. 13 (1968) 337
5) P.M. Endt and C. Van der Leun, Nucl. Phys. A214 (1973) 1
6) J.G. Van der Baan and H.G. Leighton, Nucl. Phys. A170 (1970) 607

ℓ AND j MIXING IN ^{25}Mg($\vec{\text{d}}$,p)^{26}Mg†

R.W. Babcock and P.A. Quin

University of Wisconsin, Madison, Wisconsin 53706 USA

Cross section σ and vector analyzing power iT_{11} were measured for the reaction ^{25}Mg($\vec{\text{d}}$,p)^{26}Mg at E_d = 10 MeV using the University of Wisconsin Lamb shift polarized ion source and EN tandem. The target was 1 mg/cm^2 thick and enriched to 99.2% ^{25}Mg. Beam polarization was continuously monitored by a ^{4}He polarimeter at the back of the scattering chamber. Cross section normalization was obtained by monitoring elastic scattering at $\pm 13.1°$. Particle identification with an online computer was used to eliminate background from other reactions. The experimental resolution was 50-70 keV FWHM. States in ^{26}Mg up to 6 MeV excitation were observed over the angular range 15° to 160°.

The ground state spin of ^{25}Mg is 5/2, so that in general several ℓ_n and j_n values may contribute to the transition to the same final state[1]. For (d,p) reactions, the cross section σ and the product $\sigma \cdot iT_{11}$ for each j_n transferred add incoherently within the framework of DWBA. For the final states considered here, ℓ_n = 0, j_n = 1/2 and ℓ_n = 2, j_n = 3/2, 5/2 are allowed by angular momentum and parity conservation. No appreciable ℓ_n = 4 single particle strength is expected in the range of excitation energies studied. DWBA calculations for direct reactions on nearby nuclei fit cross sections well, but give only a qualitative description of the vector analyzing power. For this reason we have used calibration curves from nearby nuclei (^{20,22}Ne, ^{24}Mg($\vec{\text{d}}$,p)) to fit mixed transitions. These calibration curves have also been used to fit mixed transitions in ^{21}Ne($\vec{\text{d}}$,p) (ref. 2). The cross section measurements were used to separate ℓ_n = 0 from ℓ_n = 2 contributions, and the vector analyzing power was then used to separate the j_n = 3/2 and j_n = 5/2 components of the ℓ_n = 2 contribution. The shape of the cross section was assumed to depend only on ℓ_n, not on j_n. DWBA calculations were used to normalize the cross section calibration curves so that absolute spectroscopic factors could be obtained. All fits were limited to the region around the stripping peak because the calibration curves are only valid in this angular region.

Figure 1 shows the results for transitions to states in ^{26}Mg at 1.81, 3.94 and 4.84 MeV. All three transitions show unambiguous mixing of ℓ_n = 0 and ℓ_n = 2. The 1.81 MeV state is populated by a predominantly j_n = 5/2 transition, with a small ℓ_n = 0 admixture. The large j_n = 5/2 component is consistent with the previous assignment of this level to the ground state rotational band[3]. The 3.94 MeV level is believed to be the second member of the one-phonon vibrational band. One might expect all three j_n values to contribute due to the complicated internal structure of this state, but we find only j_n = 1/2 and j_n = 3/2 components. This is in contrast to the first vibrational level at 2.94 MeV which could not be well fit by any combination of the allowed j_n values. The level at 4.84 MeV has approximately equal contributions from all three j_n values. The spectroscopic factors for these three states are given in table 1. The relative errors in these numbers are estimated to be 20-50%. These large uncertainties are due to fluctuations in the shape of the vector analyzing power angular distribution for transitions with unique j_n on nearby nuclei. There are two well resolved pure j_n = 5/2 transitions to 0^4 states in ^{26}Mg, the ground state and the weakly excited 3.59 MeV state. The analyzing power for these two states shows a definite Q-value dependence. Since the

$j_n = 5/2$ calibration curve was not changed with Q-value however, it fit only the 3.59 MeV transition. A more precise determination of spectroscopic factors for mixed transitions is probably only possible for heavier targets, where the Q-value and mass dependence of the vector analyzing power is small and where compound nucleus contributions do not produce appreciable fluctuations for reasonable target thickness.

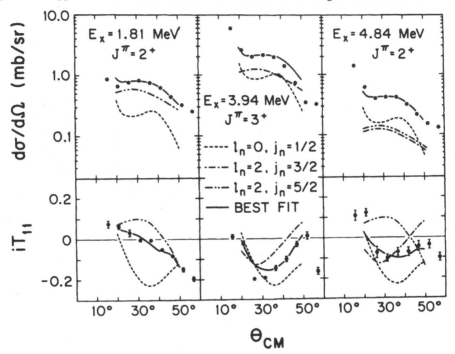

Fig. 1. Cross section and vector analyzing power fits for mixed transitions in $^{25}Mg(\vec{d},p)^{24}Mg$.

Table 1

State	J^π	Spectroscopic Factors		
		$j_n=1/2$	$j_n=3/2$	$j_n=5/2$
1.81	2^+	0.17	0	0.48
3.94	3^+	0.34	0.63	0
4.84	2^+	0.08	0.07	0.07

References

† Supported in part by the U.S. Atomic Energy Commission
1) D.C. Kocher and W. Haeberli, Phys. Rev. Lett. 23 (1969) 315
2) R. Das, A.R. Stanford, J.A. Thomson, W.H. Wong, R.W. Babcock, R.K. TenHaken and P.A. Quin, to be published
3) B. Cujec, Phys. Rev. 136 (1964) B1305

THE STUDY OF POSITIVE PARITY LEVELS IN ^{41}Ca POPULATED IN THE ^{40}Ca($\vec{d}$,p)^{41}Ca(J$^+$) REACTION

R.N. Boyd, D. Elmore and H. Clement[†]
NSRL[*], University of Rochester, Rochester, NY 14627;
W.P. Alford
University of Western Ontario[**], London, Ontario, Canada;
J.A. Kuehner and G. Jones
McMaster University[**], Hamilton, Ontario, Canada

The positive parity levels in ^{41}Ca ranging in spin up to 17/2$^+$, have been interpreted[1] as having wavefunctions consisting primarily of $f_{7/2}$ and $p_{3/2}$ neutron states coupled to ^{40}Ca(3$^-$) and ^{40}Ca(5$^-$) core states. The ^{41}Ca(α,α') experiment[2] selects the $|f_{7/2}\blacksquare^{40}$Ca(I$^-$)> part of the wavefunction. We have attempted to determine the rest of the wavefunction for many of these ^{41}Ca levels by examining differential cross sections[3] and analyzing powers produced in the ^{40}Ca($\vec{d}$,p) reaction.

The McMaster University Van de Graaff facility with its Lamb Shift polarized ion source[4] was used to produce the 11 MeV polarized deuteron beam. Typical beams on target were 140 nA, with a polarization of P_z=0.8. Proton spectra were observed using the magnetic spectrograph and two 5 cm long solid state position sensitive detectors. The resulting resolution was about 15 keV FWHM. Typical spectra with spin up and with spin down for a range of excitation energy from about 2.4 to 3.0 MeV are shown in Fig. 1. Data were taken in 10° steps from 10° to 90° laboratory angle.

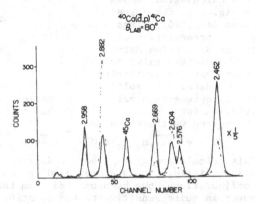

Fig. 1: Proton spectrum for ^{40}Ca($\vec{d}$,p) at E_d=11 MeV. Solid curve: spin down, dotted curve: spin up.

Differential cross section and analyzing power data were taken for about twenty ^{41}Ca levels ranging in excitation from 2.4 to 5.0 MeV. Data for the levels at 2.46(3/2$^-$), 2.88(7/2$^+$) and 3.20(9/2$^+$) MeV are shown in Fig. 2. The 2.46 MeV level is shown to indicate that the parameters used[6,7] in the (one-step) DWBA calculation represent the general features of both cross sections and analyzing powers out to an angle of 60° or so. The fit to the ground state (7/2$^-$) data[5] was somewhat better than that for the 2.46 MeV level. For the positive parity levels the calculations were made with the two-step CCBA code CHUCK[8]. Experimentally determined inelastic scattering[2] and transfer[5] strengths were used for the transitions for which they were available. The only remaining adjustable parameters were the (properly normalized)coefficients of the wavefunctions of the ^{41}Ca(J$^+$) levels.

The data for the 3.20 MeV (9/2$^+$) level are represented qualitatively by either the one-step (dashed curve, $\mathcal{J}$=0.021) or by the full calculation (solid curve). The full calculation shown involved three two-step reaction trajectories, each consisting of both transfer and inelastic scattering. Intermediate states included the ^{40}Ca(3$^-$), ^{41}Ca(7/2$^-$,0.0 MeV) and ^{41}Ca(3/2$^-$,1.95 MeV) levels. In addition a small g$_{9/2}$ one-step

672

transfer was included in the full calculation. The wavefunction for the state represented by this calculation is

$$|9/2^+> = \sqrt{.01}|n_{g9/2} \otimes 0^+> - \sqrt{.62}|n_{f7/2} \otimes 3^-> + \sqrt{.37}|n_{p3/2} \otimes 3^->.$$

Fig. 2 shows the excellent representation of even most of the details of the analyzing power, given by this calculation.

The 2.88 MeV ($7/2^+$) level is somewhat more interesting than the 3.20 MeV level in that the calculated analyzing powers for the $7/2^+$ state exhibit a large sensitivity to the assumed wavefunction. While the one-step calculation, shown in Fig. 2 as the dashed curve ($\mathcal{J}=0.026$), gives a good representation of the cross section data, its predicted analyzing power bears no similarity to the data. The solid curve represents a calculation for a state having the wavefunction

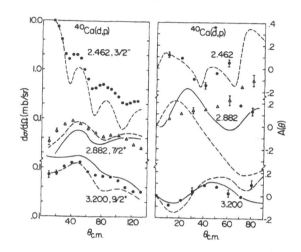

Fig. 2: Cross sections and analyzing powers to three ^{41}Ca levels. Solid curves: CCBA, dashed curves: DWBA.

$$|7/2^+> = \sqrt{.27}|n_{g7/2} \otimes 0^+> - \sqrt{.49}|n_{f7/2} \otimes 3^-> + \sqrt{.49}|n_{p3/2} \otimes 3^->.$$

This calculation gives a better data representation, but the difficulties in fitting the data around 50° probably indicate that important configurations have been omitted from the wavefunction. The analyzing power is quite sensitive to the relative phase between the $|n_{g7/2} \otimes 0^+>$ and the other pieces of the wavefunction; reversal of that phase produces an analyzing power similar to the one-step prediction.

---

†NATO Fellow

*Supported by the National Science Foundation

**Supported by the National Research Council of Canada

1)P. Goode and L. Zamick, Nucl. Phys. A129 (1969) 81

2)M.J.A. de Voigt, D. Cline and R.N. Horoshko, Phys. Rev. C10 (1974) 1798

3)R.N. Boyd, D. Elmore and H. Clement, private communication; (See also K.K. Seth, A. Saha and L. Greenwood, Phys. Rev. Lett. 31 (1973) 552

4)The source was built by General Ionex Corp., 25 Hayward St., Ipswich, Mass. 01938

5)D.C. Kocher and W. Haeberli, Nucl. Phys. A172 (1971) 652

6)F.D. Becchetti and G.W. Greenlees, Phys. Rev. 182 (1969) 1190

7)R.C. Johnson and P.J.R. Soper, Phys. Rev. C1 (1970) 976

8)P.D. Kunz, University of Colorado, 1973, unpublished

INVESTIGATION OF THE ISOSPIN THRESHOLD ANOMALY IN (D,P) REACTIONS WITH POLARIZED DEUTERONS

W. Stach, G. Graw[+], W. Kretschmer, C. Hategan[++]
Physikalisches Institut der Universität Erlangen-Nürnberg,
Erlangen,West-Germany

A systematic investigation of the isospin threshold anomaly in the excitation curves of the vector analyzing power of $(\vec{d},p)$ reactions has been performed with several target nuclei in the $A \approx 90$ and $A \approx 110$ mass region. Pronounced effects have been found for the $(\vec{d},p_0)$ reactions on ^{88}Sr, ^{90}Zr, ^{92}Mo and for the $(\vec{d},p_2)$ reaction on ^{91}Zr, whereas the effects for the $(\vec{d},p_0)$ reactions on ^{106}Cd, ^{107}Ag and ^{108}Pd are rather small.

The anomaly is assumed to originate from the isospin coupling of the 3p - neutron single particle resonance to the proton analog channel. In the observed mass region the 3p resonance is near the neutron emission threshold.

The analysis has been performed according to Lane's proposal[1,2]. The anomaly effect is described phenomenologically by a resonance term which is added to the scattering amplitudes (i.e. radialintegrals of the DWBA-code) of the l=1,j=3/2 or 1/2 proton exit channel in the following way

$$S^{j^\pi}_{l_i,j_i} = S^{DWBA}_{l_i,j_i} \ (\ 1 + RES\ ) \ .$$

The aim of our analysis is to show that

1.) The anomaly is an exit channel phenomenon.
Fig. 1) shows as an example the $(\vec{d},p_0)$ reaction on ^{92}Mo. The solid and dashed curves are DWBA calculations with and without a resonant p3/2 wave in the exit channel. The resonance parameters were adjusted to reproduce the excitation function of $\sigma(\theta)$ at $\theta=160°$ only. At the same time the other angles and observables are reproduced quite well. Moreover this adjustment yields the same parameters for the various entrance channels and confirms the anomaly as an exit channel phenomenon.

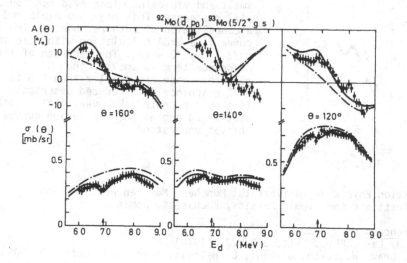

Fig. 1)

NUCLEUS	E^{thr}	R	$\emptyset$
^{88}Sr	7.5	$1.2 \pm .1$	$.8 \pm .2$
^{90}Zr	7.2	$1.16 \pm .1$	$.6 \pm .2$
^{92}Mo	7.0	$.98 \pm .1$	$.4 \pm .2$
^{91}Zr	7.4	$.86 \pm .1$	$.9 \pm .2$

2.) The mass dependence of the anomaly effect.

This may be explained by the magnitude of the resonance term at its maximum (at the threshold). This representation is rather independent of barrier penetrabilities which are included in the DWBA terms. The table shows a weak decrease with N and Z.

3.) A "j-effect" (p3/2 or p1/2 resonance ?)

The above calculations have been performed with a resonant p3/2 wave only. Fig. 2) illustrates the anomaly effect for an assumed p1/2 wave in the exit channel. The resonance term was adjusted to reproduce the cross-section data again (by multiplication of the resonance term with a factor near five). The vector analyzing power is now just opposite to the p3/2 curve, thus showing a clear "j-effect".

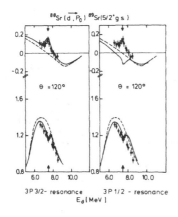

^{88}Sr($\vec{d}$,P_0) ^{89}Sr(5/2$^+$g s)

$\theta = 120°$ $\theta = 120°$

3P 3/2- resonance   3P 1/2 - resonance

E_d(MeV)

Fig. 2)

According to calculations of Fiedeldey and Frahn [3] one would expect the p1/2 resonance at the threshold in the $A \approx 110$ mass region. Therefore we have measured ($\vec{d}$,p_0) reactions on ^{106}Cd, ^{107}Ag and ^{108}Pd. Fig. 3) shows the excitation curves of $\sigma(\theta)$ and vector analyzing power $A_y(\theta)$ at $\theta=160°$. The data points of $\sigma(\theta)$ show an "anomaly structure" which can be reproduced by calculations with a p3/2 wave and also with a p1/2 wave. On the other hand the variation in $A_y(\theta)$ is very small and yields no clear evidence for a p1/2 resonance. This might be explained by a cancellation of both resonances. The solid curve is a DWBA calculation with resonant p3/2 and p1/2 waves. The strength of the resonance term is smaller than 0.1 for the p3/2 resonance and smaller than 0.5 for the p1/2 resonance in a reduced description. It should be noted, however, that at other scattering angles the excitation curves are not yet understood.

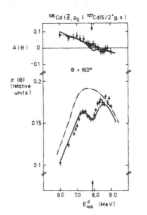

^{106}Cd($\vec{d}$,P_0) ^{107}Cd(5/2$^+$g.s.)

$A(\theta)$

$\theta = 160°$

$\sigma(\theta)$ (relative units)

E^d_{lab} (MeV)

Fig. 3)

[+] Sektion Physik der Universität München, München West-Germany
[++] Institute for Atomic Physics, Bucharest, Romania

references:
1) A.M. Lane, Phys. Lett. 33B, 274 (1970)
2) G. Graw, W. Stach, G. Gaul, C. Hategan, Phys. Rev. Lett. 30, 989 (1973)
3) H. Fiedeldey, W.E. Frahn, Ann. Phys. 19, 428 (1962)

L AND J-DEPENDENT EFFECT IN (d,p) REACTIONS

F. D. Santos

Laboratório de Física, Faculdade de Ciências, Lisboa, Portugal

It has been noted by various authors[1-3] that in (d,p) reactions above the Coulomb barrier the tensor analysing power T_{22} angular distributions at forward angles have a pronounced negative dip for $j_n = \ell_n + 1/2$ transitions and are positive or close to zero for $j_n = \ell_n - 1/2$ transitions. Another distinctive feature of the data shown in Fig. 1 is that the angle θ'_p of the first minimum in T_{22} for $j_n = \ell_n + 1/2$ increases with ℓ_n. The same ℓ_n-dependence is also recognizable when $j_n = \ell_n - 1/2$ although not so clearly. This ℓ_n and j_n-dependence which can be reproduced in DWBA calculations only by inclusion of deuteron D-state effects is interpreted in terms of a simple reaction model based on the surface localization of the transfer process.

T_{22} gives a measure of the reaction mechanism sensibility to rotations of the incident deuteron spin along the z-axis (Madison Convention system). We consider a tensor polarized deuteron beam with axial symmetry along $\vec{n}$ and fractional spin projection populations $N_+ = N_- = 1/2$ and $N_o = 0$. The most convenient orientation to measure T_{22} is to have $\vec{n}$ in the xy plane. With these specifications we get

$$\frac{d\delta}{d\phi} = -\sqrt{3} \, \sin(2\phi) \, T_{22} \qquad (1)$$

where $\delta = \sigma/\sigma_o$ is the ratio between the cross sections for a polarized and unpolarized beam and ϕ is the angle between $\vec{n}$ and the x-axis. Thus when $\vec{n}$ is rotated from the x to the y-axis the derivative has opposite sign to T_{22}. Due to the surface localization of the reaction for small momentum transfer the transfer probability is largest in the reaction plane where conservation of orbital angular momentum along the y-axis favours high values of total orbital angular momentum transfer. Thus the transfer from the deuteron D-state is favoured when $\vec{L}$, the deuteron internal orbital angular momentum is antiparallel to ℓ_n. This implies a preferencial orientation of the stripped neutron spin parallel to $\vec{\ell}_n$ when $\vec{n}$ is along the y-axis. We therefore conclude that in $j_n = \ell_n + 1/2$ transitions δ must increase with ϕ as it varies from 0 to $\pi/2$. This from eq(1) implies a negative T_{22} in accordance with the observed j_n-dependence.

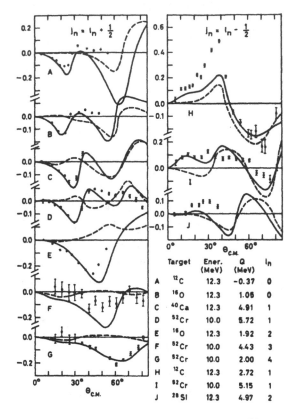

Fig. 1. T_{22} angular distributions[1-4] for (d,p) reactions. The broken and solid curves are DWBA predictions for a pure S-state deuteron and including the S and D-states.

As for the ℓ_n-dependence we find a strong correlation between the angle Θ'_p and the angle Θ_p of the differential cross section main peak. Representing by $\vec{k}'_p$ and $\vec{k}_p$ the proton asymptotic momenta corresponding to Θ'_p and Θ_p the vector

$$\vec{K}_d = \vec{k}_p - \vec{k}'_p$$

is a measure of the extra momentum at the point of transfer of neutrons stripped from the deuteron D-state. The component of $\vec{K}_d$ along the z-axis must therefore be positive which implies Θ'_p greater than Θ_p in agreement with experiment. The calculation of K_d from the experimental values of Θ'_p and Θ_p in 26 T_{22} angular distributions with[1-4] ℓ_n ranging from 0 to 4 gives a mean value of $< K_d >$av. $= 0.23 \pm 0.06$ fm^{-1} showing that it is remarkably constant. Further analysis is being pursued to determine if K_d should be identified with the deuteron asymptotic momentum $\alpha = 0.2316$ fm^{-1}.

References

1) R. C. Johnson et al. Nucl. Phys. A208(1973)221
2) L. D. Knutson et al. Phys. Rev. Lett. 31(1973)392
3) G. Delic and B. A. Robson, Nucl. Phys. A232(1974)493
4) N. Rohrig and W. Haeberli, Nucl. Phys. A206(1973)225

ON THE TWO-BODY SPIN-ORBIT USED IN DWBA CALCULATIONS

J. Raynal
Service de Physique Théorique
CEN-Saclay, BPn°2, 91190 Gif-sur-Yvette, France

In earlier publications[1,2] we used a two-body spin-orbit interaction with two opposite Yukawa form factors of range 0.55 and 0.325 respectively and its zero-range limit. This potential was obtained by comparison to the soft core potential of Bressel et al.[3] and was estimated to correspond to a slightly too strong spin-orbit interaction for the optical model. However, it appeared an error of a factor 4 in the derivation of this potential[4]. But, after this correction, the two-body spin-orbit interaction is too weak by a factor of 3 at least.

We can take as an example of two-body interaction the one derived by J. Borysowicz, H. Mc Manus and G. Bertsch which uses the sum of 4 Yukawa form factors with small ranges. The parameters are given in the Table below. When used to describe the inelastic scattering of 20.3 MeV protons on the first 2^+ of ^{90}Zr , this interaction is unable to describe the analyzing power. With the spin-orbit included or not, the cross-section is well described but the analyzing power remains negative where the experiment gives positive values. When the spin-orbit interaction is multiplied by 2, there is a small positive bump for the analyzing power around 65° but the cross-section worsens. When multiplied by 3 or 4 the analyzing power saturates but there are large effects on the cross-section.

In order to relate this increase of the spin-orbit interaction which seems to be needed, one can consider, as a first step, its zero-range limit[1]. Such a limit acts only in the relative $L = 1$ and $S = 1$ subspace of the incident and target nucleons. Consequently, it is purely $S = 1$, $T = 1$. The interaction described in the table includes also a $T = 0$ spin-orbit, which can act only in relative $L = 2$ states. This $T = 0$ component can be eliminated (using $V_{pn} = 1/2\ V_{pp}$ for the spin orbit) without changes larger than some percent even when the spin-orbit interaction is multiplied by 4 .

	range	scalar	$\sigma_1 . \sigma_2$	$\vec{L}.\vec{S}$	tensor
	.2	1323.6	−1323.6	−6701.	3884.4
V_{pp}	.4	751.9	− 751.9	− 49.	− 245.6
	.5	− 604.8	604.8	9.7	210.5
	.7	0.	0.	0.	0.
	.2	5177.1	843.3	−3604.6	−12416.6
V_{pn}	.4	1417.1	− 28.9	− 105.	145.7
	.5	−1432.2	− 74.2	17.9	− 105.4
	.7	0.	0.	0.	− 14.1

678

The strength of this zero-range limit is the strength of the finite range interaction multiplied by the fifth power of the range. The exchange term is identical to the direct term, and the strength for their sum obtained from the table is -4.7 . Results obtained with -18.8 are shown on the figures. The effect on cross-section and analyzing power is stronger than the one of finite range but is qualitatively the same. The effect of range is stronger than for the interaction which was used in ref.1 or 3. The relatively simple expression of the zero-range limit[1] allows direct comparison with some other phenomenological values which are the parameters of the optical model or the Skyrme force used for nuclear structure calculations.

The central part of this interaction is only $S = 1$, $T = 0$ and $S = 0$, $T = 1$ with zero-range strength -102.6 and -67.6 respectively. If used to compute an optical potential for a mirror nucleus, the strength of the central potential is 18 larger than the spin-orbit interaction, the form factors of these two potentials being defined from the nuclear density. The usual ratio is 7 to 10, which means that the two-body spin-orbit is 2 to 2.5 times too small.

The zero-range limit of the spin-orbit interaction is exactly the spin-orbit used in Skyrme interaction[5] except for a factor 4π in the definition of its depth. The interaction described in the Table corresponds to a Skyrme interaction with 30 MeV for its spin-orbit whereas values as 100 to 150 are commonly used. The central zero-range potentials are of the same order.

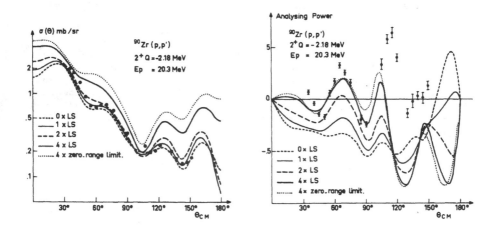

References

1) H.H. Barshall and W. Haeberli, "Polarization Phenomena in Nuclear Reactions", p.687, The University of Wisconsin Press, Madison (1971).
2) J. Raynal, "The Structure of Nuclei", Trieste Lectures 1971 (Vienna 1972).
3) C.N. Bressel et al., Nucl. Phys. A124 (1969) 624.
4) See "Note added in proof" of ref.2.
5) J.P. Blaizot and J. Raynal, Lettere al Nuovo Cimento 12 (1975) 508.

Analysing powers of the $^{19}F(\vec{d},t)^{18}F$ reaction

J.M. Nelson, O. Karban, E. Ludwig* and S. Roman
Department of Physics, University of Birmingham, England

Measurements have been made of the vector analysing powers and differential cross sections of the $(\vec{d},t)$ reaction on ^{19}F leading to states in ^{18}F below 3 MeV. The experiments were performed with the 12.3 MeV polarized beam from the Radial Ridge cyclotron incident on a 1.5 mg/cm^2 target of CaF_2. Three dE x E telescopes were placed at identical scattering angles to the left and right of the beam axis.

The elastic data were analyzed in terms of the optical model and the resultant potential was used together with a t + ^{18}O potential[1] in finite-range DWBA calculations with the code NFIMAC.[2] The predictions for the positive parity states are shown with the experimental analysing power data in fig. 1. The four states in ^{18}F between 0.94 MeV and 1.12 MeV were not resolved but only transitions to the 3^+ (0.94 MeV) and 0^+ T=1 (1.04 MeV) states were considered.

In three of the five transitions considered, the total angular momentum transferred is double-valued. The deduction of the spectroscopic factors from the straightforward comparison between theoretical and experimental cross sections is no longer possible unless it can be shown that the spectroscopic factor for one of the two transfers is negligibly small. The analysing powers are sensitive to the admixture of components carrying different angular momenta and this should provide a useful method of testing theoretically derived spectroscopic factors.

Two-body matrix elements for the s-d shell[3] were used with a version[4] of the Oak Ridge–Rochester shell model code to generate a series of spectroscopic factors for the transitions being studied. All the calculations predict the $d_{3/2}$ components to be weak in all transfers except that to the 2.52 MeV 2^+ state where it is typically 50% of the $d_{5/2}$ strength. Apart from this state, the $d_{3/2}$ component is ignored in the calculations shown.

The predictions for the 2.52 MeV state contain the sum of the $j=3/2$ and $j=5/2$ contributions in proportions specified by (a) the Kuo–Brown and (b) RIP (0.032/0.131) spectroscopic factors. In general, the fit requires the higher proportion of $d_{3/2}$ as specified by Kuo–Brown (0.076/0.154). This cannot be concluded from an analysis of the cross section data alone. The expected j–dependence is observed in the individual calculations.

The fit to the unresolved states at about 1 MeV contains the sum of the 3^+ and 0^+ predictions. It is evident that these are insufficient of themselves to explain the data and the inclusion of the 0^- transition will have to be investigated.

The agreement between predicted absolute cross sections and data is good except for the 1.70 MeV state where C^{2}S is an order of magnitude larger than the predictions.

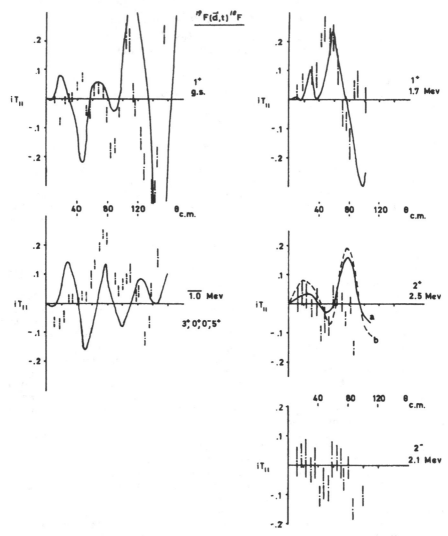

Fig. 1. Vector analysing powers and DWBA predictions for the $^{19}F(\vec{d},t)^{18}F$ reaction induced by 12.3 MeV polarized deuterons.

References

* On leave from Dept. of Physics, University of North Carolina
1) D.J.Pullen, J.R.Rook and R.Middleton, Nucl. Phys. **51** (1964) 88
2) J.M.Nelson and B.E.F.Macefield, University of Birmingham Report 74-9 (1974) unpublished
3) e.g. E.C.Halbert, J.B.McGrory, B.H.Wildenthal and S.P.Pandya, Adv. in Nucl. Phys. Vol 4 (1971) 316
4) A.P. Zuker, private communication

THE ^{207}Pb($\vec{d}$,t)^{206}Pb REACTION USING
15-MeV VECTOR POLARIZED DEUTERONS*

M. D. Kaitchuck, T. B. Clegg, W. W. Jacobs, & E. J. Ludwig
University of North Carolina, Chapel Hill, N. C. 27514 USA &
Triangle Universities Nuclear Laboratory, Durham, N. C.

The structure of ^{206}Pb has previously been studied by the ^{207}Pb(d,t)^{206}Pb [1]) and the ^{207}Pb(p,d)^{206}Pb [2]) reactions, using cross section measurements and sum rules to establish nuclear configurations. We have studied the ^{207}Pb($\vec{d}$,t)^{206}Pb reaction using 15-MeV vector-polarized deuterons to obtain cross section and vector analyzing power (VAP) distributions in 5° steps from 25° to 110° for approximately 10 levels below 3.12-MeV excitation.

Since the spin of the ground state of ^{207}Pb is $1/2^-$ the only allowed angular momentum transfer for a 0^+ state in ^{206}Pb, such as the ground state or the 1.17-MeV excited state, is $1/2^-$. For the 2^+ states, such as at 0.80 and 1.47 MeV, two transfers of different ℓ - and j-values, $p_{3/2}$ and $f_{5/2}$ are allowed, and for 4^+ states, such as at 2.93 MeV, transfers of $f_{7/2}$ and $h_{9/2}$ are allowed.

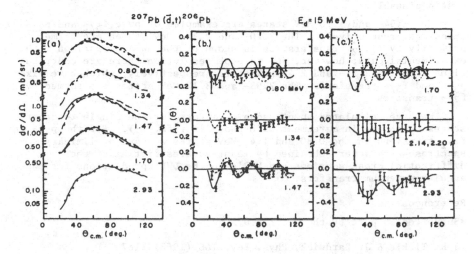

Fig. 1. Differential cross sections and vector analyzing powers for ^{207}Pb(d,t)^{206}Pb. The curves are DWBA predictions using the computer code DWUCK[3]) for the j-transfers noted. Solid curves are $j = \ell + 1/2$ transfers and dashed curves are $j = \ell - 1/2$ transfers. Final states in ^{206}Pb and angular momentum transfers shown are: 0.80, $p_{3/2}$ and $f_{5/2}$; 1.34, $f_{5/2}$; 1.47, $p_{3/2}$ and $f_{5/2}$; 1.70, $p_{1/2}$ and $p_{3/2}$; 2.20, $i_{13/2}$; and 2.93, $f_{7/2}$.

Selected experimental differential cross section and VAP distributions are shown in fig. 1. The 1.70-MeV state results from a $p_{3/2}$ transfer and illustrates that for a single ℓ transfer, the VAP distribution discriminates the j transfer more clearly than the differential cross section distribution. The DWBA predictions for the 0.80- and 1.47- MeV states (populated by a mixture of p and f waves) illustrate that the differential cross sections are sometimes superior to the VAP distributions in distinguishing the ℓ transfer. This results from the similarity in phase of the VAP distributions for different j transfers contributing to the same final state.

Both our DWBA predictions and data illustrate the Q- value dependence of the VAP for a (d,t) sub-coulomb transitions[4]. As the Q-value goes from zero toward negative values the VAP becomes increasingly negative for states formed by $j = \ell + 1/2$ transfer and increasingly positive for those where $j = \ell - 1/2$. For example, there is a change in relative position of the $p_{3/2}$ and $f_{5/2}$ DWBA predictions for the 0.80-and 1.47-MeV states. This effect is stronger still for large ℓ - value transfers as might be expected from the semi-classical description[4].

Theoretical calculations predict that for states in ^{206}Pb with low excitation energy (< 1 MeV) the VAP for all j-transfers with $j = \ell + 1/2$ will be on the average slightly positive and for all $j = \ell - 1/2$ transfers slightly negative. For states of higher excitation energy (>2 MeV), $j = \ell + 1/2$ transfers become significantly positive in VAP, and $j = \ell - 1/2$ transfers become significantly negative, so that even if the vector analyzing powers are in phase, the j-transfers are easily distinguishable.

The 1.34- and 1.70-MeV states are examples of pure $f_{5/2}$ and $p_{3/2}$ transfers, respectively. The 2.20- and 2.38- MeV states are formed primarily by $i_{13/2}$ transfers, as is shown by the DWBA comparison with experimental data; however, due to poor resolution there are contributions from the 2.14- and 2.43-MeV states respectively, which have mixed $p_{3/2}$ and $f_{5/2}$ transfers. The 2.93- and 3.12-MeV states are formed by $f_{7/2}$ transfers.

We have calculated spectroscopic factors for the single ℓ-transfer states which are in agreement with those in refs.[1,2]. For the $p_{3/2}$ and $f_{5/2}$ transfers in the 0.80 and 1.47-MeV states we have preliminary spectroscopic factors obtained by least-squares fitting of the differential cross sections and VAP's to the DWBA predictions for both angular momentum transfers.

References

*Supported in part by the U.S.E.R.D.A.

1) R. Tickle & J. Bardwick, Phys. Rev. 166 (1968) 1167
2) W. A. Lanford & G. M. Crawley, Phys. Rev. C9 (1974) 647
3) P. D. Kunz, University of Colorado, private communication
4) S. E. Vigdor, et al., Nucl. Phys. A210 (1973) 93

D-STATE EFFECTS IN ($\vec{d}$,t) REACTIONS[*]

L.D. Knutson[†] and B.P. Hichwa
University of Wisconsin, Madison, Wisconsin 53706

The effect of the deuteron D state in ($\vec{d}$,p) reactions is now quite familiar[1]). The D state has little effect on the cross section and vector analyzing power of the reaction, but has a large effect on the tensor analyzing powers. It is expected that similar D state effects should be present in all deuteron-induced single-nucleon transfer reactions. In this paper we present measurements of the tensor analyzing powers for ($\vec{d}$,t) reactions on ^{118}Sn and ^{208}Pb at energies near the Coulomb barrier. The measurements are compared with distorted-wave calculations which include D state effects. We argue that the observed tensor analyzing powers arise primarily from the D state components in the triton wave function, rather than from the deuteron D state.

In the distorted-wave Born approximation, the transition amplitude for the reaction A($\vec{d}$,t)B is given by[2])

$$T = \int \chi_d^{(-)*}(\vec{r}_d) < Ad|V_{dn}|tB > \chi_t^{(+)}(\vec{r}_t) d^3r_t d^3r_d \tag{1}$$

The matrix element in eq. (1) can be written in the form

$$< Ad|V_{dn}|tB > = < A|B > < d|V_{dn}|t > \tag{2}$$

From the Schrodinger equation for the deuteron and triton one obtains[2])

$$< d|V_{dn}|t > = \left[\frac{3\hbar^2}{4M} \nabla^2 - B_t + B_d \right] R(\vec{r}) \tag{3}$$

where

$$R(\vec{r}) = < d|t > = \int \phi_d^*(\vec{p}) \phi_t(\vec{r},\vec{p}) d^3p \tag{4}$$

Here ϕ_d and ϕ_t are the internal wave functions for the deuteron and triton respectively. The quantity $\vec{p}$ is the deuteron internal coordinate and $\vec{r}$ is the separation of the neutron from the deuteron center of mass.

From general angular momentum and parity restrictions, it is possible to show that the function R can contain only S and D state terms. Specifically we can write

$$R(\vec{r}) = -\frac{\sqrt{3}}{2} \sum_{L=0,2} \sum_{\Lambda,\sigma_d'\sigma_n} u_L(r) Y_L^{\Lambda *}(\hat{r}) <1\sigma_d',L\Lambda|1\sigma_d> < \tfrac{1}{2}\sigma_n,1\sigma_d'|\tfrac{1}{2}\sigma_t >|\chi_{\frac{1}{2}}^{\sigma_n}>$$

where $|\chi_{\frac{1}{2}}^{\sigma_n}>$ is the spin function for the transferred neutron. The tensor analyzing powers for a ($\vec{d}$,t) reaction are sensitive to the D state term in $R(\vec{r})$. This term can be non-zero if either the deuteron or triton wave function contains a D state. From the deuteron D state, one obtains terms proportional to $Y_2^m(\hat{p})$ in the integral for $R(\vec{r})$. These terms will integrate to zero unless the triton wave function depends on the angular coordinates of $\hat{p}$. Thus one expects that the contribution from the deuteron D state will be relatively small (the deuteron D state has little effect because the deuteron is a spectator in the ($\vec{d}$,t) reaction). From the triton D state wave functions[3]) one can obtain terms which are proportional to either $Y_2^m(\hat{p})$ or $Y_2^m(\hat{r})$. In this latter case the spherical harmonic can be taken outside the integral and one directly obtains an L=2 term in $R(\vec{r})$. Thus, we expect that the tensor analyzing powers for ($\vec{d}$,t) reactions are primarily sensitive to the triton configuration in which one neutron has orbital angular momentum L=2 relative to the remaining nucleons.

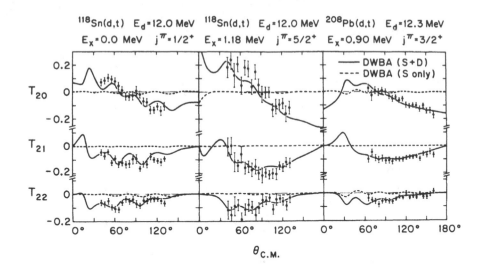

Fig. 1. Angular distributions of the three tensor analyzing powers for (d⃗,t) reactions on ^{118}Sn and ^{208}Pb. The solid curves are DWBA calculations which include D state effects, and the dashed curves are calculations in which the D state has been omitted.

In fig. 1 we present the measured tensor analyzing powers for some strong (d⃗,t) transitions. It is interesting that the angular dependence of the analyzing powers is similar to that observed for sub-Coulomb (d⃗,p) reactions, except that the signs are reversed. The dashed curves in fig. 1 show the DWBA calculations which are obtained when one includes only the L=0 term in R(r⃗). These calculations greatly underestimate the magnitude of the tensor analyzing powers. The solid curves are DWBA predictions obtained by including a D state term.

These calculations were carried out using the local-energy approximation[2]. In this approximation the D state effect depends only on the value of a single parameter, D_2. This parameter is related to the functions u_L by[2]

$$D_2 = \frac{1}{15} \int_0^\infty u_2(r)\, r^4 dr \Big/ \int_0^\infty u_0(r)\, r^2 dr \qquad (6)$$

In the calculations, D_2 was treated as an adjustable parameter. The value $D_2 = -0.24$ fm^2 provided the best fit to the measurements and was used for the calculations shown in fig. 1.

We suggest that tensor analyzing power measurements such as those presented in fig. 1 can be used to obtain new quantitative information about the D state components in the triton wave function.

References

*   Work supported in part by the U.S. Atomic Energy Commission
†   Present address: Nuclear Physics Laboratory, University of Washington, Seattle, Washington 98195

1)   L.D. Knutson, E.J. Stephenson, N. Rohrig and W. Haeberli, Phys. Rev. Lett. 31 (1973) 392
2)   F.D. Santos, Nucl. Phys. A212 (1973) 341
3)   R.G. Sachs, Nuclear theory (Addison Wesley, Cambridge, 1953) p. 180

THE TRITON D-STATE AND (d,t) REACTIONS

A. Barroso[*], A. M. Eirö[*] and F. D. Santos

Laboratório de Física, Faculdade de Ciências, Lisboa, Portugal

It is well known that the tensor analysing power T_{2q} of (d,p) reactions is particularly sensitive to the deuteron D-state. This suggests that the T_{2q} of (d,t) reactions may be a useful source of information on the triton D-state. The DWBA transition amplitude depends on the triton wave function through the overlap integral in $\vec{\rho} = \vec{r}_2 - \vec{r}_1$

$$< \chi_{1/2}^{\sigma_3} \chi_1^{\sigma_d} | \chi_{1/2}^{\sigma_t} > = \sum_{s,L=0,2} u_L(r) Y_L^\Lambda(r) (L\Lambda s\sigma|1/2\sigma_t)(1\sigma_d 1/2\sigma_x|s\sigma) \quad (1)$$

where $\chi_1^{\sigma_d}$, $\chi_{1/2}^{\sigma_t}$ are the deuteron and triton wave functions with spin projections σ_d, σ_t and $\chi_{1/2}^{\sigma_3}$ is the spin wave function of the transferred neutron. Due to spin and parity selection rules this overlap is a mixture of an S and D-state in $\vec{r} = \vec{r}_3 - (\vec{r}_1 + \vec{r}_2)/2$ formally analogous to the deuteron wave function. Thus we can apply the local energy approximation as developed in Ref. 1 to the study of finite range effects in low and medium energy (d,t) reactions. In this approximation all information on the overlap D-state is contained in the single parameter

$$D_2 = \int_0^\infty r^4 u_2(r) dr \Big/ 15 \int_0^\infty r^2 u_0(r) dr \quad (2)$$

which we have calculated using for the deuteron the Reid[2] soft core wave function and for the triton the wave function of Ref. 3 with only the dominant S-state, u_S. The inclusion of a complete D-state would be a more laborious process and since D_2 is most sensitive to the region of large internucleon distances we felt that it would be a good approximation to use a triton D-state obtained in first order perturbation theory[4]. In momentum space it is given by

$$< \vec{p},\vec{q} | D > = \left[N / (B_t - \hbar^2 p^2/M - 3\hbar^2 q^2/4M) \right] < \vec{p},\vec{q} | V_T^{12} + V_T^{23} + V_T^{31} | S > \quad (3)$$

where $\vec{p}$ and $\vec{q}$ are the momenta conjugate with $\vec{\rho}$ and $\vec{r}$, B_t the triton binding energy, V_T^{ij} the Reid soft core tensor force between nucleons i and j, | S> the triton S-state[3] and N a normalization constant which gives the total D-state probability. The overlap between the deuteron S-state and the above triton D-state shows that just the tensor force between the pairs 23 and 13 gives rise to a D-state in $\vec{r}$. This is the major contribution to the overlap D-state. The overlap between the deuteron D-state and triton S-state is likely to be small[5] and is zero in our approximation since it is proportional to the coefficient of $\left[Y_2^m(\vec{\rho})\times Y_2^\mu(\vec{r})\right]^0$ in the expansion of $u_s(\vec{r}_1,\vec{r}_2,\vec{r}_3)$. Since the triton D-state given by eq(2) depends on N our final result for D_2 is a function of the total triton D-state probability P_D. We obtain

$$D_2 = - 0.648 \left[P_D / (1 - P_D)\right]^{1/2} fm^2. \tag{4}$$

For[6] $P_D = 0.088$ this gives $D_2 = - 0.20$ fm^2 which is in good agreement with the value[7] of $D_2 = - 0.24$ fm^2 obtained from DWBA analysis of (d,t) T_{2q} angular distributions. The fact that D_2 is positive in (d,p) and negative in (d,t) is due to tensor force properties. For a spin up triton there are two spin configurations in which the overlap given by eq(1) has simultaneous contributions from the S and D-states;

$$< \chi_{1/2}^{1/2} \chi_1^0 |\chi_{1/2}^{1/2} > = - \frac{1}{\sqrt{3}} u_o(r) \frac{1}{\sqrt{4\pi}} +\sqrt{\frac{2}{15}} u_2(r)Y_2^o(\hat{r}), \tag{5}$$

$$< \chi_{1/2}^{-1/2} \chi_1^1 |\chi_{1/2}^{1/2} > =\sqrt{\frac{2}{3}} u_o(r) \frac{1}{\sqrt{4\pi}} + \frac{1}{\sqrt{15}} u_2(r)Y_2^o(\hat{r}). \tag{6}$$

By considering the relative probability of finding in the triton S-state a nucleon pair in a triplet spin up state we conclude that the configuration (5) is favoured and (6) not favoured by the tensor force. Thus u_o and u_2 have opposite signs and the triton is prolate in (5) and oblate in (6). D_2 being a measure of the triton asymmetry we can estimate the triton "intrinsic quadrupole moment". Neglecting the term involving u_2^2 we obtain $Q_2 = \pm \frac{\sqrt{2}}{3} D_2$ where the + and - signs correspond respectively to the amplitudes (5) and (6). Using the experimental value[7] of D_2 we obtain a triton "intrinsic quadrupole moment" of $|Q_2| = 0.11$ fm^2.

References

* Research sponsored in part by Instituto de Alta Cultura(proj.LF-1-II)
1) R.C. Johnson and F. D. Santos, Particles and Nuclei 2(1971)285
2) R.V. Reid, Ann. of Phys. 50(1968)411
3) A.D. Jackson, A. Lande and P. Sauer, Nucl. Phys. A156(1970)1
4) A. D. Jackson and D. O. Riska, Phys. Lett. 50B(1974)207
5) A. Barroso, A.M. Eiró and F.D. Santos, to be published
6) R.A.Brandenburg and Y.E.Kim, Phys. Lett. 49B(1974)205
7) B.P.Hichwa and L.D.Knutson,Bull. Am. Phys. Soc. 19(1974)1019 and
   private communication.

VECTOR ANALYSING POWERS FOR THE (d,α) REACTION ON ^{14}N, ^{16}O AND ^{28}Si*

E. J. Ludwig, T. B. Clegg, P. G. Ikossi and W. W. Jacobs
University of North Carolina, Chapel Hill, N. C. 27514 and
Triangle Universities Nuclear Laboratory, Durham, N. C.

Cross section and vector analysing power (VAP) distributions have been measured over an extended angular range for the three lowest-lying states of ^{12}C in the ^{14}N(d,α) ^{12}C reaction at 15 MeV, for two low-lying states of ^{14}N populated in the ^{16}O (d,α) ^{14}N reaction at 16 MeV and for six particle groups populated in the ^{28}Si(d,α) ^{26}Al reaction at 16 MeV. The measurements were made to investigate deuteron-cluster pickup in the reaction and to determine whether the VAP distributions carry information concerning the value of the total angular momentum transfer. There have been few previous studies of the VAP for (d,α) reactions.

Angular distributions for several of the groups mentioned above are shown in fig. 1. The cross-section angular distributions are forward-peaked but generally level off at angles beyond 80°. This is indicative of a substantial compound nucleus contribution at backward angles. From 20° to 30° the direct reaction contribution to the cross section for the ^{28}Si(d,α) ^{26}Al(0.42-MeV) reaction has been estimated[1]) to be approximately 70%. Hence only data at forward angles will be significant in the discussion of this reaction in terms of a direct process.

Calculations of the cross section and VAP distributions were made with the distorted wave code DWUCK, assuming either a deuteron cluster pickup from the nucleus, or the pickup of separate nucleons from the shell-model states. Both assumptions produced similar results for the cases considered. The largest differences in the calculations of the VAP occurred for the ^{14}N(d,α) ^{12}C(g.s.) reaction which are shown in fig. 1a. There are apparent differences in the VAP distributions for the residual ^{12}C states since the ground state corresponds to an L=2, J=1 transfer while the 4.43-MeV state data resemble the prediction for an L=2. J=2 transfer.

The predictions shown with the data for the ground state of ^{26}Al (fig. 1c) indicate a substantial sensitivity to the J transfer. Each curve shown with the experimental data is calculated for an L=4 transfer. There is good agreement between the solid curve (J=5) and the measured analysing powers for this state which has J^{π} = 5$^+$. Similar sensitivity to the J transfer was obtained for the transition to the 0.42 MeV-state, which appears in closest agreement with the L=2, J=3 prediction shown.

The states populated in the ^{16}O(d,α) ^{14}N reaction both have $J^{\pi}=1^+$ and involve the same transfer of L and J. There is an overall similarity in the VAP distributions although they correspond to reactions with different Q-values. Extensive DWUCK calculations have not been made; however, preliminary predictions for the VAP (fig. 1b) reproduce some of the features of the data.

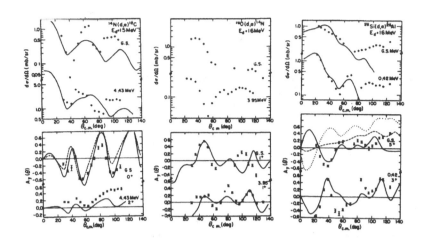

Fig. 1 a) The differential cross-section and VAP distributions for the $^{14}N(d,\alpha)$ ^{12}C(g.s. and 4.43-MeV) reaction are shown along with distorted wave calculations. The dashed line with the VAP distribution for the ^{12}C(g.s.) represents a prediction for cluster pickup while the solid line assumes the pickup of a pair of 1p nucleons from ^{14}N. b) Angular distributions for the ^{16}O (d,α) ^{14}N(g.s. and 3.95- MeV) are shown with preliminary calculations for the VAP assuming L=2, J=1. c) Angular distributions for the $^{28}Si(d,\alpha)$ ^{26}Al(q.s.) are shown with calculations assuming an L=4 transfer. The solid curve assumes J=5, the dashed curve J=4 and the dotted curve J=3. The distributions for the 0.42-MeV state are shown with an L=2, J=3 prediction.

These studies show that the VAP distributions are useful in determining the J-value transferred in a (d,α) reaction just as in single nucleon transfer reactions[2]). Cluster pickup calculations reproduce the features of the data and differ little from the two-nucleon transfer predictions for the cases studied. Finally, the apparent compound nucleus contributions to these reactions indicate that higher energies or heavier targets are desirable for detailed agreement with DWBA calculations.

References

* Work supported in part by U.S.E.R.D.A.

1) A. Richter et al., Phys. Rev. C2 (1970)1361
2) T. J. Yule and W. Haeberli, Phys. Rev. Lett. 19 (1967)756

DWBA CALCULATION FOR ^{27}Al($\vec{d}$,α)^{25}Mg REACTION

O.F.Nemets,Yu.S.Stryuk,A.M.Yasnogorodsky

Institute for Nuclear Research, Ukrainian Academy of Sciences
Kiev,USSR

The progress in the field of polarized ion sources makes it possible the polarization phenomena study in such reactions of relatively low probability as ($\vec{p}$,α) and ($\vec{d}$,α). For (p,α) reactions (or their inverse (α,p)) the domination of direct processes is manifest[1] at energy near 30 MeV, but for (d,α) reactions at energy near 10 MeV the situation is more complicated.

It is believed the structure of excitation function and the residual spin dependence of the averaged differential cross-section of ^{27}Al(d,α)^{25}Mg reaction at near 10 MeV deuteron energy provide an evidence for compound nucleus formation in the region of many overlapping levels. Under these condition, according to the statistical theory, the averaged in E_d and θ analysing power $\langle iT_{11}\rangle$ is to be zero. However, measured recently $\langle iT_{11}\rangle$[2] reaches large values which could be attributed to direct reaction contribution. So it is of interest to discuss the prediction of direct reaction theories.

In this work using the MIVPOL computer code the $T_{kq}(\theta)$ are calculated at $E_d = 13.6$ MeV within the DWBA model of deuteron pick-up from 2d-state, the depth of bound state Woods-Saxon well being chosen to account the deuteron binding energy in the target nucleus, and $a_{bs} = 0.6$ fm, $r_{bs} = 1.4$ fm, $V_{so}^{bs} = 6.5$ MeV. The parameters of deuteron optical potential are those of ref.[3]: $V = 121$ MeV, $a_v = 0.8$ fm, $r_v = 0.96$ fm; $W = 24.5$ MeV (surface absorbtion), $a_w = 0.61$ fm, $r_w = 1.33$fm; $V_{so} = 6.5$ MeV, $a_{so} = a_v$, $r_{so} = r_v$.

Predicted T_{kq} values are weakly sensitive to α-optical potential uncertainty for the sets of real well depth $V \geqslant 4V_N$ (V_N is the nucleon optical potential real well depth). That is displayed for iT_{11} in fig.1a, the numeration of curves corresponding to α-potential sets of ref.[1].

All T_{kq} show pronounced j-dependence (displayed for iT_{11} and T_{20} in fig.1b) on the transfered deuteron total angular momentum $\vec{j} = \vec{l}+\vec{s}$. That raises hopes of analysing power measurements usage to obtain the spectroscopic information (j-selection or evaluation of different j contributions) which is inaccessible in the cross-section analysis since the theoretical angular distributions for given l and different j are practically identical.

For two nucleon transfer reactions these hopes are essential since the angular momentum mixing is possible.

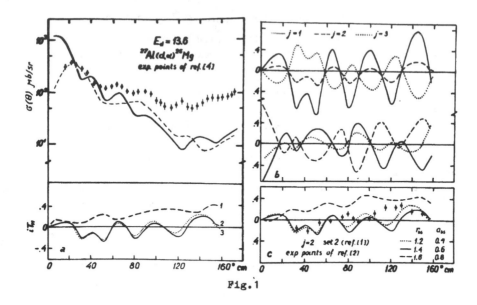

Fig.1

The detailed analysis of above questions is more expedient at higher energies since, because of possible compound nucleus formation contributions at energies under consideration, the determination of bound state well geometry parametrs from the analysis of differential cross-section is ambiguous. The effects of mentioned ambiguties on $iT_{11}(\theta)$ are shown in fig.1c.

References

1) O.F.Nemets,Yu.S.Stryuk,A.M.Yasnogorodsky, See these Proceedings, p.713
2) Y.Takeuchi,J.A.R.Griffith,O.Karban,S.Roman, Nucl.Phys. A220 (1974)589.
3) A.N.Vereshchagin et al., Izv. AN SSSR, ser.fiz. 33(1969)2064.
4) Yu.S.Stryuk,V.V.Tokarevsky, Program and Abstracts of Reports to the 23d Conference on Nuclear Spectroscopy and Nuclear Structure(Leningrad,1973), p.267.

Two nucleon transfer reactions induced by polarized deuterons

J.M. Nelson, O. Karban, E. Ludwig* and S. Roman
Department of Physics, University of Birmingham, England

The analysing powers of the $^{19}F(\vec{d},\alpha)^{17}O$ and $^{40}Ca(\vec{d},\alpha)^{38}K$ reactions have been measured at an incident deuteron energy of 12.3 MeV and are shown in fig. 1. The experimental method is described briefly in ref. 2. The ^{19}F data are of primary interest in this experiment in which ^{40}Ca forms an interesting contaminant. The experiment was performed as part of a continuing investigation of two nucleon transfer reactions on ^{19}F [1]. These reactions are characterized by the multiple values of J and L transferred. The spin transfer S is uniquely 1.

Finite-range calculations have been carried out in which all the possible (JL) transition amplitudes are correctly summed and weighted with the appropriate spectroscopic amplitudes [3]. A program has been written to calculate these amplitudes from wave functions generated by a version [4] of the Oak Ridge-Rochester shell model code. In this case, the wave functions were generated from the s-d shell matrix elements of Arima et al [5]. The two ^{17}O ground state predictions correspond to (a) J=2 only and (b) the sum of J=2 and J=3 contributions. The fit to the data suggests that the J=2 transfer is dominant although the theoretical spectroscopic amplitudes clearly indicate the importance of the J=3 transfer. The summed transfers provide better agreement between cross section data and theory than the J=2 transfer alone. A coherent summation of L=0 and L=2 transfers for J=1 is required for the 0.871 MeV transition and the fit to the data is good.

A preliminary calculation for the transition to the negative parity state at 3.05 MeV is also shown. The fit to both the analysing power and cross section data is only moderately good.

Spectroscopic amplitudes are now being calculated to allow the DWBA analysis of the ^{40}Ca data to proceed. It is interesting to note the very different character of the analysing powers of the reactions leading to the two 1^+ states.

It is clear from the calculations that have been made and the strength of the observed polarization effects that the analysing powers of these reactions are even more sensitive to the construction of the form factor, and hence the spectroscopic amplitudes, than the cross sections. This sensitivity is to be studied in further $(\vec{d},\alpha)$ reactions in the s-d shell.

References

* On leave from the Dept. of Physics, University of North Carolina
1) J.M.Nelson and W.R.Falk, Nucl. Phys. A218 (1974) 441
2) J.M.Nelson, O.Karban, E.Ludwig and S.Roman, Contribution to these proceedings, p.679
3) J.M.Nelson and B.E.F.Macefield, University of Birmingham Report 74-9 (1974) unpublished
4) A.P.Zuker, private communication
5) A.Arima, S.Cohen, R.D.Lawson and M.H.MacFarlane, Nucl. Phys. A108 (1968) 94

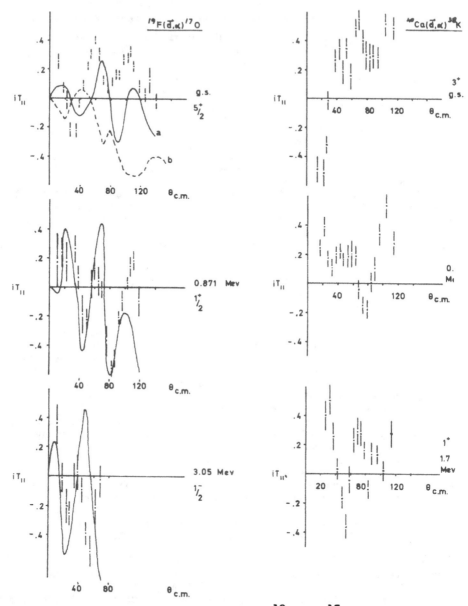

Fig. 1. Vector analysing powers of the $^{19}F(\vec{d},\alpha)^{17}O$ and $^{40}Ca(\vec{d},\alpha)^{38}K$ reactions induced by 12.3 MeV polarized deuterons. The curves are the results of DWBA predictions.

SPIN-PARITY COMBINATIONS IN s-d SHELL NUCLEI DETERMINED WITH A t_{20}-POLARIZED DEUTERON BEAM

D.O. Boerma*, W. Grüebler, V. König, P.A. Schmelzbach and R. Risler
E.T.H., Zürich, Switzerland

The analyzing power T_{20} at $\theta = 0^{\circ}$ or 180° for a (d,α) reaction on a $J^{\pi} = 0^{+}$ target nucleus: $0^{+}(d,\alpha)J_{F}^{\pi}$, characterized by the helicity amplitudes $M_{m_f}^{m_d}$, can be written as[1])

$$T_{20}(0^{\circ},180^{\circ}) = \sqrt{2} \frac{|M_1^1|^2 - |M_0^0|^2}{2|M_1^1|^2 + |M_0^0|^2} \tag{1}$$

Here m_d and m_f are the magnetic quantum numbers of the deuteron and of the final state. Due to parity conservation we have $M_0^0(\theta) = 0$ for $(-)^{J_F} = \pi$ (natural parity), so we get:

$$T_{20}(0^{\circ},180^{\circ}) = \tfrac{1}{2}\sqrt{2} \quad \text{(maximum value) for } (-)^{J_F} = \pi \tag{2}$$

$$T_{20}(0^{\circ},180^{\circ}) = -\sqrt{2} \quad \text{(minimum value) for } J_F^{\pi} = 0^{-} \tag{3}$$

$$\frac{d\sigma}{d\Omega}(0^{\circ},180^{\circ}) = 0 \qquad \text{for } J_F^{\pi} = 0^{+} \tag{4}$$

This means that if at any deuteron energy E_d, with the α-particles observed near $\theta = 0^{\circ}$ or 180° a value of T_{20} far from the maximum value is observed, then the populated level has <u>unnatural</u> parity. This rule cannot be applied in a reversed way without the introduction of a model for the reaction mechanism.

In case the reaction mechanism can be described mainly with the statistical model and J_F^{π} is unnatural the values of $|M_0^0|^2$ and $|M_1^1|^2$ have independent χ^2-distributions with two degrees of freedom around the (deuteron-) energy averaged values $<|M_0^0|^2>$ and $<|M_1^1|^2>$. Using Hauser-Feshbach type of expressions[2]) for these averaged squared amplitudes, and inserting the relations for the special Clebsch-Gordan coefficients occurring, it is possible to derive the relation: $<|M_0^0|^2> \approx 2<|M_1^1|^2>$ for $\theta = 0^{\circ}$ or 180°. With this we get the unexpected result for J_F^{π} unnatural: $<T_{20}(0^{\circ},180^{\circ})> \simeq -\tfrac{1}{4}\sqrt{2}$, with the T_{20} values distributed <u>uniformly</u> from $-\sqrt{2}$ to $\tfrac{1}{2}\sqrt{2}$ (assuming measurement at many deuteron energies). If now, at a number of deuteron energies in an E_d region where the reaction mechanism can be described by statistical theory, values of $T_{20}(0^{\circ},180^{\circ})$ (or $\frac{d\sigma}{d\Omega}(0^{\circ},180^{\circ})$) are observed all of which are compatible with (2) (or (3), or (4)) then, in an obvious way, the probability can be calculated that J_F^{π} is <u>natural</u> (or $J_F^{\pi} = 0^{-}$, or 0^{+}). Here the separation between the different E_d values should be much larger than the coherence width.

Since in practice the α particles are observed at 175°, the amplitudes M_0^1, M_0^2, etc. will give small contributions to the value of T_{20}. For an estimated effect of $|M_0^1|^2 + |M_0^2|^2 + \ldots = 10\%$ of the value of $|M_1^1|^2$, the value of $T_{20}(175^{\circ})$ for J_F^{π} natural will decrease to 0.58. The value of $T_{20}(175^{\circ})$ for $J_F^{\pi} = 0^{-}$ will increase to -1.06 due to a similar effect. The uniform distribution of T_{20} values for $(-)^{J_F} \neq \pi$ will fall off at the extreme ends.

Using similar techniques as described elsewhere[3]), values of T_{20} ($\theta_{lab} = 175^{\circ}$) have been measured for levels populated in the reactions: $^{32}S(\vec{d},\alpha)^{30}P$, $^{30}Si(\vec{d},\alpha)^{28}Al$, $^{26}Mg(\vec{d},\alpha)^{24}Na$, $^{34}S(\vec{d},\alpha)^{32}P$, $^{28}Si(\vec{d},\alpha)^{26}Al$ and $^{24}Mg(\vec{d},\alpha)^{22}Na$. In each case spectra for t_{20} positive and t_{20} negative

694

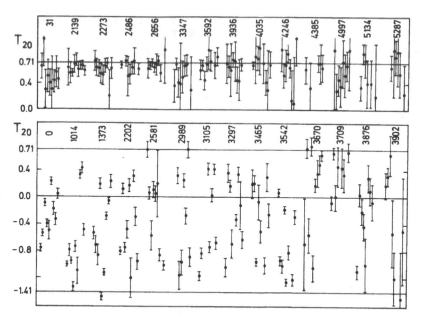

Fig. 1. Part of the measured T_{20} values for the $^{30}Si(\vec{d},\alpha)^{28}Al$ reaction to final levels, marked by the excitation energy in keV, with unnatural parity (bottom) or which have (probably) natural parity (top). Within each group of T_{20} values E_d increases from left to right.

were taken at 6-13 deuteron energies between 8 and 12 MeV. The important parts of the spectra contained α-particle peaks only, since 60-100 μm thick annular silicon surface barrier detectors (cooled to -50°C) were used, in which deuterons and protons only lost a small part of the energy, but in which the α particles were just stopped. The spectra were analysed with a computer program in which the peak positions were fixed at values calculated from the α-particle energy determined by the kinematics and the (linear) calibration. All measured cross sections were found to fluctuate, and the distribution of the T_{20} values for J_F^{π} unnatural was found to be uniform, in accordance with the statistical theory. In up to 40 cases per nucleus studied, either certain assignments of unnatural parity, or probable assignments of natural parity, $J^{\pi} = 0^+$ or $J^{\pi} = 0^-$ could be made. Combined with other spectroscopic data extracted from the literature like ℓ_n, ℓ_p values, etc., in many cases this new information led to definite J^{π} assignments. A part of the results is given in fig. 1.

References

* Present address: L.A.N., University of Groningen, The Netherlands

1) M. Simonius, in Lecture Notes in Physics 30, (Springer Verlag, Berlin, 1974) p. 38
2) W. Hauser and H. Feshbach, Phys. Rev. 87 (1952) 366
3) W. Grüebler, P.A. Schmelzbach, V. König, R. Risler, B. Jenny and D.O. Boerma, Nucl. Phys. A242 (1975) 285

USE OF THE $\vec{(d,\alpha)}$ REACTION TO DETERMINE PARITY, 0⁻ LEVELS AND t_{20} FOR THE BEAM

J.A. Kuehner, P.W. Green, G.D. Jones, D.T. Petty, J. Szücs and H.R. Weller
Tandem Accelerator Laboratory
McMaster University, Hamilton, Canada

A model-independent method for determining the parity of a level as either $\pi=(-)^J$ or $\pi=(-)^{J+1}$, for the unique identification of 0⁻ levels, and for absolute determination of the tensor polarization (t_{20}) of the beam is described.

It can be shown[1,2], using simple angular momentum coupling arguments, that for 100% tensor polarized deuterons $(t_{20}=-\sqrt{2})$ with the quantization axis along the beam direction and for a spin-zero target, the (d,α) reaction can have yield at 0° or 180° only for levels in the final nucleus which have unnatural parity $(\pi=(-)^{J+1})$. This fact allows the determination of parity for levels in odd-odd nuclei of known spin.

The method is demonstrated using the $^{16}O(d,\alpha)^{14}N$ reaction at 14 MeV. A GeO_2 target and a magnetic spectrograph were used. Fig. 1 shows the ratio of the yield for a tensor polarized beam to that for an unpolarized beam at 3° for several known levels of ^{14}N and also for the 0.72 MeV level of ^{10}B fed in the $^{12}C(d,\alpha)^{10}B$ contaminant reaction. Since the polarization was not 100% there is some yield allowed with the polarized beam for the two natural parity levels. It can be shown that the yield ratio must be $1 + \dfrac{t_{20}}{\sqrt{2}}$ for natural parity levels and $1 - \sqrt{2}\,t_{20}$ for 0⁻

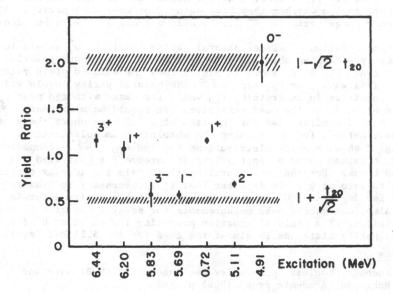

Fig. 1. The ratio of yield for a tensor polarized beam to that for an unpolarized beam is plotted for several levels in ^{14}N populated in the $^{16}O(d,\alpha)^{14}N$ reaction at 3°. One level, labelled 0.72, appeared as a contaminant from the $^{12}C(d,\alpha)^{10}B$ reaction.

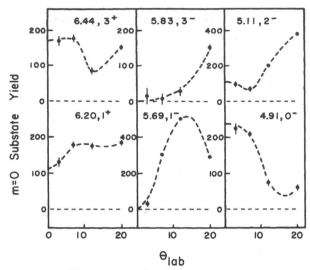

Fig. 2. The $t_{20} = -\sqrt{2}$ (m=0 substate) contribution to the yield is plotted for angles near 0° for several levels of ^{14}N populated in the $^{16}O(d,\alpha)^{14}N$ reaction at 14 MeV beam energy. Those levels for which the yield goes to zero at 0° are natural parity $(\pi=(-)^J)$.

levels. For unnatural parity levels the yield ratio must lie between these two limits. It is clear that the method allows classification of levels according to whether they have natural or unnatural parity. Also, the uniquely large ratio for 0⁻ levels allows these levels to be identified.

The observation of either natural parity levels or 0⁻ levels in this geometry allows an absolute determination of the tensor polarization parameter t_{20}. The bands shown in Fig. 1 show the allowed yield ratio fitted in this way. For t_{20} near $-\sqrt{2}$ the natural parity levels will be the most sensitive in determining t_{20} while for cases with t_{20} near $\frac{1}{\sqrt{2}}$ the 0⁻ levels will be the most sensitive. It should be mentioned that the method applies also for the case of a $^6\vec{Li}$ beam where there may not be many methods for determining the absolute beam polarization.

Fig. 2 shows angular distributions for the $t_{20} = -\sqrt{2}$ component of the beam, obtained using an appropriate difference for polarized and unpolarized beams. For the two natural parity levels the angular distributions go to zero at 0°. It is clear that it is necessary to measure within a few degrees of 0° in order to make reliable parity determinations. Also, one should make measurements for several energies to avoid the possibility of a yield fluctuation producing low yield at 0° for an unnatural parity state, as is almost the case for the 5.11 MeV level.

References

1) J.A. Kuehner, Nuclear Spin-Parity Assignments ed. N.B. Gove and R.L. Robinson (Academic Press 1966) p. 146
2) J.A. Kuehner, P.W. Green, G.D. Jones and D.T. Petty, submitted to Phys. Rev. Lett.

VECTOR ANALYSING POWERS FOR ($\vec{d}$,^{6}Li) ON ^{12}C AND ^{16}O*

W. W. Jacobs, T. B. Clegg, P. G. Ikossi and E. J. Ludwig
University of North Carolina, Chapel Hill, N. C. 27514 and
Triangle Universities Nuclear Laboratory, Durham, N. C.

Cross section and vector analysing power (VAP) angular distributions for the reactions ^{12}C($\vec{d}$,^{6}Li) ^{8}Be(g.s.) and ^{16}O($\vec{d}$,^{6}Li) ^{12}C(g.s.) have been measured using a 16-MeV purely vector polarized deuteron beam from the T.U.N.L. Lamb-shift polarized-ion source. The SiO target was 170 µg/cm^2 thick mounted on a 20 µg/cm^2 C foil. The ^{6}Li particles were identified by standard ∆E-E detector telescopes employing thin (~12µ)∆E detectors.

The results for the differential cross section and analysing power measurements are shown in fig. 1. The cross-section angular distributions are forward peaked and their shape is similar to previously measured distributions[1,2]. The analysing powers, especially for the case of ^{12}C, show a strong oscillatory behavior. An earlier investigation[1] of the energy dependence of the ^{12}C(d,^{6}Li) cross sections near 15 MeV bombarding energy exhibited evidence for a direct transfer mechanism. Hence a preliminary analysis assuming an alpha-particle pickup was attempted using the computer code DWUCK. Initial optical model parameters were taken from studies[1,2] of the (d,^{6}Li) reaction

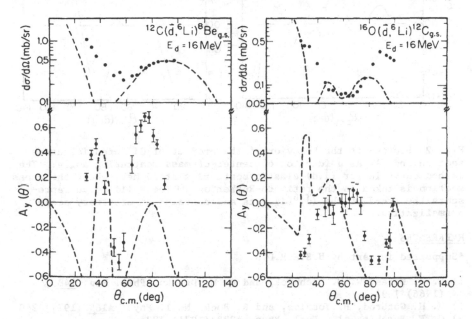

Fig. 1 Differential cross sections and analysing powers as a function of center-of-mass scattering angle for the reactions ^{12}C($\vec{d}$,^{6}Li)^{8}Be(g.s.) and ^{16}O($\vec{d}$,^{6}Li)^{12}C(g.s.). The error bars represent only statistical uncertainties. The dashed curves are preliminary DWUCK calculations.

at nearby energies, with a form factor calculated for an alpha particle bound in a 3S state. Small changes to the optical model parameters were made in an attempt to improve the fits. The addition of a small (1 MeV) ^{6}Li spin-orbit term produced a negligible effect.

In fig. 2 we compare the product of cross section and VAP data for the $(\vec{d},^6\text{Li})$ reaction with the same product for elastic $(\vec{d},d)$ scattering data taken at 15 MeV[3]). A similar comparison has been made for single-particle transfers[4]) involving L=0 in order to test the predictions of the weakly-bound projectile (WBP) model. No attempt was made in the figure to match the curves with respect to momentum transfer to the deuteron for the two different processes, although this produces a considerable effect. The similarity of shape provides some basis for considering the alpha particle as a spectator in the reaction. Although there is an arbitrary normalization of the $(\vec{d},^6\text{Li})$ data in fig. 2, the absolute VAP values indicate a slight enhancement for the $(\vec{d},^6\text{Li})$ data from ^{12}C over those previously measured for $(\vec{d},d)$, whereas for ^{16}O the values are of comparable magnitude for the two processes.

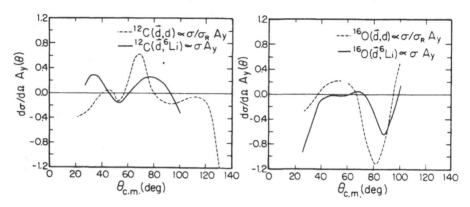

Fig. 2 Plotted is the behavior of the product of differential cross section and VAP as a function of center-of-mass scattering angle. The dashed curve is for $(\vec{d},d)$ elastic scattering at 15 MeV, where the cross section is taken as the ratio-to-Rutherford. The solid curve represents the trend of the $(\vec{d},^6\text{Li})$ data and contains an arbitrary scale normalization.

REFERENCES

*Supported in part by U.S.E.R.D.A.

1) L. J. Denes, W. W. Daehnick, and R. M. Drisko, Phys. Rev. 148 (1966) 1097.
2) H. H. Gutbrod, H. Yoshida, and R. Bock, Nucl. Phys. A165 (1971) 240.
3) C. E. Busch et al., Nucl. Phys. A233 (1974) 183.
4) S. K. Datta et al., Phys. Rev. Letters 31 (1973) 949.

ANALYZING POWER FOR THE $^{90}Zr(\vec{t},\alpha)^{89}Y$ AND $^{206,208}Pb(\vec{t},\alpha)^{205,207}Tl$ REACTIONS AT 17 MeV*

J.P. Coffin†, E.R. Flynn, R.A. Hardekopf, J.D. Sherman and J.W. Sunier
Los Alamos Scientific Laboratory, Los Alamos, New Mexico 87545, USA

Recent installation of a polarized triton source[1] at the Los Alamos Scientific Laboratory's Van de Graaff accelerator has permitted the measurement of the total spin value of proton hole states by means of the $(\vec{t},\alpha)$ reaction. This study has been carried out on targets of ^{90}Zr, ^{206}Pb, and ^{208}Pb using a 17 MeV beam of polarized tritons with an intensity of 30-50 nA on target. the reaction products were detected in a Q3D spectrometer using a one-meter-long helical proportional chamber in the focal plane.[2] The energy resolution of these experiments was 60 keV for the ^{90}Zr case (a 500 μg/cm² target), 20 keV for ^{206}Pb (150 μg/cm²), and 18 keV for ^{208}Pb (125 μg/cm²). The Q3D was operated at a solid angle of 14.2 msr, which resulted in exposure times of ∿ 10 min. per angle for the ^{90}Zr target to ∿ 1 hour for the ^{206}Pb target. Results at 10 angles were obtained for the ^{90}Zr and ^{208}Pb targets, and at four angles for the ^{206}Pb target.

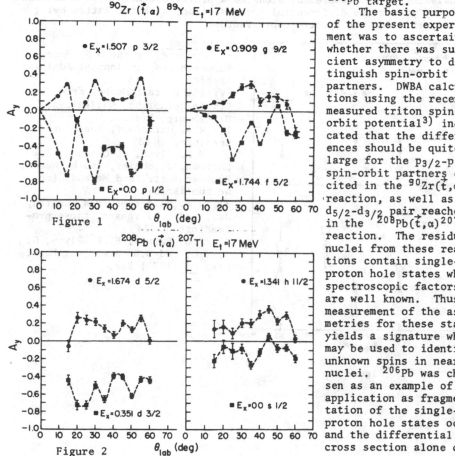

Figure 1

Figure 2

The basic purpose of the present experiment was to ascertain whether there was sufficient asymmetry to distinguish spin-orbit partners. DWBA calculations using the recently measured triton spin-orbit potential[3] indicated that the differences should be quite large for the $p_{3/2}-p_{1/2}$ spin-orbit partners excited in the $^{90}Zr(\vec{t},\alpha)$ reaction, as well as the $d_{5/2}-d_{3/2}$ pair reached in the $^{208}Pb(\vec{t},\alpha)^{207}Tl$ reaction. The residual nuclei from these reactions contain single-proton hole states whose spectroscopic factors are well known. Thus, measurement of the asymmetries for these states yields a signature which may be used to identify unknown spins in nearby nuclei. ^{206}Pb was chosen as an example of this application as fragmentation of the single-proton hole states occurs and the differential cross section alone cannot be used to assign the spin.

Figure 1 gives the value of the analyzing power A_y observed for the four prominent single-proton hole states of ^{89}Y. The dashed lines are only to connect the data points. Significant asymmetries are seen, and a clear differentiation between the 1/2⁻ and 3/2⁻ levels (both, of course, proceeding by an $\ell = 1$ proton pickup) is noted. The signs and magnitude of A_y for these latter states are in accordance with the preliminary DWBA calculations. The ^{207}Tl results are shown in Fig. 2, and again very distinctive patterns of A_y vs. θ are observed.

Figure 3 indicates the result of the ^{206}Pb$(\vec{t},\alpha)^{205}$Tl experiment at four angles. These angles represent points where large differences exist between the $d_{3/2}$ and $d_{5/2}$ A_y's. The level at 0.20 MeV is a known 3/2⁺ state, whereas the 1.14 MeV and 1.34 MeV levels were uncertain. The present data clearly indicate the total spin for these latter states is also 3/2⁺. Moreover, a tentative assignment of 11/2⁻ to the 1.48 MeV state is verified by comparison to the ^{207}Tl 11/2⁻ analyzing power.

The distinctive features of A_y as a function of angle is somewhat in contrast to the differential cross sections, which show only small variations with angular momentum transfer.[4] Indeed it would appear that the analyzing power measurement alone is a preferable method to establish both ℓ and j over the differential cross section, with the latter being necessary only for the spectroscopic factor.

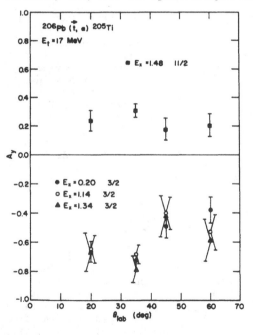

References
* Supported by the U.S. Energy Research and Development Administration
†Visiting Staff Member from Centre de Recherches Nucleaires, Strasbourg, France.
1) R.A. Hardekopf, these proceedings, p.865
2) E.R. Flynn, S. Orbesen, J. Sherman, J. Sunier, and R. Woods, Nuclear Instr. and Methods (to be published).
3) R.A. Hardekopf, P.W. Keaton, Jr., and L.R. Veeser, these proceedings, p.624
4) P.D. Barnes, E. R. Flynn, G. J. Igo, and D.D. Armstrong Phys. Rev. C1 (1970) 228.

Figure 3

THE ^{3}He (^{3}He,$\vec{p}$)^{5}Li REACTION AT 13.5 MeV †

R. Pigeon, M. Irshad, S. Sen, J. Asai and R.J. Slobodrian
Laboratoire de Physique Nucléaire, Département de Physique,
Université Laval, Québec G1K 7P4 , Canada.

The study of two nucleon transfer reactions has been confined almost exclusively to angular distributions of cross sections due to the lack of suitable polarimeters to cope with the simultaneous requirement of good energy resolution and high efficiency. Figure 1 shows the polarimetry facility consisting of Si polarimeters [1] developped at this laboratory and satisfying the above mentioned requirements. A recent addition has been the implementation of computer on-line operation for the purposes of particle identification [2]. The ^{3}He target consisted of gas at a pressure of 87 torr confined in a cell by a 2 mg cm^{-2} Havar foil window, and the polarimeters were provided with suitable gas collimators. The beam alignment was monitored using a split Faraday cup and measuring the ratio of currents of both halves. Lateral beam displacements were compensated manually with a bending magnet. The routine of data acquisition consists of two measurements with the polarimeters placed successively at opposite sides of the beam, to compensate geometrical errors. Cross sections were measured independently with a telescope using tight collimation.

Figure 2 shows the experimental results, together with curves obtained using a DWBA code [3]. This reaction is interesting because the number of participating nucleons is not large and a detailed theoretical treatment of the reaction should be possible. The symmetry of the initial state cuts in half the experimental work for the acquisition of full angular distributions. Within the usual DWBA treatment the possible (JLST) combinations are (2110), (1110) and (1101), L being odd because ^{3}He and ^{5}Li have opposite parity. The spectroscopic amplitude $S^{\frac{1}{2}}_{AB}$

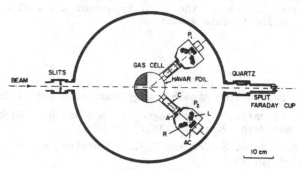

POLARIMETRY FACILITY AT
UNIVERSITE LAVAL
SCHEMATIC
P1
GAS CELL
SLITS
QUARTZ
BEAM
HAVAR FOIL
C
P2
SPLIT
FARADAY CUP
L
A
R
AC
10 cm
P1,P2 POLARIMETERS ; C: GAS COLLIMATOR
A: ANALYZERS ; L: LEFT DETECTOR ; R: RIGHT DETECTOR
AC: ANTICOINCIDENCE DETECTOR

Figure 1

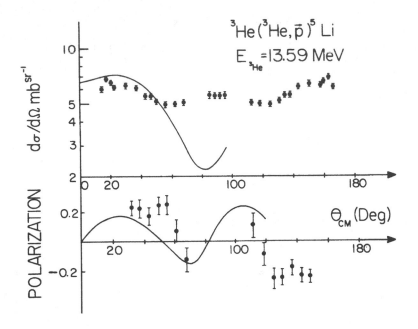

Fig. 2

reduces to a Racah coefficient, as one of the angular momenta in the 9-j symbol is zero. Figure 2 shows the sum of DWBA cross section curves for J=1 and J=2. The latter is nearly four times as large as the contribution from J=1 over the whole angular range. The polarization curve is a weighted mean of polarizations for J=1, 2 and the corresponding cross sections are used as weights. The comparison with experiment at forward angles is favourable. For large angles proper symmetrization of the elastic and reactions channels should be made and is contemplated for further analysis of these data. The implications of the transfer of a deuteron cluster instead of the usual treatment are explored in another communication to this symposium [4].

References

† Supported in part by the Atomic Energy Control Board of Canada.

1. B. Frois, J. Birchall, R. Lamontagne, R. Roy and R.J. Slobodrian, Nucl. Intr. and Meth. 96 (1971) 431.

2. M. Irshad, R. Pigeon, S. Sen and R.J. Slobodrian, contribution to this symposium, p.889

3. J.M. Nelson and B.E.F. Macefield, Oxford University Nuclear Physics Laboratory Report 18/69 (1969) unpublished.

4. J. Asai and R.J. Slobodrian, contribution to this symposium,p.703

TWO NUCLEON VERSUS DEUTERON TRANSFER IN THE ^{3}He (^{3}He, $\vec{p}$) ^{5}Li REACTION †

J. Asai and R.J. Slobodrian
Laboratoire de Physique Nucléaire, Département de Physique
Université Laval, Québec G1K 7P4, Canada.

The angular distribution of the differential cross sections and the polarizations for the ^{3}He (^{3}He, $\vec{p}$) ^{5}Li and ^{7}Li (^{3}He, $\vec{p}$) ^{9}Be at 13.5 MeV incident energy have been measured recently and are reported at this symposium [1]. The reaction mentioned in the first place has been analysed using a DWBA code for two particle transfer [2] and figure 1 summarizes the conclusions. The optical model parameters were taken from refs. 3 and 4 for the ^{3}He+^{3}He and p+^{5}Li systems respectively. Because of the identity of the target and incident particle it should be necessary to perform a proper symmetrization. However for forward scattering angles symmetrization effects are small. Discrepancies for angles larger than 50° are probably due to such effects. The ground state of ^{5}Li is a broad Breit-Wigner resonance. The shell model configuration is $(1S_{\frac{1}{2}})^4$ $(1p_{3/2})1$ and it is well separated from other levels, thus precluding ambiguities due to mixing of configuration and therefore the reaction observables are hopefully directly related to the mechanism. Denoting the transferred total, orbital and spin angular momenta by J, L and S respectively the permissible values of (J L S) are (211), (111) and (110), yet figure 1 shows that the reaction proceeds almost exclusively by (J L S) = (211). A selection of total angular momentum transfer in single nucleon transfer reactions has been observed before [5]. In the present case a suitable explanation is found by assuming that a deuteron cluster is transferred, instead of the usual general two-nucleon transfer. The ground states of ^{3}He and ^{5}Li are 1/2$^+$ and 3/2$^-$ respectively. Therefore the deuteron transfer implies that the neutron and deuteron spin be aligned antiparallel to the ^{3}He spin to close the s-shell. Finally, the orbital momentum of the proton has to be parallel to that of the deuteron to reach the 3/2$^-$ state of ^{5}Li. Consequently one obtains the value J=2. The value J=1 is not excluded by the general angular momentum coupling rules, but is not consistent with the details of the s-shell closure required in this reaction. The ^{7}Li (^{3}He, $\vec{p}$) ^{9}Be reaction can also be interpreted more adequately in terms of a deuteron transfer having thus only S=1, but the analysis is in a more preliminary stage. However correct magnitudes for the cross section have been obtained, and polarization patterns in better agreement with experiment.

TABLE 1 — OPTICAL MODEL PARAMETERS

CHANNEL	V	r	a	W	r_W	a_W	W_D	r_{WD}	a_{WD}	r_C
^{3}He+^{3}He	206.5	1.25	0.57	1.0	1.82	0.2	0.0	-	-	1.25
p+^{5}Li	59.1	1.05	0.29	0.0	1.33	0.57	10.1	1.33	0.57	1.56

Strengths are in MeV, lengths in fm.

† Work supported in part by the Atomic Energy Control Board of Canada.

704

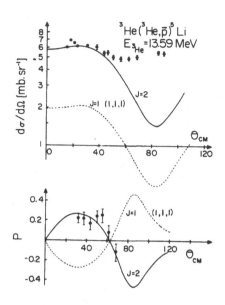

Fig. 1

References

1. M. Irshad, J. Asai, S. Sen, R. Pigeon and R.J. Slobodrian
    R. Pigeon, M. Irshad, S. Sen, J. Asai and R.J. Slobodrian
    Contributions to this symposium, p.701.

2. J.M. Nelson and B.E.F. Macefield, Oxford University Nuclear Physics
    Laboratory, Report 18/69 (1969) unpublished.

3. W.J. Roberts, E.E. Gross and F. Newman, Phys. Rev. C9 (1974) 149.

4. K. H. Bray et al. Nucl. Phys. A189 (1972) 35.

5. T.J. Yule and W. Haeberli, Nucl. Phys. A117 (1968) 1.

DIFFERENTIAL CROSS SECTION AND POLARIZATION [†]
IN THE ^{7}Li(^{3}He,p)^{9}Be g.s. REACTION AT 14 MeV

M. Irshad, J. Asai, S. Sen, R. Pigeon and R.J. Slobodrian
Laboratoire de Physique Nucléaire, Département de Physique,
Université Laval,   Québec   G1K 7P4 ,   Canada.

Measurements have been made of the proton polarization in the ^{7}Li (^{3}He, p) ^{9}Be g.s. reaction using a pair of on-line silicon polarimeters described elsewhere [1] at 14 MeV incident ^{3}He energy.   The target used was in the form of a 3.4 mg/cm^2 self-supporting foil rolled from a freshly cut natural Li. One polarimeter on each side of the beam was used and their roles were interchanged for each measurement by rotating them through 180° around the incident beam direction. Thus polarization values for each angle were obtained from a set of eight ΔE-E spectra, two for each polarimeter on either side of the beam. The associated differential cross section measurements for the same reaction were also performed. The results are shown in fig.1.

A preliminary analysis of the measurements in terms of a two nucleon transfer DWBA theory has been carried out using elastic scattering optical potentials to generate the distorted waves in the incident [2] and exit [3] channels. It is assumed that the ground state of ^{9}Be is formed by the transfer of a neutron and a proton into $1p_{3/2}$ single particle orbits and the total   nding energy is equally shared by the transferred pair. The form factor was obtained using Woods-Saxon wells expanded in harmonic oscillator basis with an oscillator parameter value of 0.5 fm^{-2}. The possible (J,L,S,T) combinations allowed by the various selection rules are (0,0,0,1), (1,0,1,0), (1,2,1,0), (2,2,0,1) and (3,2,1,0). For J=1, a coherent sum over L was performed. The calculated relative cross section plotted in fig.1 is the sum of DWBA predictions for various allowed combinations listed above. The theory predicts very different cross section and polarization patterns than those observed experimentally. The individual analysis showed that the major contribution to the predicted cross section arises from J=0 which is mainly responsible for the strong diffraction pattern peaked near zero degree. The same combination is also mainly responsible for the polarization pattern as the polarizations for various combinations were weighted according to the predicted cross sections in each case. A previous DWBA analysis of the differential cross section data for this reaction has also been reported to be unsuccessful [2]. The discrepancy between theory and experiment may be due to the neglect of other reaction mechanisms and/or to the applicability of the shell model in the case of a nucleus like ^{9}Be. The possibility of a deuteron transfer is also under consideration [4]. We are currently investigating the reaction ^{6}Li(^{3}He, p̄) ^{8}Be, with the hope of getting a better understanding of the reaction mechanism involved; ^{9}Be has a low threshold for break-up into a neutron and ^{8}Be, thus making plausible a significant clustering and a "molecular" structure consisting of two alphas bound by a neutron.

† Work supported in part by the Atomic Energy Control Board of Canada
  and by the Ministery of Education of Québec.

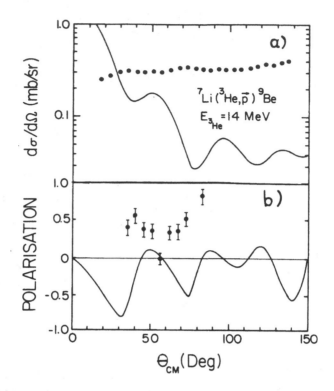

Fig. 1- a) Differential cross section. The counting statistics errors
        are smaller than the size of the plotted points. The solid
            curve represents the DWBA prediction with the optical model
            parameters of ref. [2] and ref. [3].

     b) Polarization angular distribution. The errors indicated
         are due to counting statistics only. The solid curve
         corresponds to the DWBA prediction as in 1-a).

References

1) Frois et al. Nucl. Inst. and Meth. 9(1971)431 and Irshad et al.
    4th polarization symp., p.889

2) Dixon and Edge Nucl. Phys. A156 (1970) 33.

3) Montague et al. Nucl. Phys. A199(1973) 433.

4) J. Asai and R.J. Slobodrian, contribution to this symposium, p.703.

THE POLARIZATION IN THE $^{10}\text{B}(^3\text{He}, \vec{\text{p}})^{12}\text{C}$ * 4.43 REACTION AT 10 MeV †

C.R. Lamontagne, M. Irshad, R. Pigeon and R.J. Slobodrian
Laboratoire de Physique Nucléaire, Département de Physique,
Université Laval,   Québec   G1K 7P4   Canada.

J.M. Nelson
Department of Physics, University of Birmingham,
Birmingham   B15 2TT   England.

Measurements of the differential cross section for this reaction have been carried out by Patterson et al [1] between 3.2 and 10.5 MeV. Polarization measurements were performed by Simons [2] at low energies, between 2 and 2.8 MeV. The cross section measurements indicate that beyond 8 MeV there is not evidence for resonance effects of ^{13}N, and the angular distribution patterns seem consistent with a direct reaction.

The present work has been carried out using silicon polarimeters at the polarimetry facility of Université Laval described elsewhere [3]. The analyzing power of silicon was measured recently in great detail [4] and the effective analyzing power of the polarimeters was determined from those results. The target consisted of self supported ^{10}B enriched to 92.36% of thickness approximately 400 μg cm^{-2} obtained by evaporation using an electron beam.

Routinely an asymmetry measurement consists of eight spectra, taken with two polarimeters placed at symmetrical angles with respect to the beam and also exchanging their positions. Geometrical asymmetries should thereby compensate and be considerably reduced.

Figure 1 shows the experimental results which exhibit consistency with the lower energy data [2]. The theoretical interpretation is in progress. Utilizing the current notation (JLST) for the total angular momentum, orbital, spin and isospin transfers respectively, the possible combinations are (3210) , (2210) , (1210) and (1010) , for a transition from the ^{10}B g.s. of spin, parity and isospin (3$^+$, 0) to the first excited state of ^{12}C (2$^+$, 0). The optical model parameters were obtained from the literature [5]. The figure shows DWBA curves [6] corresponding to (2210) , (3210) and (1210) , the latter two give some agreement with the measured polarization. The other (JLST) value gives also acceptable cross section angular distributions as far as shape is concerned, but fails completely to account for the polarization . The experimental results for the cross section angular distribution [1] show a backward rise not reproduced by any of the DWBA curves. The latter has been attributed in the past to heavy particle stripping [7], which is not included in the DWBA calculation and may be relevant for ^3He reactions on light nuclei.

† Supported in part by the Atomic Energy Control Board of Canada and the Ministery of Education of Québec.

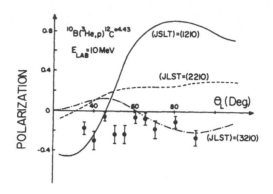

Fig. 1

References

1. J.R. Patterson, J.M. Poate and E.W. Titterton, Proc. Phys. Soc. 88 (1966) 641.

2. D.G. Simons, Phys. Rev. 155 (1967) 1132.

3. M. Irshad, J. Asai, S. Sen, R. Pigeon and R.J. Slobodrian. R. Pigeon, M. Irshad, S. Sen, J. Asai and R.J. Slobodrian, contributions to this symposium, p.889.

4. C. R. Lamontagne, B. Frois, R.J. Slobodrian, H.E. Conzett, Ch. Leemann, R. de Swiniarski, Phys. Lett. 45B(1973) 465. J.L. Duggan et al. Nucl. Phys. A151 (1970) 107.

5. N. Austern Direct Nuclear Reaction Theories, Wiley-Interscience (1969).

6. Calculated with the code of J.M. Nelson and B.E.F. Macefield, Oxford University Nuclear Physics Laboratory, Report 18169.

7. G.E. Owen, L. Madansky and S. Edwards Jr., Phys. Rev. 113 (1959) 1575.

THE j-DEPENDENCE OF ANALYSING POWER OF THE ^{58}Ni(^{3}He,d)^{59}Cu REACTION

S. Roman, A.K. Basak, J.B.A. England, O. Karban, G.C. Morrison,
J.M. Nelson and G.G. Shute[+]

The University of Birmingham, England

J-dependent effects in the cross sections of (^{3}He,d) and (^{3}He,α) reactions have been known for some time [1-3]), but such effect are usually very small and of limited usefulness in determining the spins of excited states. Differential cross sections for a number of transitions in the ^{58}Ni(^{3}He,d)^{59}Cu reaction have been determined at various energies [4-6]) showing very little structure. In particular, the measurements of Cage et al.[6]) at 33 MeV covering an angular range from 5 to 135° have found no significant j-dependence.

On the other hand the spin-orbit forces acting on a nucleon in the bound state are known to be very strong – for example from (d,p) and (d,t) studies. Since the bound state wave function is primarily responsible for the j-dependence in (d,p) reactions it is reasonable to expect effects of similar magnitude in other one-nucleon transfer reactions. This will be true for ^{3}He induced reactions, irrespective of the magnitude of the ^{3}He spin-orbit distortion, provided the reaction is a direct one. A further extension of the parallel with (d,p) reactions leads to expectation of large j-dependent effects in the analysing power of ^{3}He reactions.

In the present work a strong j-dependence of analysing power has been observed for the first time in ^{3}He reactions on a medium weight target. The j-dependence for the lp shell is reported separately.

The analysing powers of the ^{58}Ni(^{3}He,d)^{59}Cu reactions have been measured using the polarized ^{3}He beam from the University of Birmingham Radial Ridge Cyclotron. The experimental method has been described previously [7]). A selfsupporting, 99 % enriched ^{58}Ni target 2 mg/cm^{2} thick was used. The data were taken using two (left and right) sets of three ΔE x E semiconductor detector telescopes with analog particle mass identifier circuits and an acquisition system based on an IBM 1130 computer. The beam polarization was switched frequently between the "up" and "down" states with a spin precession solenoid. The beam polarization was monitored continuously using the scattering of ^{3}He by hydrogen as analyser in a downstream polarimeter.

The results for the transitions leading to the ground state and the 0.49 MeV and 0.91 MeV excited states of ^{59}Cu, whose spin assignments are known to be 3/2^{-}, $\frac{1}{2}^{-}$, and 5/2^{-} respectively are shown in fig.1 together with predictions of DWBA calculations carried out with the program NELMAC [8]).

[+] On leave from the University of Melbourne, Victoria

710

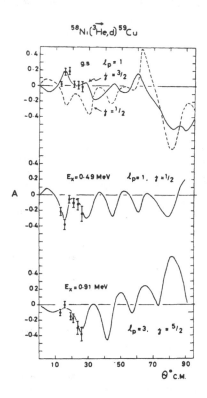

^{58}Ni(^{3}He,d)^{59}Cu

Fig. 1. Analysing power of the ^{58}Ni(^{3}He,d)^{59}Cu reaction at 33.3 MeV
incident ^{3}He energy. The curves are results of DWBA calculations for the spin values indicated.

The observed magnitude of the analysing power is large and the
j-dependence is clearly demonstrated. The j-dependent behaviour is
especially striking for the two l = 1 transitions having $j = l + \frac{1}{2}$
$l - \frac{1}{2}$ combinations, since, due to the close proximity of these states,
there is no Q-value effect which would significantly shift the
pattern of oscillations.

References
1) A.G. Blair, Phys. Rev. 140 (1965) B648
2) M.K. Brussel, D.E. Rundquist and A.I. Yavin, Phys.Rev.140 (1965) B838
3) C. Glashausser and M.E. Rickey, Phys. Rev. 154 (1967) 1033
4) G.C. Morrison and J.P. Schiffer,Isobaric Spin in Nucl. Phys.,
   Eds. J.D. Fox and D. Robson (Acad. Press, N.Y. 1966) 748
5) D.J. Pullen and B. Rosner, Phys. Rev. 170 (1968) 1034
6) M.E. Cage et al., Nucl. Phys. to be published
7) W.E. Burcham et al. Nucl. Phys. A245 (1975) 170
8) J.M. Nelson and B.E.F. Macefield, Report No 74-9, unpublished.

THE j-DEPENDENCE OF (3He,d) AND (3He,α) REACTION ANALYSING POWERS IN THE 1p SHELL

A.K. Basak, J.B.A. England, O. Karban, G.C. Morrison, J.M. Nelson,
S. Roman and G.G. Shute
Department of Physics, University of Birmingham, Birmingham B15 2TT,
England

One of the most important experimental results from polarized deuteron studies is the ability to obtain reliable spectroscopic information from measurements of the vector analysing power in transfer reactions.   With the availability of a polarized 3He beam it is of considerable interest to determine whether a similar j-dependence of the 3He analysing power exists.   Cross section measurements in the 30 MeV region have shown that the (3He,d) and (3He,α) reactions can be quantitatively described by direct reaction theories.   Therefore, j-dependence could provide a spectroscopic tool perhaps even more powerful than in reactions with polarized deuterons.

A 33 MeV polarized 3He beam with an average on-target intensity of 0.15 na and a polarization of 0.38, continuously monitored by a downstream polarimeter[1]) was incident on self-supporting 4.5 and 6.0mg/cm^2 targets of beryllium and carbon respectively.   Left-right asymmetries at 3 scattering angles were measured simultaneously using six dExE silicon counter telescopes.

The 3He analysing powers were obtained for the following final

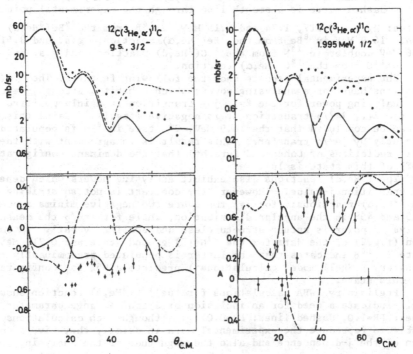

Fig.1   Cross sections and 3He analysing powers of the 12C(3He,α)11C
reaction.   DWBA predictions with a finite range parameter FR=0.0
(dashed lines) and FR=1.0 (solid lines).

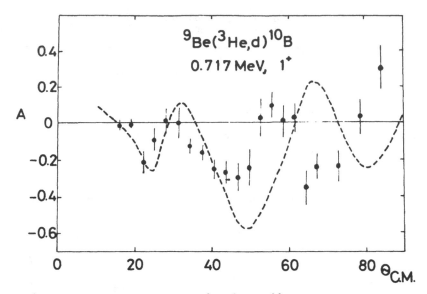

Fig.2 ^{3}He analysing power of the ^{9}Be(^{3}He,d)^{10}B reaction (full circles)
compared with that of the ^{12}(^{3}He,d)^{13}N (g.s., j=½) reaction
(dashed curve is a smooth line through the experimental points).

states: g.s., 0.717, 1.74 and 2.15 MeV in ^{10}B from the ^{9}Be(^{3}He,d) reaction; 16.9 MeV in ^{8}Be from the ^{9}Be(^{3}He,α) reaction; g.s. and 3.51 and
3.56 MeV doublet in ^{13}N from the ^{12}C(^{3}He,d) reaction, and g.s. and 1.995
MeV in ^{11}C from the ^{12}C(^{3}He,α) reaction.

The experimental results show the following features. The ^{12}C(^{3}He,α)
reactions have large analysing powers which exhibit a strong j-dependence:
the analysing power for the ℓ=1, j=½ transition is mainly positive while
for the ℓ=1, j=^{3}/$_2$ transition it is negative (fig.1). Using this feature
it can be concluded that the 16.9 MeV, 2^+ state in ^{8}Be is populated predominantly by j=^{3}/$_2$ transfer. This result is in agreement with theoretical predictions of Cohen and Kurath[2]) that the dominant configuration
(90%) of this state is 1p$_{\frac{3}{2}}$.

The (^{3}He,d) reactions also exhibit analysing powers which depend on
the transition j-value. However, the contrast is not so striking as in
the (^{3}He,α) reactions: for j=½ there are two negative minima around Θ cm
= 25 and 45° in the angular distribution, while for j=^{3}/$_2$ the measured
analysing power is rather structureless and smaller overall. A comparison (fig.2) of the data for the ^{13}N($\frac{1}{2}^-$) ground state and 0.717 MeV 1^+
state in ^{10}B indicates that the latter is populated predominantly by j=½
transfer. Shell model calculations[2]) predicts a 65% 1p$_{\frac{1}{2}}$ configuration
for this state.

Preliminary DWBA calculations for the ^{12}C(^{3}He,α) reaction shown in
fig.1 indicate a need for an inclusion of a finite range parameter (solid
lines: FR=1.0, dashed lines: FR=0.0). Although such calculations are
unable to reproduce the experimental data in detail, they give a correct
sign of the j-dependence and also the magnitude of the analysing powers.

References
1) W.E. Burcham et al., Nucl.Phys., to be published.
2) S. Cohen and D. Kurath, Nucl.Phys. A101 (1967) 1.

POLARIZATION IN (α,p) REACTIONS

O.F.Nemets,Yu.S.Stryuk,A.M.Yasnogorodsky
Institute for Nuclear Research, Ukrainian Academy of Sciences
Kiev,USSR

There are grounds to think the polarization phenomena in (α,p) reactions are associated mainly with direct reaction mechanisms. In particular, it is of interest to investigate the polarization trends, starting with the most simple and practicable model of single cluster transfer in the DWBA framework.

On the basis of differential cross-section analysis[1], in this work the calculations are performed for proton polarization in $^{28}Si(\alpha,p)^{31}P$ and $^{63}Cu(\alpha,p)^{66}Zn$ reactions at $E_\alpha = 27.2$ MeV, using the MIVPOL computer code (the polarization is equivalent to analysing power of inverse reactions $(\vec{p},\alpha)$ at proton energies near 30 MeV).

The four-parameter α-particle optical potential sets with volume absorbtion are used (see the table) which have been proposed in refs.[2,3]. The parameters of proton optical potential are those of ref.[4]. The bound state potential depths are chosen to account for the triton binding energy in the residual nucleus, the geometry parameters being selected by optimum fit to differential cross-sections.

	V MeV	W MeV	a fm	r fm
1	151	22.8	0.68	1.3
2	196	24.3	0.63	1.3
3	317	30.2	0.57	1.3
4[a]	202	28.4	0.60	1.34

a) from ref.[3]

Predicted for $^{28}Si(\alpha,p_0)^{31}P$ proton polarizations are shown in fig.1a. The numeration of the curves corresponds to the α-potential set numbering in the table. The stripping to the 4s triton state is supposed, the parameters of bound state Woods-Saxon well being $r_{bs} = 1.3$ fm, $a_{bs} = 0.5$ fm.

As it is seen from fig.1a, the high values of polarization are predicted weakly sensitive to the optical potential uncertainties when the sets of real well depth $V \gtrsim 4V_N$ are used (V_N being the nucleon optical potential real well depth), except the region $\theta \lesssim 40°$.

The proton polarization in $^{63}Cu(\alpha,p)^{66}Zn$ reaction is calculated with α-particle optical potential set 4. The stripping into 5p triton state is supposed, the parameters of bound state well being $r_{bs} = 1.45$ fm, $a_{bs} = 0.5$ fm, $V_{so}^{bs} = 6.5$ MeV.

In fig.1b it is shown the pronounced polarization dependence on

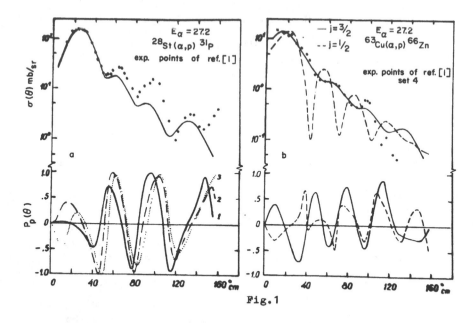

Fig. 1

transfered triton total angular momentum $\vec{j} = \vec{l} + \vec{s}$ at main peak of the differential cross-section. It is essential, that differential cross-section also shows the pronounced j-dependence.

When being collected, the experimental data will provide a mean to estimate the validity of reaction mechanism under consideration, and the extent of usefulness of polarization measurements in (α, p) reactions to obtain a spectroscopic information.

References

1) Yu.S.Stryuk,V.V.Tokarevsky, Program and Abstracts of Reports to the 20th Conference on Nuclear Spectroscopy and Nuclear Structure, part 2 (Leningrad,1970), p.279.
2) Yu.S.Stryuk,V.V.Tokarevsky, Program etc. (see ref.[1]) of the 21st Conference, part 2 (Leningrad,1971), p.85.
3) I.N.Simonov et al., Izv. AN SSSR, ser.fiz. 34(1970)1748.
4) F.G.Perey, Phys.Rev. 131(1963)754.

PION PRODUCTION WITH POLARIZED BEAMS AND TARGETS*

J.V. Noble
Department of Physics
University of Virginia, Charlottesville, VA 22901, USA

The polarized particle reactions $\vec{A}(\vec{p},\pi^+)B$ or $A(\vec{p},\pi^+)\vec{B}$ offer the possibility of differentiating between various reaction mechanisms which have been proposed to explain the presently available data on cross sections and angular distributions.[1] In a recent paper[2] I proposed $\vec{\tau}(\vec{p},\pi^+)\alpha$ as a means of distinguishing pionic stripping, Fig. 1a, for which $T_{\uparrow\downarrow}$ vanishes (up and down being defined as perpendicular to the scattering plane) from, say, the Δ_{33} mechanism, Fig. 1b, for which it seemed, <u>ab initio</u>, that $T_{\uparrow\downarrow}$ would not vanish.

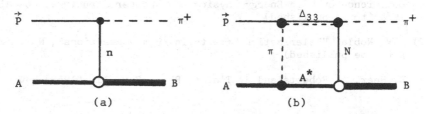

Fig. 1. Diagrammatic representation of the pionic stripping (a), and Δ_{33} (b) mechanisms.

Subsequently I realized that the inherent structure of the $\vec{\tau}(\vec{p},\pi^+)\alpha$ amplitude, imposed by considerations of rotation invariance and parity conservation, is such that $T_{\uparrow\downarrow}$ vanishes independently of dynamics, and so this reaction is not a definitive test (except, perhaps, for parity nonconservation). The difficulty arises because there are too few spins in the problem to give rise to interesting results. This suggests that we consider either targets with spin > 1/2, residual nuclei with spins > 0, or outgoing particles with spin (e.g. vector mesons or photons).

Pionic stripping into a given j-orbital predicts a definite left-right asymmetry[2] (which has been seen![3]) when the proton beam is polarized perpendicular to the scattering plane. Since the asymmetry in such reactions results from the distortion of incident and emergent waves it should be strongly mechanism-dependent as well. Calculations to examine this are proceeding. Another potentially interesting measurement . is the angular correlation between photons from the de-excitation of the residual nucleus and the outgoing pion, since various mechanisms are expected to differ in their population of the residual magnetic substates. The reaction $\vec{A}(\vec{p},\gamma)B$ should be interesting because electromagnetic theory predicts definite ratios for $\sigma_{\uparrow\uparrow}/\sigma_{\uparrow\downarrow}$, depending on the matrix elements of the current operator, which will in turn depend on reaction mechanism.

A reaction offering immediately interesting possibilities is $^2H(\vec{p},\pi^+)\vec{t}$, where the polarization of the outgoing triton is measured. The effects of d-state admixtures, spin-orbit couplings and distortions are expected to be small, and so the prediction of the pionic stripping model is that the triton should have exactly the polarization of the incident proton; whereas the Δ_{33} model predicts a substantial spin-down amplitude for the triton (taking the proton's spin as up). The enhancement of the latter near the Δ_{33} production energy would be convincing evidence for its origin.

None of the above experiments is easy, since either polarized targets, or angular correlations of sequential decays, or electromagnetic processes, or polarization-sensitive detectors are required in conjunction with weak beams (because they are polarized), and smallish cross sections. Nevertheless they would appear to be some of the most potentially rewarding intermediate energy experiments.

References

* Supported in part by the National Science Foundation

1) See, e.g., J.M. Eisenberg, in Proceedings of the Sixth International Conference on High Energy Physics and Nuclear Structure, Los Alamos, 1975 (to be published)

2) J.V. Noble, "Polarization Effects in $(\vec{p}, \pi^+)$ Reactions", Nucl. Physics A (to be published)

3) E. Heer, A. Roberts and J. Tinlot, Phys. Rev. 111 (1958) 640

Inelastic Scattering (Coupled Channels)

ELASTIC AND INELASTIC SCATTERING OF 14.2 MeV POLARIZED NEUTRONS FROM ^{12}C

R. Casparis, M. Preiswerk, B.Th. Leemann, H. Rudin, R. Wagner,
P. Zupranski*
Institut für Physik der Universität Basel

Between 22° and 152° (CM) the angular distribution of the analyzing power $A(\Theta)$ for the $^{12}C(\vec{n},n')^{12}C*$ inelastic scattering (Q = -4.43 MeV) and the neutron polarization $P(\Theta)$ for the elastic scattering have been measured. The source described in the contribution of B.Th. Leemann et al. to this conference has been used to produce polarized 14.2 MeV neutrons ($p_n \sim .5$). A time-of-flight technique allowed separation of elastically and inelastically scattered neutrons. Up to 90° a graphite cylinder served as scatterer, above 90° carbon recoils in a plastic scintillator gave an additional information to reduce the background. The measured values for $P(\Theta)$ have been corrected for multiple scattering and finite geometry using the program MULTPOL[1]. The results of the neutron polarization in the elastic scattering are in good agreement with data of Mack et al.[2] and also with data of Sené et al.[3] except at angles larger than 70° whereas the agreement with first results of a coupled channel calculation of Hodge and Tamura is poor (figure 1). The optical model parameters used herein were

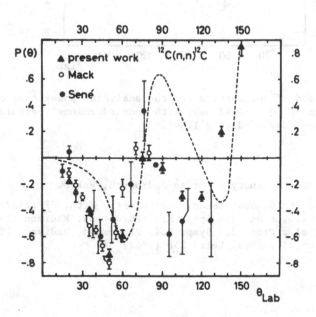

Fig.1. Comparison of our measurements of the polarization $P(\Theta)$ at 14.2 MeV with data of Mack[2] (E_n = 15.85 MeV) and Sené[3] (E_n = 14.1 MeV). The dashed line is a coupled channel calculation of Hodge and Tamura.

the same as in the analysis of the elastic and inelastic n–C differential cross sections[4] except for the radius and the diffuseness of the spin orbit potential which are .85 fm and .249 fm respectively. The ground state 0^+ and the first excited state 2^+ have been coupled. The analyzing power of the inelastic scattering is in fair correspondence with these calculations (figure 2).

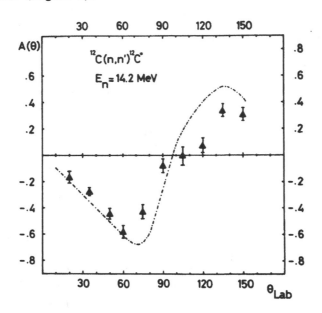

Fig.2. Comparison of our results for the analyzing power A(Θ) of the inelastic scattering (Q = −4.43 MeV) with coupled channel calculations of Hodge and Tamura (dashed–dotted line).

References

*   On leave from the Instytut Badań Jądrowych, Warsaw

1) O. Aspelund and B. Gustafsson, Nucl. Instr. Meth. 57 (1967) 197
2) G. Mack et al., Proc. 3rd Symp. Pol. Phenomena, Madison (1970) 615
3) R. Sené et al., Proc. 3rd Symp. Pol. Phenomena, Madison (1970) 611
4) G.A. Grin et al., Phys. Lett. 25b (1967) 387

MICROSCOPIC MODEL ANALYSIS OF ANALYSING POWER DATA IN (p,p') SCATTERING ON LIGHT NUCLEI

J.L.Escudié[+], E.Fabrici, M.Pignanelli and F.Resmini
Istituto Nazionale di Fisica Nucleare, Sezione di Milano, Italy

Microscopic model analysis of inelastic proton scattering has been so far only moderately successful, being plagued by uncertainties in the nucleon-nucleon (N-N) interaction to be used and by doubts as whether a direct mechanism could fully account for the data. In particular, the results for the analysing powers (A.P.) were mostly disappointing and the fits in no way comparable to those from macroscopic model analyses. The supposedly poor quality of the wave functions has been blamed for the apparent failure of the model.

In this paper we present some results obtained in a comprehensive analysis of (p,p')·A.P. data for light nuclei. This work was prompted by the discovery of an otherwise trivial error in the spin-orbit part of the direct calculation of the program DWBA70 by Raynal and Schaeffer[1]. This error was uncovered while analysing ^{18}O and ^{15}N data and its correction produced very different results at least in the direct calculation.

The results of fig. 1 were obtained with the N-N force of Borysowicz et al.[2] However other realistic forces also produce equally good results. Continous lines refer to direct calculations, dashed lines to antisymmetrized (direct plus exchange) calculations. The data were taken from the ref.s 3 to 5. The proton energy is 24.4 MeV for all nuclei, except ^{32}S data at 30.3 MeV. The agreement between the data and direct calculations is surprisingly good, while the exchange term either worsens the fit or gives no agreement at all. Transitions with the same J show a very different behaviour, which is well accounted for. As an example the 2^+ level in ^{12}C involves a $p_{3/2}$ to $p_{1/2}$ transition, while the 2^+ levels in ^{18}O and ^{20}Ne involve rearrangements in the $d_{5/2}$ and transitions from the $d_{5/2}$ to $s_{1/2}$ shells, although with different phases and amplitudes. Similar arguments hold for the 3^- levels of ^{18}O and ^{32}S.

It looks therefore like the A.P. data bear in some way the signature of the relevant wave functions, and the calculations are properly sensitive to them. More results of the same kind, not presented here, confirm this fact. One would then be tempted to say that cases of poor agreement could be rather connected with the quality of the wave function used and, moreover, that there is no evidence for important contributions from multi-step processes. It should also be noted that the fits for the 2^+, 4^+ and 3^- levels are of the same quality, and sometime better, than those currently given by the macroscopic model.

We cannot give yet a satisfactory explanation of the apparent failure of the full antisymmetrized calculation. The agreement with direct calculations looks much too systematic to be fortuitous and therefore efforts are under way to understand and possibly remove the reason of this discrepancy.

References

+   Permanent address: C.E.A., Saclay, France.
1) R.Schaeffer, Thesis, C.E.A. Report R-4000 (1970); J.Raynal, Nucl.Phys. A97 (1970) 572.
2) J.Borysowicz, H.McManus, G.Bertsch, MSU Report (1975), to be published.
3) J.L.Escudié, R.Lombard, M.Pignanelli, F.Resmini, A.Tarrats, Phys. Rev. C10 (1974) 1645.
4) H.Krug, Report C.E.A. - N-1390 (1970) 103.

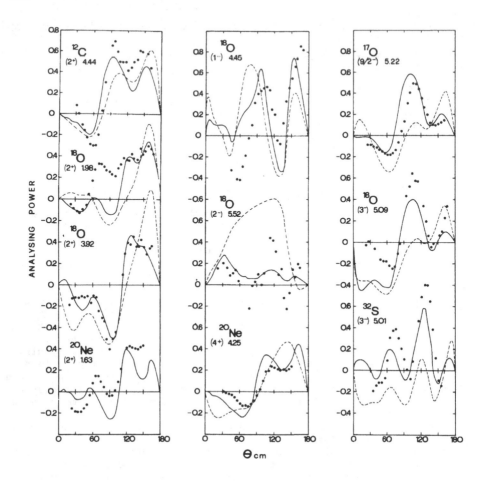

Fig. 1   Results of microscopic model calculations in fitting analysing
power data for light nuclei. See text for references to original data,
and to N-N interaction used. Continous curves refer to direct calculations, dashed ones to direct plus exchange. Wave functions are from Zu-
ker[6] for ^{17}O and ^{18}O, Cohen and Kurath for ^{12}C, Wildenthal[7] for ^{20}Ne.
For ^{32}S a simple $d_{5/2}$ to $f_{7/2}$ transition of both proton and neutron
was assumed, lacking m œe complete published wave functions.

5)   R.De Swiniarski, F.Resmini, A.D. Bacher and D.L.Hendrie, Nucl. Phys.
     to be published.
6)   A.P.Zuker, Phys. Rev. Lett., 23 (1969) 983.
7)   B.H.Wildenthal, private comunication.

ELASTIC AND INELASTIC SCATTERING OF 52 MEV VECTOR
POLARIZED DEUTERONS ON ^{12}C.

V.Bechtold, L.Friedrich, Karlsruhe Nuclear Research Center
J.Bialy,M.Junge,F.K.Schmidt,G.Strassner, University of Karlsruhe.

In the Karlsruhe polarization experiments the elastic scattering of 52 MeV deuterons on ^{12}C is used for polarization calibration. For some angles the vector analysing power is known from a double scattering experiment [1]. The angular dependence of these data was confirmed by our measurements with the vector polarized beam from C-LASKA. Furthermore, the cross section and the vector analysing power of the inelastic scattering (J=2$^+$, Q= -4,43 MeV and J=3$^-$, Q= -9,64 MeV) has been determined simultaneously.

These data were analysed by using the optical model code MAGALIE [2]. For fitting the elastic cross section and the analysing power the following optical potential was used:

$$V(r) = V_c(r) + V_o \cdot f(r) + 4i\, W_D \frac{df(r)}{dr} + V_{SO} \cdot \lambda^2 \cdot \frac{1}{\pi} \frac{1}{r} \frac{df(r)}{dr} \, \vec{\ell} \cdot \vec{o}$$

$$f(r) = \frac{1}{1 + \exp((r-R_i)/a_i)}$$

The results for parameter set b (table 1) are shown in fig. 1 (dashed line).

For the inelastic scattering DWBA calculations were performed with the parameter set b. The results are the dashed curves in fig. 1. In adjusting the DWBA cross section to the experimental data, the deformation parameters $\beta_2 = 0.55$ and $\beta_3 = 0.35$ are determined. Within this calculations a completely deformed spin-orbit potential in full Thomas form is adopted.

Parameter	r_c	V_0	r_0	a_0	W_D	r_D	a_D	V_{so}	r_{so}	a_{so}	β_2	β_3
b (opt.)	1.1	74.89	1.14	0.73	8.44	1.44	0.72	6.49	0.947	0.722		
d (CCA)	1.1	74.89	1.14	0.73	5.96	1.44	0.72	6.49	0.893	0.722	0.42	0.26

Table 1. Parameter sets for the optical model calculation and the coupled channels analysis.

Better agreement with the measured vector analysing power is achieved by coupled channels analysis (CCA), using the Karlsruhe version ot the ECIS code [3]. To get this good fit, given in fig.1 by solid lines, the parameter set d is assumed and a completely deformed spin-orbid potential too. The coupling of the inelastic states ($J=2^+$ and $J=3^-$) to the ground state leads to a reduction of the depth of the absorbing potential by 30%. Furthermore, the deformation parameters calculated by CCA differ from those of the DWBA ($\beta_2^{CCA}= 0.42$, $\beta_3^{CCA}= 0.26$). These deformation parameters are in very good agreement with those known from α- particle and electron scattering [4].

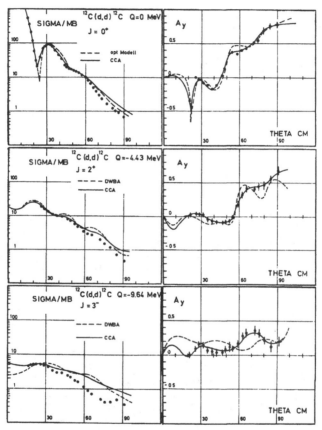

Fig. 1   Comparison between theoretical and experimental scattering data for 52 MeV deuterons on ^{12}C. The elastic cross section data are from Hinterberger et al. [5].

References

1) E. Seibt, Ch.Weddigen, Nucl.Instr.Meth. 100, 253 (1972)
2) J.Raynal, D.Ph-T/69-42 (1969)
3) G.W.Schweimer, J.Raynal, KFK-Report 1333, (1971)
4) F.Hinterberger et al., Nucl.Phys. A115, 570 (1968)
5) F.Hinterberger et al., Nucl.Phys. A111, 265 (1968

SCATTERING OF POLARIZED PROTONS FROM ^{54}Fe AT 17.7, 20.4 and 24.6 MeV*

P.J. van Hall, J.P.M.G. Melssen, H.J.N.A. Bolk, O.J. Poppema,
S.S. Klein, G.J. Nijgh,
Cyclotron Laboratory of the Eindhoven University of Technology,
Eindhoven, Netherlands

It is well known that the experimental analysing powers in the reaction ^{54}Fe($\vec{p}$,p')^{54}Fe (2$^+$, 1.41 MeV) at 18.6 MeV[1] and 19.6 MeV[2] exhibit large discrepancies with standard theoretical calculations. Raynal[3] has suggested that this may be due to an isospin dependence of the nucleon-nucleon spin-orbit force. This effect may be parametrized in a D.W.B.A. calculation using collective form factors by giving the spin-orbit potential a much larger deformation than the central potential. A factor of about 3 was needed to fit the data at 18.6 MeV. A recent analysis by Amos[4] including giant resonance effects did not solve these discrepancies either.

In order to investigate the possibility of resonance effects we have measured the analysing powers for this reaction at 17.7 MeV, 20.4 MeV and 24.6 MeV. Our results, together with the existing 18.6 MeV and 19.6 MeV data, constitute a fairly complete set. For comparison, also measurements were made for ^{56}Fe, ^{58}Ni, ^{60}Ni and ^{62}Ni on which will be reported in separate contributions. Details of the experiment can be found in the papers on ^{58}Ni and ^{60}Ni.

The results of our experiments are given in fig.1 together with optical model or D.W.B.A. calculations based on the global optical potential of Becchetti and Greenlees[5]. The D.W.B.A. code was written by B.J. Verhaar and L.D. Tolsma and included full Thomas coupling[6]. The full curves are for β(s.o.) = β(central), the dashed curves for β(s.o.) = 3 × β(central). The elastic scattering is fitted rather well though better fits may be obtained. The inelastic scattering, however, shows the well-known large discrepancies around 30 degrees and 90 degrees, which are removed only at 17.7 MeV by taking β(s.o.) = 3 × β(central).

From this study it is clear that in order to fit the data the extra deformation of the spin-orbit potential has to be energy dependent. Moreover, comparison with the results for ^{56}Fe learns that the difference between both nuclei nearly vanishes at 24.6 MeV (especially for the region around 90 degrees). This behaviour strongly suggests a resonance-like phenomenon, the description of which, however, is more complicated than the approach based on the collective model[4].

References

* Supported in part by F.O.M. - Z.W.O.
1) C. Glashausser, R. de Swiniarski, J. Thirion, A.D. Hill,
      Phys. Rev. 164 (1967) 1437
2) D.L. Hendrie, C. Glashausser, J.M. Moss, J. Thirion,
      Phys. Rev. 186 (1969) 1188
3) J. Raynal, in Structure of Nuclei, Trieste lectures (1971) p. 75
4) K. Amos and R. Smith, Nucl. Phys. A226 (1974) 519
5) F.D. Becchetti, Jr. and G.W. Greenlees, Phys. Rev. 182 (1969) 190
6) H. Sherif and J.S. Blair, Nucl. Phys. A140 (1970) 33
   B.J. Verhaar, W.C. Hermans, J. Oberski, Nucl. Phys. A195 (1972) 379

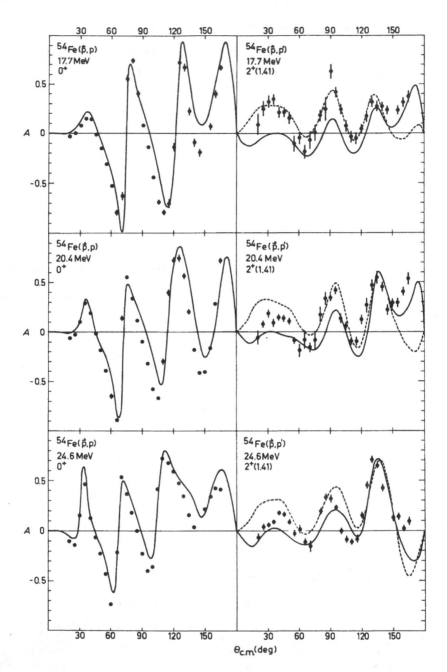

Fig.1. Analysing powers for elastic scattering by ^{54}Fe compared with predictions of the optical model and for inelastic scattering from the 2^+ level at 1.41 MeV compared with D.W.B.A. calculations.
Full curves: $\beta(s.o.) = \beta(central)$; dashed curves: $\beta(s.o.) = 3 \cdot \beta(central)$.

SCATTERING OF POLARIZED PROTONS FROM ^{56}Fe AT 17.7, 20.4 AND 24.6 MeV[*]

J.P.M.G. Melssen, O.J. Poppema, S.D. Wassenaar, S.S. Klein,
G.J. Nijgh, P.J. van Hall,
Cyclotron Laboratory of the Eindhoven University of Technology,
Eindhoven, Netherlands

With the polarized proton beam of the Eindhoven A.V.F. cyclotron analysing powers were measured in scattering from ^{56}Fe at incident energies of 17.7, 20.4 and 24.6 MeV. In this paper we report upon the elastic scattering and inelastic scattering from the first 2^+ - state (Q= - 0.85 MeV). Details of the experimental set-up are given in the companion papers of ^{58}Ni and ^{60}Ni in these proceedings

In figure 1 the measured analysing powers of the ground state and the first 2^+ state are shown for the three incident energies. The elastic scattering is compared with the predictions of the optical model potential from Becchetti and Greenlees[1]. This is a global potential and of course better agreement can be obtained with an optical model search. The agreement, however, between the experimental points and the theoretical predictions is yet quite good. So it is reasonable to use the optical model parameters of Becchetti and Greenlees for a first analysis of the 2^+ state in a D.W.B.A. calculation.

These calculations were performed with a D.W.B.A. code written by L.D. Tolsma and B.J. Verhaar, based on a formalism by Verhaar[2]. A collective form factor including a full Thomas term[3] was used. The deformation parameter of the full Thomas term was taken equal to the deformation parameter of the central part of the form factor. The results of these calculations are given in fig 1. The experimental points are well described by the theory.

The discrepancy between theory and experiment at 24.6 MeV may be caused by the misfitting of the ground state by the Becchetti-Greenlees potential at backward angles. A more detailed analysis seems to be required.

It is remarkable that no energy dependence as in the case of ^{54}Fe, described in a companion paper in these proceedings, is observed.

At forward angles, however, the experimental points tend to be higher than the theoretical predictions. This might be described by a slightly larger deformation parameter of the full Thomas term β(s.o.) $\simeq$ 1.2 β (central) in which case the deformation lengths of the full Thomas and central parts are equal.

References

* Supported in part by F.O.M. - Z.W.O.
1) F.D. Becchetti, Jr. and G.W. Greenlees, Phys. Rev. <u>182</u> (1969) 1190
2) B.J. Verhaar, W.C. Hermans, J. Oberski, Nucl. Phys. <u>A195</u> (1972) 379
3) H. Sherif and J.S. Blair, Nucl. Phys. <u>A140</u> (1970) 33

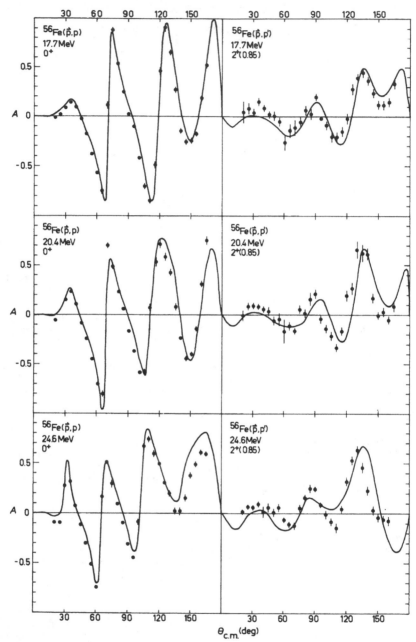

Fig 1. Analysing powers for elastic scattering by ^{56}Fe compared with predictions of the optical model and for inelastic scattering from the 2$^+$ level at 0.85 MeV compared with D.W.B.A. calculations.

SCATTERING OF POLARIZED PROTONS FROM ^{58}Ni AT 20.4 AND 24.6 MeV[*]

O.J. Poppema, S.S.Klein, S.D. Wassenaar, G.J. Nijgh,
P.J. van Hall, J.P.M.G. Melssen,
Cyclotron Laboratory of the Eindhoven University of Technology,
Eindhoven, Netherlands

In this paper we describe an experiment on the scattering by ^{58}Ni of polarized protons accelerated to 20.4 and 24.6 MeV by the Eindhoven A.V. F. cyclotron. The polarized-ion source, constructed by Van der Heide[1], is of the atomic-beam type. It delivers 3 to 5 μA protons with a polarization of about 75 per cent. A 5 keV beam is radially injected into the cyclotron by an exact copy of the Saclay trochoidal injection system[2]. Extracted beams up to 30 nA are available. About 50 per cent. of the beam is transported through a 3 mm × 3 mm aperture and focused onto a 1.5 mm × 1.5 mm spot on the target.

Scattered protons were detected by two arrays of four 3 mm Si(Li) detectors each. The "forward" array, the detectors of which were 5 degrees apart and had defining slits of one degree width, was used for the angular range from 20 to 95 degrees. The detectors of the "backward" array were 10 degrees apart and had two-degree slits. This array was used from 60 to 165 degrees.

Data were taken with this 8-detector system in 5-degree steps. The beam intensity was monitored continuously by two detectors located in the scattering chamber above and below the beam. The beam polarization was also measured continuously by scattering on ^{12}C through 52 degrees after degrading the beam energy by Al foils to 15.5 MeV (in a separate chamber downstream of the scattering chamber). Spectra were measured alternately with polarization on and off (at 20.4 MeV) or with spin up and down (other energies). The polarization was switched at intervals of about 10 seconds. The spectra from each detector were stored in corresponding alternate portions of 512 channels in a PDP-9 computer memory.

Some remarks on additional checks of the beam polarization, the angular settings, the determination of the incident energy and a description of corrections for scattering from target contaminations are made in the companion paper on ^{60}Ni in these proceedings.

The analysing powers of scattering from the ^{58}Ni 0^+ ground state, the first 2^+ excited state and (at 24.6 MeV) the 3^- excited state are shown in fig.1. They agree reasonably with optical model results obtained with the Becchetti-Greenlees[3] potential and with D.W.B.A. calculations using a collective form factor with full Thomas coupling[4]. The deformation of the spin-orbit potential was taken equal to that of the central part. We used a programme written by L.D. Tolsma and B.J. Verhaar, based on a formalism due to Verhaar[5]. Better fits have been obtained for this specific nucleus but the characteristic behaviour is not changed appreciably. A more detailed analysis will be published elsewhere.

References

[*] Supported in part by F.O.M. - Z.W.O.
1) J.A. van der Heide, Thesis, Eindhoven (1972)
2) R. Beurtey and J.M. Durand, Nucl. Instr. and Meth. 57 (1967) 313
3) F.D. Becchetti, Jr. and G.W. Greenlees, Phys. Rev. 182 (1969) 1190
4) H. Sherif and J.S. Blair, Nucl. Phys. A140 (1970) 33
5) B.J. Verhaar, W.C. Hermans, J. Oberski, Nucl. Phys. A195 (1972) 379

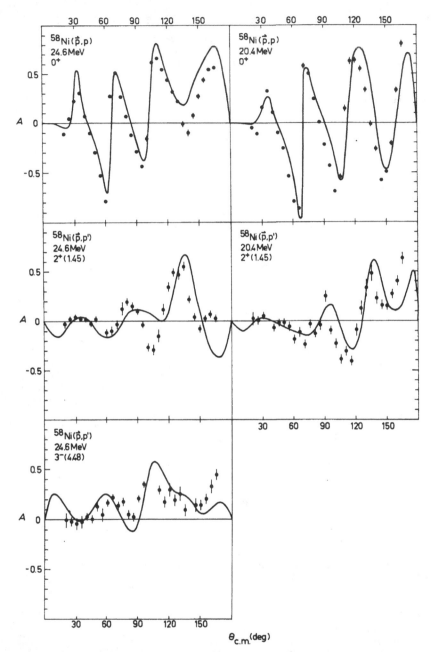

Fig 1. Analysing powers for elastic scattering by ^{58}Ni compared with predictions of the optical model and for inelastic scattering from the 2^+ level at 1.45 MeV and the 3^- level at 4.48 MeV compared with D.W.B.A. calculations.

SCATTERING OF POLARIZED PROTONS FROM ^{60}Ni AT 20.4 AND 24.6 MeV[*]

S.S. Klein, G.J. Nijgh, H.J.N.A. Bolk, P.J. van Hall,
J.P.M.G. Melssen, O.J. Poppema,
Cyclotron Laboratory of the Eindhoven University of Technology,
Eindhoven, Netherlands

In this work we report on our experiments in which 20.4 and 24.6 MeV polarized protons were scattered by a ^{60}Ni target. The experimental set-up was identical to that described in the companion paper on ^{58}Ni in these proceedings.

The energy of the proton beam was determined by comparison of the kinematic shift of the protons scattered from the hydrogen in a polyethylene target as a function of scattering angle to the excitation energy of the first excited state of the ^{12}C in the same target, assumed to be 4.433 MeV. The crossover angles were equal to within half a degree on both sides of the beam and the energy was assumed to be the average of both results.

Measurements with ^{60}Ni targets were sandwiched between similar measurements on polyethylene and mylar targets at each angular setting. From these a check on the beam polarization in addition to the measurements with the downstream polarimeter was obtained. Also the actual angle of measurement could be determined and an energy scale was provided. Finally, corrections for carbon and oxygen contaminations were easily made.

The asymmetries for scattering on ^{12}C agree with the data of Blair[1] at 20.3 MeV and Craig[2] at 24.1 MeV. The agreement with the Craig data at 20.2 MeV is qualitative only, probably due to resonance structure[1].

The actual scattering angles are consistent with nominal settings where protons scattered from hydrogen were detected. Corrections for carbon and oxygen contaminations were generally insignificant except at crossover angles.

The analysing powers of scattering from the ^{60}Ni 0^+ ground state, the first 2^+ excited state and (at 24.6 MeV) the 3^- excited state are shown in fig.1. Very good agreement is obtained with optical model and D.W.B.A. results using the Becchetti-Greenlees[3] potential and a collective form factor with full Thomas coupling[4] and equal deformations for the spin-orbit and central parts of the potential. A D.W.B.A. programme by L.D. Tolsma and B.J. Verhaar, based on a formalism due to Verhaar[5], was used. A more extensive analysis will be published elsewhere.

References

* Supported in part by F.O.M. - Z.W.O.
1) A.G. Blair et al., Phys. Rev. C1 (1970) 444
2) R.M. Craig et al., Nucl. Phys. 79 (1966) 177
3) F.D. Becchetti, Jr. and G.W. Greenlees, Phys. Rev. 182 (1969) 1190
4) H. Sherif and J.S. Blair, Nucl. Phys. A140 (1970) 33
5) B.J. Verhaar, W.C. Hermans, J. Oberski, Nucl. Phys. A195 (1972) 379

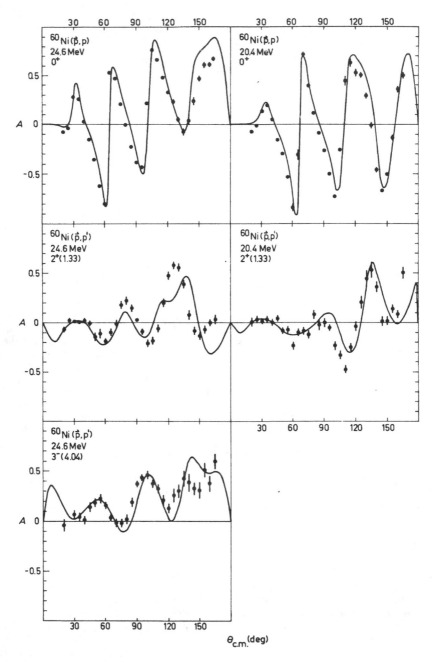

Fig 1. Analysing powers for elastic scattering by ^{60}Ni compared with predictions of the optical model and for inelastic scattering from the 2^+ level at 1.33 MeV and the 3^- level at 4.04 MeV compared with D.W.B.A. calculations.

SCATTERING OF POLARIZED PROTONS FROM ^{62}Ni AT 20.4 AND 24.6 MeV[*]

G.J. Nijgh, P.J. van Hall, H.J.N.A. Bolk, J.P.M.G. Melssen,
O.J. Poppema, S.S. Klein,
Cyclotron Laboratory of the Eindhoven University of Technology,
Eindhoven, Netherlands

In this contribution we give the results of scattering experiments with 20.4 and 24.6 MeV polarized protons using a ^{62}Ni target. The experimental situation was the same as that described in the companion paper in these proceedings on ^{58}Ni. The experimental procedure with respect to energy calibration, additional polarization checks and impurity corrections may be found in a similar companion paper on ^{60}Ni. It was again found that actual and nominal scattering angles were consistent and that contamination corrections were insignificant except at crossovers.

The analysing powers of scattering from the ^{62}Ni 0^+ ground state, the 2^+ first excited state and (at 24.6 MeV) the 3^- excited state are shown in fig.1. Very good agreement with the results of calculations is observed for elastic scattering using an optical model potential as given by Becchetti and Greenlees[1]). The analysing powers for inelastic scattering, obtained with a programme written by L.D. Tolsma and B.J. Verhaar and based on a formalism by the latter[2]) to incorporate full Thomas coupling[3]), also show a very good agreement with the experimental results. In these calculations the deformations of the spin-orbit part and of the central part of the potential were taken equal.

One may conclude that the analysing powers obtained experimentally for the Ni isotopes are generally well described by the global model used. The individual nuclides are compared with a general trend instead of obscuring the differences by parameter fitting. It looks worthwhile, however, to search for a locally better approximation slightly deviating from the Becchetti-Greenlees parametrization as a better basis for the D.B.W.A. calculations.

References

* Supported in part by F.O.M. - Z.W.O.

1) F.D. Becchetti, Jr. and G.W. Greenlees, Phys. Rev. 182 (1969) 1190
2) B.J. Verhaar, W.C. Hermans, J. Oberski, Nucl. Phys. A195 (1972) 379
3) H. Sherif and J.S. Blair, Nucl. Phys. A140 (1970) 33

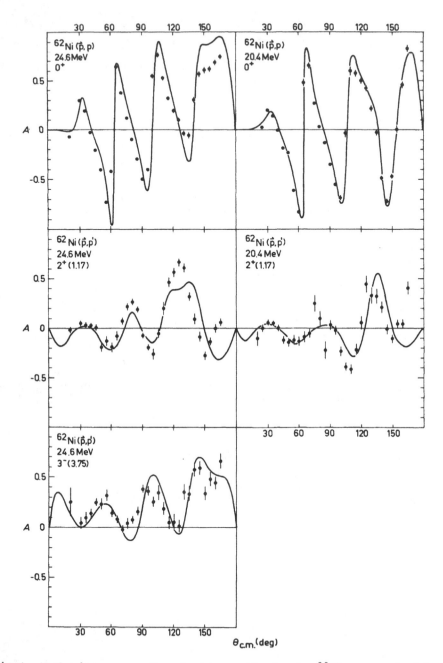

Fig 1. Analysing powers for elastic scattering by ^{62}Ni compared with predictions of the optical model and for inelastic scattering from the 2^+ level at 1.17 MeV and the 3^- level at 3.75 MeV compared with D.W.B.A. calculations.

INELASTIC SCATTERING OF 30 MeV POLARIZED PROTONS FROM ^{90,92}Zr AND ^{92}Mo

R. de Swiniarski, G. Bagieu, M. Bedjidiam[+], C.B. Fulmer[*],
J.Y. Grossiord[+], M. Massaad, J.R. Pizzi[+] and M. Gusakow[+]
Institut des Sciences Nucléaires, BP 257, 38044 Grenoble Cédex, France

Previously reported measurement of inelastic scattering of 20.3 MeV polarized protons from ^{90}Zr, ^{92}Zr and ^{92}Mo [1] have shown poor agreement between the predictions of the nuclear collective model and the analyzing power for the first 2^+ states in these nuclei, although the corresponding cross-sections for these states were well reproduced by the model. On the other hand, the analyzing power for the low-lying 3^- states in these nuclei were fairly well reproduced. Although the agreement between the analyzing power and the coupled-channels collective model calculations were somewhat improved by the introduction of the "full Thomas term" of the distorted spin-orbit potential, the overall agreement remains poor for the 2^+ analyzing power.

Since compound nucleus effect may be important at 20 MeV and therefore being partly responsable for this disagreement, a new measurement at higher energy of these data appears obvious. The data reported in this paper were taken at an energy of 30 MeV using the Grenoble cyclotron. Since the new analyzing magnet [2] recently installed was used, an overall resolution of 80 to 100 keV (FWHM) has been obtained most of the time. Up to 5 nA of analyzed polarized protons, with a polarization close to 70 % were delivered on the target.

The measurements were made using 2 or 4 telescopes comprising ΔE surface barriers detectors 1800 µm thick and E detector of 5 mm thickness,

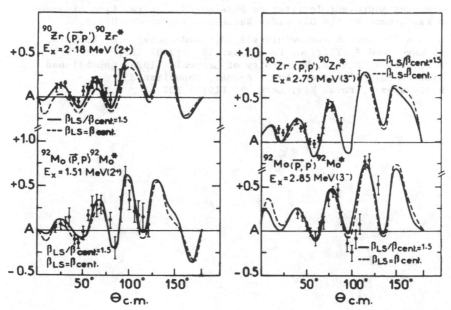

Fig. 1. Analyzing powers for the 2^+ and 3^- states of ^{90}Zr and ^{92}Mo.

all cooled by thermoelectric device to -25° C. Enriched targets obtained from ORNL were used ; they were all about 1 mg/cm^2 thick. The data were analyzed with the coupled-channels code ECIS 74 [3] using the optical model and deformation parameters of Table 1. This program includes the full Thomas term of the distorted spin-orbit potential. The figure 1 shows the good agreement obtained between the calculations and the data for the analyzing power of the low-lying 2$^+$ and 3$^-$ states in ^{90}Zr and ^{92}Mo by keeping the deformation parameters constant ($\beta_{central} = \beta_{LS}$) ; slightly improved fits were obtained by keeping the deformation lenghts constant ($\delta = \beta_{cent} r_o = \beta_{LS} r_{LS}$) while the best overall agreement, for 2$^+$ and 3$^-$ analyzing powers were obtained by increasing the ratio β_{LS}/β_{cent} to 1.5, a procedure which was often applied in the past and which has recently received a simple theoretical explanation. Preliminary analysis of ^{92}Zr data shows no particular features and equivalent fits to those presented in Fig. 1 are obtained.

Table 1. Optical model and deformation parameters

Target	V_o (MeV)	r_o (F)	a_o (F)	W_V (MeV)	W_D (MeV)	r_I (F)	a_I (F)	V_{LS} (MeV)	r_{LS} (F)	a_{LS} (F)	β_{2+}	β_{3-}
^{90}Zr	56.7	1.11	0.77	4.0	4.88	1.37	0.65	6.17	0.89	0.54	0.07	0.15
^{92}Mo	51.7	1.16	0.75	3.28	4.58	1.35	0.63	7.65	1.06	0.75	0.09	0.13

References

+ Permanent address : Institut de Physique Nucléaire, Lyon, France
* On assignment by the Oak-Ridge National Laboratory, U.S.A.

1) C. Glashausser, R. de Swiniarski, J. Goudergues, R.M. Lombard, B. Mayer and J. Thirion, Phys. Rev. 184 (1969) 1217
2) J.B. Leroux, Thesis, University of Grenoble (1973) unpublished
3) Program ECIS 74 written by J. Raynal, unpublished
4) B.J. Verhaar, Phys. Rev. Lett. 32 (1974) 307

MICROSCOPIC MODEL ANALYSIS OF 30 MeV POLARIZED PROTONS INELASTIC SCATTERING FROM ^{90}Zr AND ^{92}Mo

Dinh-Lien Pham, G. Bagieu, R. de Swiniarski and M. Massaad
Institut des Sciences Nucléaires, BP 257, 38044 Grenoble Cedex, France

Microscopic model calculations have been performed in the frame-work of the DWBA formalism, using the program DWBA 70 [1]. This program allows a complete treatment of central (S), spin-orbit (LS), and tensor (T) terms as well as full antisymmetrization of the N-N interaction, with radial form factors of the Yukawa type. The scattering amplitude is given by

$$T_A = \lambda_p T_A^{(p)} + \lambda_n T_A^{(n)}$$

where λ_p and λ_n are two normalization parameters, one for all proton and one for all neutron configurations of which the separated scattering amplitudes are respectively $T_A^{(p)}$ and $T_A^{(n)}$. In our calculations, the excited states are considered as a mixture of proton and neutron excitations. The wave functions for the ground and excited states are taken respectively to have the form

$$|\psi> = \alpha |\psi_p> + \beta |\psi_n> \qquad ; \qquad |\psi'> = \alpha |\psi'_p> + \beta |\psi'_n> .$$

For the two nuclei ^{90}Zr and ^{92}Mo (neutron closed shells), ψ and ψ' denote the simple shell model wave functions respectively for the ground and excited states. For these two nuclei, we consider the most important configuration $(2d_{5/2} - g_{9/2}^{-1})$ for the neutron amplitude. For the 2$^+$ state

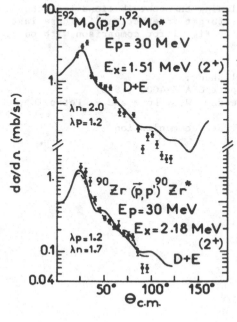

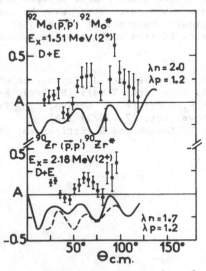

Fig. 1. Cross sections and analyzing powers for the 2$^+$ state of ^{90}Zr and ^{92}Mo. Dotted line is from ref. 3).

$$T_A^{(n)} = \beta^2 < \Psi'_n \mid V_{pn} \mid \Psi_n > = \beta^2 < (\nu d_{5/2} g_{9/2})_2 \mid V_{pn} \mid (\nu g_{9/2})_0^2 >$$

and the proton configuration is a recoupling in $g_{9/2}$. As for the nuclei ^{92}Zr, for the first 2^+ state

$$T_A^{(n)} = \beta^2 < (\nu d_{5/2})_2^2 \mid V_{pn} \mid (\nu d_{5/2})_0^2 >$$

and the proton configuration is a recoupling in $g_{9/2}$.

Fig. 1 shows the good agreement with the first 2^+ state experimental cross sections at $E_p = 30$ MeV obtained with $\alpha = (0.8)^{1/2}$, $\beta = (0.2)^{1/2}$ for ^{90}Zr and ^{92}Mo. Similar results have been obtained for ^{92}Zr with $\alpha = \beta = (2)^{-1/2}$. Optical model parameters used in these calculations are given in Table 1.

Table 1. Optical model parameters

	V_o	r_o	a_o	W_V	W_D	r_I	a_I	V_{LS}	r_{LS}	a_{LS}
	(MeV)	(F)	(F)	(MeV)	(MeV)	(F)	(F)	(MeV)	(F)	(F)
^{90}Zr	56.7	1.11	0.77	4.0	4.88	1.37	0.65	6.17	0.89	0.54
^{92}Mo	51.7	1.16	0.75	3.28	4.58	1.35	0.63	7.65	1.06	0.75

The N-N interaction used in these calculations are [2] $v_{pp}^0 = -14.5$ MeV, $v_{pp}^1 = 14.6$ MeV, $v_{pp}^{LS1} = 29.1$ MeV, $v_{pp}^{LS2} = -1490$ MeV, $v^T = -v_{pp}^{TPP} = 11.0$ MeV fm^{-2}, $v_{pn}^0 = -36.4$ MeV, $v_{pn}^1 = -2.39$ MeV, $v_{pn}^{LS1} = 20.1$ MeV and $v_{pn}^{LS2} = -752$ MeV. The range parameters are $u_S = 1.06$ fm, $u_{LS1} = 0.557$ fm, $u_{LS2} = 0.301$ fm and $u_T = 0.816$ fm.

As shown on Fig. 1, the analyzing powers don't agree with the experiment. The disagreement between calculated and experimental analyzing powers has been observed even by including the contributions from the imaginary component of the projectile-target interaction [3]. These last results (dotted line) are also shown on Fig. 1 for comparison with ours.

References

1) R. Schaeffer and J. Raynal, unpublished ;
   R. Schaeffer, Thesis (Saclay), Report CEA-R-4000 (1970)
2) R.A. Hinrichs, D. Larson, B.M. Preedom, W.G. Love and F. Petrovich, Phys. Rev. C7 (1973) 1981
3) H.V. Geramb and R. Sprickmann, private communication

STUDY OF ^{112}Cd BY INELASTIC SCATTERING OF 30 MeV POLARIZED PROTONS

R. de Swiniarski, G. Bagieu, J.Y. Grossiord,[+] A. Guichard,[+]
M. Massaad and Dinh-Lien Pham
Institut des Sciences Nucléaires, BP 257, 38044 Grenoble Cédex, France

It is known that the pure harmonic vibrational model does not explain completly the proton inelastic scattering cross-sections from several states in ^{112}Cd like the second 2^+ in this nucleus [1]. Although good fits to the data were obtained by adding to the basically two-phonon state $|2_2^+>$ some admixture of a one-phonon state, several problems in this nucleus still remain to be investigated. Moreover, recently, it has been suggested [2] that a strong two-phonon admixture should be added also to the one-phonon state of the first 2^+ in ^{112}Cd ($E_x = 0.617$ MeV (2_1^+)) to describe successfully this state. With this problem in mind, we have measured cross-sections and analyzing powers of inelastic scattering of 30 MeV protons from some low lying states in ^{112}Cd. The data taken at the Grenoble cyclotron were analyzed with the program ECIS 74 [4] by coupling the 0^+, the 2_1^+ and the 2_2^+ (1.31 MeV) using the optical model and deformation parameters of Table 1. The calculations for the data are presented on Fig. 1.

We determined for each state the mixture of one-phonon and two-phonon components in the vibrational model. In the first case, the 2_1^+ state is considered as a pure one-phonon state and the 2_2^+ state by

$$| 2_2^+ > = 0.866 \ | 1\text{-ph} > + 0.5 | 2\text{-ph} > .$$

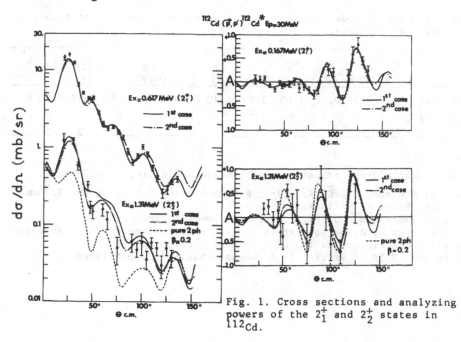

Fig. 1. Cross sections and analyzing powers of the 2_1^+ and 2_2^+ states in ^{112}Cd.

In the second case, the 2_1^+ and 2_2^+ orthogonal states are represented by

$$|2_1^+> = 0.917\,|1\text{-ph}> - 0.399\,|2\text{-ph}>$$

$$|2_2^+> = 0.399\,|1\text{-ph}> + 0.917\,|2\text{-ph}>\,.$$

The coupling parameters notation β is the same as given in ref. [1].

For the first case (solid line) we find again that the deformation parameters as well as the mixture components are similar to those obtained in ref. [1]. Nevertheless a better agreement with the data is clearly obtained for the second case but this effect is more visible on the 2_2^+ than on the 2_1^+. Finally Fig. 2 presents the excellent fit obtained for the data of the 3^- state assuming 0^+, 3^- coupling with $\beta(3^-) = 0.18$.

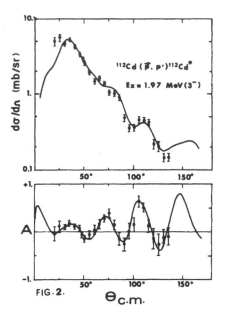

FIG. 2. $\Theta_{c.m.}$

Fig. 2. Cross section and analyzing power of the 3^- state in ^{112}Cd.

Table 1. Optical model and deformation parameters used in the coupled - channels calculations.

	V_0 (MeV)	r_0 (F)	a_0 (F)	W_V (MeV)	W_D (MeV)	r_I (F)	a_I (F)	V_{LS} (MeV)	r_{LS} (F)	a_{LS} (F)	β_2	β_{02}''
1st case	53.59	1.14	0.76	3.80	5.63	1.30	0.68	5.48	1.01	0.49	0.20	0.054
2nd case	"	"	"	"	4.50	"	"	"	"	"	"	0.040

References

+ Permanent address : Institut de Physique Nucléaire, Lyon, France

1) P.H. Stelson, J.L.C. Ford, Jr., R.L. Robinson, C.K. Wong and T. Tamura, Nucl. Phys. A119 (1968) 14
2) G.G. Shute and M.J. Fagan, private communication
3) See paper by R. de Swiniarski et al., this conference for a more complete description of the experimental section, p.735
4) Program ECIS 74 written by J. Raynal (Saclay), unpublished

ELASTIC AND INELASTIC SCATTERING OF POLARIZED DEUTERONS BY ^{112,116}Cd

A.K. Basak, J.A.R. Griffith, G. Heil*, O. Karban, S. Roman and G.Tungate
Department of Physics, University of Birmingham, England +

It was shown in ref. 1 for Fe and Ni isotopes that both elastic and inelastic vector analysing powers are sensitive to details of collective-
model description of the scattering and in particular, there is a strong
"feedback" on the elastic scattering vector analysing power (VAP) depending on the magnitude of the deformation parameter. It is therefore
desirable to study further this effect on nuclei where the mode of excitation of collective levels is better understood.
Measurements of the elastic and inelastic analysing powers were performed using both vector and tensor polarized 12.4 MeV deuterons incident
on self-supporting 2.0 mg/cm^2 targets of enriched ^{112}Cd and ^{116}Cd. A
standard experimental technique described e.g. in ref. 2 was used. The
results with the vector beam are shown for the ground state and first
2$^+$ state for both isotopes in fig. 1 and 2. As the magnitude of polarization effects decreases with rising Coulomb barrier high accuracy data
were required with statistical errors not exceeding $\pm$ 0.01 for the elastic VAP.
The experimental data show two important characteristics: The elastic
VAP, while following a regular oscillatory pattern, is "shifted" towards negative values over the whole angular region. This is a feature
which has not yet been observed in elastic scattering and cannot in
principle be reproduced by a standard optical model calculation. The
other experimental fact is that the inelastic VAP is in phase with the
elastic VAP and it would be difficult to fit both sets of data simultaneously with a coupled-channels calculation using any appreciable value
for the deformation parameter.

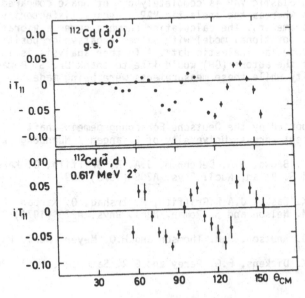

Fig.1. Elastic and inelastic scattering vector analysing power for ^{112}Cd

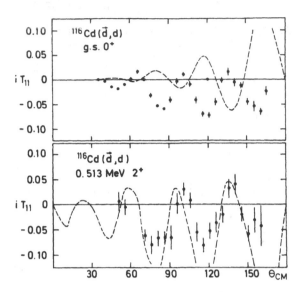

Fig. 2. Elastic and inelastic scattering vector analysing power for
^{116}Cd. Coupled channels calculation for vibrational model with β =0.20.

The latter point is demonstrated in fig. 2 where a coupled channels prediction is compared with the experimental data. Here there were used the
optical model parameters from ref. 2 and β =0.20 from ref. 4. The predicted elastic VAP is completely out of phase compared with the measured
points whereas the inelastic VAP agrees satisfactorily for the unadjus-
ted parameters. The calculation shown assumes a vibrational model for
^{116}Cd; rotational model with either negative or positive β is unable to
reproduce the inelastic data. A further analysis is in progress.
One of the autors (GH) would like to thank Dr. S. Roman for his hospitality while these measurements were being made.

References

+ Supported by the Deutsche Forschungsgemeinschaft
* Present adress: University of Erlangen / Nürnberg, W. Germany

1) R.C. Brown, A.A. Debenham, J.A. R. Griffith, O. Karban, D.C. Kocher
   and S. Roman, Nucl. Phys. A208 (1973) 589

2) A.K. Basak. J.A.R.Griffith, M. Irshad, O. Karban, E.J. Ludwig,
   J.M. Nelson and S. Roman, Nucl. Phys. A229 (1974) 219

3) L.D. Knutson, J.A. Thomson and H.O. Meyer, Nucl. Phys. A241 (1975) 36

4) J.K. Dickens, F.G. Perey and G.R. Satchler, Nucl. Phys. 73 (1965) 529

COUPLED-CHANNELS EFFECTS IN THE ^{148,152}Sm($\vec{d}$,d') REACTIONS*

F. Todd Baker[†], C. Glashausser, A. B. Robbins and E. Ventura
Department of Physics
Rutgers University, New Brunswick, New Jersey

Recent investigations[1,2]) of the inelastic scattering of vector-polarized deuterons have suggested the possibility of using the (d,d') reaction to distinguish between vibrational and rotational states in nuclei. For Fe and Ni isotopes, the relative phase of elastic and inelastic (2$^+$) iT$_{11}$ angular distributions at 12.3 MeV could be reasonably well reproduced in a coupled-channels calculation. The predictions of this phase depend critically on whether a vibrational or rotational model is used[1]).

To examine this possibility, we have used ^{148}Sm and ^{152}Sm as test cases; the transition between vibrational and rotational nuclei is known to occur between ^{148}Sm and ^{152}Sm. Sharp differences in proton scattering from these nuclei have been explained by differences in coupled-channels predictions for vibrational and rotational models[3]).

The deuteron analyzing power results at 15 MeV are shown in Fig. 1. Most of the data were taken with a position-sensitive proportional counter on the image surface of a split-pole spectrograph. The magnitudes of the analyzing

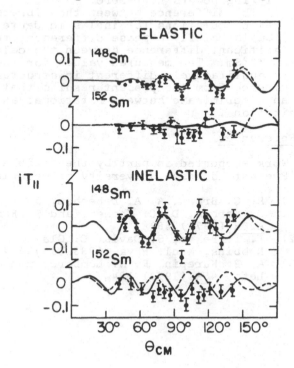

Fig. 1. Measured values of iT$_{11}$ for elastic and inelastic (2$_1^+$) scattering of 15 MeV deuterons from ^{148}Sm and ^{152}Sm. The solid curves correspond to coupled-channels calculations with a vibrational model; the dashed curves were calculated with a rotational model.

powers for ^{148}Sm are generally larger and show more pronounced structure than those for ^{152}Sm; similar differences
were observed in proton scattering.

The technique used to analyze the data was to first obtain "spherical" optical model parameters by doing a coupled-channels search on the ^{148}Sm data assuming a vibrational model for the 0^+-2^+ coupling; this optical potential was
then applied to all subsequent calculations. Differences
between the ^{148}Sm and the ^{152}Sm data should then be accounted for by model-dependence and differences in coupling
strength.

The results of the coupled-channels calculations are
shown in Fig. 1. For the weak coupling case, ^{148}Sm, there
is very little difference between vibrational and rotational
calculations; both give a good description of the data. For
the elastic scattering the rotational model does very well
in describing the ^{152}Sm data. Furthermore, due to the
stronger coupling, vibrational- and rotational-model predictions for ^{152}Sm are clearly distinguishable for both
elastic and inelastic scattering. Unfortunately neither
model provides a satisfactory fit to the inelastic scattering iT_{11} data which are negative and rather structureless
over the entire angular range measured. Attempts to improve
the fit by searching on the optical-model parameters resulted in unacceptably deep imaginary well depths; the
effect of these was simply to minimize χ^2 by making all
analyzing powers near zero.

The difference between the vibrational and rotational
model calculations for inelastic deuteron scattering from
^{152}Sm is chiefly a phase difference; this is also the most
significant difference between the calculations for ^{152}Sm
and ^{148}Sm. The measured values for the two nuclei, however,
are qualitatively different in structure. Thus there is no
evidence from the present results that deuteron scattering
can discriminate between rotational and vibrational excitations.

References

*Work supported in part by the National Science Foundation.
†Present address: University of Georgia, Athens, Ga. 30602.

1) R. C. Brown, A. A. Debenham, J. A. R. Griffith,
   O. Karban, D. C. Kocher, and S. Roman, Nucl. Phys.
   A208 (1973) 589
2) F. T. Baker, S. Davis, C. Glashausser, and A. B.
   Robbins, Nucl. Phys. A233 (1974) 409
3) A. B. Kurepin, R. M. Lombard, and J. Raynal, Phys.
   Lett. 45B (1973) 184

Giant Dipole and Quadrupole Resonances

GIANT RESONANCE EFFECTS IN THE ENERGY DEPENDENCE OF THE SPIN-FLIP
PROBABILITY FOR THE ^{12}C(p,p' $\gamma_{4.43\ MeV}$)^{12}C REACTION

R. De Leo, F. Ferrero,[+] A. Pantaleo and M. Pignanelli[⁑]
Istituto di Fisica dell'Università, Bari, Italy and
Istituto Nazionale di Fisica Nucleare, sezione di Bari

In this comunication we report an investigation of the role played,
in the case of a strong inelastic transition, by two-step processes, coming from Eλ giant resonance modes in the target nucleus.

Differential cross sections for elastic and inelastic scattering on
^{12}C have been measured at several incident energies, by using the proton
beam of the Milan University AVF cyclotron. The spin-flip probability
(SFP) for the transition to the first 2$^+$ excited state has been also determined through a proton-γ correlation, in which the 4.43 MeV deexcitation γ-ray was detected along the normal to the scattering plane. The a-
vailable data, together with some previously published result,[1] consist
of angular distributions taken at 14 proton incident energies between 15
and 40 MeV and of an excitation function, with smaller energy steps, at a
scattering angle corresponding to the backward maximum in the SFP angular
distribution.

The SFP shows a pronounced increase, resonant like, in two energy regions centered at about 20 and 28 MeV respectively. The changes in the
shape of the SFP angular distributions are instead rather limited, as
shown in the figure by data taken out and within a resonance (18.5 and
28.8 MeV). The SFP exhibits a good coherence in its energy dependence
with the elastic cross section at backward angles. In the energy regions
where bumps are found for the elastic cross section and SFP, the inelastic cross section shows only a small decrease. It is to some extent surprising that the correlation of the spin-flip part of the inelastic with
the elastic cross section is stronger than that found with the non-spin-
-flip part of the same inelastic cross section.

The increment of the SFP in the 26-35 MeV energy region over the out
resonance value, which has been estimated to be 5 and 25% for the two maxima in the angular distribution at forward and backward angles respectively, reproduces very closely the energy dependence of the E2 giant quadrupole resonance strength, as extracted from the energy dependence of
the transitions to unnatural parity 1$^+$ states[2] (continous lines in the
figure). The 20 MeV resonance is well defined only at forward angles and
should correspond to the E1 giant dipole resonance. This latter resonance
as determined by the ^{11}B(p, γ_o) reaction,[3] is however centered at an excitation energy that corresponds to about 22.6 MeV for a proton incident
on ^{12}C. A better agreement in the energy position can be found with one
of the peaks in the excitation function for the ^{11}B(p, γ_1) reaction. It
may be suggested that different intermediate states are selected, within
the giant resonance, according to the different structure of the final
state observed.

Both macroscopic coupled channels and microscopic DWBA calculations
give satisfactory descriptions to the data only out of resonances. In the
resonance regions one can reproduce, by coupled channels calculation, the
inelastic cross section or the SFP only separately by using different values of the deformation parameters.

The present work emphasizes the usefulness of spin-flip data in a
reaction mechanism study and the necessity, even for a strong collective
transition of a theoretical analysis,[4] in which giant resonances are taken esplicitly into account. Further theoretical evaluations are in pro-

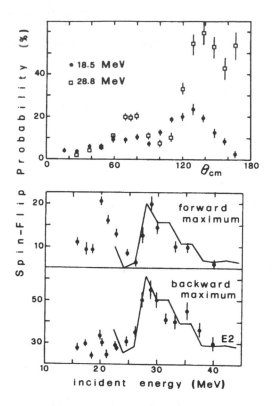

Fig.1. Upper part: angular distributions of the spin-flip probability out (18.5 MeV) and within (28.8 MeV) a resonance. Medium and lower parts: the experimental values of the maxima in the spin-flip probability at forward and backward angles plotted against the incident energy. The continous lines represent the strength of the E2 quadrupole giant resonance as given in ref. 2.

gress.

References

+   Present address: Istituto di Fisica dell'Università, Torino.
⚢   Permanent address: Istituto di Fisica dell'Università, Milano.

1) M.A.D. Wilson and L.Schecter, Phys. Rev. C4 (1971) 1103, also for previous references.
2) H.V.Geramb and R. Sprickmann, Report KFA-IKP 10/74 (Julich, 1973) p. 220 and R. Sprickmann, Report Jul-1137-KP (Julich, 1974).
3) R.G. Allas, S.S. Hanna, L. Meyer-Schutzmeister and R.E. Segel, Nucl. Phys. 58 (1964) 122.
4) H.V. Geramb, R. Sprickmann and G.L. Strobel, Nucl. Phys. A199 (1973) 545.

CALCULATED ANGULAR DISTRIBUTIONS AND POLARIZATIONS FOR THE
REACTION ^{15}N$(p,\gamma_o)^{16}$O†

D. G. Mavis, H. F. Glavish and D. C. Slater*
Department of Physics, Stanford University
Stanford, California 94305 U.S.A.

The giant dipole resonance of ^{16}O observed via the ^{15}N$(p,\gamma_o)^{16}$O and ^{15}O$(\gamma,n_o)^{16}$O reactions exhibits two dominant peaks (see fig. 1) at excitation energies of 22.3 and 24.45 MeV. These two peaks carry a major part of the E1 strength and have been interpreted as collective single particle-hole excitations generated from a particle-hole interaction acting on unperturbed single particle shell model excitations[1,2]. In terms of this model the two peaks are predicted to have quite different particle-hole configurations, being dominantly $d_{5/2}p_{3/2}^{-1}$ at 22.3 MeV and $d_{3/2}p_{3/2}^{-1}$ at 24.45 MeV. On the other hand, angular distribution and polarization measurements[3] in the ^{15}N$(p,\gamma_o)^{16}$O reaction, show that the $s_{1/2}$ and $d_{3/2}$ proton capture matrix elements (the only ones allowed for E1 radiation by angular momentum and parity conservation) have remarkably constant relative amplitudes. The calculations reported here were made to see if the simple shell model description can account for such constancy. The matrix elements $T_{\ell j}$ were determined using the formalism of Feshbach, Kerman and Lemmer[4], which yields[4,5]

$$T_{\ell j} = \langle \ell j | D_\gamma | 0 \rangle + \sum_k \langle \ell j | V_{ph} | d_k \rangle \langle d_k | D_\gamma | 0 \rangle (E - E_k + i\tfrac{1}{2}\Gamma_k)^{-1} \qquad (1)$$

where $|\ell j\rangle$ describes the continuum nucleon and the hole state of the residual mass 15 nucleus. The doorway states $|d_k\rangle$ are the collective particle hole configurations and E_k and Γ_k are their energies and widths. The particle-hole interaction V_{ph} was taken as

$$V_{ij} = -584.1(0.865 + 0.135 \underset{\sim}{\sigma}_i \cdot \underset{\sim}{\sigma}_j) \delta(\vec{r}_i - \vec{r}_j).$$

The quantity D_γ is the electric dipole operator. The unperturbed single particle wave functions were generated from a real Wood-Saxon well adjusted to correctly reproduce single particle energies.

The results of the calculation for the ^{15}N$(p,\gamma_o)^{16}$O reaction are compared with experimental data in fig. 1. The coefficients a_2 and b_2 shown at the right of fig. 1 are the $k = 2$ coefficients in the expressions for the cross section $\sigma(\theta) = A_o\{1 + \sum_{k=1} a_k P_k(\cos\theta)\}$ and analyzing power $A(\theta) = A_o \sum_{k=1} b_k P_k{}'(\cos\theta)/\sigma(\theta)$. They are related to the T-matrix elements $s_{1/2}\exp(i\theta_s)$ and $d_{3/2}\exp(i\theta_d)$ by $b_2 = \frac{1}{\sqrt{2}} s_{1/2} d_{3/2} \sin(\theta_d - \theta_s)$ and $a_2 = -0.5d_{3/2}^2 + \sqrt{2}s_{1/2}d_{3/2}\cos(\theta_d - \theta_s)$ with the normalization $s_{1/2}^2 + d_{3/2}^2 = 1$. The quadratic nature of these equations results in two sets of solutions (I and II) for the experimental matrix elements, as shown at the left in fig. 1. The calculations are consistent with solution I and are able to reproduce the observed constancy in the $s_{1/2}$ and $d_{3/2}$ amplitudes. Even the phase difference is quite well reproduced, but it should be noted that there is a small disagreement and this is responsible for a somewhat larger disagreement in the a_2 coefficient. This is because the cosine term is very sensitive to $\theta_d - \theta_s$ when this phase difference is near 90°.

The theoretical curves in fig. 1 included a small third doorway at E_x = 19.7 MeV of predominantly $s_{1/2}p_{3/2}^{-1}$ configuration. Although eq. (1) does not show doorway-doorway couplings these were actually included in the calculations but did not make a substantial difference to the quality of the fit. The doorway width is the sum of two parts $\Gamma_k = \Gamma_p + \Gamma_Q$ where Γ_p describes the damping to the continuum and can be calculated, while Γ_Q is the damping to more complicated nuclear configurations and is an adjustable parameter in the theory. However Γ_p dominates Γ_Q for a light nucleus such as ^{16}O and while the fits shown in fig. 1 have Γ_Q = 500 keV, there is little change if Γ_Q is disregarded and set equal to zero. Similar quality fits were obtained for the $^{16}O(\gamma,n_O)^{15}O$ reaction. As usual with a particle-hole model the calculated cross sections are too high.

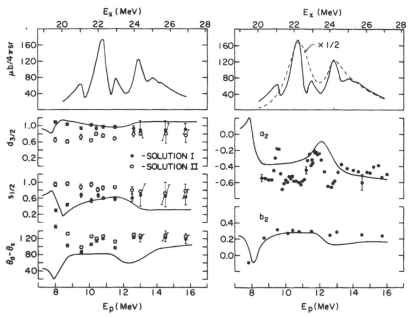

Fig. 1. Theoretical fits and experimental data for the reaction $^{15}N(p,\gamma_O)^{16}O$. The solid curves in the upper part of the figure are the experimental cross sections. The remaining solid curves are theoretical fits generated from eq. (1). The broken curve (upper right) is the cross section generated from eq. (1).

References

† Supported in part by the National Science Foundation
* Present address: LAMPF, Los Alamos Sci. Lab., Los Alamos, N.M. 87544
1) J. P. Elliot and B. H. Flowers, Proc. Roy. Soc. (London) A242 (1957) 57
2) G. E. Brown, L. Castillejo and J. A. Evans, Nucl. Phys. 22 (1961) 1
3) S. S. Hanna et al., Phys. Lett. 40B (1972) 631
4) H. Feshbach, A. K. Kerman and R. H. Lemmer, Ann. of Phys. (N.Y.) 41 (1967) 230
5) W. L. Wang and C. M. Shakin, Phys. Rev. C5 (1972) 1898

751

GIANT RESONANCE AS INTERMEDIATE STRUCTURE IN THE $^{16}O(\vec{p},p')^{16}O^*$ (2⁻, 8.88 MeV) TRANSITION

D. Lebrun, G. Perrin, J. Arvieux, M. Buenerd, P. Martin,
P. de Saintignon
Institut des Sciences Nucléaires, BP 257, 38044 Grenoble Cédex, France
R. Sprickmann and H.V. von Geramb
Institut für Kernphysik, K.F.A., Jülich, Germany

The investigation of giant resonances (GR) has been a central goal of many experimental and theoretical studies [1]. One way to get information about these GR is to study their contributions as virtual intermediate states in inelastic transitions to low lying levels in nuclei.

An analysis of differential cross sections for inelastic scattering of protons to unnatural parity state $J^\pi = 2^-$ ($E_x = +8.88$ MeV) in ^{16}O is a good tool for this purpose [2].

Since it is well known that spin dependent data are more sensitive to the details of the reaction mechanism [3], such a measurement was performed in the Institut des Sciences Nucléaires de Grenoble. Analyzing power as well as differential cross sections for the reaction $^{16}O(\vec{p},p')^{16}O^*$ (2⁻, 8.88 MeV) were extracted for 33.8, 35.8 and 36.8 MeV incident energies.

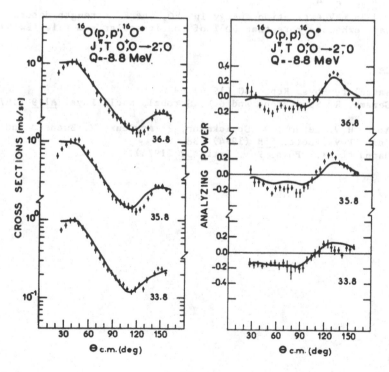

Fig. 1

Calculations were conducted using a full microscopic DWA formalism including exchange contribution due to antisymmetrisation between incident and valence target nucleus. A two step process, where resonances appear as intermediate states, was added coherently to the usual one step valence mechanism. The results are reported on Fig. 1 and Table 1.

Table 1. Coupling constants extracted from this analysis

E_p (MeV)	33.8	35.8	36.8
Y_1	5.9	5.8	4.0
Y_2	7.2	6.5	4.0
Y_3	16.7	17.8	17.0

Units of Y_λ are $10^{-3} \times MeV^{-1}$

The relevent coupling strength constants Y_λ needed to reproduce the data show evidence of excitation of the E1, E2, E3 giant resonances in ^{16}O.

Present analysis is in good agreement with photonuclear reaction data for E1 and E2 [4].

Around 35 MeV excitation energy in ^{16}O, the E3 strength determined by this way, exhausts more than 60 % of the isoscalar sum rule (EWSR) [1].

References

1) G.R. Satchler, Phys. Rep. 14C (1974) 99
2) H.V. Geramb, R. Sprickmann and G.L. Strobel, Nucl. Phys. A199 (1973) 545
3) K.A. Amos, H.V. Geramb, R. Sprickmann, J. Arvieux, M. Buenerd and G. Perrin, Phys. Lett. 55B (1974) 138
4) S.S. Hanna et al., Phys. Rev. Lett. 32 (1974) 114

E1 AND E2 STRENGTH IN ^{32}S OBSERVED WITH POLARIZED PROTON CAPTURE[*]

E. Kuhlmann[†], H. F. Glavish, J. R. Calarco, S. S. Hanna, and D. G. Mavis
Department of Physics, Stanford University
Stanford, California 94305 U.S.A.

The polarized proton capture reaction is a very powerful tool for studying the giant E1 resonance as well as M1 and E2 strength in nuclei. The reaction matrix elements for radiative capture can be studied thoroughly if not only the angular distributions, $\sigma(\theta)$, but also the analyzing powers, $A(\theta)$, of the emitted photons are observed.

We have measured $\sigma(\theta)$ and $A(\theta)$ for proton capture by ^{31}P at eleven energies between $E_p = 6$ and 11 MeV. They were analyzed according to $4\pi\sigma(\theta) = \sum_{k=0}^{4} A_k P_k(\theta)$ and $4\pi\sigma(\theta)A(\theta) = \sum_{k=1}^{4} B_k P_k^1(\theta)$. The nine experimentally determined coefficients, A_k and B_k, can be expressed in terms of the reaction matrix elements which are written, for example, for capture of a $p_{1/2}$ proton as $|p_{1/2}|\exp i\phi_{1/2}$.

If M1 radiation is neglected (for the moment) in the reaction ^{31}P$(\vec{p},\gamma_0)^{32}$S, the only allowed channels are $p_{1/2}$ and $p_{3/2}$ (for E1) and $d_{3/2}$ and $d_{5/2}$ (for E2). Higher multipoles can be neglected. The four amplitudes and the three phases relative to that for the $p_{1/2}$ channel are then determined from the seven coefficients other than A_1 and B_1. Any M1 radiation, if present, enters linearly into A_1 and B_1 and can be neglected in the remaining coefficients. Finally the coefficients A_1 and B_1 are found to be consistent with the remaining coefficients, showing, therefore, that M1 is in fact negligible in the region studied.

The results are shown in fig. 1, where the extracted amplitudes as well as the relative phases are plotted along with the 90° excitation function as observed with a PH$_3$ gas target the thickness of which corresponded to a 200-keV proton energy loss. It is remarkable that despite the pronounced structure in the yield curve the relative E1 amplitudes which account for most of the total cross-section are rather constant. An alternate solution for the reaction amplitudes is possible because the relationships between the coefficients A_k, B_k and the reaction matrix elements are quadratic. The alternate solution in this case is a trivial exchange of $p_{1/2}$ with $p_{3/2}$ and of $d_{3/2}$ with $d_{5/2}$.

If the results are expressed in the channel-spin S representation they show that both E1 and E2 capture proceed mainly via $S = 0$. This can be understood by assuming that the last four valence nucleons couple to $S = 0$, $L = 0$, and $T = 0$. Since the electric multipoles do not involve spin flip, the particle is most likely to be captured in the $S = 0$ channel. In fact the results establish that $S = 0$ capture accounts for 98% of the energy averaged E1 cross section. The similarly averaged E2 cross section is 80% $S = 0$ although there are localized regions, particularly near 9 MeV, where the $S = 1$ E2 strength is comparable to that for $S = 0$.

Also shown in fig. 1 is the observed E2 cross section. The integrated strength exhausts about 30% of the E2 sum rule. A similar amount of E2 strength can be expected in the n_0 channel. Figure 2 shows a summary of E2 strength observed in ^{32}S in low-lying resonances and capture reactions. Considering the additional strength in unobserved channels, we see that a very large amount of E2 strength is spread over a wide range. It is not clear how this strength divides into isoscalar or

isovector components. These results are consistent with the inelastic
α scattering results from Texas A & M[1] which show a compact E2 resonance
in nuclei with A ≥ 40 but no concentrated E2 strength in the 2s-1d shell
nuclei.

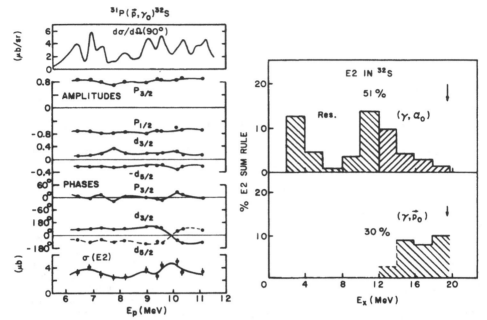

Fig. 1. The resultant E1 and E2
amplitudes and phases (relative
to the $p_{1/2}$ phase). The E2 proton capture cross section is
shown at the bottom.

Fig. 2. The E2 strength in ^{32}S in
percentage of the energy weighted
sum rule. Included are the bound
state E2 strengths[2] and the contributions from (γ, α_0)[3] and the
present work.

References

* Supported in part by the National Science Foundation
† Present address: Inst. Exp. Physik I, RUB Bochum, Germany
1) J. M. Moss et al., Phys. Rev. Lett. 34 (1975) 748
2) S. S. Hanna, Conf. on Nuclear Structure and Spectroscopy, Amsterdam,
   1974
3) E. Kuhlmann et al., Stanford Nuclear Physics Progress Report,
   1973-74, p. 21

SEARCH FOR GIANT RESONANCE EFFECTS IN $^{56}Fe(\vec{p},p)^{56}Fe$

CROSS SECTION AND POLARIZATION MEASUREMENTS

H.R. Weller[*], J. Szücs, J.A. Kuehner, G.D. Jones[†] and D.T. Petty
Tandem Accelerator Laboratory
McMaster University, Hamilton, Canada

This experiment was designed to measure the cross sections and analyzing powers of elastically scattered protons from ^{56}Fe over the region of the giant dipole resonance of ^{57}Co[1,2]. The target was enriched ^{56}Fe and was 4 mg/cm^2 thick (about 120 keV for 10 MeV protons). The polarized proton beam was provided by the McMaster polarized ion source. Typical on target beam was about 40 nA with 80% polarization.

Excitation curves were measured in 100 keV steps at eight angles from 9.3 to 16.5 MeV. Typical results are shown in Fig. 1. The giant resonance of ^{57}Co is expected to be centered around $E_p=14$ MeV. A suggestion of broad structure can be seen in the present data at $\theta_{lab}=135°$, although the physical origin of this structure requires further analysis.

Angular distributions of analyzing powers were measured at 18 angles in $\sim7.5°$ steps over the range from $\sim20°$ to $\sim150°$. The data were taken using eight detectors – four left and four right – with the proton beam being run alternately with spin up and with spin down. Measurements were made in 1 MeV intervals for $E_p=8$ to 17 MeV. An unpolarized beam was used to obtain cross section measurements at these same angles and energies. The analysis of these data began by using an optical model search code to generate a set of "initial" phase shifts at $E_p=10$ MeV. These phase shifts were then searched to minimize the χ^2 at this energy. The resulting phase shifts were used as initial phase shifts at adjacent energies. This process was repeated to include all measured energies. A typical set of data is shown for $E_p=11.0$ MeV in Fig. 2 along with the fit obtained from the phase shift analysis. The complex phase shifts at 11.0 MeV were:

	$s_{1/2}$	$p_{1/2}$	$p_{3/2}$	$d_{3/2}$	$d_{5/2}$	$f_{5/2}$	$f_{7/2}$
δ (deg)	11.6	-33.7	-28.6	-68.4	-55.3	-6.4	2.3
γ	0.53	0.92	0.68	0.25	0.60	0.72	0.59

where the complex phase shift is represented by the real part $\delta_\ell^\pm$ and the imaginary part $\eta_\ell^\pm$ (reported in terms of $\gamma_\ell^\pm = \exp(-2\eta_\ell^\pm)$.

The preliminary phase shift analysis indicates that the dominant absorption at $E_p \sim 14$ MeV is in the s and d partial waves. Although a giant dipole resonance built on the ground state of ^{57}Co must have J^π of $5/2^+$, $7/2^+$ or $9/2^+$, one built on the first excited state would have J^π of $1/2^+$, $3/2^+$ or $5/2^+$. The phase shifts which best describe the data over the energy region studied are presently being investigated. Efforts to reproduce their energy dependence by calculating the effects of broad resonance structures are also underway.

Measurements of cross sections and analyzing powers for elastically scattered protons from ^{54}Fe and ^{58}Fe have also been obtained in this energy region.

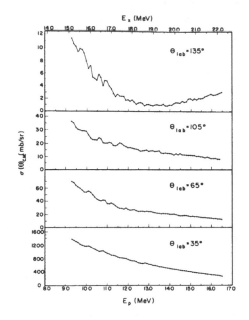

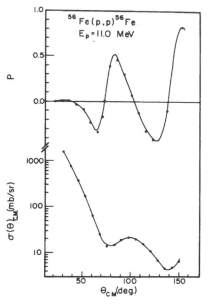

Fig. 1

The excitation curves for the
$^{56}Fe(p,p)^{56}Fe$ reaction as a function
of proton bombarding energy. The
solid lines are smooth curves drawn
through the data points.

Fig. 2

The angular distribution of cross
section and analyzing power for
the $^{56}Fe(p,p)^{56}Fe$ reaction at
E_p=11.0 MeV. The solid curve
represents the result of the fit
generated by the phase shifts
given in the text. The statistical errors associated with the
data points are less than the
size of the dots.

References

* Permanent address: University of Florida, Gainesville, Florida
† Permanent address: Oliver Lodge Laboratory, University of Liverpool,
                      U.K.

1) J. V. Maher et al., Phys. Rev. C9 (1974) 1440
2) H.R. Weller et al., Bull. Am. Phys. Soc. 19 (1974) 998

A STUDY OF THE GIANT DIPOLE RESONANCE REGION IN Co ISOTOPES
USING POLARIZED PROTON CAPTURE MEASUREMENTS

H. R. Weller,* R. A. Blue, G. Rochau and R. McBroom,
University of Florida** and Triangle Universities Nuclear
Laboratory,+ Durham, N. C.
N. R. Roberson, D. G. Rickel, C. P. Cameron and R. D. Ledford,
Duke University and TUNL and
D. R. Tilley
N. C. State University and TUNL

Previous workers[1] have measured ^{54,56}Fe(p,γ_0) cross sections in the giant dipole resonance region of ^{55}Co and ^{57}Co. We have extended this work[2] by measuring the angular distributions of the analyzing power for the ^{54,56}Fe($\vec{p},\gamma_0$) reactions through the GDR region. In addition, both cross section and analyzing power measurements have been obtained from the ^{58}Fe($\vec{p},\gamma_0$)^{59}Co reaction. The 90° yield curve for the inverse reaction is shown in Fig. 1. These three reactions, with target isospins of T= 1, 2 and 3 provide a means for examining the isospin splitting and the relative strengths of the $T_<$ and $T_>$ components of the GDR as a function of T.

Figure 1

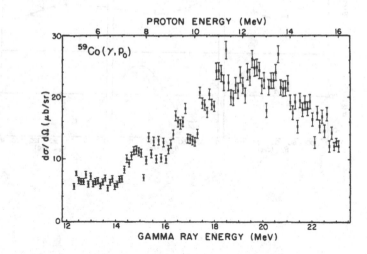

The polarized proton beam used in this work was obtained from the TUNL polarized ion source. The γ-rays were detected with our 10" x 10" NaI detector equipped with an anticoincident plastic shield. The best γ-ray resolution obtained with this system to date is 2.4% at E_γ= 21 MeV.

Typical results for the three targets are shown in Fig. 2. The curves represent fits to the cross section and cross section times analyzing power measurements using Legendre and associated Legendre polynomials respectively. The extracted coefficients for the 3 isotopes show the same energy dependence through the giant dipole resonance region. The most striking features are that 1) the a_2 coefficients decrease in value with increasing energy, changing sign at $E_\gamma \approx 18$ MeV,[1]

2) the b_2 coefficients are all negative and increase in magnitude through the dipole region from near zero to 0.2, and 3) the a_1 and b_1 coefficients are finite. These finite odd order coefficients imply multipolarities other than E1 in the radiations.

Because of the $7/2^-$ spin of the ground states of the Co isotopes, a general T-matrix solution is difficult. As a first step in the analysis, we have assumed pure E1 radiation and the T-matrix amplitudes $g_{9/2}$, $g_{7/2}$ and $d_{5/2}$. The analysis indicates that the $g_{7/2}$ strength (spin-flip) can be neglected. With the $g_{7/2}$ amplitude set to zero, we have found that there exist two solutions which describe the data. In one solution the $g_{9/2}$ amplitude is dominant while in the other, $d_{5/2}$ dominates. In both cases the dominant amplitude accounts for around 95% of the cross section at low excitation energies and varies to about 75% at the high energy side of the dipole resonance. The relative phase changes from 0° to 90° over this same energy range. The implications of these results vis a vis the isospin splitting suggested in Ref. 1 will be discussed.

Figure 2

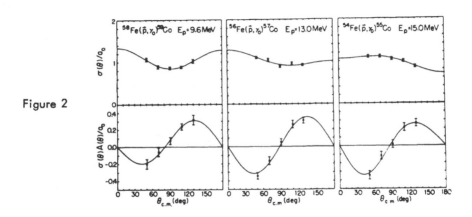

* Partially supported by Research Corporation and the National Science Foundation
** Partially supported by the Southern Regional Education Board
+ Partially supported by the U.S. Energy Research and Development Administration
1) J. V. Maher et al., Phys. Rev. C9, 1440 (1974)
2) H. R. Weller et al., Bull. Am. Phys. Soc. 19, 998 (1974)

POLARIZATION OF PHOTONEUTRONS FROM
THE THRESHOLD REGION OF ^{208}Pb*

R. J. Holt and H. E. Jackson
Argonne National Laboratory, Argonne, Illinois 60439

The ^{208}Pb nucleus should provide an ideal demonstration of a collective M1 resonance since both the lower proton and neutron orbitals ($h_{11/2}$, $i_{13/2}$) of a spin-orbit-split pair are filled while the upper orbitals($h_{9/2}$, $i_{11/2}$) are empty. A detailed theoretical study[1] has shown that there is considerable M1 strength (Γ_{γ_0} = 61 eV) at an excitation of 7.5 MeV in ^{208}Pb. Bowman et al.[2] observed seven 1$^+$ states in ^{208}Pb with a total M1 width of 51 eV, centered at 7.9 MeV and spread over a range of 700 keV. This observed fragmentation has led to an even more detailed theory.[3] However, Toohey and Jackson[4] suggested that Bowman et al. could not have uniquely determined the parities of the resonances in ^{208}Pb by observations of the photoneutron cross sections and angular distributions.

In order to define the parities of these states, we have measured the polarization of photoneutrons from the ^{208}Pb(γ,n_0) ^{207}Pb reaction at angles of 90° and 135°. This work represents the first measurement of the polarization of photoneutrons from resonances near threshold. The target (2 cm × 5 cm × 0.5 cm plate of 99.1% ^{208}Pb) was irradiated with bremsstrahlung produced by electrons, from the Argonne high-current linac, impinging on a 0.2-cm thick silver converter. The linac was operated in a mode which produced 9 MeV, energy-analyzed electron pulses with a 25-A peak current and a 4-ns width at a rate of 800 Hz. The neutron energies were determined with good resolution using a time-of-flight method. The polarization was measured by allowing the neutrons to scatter from a liquid ^{16}O analyzer (7.5 cm dia. cylinder). A neutron spin precession solenoid, designed for use with a continuous spectrum of neutrons, was used in order to measure the asymmetry.

The differential polarization for the ^{208}Pb($\gamma,\vec{n}_0$)^{207}Pb reaction is given by

$$\frac{d\vec{p}}{d\Omega} = \hat{k} \, \lambdabar_\gamma^2 \left\{ (0.38 \, a_s a_{po} \sin \Delta_{spo} + 0.27 \, a_s a_{p1} \sin \Delta_{sp1}) \right.$$
$$\left. \times \sin \theta + 0.20 \, a_s a_d \sin \Delta_{ds} \sin 2\theta \right\},$$

where a_s, a_p and a_d are the amplitudes for s,p and d-wave neutron emission and the $\Delta_{ij} = \delta_i - \delta_j$ are the phase differences. Here, $\lambdabar_\gamma$ is the photon reduced wave length and $\hat{k}$ is the unit vector whose direction is perpendicular to the reaction plane. The results for four resonances (538, 613, 651 and 846 keV) in ^{208}Pb are shown in Fig. 1. According to eq. 1, nonzero polarization at a reaction angle of 90° indicates E1-M1 interference. At an angle of 135°, polarization can also be produced by an E1 state if the s and d partial waves interfere. The observations are summarized in Table 1. These four states were previously assigned as M1 excitations. We conclude that only the 613-keV level is 1$^+$; the other three states are 1$^-$. Hence, the M1 collective resonance is not as fragmented as previously thought.

We wish to thank B. R. Mottelson for a very useful discussion of M1 collective states.

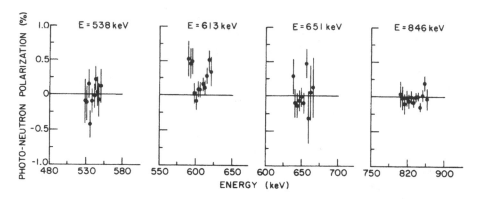

Fig. 1. The observed photoneutron polarization at 90° for four states in
^{208}Pb. Only the 613-keV level has a non-zero polarization.

Table 1. Summary of polarization measurements and the resulting
parity assignments.

E_R (keV)	Polarization $\theta = 90°$	$\theta = 135°$	J^π
538	No	Yes	1^-
613	Yes	Yes	1^+
651	No	Yes	1^-
846	No	Yes	1^-

References

1) J. D. Vergados, Phys. Lett. 36B (1971) 12.
2) C. D. Bowman, J. R. Baglan, B. L. Berman and T. W. Phillips, Phys. Rev.
   Lett. 25 (1970) 1302.
3) T. - S. H. Lee and S. Pittel, Phys. Rev. C 11 (1975) 607.
4) R. E. Toohey and H. E. Jackson, Phys. Rev. C 6 (1972) 1440.

*
Work performed under the auspices of the U.S. Energy Research and
Development Administration.

Inelastic Scattering and Compound Nucleus Effects;
Analog States and Fluctuations

A STUDY OF ANALOG RESONANCES IN ^{87}Y EXCITED BY ELASTIC AND INELASTIC SCATTERING OF POLARIZED PROTONS*

H. T. King† and D. C. Slater††
Department of Physics
Stanford University, Stanford, California 94305

Since polarization observables depend sensitively upon interference between contributing partial waves, polarized beam experiments can significantly increase the quantity and quality of information available from the study of isobaric analog resonances (IAR's). Polarization experiments can greatly reduce uncertainties in the extraction of IAR parameters and in addition shed some light on the reaction mechanisms involved in the scattering. For example, in favorable cases, the contribution to the scattering from the complicated compound nuclear (CN) states in the vicinity of the IAR can be separated from the "direct" scattering in a very nearly model-independent way[1]).

In this work IAR's in ^{87}Y were investigated by proton elastic scattering from ^{86}Sr and inelastic scattering to the 2_1^+ state of ^{86}Sr. Cross section, analyzing power, and spin-flip asymmetry[2]) data were taken on several resonances and analyzed in terms of coherent direct reaction and analog resonance contributions plus an incoherent CN contribution. The CN effects were expected to be important since all the resonances lie below or barely above the (p,n) threshold.

Compound elastic cross sections were extracted for d-wave resonances using the method of ref.[1]); good agreement with Hauser-Feshbach (HF) predictions is indicated in fig. 1(a). Unlike the elastic scattering case, the inelastic CN cross section cannot be accurately extracted in a model-independent way and a CN theory is therefore required to account for compound scattering. Fig. 1(b) shows inelastic scattering data for the 5.75-MeV $3/2^+$ resonance and fits which utilize the Engelbrecht-Weidenmüller (EW) formalism[3,4]) to compute the CN contribution. Whereas for elastic scattering HF and EW predictions are similar, for inelastic scattering they are quite different: e.g., HF theory, which disregards correlations between CN reduced widths induced by the IAR, predicts no CN contribution to the spin-flip asymmetry ΔS. The EW theory was found superior, as expected, for computing the CN contribution in the inelastic fits.

The $|\ell_j x 0_1^+\rangle$ and $|\ell_j x 2_1^+\rangle$ amplitudes for the neutron-core parent state wave functions, as obtained from the IAR partial widths for three d-wave resonances, are given in Table 1. Sums of spectroscopic factors are considerably smaller than one for the two $3/2^+$ resonances, indicating the presence of more complicated particle-core components.

Table 1. Particle-core parent state amplitudes in ^{87}Sr.

E_x(PARENT)	J^π	$lj \times 0_1^+$	$s_{1/2} \times 2_1^+$	$d_{3/2} \times 2_1^+$	$d_{5/2} \times 2_1^+$
1.78 MeV	$5/2^+$	0.70±0.03	0.17±0.04	0.10±0.06	0.45±0.15
2.68 MeV	$3/2^+$	0.27±0.03	-0.16±0.03	0.05±0.09	0.36±0.12
3.60 MeV	$3/2^+$	0.38±0.03	0.15±0.02	0.06±0.09	-0.15±0.06

References

*Supported in part by the National Science Foundation.
†Present address: Department of Physics, Rutgers University, New Brunswick, New Jersey 08903.
††Present address: LASL, Los Alamos, New Mexico 87544.

1) W. Kretschmer and G. Graw, Phys. Rev. Lett. 27 (1971) 1294
2) R. Boyd et al., Phys. Rev. Lett. 27 (1971) 1590
3) C. A. Engelbrecht and H. A. Weidenmüller, Phys. Rev. C8 (1973) 859
4) G. Graw et al., Phys. Rev. C10 (1974) 2340

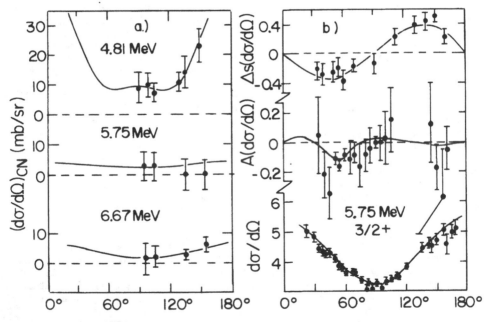

Fig. 1a. Extracted on-resonance compound elastic cross sections and Hauser-Feshbach predictions.
Fig. 1b. ^{86}Sr(p,p')^{86}Sr (2_1^+) data and fits for 5.75 MeV resonance.

INVESTIGATION OF COMPOUND NUCLEUS REACTIONS ON THE TARGET NUCLEI ^{88}SR AND ^{89}Y WITH POLARIZED PROTONS[+]

W. Kretschmer and G. Graw*
Physik. Inst. der Universität Erlangen-Nürnberg; Erlangen, West-Germany

In scattering experiments, where the energy resolution is much greater than widths and spacings of the finestructure resonances, the energy averaged cross section is a sum of the direct (DI) and compound nucleus (CN) cross section. The polarization dependent part of the cross section $\langle \sigma A \rangle$, which is independent of the CN-scattering, can be used to determine the direct part of the cross section and from that $\sigma^{CN} = \sigma_{exp} - \sigma_{DI}$[1]). We have investigated the proton induced CN-reactions on the target nuclei ^{88}Sr and ^{89}Y near the lowest isobaric analog resonances (IAR). In this energy region around 5 MeV direct inelastic reactions are strongly reduced due to the Coulomb barrier and the (p,n)-reactions proceed to 3 levels in the residual nuclei ^{88}Y and ^{89}Zr, respectively. Most partial waves have weak or medium absorption, the absorption is strong for the resonant $d_{5/2}$ partial wave at resonance energy and for the neutron p-waves. The simultaneous description of compound-elastic and (p,n)-cross sections allows a critical test of the various Hauser-Feshbach formulas[2,3]) for a situation, where only few channels are open and furthermore weak and strong absorption is present. Recently a similar test of CN-formulas has been made by Genz et al.[4]) in a many channel case ($E_p \approx 6.1$MeV near the $S_{1/2}$ IAR observed in the (p,p$_o$) and (p,n) reactions on ^{88}Sr), where the compoundelastic cross section has been extracted from a fluctuation analysis.

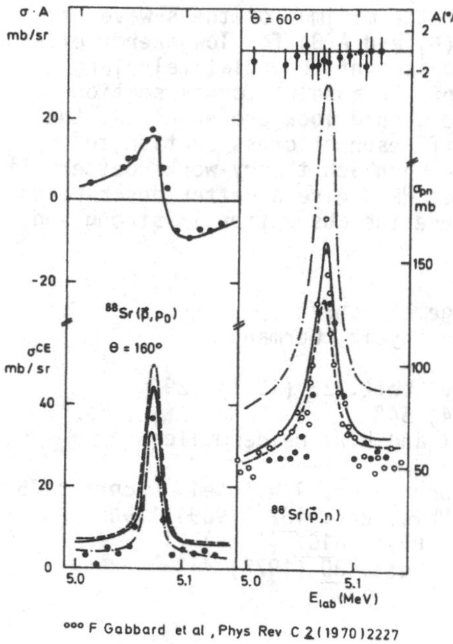

Fig. 1 and 2 show the ($\vec{p}$,p$_o$) and ($\vec{p}$,n) measurements on ^{88}Sr and ^{89}Y. The direct part of the reaction (that means the energy averaged S-matrix $\langle S_{cc'} \rangle$) is determined by $\langle \sigma A \rangle$: the off resonant S-matrix by an angular distribution and the resonant (IAR) S-matrix by an excitation function at 6 angles. Since the proton energy is several MeV below the Coulomb barrier, the optical model phase shifts are very small and rather insensitive of the details of the optical model. So we used in the case of ^{88}Sr the standard set of Becchetti, Greenlees[5]) and for ^{89}Y the potential of Johnson and Kernell[6]), but in both cases with a reduced imaginary potential of $W_D = 4$MeV. The resonance parameters for the IAR are listed in ref. 7. For the ^{88}Sr ($\vec{p}$,n) ^{88}Y reaction we obtained $A(\theta)=0$ within the experimental error of $\pm 1,5\%$ indicating the purely compound nucleus character of this process. The CN-cross sections ($\vec{p}$,p$_o$) and (p,n) are shown on the lower half of the figs. together with Hauser Feshbach

°°° F. Gabbard et al., Phys Rev C 2 (1970) 2227

Fig. 1: ^{88}Sr ($\vec{p}$,p$_o$) and ($\vec{p}$,n) near the g.s. $d_{5/2}$ IAR

766

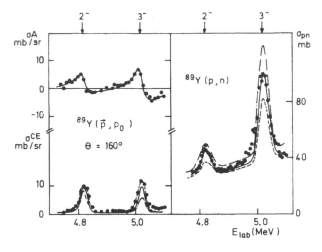

calculations in the different modifications[2,3]. The dashed-dotted curve is calculated with the original Hauser-Feshbach formula, the dashed curve with Moldauers modification for width fluctuation correction[2] and the solid curve is calculated with a formula suggested by Hofmann, Richert, Tepel, Weidenmüller[3] (HRTW) for the case of medium and strong absorption. The transmission coefficients used in these theories are determined by the proton and neutron optical potential. Since the proton potential is fixed by elastic scattering, the neutron optical potential can

(p,n): Johnson, Kernell; Nucl Phys A 107 (68) 21

——— Hofmann et al

– – – Hauser Feshbach

– · – · Moldauer (with fluctuation correction)

Fig. 2: ^{89}Y $(\vec{p},p_0)$ and (p,n) near the 2^-, 3^- IAR.

influence the distribution of the CN-decay. We used a potential suggested by Moldauer[8] who described with it the minimum of the s-wave neutron strength function for A $\approx$ 90 and $\sigma(\theta)$ and A(θ) for low energy elastic neutron scattering on ^{90}Zr. Taking into account the relatively large estimated errors in the determination of the absolute cross sections ($\pm$4mb/sr for σ^{CE} and 20% for $\sigma_{p,n}$) one should look preferably at the difference of the resonant minus the off resonant cross section to decide which modification of the Hauser Feshbach theory works better. It seems that in both cases the formulae of HRTW give a better description of the data in the resonance region, where the absorption is strong and where only few channels are open.

+ Supported by the Deutsche Forschungsgemeinschaft
* Present address: LMU München, Sektion Physik, Germany

1) W. Kretschmer and G. Graw, Phys. Rev. Lett. 27 (1971) 1294
2) P.A. Moldauer, Phys. Rev. B135 (1964) 642
3) H.M. Hofmann, J. Richert, J.W. Tepel and H.W. Weidenmüller
   Ann. Phys. 90 (1975) 403
4) H. Genz, E. Blanke, A. Richter, G. Schrieder, J.W.Tepel, preprint 75
5) F.D. Becchetti and G.W. Greenlees, Phys. Rev. 182 (1969) 1190
6) C.H. Johnson and R.L. Kernell, Nucl. Phys. A107 (1968) 21
7) G. Graw and W. Kretschmer,Phys. Rev. Lett. 30 (1973) 713
8) P.A. Moldauer, Phys. Rev. Lett. 9 (1962) 17

El ANALOGUE STATES IN ^{90}Zr STUDIED WITH POLARIZED PROTON CAPTURE[*]

J. R. Calarco, P. M. Kurjan, G. A. Fisher,[†] H. F. Glavish, D. G. Mavis
and S. S. Hanna
Department of Physics, Stanford University
Stanford, California 94305 U.S.A.

Studies of the ^{89}Y(p,γ_o) reaction[1,2] have revealed the existence of sharp, isolated 1^- analogue (IAR) resonances at E_p = 6.15 and 8.01 MeV which have been assigned isospin $T_>$ = 6 because of their narrow width and the known parent spectrum of ^{90}Y. We have studied the same reaction with polarized protons in the regions of these two IAR's in steps of 25 keV with a 30 keV (average) thick target. The angular distributions $\sigma(\theta)$ and the analyzing powers $A(\theta)$ have been analyzed according to

$$4\pi\sigma(\theta) = \sum_{k=0}^{3} A_k P_k(\theta) \quad \text{and} \quad 4\pi\sigma(\theta)A(\theta) = \sum_{k=1}^{3} B_k P_k^1(\theta).$$ The results are

shown in figs. 1 and 2 for the two resonances. The reaction amplitudes, $|s|$ and $|d|$, and their relative phase, $\delta = \phi_d - \phi_s$, have been extracted from $A_0 = |s|^2 + |d|^2$, $A_2 = -0.5|d|^2 + \sqrt{2}|s||d|\cos\delta$, and $B_2 = (1/\sqrt{2})|s||d|\sin\delta$.

In the region near the 6.15-MeV IAR the data yield two solutions: the one indicating approximately pure s-wave capture (labeled II in fig. 1) is consistent with results from (p,p_0) studies[3]; the other (labeled I) indicating dominant d-wave capture ($\sim$ 90%) is, however, not ruled out by these data.

The IAR at 8.01 MeV is not seen strongly in p,p_0[3], but is observed in $p,p_2(3/2^-)$ and $p,p_3(5/2^-)$.[3,4] suggesting the presence of configurations other than $d_{3/2}p_{1/2}^{-1}$. Furthermore, a rather pure $d_{3/2}p_{1/2}^{-1}$ configuration is observed at E_p = 7.45 MeV[3] in elastic scattering. Calculations[5] have predicted the parent configuration of the IAR at 8.01 MeV to be predominantly $s_{1/2}p_{3/2}^{-1}$. The experimental results (fig. 2), however, establish that, regardless of which of the two possible solutions (I or II) is correct, photoemission proceeds via both s- and d-waves. There are two possible interpretations of this result. The first is that the 8.01 MeV IAR is not a pure configuration but contains large admixtures of s and d particles. The second possibility is that the configuration is reasonably pure, but the coupling to the continuum involves both channels. Such a coupling is possibly expected when the widths are calculated with the doorway state model of Feshbach and coworkers[6]. In this calculation the reduced widths are proportional to $<\phi_A|V_{ph}|\psi_{\ell j}^{(+)}>$ where ϕ_A represents the IAR and $\psi_{\ell j}^{(+)}$ the particle in the continuum and the residual nucleus in a one-hole state. The particle-hole interaction V_{ph} allows a 1p-1h analogue configuration to couple to various continuum 1p'-1h' configurations which conserve total $J^\pi = 1^-$.

Therefore, the IAR at E_p = 8.01 MeV is either a rather mixed configuration or its decay is complicated by coupling to the continuum through the p-h interaction V_{ph}. Calculations are underway to attempt to resolve this question. The result will clearly be important for the interpretation of IAR's observed with non spin-zero targets and in particular the El analogue resonances.

Lastly it is interesting to speculate that if solution II is correct at 6.15 MeV and solution I turns out to be true at 8.01 MeV, then a crossover of the two solutions must occur in the region inbetween.

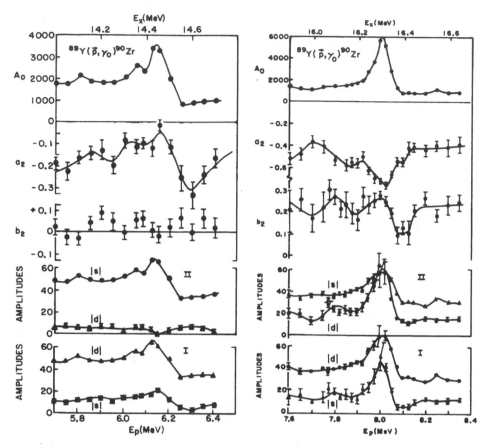

Fig. 1. The IAR at 6.15 MeV. Shown from top to bottom are: $\sigma_{tot} = A_0$, $a_2 = A_2/A_0$, $b_2 = B_2/A_0$, and the 2 solutions for the reaction amplitudes. Not shown are the relative phases $\phi_d - \phi_s$.

Fig. 2. The IAR at 8.01 MeV. Shown from top to bottom are: $\sigma_{tot} = A_0$, $a_2 = A_2/A_0$, $b_2 = B_2/A_0$, and the 2 solutions for the reaction amplitudes. Not shown are the relative phases $\phi_d - \phi_s$.

References

*   Supported in part by the National Science Foundation
†   San Francisco State University, San Francisco, Ca. 94132
1) M. Hasinoff, G. A. Fisher, and S. S. Hanna, Nucl. Phys. A216 (1973) 221
2) J. L. Black et al., Nucl. Phys. A92 (1967) 365
3) J. E. Spencer et al., Stanford Nuclear Physics Progress Report, 1970, p. 76
4) D. D. Long and J. D. Fox, Phys. Rev. 167 (1968) 1131
5) B. J. Dalton and D. Robson, Nucl. Phys. A210 (1973) 1
6) See e.g. W. L. Wang and C. M. Shakin, Phys. Rev. C5 (1972) 1898

CHANNEL CORRELATIONS IN THE COMPOUND NUCLEAR DECAY OF ANALOG RESONANCES[+]

J.H.Feist, G.Graw[++], W.Kretschmer and P.Pröschel

Physikalisches Institut der Universität Erlangen-Nürnberg,
Erlangen, West-Germany

The presence of channel correlations (CC) in the inelastic compound nuclear (CN) decay of an isobaric analog resonance (IAR)[1] is shown from a detailed analysis of cross section, analyzing power and polarization. Angular distributions (fig.1) and excitation functions (fig.2) have been measured for the $^{90}Zr(p,p')^{90}Zr(2^+,2.18$ MeV) reaction in the energy region of three isolated IAR's with $J^{\pi}=3/2^+$ and resonance energies 6.81 MeV, 7.66 MeV and 7.86 MeV. The polarization P was obtained in a "spin flip" particle-gamma coincidence[2] with a polarized beam (30 nA, P=75%).

From the analysis of $\langle\sigma A\rangle$ in the elastic scattering, the magnitudes of the optical potential and of the IAR's have been determined independently from the presence of the compound elastic scattering. {U=57 MeV, r=1.12 fm, a=0.75fm; W≈8 MeV, r=1.32 fm, a=0.59 fm; VS=6.2 MeV, r=1.01 fm, a ≃ 1.05 fm; total widths 53 keV, 36 keV and 48 keV; elastic decay widths 17 keV, 2.4 keV and 6.4 keV; mixing phases near zero}.

For the inelastic data χ^2 fits have been performed for the alternative assumptions with and without CC. The free parameters are the inelastic decay amplitudes $\Gamma_i^{1/2}$ of the

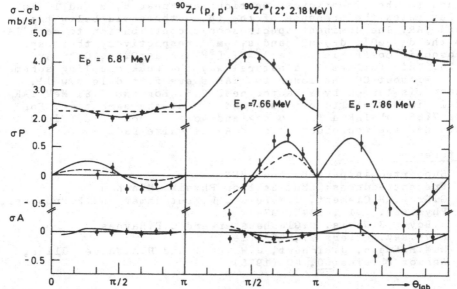

fig.1 Independent best fits with and without CC (solid and dashed lines) for three IAF's, σ^b is the calculated Hauser Feshbach cross section with $J^{\pi} \neq 3/2^+$.

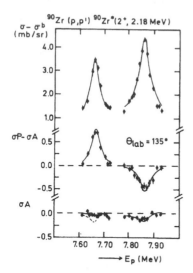

^{90}Zr (p,p') ^{90}Zr*(2$^+$, 2.18 MeV)

σ − σ^b
(mb/sr)

σP − σA

σA

θ_{lab} = 135°

E$_P$ (MeV)

fig.2 Excitation functions and fits with and without CC. For the lower resonance, a different fit is presented than in fig.1

resonant channels i=s$_{1/2}$, d$_{3/2}$, d$_{5/2}$ and g$_{7/2}$. No assumptions concerning the strength of isospin mixing have been made.

The direct part of <σ>,<σA> and <σ> has been calculated by the coherent superposition of the IAR's and the direct excitation of the 2$^+$ state (DWBA, β$_2$=0.074). The CN (Hauser Feshbach) calculation of <σ>^{CN} and <σP>^{CN} (<σA>^{CN}=0) has been done in the formalism of Hofmann et al.[3] It takes into account the presence of direct reactions and is applicable to any strength of absorption. The penetrability matrix encludes the optical model absorption in all channels and the IAR terms. They cause the enhancement of <σ>^{CN} and the channel correlation <S$^{fl}_{ab}$S$^{fl*}_{ac}$>≠0 for b≠c. The consequences of CC are <σP>^{CN}≠0 and additional contributions to <σ>^{CN} The case without CC was calculated in the same procedure but the CC terms with b≠c equated to zero subsequently.

With CC the fits to the resonances (figs.1 and 2) are excellent. For J$^\pi$=3/2$^+$ the ratios of the compound to the direct cross sections are near 6, 3 and 1. This illustrates the presence of the (p,n) threshold below the second IAR. The dominant spectroscopic contribution to each IAR is the d$_{3/2}$⊗0$^+$, d$_{5/2}$⊗2$^+$ and s$_{1/2}$⊗2$^+$,respectively, their spectroscopic coefficients (a$_i$)^2=Γ$_i$/Γ$_i^{sp}$ are a^2=0.34, 0.76 and 0.26. This follows in a natural way the weak coupling scheme.

Without CC the lower two IAR's are fitted less well, the χ^2 is larger by a factor near 2.5. For the 6.81 MeV IAR it is impossible to fit <σP> and <σ> at the same time. For the 7.66 MeV IAR a fit to <σ> and <σP> is possible (fig.2), but then the reproduction of <σA> is quite bad.

References

† Supported in part by the BMFT
†† Present address: LMU, Sektion Physik, München
1) G.Graw, H.Clement, J.H.Feist, W.Kretschmer and P.Pröschel, Phys.Rev. C.10, 2340(1974)
2) R.Boyd, S.Davis, C.Glashausser and C.F.Haynes Phys.Rev.Letters 27, 1590(1971)
3) H.M.Hofmann, J.Richert, J.W.Tepel and H.A.Weidenmüller, Ann.of Physics 90, 403(1975)

CORRELATIONS AMONG COMPOUND-NUCLEUS WIDTHS NEAR AN ANALOGUE RESONANCE

R. Albrecht, R.-J. Demond, J.P. Wurm

Max-Planck-Institut für Kernphysik, Heidelberg, Germany.

One of the basic assumptions of the statistical model of nuclear reactions is that the decay amplitudes to different channels are uncorrelated. This assumption is expected to be no longer valid in the presence of direct reactions. First experimental evidence for channel-channel correlations (CCC) in compound nuclear (CN) scattering has been found from a study of the inelastic scattering of protons at isobaric analogue resonances (IAR)[1-4].

We have measured (p,p'γ)-angular correlations in the vicinity of the $3/2^+$ IAR at E_p = 7.64 MeV in ^{90}Zr + p. Excitation functions for 12 p'γ angular combinations are shown in Fig. 1 for scattering via J^π = $3/2^+$ IAR and CN resonances. Three different processes contribute to the observables as cross section $d\sigma/d\Omega$ and angular correlation $d^2\sigma/d\Omega_pd\Omega\gamma$: 1. CN scattering via resonances with $J^\pi \neq 3/2^+$. This part is not affected by the IAR and is fixed by fitting the off-resonance angular correlation using conventional Hauser-Feshbach (HF) theory (and is subtracted in Fig.1). 2. The direct scattering via IAR. 3. CN scattering via resonances with J^π = $3/2^+$. In this part the contribution due to the optical model absorption is enhanced by the flow through the direct channels of the IAR.

Using the conventional HF formula one expects no CCC originating from the third part and hence no bilinear terms of the fluctuating S-matrix $\langle S_{cc}^{fl} S_{cc}^{fl*} \rangle$ with $c' = c''$. The simplest formula that takes into account CCC is Satchler's non-eigenbasis representation of CN scattering.

$$\sigma^{cN} \propto \langle S_{cc'}^{fl} S_{cc''}^{fl*} \rangle = P_{cc'} P_{c'c''} / \Sigma_\alpha P_{c\alpha c}$$

The transmission matrices

$$P_{cc'} = \delta_{cc'} T_c^0 + \frac{e^{i(\phi_c - \phi_{c'})} \Gamma_c^{1/2} \Gamma_{c'}^{1/2}}{(E - E_R)^2 + \Gamma^2/4} \cdot \Gamma_{cc'}^+$$

contain the usual optical model transmission coefficients T_c^0 and the decay amplitudes of the IAR to the channel c $\Gamma_c^{1/2}$ with phases ϕ_c. E_R and Γ are the energy and the total width of the IAR, respectively. The complex quantity $\Gamma_{cc'}^+$, is a channel dependent spreading width which goes over into the usual Γ^+ for zero optical model absorption η_c.

$$\Gamma_{cc'}^+ = 1/2 \Gamma (e^{-2\eta_c} + e^{-2\eta_{c'}}) - \Gamma^+ - i (E - E_R)(e^{-2\eta_c} - e^{-2\eta_{c'}}) \; ; \; \Gamma^+ = \Sigma_\alpha \Gamma_\alpha$$

We tried to fit the experimental data with both the conventional HF formula and the Satchler ansatz. Magnitudes and signs of the decay amplitudes to the inelastic 2^+ channel were varied while E_R, Γ, Γ^+ and the elastic width Γ_{el} were taken from independent experiments. The HF formula (dashed line in Fig.1) is not able to reproduce the special features of the on resonance angular correlation while with the Satchler ansatz the angular correlation is fitted well (solid line in Fig.1). With the best-fit decay amplitudes $\Gamma^{1/2}$(s1/2,d3/2,d5/2,g7/2)=(2.0,0.5-1.2,0.2) keV$^{1/2}$ and the Satchler ansatz the inelastic polarization of (+19±5)% of our double scattering experiment[1] and that of the Erlangen group[3] is well reproduced while only half of the measured polarization can be described using the conventional HF formula.

The failure to explain the special features of the angular correlation data in ^{90}Zr+p without CCC gives an additional evidence that the CN decay amplitudes to different channels are strongly correlated in the

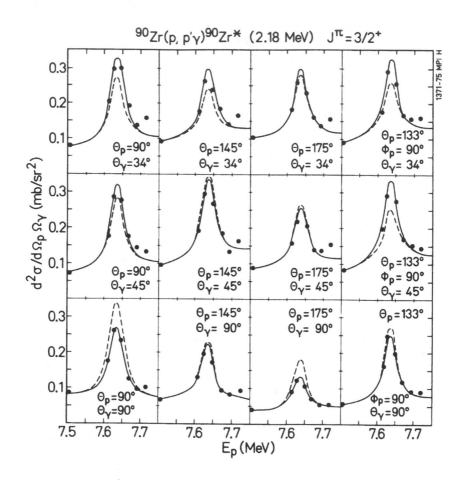

^{90}Zr(p, p'γ)^{90}Zr* (2.18 MeV) $J^{\pi} = 3/2^+$

vicinity of an IAR. A theoretical approach covering the effects of CCC as well as other modifications due to strong absorption has recently been given by Weidenmüller et al.[5].

References

1) R.Albrecht, K.Mudersbach, J.P.Wurm and V.Zoran,
   Proceedings of the International Conference on Nuclear Physics,
   Munich 1973 (North-Holland, Amsterdam 1973) p.577

2) S.Davis, C.Glashausser, A.B.Robbins, G.Bissinger, R.Albrecht and
   J.P.Wurm, Phys.Rev.Lett. 34 (1975) 215

3) G.Graw, H.Clement, J.H.Feist, W.Kretschmer, and P.Pröschel,
   Phys.Rev. C10 (1974) 2340

4) E.Abramson, R.A.Eisenstein, I.Plesser, Z.Vager and J.P.Wurm,
   Nucl.Phys. A144 (1970) 321.

5) H.M. Hofmann, J.Richert, J.W.Tepel and H.A.Weidenmüller,
   Ann.Phys. 90, 2 (1975) 403

ON THE ENHANCED POLARIZATION DUE TO CHANNEL - CHANNEL CORRELATIONS

A.Berinde, C.Deberth, G.Vlăducă and V.Zoran

Institute for Atomic Physics, Bucharest, Romania

As it was shown in ref.[1-4] , the polarization of the inelastically scattered protons on an IAR contains an appreciable contribution from the channel - channel correlations (CCC) between the fluctuating parts of the S-matrix elements. While a complete account of the corelations induced by the doorway - state mechanism is still missing, several approximate expressions for calculating $\langle S_{cc'}^{fl} S_{cc''}^{fl*} \rangle$ have been proposed. The theory of Abramson et al.[5], labeled hereafter with A, is based on the assumption of the fine structure coupling to the continuum only through the IAR. In theory B of Kawai, Kerman and Mc Voy[6] a simple expression is derived in the limit of many directly coupled open channels. A more complete treatment is given by the theory C1 of Engelbrecht and Weidenmüller[7]. In the limit of many directly coupled channels this theory reduces to the expression B, with an additional term due to the neglect of the mutual level repulsion[8]. This case will be referred as C2. All these theories have been deduced in the limit of small absorption coefficients. Up to now only the theories A and B were used in fitting the experimental data. No absorption was considered in the resonant channels. Within this assumption, the observables $\langle S_{cc'} S_{cc''}^{*} \rangle$ are enhanced independently of the indices c, c' , c" which gives a complete correlation of the fluctuations of the S-matrix elements between the resonant channels in both theories A and B [3,8].

The aim of this note is to test the independence of the enhancement on the channel indices and to compare its relative magnitude for all the mentioned theories. In the calculations the absorption was not neglected. The numerical study was performed for the d3/2 IAR at E_p = 7.65 MeV in ^{90}Zr (p,p2) 2.18 MeV, 2+ . The optical model parameters for protons are taken as in ref.[2],for neutrons the set A from ref.[10]. The resonance parameteres are taken from a preliminary version of ref.[3] and are close to the published ones. The fine structure enhancement may be characterized by the quantity:

$$\alpha_{c'c''} = \langle S_{cc'}^{fl} S_{cc''}^{fl*} \rangle / \langle S_{cc'}^{res} \rangle \langle S_{cc''}^{res*} \rangle$$

and its average value at the resonance energy for all non-diagonal indices is given in Table 1.

TABLE 1

Theory	A	B	C1	C2
$\overline{\alpha}_{c'c''}$	1.28 ± 0.04	2.03 ± 0.05	1.51 ± 0.10	2.77 ± 0.07

All the individual values $\alpha_{c'\,c''}$ are contained within
the limits given in the table, which shows that for all formalisms the
enhancement remains relatively channel independent even at moderate
absorptions. This is not the case for the diagonal terms $\alpha_{c'\,c'}$ which
illustrates the correlation damping due to the absorption. On the other
hand $\alpha_{c'\,c''}$ is strongly model dependent and consequently, when considering absorption, also the spectroscopic information which can be
extracted from polarization measurements. An analysis with all these
theories of the published experimental data in the region A=9o in view
of extracting spectroscopic information is in progress.

References

1) R.Albrecht, K.Mudersbach, J.P.Wurm and V.Zoran, in Proceedings
   of the International Conference on Nuclear Physics, München 1973,
   ed.by J.de Boer and H.J.Mang (North Holland, Amsterdam 1973) p.577.

2) R.Albrecht, K.Mudersbach, J.P.Wurm and V.Zoran, Rev.Roum.Phys.19
   (1974) 823.

3) G.Graw, H.Clement, J.H.Feist, W.Kretschmer and P.Pröschel,
   Phys.Rev. C1o (1974) 234o.

4) S.Davis, C.Glashausser, A.B.Robbins, G.Bissinger, R.Albrecht and
   J.P.Wurm, Phys.Rev.Lett.34 (1975) 215.

5) E.Abramson, R.A.Eisenstein, I.P.Flesser, Z.Vager and J.P.Wurm,
   Nucl.Phys. A 144 (197o) 321.

6) M.Kawai, A.K.Kerman and K.W.McVoy, Ann.Phys. 75 (1973) 156

7) C.A.Engelbrecht and H.A.Weidenmüller, Phys.Rev. C8 (1973) 859.

8) H.L.harney, Preprint MPI Heidelberg - 1974 - V1

9) K.P.Lieb, J.J.Kent and C.F.Moore, Phys.Rev. 175 (1958) 1482

1o) U.Jahnke, Ph.D.thesis, University of Berlin (197o).

ANALOGUE RESONANCES IN PROTON INELASTIC SCATTERING FROM ^{126}Te

W.P.Th.M. van Eeghem and B.J. Verhaar

Department of Physics, Technische Hogeschool, Eindhoven, Netherlands

The study of elastic and inelastic proton scattering via isobaric analogue resonances yields valuable and reliable nuclear structure information: in addition to the angular momentum quantum numbers of the parent state, the set of coefficients of fractional parentage (cfp) may be obtained, relating the parent state to the target ground and excited states. Using the Utrecht tandem Van de Graaff accelerator in combination with the Eindhoven polarized proton source, we measured[1] the cross section σ and analyzing power A as a function of the cm proton energy E in the range 9.65-10.27 MeV for the reaction ^{126}Te($\vec{p},p_1$)^{126}Te* (E_x=0.667 MeV) at the cm scattering angles 140° and 160°, to supplement previous elastic scattering experiments, carried out in particular by our group.

The analysis contains some new features:

a) The separation of the analysis in two parts. In the first part the energy dependence of σ and σA is written as a DWBA background term and a linear combination of resonance functions $r_\lambda^2(E)$ and $r_\lambda(E)\cos\delta_\lambda(E)$, where r_λ and δ_λ are the amplitude and phase of $\Gamma_{\lambda 1j}^{\frac{1}{2}}/(E_\lambda-E-\frac{1}{2}i\Gamma_\lambda)$ for resonance λ, known from elastic scattering. For σ and σA at each scattering angle the coefficients (two per resonance) in the linear combination can be obtained by a linear χ^2 method, which also yields the corresponding correlation matrix. In the second part of the analysis the coefficients thus determined are confronted with theoretical linear and quadratic expressions in the unknown inelastic partial width amplitudes $\Gamma_{\lambda 1'j'}^{\frac{1}{2}}$ by a χ^2 procedure taking into account the correlations calculated in the first step. This separation is advantageous if a sufficient number of data points per resonance is available at the same scattering angle.

b) Subsidiary conditions on the $\Gamma_{\lambda 1'j'}^{\frac{1}{2}}$ or the corresponding cfp's $\beta_{\lambda 1'j'}$ are taken into account in the second step of the analysis. From the normalization of the parent state we have

$$\beta_{\lambda 1j}^2 + \sum_{1'j'} \beta_{\lambda 1'j'}^2 - 1 \leq 0, \tag{1}$$

and from the orthogonality of the parent states λ' and λ of the same J^π, making use of Schwarz's inequality

$$(\beta_{\lambda'1j}\beta_{\lambda 1j} + \sum_{1'j'} \beta_{\lambda'1'j'}\beta_{\lambda 1'j'})^2 \tag{2}$$
$$- (1 - \beta_{\lambda'1j}^2 - \sum_{1'j'}\beta_{\lambda'1'j'}^2)(1 - \beta_{\lambda 1j}^2 - \sum_{1'j'}\beta_{\lambda 1'j'}^2) \leq 0.$$

The Σ is restricted to decay to the first excited state. To the χ^2 function of the second step we add $p \sum_n G_n^2$ where G_n for $n \leq$ the number of resonances is equal to the left-hand side of eq. (1) if this is positive and zero otherwise. Each of the remaining G_n functions is similarly defined for a combination of two resonances with equal J^π in terms of eq. (2). In the search code a large value for p is chosen to ensure that only solutions are found which satisfy the two kinds of subsidiary conditions.

The σ and σA values measured are presented below as a function of proton energy. The background amplitudes were calculated using a DWBA

E(MeV) J$^\pi$		$\beta_{\lambda l'j'}$					
		$p_{1/2}$	$p_{3/2}$	$f_{5/2}$	$f_{7/2}$	$h_{9/2}$	$h_{11/2}$
9.705	A	0.42±0.09	−0.12±0.10	−0.03±0.23	−0.08±0.11		
3/2$^-$	B	−0.12±0.12	0.54±0.17	0.34±0.15	−0.25±0.09		
9.740	A		−0.11±0.06	−0.36±0.10	−0.03±0.08	−0.6 ±0.3	0.06±0.20
7/2$^-$	B		−0.21±0.03	0.01±0.12	−0.09±0.10	0.85±0.07	−0.30±0.15
9.847	A		−0.30±0.04	−0.32±0.18	−0.33±0.10	0.4 ±0.3	−0.6 ±0.2
7/2$^-$	B		−0.27±0.04	0.15±0.16	−0.43±0.08	−0.40±0.17	−0.4 ±0.2
9.903	A	−0.14±0.10	0.22±0.13	−0.56±0.15	−0.73±0.07		
3/2$^-$	B	−0.15±0.10	0.11±0.13	−0.70±0.11	−0.63±0.11		
10.150	A		0.46±0.05	−0.66±0.08			
1/2$^-$	B		0.46±0.05	−0.72±0.08			

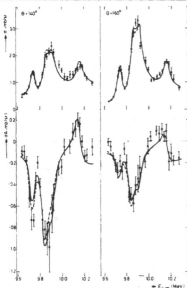

code[2] including deformation of both real and imaginary central potentials, as well as Coulomb and spin-orbit potentials. Deformation of the spin-orbit potential turned out to have little effect on the amplitudes at the energies used in this investigation. An additional advantage of the previously mentioned separation a) is the possibility to determine the deformation parameter β_2 multiplying the DWBA amplitudes by minimizing the total χ^2 value as a function of β_2 in the relatively simple first step only. The optimum β_2 value turned out to be 0.18 in good agreement with the value 0.163 from Coulomb excitation[3]. We found two sets of partial width amplitudes and corresponding cfp's. The latter are listed in the table. In the figure the solid (dashed) curves show σ and σA calculated for set A (B).

References

1) W.P.Th.M. van Eeghem, Thesis Technische Hogeschool Eindhoven, 1975
2) B.J. Verhaar, W.C. Hermans and J. Oberski, Nucl. Phys. A195 (1972) 379
3) P.H. Stelson and L. Grodzins, Nucl. Data A1 (1965) 31

NUCLEAR STRUCTURE STUDIES OF N=83 NUCLEI FROM POLARIZED PROTON SCATTERING NEAR ISOBARIC ANALOG RESONANCES†

H. Clement* and G. Graw**

Tandemlabor der Univ. Erlangen-Nürnberg, D-852 Erlangen, Germany

The quantitative analysis of the elastic and inelastic scattering of polarized protons in the region of isobaric analog resonances provides the unambiguous extraction of their decay amplitudes which, in turn, contain direct information on the wave functions of the parent states.[1] In the core-coupling expansion those are represented by

$$|\lambda>_{J\pi} = \sum_{i,\ell j} a^{\lambda}_{i,\ell j} |n_{\ell j} \otimes C^i(I)>_{J\pi}$$

The configurations with the N=82 core C^i in the ground state or in an excited state are isobaric analog to the resonant elastic or inelastic proton scattering channels; they are observed with decay amplitudes proportional to the spectroscopic amplitudes $a^{\lambda}_{i,\ell j}$.

For the nuclei ^{138}Ba, ^{140}Ce and ^{144}Sm, the energy excitation functions of differential cross section and analyzing power have been measured from about 9 - 12.7 MeV at several angles for the elastic scattering and the inelastic scattering to the first excited 2^+-state (Fig. 1).

From the analysis of the elastic scattering 17 - 19 levels per nucleus have been identified and their resonance parameters including spectroscopic factors ($=\{a^{\lambda}_{0,J\pi}\}^2$) extracted.

From the analysis of the inelastic scattering, the amplitudes of the configurations with a $p_{1/2}$, $p_{3/2}$, $f_{5/2}$ or $f_{7/2}$ single particle neutron coupled to the core in the 2^+_1 excited state have been obtained for the 8 - 11 lowest levels in each nucleus.

The extracted matrices of spectroscopic amplitudes have been found to be fully consistent with the sumrules due to completeness and orthonormality. There is a qualitative agreement with unified model calculations[2], especially in the signs. The $|f_{7/2} \otimes 2^+>$ configuration, however, was found to be

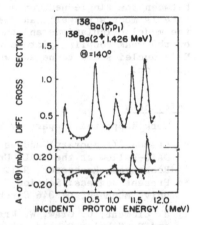

Fig. 1: Inelastic scattering to the first excited 2^+-state in ^{138}Ba. The dashed curves show the pure DWBA-result, the solid curves the full calculation with the best fitting set of resonance parameters.

less dominant than predicted by the model calculation.

For most levels having small single particle strength the configurations with the core in the 2^+_1-state turned out to be the dominant pieces of the wave function. For some of them, however, also those are not strongly enough excited to exhaust the main strength of the wave function and configurations with higher core excitation must be expected to contribute significantly.

The average behavior of the structure of the levels up to an excitation energy of about 2 MeV is shown in fig. 2. Here the average strengths of the configurations $|n_{\ell j} \blacksquare C^1(I) >$ are defined as $\overline{S}_{\ell j}^i = 1/n \overline{\sum_\lambda \{a_{\lambda, \ell j}^i\}^2}$ with the sum running over all appropriate levels and taking the average of all three nuclei; n stands for the number of possibilities of coupling the neutron total angular momentum j to the core spin I_i. The completeness relation then gives unity as an upper limit for $\overline{S}_{\ell j}^i$.

For the single particle configurations $|n_{\ell J} \blacksquare 0^+_{g.s.}>$ the experimental results (full bars) decrease with regard to the shell model spin sequence as expected from a simple shell model picture. The unified model calculations (open bars) exhibit the same overall behavior, but in general predict larger strengths than we observe.

In the case of the configurations $|n_{\ell j} \blacksquare 2^+_1 >$ the weak-coupling picture favors those having the neutron in the $f_{7/2}$ orbit to be by far the most dominant ones, as the unified model calculations demonstrate. From the analyses of the data we get average strengths for the 2^+-core configurations (with the neutron in either $p_{1/2}$, $p_{3/2}$, $f_{5/2}$ or $f_{7/2}$ orbitals), which in total are of nearly the same amount as the model calculations predict. This total strength of 2^+-core excitation, however, is spread out fairly evenly over all neutron orbitals leading to a far less obvious dominance of the $|f_{7/2} \blacksquare 2^+_1 >$ configurations.

This implies that the interaction between the single neutron and the N=82 core is much stronger than assumed in the model calculations and the validity of the weak coupling picture for the N=83 nuclei seems to be somewhat in question.

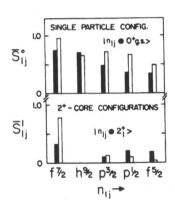

References

† Work supported in part by BMFT

* Present (temporary) address:
NATO-Fellow at the NSRL, Univ. of
Rochester, Rochester, N.Y. 14627
** Present address: Sektion Physik der
Univ. München, D-8046 Garching

1) H. Clement, G. Graw, W. Kretschmer
and P. Schulze-Döbold, Phys. Rev.
Lett. 27 (1971) 526; Int. Conf. Nucl.
Sruct., Amsterdam 1974, Proc. Vol.
1, 224; see also R. N. Boyd, R. Arking, J. C. Lombardi, A.B. Robbins,
S. Yoshida, D. C. Slater, H. T. King
and R. Avida, Nucl. Phys. A228 (1974)
253
2) G. Vanden Berghe, K. Heyde, M.Waro-
quier, Phys. Lett. 38B (1972) 467
and private communication

Fig. 2: Distribution of the strengths of the particle-core-coupling configurations over the single particle orbitals $n_{\ell j}$, averaged over the nuclei ^{139}Ba, ^{141}Ce and ^{145}Sm. Full bars: experimental results, open bars: unified model calculations[2]

ANALYZING POWER IN THE ^{206}Pb($\vec{p}$,p$_0$)^{206}Pb REACTION NEAR THE $3p_{1/2}$ ISOBARIC ANALOG RESONANCE

M. P. Baker,[†] T. A. Trainor, J. S. Blair, J.G. Cramer, and W.G. Weitkamp
University of Washington, Seattle, Washington, USA*

Recent S-matrix theoretical predictions[1] indicate that, under a few very general assumptions, the following relation exists among the parameters of an isoboric analog resonance (IAR):

$$\Gamma \geq 2 \; \overline{\cos \; (2\phi_c)} \; \sum_c \Gamma_c \tag{1}$$

where Γ_c and ϕ_c are the partial width and resonance mixing phase, respectively of the IAR in the channel c and Γ is the total width of the IAR. The bar in expression (1) is necessary since it has been assumed that the ϕ_c are approximately equal and the average value of cos ($2\phi_c$) has been extracted from the sum. For IAR where $\Gamma < 2\sum_c \Gamma_c$, expression (1) yields a lower limit on the value of $2\phi_c$. Previous differential cross-section measurements[2] of elastic and inelastic proton scattering on ^{206}Pb give $\Gamma/2\sum_c \Gamma_c \simeq 0.74$ for the $3p_{1/2}$ IAR in ^{207}Bi which implies that $2\phi \gtrsim 42^o$. This result is surprising in that IAR measurements near other closed shells[3] (A = 90 and 140) yield much smaller values for the resonance mixing phase in the elastic channel (ϕ_R).

In order to experimentally confirm this prediction for ϕ_R, analyzing-power excitation functions have been obtained for elastic proton scattering on ^{206}Pb in the vicinity of the $3p_{1/2}$ IAR. An accurate determination of both ϕ_R and the elastic partial width (Γ_p) can be made from such analyzing power measurements since the off-resonance analyzing power is very small. The excitation functions were obtained at 120, 150 and 165° for proton bombarding energies between 11.0 and 13.6 MeV in 50 or 100-keV steps. The target was isotopically enriched, self-supporting ^{206}Pb approximately 400 μg/cm^2 thick. Left-right asymmetries were measured with symmetric Si(Li) detector pairs with acceptance angles of ±2°.

The excitation functions obtained are shown in fig. 1. The target thickness for incident protons in this energy range is about 6 keV. The statistical uncertainties are generally 0.01 or less. Structure due to the $3p_{1/2}$ resonance at 12.2 MeV is clearly seen at all three angles, with the most distinctive resonance effect observed at 120°. The curves in fig. 1 are calculations of the analyzing power incorporating the optical model to generate the potential scattering background and a Breit-Wigner term to describe the resonance. The total width of the resonance was taken from the results of the differential cross-section measurements for proton inelastic scattering[2]. The resonance energy (E$_R$) and Γ_p were adjusted to give a reasonable fit to the data although no formal search was made. Two curves are shown for each angle, one with ϕ_R set equal to 0°, the other with ϕ_R = + 15°. The 120° data are the most sensitive to changes in the overall phase of the $p_{1/2}$ amplitude. This overall phase, however, depends not only upon ϕ_R, but also upon the real part of the $p_{1/2}$ optical model phase shift ($2\lambda_{p1/2}$). For the optical model parameters used here $2\lambda_{p1/2} = 7^o$. Since this phase is quite small, variations in optical model parameters are not expected to cause large changes in its magnitude. The resonance parameters obtained from this procedure are also given in fig. 1.

The $\phi_R = 0°$ curve certainly gives a superior fit to the data at 120°, but as shown in fig. 1 the Γ_p required is 18 keV; a value 50% larger than that determined in the previous study.[2] Since the inelastic scattering cross sections are proportional to the product $\Gamma_p \Gamma_{p'}$ (where $\Gamma_{p'}$ is the inelastic scattering partial width), the previously extracted $\Gamma_{p'}$ are also 50% too large. If the present values of Γ_p and $\Gamma_{p'}$ are substituted into eg. (1), the discrepancy between the theoretical prediction for ϕ_R and the experimental result no longer pertains, i.e., $\Gamma/2 \sum \Gamma_c \cong 1$. In addition, the value determined for Γ_p in the present work brings the spectroscopic factor deduced from IAR theory into agreement with that derived from a recent DWBA analysis for the analogous ^{207}Pb (p,d) ^{206}Pb reaction.[4]

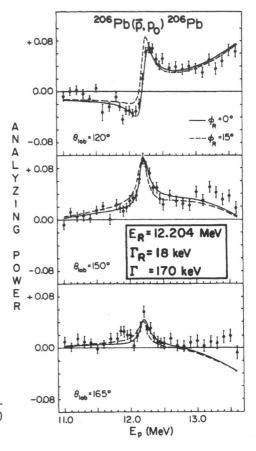

Fig. 1 Analyzing power excitation functions for ^{206}Pb$(\vec{p},p_0)^{206}$Pb near the $3p_{1/2}$ IAR.

REFERENCES

† Present Address: Los Alamos Scientific Laboratory, University of California, Los Alamos, New Mexico
* Supported in part by ERDA
1. P. von Brentano in "Proceedings of the Europhysics Study Conference on Intermediate Processes in Nuclear Reactions", ed. N. Cindro (Springer-Verlag, New York, 1973) p. 267.
2. C. D. Kavaloski, J. S. Lilley, Patrick Richard and Nelson Stein, Phys. Rev. Letts. 16, 807 (1966).
3. G. Graw in "Polarization Phenomena in Nuclear Reactions", ed. H. H. Barschall and W. Haeberl: (University of Wisconsin Press, Madison, Wisconsin, 1972) p. 179.
4. W. A. Lanford and G. M. Crawley, Phys. Rev. C9, 646 (1974).

ANALYZING POWER FOR PROTON-ELASTIC SCATTERING FROM ^{208}Pb
NEAR THE LOW-LYING ISOBARIC ANALOG RESONANCES

M. P. Baker,[†] J. S. Blair, J. G. Cramer, E. Preikschat, and W. Weitkamp
University of Washington, Seattle, Washington, USA*

The isobaric analogs of the low-lying states of ^{209}Pb, seen as resonances in elastic and inelastic scattering of protons from ^{208}Pb, have
been shown to involve relatively simple configurations,[1,2] This has led
to interest in making precise comparisons between experimental and theoretical values of the elastic partial widths (Γ_p) of these states. Such
comparisons provide information on the extent to which the shell model
can reliably predict the details of nuclear structure in this region.
It has been difficult, however, to extract unique values for the Γ_p from
the available differential cross-section data because the total widths
of these isobaric anolog resonances (IAR) are approximately the same as
the spacings between them. This has resulted in a large variety of extracted Γ_p with large uncertainties. These ambiguities can, in principle,
be resolved by measurement of more than one observable near IAR.

We have, therefore, simultaneously measured analyzing power (A) and
differential cross section (σ) excitation functions for elastic proton
scattering from ^{208}Pb over the energy range 14-18 MeV in which the low-
lying IAR are observed. Left-right asymmetries produced by the polarized
proton beam scattering from an isotopic ^{208}Pb target were measured at
scattering angles of 135, 150, and 165° using Si(Li) detectors with acceptance angles of ±2°. The previous elastic scattering results have
revealed the presence of 6 even-parity IAR.[2] The excitation functions
obtained for the elastic scattering are shown in fig. 1. The target
thickness was 15-20 keV and the data were obtained in either 25 or 50 keV
steps. The statistical uncertainties are typically 0.01. Structure due
to all 6 is clearly seen, particularly in the excitation functions of A.

As an initial step in the analysis, the potential scattering background was parameterized and the resonance parameters extracted by a χ^2
minimization fitting procedure. The Γ_p^{exp} determined from this search
procedure are given in table 1; the results quoted are averages of values
extracted at each of the three scattering angles.

As a consistency check on the appropriateness of a parameterized
potential scattering background, an optical model search was also performed. Breit-Wigner terms, to describe the IAR, were added to the usual S-matrix background terms and the resonance parameters were held fixed
at the values obtained from the parameterized fit. Starting with a global parameter set,[3] the optical model parameters were allowed to vary in
order to fit the data for A and σ simultaneously. The fit obtained is
shown in fig. 1, along with the fits determined by parameterizing the
potential scattering backgrounds at each angle independently.

Having obtained an adequate phenomenological description of the
elastic scattering process, calculations, based on the theory of Bund and
Blair,[4] were made for the single particle widths (Γ_p^{th}) of these IAR using
the code ANALOG.[5] The results of these calculations are also given in
table 1. The ratio ($\Gamma_p^{exp}/\Gamma_p^{th}$) is then the spectroscopic factor (SF) for

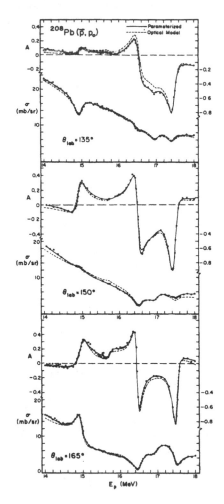

Table 1. Experimental and theoretical elastic scattering partial widths and spectroscopic factors for the IAR of ^{209}Pb.

$n_{\ell j}$	E_R (MeV)	Γ_p^{exp} (keV)	Γ_p^{th} (keV)	S(IAR)	S(d,p) (±10%)
$g_{9/2}$	14.92	21 ± 2	20.4	1.04 ± 0.10	0.91
$i_{11/2}$	15.72	2.0 ± 0.3	2.4	0.84 ± 0.13	1.2
$d_{5/2}$	16.50	50 ± 2	50.5	1.00 ± 0.04	0.89
$s_{1/2}$	16.96	49 ± 4	54.0	0.91 ± 0.07	0.91
$g_{7/2}$	17.43	30 ± 4	33.8	0.90 ± 0.12	1.1
$d_{3/2}$	17.47	47 ± 7	51.1	0.93 ± 0.14	0.97

a given state. The SF obtained from a recent DWBA analysis of the ^{208}Pb(d,p) ^{209}Pb reaction[6] at 12.3 MeV are also shown in table 1 for comparison. Unfortunately the agreement in magnitude is illusory since the bound state geometry in the (d,p)analysis is much larger than in our analysis.

In conclusion, the disagreement in the literature over the values of Γ_p for the IAR of the low-lying states of ^{208}Pb has been resolved. The Γ_p determined in this study, when compared to theoretical calculations, give SF equal to unity within experimental uncertainties. These SF for the IAR agree relatively but not in magnitude with those determined from (d,p) reactions to the parent states.

Fig. 1. Analyzing power and differential cross-section excitation functions for ^{208}Pb($\vec{p}$,p)^{208}Pb. The curves are theoretical fits to the data with the direct elastic scattering generated from the optical model or parametrized by polynomials in energy.

REFERENCES

† Present address: Los Alamos Scientific Laboratory, University of California, Los Alamos, NM 87545.
* Supported in part by ERDA.
1) W. R. Wharton, P. von Brentano, W. K. Dawson, and P. Richard, Phys. Rev. *176*, 1424 (1968).
2) S. A. A. Zaidi, J. L. Parish, J. G. Kulleck, C. Fred Moore, and P. von Brentano, Phys. Rev. *165*, 1312 (1968).
3) F. D. Becchetti, Jr., and G. W. Greenlees, Phys. Rev. *182*, 1190 (1969).
4) G. W. Bund and J. S. Blair, Nucl. Phys. *A144*, 384 (1970).
5) Gerhard Bund, Ph.D. Thesis, University of Washington, unpublished (1968).
6) R. F. Casten, E. Cosman, E. R. Flynn, Ole Hansen, P. W. Keaton, N. Stein, and R. Stock, Nucl. Phys. *A202*, 161 (1973).

ANALYZING POWER FLUCTUATIONS IN
THE ^{19}F($\vec{p}$,p) and ^{19}F($\vec{p}$,α) REACTIONS*

J. Eng, C. Glashausser, H. T. King,
A. B. Robbins, and E. Ventura
Department of Physics
Rutgers University, New Brunswick, New Jersey

Recent investigation of the analyzing power A_y for elastic and inelastic scattering from ^{26}Mg and ^{27}Al has revealed evidence for intermediate structure in these reactions[1]. The incident proton energies, 6 to 12 MeV, had considerable overlap with the incident proton energies at which the giant dipole resonance (GDR) is observed in (p,γ) reactions. Indeed, the structure observed in the excitation functions for the ^{27}Al($\vec{p}$,p')^{27}Al analyzing powers showed a surprising qualitative resemblance to the structure previously observed[2] in the ^{27}Al(p,γ)^{28}Si cross-section excitation function. While the compound states contributing to the GDR should be excited in (p,p) and (p,p') reactions, it is not expected that they would be preferentially excited as they are in the capture reaction.

The ^{19}F(p,γ$_0$)^{20}Ne cross-section excitation function reveals striking structure[3], as shown in Fig. 1; in addition the γ$_0$ and γ$_1$ channels are strongly correlated. Thus ^{19}F is an excellent target with which to search for evidence of anomalous correlation between the (p,γ) and (p,p') reactions. We have studied the analyzing power as a function of incident proton energy between 4.0 and 10.0 MeV at 140° and 160° for scattering from ^{19}F. Data taken in 50 keV steps at 140° for elastic scattering and for inelastic scattering to the fifth excited state at 1.55 MeV (3/2$^+$) are shown in the figure. Also illustrated are analyzing powers for the ^{19}F(p,α$_0$)^{16}O reaction over the same energy range. Compound states corresponding to the predominantly T = 1 GDR in ^{20}Ne should not contribute to the (p,α$_0$) reaction.

The structure observed in the excitation functions of A_y is qualitatively consistent with the structure expected from Ericson fluctuations. The fluctuations are most pronounced in the (p,α) data where the number N of independent reaction channels is smallest; the fluctuations are smallest in p$_5$ where N is largest. There are no strong correlations between the analyzing powers for the different reaction channels, or between the analyzing power and the cross section for a particular reaction channel.

The pronounced structure observed in the ^{19}F(p,γ) cross section is not visible in any of the elastic or inelastic excitation functions. The (p,α$_0$) analyzing power excitation function shows strong oscillations but these are not correlated with the (p,γ) cross sections. Thus there is no evidence from this work for any anomalous correspondence between the capture cross section and the analyzing power in other reaction channels.

In addition, there is no evidence for non-statistical

784

structure. However, since a "bump" must be about 15Γ wide in order to be detected on the basis of tests used for ^{26}Mg and ^{27}Aℓ, it would be difficult to distinguish between such a bump and a change in the direct reaction background.

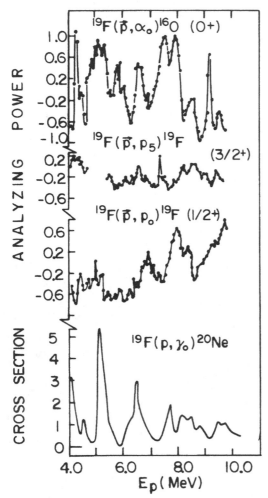

Fig. 1. Measured values of the analyzing power as a function of incident proton energy E_p at 140° for the reactions ^{19}F($\vec{p}$,α)^{16}O (ground state), ^{19}F($\vec{p}$,p')^{19}F* (1.55 MeV), and ^{19}F($\vec{p}$,p)^{19}F. Error bars are generally about as large as the data points; the solid lines are drawn through data points taken every 50 keV. The cross section excitation function for the ^{19}F(p,γ_0)^{20}Ne (ground state) reaction is taken from Ref. 3.

References

*Work supported in part by the National Science Foundation.

1) C. Glashausser, A. B. Robbins, E. Ventura, F. T. Baker, J. Eng, and R. Kaita, to be published and contribution to this conference, p.787
2) P. P. Singh et al., Nucl. Phys. 65 (1965) 577
3) R. E. Segel et al., Nucl. Phys. A93 (1967) 36

PROBABILITY DISTRIBUTIONS FOR COMPOUND NUCLEAR FLUCTUATIONS IN VECTOR ANALYZING POWERS*

R. F. Haglund, Jr., J. M. Bowen and W. J. Thompson
University of North Carolina, Chapel Hill, N.C. 27514 USA
and Triangle Universities Nuclear Laboratory, Durham, N.C.

Compound nuclear (CN) fluctuations in the cross section $\sigma(E,\theta)$ are frequently analyzed by writing $\sigma(E,\theta)$ as a sum of "basic cross sections" $\sigma_\alpha(E,\theta)$, where α labels channel spins.[1] However, the assumptions underlying this method may be invalid for projectiles with spin; moreover, the formalism does not generalize to analyzing powers.

We have tested an alternate approach, in which the helicity matrix elements are treated as independent variables in a statistical analysis. Semi-realistic calculations of fluctuations in cross section $\sigma(E,\theta)$ and analyzing power cross section $\sigma A_y(E,\theta)$ were made for proton elastic scattering on ^{26}Mg, for $5 \leq E_p \leq 10$ MeV and $90^\circ \leq \theta_{cm} \leq 165^\circ$. The excitation function step size was 25 keV, yielding 200 points at each angle.

For spin 1/2 elastic scattering with overlapping CN levels, the helicity matrix elements are

$$f(E,\theta) = f_D + f_{CN} = f_D + \sum_{J,L} e^{2i\omega_L} (J+\tfrac{1}{2}) P_L(\cos\theta) \sum_\lambda \frac{\Gamma_{\lambda J}}{E - E_\lambda + i\Gamma/2} \quad (1)$$

$$g(E,\theta) = g_D + g_{CN} = g_D + \sum_{J,L} e^{2i\omega_L} (-)^{J-L-\frac{1}{2}} P_L^1(\cos\theta) \sum_\lambda \frac{\Gamma_{\lambda J}}{E - E_\lambda + i\Gamma/2} \quad (2)$$

In this representation, the cross section and vector analyzing power are

$$\sigma(E,\theta) = \frac{1}{k^2}\left(|f|^2 + |g|^2\right) \quad (3)$$

$$\sigma A_y(E,\theta) = \frac{2}{k^2} \operatorname{Im}(fg^*) \quad (4)$$

The direct reaction terms f_D and g_D are obtained by optical model calculation. The ω_L are relative Coulomb phases. The index λ labels the CN levels; Γ is the coherence width. The complex partial widths $\Gamma_{\lambda J}$ are generated by a Tausworthe pseudo-random algorithm.[2] In the random phase approximation (RPA), the energy averages $<f_{CN}>$, $<g_{CN}>$ and $<f_{CN}g_{CN}^*>$ vanish, while $<|f_{CN}|^2>$ and $<|g_{CN}|^2>$ are finite.[3]

Calculations show that $f_{CN}(E)$ and $g_{CN}(E)$ are random and independent at the 10% confidence level. The choice of these quantities as the independent variables fixes the form of the probability density function for σ and σA_y.[4] Figure 1 shows typical results from an analysis of synthetic excitation functions.

For deuteron elastic scattering, $\sigma(E,\theta)$ and $\sigma T_{kq}(E,\theta)$ are quadratic functions of five helicity amplitudes. Thus, the analysis would be

expected to proceed similarly, with the difference that the fluctuating
quantities will have more degrees of freedom

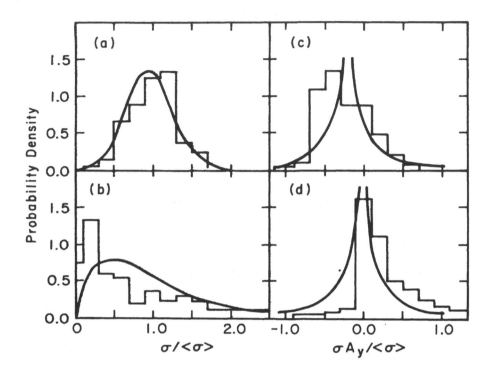

Fig. 1. Histograms are shown for the following distributions at 150°:
(a) Differential cross section with direct-reaction fraction y_d=.9.
(b) Differential cross section with y_d=0. (c) Analyzing power cross
section σA_y with y_d=.9. (d) Analyzing power cross section with y_d=0.
Probability density functions are shown as solid curves. For the
cross section, y_d is used as a fitting parameter. For A_y, the mean of
the distribution is proportional to y_d, and the fitting parameter is s_1,
the relative CN spin-flip fraction.

References

* Work supported in part by U.S.E.R.D.A.

1) J. P. Bondorf and R. B. Leachman, Mat. Fys. Medd. Dan. Vid. Selsk.
    34 (1965) 19
2) John RB. Whittlesey, Comm. A. C. M. 11 (1968) 641
3) E. Kujawski and T. J. Krieger, Phys. Lett. 27B (1968) 132
4) William C. Guenther, J. Am. Stat. Assn. 5 (1964) 57; J. Wishart and
    M. S. Bartlett, Proc. Camb. Phil. Soc. 28 (1932) 455

EVIDENCE FOR NON-STATISTICAL STRUCTURE IN THE INELASTIC SCATTERING OF POLARIZED PROTONS FROM ^{26}Mg[*]

C. Glashausser, A. B. Robbins, E. Ventura, F. T. Baker,[†]
J. Eng and R. Kaita
Department of Physics
Rutgers University, New Brunswick, New Jersey 08903

The analyzing power A_y for inelastic scattering from ^{26}Mg has been measured as a function of incident proton energy E_p in 50 keV steps from 5.5 to 9.5 MeV for a variety of scattering angles. Statistical analysis[1] of these data reveals evidence for intermediate structure in the excitation functions of A_y, although previous analysis of cross-section data indicated no deviations from the statistical model[2]).

The qualitative features that indicate intermediate structure are evident in Fig. 1. The actual data for scattering to the second excited state (Fig. 1B) show characteristic Ericson fluctuations; a coherence width Γ of about 50 keV has been previously determined[2]). Considerable structure remains when these fluctuations have been smoothed as in Fig. 1A where each data point represents an average over 150 keV. Large bumps, about 750 keV wide, are evident at 8.8 MeV in p_2 and at 6.8 and 8.5 MeV in p_1. Angular distributions at several energies averaged over 250 keV are shown in Fig. 2. The most dramatic changes in the shapes

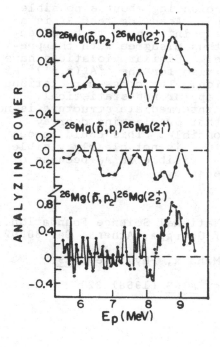

Fig. 1. Measured values of A_y for inelastic scattering to the 2^+_1 (1.81 MeV) and 2^+_2 (2.94 MeV) states of ^{26}Mg at 140°. The error bars are generally smaller than the data points. Actual measured values are shown in Fig. 1B; smooth lines have been drawn through data points averaged over 150 keV in Fig. 1A.

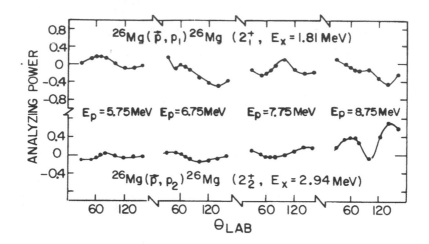

Fig. 2. Angular distributions of A_y for inelastic scattering from ^{26}Mg. Each data point represents an average over 250 keV.

of the angular distributions occur over the energies of these bumps.

The statistical tests of Baudinet-Robinet and Mahaux[1] show that there is less than a 1% chance that the data points for either p_1 or p_2 in Fig. 1A constitute a series of independent numbers randomly fluctuating about a possible smooth direct-reaction background. Assuming that 3Γ is a sufficient interval for the data points shown to be effectively independent, these data thus disagree with the predictions of the statistical model. Similar violations have been found at other angles, and also for the ^{27}Al$(p,p')^{27}$Al reaction in the same energy region. The cross-correlations between different channels also appear non-statistical.

The nature of the apparent intermediate structure is not yet clear, but it seems likely that overlapping doorway states are at least partly responsible. This could account for the fact that unusual structure is not clearly visible in the cross sections; it becomes visible in A_y because of the sensitivity of A_y to interference.

References

*Work supported in part by the National Science Foundation.
†Present address: University of Georgia, Athens, Ga. 30602.

1) Y. Baudinet-Robinet and C. Mahaux, Phys. Rev. C9 (1974) 723
2) O. Häuser et al., Nucl. Phys. A109 (1968) 329

INVESTIGATION OF ERICSON FLUCTUATIONS BY ^{27}Al(p,α)^{24}Mg

K. Imhof, G. Heil, G. Klier, M. Wangler

Physik. Inst. der Universität Erlangen-Nürnberg; Erlangen, West-Germany[+]

The reaction ^{27}Al($\vec{p}$,α)^{24}Mg have been measured to the first five states in ^{24}Mg from 9.20 to 11.45 MeV in steps of 25 keV and from $(\theta)_{lab}$=50-180° in steps of 10 degrees. Figure 1 shows the analysing power A(θ) for (p,α$_o$) over the range of energy. The error is about $\pm$ .05.

The differential cross section σ(θ) as well as σ(θ)A(θ) for the ground state transition have been used to extract the coherence width Γ. The mean values arge $\bar{\Gamma}$(σ)=38.6 $\pm$ 6.6 keV for σ(θ) using a "sliding averaging interval" and $\bar{\Gamma}$(σA)=39.4 $\pm$ 8.0 keV for σ(θ)A(θ). It isn't necessary for A to use an averaging interval, which has been shown by Haglund and Thompson[1] by means of synthetic excitation functions.

The amount of the direct process y_D on ^{27}Al(p,α) in this region of

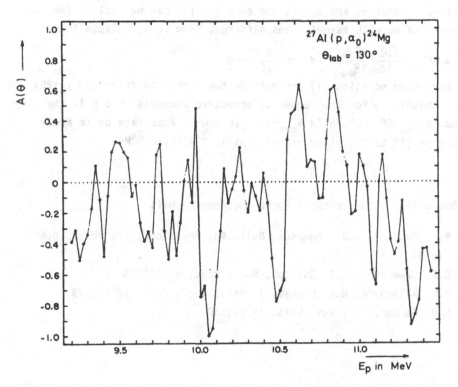

Fig. 1 : Excitation function for the analysing power A(θ).

790

energy is small[2]. But the "histogram-method" or the autocorrelation function

$$C(\mathcal{E}=0,\theta) = \frac{1-y_D^2}{N}$$ are not sensitive enough to duduce small values for

y_D. By using a polarized beam it is easy to determine a pure compound process, because in this case the energy averaged value $\langle \mathcal{O} A\rangle = 0$. As one can see in figure 1 this isn't the case for the (p,α_o)-reaction. Therefore a formular has been applied, which we have developed from the scattering of protons on nuclei with spin 0:

$$C_{\mathcal{O}A}(\mathcal{E}=0,\theta) \approx \frac{1}{y_D^2} - 1 \tag{1}$$

$$\text{with } C_{\mathcal{O}A}(\mathcal{E},\theta) = \frac{\langle \mathcal{O}A(E,\theta)\cdot \mathcal{O}A(E+\mathcal{E},\theta)\rangle}{\langle \mathcal{O}A(E,\theta)\rangle \langle \mathcal{O}A(E+\mathcal{E},\theta)\rangle} - 1 \tag{2}$$

In scattering amplitude formalism $\langle (\mathcal{O}A)^2\rangle$ is a sum of products of basis cross sections. By assuming that $\langle \mathcal{O}A\rangle = (\mathcal{O}A)$ direct and all basic cross section quantities are equal, the equation (1) can be applied for more general cases with target spins different from zero, because then

$$C_{\mathcal{O}A}(0) = \frac{\langle (\mathcal{O}A)^2\rangle}{(\mathcal{O}A)^2_{direct}} - 1 \approx \frac{1}{y_D^2} - 1$$

We have used equation (1) and got for the direct contribution $y_D=25\%$ and with this value for the "number of effective channels" $N=5.5$ in the range from $(\theta)_{lab}=50-150°$ while $N_{max}=6.0$. This will encourage us to apply equation (1) to the other transitions in $^{27}Al(p,\alpha)^{24}Mg$.

+ Supported by the Deutsche Forschungsgemeinschaft

1) R.F. Haglund, W.J. Thompson, Bull. Am. Phys.Soc., II, vol.20,p.693 (1975)
2) G.P. Lawerence, A.R. Quinton, Nucl. Phys. 65(65)275
B.W. Allardyce, W.R. Graham, I. Hall, Nucl. Phys. 52(64)229
G.M. Temmer, Phys.Rev. Lett. 12(64)330

STATISTICAL FLUCTUATIONS OF CROSS SECTIONS AND ANALYZING POWERS
OF ^{88}Sr(p,p$_o$) AND ^{90}Zr(p,p$_o$) AROUND E$_p$ = 12.5 MeV

H. Paetz gen. Schieck and G. Berg [+],
Institut für Kernphysik der Universität zu Köln, 5 Köln 41, Germany
and
G. G. Ohlsen and C. E. Moss,
Los Alamos Scientific Laboratory, University of California,
Los Alamos, New Mexico 87544, USA

Recently, statistical fluctuations in proton induced reactions have found renewed interest by the inclusion of isospin-dependent effects in the statistical model [1]. Two independent sets of compound levels, $T^<$ and $T^>$ states can be excited by proton bombardment whereas isospin selection rules prevent the decay of $T^>$ states via (p,α) reactions. The different coherence widths found for (p,p) and (p,α) reactions on ^{31}P were interpreted as a consequence of isospin selection rules [2]. The availability of high intensity polarized ion sources allows investigations of fluctuations in excitation functions of analyzing powers in sufficiently small energy steps.

Around the cross section minimum of proton elastic scattering from ^{88}Sr and ^{90}Zr at $\theta = 175^o$ around E$_p$ = 12.5 MeV [3][4] strong statistical fluctuations have been observed with markedly different coherence widths for the two target nuclei. This difference has been explained by assuming a coexistence of $T^<$ and $T^>$ fluctuations in the scattering cross section but with $R = \sigma^>/\sigma^<$ significantly >1 for ^{88}Sr and <1 for ^{90}Zr. Therefore in ^{88}Sr(p,p$_o$) one observes resolved $T^>$-fluctuations with $\Gamma \approx$ 20 keV whereas for ^{90}Zr(p,p$_1$) experimentally averaged $T^<$-fluctuations with $\Gamma_{exp} \approx 10$ keV dominate ($\Gamma_{coh} \approx 200$ eV). Level density calculations showed the validity of the Ericson condition $\Gamma/D > 2$ for both types of states.

Fig. 1 shows excitation functions in 10 keV steps of cross-sections, analyzing powers and "differential" analyzing powers as measured and averaged with a 20 keV interval. The data were measured using the polarized beam from the LASL Lambshift polarized ion source with a typical proton polarization of P = -0.88 and intensities close to 200 nA on target after acceleration by the FN Tandem accelerator. The polarization was regularly determined from the quench ratio which in turn was checked with ^{4}He($\vec{p}$,p) scattering). Four surface barrier detectors were used in the "supercube" scattering chamber at the two scattering angles and each measurement was performed with both spin up and spin down. The targets were a self-supporting foil of ^{90}Zr and ^{88}Sr CO$_3$ evaporated on a thin carbon backing such that the target-plus-beam-resolution was about 10 keV at 12.5 MeV.

The data show strong fluctuations for both nuclei, especially for $d\sigma/d\Omega \cdot A$. The analyzing power data more clearly than $d\sigma/d\Omega$ exhibit a fine-structure with the width of the experimental resolution superimposed on a coarse structure of $\Gamma > 25$ keV. Due to increasing direct contributions and increasing number of effective spin channels the cross sections at 140° are almost structureless whereas the analyzing powers still fluctuate strongly indicating a smaller influence of direct contributions and a reduction of the number of effective spin channels by the incident beam polarization as compared to the cross section [6].

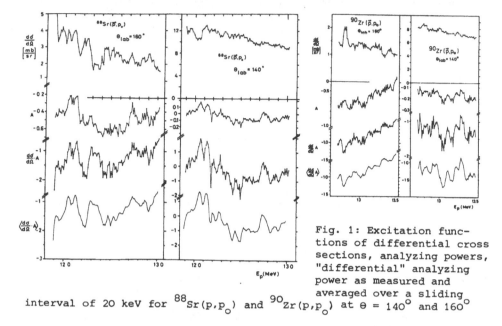

Fig. 1: Excitation functions of differential cross sections, analyzing powers, "differential" analyzing power as measured and averaged over a sliding interval of 20 keV for ^{88}Sr(p,p_o) and ^{90}Zr(p,p_o) at $\theta = 140°$ and $160°$

From the data of Fig. 1 we conclude that the analyzing power measurements confirm in a sensitive way the idea that in ^{88}Sr(p,p_o) and ^{90}Zr(p,p_o) we observe two classes of Ericson fluctuations from separately overlapping $T^<$ and $T^>$ levels in the respective compound nuclei at about 20 MeV excitation energy.

References

* Work performed under the auspices of the U.S.A.E.C. and supported in part by the Bundesministerium für Forschung und Technologie
+ Present address: Department of Physics, University of Wisconsin, Madison, Wisconsin 53701, USA

1) D. Robson, A. Richter and H. L. Harney, Phys. Rev. C8 (1973) 153
2) M. Kildir and J. R. Huizenga, Phys. Rev. C8 (1973) 1965
3) K. Schulte, G. Berg, P. von Brentano and H. Paetz gen. Schieck, Nucl. Phys. A241 (1975) 272
4) G. Berg, H. Paetz gen. Schieck, U. Scharfer, K. Schulte and P. von Brentano, Proc. Int. Conf. on Nuclear Physics, München 1973, ed. de Boer and H. J. Mang (North Holland, Amsterdam, 1973) p. 530
5) G. Berg, W. Kühn, H. Paetz gen. Schieck, K. Schulte and P. von Brentano, to be published
6) G. Latzel, G. Berg and H. Paetz gen. Schieck, Fourth Polarization Symposium, p.793

STATISTICAL ANALYSIS OF ERICSON FLUCTUATIONS IN ANALYZING POWER EXCITATION FUNCTIONS *

G. Latzel, G. Berg [+] and H. Paetz gen. Schieck
Institut für Kernphysik der Universität zu Köln
D 5 Köln 41, Germany

In another contribution to this conference Ericson fluctuations in experimentally obtained analyzing power excitation functions of polarized protons scattered from ^{90}Zr and ^{88}Sr have been discussed [1].

In the case of cross section fluctuations the data can be analyzed in terms of the autocorrelation function [2]

$$C(\varepsilon) = \frac{\langle \sigma(E+\varepsilon)\sigma(E)\rangle - \langle\sigma(E)\rangle^2}{\langle\sigma(E)\rangle^2} = \frac{1}{N_{eff}} \cdot \frac{\Gamma^2}{\Gamma^2+\varepsilon^2} \cdot (1-y_D^2)$$

with N_{eff} = number of effective spin channels, Γ = coherence width of overlapping compound levels and $y_D = \sigma^{DI}/(\sigma^{DI}+\langle\sigma^{CN}\rangle)$ the direct contribution to the average cross section. Thus a knowledge of two of the three quantities N_{eff}, Γ or y_D would allow to determine the third one from $C(\varepsilon)$. In the case of a more general observable like the differential analyzing power $A \cdot d\sigma/d\Omega$ normally angular momentum coupling has to be taken into account explicitly and no simple relationship between the autocorrelation $C(\varepsilon)$ of this observable, y_D and Γ will result. From the formal similarity of the autocorrelation function $C(\varepsilon)$ for the cross section [3] and differential analyzing powers [4] with the approximations made by Ericson [3] (all particle spins small compared to typical angular momenta and not all zero) follows a similar proportionality $C(\varepsilon) \sim \Gamma^2/(\Gamma^2+\varepsilon^2)$ for the differential analyzing power. Lambert and Dumazet [4] showed that the normalized variance $C(\varepsilon=0)$ is larger for the differential analyzing power than for the differential cross section for all scattering angles. This is equivalent to a lower number of spin channels for any angle as is to be expected from physical reasons for a polarized beam. The same conclusion was drawn by Kujawski and Krieger [5] by considering the relative variance of the polarized cross section (defined as $d\sigma_{pol}/d\Omega = d\sigma_o/d\Omega(1+A_y \cdot P_y)$):

$$C(0)(d\sigma_{pol}/d\Omega) = C(0)(d\sigma_o/d\Omega) + \frac{\langle[P_y A_y(d\sigma_o/d\Omega)]^2\rangle}{\langle d\sigma_{pol}/d\Omega\rangle^2}$$

A fluctuation analysis of analyzing powers can therefore be useful to extract the coherence width Γ in cases where cross sections will not show fluctuations due to large direct contributions or higher number of effective spin channels [1], especially with available polarized beams of sufficiently high intensity.

Another major advantage is the possibility of extracting the direct contribution using the fact that $\langle d\sigma/d\Omega \cdot A_y\rangle$ is a purely direct quantity even in the presence of channel-channel-correlations.

We have analyzed the differential analyzing powers of ^{88}Sr$(\vec{p},p_o)$ and ^{90}Zr$(\vec{p},p_o)$ by performing an autocorrelation analysis on $d\sigma_{pol}/d\Omega$ and $d\sigma_o/d\Omega$ with the results shown in table 1.

Table 1

	θ_{lab}	^{88}Sr(p,p$_0$)			^{90}Zr(p,p$_0$)			^{90}Zr(p,p$_1$)
		140°	160°	175° [a]	140°	160°	175° [a]	175° [a]
C(o)	$d\sigma_{pol}/d\Omega$	0.004	0.028	--	0.002	0.012	--	--
	$d\sigma_o/d\Omega$	0.003	0.025	0.120	0.001	0.013	0.050	0.070
Γ (keV)	$d\sigma'_{pol}/d\Omega$	24	28	--	25	24	--	--
	$d\sigma_o/d\Omega$	19	19	22	22	22	17	10

[a] From reference [6]

The numbers of table 1 confirm the general idea that fluctuations of the differential analyzing powers are "stronger" than of the differential cross sections even when the relatively large FRD-errors of about 30 % are considered. The width Γ as determined by the autocorrelation analysis of the elastic scattering is the width of the $T^>$-component of $d\sigma/d\Omega$ whereas ^{90}Zr(p,p$_1$) shows a width which is equal to the target + beam energy-resolution and can be understood as the width of averaged $T^<$ compound states (see [1]).

References

* Supported in part by the Bundesministerium für Forschung und Technologie
+ Present address: Department of Physics, University of Wisconsin, Madison, Wisconsin 53701, USA

1) H. Paetz gen. Schieck, G. Berg, G. G. Ohlsen and C. E. Moss, Fourth Polarization Symposium, p.791
2) M. G. Braga-Marcazzan and L. Milazzo Colli, Progr. Nucl. Phys. **11** (1970) 145
3) T. Ericson, Annals of Physics **23** (1963) 390
4) M. Lambert and G. Dumazet, Nucl. Phys. **83** (1966) 181
5) E. Kujawski and T. J. Krieger, Phys. Lett. **27B** (1968) 132
6) G. Berg, W. Kühn, H. Paetz gen. Schieck, K. Schulte and P. von Brentano, to be published

A STUDY OF FLUCTUATIONS IN THE CROSS-SECTION AND THE ANALYZING POWER OF 28SI($\vec{D}$,D)- AND 28SI($\vec{D}$,P)-REACTIONS

R. Henneck (University of Erlangen-Nürnberg, Erlangen, W.Germany)
T.B. Clegg, R.F. Haglund, E.J.Ludwig (University of North Carolina, Chapel Hill and TUNL, Duke University, N.C., USA)
G. Graw (University of München, München, W. Germany)

The statistical model as developed by Ericson[1], Brink and Stephen[2] has been applied successfully to many cases, where either low-spin particles interact or the reaction mechanism is known from other sources. In most cases however, this method displays a severe ambiguity between the "number of effectively contributing channels N" and the "direct interaction contribution y_D".This ambiguity might be removed by the additional measurement of the analyzing power in a polarization experiment. In the case of elastic proton scattering from O-spin targetnuclei e.g. an analysis of the cross-section excitation function and the analyzing power excitation function together results in an unambiguous determination of y_D:

$$\frac{R(\sigma,\varepsilon=0) + R(\sigma A,\varepsilon=0)}{\langle\sigma\rangle^2} = 1 - y_D^2 \qquad (a)$$

where $R(Q,\varepsilon)$ is the relative autocorrelation function

$$R(Q,\varepsilon) = \langle Q(E+\varepsilon)Q(E)\rangle - \langle Q(E+\varepsilon)\rangle\langle Q(E)\rangle \qquad (b)$$

of the quantity Q.
This formula encourages to go a step further and to investigate deuteron-induced reactions from O-spin target-nuclei.
We have studied several 28Si($\vec{d}$,d)- and 28Si($\vec{d}$,p)-reactions with the purely vector-polarized deuteron beam of the TUNL Lambshift source. We measured 18 excitation functions of the differential cross-section $\sigma(\theta)$ and the vector-analyzing power $A(\theta)$ within the energy-range $6.00 \leqslant E_d \leqslant 10.46$ MeV in steps of 20 keV. The experimental energy resolution was also about 20 keV. The figure shows $\sigma(\theta)$ and $\sigma(\theta)A(\theta)$ for the elastic deuteron scattering and the (d,p_1)- and (d,p_2)-reactions at θ_{lab} = 150°.

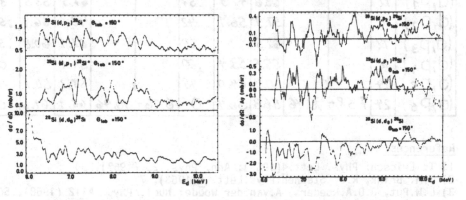

The analysis of the $\sigma(\theta)$-excitation functions was performed via the normalized autocorrelation function $C(\varepsilon)$[1], whereas for the (σA)-data the relative autocorrelation function $R(\sigma A,\varepsilon)$ from equ. b was used. Corrections for finite-range-of-data and experimental energy resolution effects have been applied[3]. As the gross-structures in the averaged background of $\sigma(\theta)$ have a strong influence on the behaviour of $C(\varepsilon)$, the coherence width $\Gamma(\sigma)$ and $C(\varepsilon=0)$ could only be extracted from the trend-reduced data[3]. For the (σA)-data however, a trend-reduction turned out to be not necessary. This feature has been shown by Haglund and Thompson[4] by means of synthetic excitation function studies.

The results of this analysis are listed in the table for all 9 reaction channels at θ_{lab} = 130° and θ_{lab} = 150°. The coherence widths $\Gamma(\sigma)$ and $\Gamma(\sigma A)$ from the analysis over the whole energy-range ΔE have similar values, the relative statistical errors being about 15%. The average over all 18 excitation functions $\overline{\Gamma(\sigma)}$ = 61keV was found to be somewhat larger than $\overline{\Gamma(\sigma A)}$ = 55 keV, where the latter average was taken over 15 (instead of 18) excitation functions. An equality has been predicted by Lambert and Dumazet[5]:
The direct interaction contribution y_D has been extracted using the assumption $N = N_{max}$ from the autocorrelation function $C(\varepsilon=0)$ and also from the cross-section distributions[2]. The results of both methods agree within 10%, therefore only y_D-obtained from $C(\varepsilon=0)$ is given in the table. From the assumption $N = N_{max}$ it is clear that these values can only be regarded as estimates and no errors are given.
A determination of y_D and N at the same time with the cross-section distribution method is only feasible for small values of N and proved to be impossible.

reaction	N_{max}	θ_{LAB}=130°				θ_{LAB}=150°			
		ΔE(MeV)	$\Gamma(\sigma)$	$\Gamma(\sigma A)$	y_D	ΔE(MeV)	$\Gamma(\sigma)$	$\Gamma(\sigma A)$	y_D
(d,d_0)	4	7.58 – 10.46	58.4	56.6	.97	6.00 – 10.46	54.9	51.3	.89
(d,d_1)	22	7.58 – 9.84	64.9	43.4	.77	6.26 – 9.18	66.7	50.0	.68
(d,p_0)	6	7.58 – 10.46	62.7		.83	6.26 – 10.46	82.8	61.2	.24
(d,p_1)	12		63.6	42.9	.57		67.9	53.5	.34
(d,p_2)	18		57.9	56.1	.42		43.6	66.4	.52
(d,p_3)	12		62.5		.65		62.1	60.8	.53
(d,p_4)	18		55.5	53.4	.60		54.4		0
(d,p_5)	24		60.9	64.0	.45		58.3	50.1	0
(d,p_6)	24	7.58 – 10.46	56.3	56.4	0	6.26 – 10.46	68.9	61.6	0

References:

1) T. Ericson: Phys. Lett.4(1963); Ann. Phys. 23 (1963),390
2) D.M. Brink, R.O. Stephen:Phys.Lett. 5(1963), 77
3) L.W.Put, J.D.A.Roeders, A.van der Woude: Nucl. Phys. A112 (1968), 561
   J.D.A. Roeders: Ph.D. thesis, University of Groningen
   A.van der Woude: Nucl. Phys. 80(1966), 14
4) R.F. Haglund, W.J. Thompson: Bull. Am.Phys. Soc.,II, Vol.20,p.693 (75)
5) M.Lambert, G. Dumazet:Nucl. Phys. 83(1966), 181

Particle-γ Angular Correlations
and New Applications of Polarized Beams

SPIN-FLIP PROBABILITY AND SPIN-FLIP ASYMMETRY IN THE REACTION $^{12}C(\vec{n},n')^{12}C^*(2^+, 4.44$ MeV) AT 15.85 MeV

M. Thumm, G. Mertens, H. Lesiecki, G. Mack and K. Schmidt
Physikalisches Institut der Universität Tübingen,
D-7400 Tübingen, Morgenstelle, Germany

In recent years there has been a considerable amount of research carried out in order to investigate the spin dependence of the effective two-nucleon interaction and the reaction mechanism in inelastic scattering of nucleons from nuclei. With the z-axis chosen along the normal to the scattering plane[+] the inelastic observables may be expressed in terms of the incident and the outgoing spin projections of the nucleon by

$$d\sigma(\theta)/d\Omega = \tfrac{1}{2}(\sigma_{++}(\theta) + \sigma_{--}(\theta) + \sigma_{+-}(\theta) + \sigma_{-+}(\theta)) \equiv \tfrac{1}{2}\sigma(\theta)$$

$$p_z(\theta) = (\sigma_{++}(\theta) + \sigma_{+-}(\theta) - \sigma_{-+}(\theta) - \sigma_{--}(\theta))/\sigma(\theta)$$

$$A_z(\theta) = (\sigma_{++}(\theta) + \sigma_{-+}(\theta) - \sigma_{+-}(\theta) - \sigma_{--}(\theta))/\sigma(\theta)$$

$$S(\theta) = (\sigma_{+-}(\theta) + \sigma_{-+}(\theta))/\sigma(\theta)$$

$$\Delta S(\theta)/S(\theta) = (\sigma_{-+}(\theta) - \sigma_{+-}(\theta))/(\sigma_{-+}(\theta) + \sigma_{+-}(\theta)) \equiv \tfrac{1}{2}(A_z(\theta) - p_z(\theta))/S(\theta)$$

From these equations it is obvious that polarization p_z, analyzing power A_z, spin-flip probability (SFP) S and spin-flip asymmetry (SFA) $\Delta S/S$ are more sensitive to details of a reaction theory than is the differential cross section $d\sigma/d\Omega$.

For 0^+ to 1^+ and 0^+ to 2^+ excitations the SFP can be determined with an unpolarized beam by measuring the absolute angular correlation between the inelastically scattered nucleons and the subsequent deexcitation γ-rays emitted perpendicular to the scattering plane. The SFA and thus the difference between A_z and p_z may be deduced by measuring the SFP using a polarized beam[1]. The proton SFA at isobaric analog resonances was found to be large and sensitive to the structure of the resonances[1]. In the $^{12}C(\vec{p},p')^{12}C^*(4.44$ MeV) scattering at 30 and 50 MeV[2,3], where the direct reaction mechanism is dominant, also an analyzing power - polarization inequality was observed and attributed according to Sherif[4] to the interference of effective interactions which are even and odd under time reversal.

In order to continue our research on the inelastic neutron scattering from ^{12}C [5,6] we have measured $S(\theta)$ and $\Delta S(\theta)/S(\theta)$ in the reaction $^{12}C(\vec{n},n')^{12}C^*(4.44$ MeV) at 15.85 MeV. At this energy a compound nucleus resonance in the elastic cross section and to a lesser extent in the inelastic cross section has been observed[7].

Neutrons with a primary polarization $p_B = -0.135$ [8,9] were produced in the $^3H(d,\vec{n})^4He$ reaction. The sign of p_B was changed between alternative runs by means of a superconducting spin-precession solenoid. Neutron time-of-flight technique with carbon recoil detection in a plastic scintillator and fast n-γ coincidence technique in spin-flip geometry allowed a clear separation of elastically and inelastically scattered neutrons[5].

<u>Spin-flip probability</u>. The SFP is proportional to the sum of the n'-$\gamma_\perp$ coincidences from a pair of runs with primary polarization up and down. It is independent of p_B. The values of S have been corrected for non-spin-flip contributions due to off-z-axis γ-rays. The results are shown in fig. 1 together with the proton SFP data of Wilson and Schecter[10] at 15.9 MeV. The relatively small values of

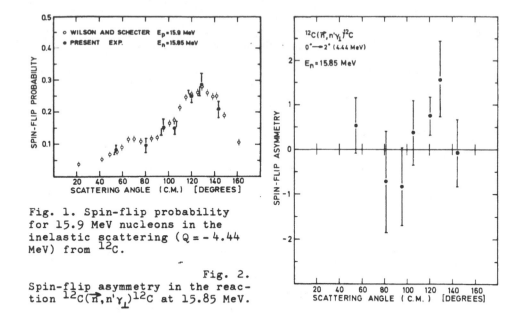

Fig. 1. Spin-flip probability for 15.9 MeV nucleons in the inelastic scattering ($Q = -4.44$ MeV) from ^{12}C.

Fig. 2. Spin-flip asymmetry in the reaction $^{12}C(\vec{n},n'\gamma_\perp)^{12}C$ at 15.85 MeV.

S at forward angles and the good agreement between the neutron and proton SFP angular distributions indicate that at this energy the reaction mechanisms for neutron and proton inelastic scattering are very similar, Coulomb effects are small and compound nucleus resonances are not dominating.

Spin-flip asymmetry. The SFA is proportional to the asymmetry in the number of n'-$\gamma_\perp$ coincidences measured in a pair of runs with primary polarization up and down. The results are given in fig. 2. In spite of the poor statistics the data seem to indicate some difference between A_z and p_z in the vicinity of 130°.

References

+) This choice of the z-axis is not consistent with the Madison Convention but it is necessary for a convenient description of spin-flip.

1) R. Boyd et al., Phys. Rev. Lett. 27 (1971) 1590 and 29 (1972) 955
2) M. Ahmed, J. Lowe, P.M. Rolph and V. Hnizdo, RHEL/R-187 (1969)
3) T. Hasegawa et al., in Proc. Int. Conf. on Nuclear Physics, Eds. J. de Boer and H.J. Mang, München 1973, Vol. 1, p. 409
4) H. Sherif, Can. J. Phys. 49 (1971) 983
5) G. Mertens, M. Thumm and H.V. Geramb, Nucl. Phys. A232 (1974) 472
6) M. Thumm, G. Mertens and G. Mack, Z. Naturf. 28a (1973) 1223
7) F. Boreli, B.B. Kinsey and P.N. Shrivastava, Phys. Rev. 174 (1968) 1147
8) G. Hentschel, G. Mack and G. Mertens, Z. Naturf. 23a (1968) 1401
9) W. Tornow, G. Mack, G. Mertens and H. Spiegelhauer, Phys. Lett. 44B (1973) 53
10) M.A.D. Wilson and L. Schecter, Phys. Rev. C4 (1971) 1103

POLARIZATION EFFECTS IN DE-EXCITATION
GAMMA-RAYS FOLLOWING ($\vec{p}$,p') REACTIONS*

H. T. King, C. Glashausser, M. Hass[†], A. B. Robbins
and E. Ventura
Department of Physics
Rutgers University, New Brunswick, New Jersey

In this work de-excitation γ-rays following ($\vec{p}$,p') reactions were studied without detection of the outgoing protons. Such an experiment was first reported by Glavish et al.[1]. Transversely polarized protons from the Rutgers FN tandem and atomic beam polarized source were used to bombard targets of ^{26}Mg and ^{92}Mo; Ge(Li) detectors located symmetrically on either side of the beam, in a plane perpendicular to the beam polarization, were used to measure the left-right asymmetry of de-excitation γ-rays following inelastic scattering.

The general expression for the left-right γ-ray asymmetry is given by

$$A_\gamma \, d\sigma/d\Omega_\gamma = \sum_{\substack{K \geq 2 \\ \text{even}}} B_K \, \rho_{K1} \, P_K^1(\theta_\gamma)$$

where B_K depends only on the two states involved in the γ-decay and ρ_{KQ} is the spherical tensor form of the density matrix for the γ-decaying state. If the decaying state is formed directly, i.e. not by feeding from other states, ρ_{KQ} depends in a straightforward way on the transition matrix elements for the reaction that excites the state. Indirect formation of the state is expected to decrease the observed γ-ray asymmetry.

The elements ρ_{K1} are non-zero only if there is interference between different proton partial waves in the entrance channel, and therefore no γ-ray asymmetry should be observed for the case of scattering through an isolated resonance in the compound nucleus. If the compound nuclear (CN) levels overlap one another, the γ-ray asymmetry is expected under high resolution to fluctuate about an average value which is zero in the absence of coherent processes such as direct reactions.

Fig. 1 shows results for $^{26}Mg(\vec{p},p')^{26}Mg$ ($2_1^+ \rightarrow 0_1^+$) and ($2_2^+ \rightarrow 2_1^+$). The asymmetries are generally quite small - less than 10% for the $2_1^+ \rightarrow 0_1^+$ transition, and less than 5% for the $2_2^+ \rightarrow 2_1^+$ transition. Ericson fluctuations are apparent, but there is no clear structure of a coarser nature as has been seen in Mg($\vec{p}$,p') particle work[2].

For the $^{92}Mo(p,p')^{92}Mo$ ($2_1^+ \rightarrow 0_1^+$) reaction, where numerous strong analog resonances are superimposed on a large CN background, no γ-ray asymmetry is observed on or off resonance below $E_p \sim 7.5$ MeV. As expected no asymmetry arises from either the isolated resonances or the energy-averaged CN background. An asymmetry is seen only at higher energies

where direct reactions begin to contribute and the analog
resonances begin to overlap.

References

*Work supported in part by the National Science Foundation.
†On leave from the Weizmann Institute, Rehovot, Israel.

1) H. F. Glavish et al., Bull. Am. Phys. Soc. 20 (1975) 85
2) C. Glashausser et al., Bull. Am. Phys. Soc. 19 (1974)
   1012

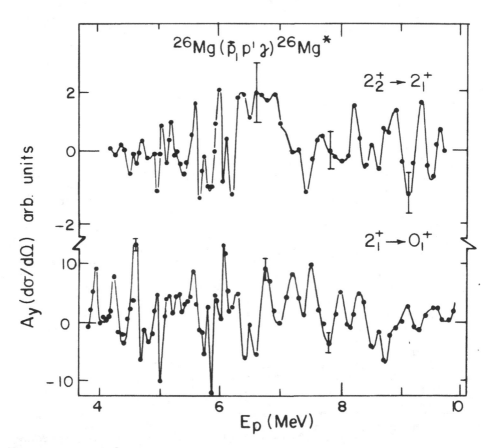

Fig. 1. Gamma-ray asymmetries at 60° for two transitions
in ²⁶Mg following inelastic polarized proton scattering.
Outgoing protons were not detected.

MULTISTEP PROCESSES IN THE REACTION ^{24}Mg(d,d')^{24}Mg
DEMONSTRATED BY PARTICLE-γ ANGULAR CORRELATIONS

U. Scheib, M. Berg, W. Eyrich, A. Hofmann,
S. Schneider, and F. Vogler

Physikalisches Institut der Universität Erlangen-
Nürnberg, Erlangen, Germany

Polarization phenomena studied either by polarized projectiles and/or by particle-γ-angular correlation experiments
provide more information about the reaction mechanism than
the analysis of the differential cross section alone. To analyse cross section data often DWBA or CCBA calculations are
used. While in the DWBA approach only the direct coupling of
the final state to the target ground state is considered, the
CCBA calculations include multistep processes too. It is the
purpose of this note to show that the comparison of such calculations with particle-γ-correlation data provides a semiquantitative measure of the importance of multistep processes.

As an example we have investigated the scattering of 10 MeV
deuterons on ^{24}Mg. The cross sections of the elastic and inelastic scattering to the first excited state as well as the
d_1-γ "in plane" angular correlation were measured absolutely.
The scattered particles were observed between 25° and 95° in
steps of 5°. To detect the coincident γ-rays a Ge(Li)-Detector
located at 6 different positions in the reaction plane was
used.

The correlation function describing the experiment is of the
form

$$W(\vartheta_\gamma = \frac{\pi}{2}, \varphi_\gamma) = A + B \sin^2(\varphi_\gamma - \varphi_1) + C \sin^2 2(\varphi_\gamma - \varphi_2)$$

where A, B, C, φ_1 and φ_2 depend on the scattering angle. The
z-axis is chosen perpendicular to the reaction plane. The parameter A is proportional to the polarization of the excited
nucleus, B is proportional to the half-spinflip probability.
C squared is proportional to the product $|\varrho_{2\,2}| \cdot |\varrho_{-2-2}|$, where
$\varrho_{M\,M'}$ is the density matrix describing the residual nuclear
excited state. The phases φ_1 and φ_2 are proportional to the

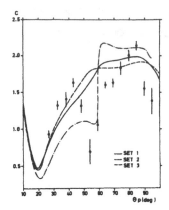 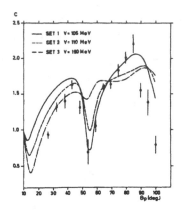

Fig. 1: Comparison of the parameter C with DWBA-calculations
(left side) and CC-calculations (right side)

phases of ϱ_{1-1} and ϱ_{2-2} respectively. In the forward angular
region the half-spinflip probability is found to be small, which
means that B is negligible. So only three parameters remain in
the correlation function, which can be extracted from the experimental data for each scattering angle.

To analyse the experimental results DWBA and CC-calculations
have been done. Fig. 1 shows the comparison of the DWBA- and
CC-calculations with the experimental values of C. The potential sets of the different calculations have been obtained
by fitting the elastic and inelastic cross sections. While all
calculations describe the differential cross section quite
well the DWBA-calculations are not able to reproduce the
structure of C. The CC-calculations however give a good
agreement with the data for potentials with a depth of about
100 MeV.

We conclude that DWBA is not able to describe the reaction
$^{24}Mg(d,d_1)^{24}Mg$ although it leads to a good agreement with
the cross section data. From our angular correlation data
analyses it is obvious, that multistep processes which are
included in the CC-calculations play an important role in
this reaction.

APPLICATION OF MULTI-STEP PROCESSES TO NUCLEAR STRUCTURE STUDIES: THE ^{28}Si($\vec{d}$,pγ) REACTION

H. Clement[+] and R.N. Boyd
NSRL*, University of Rochester, Rochester, NY 14627;
C.R. Gould
North Carolina State University and TUNL**, Durham, NC 27706;
T.B. Clegg
University of North Carolina and TUNL**, Durham, NC 27706

In a previous study[1] proton-gamma ray angular correlations (PGAC) were studied for the ^{28}Si(d,pγ)^{29}Si(5/2$^+$,2.03 MeV) reaction at particle scattering angles near the stripping peak. The correlations were found to be sensitive to the excited core configurations existing in the state populated by the reaction. However in that study some questions still remained. Thus we have performed the ^{28}Si($\vec{d}$,pγ) reaction to try to resolve these ambiguities. Here we present results for two of the ^{29}Si levels examined.

The Van de Graaff accelerator and polarized ion source[2] of TUNL were used to produce the 7.6 MeV vector polarized (P_z= 0.60) deuteron beam. Typical beams on the 500 μg/cm^2 natural Si target were 30 nA. Five NaI(Tl) gamma ray detectors were operated in coincidence with a particle detector located near the ℓ=2 stripping peak. All detectors were in the horizontal plane and the beam polarization axis was normal to this plane.

The polarized PGAC data, defined as correlation yield with spin up minus that with spin down, for four particle scattering angles are shown, in Fig. 1, for the ground state decay of the ^{29}Si(2.03 MeV) level. The dashed-dotted curve represents a one-step DWBA calculation. It was shown in ref. 1 to give a reasonable representation of cross section and PGAC data. However Fig. 1 shows that some of the gross features of the polarized PGAC data, most notably those taken at particle

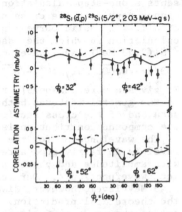

Fig. 1: Polarized PGAC for ^{28}Si($\vec{d}$,pγ)^{29}Si(2.03 MeV). Solid curves: CCBA, dashed-dotted curves: DWBA.

detector angles of 42° and 52°, are not represented by this calculation.

The two-step CCBA code CHUCK[3] was used to calculate the cross section and correlations. Each two-step reaction trajectory involved both inelastic scattering and transfer, with the ^{29}Si(2$^+_1$), ^{29}Si(g.s.), or ^{29}Si(3/2$^+$,1.27 MeV) levels as the intermediate state. The solid curve in Fig. 1 shows the results of a calculation in which the assumed wavefunction for the ^{29}Si(2.03 MeV) level was

$$|5/2\rangle = \sqrt{0.14}\,|n_{d_{5/2}} \otimes 0^+\rangle + \sqrt{0.64}\,|n_{s_{1/2}} \otimes 2^+\rangle - \sqrt{0.21}\,|n_{d_{3/2}} \otimes 2^+\rangle.$$

The ^{28}Si(0$^+$) and ^{28}Si(2$^+$) cores were assumed to be members of an oblate[4] rotational band. This CCBA calculation gave only a slightly better representation of the cross section and PGAC data than did the one-step calculation. However, Fig. 1 shows that it also reproduces the features of the polarized PGAC data quite well, unlike the one-step calculation.

The cross section and polarized PGAC data for the ground state decay

806

of the ^{29}Si(3/2$^+$,2.43 MeV) level are shown
in Fig. 2. While the cross section data
shown[4,5] were taken at an incident deuteron energy of 9.0 MeV, ref. 4 shows that
little change in the general features of
this cross section occurs between 9 and 11
MeV and our cross sections at 7.6 MeV were
very similar to the 9.0 MeV data. This
state has[6] a one-step spectroscopic
factor of essentially zero. At a deuteron
energy of 7.6 MeV, the compound nuclear
cross section is about half of that observed. However for energy averaged data
the compound contribution does not interfere with the direct part of the scattering. Then the polarized PGAC as defined
above should be free of compound effects.
The dashed-dotted curve in Fig. 2 represents a one-step calculation; it clearly
cannot represent the cross section data
at all. The solid curve represents a
calculation in which the assumed
^{29}Si(2.43 MeV) level wavefunction was

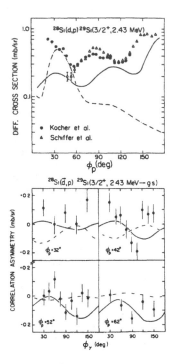

$$|3/2> = \sqrt{0.005}\,|n_{d_{3/2}} \otimes 0^+> - \sqrt{0.995}\,|n_{d_{3/2}} \otimes 2^+>.$$

It gives a more reasonable representation
of the general features both of the PGAC
data and of the cross section data after
the compound contribution is accounted for.

The wavefunction for the ^{29}Si(2.03 MeV)
level is in fairly good agreement both
with amplitudes and phases of the theoretical wavefunction[7]. Our result for the
2.43 MeV level, however, disagrees with
the theoretical prediction. The coeffici-
ent of the $|n_{d_{3/2}} \otimes 0^+>$ piece of the

Fig. 2: Polarized PGAC and
cross section for
^{28}Si($\vec{d}$,pγ)^{29}Si(2.43 MeV).
Solid curve: CCBA, dashed-
dotted curve: DWBA.

theoretical wavefunction (α=.15) is obviously much larger than the experimental result[6]. Furthermore our calculations indicate that the
principle excited core piece of the wavefunction is $|n_{d_{3/2}} \otimes 2^+>$, while
that of the theoretical wavefunction is $|n_{s_{1/2}} \otimes 2^+>$.

†NATO Fellow
 *Supported by the National Science Foundation
**Supported by the Energy Research and Development Administration
1)R.N. Boyd, J. Kaminstein, H. Clement and R. Arking, to be published in
   Phys. Rev.
2)T.B. Clegg, G.A. Bissinger and T.A. Trainor, Nucl. Instr. & Meth. 120
   (1974) 445
3)P.D. Kunz, University of Colorado, 1973, unpublished
4)J.P. Schiffer, L.L. Lee, Jr., A. Marinov and C. Mayer-Boricke, Phys.
   Rev. 147 (1966) 829
5)D.C. Kocher, P.J. Bjorkholm and W. Haeberli, Nucl. Phys. A172(1971)663
6)M.C. Mermaz, C.A. Whitten, Jr., J.W. Champlin, A.J. Howard and D.A.
   Bromley, Phys. Rev. C4 (1971) 1778
7)B. Castel, K.W.C. Stewart and M. Harvey, Can. Journ. Phys. 48 (1970)
   1490

POLARIZED BEAM ANGULAR CORRELATION MEASUREMENTS IN ^{41}Ca[*]

C. R. Gould and D. R. Tilley
North Carolina State University and TUNL[+]
C. Cameron, R. D. Ledford and N. R. Roberson
Duke University and TUNL[+]
T. B. Clegg
University of North Carolina and TUNL[+]

The structure of the two $3/2^-$ states in ^{41}Ca at 1943 and 2463 keV has long been of interest.[1,2] The simple shell model predicts just one $3/2^-$ state, having the configuration $p_{3/2}$ coupled to the ^{40}Ca core. Both the 1943 and 2463 keV states are strongly populated in the ^{40}Ca$(d,p)^{41}$Ca reaction however, indicating significant single particle configurations in each level. The γ-ray decay properties of these levels are very different though. The 1943 keV level decays to the $7/2^-$ ground state with a strong BE2 transition of 65 e^2fm^4. The 2463 keV level has no observable ground state decay[3] (BE2 $<$ 0.06 e^2fm^4) but instead decays solely to the 1943 keV level via a low energy transition.

In this note we report measurements of the mixing ratios δ, of the decays of these $3/2^-$ states. We utilize the technique of particle γ-ray angular correlations induced by tensor polarized beams of deuterons in a collinear geometry. These mixing ratios have not previously been accessible to measurement in the ^{40}Ca$(d,p\gamma)^{41}$Ca reaction[4,5] because of the non zero spin of the deuteron. The theoretical correlation formula in this case depends on both the mixing ratio and the unknown ratio of the population of the $|m| = 3/2$ and $|m| = 1/2$ magnetic substates. Since the experimental correlation for the decay from a $j = 3/2$ state yields only one number (the coefficient of the second order Legendre polynomial), it is clear that, in general, an unpolarized beam correlation experiment cannot provide unique solutions for the mixing ratio. By combining the results obtained with deuteron beams having two different tensor polarizations, a correlation can be derived which depends only on the mixing ratio of the decay of interest and which can be analyzed without invoking any unknown population parameter in the fitting procedure.[6,7]

The experiments were performed with 30nA beams of 4.04 MeV tensor polarized deuterons from the TUNL Lamb shift polarized ion source.[8] The protons from the ^{40}Ca$(d,p\gamma)^{41}$Ca reaction were detected at $\sim$180° in a 500 μm annular silicon detector. The coincident γ-rays were detected at angles 30°, 45°, 60° and 90° in four integral NaI(Tℓ) detectors. Measurements were alternately made with beams of polarizations $p_z = p_{zz} = 0.69 \pm 0.02$ and $p_z \sim 0$, $p_{zz} = -1.38 \pm 0.04$ respectively. These polarizations were measured by the quench ratio method[8] and equal amounts of beam on target were accumulated in each polarization state.

The correlation results for the 2463 $\rightarrow$ 1943 transition are shown in Fig. 1. The derived correlation[7]

$$W(\theta) = W_\alpha(\theta) - \frac{2 + p_{zz}^\alpha}{2 + p_{zz}^\beta} \; W_\beta(\theta)$$

is shown with open circles. ($W_{\alpha,\beta}(\theta)$ is the experimental correlation obtained with the incident beam having tensor polarization $p_{zz}^{\alpha,\beta}$). On the right, the plot of χ^2 versus arctan δ indicates there are two

808

possible solutions for the mixing ratio: $\delta = -0.11 \pm 0.07$ or -2.6 ± 0.5. The smaller value is within two standard deviations of zero and is consistent with pure M1 for the transition. The larger value implies the 2463 → 1943 decay is mostly E2. This seems less reasonable for such a low energy decay but cannot be ruled out on the basis of transition strength systematics. The lifetime of the 2463 keV level is known[3] only to be greater than 1.5 ps. The lack of broadening of the coincident time peak in our work indicates $\tau < 5$ ns for this level.

Our correlation results for the 1943 → 0, $3/2^- \to 7/2^-$ transition indicate $\delta = -0.17 \pm 0.14$ or $\delta > 2$. The small value is again within two standard deviations of zero and is consistent with the expected value of $\delta = 0$ for a pure E2 transition. This non zero result may indicate the presence of a small systematic error in these data.

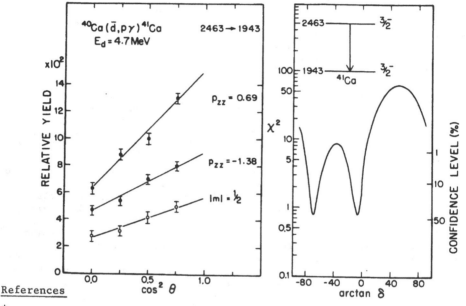

References

* Work supported in part by USERDA.
+ Triangle Universities Nuclear Laboratory, Duke Station, Durham, N. V. 27707, U.S.A.

[1] W. J. Gerace and A. M. Green, Nucl. Phys. A93 (1967) 110.
[2] P. Fedeman, G. Greek and E. Osnes, Nucl. Phys. A135 (1969) 545.
[3] H. Laurent, S. Fortier, J. P. Schapira, R. S. Blake, F. Picard, J. Dalmas, Nucl. Phys. A164 (1971) 279.
[4] G. Johnson, R. S. Blake, H. Laurent, F. Picard, and J. P. Schapira, Nucl. Phys. A143 (1970) 562).
[5] L. C. McIntyre, Phys. Rev. C9 (1974) 200.
[6] J. D. McCullen and R. G. Seyler, Nucl. Phys. A139 (1969) 203.
[7] C. R. Gould, R. O. Nelson, J. R. Williams, D. R. Tilley, J. D. Hutton, N. R. Roberson, C. E. Busch and T. B. Clegg, Phys. Rev. Lett. 30 (1973) 298.
[8] T. B. Clegg, G. A. Bissinger and T. A. Trainer, Nucl. Inst. & Meth. 120 (1974) 445.

Angular Correlation Measurement for $^{13}C(^{3}He,\alpha,\gamma)^{12}C$(4.43 MeV) Reaction

T. Fujisawa, H. Kamitsubo, T. Wada, M. Koike*
Y. Tagishi** and T. Kanai***
The Institute of Physical and Chemical Research,
Wako-shi, Saitama 351, Japan

Angular correlations of α particles leading to the first excited state of ^{12}C with following gamma rays were measured for $^{13}C(^{3}He,\alpha)^{12}C*$ reactions at the bombarding energy of 29.2 MeV. Particle distributions in coincidence with gamma rays emitted parpendicularly to the reaction plane will give information on the reaction mechanism. An example of this technique is so-called spin-flip measurement in inelastic scattering of spin non-zero particles from even nuclei[1]. The purpose of the present experiment is to apply this technique to the transfer reactions and to see how it is useful to study the reaction mechanism.

The symmetry property of the reaction matrix[2] or the Bohr theorem [3] give the relation in reaction A(a,b)B as

$$M_A - M_B + m_a - m_b = even/odd \qquad as \quad \Delta\pi = even/odd \qquad (1)$$

where M_A and M_B are z components of the spins of nuclear states A and B, m_a and m_b are those of particles a and b, respectively, and $\Delta\pi(=\pm1)$ is the overall parity change in the reaction. From Eq. (1) there is a simple spin relation in the reaction $^{13}C(^{3}He,\alpha)^{12}C*$(4.43MeV,$2^+$) and Table 1 gives the relation.

Ions of ^{3}He were accelerated at the IPCR Cyclotron. Energy spread of the beam was about 0.1%, its horizontal divergence was within 1 deg. and a size of the beam spot on the target was 2mm x 2mm. Beam intensity was kept below 10 nA in order to reduce the chance coincidence. The target was a self-supported foil of ^{13}C and the thickness was about 700 μg/cm^2 and purity of ^{13}C in the target was 90%. Emitted particles were detected by a surface barrier detector of 2000μm thick and gamma rays were measured with a 7.6cm x 7.6cm NaI(Tl) crystal mounted on a 56AVP photo-multiplier. The gamma detector was placed perpendicular to the reaction plane and well shielded with a lead cylinder. The solid angles of both detectors were large so that the coincidence efficiency was much improved.

The single spectra of α particles were measured at the same time as the coincident ones with 4.43 MeV gamma rays. Then a ratio of coincident yields to total ones was obtained for α particles, α_1, going to the first 2^+ state of ^{12}C. In case of the inelastic scattering this ratio corresponds

to the spin-flip probability.   Fig.1 shows the angular dependence of this
ratio and the angular distribution of α_1 obtained in the present work.   As
can be seen in the figure, the ratio has a weak angular dependence and
increases at backward angles.   Because of a large solid angle of the
particle detector, angular uncertainty is about $\pm$ 5 deg.

The ratio should be very sensitive to the distorting potential, to
spin dependence of the interaction which causes the transfer of the nucleon
and to the reaction mechanism such as two-step process. DWBA analysis[4] is
now in progress.

References
* permanent adress; Inst. for Nucl. Study, Univ. of Tokyo
**permanent adress; Dept. of Phys., Tsukuba Univ.
***permanent adress;Natl. Inst. of Radiological Sci.

1) F.H.Schmidt, R.E.Brown, J.B.Gerhart and W.A.Kalosinski: Nucl.Phys.52
   (1964) 353
   S.Kobayashi, S.Motonaga, Y.Chiba, K.Katori, A.Stricker, T.Fujisawa and
   T.Wada,   J. Phys. Soc. of Japan 29(1970) 1

2) F.Rybicki, T.Tamura and G.R.Satchler, Nucl. Phys. A146(1970) 659

3) A.Bohr, Nucl. Phys. 10(1959) 486

4) E.M.Kellogg and R.W.Zurmuhle, Phys. Rev. 152(1966) 890

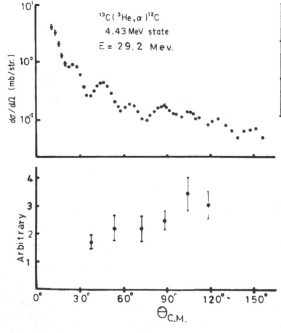

spin direction	M_B	γ-ray $\perp$
parallel	0 or ± 2	not emitted
anti-parallel	± 1	emitted

Table 1

Fig.1 Angular correlations
for $^{13}C(^{3}He,\alpha\gamma)^{12}C$ and differential cross sections of
$^{13}C(^{3}He,\alpha)^{12}C*(4.43$ MeV$)$.

^{24}Mg$(\alpha,\alpha'\gamma)$ ANGULAR CORRELATION MEASUREMENTS
AT 104 MEV AS A TEST FOR THE QUADRUPOLE DEFORMATION

W. Eyrich, M. Berg, A. Hofmann, H. Rebel[*], U. Scheib,
S. Schneider, and F. Vogler
Physikalisches Institut der Universität Erlangen, Germany

In particle-γ angular correlations interference terms appear
which are composed of the reaction amplitudes, referring to the
different magnetic substates of the residual excited state.
(Similar interference terms, though referring to different projectile substates, contribute to the analysing power in reactions with polarized beams.) Therefore correlation experiments
as well as experiments with polarized projectiles provide more
insight into the reaction mechanism than differential cross
section measurements.

Analyses of scattering cross sections of 104 MeV α-partic-
les from sd-shell nuclei have shown [1] that coupled channels
analyses on the basis of the rotational model allow in principle the extraction of the sign and magnitude of the deformation parameters. From $(\alpha,\alpha_1\gamma)$ angular correlation measurements
on even-even nuclei the reaction amplitudes can be determined
separately (see for example ref. [2]). Since these amplitudes are
more sensitive to assumptions of the reaction model than the differential cross section [2], one expects that especially variations of deformation parameters yield more pronounced effects
in the reaction amplitudes than in the differential cross section.
Therefore $(\alpha,\alpha_1\gamma)$ measurements on ^{24}Mg have been done at the
Karlsruhe cyclotron at a bombarding energy of 104 MeV. The partic-
le-γ-coincidence measurements were performed in "in-plane" geometry by use of four Si(Li) detectors for particle detection and
two Ge(Li) γ-detectors. The considerable γ-ray background, due
to the high incident energy, could be sufficiently reduced. Double
differential cross sections were measured at 14 different particle
scattering angles ($8^{\circ} \leqslant \emptyset_{\alpha,\text{C.M.}} \leqslant 40^{\circ}$) and at 6 different positions of the γ-detector; i.e. 84 points of the correlation func-
tion have been determined.

From the angular correlation function and the differential

cross section one can extract the absolute squares of the reaction amplitudes X_0, X_2, X_{-2} and the relative phase between X_2 and X_{-2} as a function of the particle scattering angle (the indices refer to the magnetic substates of the residual excited state; the z-axis is chosen perpendicular to the scattering plane).

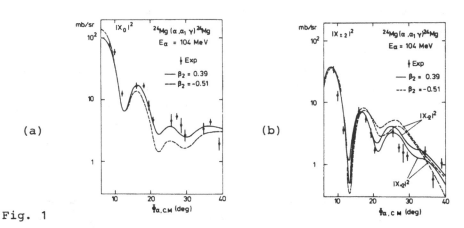

(a)   (b)

Fig. 1

Experimental values of the absolute squares of the reaction amplitudes $\left|X_0\right|^2$ (a) and $\left|X_{\pm 2}\right|^2$ (b) compared with coupled channels calculations for prolate and oblate deformation.

In Fig. 1 (a) the experimental value of $\left|X_0\right|^2$ are shown as a function of the scattering angle, and are compared with coupled channels calculations for prolate and oblate deformation. The parameters used are the best fit values for the differential cross section of ref. [1] for prolate and oblate deformations, respectively.

Fig. 1 (b) shows the corresponding calculations for $\left|X_2\right|^2$ and $\left|X_{-2}\right|^2$. The experimental value of $\left|X_2\right|^2$ and $\left|X_{-2}\right|^2$ turned out to be equal within the experimental errors. From both figures it is obvious that only the assumption of a prolate deformation is in agreement with the experiment.

*) Kernforschungszentrum Karlsruhe, Zyklotronlabor

1) H. Rebel et al., Nucl. Phys. A 182 (1972) 669
2) H. Wagner, A. Hofmann, and F. Vogler,
   Phys. Lett. 47 B (1973) 497

Particle-particle angular correlations as a tool for
analyzing nuclear polarization.

W.C. Hermans, M.A.A. Sonnemans, J.C. Waal, R. van Dantzig
(IKO, Amsterdam), B.J. Verhaar (T.H. Eindhoven)

In a nuclear reaction A(a,b)B both the reaction-
product b and the residual nucleus B will in general be
polarized. Direct measurement of the polarization of the
light particle b is, though difficult, often possible;
this is usually not the case for the residual nucleus B.
If, however, B betrays itself by decaying afterwards,
information concerning its polarization can be obtained
by studying the angular distribution of the decay [1].
In most cases, studied so far the decay is a γ-emission.
The spin of the γ often complicates the analysis. A
much cleaner case arises if B decays by α-particle
emission, leaving behind a spin-zero nucleus. In practice
these cases are restricted to lighter nuclei, e.g. B=^{8}Be,
^{12}C etc. Even then it is not, in general, possible to
obtain complete information on the polarization of B.
This would either require a full knowledge of the polarizations of a and b, or these particles should be spinless.
In experiments with unpolarized beams and non-determined
polarization of the reaction product, one still obtains partial, "summed" information on the population of
the magnetic substates of B, which can serve to test
ideas on nuclear structure or reaction mechanisms.

A good example to apply these considerations to is
the reaction ^{9}Be(d,t) ^{8}Be(16.9 MeV, 2^{+})$\rightarrow \alpha+\alpha$. The excited state of ^{8}Be is a well-defined resonance. Angular
correlations are most adequately described in the so-
called recoil centre of mass (RCM) coordinate system,
in which the decaying nucleus is at rest. Angular
correlation patterns for a close grid of $(\theta_\alpha, \phi_\alpha)$-values
in the RCM system for a number of triton scattering
angles θ_t were measured with the BOL-system at IKO [2].
The angular correlation function can be expanded:

$$W(\theta_t; \theta_\alpha, \phi_\alpha) = \sum_{JM} c_{JM}(\theta_t) Y_{JM}(\theta_\alpha, \phi_\alpha).$$

Various symmetries restrict the summations to J=0,2,4
and 0$\leq$M$\leq$J, M even. The coefficients have been extracted
from experiment by a least squares fit. As examples,
c_{20} and $|c_{22}|$ are shown in figures 1,2. The angular
dependence of $|c_{22}|$ is similar to that of the angular
distribution (c_{00}), the magnitude is 1/5. The coefficients c_{4M} are much smaller, though different from zero.

The coefficients can be expressed in terms of the
T-matrix elements of the pick-up reaction and in this
way the vast amount of data points can be confronted
with theory. If one assumes a one-step DWBA mechanism
for the pick-up process, interesting conclusions can
be drawn about the state in ^{9}Be from which the neutron
is picked-up. If this state is assumed to be a pure
$p_{3/2}$ or $p_{1/2}$ single particle state, the angular

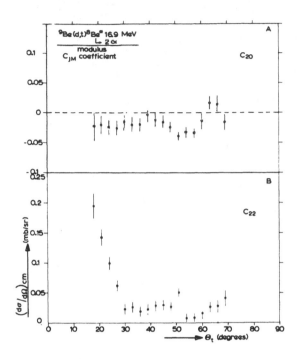

^{9}Be(d,t)^{8}Be* 16.9 MeV
$\rightarrow$ 2α
modulus
c_{JM} coefficient

correlation pattern has been shown to be isotropic [3]. If the $p_{3/2}$ and $p_{1/2}$ states occur mixed, still the c_{4M} (M=0,2,4) should be zero. The experiment clearly shows non-isotropy, indicating a mixture of $p_{3/2}$ and $p_{1/2}$, and possibly a small admixture of $f_{7/2}$ or other states. Besides, a deviation from the one-step description could complicate the picture.

1) J.G. Cramer, W.W. Eidson, Nucl. Phys. 55 (1964) 593
2) M.A.A. Sonnemans, Multiparticle reactions induced
   by deuterons on Be, Thesis, Amsterdam 1974
3) W.C. Hermans, c.s., Proceedings Int. Conf. on Nuclear
   Physics, Munich 1973.

STUDY OF UNBOUND STATES IN ^{41}Sc[*]

P. Viatte, S. Micek[†], R. Müller, J. Lang, J. Unternährer,
C.M. Teodorescu[††] and L. Jarczyk[†]
Laboratorium für Kernphysik, Eidg. Techn. Hochschule
8049 Zürich, Switzerland

The decay angular distribution gives for particle unstable states an indication of the degree of polarization, and thus of the formation mechanism and spin of the short-lived decaying nucleus. Particle-particle correlation experiments therefore present an alternative tool to usual polarization measurements for the investigation of unbound states.

Measurements of stripping reactions to unbound states and their analyses in the framework of the DWBA have become a matter of increasing interest in recent years. This is not only because of the information obtained regarding the structure of comparatively high excited states, but especially because a comparison of elastic scattering and stripping reaction checks the accuracy of the DWBA analysis[1].

In this experiment levels in ^{41}Sc were investigated by proton-neutron correlation measurements of the ^{40}Ca(d,n)^{41}Sc*(p)^{40}Ca reaction performed at an incident deuteron energy of 11 MeV. Coincident events between the outgoing neutrons detected in a NE213 liquid scintillator and protons observed with a surface barrier detector were recorded[2]. In the spectra of the outgoing particles about 25 resonances corresponding to proton unbound levels in ^{41}Sc were observed. For seven levels the angular correlation $d\sigma^2/d\Omega_n d\Omega_p$ was determined for fixed neutron angle (25°) and various proton angles.

The experimental data could be interpreted successfully in the framework of the DWBA. The calculations were carried out with the code VENUS[3] modified for the treatment of resonant states using the method of Vincent and Fortune[1] and extended to permit calculation of the particle-particle correlation function.

By comparison of the shapes of the measured and calculated p-n correlation curves it was possible to assign spin and parities to 7 of the observed states in ^{41}Sc in an excitation energy range from 3 to 8 MeV. The ratio of measured to calculated cross sections yielded the resonance widths, which are for all cases in excellent agreement with those obtained from measurements of the elastic proton scattering[4]. Examples are given in figure 1.

The assumption that the studied states in ^{41}Sc are populated through a stripping process appears therefore to be adequat and this mechanism seems to be accurately described by the DWBA formalism.

References

* Supported by the Schweizerischer Nationalfonds
† On leave from Jagellonian University, Cracow, Poland
††On leave from Institute for Atomic Physics, Bucarest, Romania

1)H.T. Fortune, C.M. Vincent, Phys. Rev. 185 (1969) 1401
2)J. Unternährer, R. Müller (submitted to Nucl. Instr. and Meth.)
3)T.Tamura, W.R. Coker, F. Rybicki, Comp. Phys. Comm. 2 (1971) 94
4)P.M. Endt and C. Van der Leun, Nucl. Phys. A214 (1973) 481
  W. Mittig, Thesis, University of Paris-Sud, 1971 (unpublished)

816

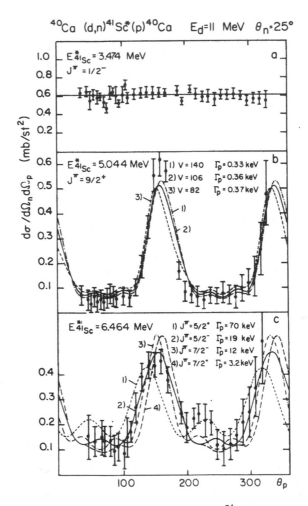

Fig. 1. The proton-neutron correlations $d\sigma^2/d\Omega_n d\Omega_p$ for 3 states in ^{41}Sc, studied by the reaction ^{40}Ca(d,n)^{41}Sc*(p)^{40}Ca at Ed = 11 MeV are plotted as a function of the proton angle (for fixed θ_n). The 3 curves in b show that the optical model parameters are uncritical. In figure c, different spin-parity assumptions are used for the DWBA calculations, curve 2 giving the best χ^2-value.

ON THE POLARIZATION OF INTERMEDIATE PARTICLES IN SEQUENTIAL DECAY

Albrecht Lindner

1. Institut für Theoretische Physik der Universität Hamburg

It will be shown for two-step-processes $a + b \to \vec{c} + d \to c_1 + c_2 + d$, that the even polarization tensors of particles c are intimately connected with the angular correlation of the final products. This yields a new method to measure the polarization of reaction products.

If we describe the directions of propagation in the c.m. system by Ω_i and Ω_f for the reaction $a + b \to c + d$ and by Ω for the relative motion $c_1 - c_2$, the cross section can be expanded in terms of Biedenharn's triple correlation functions[1]:

$$\frac{d^5\sigma}{d\Omega_f\, d\Omega\, dE} = \frac{\pi}{4\, k_i^2\, \hat{s}_a^2\, \hat{s}_b^2} \sum_{n_i, n_f, n} \hat{n}_i\, \hat{n}_f\, \hat{n}\; B_{n_i n_f n}\; P_{n_i n_f n}\left(\Omega_i, \Omega_f, \Omega\right). \quad (1)$$

(This eq. generalizes the well-known expansion in terms of Legendre polynomials applicable for two-particle reactions to the case of three-particle reactions). The coefficients B can be determined by experiment[2].

From the coefficients B we can derive the polarization of the intermediate particles c. This is especially simple if we express the polarization in terms of spherical invariants as proposed by the author[2]

$$t^{(n_c)}(\Omega_c) = \sum_{n_f, n_i} \hat{n}_f\, \hat{n}_i\, \hat{n}_c\; b_{n_f n_i n_c}\; P_{n_f n_i n_c}\left(\Omega_f, \Omega_i, \Omega_c\right), \quad (2)$$

but is also possible in terms of the polarization tensors recommended by the Madison convention.

$$t^{(n_c)}_{\nu_c} = \frac{\hat{n}_c}{\sqrt{4\pi}} \int d\Omega_c\; t^{(n_c)}(\Omega_c)\; Y^{(n_c)}_{\nu_c}(\Omega_c) \quad (3)$$

$$= i^{-n_c} \sum_{n_f, n_i} (-)^{\frac{n_f + n_i}{2}} \begin{pmatrix} n_f & n_i & n_c \\ 0 & \nu_c & -\nu_c \end{pmatrix} \sqrt{\frac{(n_i - \nu_c)!}{(n_i + \nu_c)!}}\; \hat{n}_f\, \hat{n}_i\, \hat{n}_c\; b_{n_f n_i n_c}\, P^{\nu_c}_{n_i}(\cos\theta).$$

It is the aim of this note to show the connection between the real coefficients b and B.

In the case of a two-step-process $a + b \to c + d \to c_1 + c_2 + d$, the transition matrix factorizes into a transition matrix of the process $a + b \to c + d$ and an amplitude A of the decay process $c \to c_1 + c_2$:

$$\left\langle \left(E_f l_f\left((E\, 1(s_1 s_2) s) s_c\, s_d\right) s_f\right) \mathcal{F}\, \middle|\, T\, \middle|\, \left(E_i l_i (s_a s_b) s_i\right) \mathcal{F} \right\rangle \quad (4)$$

$$= \left\langle \left(E\, 1(s_1 s_2)\, s\right) s_c \middle| A \middle| s_c \right\rangle \left\langle \left(E_f l_f (s_c s_d) s_f\right) \mathcal{F}\, \middle|\, T\, \middle|\, \left(E_i l_i (s_a s_b) s_i\right) \mathcal{F} \right\rangle.$$

Comparing the expressions for B and b in this channel spin representation - cf.[1,2] - we find the one-to-one correspondence

$$B_{n_i n_f n} = b_{n_f n_i n} \times \left(\frac{2 k_i \hat{s}_a \hat{s}_b}{k_f}\right)^2 \hat{s}_c \sum_{s,l,l'} (-)^{s+s} \hat{l} \hat{l}' \begin{pmatrix} l\,l' & n \\ o\,o & o \end{pmatrix} \begin{Bmatrix} l\,l'\,n \\ s_c s_c s \end{Bmatrix} \times \quad (5)$$

$$\times \left\langle (l_{(s_1 s_2)} s) s_c \middle| A \middle| s_c \right\rangle^* \left\langle (l'_{(s_1 s_2)} s) s_c \middle| A \middle| s_c \right\rangle.$$

Here, the index n as well as the sum $n_f + n_i$ must be even if parity is conserved. Therefore, all polarization tensors of even rank are connected with the angular correlation in the cross section.

In order to evaluate the polarization tensors with the help of the angular correlation one apparently needs the break-up amplitudes A. Affairs are most simple if there exists only one such amplitude. In this case only a normalization constant is missing, which anyhow must be introduced if the cross section has not been measured absolutely. This simplification appears if the spins of particles c_1 and c_2 only allow their sum $s < 1/2$ (or $s = 1$ but $\pi_1 \pi_2 \pi_c = (-1)^{s_c}$ such that $l=l'=s$). As examples I may mention the reaction $d + \alpha \rightarrow \alpha + n + p$ proceeding via ^{5}He$(3/2^-)$ or via ^{5}Li$(3/2^-)$ or reactions proceeding via ^{8}Be$^*(2^+,4^+)$ which have been measured by some authors[3,4] in complete experiments.

In other cases, however, one cannot avoid to evaluate the break-up amplitudes of the decay $c \rightarrow c_1 + c_2$. This is not surprising since one compares the processes $a + b \rightarrow c + d$ and $a + b \rightarrow c_1 + c_2 + d$.

Finally it should be mentioned that eq.(5) can be used not only to infer new polarization data from the angular correlation but also to extract angular correlations in sequential decay processes from calculations of two-particle reaction theories. This has been done e.g. by P. Heiss[5].

I gratefully acknowledge fruitful discussions with P. Heiss.

References

1) A. Lindner, Nucl. Phys. A199 (1973) 110
2) A. Lindner, Nucl. Phys. A230 (1974) 477
3) T. Tanabe, J. Phys. Soc. Japan 25 (1968) 21
   R.E. Warner and R.W. Bercaw, Nucl. Phys. A109 (1968) 205
   P.A. Assimakopoulos, E. Beardsworth, D.P. Boyd and P.F. Donovan, Nucl. Phys. A144 (1970) 272
   T. Rausch, H. Zell, D. Wallenwein and W.v.Witsch, Nucl. Phys. A222 (1974) 429
   E. Hourany, H. Nakamura, F. Takéutchi and T. Yuasa, Nucl. Phys. A222 (1974) 537
   K. Prescher (Univ. Köln), to be published
4) J.D. Bronson, W.D. Simpson, W.R. Jackson and G.C. Phillips, Nucl. Phys. 68 (1965) 241
   J.M. Lambert, P.A. Treado, D. Haddad, R.A. Moyle and J.C. Sessler, Phys. Rev. Lett. 27 (1971) 820
   P.A. Treado, J.M. Lambert, V.E. Alessi and R.J. Kane, Nucl. Phys. A198 (1972) 21
5) P. Heiss, Z. Physik A272 (1975) 267

HYPERFINE STUDIES BY USE OF POLARIZED REACTIONS[*]

T. Minamisono, J. W. Hugg, D. G. Mavis, T. K. Saylor, S. M. Lazarus
H. F. Glavish and S. S. Hanna
Department of Physics, Stanford University
Stanford, California 94305 U.S.A.

A new method for studying hyperfine interactions has been developed in which polarized nuclei are produced from nuclear reactions initiated with fast polarized beams of protons and deuterons. We have demonstrated that in these reactions the net polarization transferred to all residual nuclei is large enough to make possible quantitative studies of nuclear polarization and moments.

As examples of the process, the β-emitting nuclei ^{8}Li, ^{12}B and ^{29}P have been produced and polarized by $(\vec{d},p)$ and $(\vec{d},n)$ reactions with the deuteron beam polarized perpendicular to the beam direction. The targets were Li metal, ZrB_2 and Si crystals, respectively, thick enough (100 mg/cm^2) to stop both the beam and all the recoil nuclei. The polarization transferred to the recoil nuclei was measured by detecting the resulting β asymmetries. Since all recoil nuclei are used, high β counting rates of $10^3 \rightarrow 10^4$ cps were obtained in two particle telescopes placed at 0 and 180° to the polarization direction. The incident beam was pulsed and β particles were counted only during the period between pulses. To eliminate possible instrumental asymmetries, counts were accumulated with the beam polarization alternately on and off. The experimental arrangement in which the measurements were made is shown in fig. 1, and the results are plotted in fig. 2. It is seen that for $^7\text{Li}(\vec{d},p)^8\text{Li}$ and $^{11}\text{B}(\vec{d},p)^{12}\text{B}$ the net polarization transferred to all residual nuclei (with a beam polarization $P_z = 0.6$) produces an up-down asymmetry of about 9% in the β-decay, which is fairly independent of energy for the thick targets used. It is interesting that in the case of $^{28}\text{Si}(\vec{d},n)^{29}\text{P}$ the polarization transferred is opposite (and smaller) than in the other examples. This polarization transfer mechanism has also been observed in the reactions $^{31}\text{P}(\vec{p},n)^{31}\text{S}$ and $^{39}\text{K}(\vec{p},n)^{39}\text{Ca}$ in thick targets of P_4S_7 and $KCaF_3$.

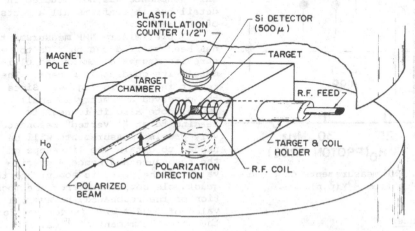

Fig. 1. Experimental set-up used to detect β asymmetries and measure NMR of polarized nuclei produced in reactions with polarized beams.

820

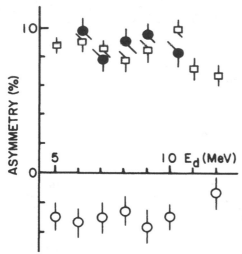

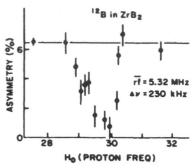

Fig. 3. NMR measurements on polarized ^{12}B from ^{11}B($\vec{\text{d}}$,p)^{12}B.

Fig. 2. β-asymmetry from ^{7}Li($\vec{\text{d}}$,p)^{8}Li (closed circles), ^{11}B($\vec{\text{d}}$,p)^{12}B (squares) and ^{28}Si($\vec{\text{d}}$,n)^{29}P (open circles).

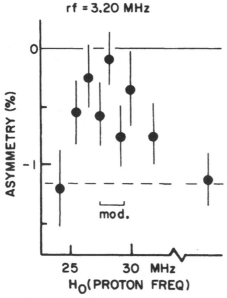

Fig. 4. NMR measurements on polarized ^{39}Ca from ^{39}K($\vec{\text{p}}$,n)^{39}Ca.

With the proton beam measurements could be made with spin up and down. The polarized nuclei produced by this technique have been used to study nuclear-spin relaxation phenomena and to measure nuclear moments by means of NMR detection[1] in the setup of fig. 1. The study made of ^{8}Li is reported in a separate contribution to this conference. In the case of ^{12}B for which the nuclear moments are known, an NMR measurement was carried out as a test of the method. The resonance obtained is shown in fig. 3 where it can be seen that the β-asymmetry, i.e. polarization of ^{12}B, is almost completely destroyed at resonance. This resonance was not studied in detail but it confirms all aspects of the method.

A preliminary NMR measurement has been carried out on ^{39}Ca to measure its magnetic moment and ultimately its quadrupole moment. The result is shown in fig. 4. Since the observed asymmetry is small ($\approx 1\%$, note also it is negative leading to an "inverted" resonance) very precise measurements will be needed to obtain the lineshape of the resonance and hence a precise value of the magnetic moment and the quadrupole coupling. From the location of the resonance, however, a value of 0.94 ± 0.10 is derived for the magnetic moment of ^{39}Ca.

References

* Supported in part by the National Science Foundation
1) K. Sugimoto et al., Phys. Lett. <u>18</u> (1965) 38

MEASUREMENT OF THE QUADRUPOLE MOMENT OF ^{8}Li BY USE OF A POLARIZED BEAM AND NMR DETECTION*

J. W. Hugg, T. Minamisono, D. G. Mavis, T. K. Saylor, S. M. Lazarus,
H. F. Glavish and S. S. Hanna
Department of Physics, Stanford University
Stanford, California 94305 U.S.A.

In a separate contribution to this conference (T. Minamisono et al.) the method of producing polarized nuclei with polarized beams is described. In this paper we report an application of the method to a measurement of the quadrupole moment of ^{8}Li.

The ^{8}Li nuclei ($I^\pi = 2^+$, $T_{1/2} = 0.84$ sec) were produced and polarized by initiating the reaction ^{7}Li(d,p)^{8}Li with vector polarized deuterons ($P_z = 0.6$) from the Stanford FN tandem Van de Graaff accelerator. Since recoil angle selection was not required, a thick target of LiIO$_3$ (hexagonal single crystal) was used which produced a β-particle counting rate from ^{8}Li as high as 10^3 per sec for an incident beam current of only 10 nA. The polarization of the deuteron beam was transverse to the beam direction, and a holding field H$_0$ of 1.63 kG was applied along the polarization direction [see fig. 1(a) for measurements of the polarization vs. holding field]. The β-particles from the ^{8}Li decay were detected by two particle telescopes placed at 0° and 180° to the polarization direction. The LiIO$_3$ target was exposed to the beam for periods of 1 sec every 4 sec and β-particles were counted only with the beam off. The polarization of the deuteron beam was set equal to zero on alternate cycles, so that the absolute β-particle asymmetry could be determined independent of instrumental asymmetries.

The asymmetry in the β-decay from polarized ^{8}Li was measured as a function of time following each pulse so as to measure the spin-lattice relaxation time T$_1$. Typical results for E$_d$ = 9 MeV are shown in fig. 1(b). The initial asymmetry was determined to be 0.035 ± 0.002 with a relaxation time of T$_1$ ≈ 11 sec which is much longer than the half-life of ^{8}Li. The large value of T$_1$ demonstrated the usefulness of the LiIO$_3$ crystal for recoil implantation. The initial ^{8}Li polarization measured in LiIO$_3$ was about 1/3 of that measured in Li metal. This suggests that 1/3 of the recoils find substitutional Li sites, where the electric field gradient is known to be axially symmetric in a direction parallel to the crystal c-axis[1]. To measure the quadrupole moment of ^{8}Li[2] an rf magnetic field was applied perpendicular to the holding field which served as the NMR field H$_0$. If a "small" quadrupole interaction is present in addition to the "large" magnetic interaction with H$_0$, one expects to observe four resonances of slightly different frequencies. (See fig. 2.) Each resonance line was detected by scanning with the rf frequency, while the other three transitions were saturated by modulated rf signals. The four NMR lines observed for the case where the crystal c-axis was parallel to H$_0$ are shown in fig. 2. The solid curves in fig. 2 are the best fits which take into account the intrinsic Gaussian line shape and the level of the rf intensity. At the center of each resonance it is seen that the polarization is completely destroyed when the other transitions are saturated. This result shows that the nuclear spin is I = 2, that the field gradient is unique, and that the ^{8}Li nuclei which keep their polarization sit in equivalent sites in the LiIO$_3$ crystal. The intrinsic line broadening obtained from the line-shape analysis was consistent with the

822

dipolar broadening estimated for ^{8}Li in this crystal. Thus, broadening produced by radiation damage was small, although a holding field of about 1 kG was necessary to decouple the time-dependent interactions due presumably to radiation damage. [See fig. 1(a).] The electric quadrupole coupling constant of ^{8}Li in LiIO$_3$, as determined from the resonant frequencies in fig. 2, is $|eqQ| = 29.2 \pm 0.8$ kHz. Using the known[1] coupling constant of ^{7}Li, we obtain the ratio of ground state quadrupole moments

$$\left| Q(^8\text{Li})/Q(^7\text{Li}) \right| = 0.66 \pm 0.05$$

in fair agreement with the result obtained using polarized neutron capture by Ackerman et al.[2].

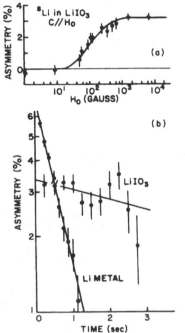

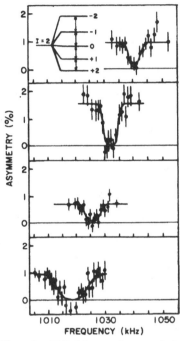

Fig. 1. (a) Polarization of ^{8}Li vs. holding field H_0; (b) Polarization vs. time in metal and LiIO$_3$.

Fig. 2. NMR lines observed for polarized ^{8}Li in LiIO$_3$.

References

* Supported in part by the National Science Foundation
1) V. M. Sarnatskii, V. A. Shutilov, T. D. Levitskaya, B. I. Kidyarov, and P. L. Mitnitskii, Sov. Phys. Solid State 13 (1972) 2021
2) H. Ackerman, D. Dubbers, M. Grupp, P. Hertjans and H. -J. Stockman, Phys. Lett. 52B (1974) 54

Polarized Heavy Ions

ELASTIC SCATTERING OF VECTOR POLARIZED ^{6}LI ON ^{4}HE

P.Egelhof, J.Barrette[+)], P.Braun-Munzinger, D.Fick. C.K.Gelbke, D.Kassen
and W.Weiss, Max-Planck-Institut für Kernphysik, Heidelberg

E.Steffens, I.Institut für Experimentalphysik der Universität Hamburg

The elastic scattering of ^{6}Li from ^{4}He has been subject of recent investigations[1,2]. In all cases the observed angular distributions of the
differential cross section show pronounced structures at intermediate and
backward angles. These structures have been interpreted[1,2] as due to the
elastic transfer of a deuteron between the two ^{4}He-cores. In order to obtain a more detailed information on the reaction mechanism a study of the
vector analyzing power for this reaction had been started using the polarized ^{6}Li beam of the Heidelberg EN-tandem[3]. The experiment has been performed at 22.8 MeV beam energy and scattering angles of $35 < \Theta_{cm} < 150^\circ$.
As ^{4}He target a gas cell was used which was specifically constructed to
allow for measurements of the left-right asymmetry. The entrance and exit
windows consisted of Havar-foils of 2.1 mg/cm^2. The target has been
operated at pressures around 500 torr ^{4}He corresponding to an effective
target thickness between 30 and 80 µg/cm^2. Taking into account the energy
loss in the entrance foil and in the gas target the beam energy at the
active volume of the gas cell is $E_{lab} \approx 21.3$ MeV. The scattered ^{6}Li- and
the rescattered ^{4}He-nuclei have been detected with ΔE-E telescopes.
Relative and absolute normalization of the data has been obtained by
measuring the Rutherford scattering of ^{6}Li from a 3.6% Kr-admixture to
the ^{4}He-gas. The absolute cross-sections are accurate up to 20%. The
asymmetry for vector polarized ^{6}Li has been obtained by successively
measuring with a polarized and unpolarized beam. Beam polarization has
been checked to be constant throughout the experiment. The data are
presented in Fig.1. As expected the angular distributions of the differential cross-section exhibits strong oscillations. Pronounced structures
are also observed in the asymmetry, demonstrating that there are strong
polarization effects in the elastic scattering of ^{6}Li on ^{4}He. The
structures observed in the asymmetry can probably not be explained in a
simple diffraction model since they are not proportional to the derivative of the angular distribution of the differential cross-section.
Detailed calculations of the angular distributions of the differential
cross-section and the analyzing power are planned to settle the question
of elastic transfer in this reaction.Finally two points should be noted:

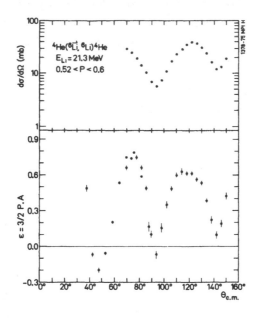

Fig.1 Angular distributions of the differential cross-section and of the
asymmetry observed in the elastic scattering of vector polarized
^{6}Li on ^{4}He at 21.3 MeV beam energy.

i)   The maximum of the asymmetry $\varepsilon = 0.79\pm0.01$ at $\theta_{cm} = 75.8^\circ$ is the high-
     est value of ε observed up to now with the vector polarized ^{6}Li-beam
     of the Heidelberg EN-tandem. This gives a lower limit of the beam
     polarization of P > 0.52.

ii) Because of the plateau of the asymmetry between scattering angles
     $110^\circ < \theta_{cm} < 125^\circ$ the elastic ^{6}Li-^{4}He scattering may be a suitable
     monitor for the vector polarization of ^{6}Li.

+)   supported by the NRC of Canada

1) H.Bohlen,N.Marquardt,W.von Oertzen,Ph.Gorodetzky,Nucl.Phys.A179(1972)504

2) M.Bernas,R.DeVries,B.G.Harvey,D.Hendrie,J.Mahoney,J.Sherman,J.Steyaert,
   M.S.Zisman,Nucl.Phys. A242 (1975) 149

3) E.Steffens,H.Ebinghaus,F.Fiedler,K.Bethge,G.Engelhardt,R.Schäfer,W.Weiss,
   D.Fick,Nucl.Instr.Meth124(1975)601 and contribution to this conference,
   p. 871.

ELASTIC SCATTERING OF VECTOR POLARIZED ^{6}LI

W.Weiss, P.Egelhof, K.D.Hildenbrand[+], D.Kassen, M.Makowska-Rzeszutko[++]

and D.Fick, Max-Planck-Institut für Kernphysik, Heidelberg

E.Steffens, I. Institut für Experimentalphysik der Universität, Hamburg

Using the vector polarized ^{6}Li-beam of the Heidelberg EN-Tandem[1] the elastic scattering of ^{6}Li on ^{12}C, ^{16}O, ^{28}Si and ^{58}Ni has been investigated at a bombarding energy of E_{Li} = 22.8 MeV. The structure of the observed angular distributions of σ/σ_R depends in heavy ion scattering critically on E/E_{coul}. Therefore the observed angular distributions of σ/σ_R display for ^{12}C ($E/E_{coul} \approx 3.5$) an oscillating pattern (fig.1) which looks like Frauenhofer diffraction scattering. The angular distribution of σ/σ_R for ^{58}Ni ($E/E_{coul} \approx 1.4$) is without any structure (fig.1) and looks like Fresnel diffraction scattering. The analyzing power had been obtained by measuring successively with a polarized and unpolarized beam (high frequency transition of the source[1] on and off respectively). The polarization of the beam is known to be between 0.52 and 0.6 . The lower bound is deduced from the highest up to now with this source observed asymmetry in the ^{6}Li-^{4}He elastic scattering[2]. The higher bound is estimated from the source parameters itself[1]. The polarization was monitored after some couple of runs either with the ^{1}H(^{6}Li,^{4}He)^{3}He reaction or, in a later stage of the experiment using the elastic scattering of ^{6}Li on ^{12}C. Around θ_{cm} = 44.5° and E_{Li} = 22.8 MeV a flat maximum of the asymmetry connected with a fairly high cross section had been observed (fig.1). For the display of the data the polarization P of the beam has been assumed to be 0.577 exactly. For this value the observed asymmetry is just equal to $< iT_{11} >$. For the scattering of vector polarized ^{6}Li on ^{12}C (Frauenhofer scattering) the observed analyzing power has an oscillating structure and reaches up to 60% of its maximum value. On the other hand the observed angular distribuiton of the analyzing power for ^{58}Ni (Fresnel scattering) is without any structure and the maximum observed values are very small. The observed angular distributions of σ/σ_R and $< iT_{11} >$ for ^{16}O and ^{28}Si which will be shown in another contribution[3] confirm this trend: if the mass and the charge of the target nucleus increases the angular distribution of σ/σ_R and $< iT_{11} >$ become more and more structureless and the observed analyzing powers become smaller and smaller.

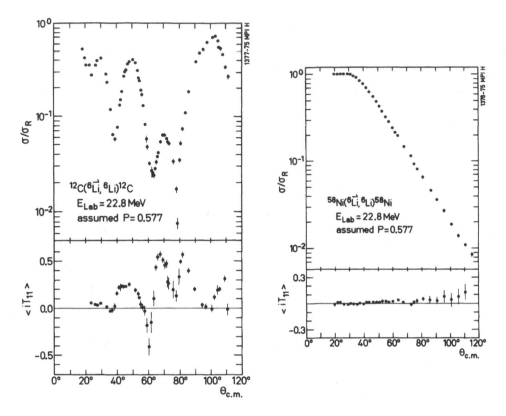

Fig.1 Differential cross section (σ/σ_R) and analyzing power for the elastic scattering of vector polarized ^{6}Li on ^{12}C (left part) and ^{58}Ni (right part). E_{Li} = 22.8 MeV.

+) present address: Gesellschaft für Schwerionenforschung, Darmstadt

++) on leave of absence from Nuclear Research Institute, Krakow

1) E.Steffens, H.Ebinghaus, F.Fiedler, K.Bethge, G.Engelhardt, R.Schäfer, W.Weiss.and D.Fick, Nucl.Instr.Meth. 124 (1975) 601 and contribution to this conference, p.871

2) P.Egelhof, J.Barrette, P.Braun-Munzinger, D.Fick, C.K.Gelbke, D.Kassen, W.Weiss and E.Steffens, contribution to this conference, p.825

3) D.Fick, Polarized Heavy Ions, contribution to this conference, p.357

STUDY OF POLARIZED ^{6}Li ELASTIC SCATTERINGS ON ^{12}C AND ^{58}Ni NUCLEI

WITH THE FOLDED (L·S) POTENTIAL

K.-I. Kubo and H. Amakawa

Department of Physics, University of Tokyo, Tokyo, Japan

The optical potential of heavy-ion scattering is not yet well established. Particularly there exists very little experimental information about its spin-dependence. It has not been tested whether the one-body spin-orbit coupling type potential is useful for interpretation of polarization phenomena in heavy-ion collisions. In our previous work (ref. 1), the gross properties of spin-orbit potential calculated by the folding method has been investigated.

Very recently the elastic scattering experiments have been done (ref. 2) using the polarized ^{6}Li beam of 22. 8 MeV on several target nuclei from ^{12}C to ^{58}Ni and the asymmetries have been measured. They show large asymmetry probability for the lighter targets ($\sim$60% for ^{12}C) and strong angular dependence, but small probability for the heavier targets.

In this paper, the success of theoretical attempt to investigate the polarized ^{6}Li scatterings on ^{12}C and ^{58}Ni nuclei using one-body (L·S) potential is reported. The (L·S) potential was calculated by the folding method. The d + α cluster model was used for ^{6}Li nucleus. The relative wave function was obtained by solving the Woods-Saxon potential problem. For the original (L·S) potential of ^{12}C + d and ^{58}Ni + d scatterings, Thomas form is used and the parameters are Vs=7. 42 (9. 6), r$_0$=0. 9 (0. 87) and a=0. 18 (0. 48) for ^{12}C (^{58}Ni), respectively. These parameters have been found to reproduce the experimental analyzing power as well as the elastic scattering cross section in ref. 3 for ^{12}C and in ref. 4 for ^{58}Ni, respectively.

The elastic scattering cross sections and the asymmetries produced by the vector type analyzing power were calculated using those folded (L·S) potentials, but with researching the central potential parameters, which had been found originally without spin-orbit potential, to make the x-square minimum. The absolute value of experimental cross section for the ^{12}C + ^{6}Li elastic scattering is not measured and then the cross section at θcm=50° is normalized tentatively to 0. 325 in Rutherford ratio by comparing the data with other experimental results reported in ref. 5. The beam polarization is 0.55 on the average. The guess values for parameter search are employed from ref's. 6 and 7, which can well reproduce the transfer reaction data as well as the elastic scattering data at 32 MeV for ^{12}C and at 28 MeV for ^{58}Ni, respectively. Those guess values and the parameters converged are given in Table 1.

In Fig. 1, the results obtained are shown by solid lines compared with the observed data (solid circles). For the ^{12}C + ^{6}Li elastic scattering, the present calculation does not give a complete fit to the experimental data in angular region θcm$\gtrsim$80°. The asymmetry calculated show a very well fit to the data in the angular region where the elastic scattering is well reproduced. For the ^{58}Ni + ^{6}Li scattering, the experimental results are reproduced by calculations very well.

The present success of the folding potential method for one-body (L·S) potential may indicate us a guide to investigate the spin-dependent heavy-ion scattering potentials on the basis of light-ion scattering analyses.

The authors thank Dr. D. Fick and his colleagues for informative communications. This work is supported by the Ito Science Fundation.

Table 1

		v	r_o	a	W_v	r_o'	a'	r_c
$^{12}C + ^6Li$	guess (32 MeV)[6]	173.2	1.208	.802	8.9	2.17	.945	(1.3)
	converged	188.6	1.15	.798	6.43	2.30	.958	(1.3)
$^{58}Ni + ^6Li$	guess (28 MeV)[7]	152.0	1.41	.69	10.3	1.81	.56	(1.41)
	converged	133.9	1.51	.541	9.76	1.72	1.11	(1.41)

$R_o = r_o \cdot (\text{target})^{\frac{1}{3}}$ is used.

The Coulomb parameter r_c is not changed in course of search.

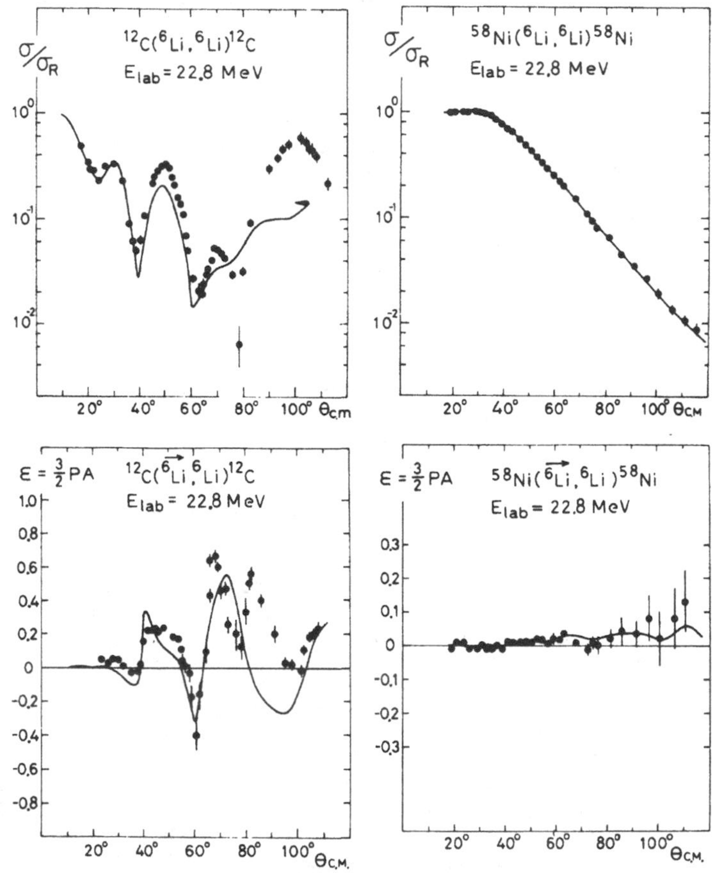

Fig. 1.
The comparisons of calculated results (solid lines) to the experimental data for the elastic scattering (upper part) and asymmetry (lower part) of the polarized ^6Li on ^{12}C and ^{58}Ni nuclei.

References:
1) H. Amakawa and K.-I. Kubo, Proc. Int. Symp. on Cluster Structure and Heavy-Ion Reactions, Tokyo (1975), to be published.
2) D. Fick et al., private communication and to be reported in this conference, p.827
3) J. A. R. Griffith et al., Nucl. Phys. A167 (1971) 87.
4) ibid. A146 (1970) 193.
5) P. K. Bindal et al., Phys. Rev. C9 (1974) 2154.
6) P. Schumacher et al., Nucl. Phys. A212 (1973) 573.
7) R. M. DeVries et al., Phys. Letters 55B (1975) 33.

POLARIZATION IN HEAVY-ION SCATTERING*

W. J. Thompson, Univ. of North Carolina, Chapel Hill, N. C. USA and Triangle Universities Nuclear Laboratory, Durham, N. C.

Elastic-scattering polarization P and inelastic-scattering spin-flip probabilities P_1 are calculated in a model[1] for the spin-orbit interaction. The model considers a single valence orbit bound to a spin-zero core. Both ^{13}C and ^{15}N can be so considered, since their $1p_{\frac{1}{2}}$ spectroscopic factors S are large[2]). The single-nucleon spin-orbit interaction with the target nucleus V_{Nso} is averaged over the nucleon motion in the $1p_{\frac{1}{2}}$ orbital to give the projectile-nucleus spin-orbit interaction V_{so} (R).

The global parameterization[3] of V_{Nso} is well approximated by

$$V_{Nso}(\vec{r}) = - V_{Ns} r^4 \exp[-r^2/C_{Ns}^2] \vec{\ell} \cdot \vec{s} \tag{1}$$

The dependences of V_{NS} and C_{NS} on the target mass number A_T are adequately approximated by

$$V_{Ns} = 15 A_T^{1/3}/C_{Ns}^7 \text{ MeV.fm}^{-4} \quad , \quad C_{Ns} = 0.70 A_T^{1/3} (1 - 0.55 A_T^{-2/3} + 0.4 A_T) \text{ fm} \tag{2}$$

With the 1p orbit described by an oscillator wave function, the folding-model, heavy-ion, spin-orbit potential is

$$V_{SO}(\vec{R}) = \frac{-SV_{Ns}}{A-1} \left(\frac{A}{A-1} \right)^4 \frac{a^{5/2} C_{Ns}^9}{12 C_N^5} \Big(105 a^2 + 10 a(10 - 28 a + 21 a^2)(R/C_{Ns})^2$$

$$+ 4(1-a)^2 (3 - 14 a + 21 a^2)(R/C_{Ns})^4 \tag{3}$$

$$+ 8a(1-a)^4 (R/C_{Ns})^6 \Big) \exp[-R^2/C_s^2] \vec{L} \cdot \vec{S}$$

Here

$$\tag{4}$$

$$a = (1 + [AC_{Ns}/((A-1)C_N)]^2)^{-1}, \quad C_s^2 = C_{Ns}^2 + ((A-1)C_N/A)^2, \quad C_N = \sqrt{(2/5)} \cdot R_{ms}$$

where R_{ms} is the RMS radius of the 1p orbit.

There are preliminary spin-flip data on $^{60}Ni(^{13}C, {}^{13}C)^{60}Ni$ at 60 MeV (lab)[4]). The nucleon spin-orbit parameters from eq. (2) are Set 1 of table 1, S=0.61 (ref.2) and R_{ms} =2.75 fm (r_{ms} = 1.17 fm). Elastic and inelastic scattering are calculated in the coupled-channels formalism, with the 1.33-MeV state described as a vibrational state with β=0.18. Optical-model parameters, chosen[4]) to fit elastic scattering data are from Set 1 in table 1. The values of P(θ) shown in fig. 1 are small at all θ where σ(θ) is significant. The inelastic-spin-flip probability is miniscule (P_1 (θ)<10^{-4} at all θ), and is much smaller than the upper limit P_1 <0.01 (ref.4). A maximum P_1 of one half this limit is obtained at the forward inelastic-scattering angles (θ <20°) if parameter Set 2 in table 1 is used. This does not significantly affect σ(θ) but makes P (θ) large in magnitude.

TABLE 1. Optical-model parameters for $^{13}C+^{60}Ni$.
(V_O=100 MeV, $R_O=R_W$=7.04 fm, $a_O=a_W$=0.6 fm, β=0.18)

	W_S(MeV)	C_{ns}(fm)	r_{ms}(fm)
Set 1	40.0	2.66	1.17
Set 2	10.0	3.28	1.44

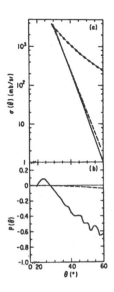

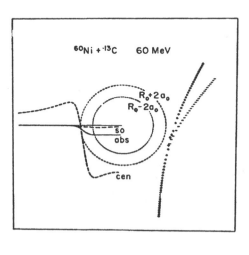

Fig. 1. Elastic-scattering for $^{60}Ni(^{13}C, ^{13}C)$ ^{60}Ni at 60 MeV vs c. m. angle. (a) Differential cross sections for, Rutherford scattering (dash-and-dot curve), Set 1 (dash curve), and Set 2 (solid curve). (b) Polarizations for Set 1 (dashed curve) and Set 2 (solid curve). (c) Classical trajectories for J=L+1/2 (+) and J=L-1/2 (-) (L=34.8) for Set 2 real potential. Central (cen), absorption (abs) and spin-orbit splitting (so) potentials are shown. The scattering angles of 50° (+) and 61° (-), and the rapid decrease of $\sigma(\theta)$ with θ, produce large negative $P(\theta)$.

References

*Supported in part by the U.S.E.R.D.A.

1) W. J. Thompson, Proc. Int. Conf. on Reactions Between Complex Nuclei (North-Holland, Amsterdam, 1974) 1 14; Bull. Am. Phys. Soc. 19 (1974) 1015
2) S. Cohen & D. Kurath, Nucl. Phys. A101 (1967) 1
3) F. D. Becchetti & G. W. Greenlees, Phys. Rev. 182 (1969) 1190
4) C. Chasman, P. D. Bond & K. W. Jones, Bull. Am. Phys. Soc. 20 (1975) 55; P. D. Bond, private communication

Production of Polarized Particles

THE HAMBURG POLARIZED NEUTRON BEAM *

R. Fischer, F. Kienle, H.O. Klages, R. Maschuw, R. Schrader
P. Suhr and B. Zeitnitz

II. Institut für Experimentalphysik, Universität Hamburg

The deuteron beam of the Hamburg Isochronous Cyclotron is used to produce polarized neutrons via the reactions D (d,n) ^{3}He and T (d,n) ^{4}He.

The gastarget system consists of a cylindrical steel cell with molybdenium windows and an uranium oven which serves as a gas reservoir.
A lot of security problems have been solved for the purpose of tritium use.
A 2π - shield build up in layers of heavy metal, copper, iron and paraffine gives a well collimated (10^{-4}sr) neutron beam at $\theta_{lab} = 30^\circ$.
A superconducting solenoid for the rotation of the neutron spin is part of the shield. The distance between the target and the place, where the neutron beam can be used for experiments, is appr. 200 cm.
Typical data at 14.5 MeV deuteron energy are: deuteron beam I_d=1-2 μA, gas pressure P_D=3 atm and neutron yield Y_n=10^4sec^{-1}.

The first experiment performed was the determination of the source polarization for the D $(d,n)^3$He reaction at the energy range available.
For this purpose a liquid-^{4}He-polarimeter has been used. Figure 1 shows the results of the measurement with different sets of n-α-phases.

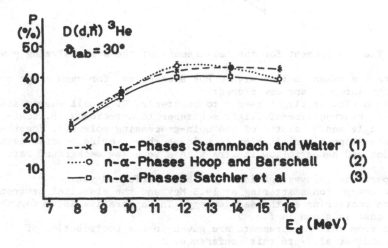

Fig.1 The neutron beam polarization at various energies

* Supported by the Bundesminister für Forschung und Technologie.

The experimental setup to investigate the analyzing power of the neutron scattering on light nuclei is shown in figure 2 in a simple view.

Up to 8 neutron detectors can be placed around a scatterer, which is performed as a scintillator. For the scattering on protons and deuterons commercial materials like NE 213 or NE 230 can be used.
The n-^{3}He-scattering can be done with a liquid-^{3}He-scintillator which has been developed in Hamburg.

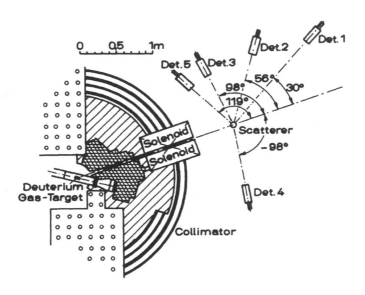

Fig. 2 The arrangement for the measurement of the n-d-analyzing power.

The data are taken on-line with a PDP 9 computer, for each event the following informations are stored:
1. neutron time-of flight target to scatterer, 2. recoil energy scatterer, 3. neutron time-of-flight scatterer to detector x, 4. scattering angle and 5. status of the spin-precessing solenoid. Together with puls-shape-discrimination techniques they provide precise data reduction and hence follow spectra with a very low background rate.

Two experimental investigations have been performed yet:
The neutron-proton scattering at 14.5 MeV and the elastical neutron-deuteron scattering at the same energy. The apparative setup for the latter case is shown in fig. 2.
Results from this measurements are given in the contribution of
R. Behrendt et al.[4] to this conference.

1) Th. Stammbach and R.L. Walter - Nucl.Phys. A 180 (1972), 225
2) B. Hoop and H.H. Barshall - Nucl.Phys. 83 (1966), 65
3) G.R. Satchler et al. - Nucl.Phys. A 112 (1968), 1
4) R. Behrendt et al. - Contribution to this conference, p.441

Performance of the IPCR Polarized Ion-Source

S. Motonaga, T. Fujisawa, M. Hemmi, H. Takebe,
K. Ikegami, and Y. Yamazaki*
The Institute of Physical and Chemical Research,
Wako-shi, Saitama, 351, Japan

We constructed a conventional polarized ion source based on the method of magnetic separation of ground state atoms[1]. A schematic diagram of the polarized ion source is shown in Fig. 1. It consists of the following four sections: 1) Dissociator 2) Separation magnet 3) Radio-frequency transition unit, and 4) Strong field ionizer.

The hydrogen or deuterium molecules are dissociated in a pylex bottle of 35 mm in diameter by an rf discharge at a frequency of 20 MHz. The dissipated rf power in the dissociator is 250 watts. The pylex bottle is cooled by air blast. Dissociated atoms difuse into a sextupole magnet through a single nozzle of inside-diameter of 3.5 mm and 10 mm long.

For separation we use the sextupole magnet having an axial length of 35 cm, an entrance aperture of 6.5 mm, and an exit aperture of 13.5 mm. The magnet produces the maximum field strength of 7.2 kG at the pole tip.

The atomic beam polarized with respect to the electron spin passes through a radio-frequency transition region. The transitions are carried out by the adiabatic passage method.[2]

The polarized atoms are ionized by electron impact in a strong field ionizer which is similar to that described by Glavish[3]. A magnetic field of about 1500 gauss is applied parallel to the beam axis.

Typical pressures in the different sections of the ion source were measured under the normal operating condition in which the gas flow rate from the nozzle was at about 0.25 Torr liter/sec : In the dissociator $p=2 \times 10^{-5}$ Torr, in the sextupole magnet $p=6 \times 10^{-6}$ Torr, and in the ionizer $p=4 \times 10^{-7}$ Torr.

In order to obtain an intense and well-collimated atomic beam from the dissociator, we examined a multicapillary nozzle and a single canal nozzle, and we chose a 3.5 mm diameter single canal nozzle.

The atomic beam intensity at the ionizer entrance behind the sextupole magnet was found to be 3×10^{15} atoms/cm^2.sec using an ionization gauge. The focusing and separating effects for the electron spin states were observed by taking pictures on MoO_3 targets. The beam from 3.5 mm nozzle hole was of 8 mm in diameter at the ionizer entrance 35 cm behind the separating magnet.

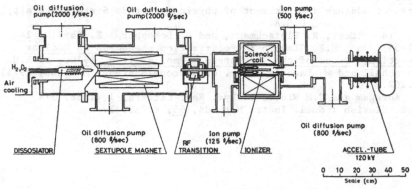

Fig. 1. Schematic diagram of the IPCR polarized ion source.

The polarized beam intensity was determined by the difference in current reading at the end of the accelerating tube placed at 60 cm behind the extracting electrode in the ionizer when the sextupole magnet was turned off and on. The intensity was 1 μA in the optimum operation. For several hours of operation the beam intensity of 0.6 μA was maintained stably.

The tensor polarization of the polarized deuteron beam was measured by the neutron anisotropy $N(0°)/N(90°)$ in the $T(d,n)$ ^{4}He reaction. For a frequency of 380 MHz the static field was varied in the region from 40 to 150 gauss. The results are shown in Fig.2, in which the values of the tensor polarization are 0.516 ± 0.058 for 2 to 6 transition, and -0.562 ± 0.071 for 3 to 5 transition.

The efficiency of transition was also examined by varying the strength of the rf rotating field B_1 from 0.21 to 2.3 gauss at 380 MHz. However, as shown in Fig.3, remarkable increase in the anisotropy was not observed.

The polarization values obtained are smaller than those theoretically expected. This is mainly due to the partial pressure of the deuterium gas in the ionizing region.

On the basis of the present study a new polarized ion source has been designed to be installed in the cyclotron.

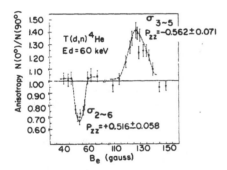

Fig. 2. Neutron anisotropy in $T(d,n)$ ^{4}He reaction with polarized deuterons.

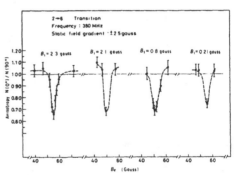

Fig. 3. Neutron anisotropies produced by different rotating fields.

References
* Present address : Department of physics, Florida State University, U.S.A.
1) A. Clausnitzer, R. Fleischmann, and H. Schopper : Z. Physik, 144 (1956) 336 ; H.F. Glavish : "Polarization phenomena in Nuclear Reaction," Proc. of the 3rd Intern. Symp. Madison, 1970., ed H.H. Barschall and W. Haebeli (Univ. of Wisconsin Press, Madison, 1971), p.267.
2) A. Abragam and J.M Winter : Phys. Rev. Letters, 1,(1958) 374
3) H.G. Glavish : Nucl. Instr. Methods, 65, (1968) 1.

THE BONN ATOMIC BEAM POLARIZED ION SOURCE[+]

F. Barz, E. Dreesen, W. Hammon, H.H. Hansen,
S. Penselin, A. Scholzen, W. Schumacher
Institut für Angewandte Physik
Universität Bonn, Bonn, W.-Germany

A polarized positive ion source is installed at the Bonn isochronous cyclotron for axial injection of polarized particles [1].

In a 27 MHz discharge H or D atoms are formed in a pyrex discharge tube. The copper nozzle of this tube is cooled by flowing liquid nitrogen. The sextupole magnet has a length of 220 mm and a constant distance between opposite pole pieces of 11,5 mm. Nuclear polarization is increased by use of a system of rf transitions. Transitions $1 \rightarrow 3$ and $2 \rightarrow 4$ are used for H and transitions $3 \rightarrow 5$, $2 \rightarrow 6$ and $1 \rightarrow 4$ are used for D. Contributions of vector polarization in the case of D can be averaged to zero if the transition $1 \rightarrow 4$ is switched on and off for equal times.

The ANAC strong field ionizer is used for ionization of the neutral beam. We now use a variation of this ionizer with a solid electron beam and a conical ionization column.

The maximum ion beam intensity of the source was 6.5 µA for protons and 8 µA for deuterons (these intensities arise only from the atomic beam). At the moment these high currents cannot be reproduced, but typical currents of 1.5 - 2 µA for protons and 2 - 2.5 µA for deuterons are available at the end of the source.

About 1/3 of all ions are in a phase space of o.3 π cm rad $\sqrt{eV}$, about 2/3 in a phase of 0.9 π cm rad $\sqrt{eV}$. The total phase space (100 % of the ions) is about 2.5 π cm rad $\sqrt{eV}$.

The tensor polarization P_{zz} of the deuteron beam was tested with the $T(d,n)^4He$ reaction. The best value observed for P_{zz} was 0.79 $\pm$ 0.02 with no blocking of the central region of the atomic beam in the sextupole magnet. By blocking a central region with a diameter of 3.5 mm at the end of the sextupole magnet a higher value for P_{zz} of 0.85 $\pm$ 0.02 was observed.

The transition probabilities for the rf transitions were measured. For the transitions $1 \rightarrow 3$ for H and $1 \rightarrow 4$ for D transition probabilities are higher than 0.95. The transition $2 \rightarrow 6$ for D is induced with a probability of 0.993 $\pm$ 0.003 and the transition $3 \rightarrow 5$ for D with a probability of 0.985 $\pm$ 0.005.

References

[+] Supported by the "Bundesministerium für Forschung und Technologie"

1) W. Schumacher et al., Nucl. Instr.Meth., to be published

A GROUND STATE ATOMIC BEAM POLARIZED ION SOURCE FOR THE HIGH VOLTAGE
TERMINAL OF THE OHIO STATE 7 MV PRESSURIZED VAN DE GRAAFF ACCELERATOR*

T.R. Donoghue, W.S. McEver[†], H. Paetz gen. Schieck[‡], J.C. Volkers,
C.E. Busch[††], Sr. Mary A. Doyle, L. Dries, and J.L. Regner
Department of Physics, The Ohio State University, Columbus, Ohio 43210

A ground state atomic beam polarized ion source has been constructed for operation within the high voltage terminal of a 7 MV
pressurized Van de Graaff accelerator. This environment is quite
unusual for a polarized source and, as such, places many special constraints on its design. For instance, the source must be very compact
(length < 1.4 m) and lightweight (as it is supported solely by the
glass accelerator tube), it must operate on less than 4 kW of power
(limited by the alternator in the h.v. terminal), and it must withstand
an external pressure of 225 psi. The most formidable problem we encountered was in satisfying our vacuum demands as quick reflection
shows that conventional vacuum systems are totally inadequate. A
custom designed system utilizing a rod-type titanium sublimator together with an integral ion pump admirably meets our requirements of high
pumping speeds (2200 ℓ/sec at 10^{-5} torr for H_2) at the dissociator, as
well as of high ultimate vacuums (~ 10^{-8} torr) under light gas load
conditions such as in the ionizer chamber. McEver et al.[1]) describe
this pumping system more fully elsewhere.

A plan view of the source is shown in the figure. Although the
general configuration of the source
is basically the same as presented[2])
at the Madison Symposium, a number of
important design changes have been
made particularly in the ion beam
portion of the source. The source is
a standard atomic beam type, consisting briefly of the following. The H_2
(or D_2) gas is dissociated in a double
walled freon cooled pyrex bottle. The
atomic beam exits through a 2 x 2 mm
diameter aperture and passes into a
short (13.3 cm long) sextupole magnet
which serves to separate the atomic
spin states. The retained components
of this beam then pass into an adiabatic fast-passage RF transitions
region. Here we effect transitions
between states 3 and 5 of deuterium
using medium field transitions (355
Mhz, B_0 = 104 ± 7 G). This is followed by the weak field transition for
deuterium (8 Mhz, 8.5 G) and for hydrogen (12 Mhz, 8.5 G). With these,
theoretical maximum polarizations are
p_{zz} = -1.0 for deuterons and p_y = -1.0
for protons, although we expect ~ 0.8
of these values because of incomplete
spin state separation in our sextupole.
The prepared atomic beam passes from

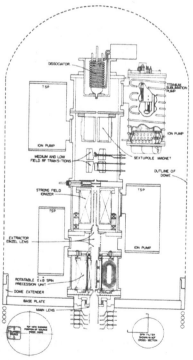

the RF transitions into a strong field solenoidal ionizer of the Glavish type and the beam extracted from this is focused by an einzel lens to a cross over in the center of an $\vec{E} \times \vec{B}$ spin precession unit. The latter can be rotated about the beam axis permitting the spin orientation axis to be rotated to any direction in space at the reaction target. To conserve power, we must use permanent magnets wherever possible (sextupole, RF transitions, ion pumps) and operate the ionizer at a lower B field than optimum and additionally use a considerable amount of solid state electronics, including a transistorized 50 watt 355 Mhz oscillator. Although the electronics are protected against voltage transients associated with the h.v. breakdown of the accelerator terminal, and additionally electronics are contained within a double-shielded cage, we still have some problems in this area to be cured. The dissociator, RF oscillators, ionizer and E x B spin unit are freon cooled from without the accelerator. We monitor the status of most of the electronic modules in the high voltage terminal using digital voltmeters whose contents are relayed to the control console via closed circuit TV.

Operationally, we measure 150-200 nA of mass analyzed beam out of the source at a test bench accelerator. With the source inside the Van de Graaff, we measure[3] for protons p_y = 0.65 with I = 60 nA on a reaction target (2.5 mm square beam size) and for deuterons, p_{zz} = 0.80 and I = 30 nA. These performance figures are reasonably good, considering the numerous compromises we had to resign ourselves to. The source does require constant maintenance mainly for electronic failures and for pumps. The latter must be rejuvenated every 10-14 days, including titanium replacement. We anticipate operating the source inside the Van de Graaff in approximately three month intervals. To facilitate this, the source and the associated electronics are each arranged as a single package unit that can be hoisted into place in a matter of hours. We anticipate increased beam intensities with several changes in accelerator beam optics and more flexibility in the spin reversal with several additional RF transitions.

References

*   Work supported in part by the U.S. National Science Foundation.
†   Now at Intertechnology Corporation, Warrentown, Va.
‡   Now at Institut für Kernphysik der Universitat Köln, Köln, West Germany.
††   Now at Scientific Atlanta, Atlanta, Georgia.

1) W.S. McEver, J.C. Volkers, and T.R. Donoghue, Nuclear Instr. and Methods, 127, (1975) (in press).
2) H. Paetz. gen. Schieck, C.E. Busch, J.A. Keane, and T.R. Donoghue, in Polarization Phenomena in Nuclear Reactions Madison, edited by H.H. Barschall and W. Haeberli (University of Wisconsin Press, Madison, 1971) p. 810.
3) T.R. Donoghue et al., in Proceedings of the Second Symposium on Ion Sources and Formation of Beams, Berkeley, 1974, edited by J. Orth, available as LBL-3399 (1974) p. IV-8-1.

THE SOURCE FOR POLARIZED DEUTERONS AND PROTONS
AT THE ETHZ TANDEM ACCELERATOR

R. Risler, W. Grüebler, V. König, P.A. Schmelzbach, B. Jenny
and W.G. Weitkamp
Laboratorium für Kernphysik, Eidg. Techn. Hochschule
8049 Zürich, Switzerland

The construction of the polarized ion source at the ETHZ is based
on an older version which was in operation up to 1973[1]. The new source
is also of the atomic beam type. In a hairpin shaped pyrex tube, deuterium or hydrogen is dissociated and formed into an atomic beam by a 2.5mm
nozzle and two diaphragms of 3.0 and 5.0mm diameter[2].

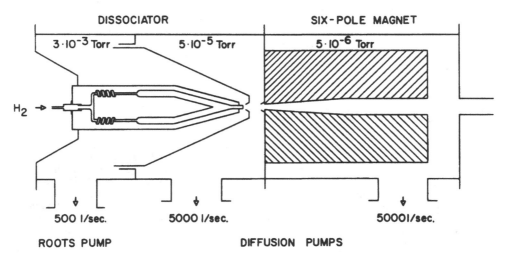

Fig. 1. The atomic beam apparatus. The diagram is drawn roughly to scale.
Typically 2.5 bar cm^3/sec deuterium or hydrogen gas is let into the
system.

Fig.1 shows the main features of the atomic beam apparatus. The separation
of the spin states takes place in the field of a tapered sextupole magnet
50 cm in length. With the possibility of a weak-field and two strong-
field RF transitions all necessary polarization states of the deuteron
beam can be obtained. The weak-field transition magnet can also be tuned
in order to produce a negatively polarized proton beam. The 2-4 microwave transition for the positive polarization will be installed in the
near future.

The strong-field ionizer is housed in a rectangular vacuum tank,
(cf. fig. 2) which is pumped by a 5000 l/sec diffusion pump and two
900 l/sec orbitron pumps. The background pressure is in the order of
$2 \cdot 10^{-7}$ torr. After extraction from the ionization region, the positive
beam is accelerated to 5 keV where the charge exchange in sodium vapour
takes place. After acceleration to ground potential (60 keV) the beam
polarization axis can be oriented in any desired direction by a Wien
filter.

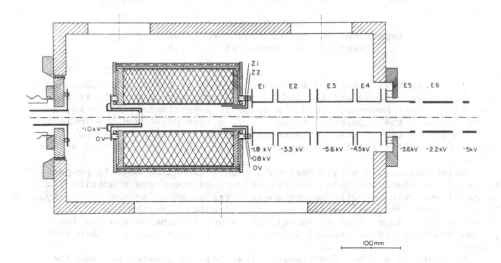

Fig. 2. Diagram of the ionizer. The beam enters from the left. At the beginning of the solenoid a tungsten filament and an accelerating grid produce the electron current for ionization. The potentials indicated are typical operating values.

The source has been running at the tandem accelerator for over 1500 hours now. Typical beam currents on target are 50 nA for deuterons and 30 nA for protons. The polarization of the deuteron beam typically reaches 87 % of the theoretical value ($p_z = \pm 0.57$, $p_{zz} = \pm 0.87$), for protons 80 % are obtained.

The beam is exceptionally stable. The observed variations stay within 1 % over a whole measuring period of 24 hours.

References

1) W. Grüebler, V. König and P.A. Schmelzbach, Nucl. Instr. and Meth. 86 (1970) 127
2) R. Risler, W. Grüebler, V. König and P.A. Schmelzbach, Nucl. Instr. and Meth. 121 (1974) 425

ACHROMATIC MAGNETIC FOCUSING OF AN ATOMIC BEAM*

H. F. Glavish
Department of Physics, Stanford University
Stanford, California 94305 U.S.A.

In atomic beam polarized ion sources, such as the type produced by ANAC[1], rf transitions are induced in the atomic beam before it is ionized in a strong magnetic field. A space of approximately 25 cm is required between the separating magnet (e.g. a sextupole) and the ionizer to incorporate the rf transition units. In this region the neutral atomic beam drifts freely and diverges, thus reducing the density at the ionizer.

Unfortunately, a single sextupole magnet cannot be made to properly focus all of the atoms into the ionizer because the focus condition for atoms of one velocity will not be correct for atoms of another velocity. In other words one must average over the velocity distribution of the atoms which is approximately Maxwellian. Also, in considering the ion yield, the extended ionization volume at the ionizer must be taken into account.

However, if a short "compressor" sextupole is located between the rf transition units and the ionizer it is possible to greatly enhance the atomic beam density at the ionizer. This is because there are now two, separated sextupoles which are able to produce an achromatic focusing condition. The situation is shown schematically in fig. 1. Slow atoms emerging from the first sextupole diverge rapidly and enter the compressor sextupole at a large radius. As they are slow they spend substantial time in the compressor and become strongly focused towards the axis. On the other hand, fast atoms diverge less from the first sextupole, spend less time in the compressor, and are at a smaller radius. Therefore, they are not focused so strongly by the compressor.

The results of detailed computer calculations, which average over a Maxwell velocity distribution and the extended ionization volume, are summarized in the table below. The pole tip field is B_m, while ℓ_i, a_i, b_i are the length, entrance aperture, and exit aperture respectively, of the separating sextupole (i = 1) and compressor (i = 2). The ion yield, in arbitrary units, applies to protons and takes into account states

	B_m (kG)	ℓ_1 (cm)	a_1 (cm)	b_1 (cm)	ℓ_2 (cm)	a_2 (cm)	b_2 (cm)	Ion Yield
I	8	35	0.35	0.70		None		1.0
II	8	35	0.35	0.76	10	0.8	0.8	2.3
III	8	35	0.35	0.76	10	0.8	0.8	2.9*
IV	12	40	0.35	1.75		None		2.0
V	12	40	0.35	1.75	10	1.0	1.0	3.6
VI	12	40	0.35	1.75	10	1.0	1.0	4.4*
VII	12	15	0.35	0.88	10	0.8	0.8	6.3*
VIII	12	40	0.35	1.75	40	1.2	0.6	2.1

* Ionization length 30 cm.

rejected by the compressor after rf transitions are induced. Case I
applied to the present ANAC polarized sources[1]. Case VIII has a long
second sextupole which does not enable very achromatic focusing to be
achieved but is interesting in that deuterons in pure magnetic substates,
either 1, 0, or -1, can be prepared. The ion yields indicated in the
table apply to an ionization region of 20 cm length, unless otherwise
indicated. The table shows that substantial gains should be achieved
with addition of a compressor sextupole. In addition to benefiting
atomic beam polarized ion sources, the improved atomic beam density would
make polarized atomic beam targets feasible.

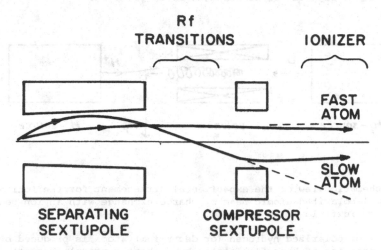

Fig. 1. Schematic illustration of achromatic focusing.

References

* Supported in part by the National Science Foundation
1) H. F. Glavish, Proceedings of the Symposium on Ion Sources and
   Formation of Ion Beams (Brookhaven National Laboratory, 1971)
   ed. Th. J. M. Sluyters, p. 207

IONIZATION OF A POLARIZED ATOMIC BEAM
BY CHARGE EXCHANGE REACTIONS[+]

W. Hammon, A. Weinig
Institut für Angewandte Physik
Universität Bonn, Bonn, W.-Germany

In atomic beam sources the polarized atomic beam is ionized by electron impact. A higher ionization efficiency is expected if charge exchange reactions with their typical high cross sections are used [1]. Our first experiment was a test for an employment of charge exchange reactions for the production of polarized protons and deuterons from the atomic beam in a magnetic field of 2 kG [2].

In a second experimental arrangement we now use a superconducting solenoid with a magnetic field strength of up to 75 kG:

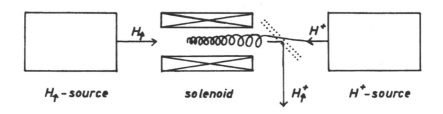

$H_\uparrow$-source   solenoid $H_\uparrow^+$ H^+-source

Fig. 1. Schematic view of the experimental arrangement for the ionization of a thermal polarized atomic beam by charge exchange with an ion beam in a high magnetic field of up to 75 kG.

A beam of polarized hydrogen (or deuterium) atoms is produced by a spatial separation of hyperfine states by a short sextupole magnet ($H_\uparrow$ -source). This thermal neutral beam is focussed into the ionization volume in the solenoid. An unpolarized proton (or deuteron) beam with an energy of 10 to 20 keV is produced in a duoplasmatron source. The path length of these fast ions is increased by a suitable injection into the ionization volume where these ions spiral around the axis of the magnetic field. In the high magnetic field the thermal atomic beam and the fast ion beam can overlap. The ions produced by charge exchange from the atomic beam are extracted by a system of electrostatic lenses. An electrostatic mirror provides energy separation of the extracted beam.

Using the reaction $H_\uparrow + H^+ \longrightarrow H_\uparrow^+ + H$ in a first test we could extract 1 μA protons from the atomic beam. The efficiency of this process is about $1 \cdot 10^{-3}$ and can be increased.

References

+ Supported by the "Landesamt für Forschung des Landes Nordrhein-Westfalen", Düsseldorf, W.-Germany

1) R. Beurtey and M. Borghini, Journal de Physique C2 (1969) 56
2) W. Hammon, S. Penselin, A. Scholzen and H. Simm, Nucl. Instr. Meth. 125 (1975) 571

FIRST ACCELERATION OF A POLARIZED PROTON BEAM TO 590 MeV WITH THE SIN ISOCHRONOUS RING CYCLOTRON

G. Heidenreich, University Basel, Basel, Switzerland

M. Daum, G.H. Eaton, U. Rohrer, E. Steiner
Swiss Instiute for Nuclear Research, SIN, 5234 Villigen, Switzerland

The SIN isochronous ring cyclotron accelerates a proton beam from the 72 MeV injector cyclotron to 590 MeV kinetic energy. An atomic beam type polarized ion source has been installed below the axial injection system of the injector cyclotron. From this ion source for the first time a polarized proton beam has been accelerated to full energy of 590 MeV. The polarization of this high energy 1 nA beam was measured in the following way: The polarized proton beam was transported to the external thin target station containing a 1 g/cm^2 carbon target. The protons scattered from this target at 8° into a secondary beam line were momentum analyzed to

$$\frac{\Delta p}{p} = 3.4 \text{ \%0}$$

corresponding to about ± 2.0 MeV energy spread. The elastically scattered protons were selected using a slit system and transported to a scintillation counter telescope at the end of the beam line. The polarization of the proton beam extracted from the ring accelerator was determined from measurements of the scattered proton intensity at 8° from the thin target for the two directions of the beam polarization and also with the unpolarized beam. As a monitor of the beam intensity in front of the scattering target, a 10 mg/cm^2 thick aluminium foil installed upstream from this target was used in conjunction with a scintillation counter telescope defining a mean scattering angle of 40° in the vertical plane i.e. parallel or antiparallel to the polarization vector of the incident proton beam. The results of the measurement are listed in table 1.

Table 1:

Spin direction	Monitor coincid.	trans.	P.A.
up (↑)	10^5	S.F.	-0.22 ± 0.005
down (↓)	10^5	W.F.	0.15 ± 0.005

Table 1: Results of the polarization measurement of the SIN 590 MeV polarized proton beam (only statistic errors).

Thus assuming a carbon analyzing power[1,2,3] at this energy and angle

$$A = 0.41 ± 0.04$$

one receives a polarization of

$$P = 0.52 ± 0.06$$

measured with the strong field transition in the polarized ion source.

This measured polarization ðf the 590 MeV beam is in fairly good agreement with earlier measurements for the 30 MeV polarized protons extracted from the injector machine. The value obtained is at the present below expectation because of the insufficient time available for optimization of the polarized ion source, and in particular for the weak field and strong field transition parameters. It is hoped to realize the design polarization of this source of 80 % at a later date.

<u>Refereences:</u>

[1]) R. Hess et al.; private communication, 1975.
[2]) M.G. Meshcheriakov et al.; JETP 4, 337 (1957).
[3]) L.S. Azhgirey et al.; Nucl. Phys. <u>43</u>, 213 (1963).

849

PRESENT STATUS OF THE KARLSRUHE POLARIZED
ION SOURCE

V.Bechtold, L.Friedrich, Karlsruhe Nuclear Research Center
D.Finken, G.Strassner, P.Ziegler, University of Karlsruhe.

The Lambshift source C-LASKA producing polarized deuterons from
charge exchange of metastables with iodine [1) is now in operation at the
Karlsruhe isochronous cyclotron. The source is installed in the basement
of the experimental hall, separated from the cyclotron by 3 m of shielding. The distance between the axial injection system and the source is
11 m. Thus the strayfield of the cyclotron is eluded. In addition
the source can be handled when the cyclotron is in operation.
The source delivers 0.8 µA vector polarized deuterons with an emittance
of 1.1 cm rad (eV)$^{1/2}$. Of this, 5% could be extracted from the cyclotron.
The vector-polarization of the 52 MeV beam was calibrated by elastic
scattering on ^{12}C. The analysing power is known from a double scattering
experiment [2). After some improvements Py = 0.45 ± 0.014 averaged over a
run of 60 h was measured. P_{vv} is estimated to be less than 0.01.

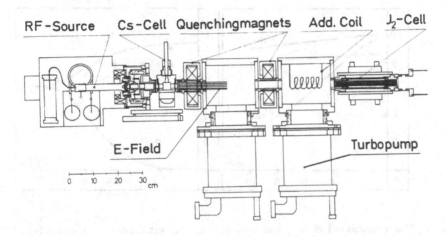

RF-Source   Cs-Cell Quenchingmagnets   Add. Coil J_2-Cell

E-Field   Turbopump

0   10   20   30 cm

Fig. 1   The Ion Source C-LASKA

Instead of oil diffusion pumps now turbomolecular pumps are used. In this
way a better stability of the beam intensity and a longer lifetime of the
ion source bottle has been achieved. To protect the bearings and the
oil of the pumps from iodine a low iodine pressure in the forevacuum is
maintened by liquid nitrogen traps.
A transverse magnetic field of 150 Oe can be produced in the iodine cell
to get transversely vector polarized deuterons. This field is compensated
electrically to avoid quenching of the metastables and broadening of the
beam. For this purpose the cell (18 mm diameter and 200 mm length) has
been sliced into 24 segments which are operated at different potentials.

The polarization of the metastable atoms is rotated to transverse direction by superposing the outgoing field of the second quenching magnet and
the stray field of the iodine cell. In the case of weak field in the
quenching magnet an additional coil was used for optimum rotation. The
iron shielding of the two quenching magnets was developed for optimum
tensor polarization $P_{yy} = 0.73$ as reported earlier[3]. In this case all
metastable ß-states are quenched. For a beam with maximum vector polarization and zero tensor polarization a high transmission $w_{2\beta} = w_{2\alpha}$ in the
second quenching magnet is necessary. Whereas the transmission $w_{2\alpha}$ does
not strongly depend on the distance from the beam axis, the transmission
$w_{2\beta}$ becomes very low at the periphery of the beam, even for low magnetic
field and no electrical field (Fig. 2). Therefore the magnetic field in
the second quenching magnet has to be reduced as far as possible without
affecting the zero crossing of the field between the two quenching
magnets. Fig. 3 shows the measured vector polarization as a function of
the magnetic field in the second quenching magnet. Assuming constant
beam density the curves P_y and P_{yy} are calculated from w_α and w_β. P_{yy}
is smaller than 0.01 if the second quenching magnet is operated at 100 Oe.

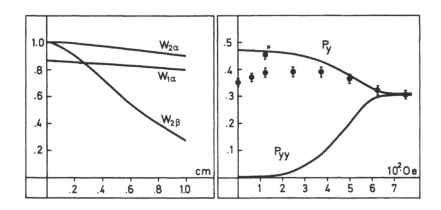

Fig. 2   The transmission $w_{2\alpha}$ and $w_{2\beta}$ of the metastable state calculated
         as a function of the beam radius in the 2nd quenching magnet at
         400 Oe maximum. In addition $w_{1\alpha}$ in the 1st quenching magnet is
         shown for 600 Oe and a deflection E-field of 15 V/cm.

Fig. 3   The measured vector polarization as a function of the maximum
         field strength in the 2nd quenching magnet. P_y and P_{yy} are calcu-
         lated. The x point represents the polarization achieved after
         final adjustment averaged over 60 h.

References

1) L.D.Knutson, Phys.Rev. A2 (1970)1878
   H.Brückmann, D.Finken, L.Friedrich, Nucl.Instr.Meth. 87(1970)155

2) E.Seibt, Ch.Weddigen, Nucl.Instr.Meth. 100(1972)253

3) V.Bechtold, H.Brückmann, D.Finken, L.Friedrich, K.Hamdi, G.Strassner,
   Proceedings of the second conference on ion sources, Vienna, Sept.72,
   498

THE LAMBSHIFT SOURCE AND 1,2 MEV-TANDEM-FACILITY IN GIESSEN, GERMANY [+]

W. Arnold, H. Berg, G. Clausnitzer, H.H. Krause, J. Ulbricht

Institut für Kernphysik, Strahlenzentrum
Universität Giessen, Germany

A Lambshift source was constructed and connected to a small selfbuilt pressurized 1,2 MeV-Tandem-Accelerator.

The source design follows the general principles (1). The main feature is a compact construction (1,6 m long) in one rectangular vacuum chamber, which is divided internally into three sections, pumped by three 6000 l/s-oil diffusion pumps. Pumps and baffles are maintained at ground potential, the vacuum chamber is insulated for injection potentials up to 20 keV. Fig. 1 gives a schematic view.

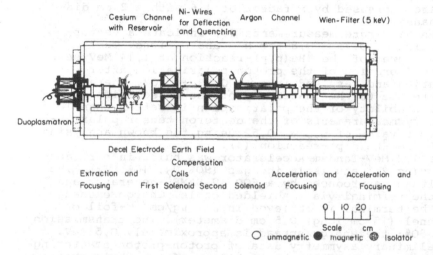

Fig. 1: Schematic of the Lambshift-Source

Positive ions are produced in a single aperture duoplasmatron ion source, which was rebuilt (2 mm distance between anode and zwischenelectrode) and equipped with a cylindrical expansion cup with 1 cm diameter, 1 cm depth. H^+-yields of 80% can be achieved under optimum conditions. The extraction system consists of a usual cone shaped electrode, einzel-lens focussing and a decelleration system in front of the cesium channel. An extraction potential between 6 and 8 kV results in optimum metastable production in a charge exchange collision with cesium. The length of the cesium channel is 10 cm, the diameter can be changed between 1 and 2 cm using different inserts; the consumption rate is 1-2 g/24 hours.

Two iron shielded quenching coils (8 cm long) generate the magnetic field for selective quenching of metastable atoms. A Sona transition region (2) is used between the two coils. The vector polarization can be switched to zero with minimal influence on the ion beam using a small transverse magnetic field superimposed on the transition region. The selective ionisation to negative ions takes place in a 28 cm long, 2 cm diameter argon cell. A longitudinal magnetic guide field generates the quantization axis, which can be rotated by the following Wien filter arrangement acting on a 5 keV polarized beam. All units are mounted on side flanges, which allows a closed spacing while maintaining high pumping speeds.

The extracted polarized negative ion current is a strong function of the arc current in the duoplasmatron. Continuous operation with 8 A allows a H^--current of 600 nA (D^-: 1000 nA) with a 1 cm diameter cesium channel; both values are increased by a factor of 1,6 with a 2 cm diameter channel.

Quenching rate measurements (3) indicate a vector-polarization of 0,75 - 0,8 for H^-- and D^--ions. The analysing power of the ^{4}He(p,p)-reaction at 1,14 MeV served for the calibration of the proton polarisation after acceleration and resulted in $p_z = 0,69 \pm 0,02$, where the errors are due to the inaccuracy of the phase shift analysis; the stability of the polarisation is better than 10^{-2}. Preliminary measurements of the deuteron tensor polarization gave a value of $p_{zz} = -0,52$ using the known analysing power of the ^{3}He(d,p)-reaction (4).

The 1,2 MeV-Tandem-Accelerator was built in our laboratory using 2,5 atü isolating gas (90% N_2, 10% SF_6). The high voltage is produced by a 600 KV SAMES-generator and fed to the terminal via a shielded cable. Charge exchange inside the terminal is achieved in a 2 µg/cm^2 C-foil or a gas channel (15 cm long, 0,5 cm diameter). The transmission is 70 - 80%; the energy spread is approximately 0,5 keV.

Preliminary asymmetry data of proton-proton scattering at 1 MeV gave insight in the stability of the whole system; apparative asymmetries are smaller than 10^{-3}. Target currents of 400 - 500 nA of protons have been used for a (p,γ)-experiment and deuteron target currents up to 700 nA were achieved after a 2 mm diameter aperture.

References:
+ Supported in part by the Bundesministerium für Forschung und Technologie and Deutsche Forschungsgemeinschaft.

(1) W. Haeberli, Ann.Rev.Nucl.Sci. 17 (1967) 373
(2) P.G. Sona, Energia Nucleare 14 (1967) 295
(3) G.G. Ohlsen, et al., Phys.Rev.Letters 27 (1971) 599
(4) L. Brown, et al., Nucl.Phys. 79 (1966) 459

THE MUNICH POLARIZED ION SOURCE

D. Ehrlich, R. Frick, P. Schiemenz
Universität München, Abteilung Kernphysik
8046 Garching, Am Coulombwall 1

A lambshifttype ion source was built for the Munich MP-Tandem. Protons or Deuterons are produced in a RF-source using inductive coupling of the RF-energy to the plasma. Typical operation is at 300 Watts forward power and less than 3 Watts reflected power.

The subsequent lens system consists of an einzel lens and a deceleration lens. The waist in the extraction region which is approximately 2 mm in diameter at an energy of 3.5 keV in case of protons is focussed to a waist of 24 mm diameter at an energy of 0.4 keV at the position of the Sona zero field crossing.

The Cesium cell having apertures of 34 mm in diameter on both sides is constructed for minimum output of Cesium into the vacuum system. Cesium atoms which are evaporated at 150°C from the oven cannot escape directly through the apertures. They are condensed at the top of the cell at 40°C and refilled into the oven. With 50 gr of Cesium the cell was operated for more than 200 hours.

Polarization is produced by a Sona type magnetic field and electric quenching and deflection fields. The solenoid arrangement contains no iron and magnetic field gradients are below 1 Gauß/cm in the zerofield region.

The beam is ionized in an Argon cell with entrance and exit apertures of 37 and 51 mm in diameter which is pumped differentially. Weak or strong magnetic fields are provided in the ionization region.

The polarized beam is accelerated behind the argon cell to 2.5 keV in case of Hydrogen and focussed into a Faraday cup. In case of Deuterium all beam energies in the source are doubled. Beam transportation vacuum is approximately $1 \cdot 10^{-5}$ Torr throughout the source.

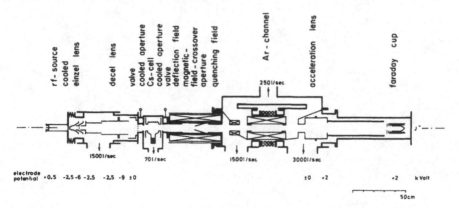

Fig. 1 General view of the Munich polarized ion source

The operation conditions of the polarized source have been optimized and the following values for beam intensity and polarization have been obtained

| | Protons | Deuterons |

Protons

$J^- = 0.7\ \mu A,\ p = 0.85$

Deuterons

$J^- = 0.8\ \mu A,\ p_z = 0.45$

$$p_{zz} \simeq 0$$

The polarization was calculated from quenching ratios. Typical values in case of Hydrogen were 760 nA with Argon and quenching field equal to 10 V/cm, 160 nA with Argon and quenching field equal to 100 V/cm, and 60 nA without Argon and quenching field equal to 100 V/cm.

The spinhandling and beam transportation system to the tandem accelerator is tested at the present time.

THE ERLANGEN LAMB SHIFT POLARIZED ION SOURCE[+]

J.H. Feist, B. Granz*, G. Graw**, H. Löh, H. Schultz***, H. Treiber****
University Erlangen-Nürnberg, West-Germany

A Lamb Shift Polarized Ion Source was installed at the Erlangen HVEC EN-Tandem in May 1974. Scince then the source worked for 2500 hours at the accelerator without any severe failure. The available current at the target is 40 - 60 nA with a polarization between 70 - 80% of the theoretical values.

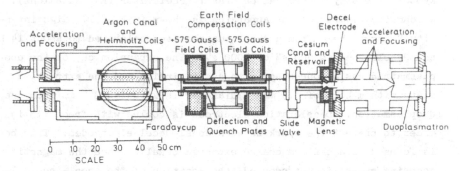

The figure shows a schematic drawing of the source. The positive proton or deuteron beam is extracted from a duoplasmatron and focussed by a 3 elektrode system into the caesium canal. The properties of the duoplasmatron limit this current to 0.5 mA. The caesium canal (116mm long, 10mm Ø) is operated at 150°C (Reservoir 115°C). The consumption of caesium is 20g/400h.

The polarization of the beam is produced with the sudden field reversal method. At the zero transition a gradient of 1.5 Oe/cm is used, the earth magnetic field can be compensated to less than 20 mOe. The compensation coils can be used also to destroy the polarization without changing the beam quality.

Negativ ions are produced in an argon gas canal (160mm long, 15mm Ø) with a magnetic field up to 200 Oe for the production of vector- or tensor polarized ions. A system of Helmholtz coils again compensates the earth magnetic field.

The source is operated at a (-50 kV) potential. The components of the spin-handling system are a 36° bending magnet, a Wienfilter and a Solenoid. The transmission of the polarized part of the beam from the first cup to the target is at most 40%, for the unpolarized part 10%.

First measurements for tensorpolarisation showed for a "purely" vector polarized beam a tensorpolarisation of 5% and vice versa.

[+] Work supported in part by the German BMFT and DFG
present address:
*   CERN, Genf, Switzerland
**   LMU München, Sektion Physik, Germany
***   Fa. Triumph, Nürnberg, Germany
****   Fa. Agfa, München, Germany

LAMB SHIFT TYPE POLARIZED ION SOURCE OF KYOTO UNIVERSITY

H. Sakaguchi and S. Kobayashi

Department of Physics, Kyoto University, Kyoto Japan

A Lamb-shift polarized ion source is under construction for the Kyoto University tandem Van de Graaff Accelerator (EN equivalent). A schmatic diagram of the apparatus is shown in Fig. 1. High intensity H^- or D^- beam from a duoplasmatron is accelerated to about 14 kV and decelerated to 500 eV or 1000 eV. The expansion cup of the duoplasmatron has a cone-shape with a slit inside and the extraction plate with a hole of 6.5 mm in diameter is situated 3.5 mm from the duoplasmatron. Deceleration electrode is cooled with freon 113 in order to prevent the spark between accel-decel electrodes. This beam is focused to a cesicum charge exchange canal. Since the magnetic focusing system is not adopted, the position of the accel-decel-electrodes is very critical. Following the cecium canal are deflection plates to remove charged components from the beam and the beam enters into the polarization region.

To polarize the metastable component of the beam, the method of sudden field-reversal as suggested by Sona [1] is used. 8 solenoid coils produce the required adiabatic field change and the rapid change of field sign for the metastable atom. The power supply of these coils are remote controled to choose the polarization mode of operation.

The polarized metastable atoms are selectively ionized in a argon charge exchange canal. And the emerged polarized H^- or D^- ions are pre-accelerated to inject into the accelerator. Prior to the injection the polarization axis of the beam is changed by a Wien-Filter to suit the experimental requirements. The polarized source is operated at -50 kV to -80 kV with respect to ground. Freon gas heat exchanger is used to cool the water bath at the high potential side. The ion souce is still under developments and testing.

[1] P. G. Sona, Energia Nucleare 14 (1967) 295.

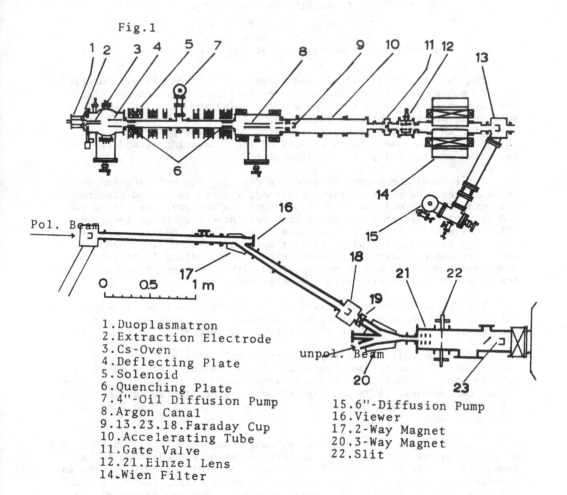

Fig.1

Pol. Beam

0   0.5   1 m

unpol. Beam

1.Duoplasmatron
2.Extraction Electrode
3.Cs-Oven
4.Deflecting Plate
5.Solenoid
6.Quenching Plate
7.4"-Oil Diffusion Pump
8.Argon Canal
9.13.23.18.Faraday Cup
10.Accelerating Tube
11.Gate Valve
12.21.Einzel Lens
14.Wien Filter

15.6"-Diffusion Pump
16.Viewer
17.2-Way Magnet
20.3-Way Magnet
22.Slit

THE UNIVERSITY OF WASHINGTON POLARIZED ION SOURCE*

T.A. Trainor and W.B. Ingalls
Nuclear Physics Laboratory
University of Washington, Seattle, Washington, USA

The UW Lamb-shift source[1] was installed in March, 1971. Polarization is achieved with the Sona magnetic-field-crossover scheme. Experience gained in operating this ion source during the past four years as well as information obtained from other installations have provided the basis for an extensive redesign completed during the past year. The design effort has been timely in view of the increasing demand for more beam time and intensity at this laboratory.

The revised ion source is shown in fig. 1. The existing support frame, light-pipe remote control system[2], vacuum system and electronics are retained with only minor modifications. A new positive ion source and extraction system and a new cesium cell have been installed. The 575 G coils and associated drift-tube sections were modified about two years ago in connection with the fast spin-flip system[3]. Other components are under construction or undergoing tests before installation. Completed components are being installed without interrupting the research program associated with the polarized source.

The duoplasmatron uses a symmetric, 70°-cone accel-decel geometry with accel and decel electrodes movable axially and transversely[4]. This ion source delivers 15-20 mA with 11-12 A arc current.

The cesium cell is a one-piece copper unit with a single, 150 W, thermostatically controlled heater and freon cooling. It is found advisable to wash the outside of the cell after each source opening to insure stable performance. A 25 g charge of cesium generally lasts 4-6 months. The cell operates typically at 110° C. The canal is 12.5 cm long and 1.25 cm diameter.

The source presently operates without a magnetic lens. In this mode of operation the movable extraction electrodes have been found to be invaluable. This is because slight changes in cesium density in the extraction region alter the focal properties and steering there. During the first day of operation the electrodes must be moved frequently to maintain constant output. After this time an equilibrium condition exists and only daily adjustment is necessary. In addition, a freon-cooled baffle has been placed between the cesium cell and extraction region to provide cesium trapping which would otherwise be provided by cooled magnetic lens elements.

A new vacuum enclosure between 575 G argon coils provides needed argon pumping between cesium and argon cells and allows installation of a valve to isolate the cesium region for servicing.

All electrostatic lens elements are twice the diameter used in the original design[1]. The Einzel lens following the argon cell is partially screened to improve pumping.

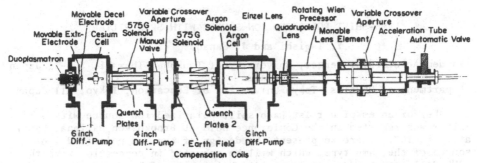

Fig. 1. University of Washington polarized ion source elevation.

The rotating Wien precessor is capable of rotating 1 kV deuteron spins through 150° in the plane of the magnet pole pieces. The precessor is 30 cm diam by 30 cm long. Field uniformity is about 5/10,000 over a 1.5 cm diam beam cross section. A singlet lens preceding the precessor compensates for a cylindrical lens component.

The acceleration tube follows the Los Alamos design. The movable lens element allows steering to position a beam waist at the small cross-over aperture.

Source operation at present is quite stable and reproducible. Proton beams are typically 20-40 nA with $p_z = 0.65$. Deuteron beams are typically 40-60 nA with $p_z = (0.7 - 0.8) \times 0.67$ and $p_{zz} \sim 0.01 - 0.02$. Variation of beam intensity is due in large part to variation of accelerator transmission with voltage and condition of stripper foils. Transmission from source to target is typically 30%. We have recently confirmed that production of a deuteron beam with pure tensor polarization is possible by application of the fast spin-flip system[3] in a manner uncorrelated with the data-acquisition. Measured values of p_z in this mode were less than the statistical accuracy of 0.003 in the measurement.

References

*   Supported in part by USERDA
1) H. Fauska et al., Nucl. Phys. Lab. Annual Report, Univ. of Washington (1971), p. 6 and subsequent reports
2) Ibid., p. 29
3) E.G. Adelberger, M.D. Cooper and H.F. Swanson, Nucl. Phys. Lab. Annual Report, Univ. of Wash. (1973), p. 1
4) W.B. Ingalls, T.A. Trainor, and Staff, Nucl. Phys. Lab. Annual Report, Univ. of Wash. (1975), p. 18

LAMB-SHIFT POLARIZED ION SOURCE WITH SPIN FILTER

Y. Tagishi and J. Sanada

Tandem Accelerator Center, University of Tsukuba, Ibaraki, 300-31 Japan
and
Department of Physics, Tokyo University of Education, Tokyo, 211 Japan

Performances of our 1st Lamb-shift polarized ion source with spin filter was reported in the Conference held at Brookhaven National Laboratory in 1971. Here we present the design of the 2nd polarized ion source of the same type, which will be operated in connection with the 12UD Pelletron tandem accelerator of University of Tsukuba.

We have tried to make some improvements in this 2nd polarized ion source in referring experiences on our 1st source and reported results of sources at other laboratories such as Los Alamos Scientific Lab. and Triangle Universities.

Fig. 1 shows the general layout of the source. Positive ions are produced in a Duoplasmatron. 10 A of the arc current is designed. The extraction electrode is a flat stainless steel plate, which can be moved in any direction without braking the vacuum. The Cs oven is made of stainless steel. It has a reservoir of Cs metal and a charge exchange canal whose inside diameter is 1 cm and length is 10 cm. The reservoir can be sealed off by a tapered stainless steel rod which is operated from the outside of the vacuum tank.

A stainless steel vacuum tank which contains the duoplasmatron, extraction system with a magnetic lens and the cesium cell is separated from the vacuum tank which contains remainders of the source. These two tanks have a hole, respectively, and are connected by a small bellows. The size of holes is just large enough to permit the passage of the beam. A small gate valve is equipped to a hole so that evacuation of each tank can be performed independently. Such constitution increases the feasibility for alignment of the system.

The RF cavity is a cylinder made of brass with 14.66 cm inside diameter and 15 cm length and has similar feature to that use in the source at Los Alamos. The Q of the cavity is 3700. The RF power of 1610 MHz is fed from an RF power source of 2.25 W maximum output. We can select RF power level between two levels: the lower for ordinary operation and the higher for full quenching of the metastable atoms.

To produce the uniform magnetic field for the spin filter an iron-shielded solenoid is made. This solenoid coils are composed of a main coil, two correction coils, two end coils and a modulation coil, with which the magnetic field can be modulated by $\pm$ 30 G. The solenoid system is placed outside the vacuum tank to reduce troubles for vacuum.

Argon cell is a cylindrical stainless steel pipe of 4 cm inside diameter and 30 cm length. To establish a quantization axis for the nuclear spin an other uniform magnetic field is provided in the argon region.

Three different types of vacuum pumps are used. Specifications of these pumps are described in the figure caption of Fig. 1.

The polarized source described above is insulated from the earth potential to 100 kV.

At present all components are delivered and we are assemling part by part.

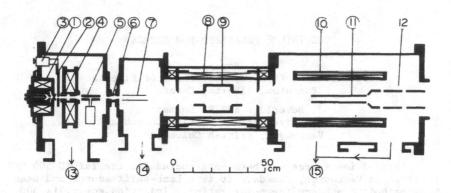

Fig.1. Schematic cross section of the polarized ion source.

1. duoplasmatron, 2. extraction electrode, 3. x,y,z goniometer,
4. magnetic lens, 5. Cs cell, 6. bellows, 7. deflection plates,
8. 575 G solenoid coil, 9. RF cavity, 10. argon solenoid,
11. argon cell, 12. focussing electrode, 13. to oil diffusion pump
(2400 l/s unbaffled), 14. to ion pump (1000 l/s H_2), 15. to cryopump
(4000 l/s Ar).

THE TRIUMF POLARIZED ION SOURCE

G. Roy
Department of Physics, University of Alberta
Edmonton, Alberta, Canada

J. Beveridge and P. Bosman
TRIUMF, University of British Columbia
Vancouver, British Columbia

A polarized ion source has been constructed for the TRIUMF 500 MeV H^- cyclotron in Vancouver, Canada. It is a Lamb-shift source and uses the Sona method to enhance the polarization. Injection energy is 300 kilovolts and the beam will be injected axially into the cyclotron.

The source is modular in construction and has four 2000 litre/second diffusion pumps. The duoplasmatron and accel-decel system are based on the Los Alamos design, as is the gap lens which injects the beam into the accelerator tube. A small analyzing magnet is placed just after this lens in order to aid in tuning the ion source. The source electronics are predominantly of our manufacture and are controlled via a light-link and Camac equipment.

The maximum output of the source has been 520 nanoamperes of H^- beam, with a polarization of 75% as calculated from atomic substate population measurements. More typical currents are 300 - 400 nanoamperes. No deuteron beam has been attempted at this time.

The spin precessor is a Wien filter and operates on the 300 kilovolt beam. The injection line (~100 feet long) is completely electrostatic. A steel pipe surrounds the injection line to reduce the stray field from the cyclotron magnet.

It is hoped to inject polarized beam into the cyclotron by August, 1975.

ALTERNATING POLARIZATION AT 1000 Hz IN A LAMB-SHIFT H SOURCE*

Joseph L. McKibben and James M. Potter
Los Alamos Scientific Laboratory, Los Alamos, New Mexico 87545, USA

Preliminary results of a search for parity violation in the scattering of longitudinally polarized protons on hydrogen at 15 MeV has already been reported[1]. This experiment requires a means of rapidly alternating the sign of the polarization so that the tiny effect can be sorted from the noise in the beam through the tandem by use of a lock-in amplifier. Reversal in sign had previously been accomplished by reversing the currents to both the spin filter and argon fields[2]; however, this process takes seconds because of the inductance in the coil of the spin-filter.

In 1972 R. R. Lewis[3] suggested that it was only necessary to alternate the argon field of only 0.6 mT in order to alternate the spin, provided a transverse field is present at the zero in the longitudinal field to adiabatically guide the spin around. While this did provide an effective method of alternating the spin, it was accompanied by an alternating displacement of the beam that gave a false experimental result.

Late in 1973 R. E. Mischke[4] called our attention to a reversal system in which the field about the region where the H$^-$ ions are formed is not alternated. With the argon field opposite in sign to the spin filter, the spin of the beam is alternated as a transverse magnetic field is turned on and off. The system is related to the Sona method of polarization in that $(m_j=+1/2, m_I=+1/2)$-metastable-hydrogen atoms will pass through the zero without reversal of spin provided no transverse field is present; however, the presence of only a 0.22 mT field is sufficient to flip the spin of the atoms. The rotation appears to involve adiabatic following of the field near 0° and 180°, but pure precession near 90°.

The basic apparatus installed in our source between the spin filter and the argon cell for producing alternating reversal is depicted in Fig. 1. The field coil (T) is shaped so the transverse field merges smoothly into the longitudinal field. Some of the coils used to produce the longitudinal field are shown in cross section. Stray magnetic fields are kept out of the reversal and argon regions by magnetic shields. Curves L and LR shown below the figure give the optimum fields on each side of the zero, note L is opposite to LR. The optimum value of the transverse field was also experimentally determined and is given by curve T. In operation a transistor circuit turns on and off the field at 1000 Hz using a square wave synchronized to the lock-in amplifier sensing the parity-violating asymmetry.

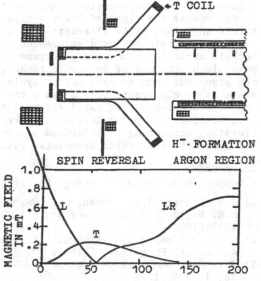

Fig. 1.   (length in mm)

The magnitude of the polarization alternated was measured several times, usually in the scattering chamber of the experiment. The left-right (or up-down) asymmetry of the scattered protons was detected by scintillators on opposite sides of the chamber. The difference in the current out of the two photomultipliers viewing the scintillators is proportional to the polarization. The spin is precessed normal to the scattering plane for the measurement. The protons are scattered with a fairly thick carbon foil since the analyzing power of hydrogen is very small. Using the slow but thorough reversing system and the polarization value measured from quenching[2], the analyzing power of the foil was found to be 10%. Returning to 1000 Hz reversal, the magnetic fields were easily optimized while watching the meter on the lock-in amplifier connected to the photomultipliers; this was readily possible because the scattering event rate was very high. This process led to the field values given in Fig. 1 and polarization reduced from that of slow reversal by only 6%.

The more accurately that the intensities in the two phases can be kept equal in the source, the more accurately the parity effect can be measured in the scattering chamber. The chief cause for the intensities being unequal is the Lorentz force that quenches some of the metastable atoms as they pass through the transverse field at $v/c \sim 10^{-3}$. We compensate this Lorentz force with an electric field that is turned on and off with the magnetic field; one of the electrodes is depicted in Fig. 1. We have been able to hold the two phases equal to one part in 10^4 by adjusting the compensation. Recently L. B. Sorensen has installed feedback from the beam current monitor and has the equality to one part in 10^5.

The most troublesome error in the parity experiment comes from beam displacement synchronized with polarization reversal. While this is much reduced in the new system where H^- ions are mostly formed in a constant field; unfortunately, some argon is present in the reversal region. On the other hand, if the Lorentz force were perfectly compensated by the electric field, no beam displacement would occur. We are improving the accuracy of the compensation. Another trick we have used for some time is to reverse the signs of the two transverse fields each time they are turned on. This still alternates the polarization but transforms 1000 Hz displacement into 500 Hz displacement that tends to get ignored by the lock-in amplifier. Moreover, a tremendous increase in pumping speed is possible by the use of cryogenic techniques. We plan to install soon a new argon cell cooled by a helium-cycle refrigerator at 20° K. These measures should practically elimate the beam displacement problem.

The development of rapid reversal tehcniques in the LASL source has progressed to the point that we believe the experimental error in parity-violating scattering of polarized protons in hydrogen will be determined solely by the statistics associated with events to well below 1×10^{-7}.

References

* Supported by the U. S. Energy Research and Development Administration

1) J.M. Potter, J.D. Bowman, C.F. Hwang , J.L. McKibben, R.E. Mischke, E.E. Nagle, P.G. Debrunner and L.B. Sorensen, Phys. Rev. Letters 33, 1307 (1974).
2) G. G. Ohlsen, J.L. McKibben, G.P. Lawrence, P.W. Keaton, Jr. and D.D. Armstrong, Phys. Rev. Letters 27, 599 (1971).
3) J.L. McKibben, R.A. Hardekopf and R.R. Lewis, Bull. Am. Phys. Soc. 18, 4, 618 (1973).
4) R.E. Mischke, private communication.

OPERATION OF THE LASL POLARIZED TRITON SOURCE*

R. A. Hardekopf

Los Alamos Scientific Laboratory, Los Alamos, New Mexico 87545 USA

A Lamb-shift polarized ion source, specifically designed for use with tritium has been constructed for the Los Alamos Scientific Laboratory tandem accelerator. This report describes briefly some of the more important design features and the initial operation. The first data obtained using this source are reported elsewhere in these proceedings.

Type and Location. A Lamb-shift source was chosen because of our experience with this type and because of its relatively low gas consumption. A newly designed spin-filter[1], about half the length of the original model, has been incorporated. This device permits an 80-90% polarized triton beam to be produced with its polarization absolutely known via the quench ratio[2]. The source is oriented vertically, as shown in fig. 1, with injection of the 60 kV beam into the tandem by an electrostatic mirror located 115 cm from the end of the accelerator. This arrangement provides vertical spin orientation at the target without the need for a spin precessor.

Ion Source. A high-current duoplasmatron, similar to the one used on the LASL polarized p-d source, produces a positive triton beam. Longitudinal and transverse adjustment of the extraction electrode and transverse adjustment of the intermediate electrode are possible during source operation. The arc requires about 60 atm-cm^3/hr of tritium gas at a pressure of about 0.2 Torr. A 500 ℓ/sec turbomolecular pump backed by a 70 ℓ/sec mercury diffusion pump with a forepressure tolerance of 35 Torr allow recirculation of this gas back into the arc region. Fig. 2 is a schematic of the pumping system, including the charcoal trap used for collecting the tritium gas after source shut-down. The inventory of T$_2$ present in the source during operation is only about 13 cm^3 (35 Curies).

Cesium Region. Following the suggestion of Bacal et al.[3], we have developed a cesium canal that recirculates cesium via capillary action in a stainless steel mesh. After initial saturation of the mesh from a heated reservoir, the ends of the canal are cooled to about 30°C, at which temperature cesium losses out the ends are minimal. An automatic controller maintains the center of the canal at about 100°C.

An important feature of this source is the ability to valve off the cesium region from both the

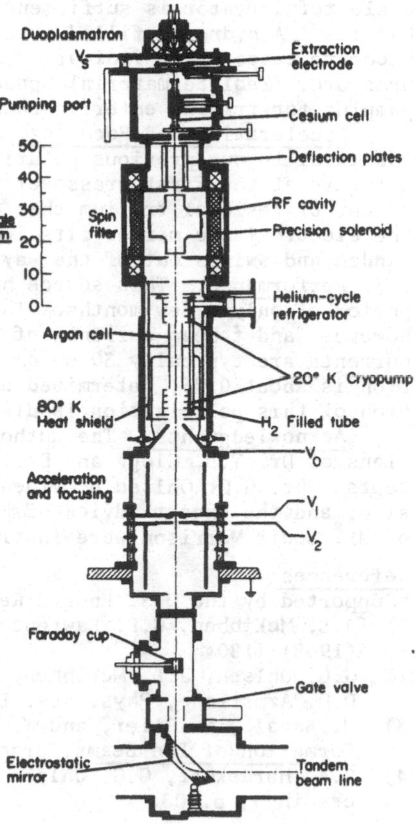

Fig. 1

duoplasmatron and argon regions. This allows filament changes and other minor maintenance without contamination of the cesium. The canals themselves are inexpensive to make and easily replaced if necessary.

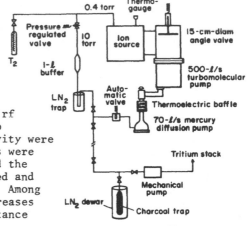

Fig. 2

Spin Filter. We have made several important changes to this device. The active length of the rf cavity was decreased from 15 cm to 7.5 cm, the "end pipes" of the cavity were DC isolated, the deflection plates were moved into the magnetic field, and the magnetic field coils were shortened and placed outside the vacuum jacket. Among other advantages, this design decreases the cesium cell to argon cell distance to 60 cm.

Argon Region. Cryopumping of the argon at ∼20°K provides high pumping speed, an extremely clean vacuum, and containment of any tritium entering this region. A 1 watt helium-cycle refrigerator is sufficient to cool an upper and lower set of baffles. A hydrogen-filled tube acts as a heat pipe and also provides accurate temperature indication by monitoring the hydrogen vapor pressure. Zeolite material bonded to a cryopanel provides additional pumping for tritium entering this region.

Acceleration and Focusing. The two-gap acceleration system used is identical to our previous polarized source design. A 1.8 mm diameter aperture at the first crossover provides clean optics and reduces the amount of residual tritium that can enter the accelerator vacuum system. The electrostatic mirror fits into the existing pump box on the FN tandem and swings out of the way when not in use.

Performance. This source has operated on the accelerator for a period of only a few months. The performance has been quite reliable, however, and $\vec{t}$ beam currents of 200 nA at the source are common. Target currents are typically 30-60 nA at this time, and triton beam polarization is about 0.85, determined on-line to better than 0.01. The calibration of this polarization is discussed elsewhere.[4]

Acknowledgments. The author gratefully acknowledges the contributions of Dr. T.B. Clegg and Dr. P.W. Keaton to the initial source concepts. Dr. G.G. Ohlsen provided considerable help in the spin-filter design, and the expert advice of Dr. J.L. McKibben and technical assistance of Mr. Louis Morrison were instrumental at all phases of this project.

References
*Supported by the U.S. Energy Research and Development Administration.
1) J.L. McKibben, G.P. Lawrence and G.G. Ohlsen, Phys. Rev. Letters 20 (1968) 1180
2) G.G. Ohlsen, J.L. McKibben, G.P. Lawrence, P.W. Keaton, Jr., and D.D. Armstrong, Phys. Rev. Letters 27 (1971) 599
3) M. Bacal , N. Olier, and W. Reichelt, Symposium on Ion Sources and Formation of Ion Beams, Brookhaven, 1971 (post deadline paper)
4) R.A. Hardekopf, G.G. Ohlsen, R.V. Poore, and N. Jarmie, these proceedings, p.903.

CONSTRUCTION AND OPERATION OF A POLARIZED ^{3}He ION SOURCE

K. Allenby, G.H. Guest, W.C. Hardy, O. Karban and W.B. Powell
Department of Physics, University of Birmingham, B15 2TT, England

There is little doubt that the use of accelerated beams of particles from polarized ion sources is experimentally most convenient for a study of reaction analysing powers. However, in the case of ^{3}He particles no such beam has been available to the experimentalists. This fact stimulated experimental and theoretical study of the physical principles of a source of polarized ^{3}He ions and its eventual construction at the Radial Ridge Cyclotron.

The basic ideas of the source were described in detail in ref. 1 and can be briefly summarized as follows: when a beam of doubly charged ^{3}He^{++} ions is incident on a thin gas target a part of it experiences electron capture to the ground (1S) or metastable (2S) state of the singly charged ^{3}He$^+$ ion. After Zeeman splitting and consequent quenching of the bottom metastable electron state the remaining metastable fraction carries a 50% nuclear polarization. To separate the latter from the unpolarized 1S fraction the beam passes through another gas target in which electron stripping from the metastable state is strongly enhanced with respect to the ground state. The resulting ^{3}He^{++} beam, originating mostly from the polarized metastable fraction is then ready for acceleration.

The feasibility of constructing a polarized ^{3}He source on such principles depends on cross sections of two atomic processes: a formation of the metastable ^{3}He^{++} ions, and selective ionisation of the metastable state. When this work was commenced there was no experimental information on either of these processes, therefore experiments were undertaken to test the basic ideas. The results obtained can be briefly summarised as follows:

(1) about 85% of the resulting doubly charged beam is formed from the 50% polarized metastable fraction;
(2) about 0.075% of the original ^{3}He^{++} beam can be converted into polarized beam ready for injection into the cyclotron;
(3) air may be used in both gas targets without a significant loss in both intensity and polarization;
(4) the optimum energy for the atomic processes is close to the required injection energy of 30 keV.

The source utilizes axial injection, and duringoperation is mounted directly above the cyclotron magnet. The design allows the source to be easily removed, so it is interchangeable with the polarized deuteron source.

The polarized ^{3}He source has been in operation for about three months, during which time a large amount of data on nuclear scattering and reactions was accumulated. Typical source characteristics were the following: on target intensity, 0.15 na; beam polarization, 0.38. The polarization was measured and monitored using the p - ^{3}He scattering, and typical asymmetries of recoil protons from a polyethylene target at 27° (lab) were 0.06. An absolute calibration was obtained using the ^{3}He - ^{4}He scattering at 13 MeV giving a systematic error on the beam polarization $P_{beam} = 0.38 \begin{subarray}{l} + 0.02 \\ - 0.04 \end{subarray}$. The polarization was switched automatically between positive and negative values, by means of a precessor coil producing a transverse magnetic field to rotatethe electron spin

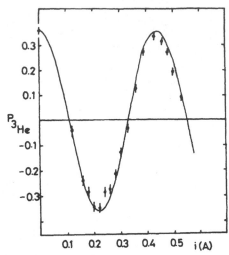

fig. 1
Beam Polarization as a function
of spin precessor coil current.

by 180°. Results obtained with
this spin precessor are shown in
fig. 1.

During an initial test run,
measurements were made of the beam
polarization as a function of the
axial magnetic field in the radio-
frequency cavity (1). These results have been explained by considering in detail the quenching
process, including the effect of
resonance saturation, by solving
exactly the coupled equations for
the magnetic substates of the $2S_{\frac{1}{2}}$
and $2P_{\frac{1}{2}}$ levels. By fitting the
calculated curve to the experimental points, it was deduced that the
amplitude of the quenching field
during normal operation of the
source is 50 ± 5 V/mm, fig. 2.
The broad plateau evident in the
figure guarantees that possible
fluctuations in the magnetic field
will not affect the beam polarization.

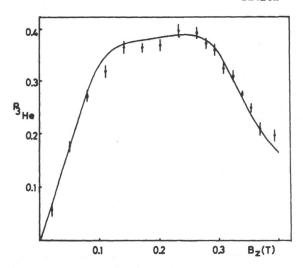

fig. 2
Beam Polarization as a function of axial
magnetic field in radiofrequency cavity.

References
(1) W.E. Burcham, O. Karban, S. Oh and W.B. Powell, Nucl.Instr. 116
     (1974) 1
(2) O. Karban, S.W. Oh and W.B. Powell, Phys.Rev.Lett., 31 (1973) 109

SOME ASPECTS OF A POLARIZED ^{3}He SOURCE PROJECT

O. Karban, S. Oh[+] and W.B. Powell
Department of Physics, University of Birmingham, Birmingham B15 2TT,
England

The finding of a suitable polarization principle[1]) and of an
efficient charge changing process[2]) have already been described in some
detail, while the operation of the Birmingham polarized ^{3}He source is
briefly described in a contribution to this Conference.   It is the purpose of the present paper to discuss a few of the other problem areas,
most of which are indicated in the schematic diagram of fig.1.

Early in the project there was an investigation into possible ^{3}He
depolarization during injection and acceleration in the magnetic field of
the cyclotron.   However, after noting that the spin precession/cyclotron
frequency ratio   for ^{3}He^{++} is only a little less favourable than for
protons (3.2 compared with 2.8 for protons) and bearing in mind that proton depolarization is not generally observed it was established without
much difficulty that measurable depolarization effects were unlikely.
Operation of the source has confirmed this.

Potentially more serious difficulties arise from the knowledge that
at the low injection energy of 32 keV background gas could cause beam
loss by electron attachment to the ^{3}He^{++} ions.   The concern was with the
axial injection hole which is 1 m long, narrow, and full of quadrupole
lenses and their associated components;   pumping is only at the ends of
the hole which has a low pumping speed.   Beam loss should begin at
pressures of about 10^{-5} Torr.   Only when final tests were made did it
become clear that the outgassing rates were sufficiently low and the
problem had been overcome.

A further uncertainty only completely resolved with the final tests,
concerned the criticality of magnetic fields in the region between the
gas canals.   The need to minimise the quenching effects of radial magnetic field components is well known, requiring as it does a gentle change
when going from a high to a low field region.   In addition, at low
fields, depolarization is more likely because the adiabaticity requirement
is harder to meet – again, if there is change, it must be gentle.   In
the present design there is an extra hazard due to the inevitable penetration of the stray fields of the cyclotron (~0.03 T) into the low
field region where a gentle change to about 0.001 T is required.   This
penetration comes about because it is necessary to break the magnetic
shield (as indicated in fig.1) in order to apply the 1 kV decelerating
potential required for energy selection.   A big effort was put into reducing this problem although it is now known that these fields are less
critical than was at first supposed.

Several difficulties arose in the design of the 10 GHz cavity
(cylindrical, 22 mm dia x 13 mm high) used to quench the bottom metastable
state.   For example, ways had to be found of keeping the Q-value high in
spite of the 10 mm diameter beam holes at each end[3]);   however, the chief
problem anticipated – unwanted ionization and electron storage in the
crossed E and B fields – was never encountered.

Of all the factors perhaps the most critical concerned the need for
total elimination of the large unpolarized background of ^{3}He^{++} which comes
unchanged from the ion source, without affecting the polarized ^{3}He^{++} (the
ratio between the two is about 1000: 1).   The scheme to achieve this
involved adding 1 keV of energy to the polarized ions only be applying
+1 kV to the second gas canal   (this happens because the polarized ions

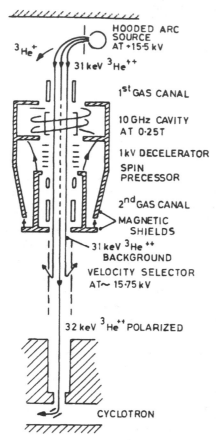

HOODED ARC
SOURCE
AT +15·5 kV

^{3}He$^+$

31 keV ^{3}He^{++}

1st GAS CANAL

10 GHz CAVITY
AT 0·25 T

1 kV DECELERATOR
SPIN
PRECESSOR

2nd GAS CANAL

MAGNETIC
SHIELDS

31 keV ^{3}He^{++}
BACKGROUND

VELOCITY SELECTOR
AT~ 15·75 kV

32 keV ^{3}He^{++} POLARIZED

CYCLOTRON

approach the second gas canal in the
singly charged state while the background ions are doubly charged); both
beams then enter an energy selector
and are decelerated by a voltage
slightly more positive than that of
the arc source so that only the polarized beam can pass through. The
difficulties in this scheme are compounded, for on the one hand the energy difference between the two beams
cannot be allowed to exceed about 1keV
or the decelerating electric fields
would cause significant quenching, and
on the other hand the small energy
difference means that (i) to eliminate
the background beam the polarized ions
must have their energy decreased to
only a few hundred electron volts at
the centre of the energy selector when
disturbances due to stray fields might
be expected to cause severe beam
losses, and (ii) the energy differences
permitted in the ion source are also
very small - about 100 eV or so. In
view of these problems it was aston-
ishing to discover how well the energy
selector worked in practice. The
central electrode (at approximately
+15.75 kV) only had to be increased by
about 100 V for conditions to change
from full background to polarized beam
only.

It will be seen that the polarized
^{3}He source, while seemingly at the
beginning to be full of difficulties
has, in operation, turned out to be
stable and remarkably uncritical.

Fig.1   Some of the problem areas of
        the polarized ^{3}He source.

References
+ Now at the University of Manitoba
1) O. Karban, S.W. Oh and W.B. Powell, Phys.Rev.Lett. 31(1973) 109.
2) W.E. Burcham, O. Karban, S. Oh and W.B. Powell, Nucl.Inst. 116(1974) 1
3) D.W. Bennett, unpublished project.

THE SOURCE FOR POLARIZED 6-LITHIUM IONS AT THE HEIDELBERG EN-TANDEM

E.Steffens and H.Ebinghaus, I. Institut für Experimentalphysik, Hamburg

K.Bethge and F.Fiedler, Phys. Inst. der Universität Heidelberg

P.Egelhof, K.D.Hildenbrand, D.Kassen, R.Schäfer, W.Weiss and D.Fick

Max-Planck-Institut für Kernphysik, Heidelberg

The source was developed at the University of Hamburg[1,2,3] and installed at the Heidelberg EN-Tandem[4]. The principle is that of an atomic beam source, but the usually used electron bombardement ionizer is replaced by a surface ionizier with an efficiency close to one. The outline of the source is displayed in fig.1. An intense atomic beam is formed by evaporating ^{6}Li from an oven through a Laval nozzle (0.5 mm $\emptyset$). After that the core of the beam is skimmed by a conical and a plane heated orifice (1 and 1.5 mm $\emptyset$ respectively). The beam is polarized with respect to the elctron spin by passing a permanent 6-pole magnet. Than it enters a weak field transition. The transition unit, which had an efficiency of 90% [4] has been redesigned in some parts and reaches now full efficiency. The surface ionizer consists of a 10x25 mm^2, 0.1 thick tungsten strip heated to about 1000° C. A monolayer of oxygen which increases the work function of tungsten to values higher than the ionization energy of lithium[5] is provided by an oxygen leak near the tungsten strip. In this way a nearly 100% efficiency of the ionizer is achieved. Since the surface absorption time is known to be 20 to 100 ms for temperatures around 1000°C [6] a strong magnetic field (B/B_c=4) has to be provided for the ionization region in order to avoid depolarization effects and to define the direction of polarization. The current of the ^{6}Li-ions leaving the ionizer is typically 20 µA. The ion beam is formed by a "Whenelt-type" electrostatic lense system. The beam is further accelerated to 13 keV and focused into the canal (1 cm $\emptyset$, 10 cm long) of a potassium charge exchange cell at -11 keV. The choice of potassium with a maximum charge exchange rate of 2.1% at 10 keV instead of cesium with 4.5% at 5 keV [7] is a compromise in favour of easy handling of the source. A solenoid at high potential produces a strong magnetic field (B/B_c=7) to avoid depolarization during the charge exchange process. The source is installed in direction of the axis of the EN-tandem 4m apart from the HVEC-source box. The beam transport system consists of two quadrupole dupletts and a Wien-filter between them for rotating the spin

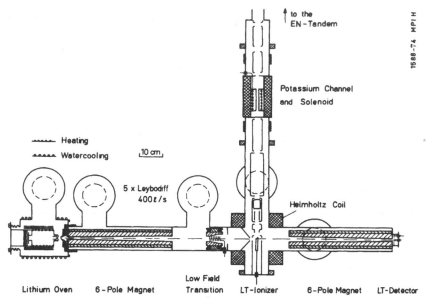

Fig.1 Source for a vector polarized ^{6}Li$^-$-beam at the Heidelberg EN-Tandem

of the ions perpendicular to the beam axis. Negative currents up to 240nA
are measured in front of the source box and about 80% is transported to
the low energy cup. The transmission rates through the tandem are now up
to 30%, since it is possible to move the stripper canal out of the beam
if the foil stripper is used. An analyzed current of about 100 nA (180 nA
max.) ^{6}Li^{3+} and a current of about 80 nA (150 nA max.) on the target can
be achieved. The polarization is not known yet exactly. Since the weak
field transition has a full efficiency now one can assume that the beam
will be purely vector polarized. From the largest asymmetry observed up
to now[8] one gets as a lower limit for the beam polarization $P \geq 0.52$.
The further development of the source is devoted to the production of
tensor polarized beams of ^{6}Li and other alkalides (^{7}Li, ^{23}Na).

1) H.Ebinghaus,U.Holm,H.V.Klapdor,H.Neuert, Z.Physik 199 (1967) 68

2) U.Holm,E.Steffens,H.Albrecht,H.Ebinghaus,H.Neuert,Z.Physik 233(1970)415

3) U.Holm,H.Ebinghaus, Nucl.Instr.Meth. 95 (1971)39

4) E.Steffens et al. Nucl.Instr.Meth. 124 (1975) 601

5) S.Datz,E.H.Taylor, J.chem.Phys. 25 (1956) 389 and references therein

6) M.Kaminsky, Ann. Phys. 18 (1966) 53

7) H.Ebinghaus et al. Ann.Rep. 1972 I. Inst. f. Exp.-Phys., Hamburg

8) P.Egelhof et al. contribution to this conference, p.825

POPULATION OF AN ATOMIC SUBSTATE BY THE WEAK FIELD
TRANSITION AND VECTOR POLARIZATION

D. Kassen and P. Egelhof (Max-Planck-Institut für Kernphysik, Heidelberg)

E. Steffens (I. Institut für Experimentalphysik der Universität Hamburg)

The Heidelberg source for polarized lithium ions[1] offers the possibility of measuring the population number N_4 of the 4th hfs component ($F = 3/2$, $m_F = -3/2$). This can be done by analyzing the electronic polarization of the atomic beam by a second 6-pole magnet[2]. A small portion of the atomic beam passes the ionizer strip and reaches the Langmuir-Taylor detector (see ref.1,Fig.1; ion current about 10 nA). Therefore the measurement of N_4 is possible during the normal operation of the source. The proper operation of the weak field transition, which is assumed to transfer m_F into $-m_F$ components, is indicated by a fall-off of the detector current to 2/3, when switching on the weak field transition. This means that the population number N_4 was changed by the transition from zero to 1/3.

Such measurements have been done as a function of the static magnetic field inside the transition simultaneously to measurements of the asymmetry in a nuclear reaction. The frequency of the weak field transition was 6.00 MHz. The gradient was about 1.2 Gs/cm and had positive[3] sign. The results are shown in Fig.1. ε is the asymmetry measured in the elastic scattering on ^{4}He[4] by a single telescope. C is a normalization factor to make the two guide lines for ε and $C \cdot N_4$ coincide in the center of the transition.

Both quantities have a plateau, but that of $C \cdot N_4$ is broader. There is no one-to-one correspondence between them if one compares the rise and fall region. This is in contrast to results[5] obtained by applying the Majorana formula which gives P_z, P_{zz} and N_4 in terms of a single parameter r , a fictitious transition probability.

Apart from this N_4 is a useful quantity to adjust the static field of the transition, if one fixes the gradient and oscillating field to their optimum values, and for controlling the transition during operation of the source.

874

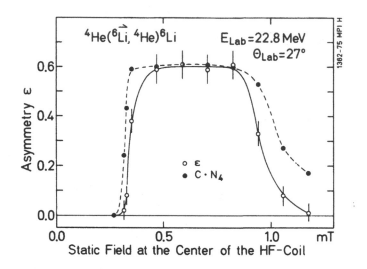

Fig. 1 Observed asymmetry ε and hfs-population number N_4 as a function
of the static magnetic field in a weak field transition.

References

1) E.Steffens, H.Ebinghaus, F.Fiedler, K.Bethge, G.Engelhardt, R.Schäfer,
    W.Weiss, D.Fick, Nucl.Instr.Meth. 124 (1975) 601 and contribution
    to this conference, p. 871
2) U.Holm and H.Ebinghaus, Nucl.Instr.Meth. 95 (1971) 39
3) S.Oh, Nucl.Instr.Meth. 82 (1970) 189
4) P.Egelhof, J.Barrette, P.Braun-Munzinger, D.Fick, C.K.Gelbke, D.Kassen,
    W.Weiss, E.Steffens, contribution to this conference.
5) K.Jeltsch, P.Huber, A.Janett, H.R.Strieber, Helv.Phys.Act. 43(1970)279.

A HIGHLY POLARIZED LITHIUM-6 TARGET

J. Ulbricht, F. Wittchow, U. Holm, K. D. Stahl and H. Ebinghaus

I. Institut für Experimentalphysik der Universität Hamburg

Experiments on the production of a polarized target by storing a polarized lithium beam on moving target backings showed too low relaxation times[1]. In this work a surface ionizer has been used to store the polarized particles. The desorption time is some 10 to 100 ms[2], and surface ionization conserves polarization[3]. A polarized lithium beam equivalent to 5μA from a conventional atomic beam source[1] hits a heated oxygenated tungsten tape placed in a strong magnetic field of 50 mT. The adsorbed lithium-6 ions were exposed to 20μA deuterons of 425 keV. The tensor polarization P_{zz} was calculated from the asymmetry of the α-particles from the reaction $^6Li(d,\alpha)\alpha$ using the known analyzing powers of the reaction $^6Li(\vec{d},\alpha)\alpha^{4,5}$[4,5]. Fig. 1b shows the experimental points of P_{zz}. In comparison with these values the solid curve was calculated from the efficiency of the hf-transitions given in fig. 1a.

Maximum values for P_{zz} are + o,81 $\pm$ 0,06 and - 0,73 $\pm$ 0,11.

Fig. 2 illustrates the nearly linear increase of P_{zz} and the nearly quadratic decrease of the counting rate N with temperature. The product $P_{zz}^2 \cdot N$ seems to be a constant. The target thickness was $3 \cdot 10^{14}$ lithium ions per cm^2. Further investigations are planned with ^6Li, ^7Li and other alkalis.

[1] H. Ebinghaus, E. Steffens, J. Ulbricht, F. Wittchow,
Nucl. Instr. Meth. 125(1975)73

[2] H. L. Daley, A. Y. Yahiku, J. Perel,
J. Chem. Phys. 52(1970)3577

[3] U. Holm, E. Steffens, H. Ebinghaus, H. Neuert,
Z. Physik 233(1970)415

[4] R. Neff, P. Huber, H. P. Nägele, H. Rudin, F. Seiler,
Helv. Phys. Acta 44(1971)679

[5] D. Fick, Z. Physik 237(1970)131

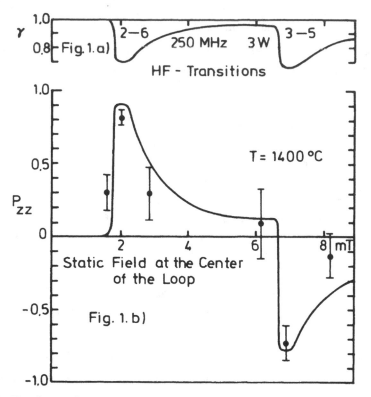

Fig.1a) The intensity of the atomic beam behind an analyzing sextupole
as a function of the static magnetic field in the hf-transition.
Fig.1b) The expected function P_{zz} via the static field calculated from
Fig.1a) and the ecperimental points from the reaction $^6\vec{Li}(d,\alpha)\alpha$.

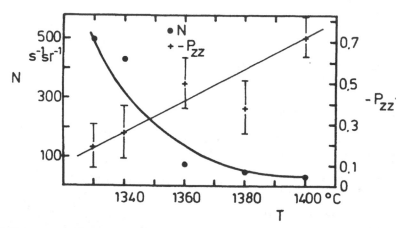

Fig.2) Mean counting rate N of the detectors and the negative tensor polarization $-P_{zz}$ of the ^{6}Li target against the temperature T of the tungsten.

GaAs: A NEW SOURCE OF POLARIZED ELECTRONS

D.T. Pierce[+], and F. Meier
Laboratorium für Festkörperphysik
Eidgenössische Technische Hochschule Zürich,
CH-8049 Zürich, Switzerland

By proper surface treatment, p-doped GaAs can be converted into a photoemitter of extremely high quantum efficiency Y. Scheer and van Laar obtained Y ≳ 0.1 electron per incident photon above threshold.[1] In this paper we show that when using circularly polarized light of suitable frequency instead of unpolarized light, the electrons emitted from GaAs are spin polarized up to 50 %.[2] Due to the simplicity of the device, GaAs may find wide application as a source of polarized electrons.

Technically, the high quantum efficiency is obtained by covering a clean surface of p-doped GaAs[3] with a $\sim$ 10 Å thick layer of Cs_2O[4]. Then a peculiar situation is reached where the vacuum level lies below the conduction band of the bulk GaAs, a phenomenon called negative electron affinity. Neglecting any band bending effects near the surface, it implies that all electrons in the conduction band have enough energy to escape into the vacuum. Then, the escape depth is determined by the recombination probability from the conduction band into the valence band. The corresponding escape depth is $\sim$ 1 to 10 μ, i.e. orders of magnitude larger than for most photoemitters. This explains the high quantum efficiency.

As a semiconductor, GaAs possesses a band gap E_G separating the fully occupied valence band from the conduction band; it is located at the center of the Brillouin zone, i.e. at the Γ - point, see insert Fig. 1. The band gap is $E_G \sim$ 1.4 eV. At Γ , the wave functions are atom-like and therefore the optical transition probabilities can be borrowed from atomic physics.[5] They are such that when GaAs is irradiated by circularly polarized light $h\nu$ = E_G, there will be a majority of one spin direction of the photoexcited electrons in the conduction band. This process is called optical pumping. A crucial requirement for getting a polarization in the conduction band is that the valence band is split by spin orbit interaction (into a $p_{3/2}$ and $p_{1/2}$ band in case of GaAs). At $h\nu$ = E_G only transitions from the $p_{3/2}$ band (4-fold degenerate at Γ) occur.

Neglecting any spin dependent scattering during photoemission, theory predicts 50 % polarization at $h\nu$ = 1.4 eV. Measurements made at 4.2 K showed that this value can be experimentally obtained within a few percent, see Fig. 1.

Measurements made at 77 K gave the same P as at 4.2 K. However, at room temperature P is reduced to roughly half its value at low temperatures.

When the light energy is only slightly increased above E_G, the polarization drops sharply. The main reason is that then transitions from the split-off valence band into the conduction band become possible. Again using atomic transition probabilities, one easily finds that P should go to zero, in agreement with experiment.

The structure around hν = 3 eV reflects special band structure properties of GaAs and is well understood.[8]

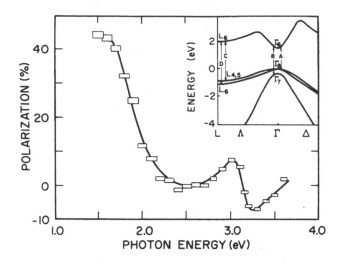

Fig.1. Polarization of photoemitted electrons as function of light energy. Insert: Band structure of GaAs: Gap $E_G = E(\Gamma_6) - E(\Gamma_8)$ Γ_7 is the top of the split-off $p_{1/2}$ band.

References

+ Present address: National Bureau of Standards, Surface and Electron Physics Section, Washington, D.C.20234
1) J.J. Scheer and J. van Laar, Solid State Commun. 3, 189 (1965)
2) The polarization is defined as P = (N↑ - N↓)/(N↑ + N↓); the quantization direction is the direction of propagation of the light.
3) Our GaAs crystals were doped with 1.3×10^{19} cm^{-3} Zn.
4) L.W. James, G.B. Antypas, J. Edgecumbe, R.L. Moon, and R.L. Bell, J.Appl.Phys. 42, 4976 (1971)
5) see e.g. H.G. Kuhn, Atomic Physics, p. 211 ff. (Longman, Green & Co. Ltd. (1969))
6) D.T. Pierce, F. Meier, and P. Zürcher, Phys. Lett. 51A, 465 (1975)

Measurements Techniques

A MOTT-SCHWINGER SCATTERING POLARIMETER FOR MeV NEUTRONS

AND POLARIZATION IN THE ^{2}H(d,n)^{3}He REACTION

R.B. Galloway and R. Martinez Lugo
Physics Department, University of Edinburgh, Scotland

Small angle scattering systems employed hitherto in the determination of the polarization of neutrons in the MeV region[1-3] have been
designed primarily to investigate small angle scattering theory[4].
Consequently they have particularly fine collimation of the neutron beam
incident on the scatterer, very small spread in scattering angle
accepted by the neutron detectors and necessarily very low efficiency
of data collection. The particular attraction of an efficient Mott-
Schwinger scattering polarimeter is that the neutron polarization, Pn,
is determined directly from experimental observables,

$$Pn = \frac{\sigma(\theta,0) - \sigma(\theta,\pi)}{(k\sigma_T\gamma \cot\theta/2)/\pi}$$ in the nomenclature of ref.[4]. It has

been suggested[5] that it should be possible to construct a polarimeter
for neutrons of a few MeV energy based on small angle Mott-Schwinger

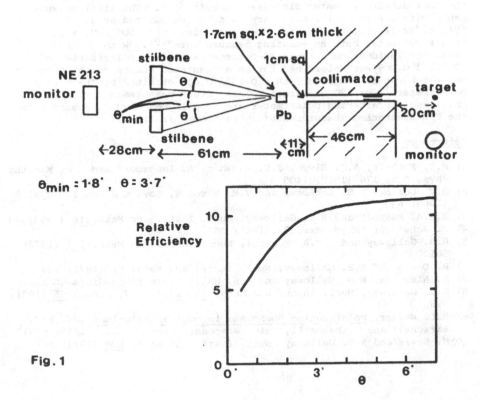

Fig. 1

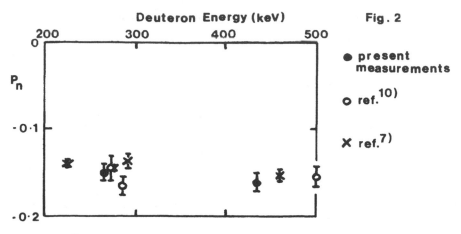

Fig. 2

● present measurements

○ ref.[10]

✕ ref.[7]

scattering[4] which has an efficiency of data collection comparable to that attainable in a conventional ^{4}He scattering system employing a gas scintillator. Such a polarimeter has been constructed, as outlined in fig. 1. It has a data collection efficiency of about 30 times that of the Mott-Schwinger system discussed in refs.[3,5]. The first measurements with the new polarimeter are on neutrons emitted at 45° from the ^{2}H(d,n)^{3}He reaction for deuteron energies less than 500 keV, a region of discrepancies between existing measurements[8,9]. With 340 keV deuterons incident on a thick Ti-D target a neutron polarization of -0.158 ± 0.009 was determined for an accumulated charge on the target of 46C which can be compared with - 0.152± 0.009 obtained by ^{4}He scattering polarimeter[7] for 12C on target. Thin target measurements with the small angle scattering polarimeter are also in excellent agreement with the ^{4}He scattering measurements[7,10] as shown in fig. 2.

References

1) F.T. Kuchnir, A.J. Elwyn, J.E. Monahan, A. Langsdorf and F.P. Mooring, Phys. Rev. 176 (1969) 1405
2) G.V. Gorlov, N.S. Lebedeva and V.M. Morozov, Soviet J. Nucl. Phys. 8 (1969) 630
3) R.M.A. Maayouf and R.B. Galloway, Nucl. Instr. and Meth.118 (1974) 343
4) J. Schwinger, Phys. Rev. 73 (1948) 407
5) R.B. Galloway and R.M.A. Maayouf, Nucl. Instr. and Meth. 105 (1972) 561
6) H. Davie and R.B. Galloway, Nucl. Instr. and Meth. 92 (1971) 547
7) A. Alsoraya, R.B. Galloway and A.S. Hall, these proceedings, p.520
8) R.B. Galloway, Nucl. Instr. and Meth. 92 (1971) 537, errata 95 (1971) 393
9) R.L. Walter, Polarization Phenomena in Nuclear Reactions (Ed. H.H. Barschall and W. Haeberli, Univ. Wisconsin Press Madison 1970) p.317
10) H. Davie and R.B. Galloway, Nucl. Instr. and Meth. 108 (1973) 581

COMPUTER SIMULATION OF MULTIPLE SCATTERING
EFFECTS IN 14 MeV n-p ANALYZING POWER MEASUREMENTS

J.C. Duder, A. Chisholm, R. Garrett and J.E. Brock
Physics Dept, University of Auckland, Private Bag, Auckland, New Zealand

We are presently measuring the analyzing power of 14 MeV neutrons elastically scattered by protons in the liquid scintillator NE213. Our detection method is to observe the pulse height spectrum of recoil nuclei in the scintillator when gated by neutrons scattered into left and right side detectors. While other sources of background may be removed in a well planned experiment, multiply scattered neutrons are not eliminated by the gating, and the detailed shape of the multiply scattered background under the n-p single-scattering peak remains difficult to determine experimentally. Moreover, this background may be different when gated by left or right side detectors since n-C elastic scattering may occur with large analyzing powers. There is thus the possibility of a systematic error in determining the n-p analyzing power. Leemann et al[1] have measured significantly different 14.2 MeV n-p analyzing powers at $\theta_{cm} = 90^{\circ}$ with scatterers of diameter 3.81 cm and 2 cm, and propose that multiple scattering events involving carbon may be the cause of this discrepancy. Yet Jones and Brooks[2], with a 1.5 cm^3 scatterer at 21.6 MeV obtained an n-p analyzing power consistent with the larger of Leemann's scatterers.

A Monte Carlo organic scintillator detection efficiency code[3] has been modified in order to explore such questions. In the present simulation, a 100% polarized 14 MeV collimated neutron beam is incident on the curved face of a cylinder of NE213. Dimensions of 10.16 cm, 5.08 cm or 2.54 cm in diameter by 15.24 cm or 7.62 cm long have been used. Coaxial with and 40 cm from the NE213 is an array of side detectors. Each side detector spans 8° in polar angle and 14° in azimuth. For every neutron which scatters into a particular side detector, account is kept of the n-p and n-C elastic and inelastic events it underwent, and of the light consequently generated in the NE213. In this way one obtains, for each side detector, various gated singles spectra, a gated multiples spectrum, and on adding these a gated totals spectrum simulating an experimental spectrum. Stanton's cross section data[3] are used, and the analyzing power of n-p scattering is assumed to be zero. The analyzing power of n-C elastic scattering is taken to be $A_z = -1$ at all energies and angles, an expedient which strongly enhances multiple scattering asymmetries. By fitting an ungated simulated spectrum to a corresponding ungated experimental spectrum, a parameter describing the overall detector resolution has been determined. This is the one photo-electron level[3] and it was found to be 12 keV electron equivalent. The fit was sufficiently good that it is inappropriate to include light attenuation effects[4] explicitly. Edge effects[5] are negligible. Simulated coincidence-gated spectra and corresponding experimental spectra are shown in fig. 1. We deduce that:

1) the multiply scattered background in gated spectra is approximately proportional to the linear dimension of the scatterer in the scattering plane, and is almost independent of its length perpendicular to the plane;
2) a horizontal extrapolation of the multiply scattered background on

the low energy side of the n-p singles peak is a good estimate of the multiply scattered background under the peak;

3) the large carbon analyzing power assumed in the simulation induces, in the n-p peak, a spurious asymmetry of typically (-1.8±0.2)% at $\theta_{lab} = 50°$ for a 5.08 cm diameter scatterer. This is an extreme upper limit which would not arise in practice; the spurious asymmetry in a real n-p experiment may well be an order of magnitude less. Further simulation with more realistic carbon analyzing powers is planned.

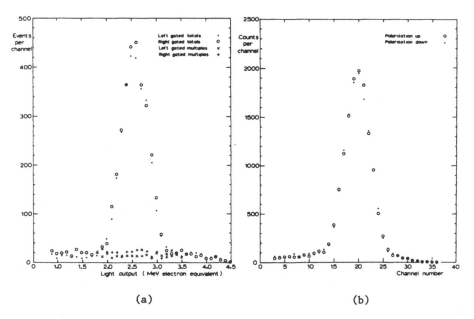

(a)   (b)

Fig. 1. Coincidence-gated recoil spectra for a 5.08 cm diameter scatterer and $\theta_{lab} = 50°$.
(a) Simulated spectra. In channels where multiple events are not explicitly indicated the total consists entirely of multiples.
(b) Experimental spectra. Note that coincidence gated events are subject to a time-of-flight requirement which excludes some of the large recoil-energy events allowed in the simulation.

References

1. B. Leemann, R. Casparis, M. Preiswerk, H. Rudin, R. Wagner and P. Zuprański, Helv. Phys. Acta 47 (1974) 479
2. D.T.L. Jones and F.D. Brooks, Nucl. Phys. A222 (1974) 79
3. N.R. Stanton, A Monte-Carlo program for calculating neutron detection efficiencies in plastic scintillator, Ohio State University Preprint COO-1545-92 (Feb 1971)
4. R. De Leo, G. D'Erasmo, A. Pantaleo and G. Russo, Nucl. Instr. Meth. 119 (1974) 559
5. J.W. Watson and R.G. Graves, Nucl. Instr. Meth. 117 (1974) 541

A NEW TYPE OF HELIUM RECOIL POLARIMETER FOR FAST NEUTRONS[*]

C.P. Sikkema

Laboratorium voor Algemene Natuurkunde der Rijksuniversiteit
Groningen, Groningen, the Netherlands

The polarization of fast neutrons up to about 20 MeV energy is deduced most frequently from the azimuthal or left-right asymmetry in n-
^{4}He scattering. When the asymmetry is measured by detecting the scattered neutrons, an important source of error is a background caused by
stray neutrons entering the neutron detector. This problem is avoided if
one uses only the recoil α-particles for the asymmetry measurement. With
previously employed methods of this type, the neutrons traverse a volume
of gaseous helium, in which slits or tubes are mounted to confine the directions of the α-particles. In most cases proportional counters are
used as detectors. However, the employed diaphragms also limit the effective helium volume and therefore the efficiency is low. Moreover,
geometrical asymmetries are introduced. Alternatively, cloud chambers
have been applied for measuring the directions of the recoil particles.
The application of this technique, however, is seriously restricted by
the long time required for the measurement of the photographed tracks.
To overcome these problems, we have developed a multiwire proportional
chamber or "drift chamber" technique for determining the directions of
the recoil α-particles electronically. Since there are no obstacles like
slits or tubes in the helium, the efficiency is good and there are no
geometrical asymmetries. The electrode arrangement is shown schematically
in fig. 1. A collimated neutron beam traverses the middle part of the
region with a homogeneous electric field. The beam is directed parallel
to the electrodes and perpendicular to the wires. From the chamber three
types of direction-sensitive pulses are deduced, which we call the D-
pulse, the D'-pulse and the F-pulse respectively. The height of the D-
pulse is proportional to the charge collected on the wires and therefore
to the particle energy E_α. We have the relation $E_\alpha = E_{\alpha 0} \cos^2\theta$, θ being
the scattering angle and $E_{\alpha 0}$ the maximum recoil energy, which is proportional to E_n. We assume the neutrons to be monoenergetic and therefore
the D-pulse not only fixes E_α and the particle range R but also θ and the
track component R_y. The D'-pulse is obtained by differentiating the D-
pulse. For a given D-pulse height, the height of the D'-pulse only depends
on the risetime and therefore on $|R_z|$, the absolute magnitude of the
track component R_z. The height of the F-pulse is, for a given height of
the D-pulse, approximately proportional to $x_{n+1} - x_n$, the difference in
position of the discharges on two successive collecting wires. The cen-
tres of the discharges are the intersections between the wires and the
projection of the particle track on the plane of the wires (cf. fig. 1).
We therefore have the relation $(x_{n+1} - x_n)/d = R_x/R_y$, d being the distance between two successive wires. Hence for a given height of the D-
pulse, the height of the F-pulse is approximately proportional to the
track component R_x. The F-pulse is generated by processing pulses, induced on insulated sections of the lower cathode plate, by means of an
analog computer. The algebraic processing method is based on a theorem
which we described in a previous paper[1]. This paper also contains a full
description of the system. An asymmetry measurement consists of recording the three-parameter correlation between the heights of the pulses D,
D' and F. From this correlation, distributions of the pulse height F are
taken for single D-pulse height channels and D'-pulse height channels ex-

886

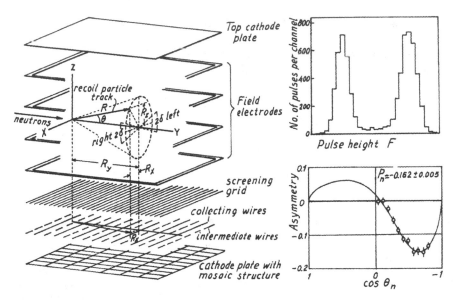

Fig. 1 (left), geometry of the drift chamber and the method of asymmetry measurement (schematic). Fig. 2 (top right), typical distribution of the pulse height F for a given D-pulse channel and a threshold applied to the D'-pulse. Fig. 3 (bottom right), asymmetries deduced from histograms of the type shown in fig. 2, plotted as a function of cos θ_n, θ_n being the neutron scattering angle in the c.m.s. system. Neutrons from the reaction $^2H(d,n)^3He$, for E_d = 268 keV and Θ_n = 48°.

ceeding a given threshold. (fig. 2). The F-pulse distributions therefore correspond to tracks of recoil particles with a given E_α and scattering angle θ, which are lying on a cone with the axis parallel to the neutron beam and half top angle θ. Moreover the threshold on the D'-pulse puts an upper limit to $|R_z|$. Hence the F-pulse spectra correspond to recoil α-particles scattered at a given angle θ to the left and to the right into equal azimuthal intervals 2δ centered in the plane of the electrodes (fig 1). This is properly reflected by the two peaks. From the numbers of pulses in the peaks the asymmetry is calculated in the well-known way. The method simultaneously yields asymmetries for a range of angles θ. A typical result is shown in fig. 3. The neutron polarization is determined by fitting the measured asymmetries with a function $P_n P_{He}(\theta)$, the analyzing power $P_{He}(\theta)$ being calculated using known phase shifts[2]. The application of the polarimeter for neutrons from the reaction $^2H(d,n)^3He$ has been described in a separate contribution[3].

References
* Supported by the Netherlands Foundation for Fundamental Research of Matter (FOM).
1) C.P. Sikkema, Nucl. Inst. 122 (1974) 415
2) B. Hoop and H.H. Barschall, Nucl. Phys. 83 (1966) 65
   Th. Stammbach and R.L. Walter, Nucl. Phys. A180 (1972) 225
3) C.P. Sikkema and S.P. Steendam, these Proceedings, p.518.

Absolute Neutron Polarization from Measurements of $A_{zz}(0^\circ)$ in the T($\vec{d}$,n)⁴He Reaction*

P.W. Lisowski and R.L. Walter, Los Alamos Scientific
Laboratory and Triangle Universities Nuclear Laboratory
and
R.A. Hardekopf and G.G. Ohlsen
Los Alamos Scientific Laboratory, Los Alamos, N.M. 87545, USA

In a contribution to the Third Polarization Symposium, Ohlsen, Keaton, and Gammel[1] reported that certain properties of the spin structure of the T($\vec{d}$,n)⁴He reaction permitted a calculation of the longitudinal neutron polarization from the zero-degree analyzing power $A_{zz}(0^\circ)$ and the deuteron beam polarization. In this contribution we report measurements of $A_{zz}(0^\circ)$ for the T($\vec{d}$,n)⁴He reaction which were made in preparation for a calibration of the ⁴He(n,n)⁴He analyzing power above 20 MeV.[2] In addition we compare our results to those for the mirror reaction ³He($\vec{d}$,p)⁴He.[3] Finally we show values of the neutron polarization calculated from our analyzing power data and discuss the usefulness of the T($\vec{d}$,n)⁴He reaction as a calibrated source of polarized neutrons.

For reactions with the spin structure 1/2 + 1 → 1/2 + 0 such as T($\vec{d}$,n)⁴He, the longitudinal polarization transfer coefficient $K_z^{z'}(0^\circ)$ is related to the analyzing power $A_{zz}(0^\circ)$ by the equation

$$K_z^{z'}(0^\circ) = \frac{2}{3}\left[1 + \frac{1}{2}A_{zz}(0^\circ)\right]. \qquad (1)$$

Using the above expression, the polarization $p_{z'}$ of the outgoing neutrons may be calculated from the deuteron vector and tensor beam polarizations p_z and p_{zz} to give

$$p_{z'} = \frac{[1+\frac{1}{2}A_{zz}(0^\circ)]\,p_z}{[1+\frac{1}{2}A_{zz}(0^\circ)\,p_{zz}]} \qquad (2)$$

Figure 1 shows the neutron polarization plotted as a function of A_{zz} for several values of polarization of an $m_I = 1$ deuteron beam, i.e. one with $p_z = p_{zz} = p_Q$. For A_{zz} values between -1 and -2 the neutron polarization depends quite strongly on p_Q for realistic values of beam polarization.

The measurements reported here were performed using the Lamb-shift ion source and the accelerator facility of the Los Alamos Scientific Laboratory. In a separate experiment the quench-ratio method for measuring the polarization of the deuteron beam was verified to be correct to ± 0.01 by scattering from ⁴He at 4.57 MeV and 58° cm. Under these conditions A_{yy} for ⁴He($\vec{d}$,d)⁴He is known[4] from phase shift calculations to be -1.0. The T($\vec{d}$,n)⁴He zero degree analyzing power was then determined for 12 energies from 3.5 to 12.8 MeV from measurements of the relative neutron yield in a liquid helium scintillator for deuteron beams with different m_I values.

The results are presented in fig. 2. Also shown are the earlier T($\vec{d}$,n)⁴He data of Broste et al.[5] and a smooth curve which represents $A_{zz}(0^\circ)$ for the mirror reaction ³He($\vec{d}$,p)⁴He[3]. The $A_{zz}(0^\circ)$ values for the two reactions approach each other at both high and low energies, but differ significantly in the energy region near 5 MeV. Shown in Fig. 3

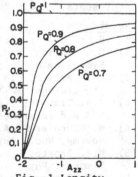

Fig. 1 Longitudinal neutron polarization as a function of A_{zz} for different p_Q values.

are values of the longitudinal neutron polarization p_z' which were calculated from the measured $A_{zz}(0^O)$ using a typical value for p_Q of 0.8. The errors associated with p_z' were compounded from both ΔA_{zz} and Δp_Q and reflect the present uncertainty with which absolute neutron polarization values have been determined.

Utilizing the $T(\vec{d},n)^4He$ reaction as a polarized neutron source has the disadvantage of requiring a dipole magnet to precess the neutron spin into a transverse orientation. In addition, the figure of merit is lower than that for similar transverse polarization transfer reactions[6,7]. However, considering the fact that longitudinal polarization transfer in the $T(\vec{d},n)^4He$ reaction provides a calibrated source of polarized neutrons, it is a valuable tool as a calibration standard for neutron polarization studies.

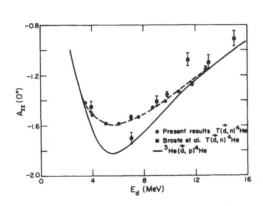

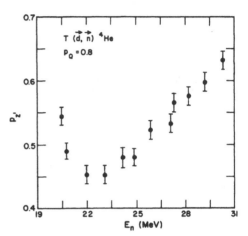

Fig. 2 $A_{zz}(0^O)$ for $T(\vec{d},n)^4He$ and $^3He(\vec{d},p)^4He$. The dashed curve is a guide to the eye. The solid curve represents $^3He(\vec{d},p)^4He$ results.

Fig. 3 Values of longitudinal neutron polarization calculated from the present A_{zz} data.

References

* Supported by the U.S. Energy Research and Development Administration
1) G.G. Ohlsen, P.W. Keaton and J.L. Gammel, Proc. 3rd International Symposium on Polarization Phenomena in Nuclear Reactions, p.512
2) R.L. Walter, P.W. Lisowski, G. G. Ohlsen and R.A. Hardekopf, these proceedings, p.536
3) T.A. Trainor, T.B. Clegg and P.W. Lisowski, Nucl. Phys. A220 (1974) 533 and references therein.
4) W. Grüebler, P.A. Schmelzbach, V. König, R. Risler, B. Jenny and D. Boerma, Nucl. Phys. A242 (1975) 285
5) W.B. Broste, G.P. Lawrence, J.L. McKibben, G.G. Ohlsen and J.E. Simmons, Phys. Rev. Letters 25 (1970) 1040
6) J.E. Simmons, W.B. Broste, T.R. Donoghue, R.C. Haight, and J.C. Martin, Nucl. Instr. and Methods 106 (1973) 477
7) G.G. Ohlsen, R.A. Hardekopf, R.L. Walter and P.W. Lisowski, these proceedings, p.558

A HIGH EFFICIENCY GENERAL PURPOSE ON-LINE SILICON POLARIMETER SYSTEM †

M. Irshad,   S. Sen,   R. Pigeon   and   R.J. Slobodrian,
Laboratoire de Physique Nucléaire, Département de Physique,
Université Laval,   Québec   G1K 7P4 ,   Canada.

Polarimetry based on double scattering techniques usually use carbon or helium on polarization analysers. But this limits the polarimeter efficiency when high resolution in energy is sought and vice-versa. The polarimeter system we have developed overcomes these limitations by using a silicon analyser which acts as a ΔE detector at the same time. Basically the polarimeter consists of a 1mm thick silicon ΔE detector and two silicon stopping detectors for detecting L and R scattered particles at 28°, and another detector behind the analyser used in anticoincidence to eliminate the unscattered particles as shown in fig. 1. The basic design of the polarimeter has already been described [1]. We have devised a system to use the polarimeters (one on each side of the beam) with a PDP-9 computer on-line. Three 400-channel pulse height analysers each with its memory divided in halves are used to record $\Delta E_{1,2}$, $E_{1L,R}$ and $E_{2L,R}$ pulses respectively. The ADC outputs corresponding to these signals give the energy deposited by the reaction products in ΔE detectors and their total energy respectively. The primary data are stored on-line and written on a magnetic tape as records consisting of 133 events, each event being defined as three channel conversions in coincidence from three ADC's. The basic logic is to record a new event whenever there is an acceptable signal in any one of the ΔE detectors, ADC1 simultaneously with a signal in any of E detectors, ADC2 or ADC3. This event corresponds to a new channel conversion only in one of the E ADC's depending on ΔE signal location in the first ADC, the previous channel conversion being repeated for the other E ADC. The separation of these events into those due to protons and to other particles, for each detector telescope, is done setting energy windows on ΔE and E channel conversions to match with the energy loss in the analyser detectors and the total energy of these particles, as determined from the reaction kinematics. Thus this technique enables one to identify the reaction products and obtain their energy spectra without using analog particle identifying circuits. This type of polarimeter with an on-line particle identification system can play an important role in polarization measurements involving low count rates, as generally is the case in double scattering, triple scattering and in simultaneous measurements of asymmetries when more than one type of particles is produced in a nuclear reaction. The system described here is easily adaptable to handling four polarimeters simultaneously without changes. With larger ADC's it can handle easily eight to sixteen polarimeters.

It is a pleasure to acknowledge the most able assistance of Mr Richard Bertrand in implementing the interface for the specific application to the polarimeter system, and to Mr Laurent Pouliot for his superb work in constructing all the mechanical parts of the system.

† Work supported in part by the Atomic Energy Control Board of Canada and the ministery of Education of Québec.

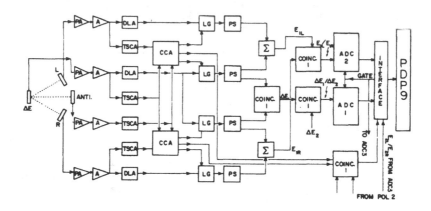

Fig. 1   Block diagram of the electronics for one polarimeter. The
          symbols used are :   ΔE-analyser silicon detector, E_L - left
          stopping silicon detector, E_R - right stopping silicon detector,
          ANTI - anticoincidence silicon detector, PA-Preamplifier,
          A-amplifier, DLA-delay amplifier, TSCA-timing single channel
          analyser, CCA-coincidence - anticoincidence circuit, LG-linear
          gate, PS-Pulse stretcher, Σ-sum circuit, COINC 1 - coincidence
          circuit set to coincidence requirement 1.

References :

1)   B. Frois, J. Birchall, R. Lamontagne, R. Roy and R.J. Slobodrian,
      Nucl. Inst. and Meth. 9(1971)431 and references therein.

ABOUT THE POSSIBILITY OF MEASURING POLARIZATION OF PROTONS EMITTED IN NUCLEAR REACTIONS AT BACKWARD ANGLES USING RING GEOMETRY

M.I.Krivopustov, I.V.Sizov, H.Oehler

Laboratory of Neutron Physics, Joint Institute for Nuclear Research, Dubna, USSR.

In papers[1,2] the polarimeter in which the analyser is in the form of a ring was discussed. This ensures the high efficiency of experiments on polarization. This method is used to measure the proton polarization in the reaction $^{12}C(^{3}He, p)^{14}N$ (ground state)[3]. Recently the ring-polarimeter method was used to study the reaction product polarization in papers[4-7]. However, untill now it was not employed for measuring the polarization of particles flying at backward angles.

In the present report an extension for the construction of ring-polarimeter for measuring the polarization at backward angles was carried out. This was used to measure the proton polarization from the reaction $^{12}C(^{3}He, p)^{14}N$ (g.s.) for $\theta > 90°$. Schematic diagram of the experimental technique is shown in fig. 1. Specially, the polarimeter needs the use of a coaxial semiconductor detector, it is necessary to use a tube in order to protect the detectors from the incident beam. Also strict requirements for exact polarimeter details and focusing of the beam were taken into consideration.

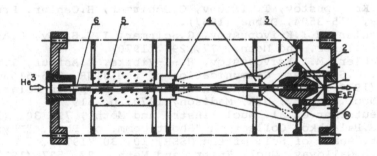

Fig. 1. Schematic diagram of polarimeter with ring analyser for measuring nucleon polarization from nuclear reaction at backward angles. 1 - target, 2 - aluminium foil, 3 - analyser target, 4 - annular semiconductor detector, 5 - lead sheet.

The method gives the asymmetry η and polarization $\vec{P}$, the analysis of results gives also the dependence on these values on physical and geometrical parameters which remains analogical to the case of measuring polarization for angles $\theta > 90°$. It is reported in details in papers[1,2]. The experimental results for proton polarization obtained from $^{12}C(^{3}He, p)^{14}N$ for $\theta > 90°$ are given in the table.

E ^{3}He (keV)	angle θ (degrees)	analyser	asymmetry η	polarization $\vec{P}$
2950	130±3	C	0.01±0.03	-0.02±0.06
3520	135±3	Au	-0.01±0.02	-
3520	135±3	C	0.27±0.04	-0.56±0.06

The reliability of measurements and calculations of a geometrical factor and average polarizing power has been controlled by experiments with gold analyser as well as by a comparison with the results obtained by the left-right asymmetry.

The results of first measurements with the new polarimeter show that, although there are difficulties in the method and technique of ring-polarimeter, it can be successfully used to measure the polarization for all range of angles $0° - 180°$.

References

1. M.I.Krivopustov, I.V.Sizov, G.Schirmer, H.Oehler. Preprint JINR, P15-3504, Dubna (1967).
2. H.Oehler, M.I.Krivopustov, G.Scirmer, I.V.Sizov, F.Asfour. Nucl, Instr. and Meth., 77, 293 (1970).
3. H.Oehler, M.I.Krivopustov, H.-I-Vibike, F.Asfour, I.V.Sizov, G.Schirmer. Communication of the JINR, P15-5156, Dubna (1970). Proc. of the Third Intern. Symp. on Polar.Phen. in Nucl. React., USA, Madison, 1970, p. 619.
4. M.Deutscher et al. Nucl. Instr. and Meth., 71, 301 (1969).
5. Yu.G.Balashko. Collection "Short Comm. of Phys. "Phys. Inst. Acad. of Sci. of the USSR, 10, 30 (1974).
6. G.R.B.Gallovay. Nucl. Instr. and Meth., 92, 537 (1971).
7. M.V.Artsimóvich et al. All-Union XXIV Conf. on Nucl. Spectr. and Struct. of Atomic Nucleus (Kharkov 1974). Leningrad, "Nauka" 1974, p. 357.

AN OPTIMAL METHOD OF MEASURING ANALYSING POWERS WITH THE POLARIZED BEAM FROM AN ATOMIC BEAM SOURCE AT A TANDEM ACCELERATOR

V. König, W. Grüebler, P.A. Schmelzbach

Laboratorium für Kernphysik, Eidg. Techn. Hochschule,

8049 Zürich, Switzerland

In the target coordinate system shown in fig.1, the cross section for a scattering or reaction with polarized deuterons can be written as

$$\sigma(\theta) = \sigma_o(\theta) \cdot \{1 + \sqrt{2}\sin\alpha \cdot \cos\beta \cdot \hat{t}_{10} \cdot iT_{11} + \frac{1}{2}(3\cos^2\alpha - 1) \cdot \hat{t}_{20} \cdot T_{20}$$
$$+ \sqrt{\tfrac{3}{2}}\sin 2\alpha \cdot \sin\beta \cdot \hat{t}_{20} \cdot T_{21} - \sqrt{\tfrac{3}{2}}\sin^2\alpha \cdot \cos 2\beta \cdot \hat{t}_{20} \cdot T_{22}\}$$

with source parameters $\hat{t}_{10}$ and $\hat{t}_{20}$, the spin-tensor moments T_{kq} according to the Madison convention and the spin direction angles α and β given in fig.1.

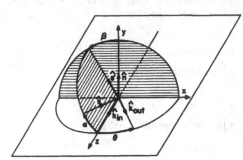

Fig.1: Target coordinate system in which the polarization of the incident beam is described: z is along the incident momentum $\hat{k}_{in}$, y is along $n = \hat{k}_{in} \times \hat{k}_{out}$. The spin-alignment axis $\hat{s}$ makes an angle α with $\hat{k}_{in}$, the projection of $\hat{s}$ onto the xy-plane makes an angle β with $\hat{n}$.

In the method described in the following the four analysing powers T_{qk} are measured in four runs separately. For each run α and β are chosen in such a way that the wanted T_{qk} can be extracted easily from the measured counting rates. A spin rotation device (for instance a Wien filter which can be rotated around the beam axis) is needed to set the angles α and β to the desired values and the two detectors for each scattering angle θ, left and right in the horizontal plane at angles β and $\beta + \pi$ respectively. For the determination of the unpolarized cross section it is necessary to be able to change between positive and negative beam polarization in short intervals and to collect the spectra for both polarizations separately. The best way to do so is to have three RF-transitions as shown in table 1, changing between states b and c for tensor measurements and between d and e for vector measurements. However, the RF-transitions should be tuned carefully, in order to get the same absolute value of the polarization for both signs.

Table 1: Configuration of RF-transitions

	2 → 6	3 → 5	weak field	$\hat{t}_{10}$	$\hat{t}_{20}$
a	off	off	off	0	0
b	off	on	on	$-1/\sqrt{6}$	$+1/\sqrt{2}$
c	off	on	off	$+1/\sqrt{6}$	$-1/\sqrt{2}$
d	on	on	off	$+\sqrt{2/3}$	0
e	off	off	on	$-\sqrt{2/3}$	0

The advantages of this method are the following:
a) Since only two detectors in one plane are used, it is easier to measure at extreme forward and backward angles, or to measure more scattering angles Θ simultaneously than with a device with four detectors in two planes.
b) The measurement is totally independent of the ratio of the solid angles of both detectors.
c) For an accuracy of the T_{qk} of 0.01, a determination of the spin direction angles α and β to only $\pm 4^\circ$ is sufficient. However, the same accuracy for T_{21} requires an accuracy of β better than $\pm 2^\circ$.
d) The statistical accuracy of the measured T_{qk} is better for a measurement changing the sign of the beam polarization than for a measurement comparing only the polarized and unpolarized beams.

With the two detectors and the positive and negative beam polarizations four counting rates N_L^+, N_L^-, N_R^+, N_R^- are obtained from which the ratios L and R – for the left and right detectors respectively – can be calculated, both being independent of the solid angles:

$$L = \frac{N_L^+ - N_L^-}{N_L^+ + N_L^-} = +\sqrt{2}\sin\alpha\cdot\cos\beta\cdot\hat{t}_{10}\cdot iT_{11} + \tfrac{1}{2}(3\cos^2\alpha-1)\cdot\hat{t}_{20}\cdot T_{20}$$
$$+\sqrt{\tfrac{3}{2}}\sin 2\alpha\cdot\sin\beta\cdot\hat{t}_{20}\cdot T_{21} - \sqrt{\tfrac{3}{2}}\sin^2\alpha\cdot\cos 2\beta\cdot\hat{t}_{20}\cdot T_{22}$$

$$R = \frac{N_R^+ - N_R^-}{N_R^+ + N_R^-} = -\sqrt{2}\sin\alpha\cdot\cos\beta\cdot\hat{t}_{10}\cdot iT_{11} + \tfrac{1}{2}(3\cos^2\alpha-1)\cdot\hat{t}_{20}\cdot T_{20}$$
$$-\sqrt{\tfrac{3}{2}}\sin 2\alpha\cdot\sin\beta\cdot\hat{t}_{20}\cdot T_{21} - \sqrt{\tfrac{3}{2}}\sin^2\alpha\cdot\cos 2\beta\cdot\hat{t}_{20}\cdot T_{22}$$

With a suitable choice of α and β it follows:

$\alpha = 0^\circ$:
$$T_{20} = \frac{1}{\hat{t}_{20}}\cdot\frac{L+R}{2}$$

$\alpha = 90^\circ$ $\beta = 90^\circ$:
$$T_{22} = \frac{1}{\sqrt{3/2}\hat{t}_{20}}\cdot\frac{L+R}{2} + \frac{1}{\sqrt{6}}\cdot T_{20} \qquad \text{(with known } T_{20}\text{)}$$

$\alpha = 45^\circ$ $\beta = 90^\circ$:
$$T_{21} = \frac{1}{\sqrt{3/2}\hat{t}_{20}}\cdot\frac{L-R}{2}$$

$\alpha = 90^\circ$ $\beta = 0^\circ$:
$$iT_{11} = \frac{1}{\sqrt{2}\hat{t}_{10}}\cdot\frac{L-R}{2} \qquad \begin{array}{l}\text{(with purely vector}\\\text{polarized beam)}\end{array}$$

If the spin direction angles α and β deviate from the correct values by $\Delta\alpha$ and $\Delta\beta$, the following errors for the T_{qk} are obtained:

$$\Delta T_{20} = \tfrac{3}{2}\Delta\alpha^2\cdot T_{20} + \sqrt{\tfrac{3}{2}}\cos 2\beta\cdot\Delta\alpha^2\cdot T_{22}$$
$$\Delta T_{22} = (\tfrac{1}{2}\Delta\alpha^2 + 2\Delta\beta^2)\cdot T_{22} - \sqrt{\tfrac{3}{2}}\Delta\alpha^2\cdot T_{20}$$
$$\Delta T_{21} = (2\Delta\alpha^2 + \tfrac{1}{2}\Delta\beta^2)\cdot T_{21} + \tfrac{\sqrt{2}}{3}(\Delta\beta + \Delta\alpha\Delta\beta)\cdot iT_{11}$$
$$\Delta iT_{11} = \tfrac{1}{2}(\Delta\alpha^2 + \Delta\beta^2)\cdot iT_{11}$$

All terms in $\Delta\alpha$ are quadratic and therefore in the worst case, i.e., for all T_{qk} maximum, an error of about 0.01 is obtained for a value of $\Delta\alpha = 4^\circ$. This means that a determination of $\alpha \pm 4^\circ$ is sufficient for an accuracy in the T_{qk} of 0.01. The same is true for the angle β, except for T_{21}: in this case there exists a term with $\Delta\beta\cdot iT_{11}$ which is only linear in $\Delta\beta$, and therefore requires $\Delta\beta < 2^\circ$ for an error smaller than 0.01 in T_{21}.

AN OPTIMAL METHOD OF MEASURING ANALYSING POWERS WITH THE POLARIZED BEAM FROM AN ATOMIC BEAM SOURCE AT A CYCLOTRON

V. König, W. Grüebler, P.A. Schmelzbach
Laboratorium für Kernphysik, Eidg. Techn. Hochschule,
8049 Zürich, Switzerland

The spin direction of a beam coming from a cyclotron is fixed by the magnetic field of the machine to $\alpha = 90^\circ$ in the vertical direction. This implies that it is impossible to measure T_{21}, and that the angle β is fixed for all directions around the beam axis. (For the definition of the spin direction angles α and β see fig. 1 of the preceding paper).

In this case it is more convenient to change from spherical to cartesian coordinates in the target coordinate system. With $\alpha = 90^\circ$ the general formula for a scattering or reaction with polarized deuterons will reduce to

$$\sigma(\theta) = \sigma_o(\theta) \cdot \{1 + \tfrac{3}{2}\cos\beta \cdot \hat{p}_z \cdot A_y + \tfrac{1}{2}\sin^2\beta \cdot \hat{p}_{zz} \cdot A_{xx} + \tfrac{1}{2}\cos^2\beta \cdot \hat{p}_{zz} \cdot A_{yy}\}$$

with source parameters $\hat{p}_z$ and $\hat{p}_{zz}$ and analysing powers A_y, A_{xx}, A_{yy}.

Hence, one should measure A_y and A_{yy} in the horizontal plane ($\beta = 0^\circ$) and A_{xx} in the vertical plane ($\beta = 90^\circ$). To do this simultaneously, it is necessary to have a complicated system of detectors in two planes at the same scattering angle θ. Measurements at extreme forward and backward angles would be difficult with such a system. It is therefore more convenient to have a scattering chamber which can be rotated by 90° around the beam axis with two detectors to the left and right at the same scattering angle θ. With the configuration of the RF-transitions given in table 1 of the preceding paper one can apply the measuring technique given there, with the advantages of this method.

The measurement of the analysing powers A_y, A_{yy} and A_{xx} is taken in two runs:

a) Scattering chamber horizontal, $\beta = 0^\circ$:

The detectors are left and right, and with the positive and negative beam polarizations one gets four counting rates N_L^+, N_L^-, N_R^+ and N_R^- from which one can calculate the ratios L and R which are independent of the values of the solid angles of the detectors:

$$L = \frac{N_L^+ - N_L^-}{N_L^+ + N_L^-} = + \tfrac{3}{2}\hat{p}_z \cdot A_y + \tfrac{1}{2}\hat{p}_{zz} \cdot A_{yy}$$

$$R = \frac{N_R^+ - N_R^-}{N_R^+ + N_R^-} = - \tfrac{3}{2}\hat{p}_z \cdot A_y + \tfrac{1}{2}\hat{p}_{zz} \cdot A_{yy}$$

From these ratios follow:

$$A_{yy} = \frac{1}{\hat{p}_{zz}} \cdot (L + R)$$

$$A_y = \frac{1}{3\hat{p}_z} \cdot (L - R)$$

b)Scattering chamber vertical, $\beta = 90^\circ$:

The detectors are up and down and one can calculate from the four counting rates N_U^+, N_U^-, N_D^+ and N_D^- the ratios U and D:

$$U = \frac{N_U^+ - N_U^-}{N_U^+ + N_U^-} = \tfrac{1}{2}\hat{p}_{zz} \cdot A_{xx}$$

$$D = \frac{N_D^+ - N_D^-}{N_D^+ + N_D^-} = \tfrac{1}{2}\,\hat{p}_{zz} \cdot A_{xx}$$

and :

$$A_{xx} = \frac{1}{\hat{p}_{zz}} \cdot (U + D)$$

A possible deviation $\Delta\beta$ from the correct value of β gives the following errors in the measured analysing powers:

$$\Delta A_{yy} = (A_{yy} - A_{xx}) \cdot \Delta\beta^2$$

$$\Delta A_{xx} = (A_{xx} - A_{yy}) \cdot \Delta\beta^2$$

$$\Delta A_y = \tfrac{1}{2}\,A_y \cdot \Delta\beta^2$$

All these errors are quadratic in $\Delta\beta$ and therefore the influence of an incorrect setting of β is very small. Further the result is totally independent of the ratio of the solid angles of both detectors.

Normally the measurement of A_y is done with a purely vector polarized beam, because of the higher value of $\hat{p}_{zz}$ in this case. On the other hand, using the mixed vector and tensor polarized beam in which case $\hat{p}_z = \tfrac{1}{3}\hat{p}_{zz}$, one gets the same statistical accuracy for A_y and A_{yy}, since the factor 3 in the vector term of the scattering formula cancels the factor $\tfrac{1}{3}$ for the smaller vector polarization in the beam.

EXTREME VALUES OF THE ANALYZING POWER FOR SPIN-1 POLARIZATION*

F. Seiler⁺, F. N. Rad and H. E. Conzett
Lawrence Berkeley Laboratory, University of California
Berkeley, California 94720

The large experimental values of the components A_y and A_{yy} of the analyzing power for deuterons in several processes suggest the possibility of extreme values $A_y = \pm 1$ and $A_{yy} = 1$. Criteria to judge maximum possible polarization states for an ensemble of polarized spin-1 particles produced in a nuclear reaction have been given by Lakin[1] and Minnaert[2]. They are based on the fact that the density matrix is positive semidefinite. Its expansion in terms of tensor operators τ_{kq} imposes conditions on the tensor moments t_{kq}. Due to time reversal invariance, identical limitations apply for the polarization efficiencies T_{kq}. They are particularly simple in a transverse coordinate system S' with the z'-axis perpendicular to the reaction plane.

$$\delta^2 = \frac{1}{2} \left[(T'_{10})^2 + (T'_{20})^2 + 2|T'_{22}|^2 \right] \leqslant 1. \tag{1}$$

$$\varepsilon^2 = (T'_{20} + \sqrt{2})^2 - 3(T'_{10})^2 - 6|T'_{22}|^2 \geqslant 0. \tag{2}$$

The observables T'_{kq} in terms of those defined by the Madison Convention are

$$T'_{10} = \sqrt{2} \; iT_{11} = \frac{1}{2} \sqrt{6} \; A_y,$$

$$T'_{20} = -\frac{1}{2} (T_{20} + \sqrt{6} \; T_{22}) = \frac{1}{2} \sqrt{2} \; A_{yy}, \tag{3}$$

$$T'_{22} = \frac{1}{4} (\sqrt{6} \; T_{20} - 2T_{22}) - iT_{21} = \frac{1}{6} \sqrt{3} \left[(A_{zz} - A_{xx}) + i2A_{xz} \right].$$

In the space $(T'_{10}, \sqrt{2}|T'_{22}|, T'_{20})$, $\delta = 1$ defines a sphere and $\varepsilon = 0$ an inscribed cone (figure). Points with $A_{yy} = 1$ lie on the base of the Lakin cone $(T'_{20} = 1/2 \sqrt{2})$. Only there, values of $A_y = \pm 1$ can be attained (points A and B), and then only if $|T'_{22}| = 0$. Thus

$$A_{yy} = 1, \qquad A_{xx} = A_{zz} = -\frac{1}{2}, \qquad A_{xz} = 0. \tag{4}$$

Consequently $A_{yy} = 1$ is a prerequisite for $A_y = \pm 1$, and the other efficiencies are numerically determined. This was shown for $^4\text{He}(\vec{d},d)^4\text{He}$ by Grüebler et al.[3], using the properties of that particular M-matrix. From the derivation here, it is obvious that eqs.(4) are always valid. In several reactions, regions with large values of A_y have been found[4], where eqs. (4) are nearly satisfied, but the data density is not good enough to allow any conclusions. Also eqs. (4) are necessary but not sufficient conditions for $A_y = \pm 1$, since they are satisfied by any point on the line AB. In order to prove the existence of any extreme value, the fulfillment of the relevant conditions on the M-matrix has to be established. However these conditions, given for several reactions in papers contributed to this conference, are not necessarily independent. For $A_y = \pm 1$ for instance, eq. (2) for $\varepsilon^2 = 0$ directly imposes the requirement $A_{yy} = 1$.

It also should be noted that it is the direction perpendicular to the scattering plane (y-axis) which yields the large values of the analyzing power. The Madison Convention, generally used for the description of both the M-matrix and the experimental data, does not use this preferred direction as the quantization axis. An investigation of the conditions on the M-matrix in a transverse coordinate system is in progress.

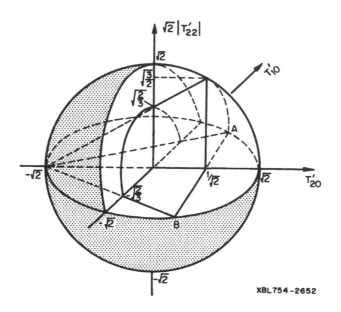

XBL754-2652

References

* Work performed under the auspices of the U. S. Energy Research and Development Administration.
+ On leave of absence from the University of Basel, Switzerland.

1) W. Lakin, Phys. Rev. 98 (1955) 139.
2) P. Minnaert, Phys. Rev. Lett. 16 (1966) 672.
3) W. Grüebler, P. A. Schmelzbach,V. König, R. Risler, B. Jenny and D. Boerma, Nucl. Phys. A242 (1975) 285.
4) F. Seiler, F. N. Rad and H. E. Conzett, Lawrence Berkeley Laboratory Report, LBL-3496.

ABSOLUTE CALIBRATION OF THE ANALYSING POWER T_{20} AT 0° FOR THE ^{3}He($\vec{d}$,p)^{4}He REACTION

P.A. Schmelzbach, W. Grüebler, V. König, R. Risler, D.O. Boerma
and B. Jenny
Laboratorium für Kernphysik, Eidg. Techn. Hochschule,
8049 Zürich, Switzerland

In previous works[1,2] the ^{3}He($\vec{d}$,p)^{4}He reaction at 0° has been proposed as an analyser for the continuous monitoring of the tensor polarization of polarized deuteron beams. The absolute calibration of the analysing power T_{20} for this reaction with the use of the ^{16}O($\vec{d}$,α_1)^{14}N* reaction has been discussed in ref.2). During the last years a ^{3}He-polarimeter has been used sucessfully for all experiments performed with the ETH tensor polarized deuteron beam. However, since the installation of a new, more powerful polarized ion source allows more precise measurements a new calibration of the polarimeter was needed between 1.0 and 12 MeV.

The polarimeter has been calibrated and the values for the analysing power of the ^{3}He($\vec{d}$,p)^{4}He reaction at the corresponding energies have been deduced from this calibration. Consequently, because of the thick ^{3}He target used in the polarimeter, a large energy spread had to be taken into account at small deuteron energies.

In a first experiment, the polarimeter has been calibrated absolutely with the help of the ^{16}O($\vec{d}$,α_1)^{14}N* reaction. The experimental method was the same as in ref.1). However the ^{16}O gas cell had nickel windows 1.25 μm thick and the pressure was reduced to 0.8 atm.

Many angle-energy combinations have been investigated between 5.5 and 10 MeV. The most satisfying spectra have been obtained at 7.20 MeV and θ_{lab}= 35° and 50°. The linearly substracted background was of the order of 10%. The corresponding energy at the entrance in the polarimeter is 7.07 MeV and 6.69 MeV at the middle of the ^{3}He-cell. Under these conditions, the analysing power was found to be -1.2287±0.0072. The absolute error arising from incorrect background substraction is estimated to be smaller than 1%.

In a second experiment, the analysing power of the polarimeter was determined between 1.60 and 11.83 MeV using the value -1.2287 at 7.07 MeV as reference. For the determination of the analysing power between 4.63 and 11.83 MeV, the mean value of the polarization measured at 7.07 MeV before and after each run was used. When a difference larger than the statistical error was observed, the procedure was repeated. The analysing power of the polarimeter was determined in the same way between 2.67 and 4.63 MeV with beam polarization measurement at 4.63 MeV at which energy the analysing power was found in the previous part of the experiment, and between 1.60 and 2.67 MeV with the value measured at 2.67 MeV as reference value. The results as a function of the mean deuteron energy in the ^{3}He-cell are shown in fig.1.

It has been recently shown that the analysing power A_{yy} of the ^{4}He($\vec{d}$,d)^{4}He scattering can be used for the absolute calibration of the tensor polarization t_{20} of a deuteron beam[3]. Using the measurements of A_{yy} of ref.3) around 4.6 and 11.9 MeV, one can, by setting the maximum of $A_{yy}(\theta,E)$ equal to 1, calibrate the polarimeter absolutely with an accuracy of .5%. The results are shown as triangles in the fig.1 at the corresponding energies in the ^{3}He-cell. The agreement with the calibration using the $^{16}O(\vec{d},\alpha_1)^{14}N^*$ reaction is excellent.

Recently·Trainor and Clegg[4] habe published an absolute calibration of A_{zz} between 6.6 and 15.8 MeV for the ^{3}He$(\vec{d},p)^4$He reaction at 0°. They also used the $^{16}O(\vec{d},\alpha_1)^{14}N^*$ reaction as standard but at 9.8 MeV deuteron energy. Their results expressed in term of the spherical tensor T_{20} are also shown in fig.1. The agreement between both measurements in the overlapping region is excellent.

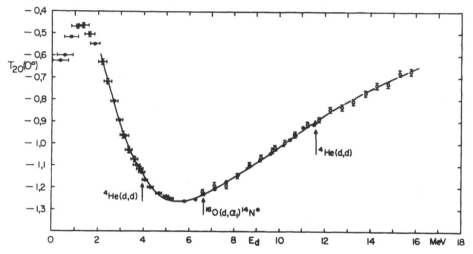

Fig.1. The analysing power T_{20} of the ^{3}He$(\vec{d},p)^4$He reaction at 0° up to 15.79 MeV. The square represents the absolute measurements at 6.69 MeV. The triangles at 3.92 and 11.56 MeV are results obtained from the calibration with the ^{4}He$(\vec{d},d)^4$He scattering. The open circles are the results of ref.4). The curve is calculated with the parameters given in the text.

Parts of the excitation function can be very well reproduced by second order expansions with the following parameters:

$2.18 \leqslant E \leqslant 5.83$ MeV : $T_{20}(0^\circ,E) = 0.4549 - 0.6323 E + 0.0583 E^2$

$5.83 \leqslant E \leqslant 11.60$ MeV : $T_{20}(0^\circ,E) = -1.4507 + 0.0139 E + 0.0029 E^2$

$10.31 \leqslant E \leqslant 15.79$ MeV : $T_{20}(0^\circ,E) = -2.1178 + 0.1401 E - 0.0031 E^2$

A comparison of these new results with the results of ref.1) shows that the analysing power measurements based on the calibration of ref.1) are about 2.5% to large.

References

1) W. Grüebler, V. König, A. Ruh, R.E. White, P.A. Schmelzbach, R. Risler and P. Marmier, Nucl. Phys. A165 (1971) 505
2) V. König, W. Grüebler, A. Ruh, R.E. White, P.A. Schmelzbach, R. Risler and P. Marmier, Nucl. Phys. A166 (1971) 393
3) W.Grüebler, B.Jenny, P.A. Schmelzbach, V. König, R. Risler and D.O. Boerma, Nucl. Phys. A242 (1975) 285
4) T.A. Trainor, T.B. Clegg and P.W. Lisowski, Nucl. Phys. A220 (1974) 533

A POLARIMETER FOR 2-14 MeV VECTOR-POLARIZED DEUTERON BEAMS[*]

R.R. Cadmus, Jr. and W. Haeberli
Department of Physics, University of Wisconsin
Madison, Wisconsin, 53706, U.S.A.

A polarimeter for use primarily with 2-14 MeV vector-polarized deuterons has been constructed and calibrated. Elastic scattering of deuterons from ^{4}He is used as the analyzing reaction. The only scattering angles for which the analyzing power of ^{4}He$(\vec{d},d)^4$He remains large over the entire energy range of interest are those near θ_{cm}=150° (θ_{lab}=126°). Because the energy of the deuterons scattered at this angle is inconveniently low, the corresponding α particles were observed at a lab angle of 15°. The polarimeter is contained in a small scattering chamber attached to the beam exit port of the main scattering chamber. The polarimeter chamber is filled with ^{4}He at a pressure of about 0.5 atm, and is separated from the main scattering chamber by a 2.5 μm thick Havar window. After passing through this window, the beam is collimated by a pair of slits 1.3 mm wide and 7.4 mm high. These slits are 38.1 mm and 79.4 mm from the center of the polarimeter. A small ionization chamber located between these slits serves as a beam intensity monitor. Two detector assemblies are located symmetrically about the beam axis at a scattering angle of 15.1°. Each assembly consists of two slits to define the scattering angle, and a single 50 μm thick partially depleted surface barrier detector. The front slit is 1.3 mm wide and is located 21.7 mm from the center of the polarimeter. The second slit is 50.5 mm behind the first and is 2.5 mm wide. Both slits are about 12.6 mm high.

For incident deuteron energies above about 6 MeV, the α particles lose more energy in the thin detector than do the deuterons scattered at 15°, and the α peak therefore lies above the deuteron peak in the pulse height spectrum. The background under the peak is 1-2% of the peak area. Between 2 and 6 MeV the α peak lies below the deuteron peak in the spectrum, but the gas pressure and detector depletion depth can be adjusted to separate the peaks completely, although the background is 2-5% in this case. At 10 MeV the polarimeter counting rate (left plus right) is about 150 counts per nC of beam after the collimating slits.

The polarimeter was calibrated by making simultaneous observations of the asymmetries in the polarimeter and in a good-geometry ^{4}He$(\vec{d},d)^4$He experiment in the main scattering chamber. The polarimeter analyzing power was then determined from ^{4}He$(\vec{d},d)^4$He analyzing powers of refs.[1-5]). The measurements, as well as the resulting smoothed calibration curve, are shown in the lower half of fig. 1. The uncertainties shown include the uncertainties in overall normalization. The upper half of fig. 1 shows ^{4}He$(\vec{d},d)^4$He analyzing powers calculated from the phase shifts of refs.[2,6,7]). The measurements are quite similar to the calculations.

This polarimeter has been used in numerous experiments and has performed well. Its efficiency is sufficient for the experiments done at this laboratory, but a substantial increase in efficiency with only a slight decrease in analyzing power could be achieved by altering the slit dimensions.

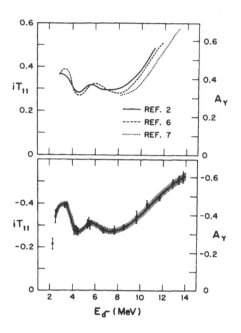

Fig. 1. Measured polarimeter analyzing power (lower frame) and phase shift calculations of the ^{4}He($\vec{d}$,d)^{4}He analyzing power at θ_{cm}=150° (upper frame).

References

*   Supported in part by the U.S. Atomic Energy Commission
1) A. Trier and W. Haeberli, Phys. Rev. Lett. 18 (1967) 915. The analyzing powers from this reference were corrected for a 50 keV error in the beam energy calibration (see ref.[2])
2) L.G. Keller and W. Haeberli, Nucl. Phys. A156 (1970) 465
3) W. Grüebler, V. König, P.A. Schmelzbach and P. Marmier, Nucl. Phys. A134 (1969) 686. The values of iT_{11} were renormalized as suggested in V. König, W. Grüebler, A. Ruh, R.E. White, P.A. Schmelzbach, R. Risler and P. Marmier, Nucl. Phys. A166 (1971) 393
4) G.G. Ohlsen, P.A. Lovoi, G.C. Salzman, U. Meyer-Berkhout, C.K. Mitchell and W. Grüebler, Phys. Rev. C8 (1973) 1262
5) C.C. Chang, H.F. Glavish, R. Avida and R.N. Boyd, Nucl. Phys. A212 (1973) 189. The results of this reference were corrected for the fact that the analyzing powers reported are values of A_y and not of iT_{11}
6) P.A. Schmelzbach, W. Grüebler, V. König and P. Marmier, Nucl. Phys. A184 (1972) 193
7) W. Grüebler, P.A. Schmelzbach, V. König, R. Risler and D. Boerma, Nucl. Phys. A242 (1975) 265

CALIBRATION OF A POLARIZED TRITON BEAM*

R. A. Hardekopf, G. G. Ohlsen, R. V. Poore and N. Jarmie
Los Alamos Scientific Laboratory, University of California
Los Alamos, New Mexico 87545

The new LASL polarized triton source[1] produces a beam whose polarization may be determined by the quench-ratio method[2]. Briefly, this method involves altering conditions in the nuclear spin-filter in such a way that the unpolarized component of the current can be measured. The beam polarization is then $1-1/Q$ where Q is the ratio of the normal to the quenched beam current. This measurement is made periodically on the accelerated beam during experimental runs. We usually observe a variation in the beam polarization less than 0.005 for extended periods.

The accuracy of the quench-ratio polarization determination for protons is discussed at length in ref. 2). Calibrations with protons and deuterons have demonstrated the accuracy to 1/2% on many occasions. Nevertheless, we thought it desirable to check the triton polarization if a suitable analyzing reaction could be found.

The existence of several $A_y = 1.0$ points was first proved[4] and experimentally located[2,5] in ${}^4\text{He}(\vec{p},p){}^4\text{He}$ elastic scattering. This provided an absolute calibration point for polarized proton beams. Analyses of ${}^4\text{He}(t,t){}^4\text{He}$ elastic scattering has indicated that large triton polarization is produced in this system[6,7], and Keaton et al.[8] have measured a maximum polarization of 0.92 near 12.25 MeV in a double-scattering experiment.

In searching for a possible $A_y = \pm 1.0$ point in this system, we use the well known quadratic relation for spin-1/2 on spin-0 scattering

$$(A_y)^2 + (K_x^{x'})^2 + (K_z^{x'})^2 = 1.0,$$

and plot contours of $K_x^{x'}$ vs $K_z^{x'}$. Predictions for these observables are available from the R-Matrix analysis of Dodder and Hale[7]. Figure 1 shows the calculated contours at 11.0 and 12.0 MeV for ${}^4\text{He}(t,t){}^4\text{He}$ scattering which indicate that at some energy between 11 and 12 MeV, a contour must pass through the origin. From the quadratic relation, when $K_x^{x'} = K_z^{x'} = 0$, we must have $A_y = \pm 1$. The cross marks on the contours are drawn at 5° c.m. intervals to give an estimate of the angle where "crossing" occurs. The sign of A_y is found to be negative in this angular region.

On the first experimental run with a polarized triton beam, we obtained data at several angles and energies near the expected minimum in A_y. Figure 2 shows, on an expanded scale, the analyzing powers measured using the quench-ratio method for determination of the beam

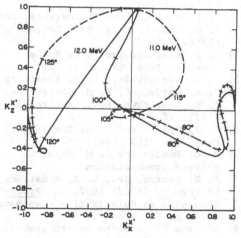

Fig. 1

polarization. The lines through the data are to guide the eye. These experimental results show that A_y is near −1.0 over a fairly wide energy range, from 10.8 to 11.4 MeV, with the most likely minimum at 11.1 MeV, θ_{cm} = 95°. This angle corresponds to a laboratory angle of 56.2° for elastic tritons or 42.5° for recoil alpha particles.

The data show the double-humps predicted by the calculations[7] and also evidenced by the loop in the 11.0 MeV K^{x}_{x} vs K^{x}_{z} contour of fig. 1. It is possible that A_y reaches −1.0 at several points in this vicinity because of this feature, and a more extensive analysis using the data just obtained will check this.

In conclusion, we have located minima in ${}^4\mathrm{He}(\vec{t},t){}^4\mathrm{He}$ analyzing power at which A_y = −1.0 and have calibrated the quench-ratio method for determination of the triton beam polarization to an accuracy of at least ± 0.005.

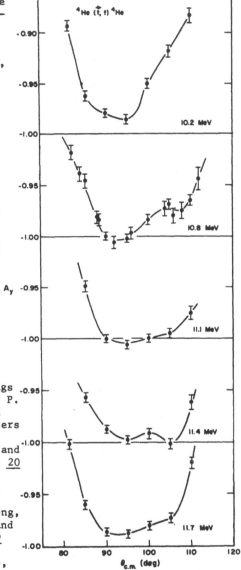

Fig. 2

References

* Work performed under the auspices of the U.S. ERDA
1) R. A. Hardekopf, these proceedings
2) G. G. Ohlsen, J. L. McKibben, G. P. Lawrence, P. W. Keaton, Jr., and D. D. Armstrong, Phys. Rev. Letters 27 (1971) 599
3) J. L. McKibben, G. P. Lawrence, and G. G. Ohlsen, Phys. Rev. Letters 20 (1968) 1180
4) G. R. Plattner and A. D. Bacher, Phys. Letters 36B (1971) 211
5) P. W. Keaton, Jr., D. D. Armstrong, R. A. Hardekopf, P. M. Kurjan, and Y. K. Lee, Phys. Rev. Letters 29 (1972) 880
6) R. J. Spiger and T. A. Tombrello, Phys. Rev. 163 (1967) 964
7) D. C. Dodder and G. M. Hale, private communication
8) P. W. Keaton, Jr., D. D. Armstrong, and L. R. Veeser, Phys. Rev. Letters 20 (1968) 1932, and Proceedings of the 3rd International Symposium on Polarization Phenomena in Nuclear Reactions, Madison, 1971, p. 677 and 680
9) K^{x}_{x} and K^{x}_{z} are the modern symbols for the familiar Wolfenstein parameters R and A, respectively.

EXPERIMENTAL EXAMINATION OF THE ^{3}HE OPTICAL PUMPING THEORY

R. Beckmann, U. Holm and B. Lindner
I. Institut für Experimentalphysik der Universität Hamburg, Zyklotron,
2ooo Hamburg 5o, Luruper Chaussee 149

The nuclear orientation of ^{3}He by optical pumping is a well known and often used technique [1]. Timsit et al. [2] have derived an equation for the attainable ^{3}He polarization as a function of the experimental parameters. But there is a slight difference between the theoretical predictions and their measurements [3] of the equilibrium nuclear polarization in dependence on the pumping light intensity. The conformity of theory and experiment however is very important because the fundamental assumptions of this theory enter the determination of the ^{3}He target nuclear polarization. Therefore we have examined this connexion once more.

The experiments reported here were carried out in a conventional optical pumping apparatus [4]. Changes in the amount of pumping light transmitted through the cell during the adiabatic rotation of the uniform magnetic field were used to determine the ^{3}He polarization. The pumping light intensity and the light detection system together were constant to 2 %. Under the assumptions that all relaxation processes are independent from one another and that the ^{3}He gas is in statistical equilibrium Timsit et al. [2] calculated a formula which we give in an abbreviated form:

$$(1) \qquad dP/dt = - P/T_g + A(I) \, T(P)$$

Here P is the polarization and T_g is a relaxation time describing all angular momentum relaxation processes. A(I) is proportional to the pumping light intensity I absorbed at $P=o$. T(P) is given by $T(P) = R_+(1-P)^2(3.63-2.76P)-R_-(1+P)^2(3.63+2.76P)-R_o(1-P^2)1.26P$ where the parameters R_+, R_-, R_o give the fractions of right circular, left circular and linear polarization of the pumping light. The values of these parameters depend on the geometry of the experimental arrangement. We found $R_+ = o.81$, $R_- = o.13$, $R_o = o.06$. To compute $P_o(I)$ from equation (1) the quantities T_g and A(I) must be known.

The relaxation time T_g has been determined by measuring P(t) as a function of time t with the pumping light cut off (A=o). In agreement with the upper formula we found P(t) to be exponential. The derived value of T_g was (86 ± 5)s. For a given light intensity I_o the value $A(I_o)$ was determined in the following way: we computed P(t) by numerical integration of (1) for the "pumping up" process from $P=o$ to the equilibrium polarization P_o. As a result we have got an approximately linear dependence between $\ln(P_o-P)$ and t. If equation (1) is true the one parameter $A(I_o)$ determines two quantities: the equilibrium polarization P_o and the slope $d \ln(P-P_o)/dt$. Fig. 2 shows the fit to the measurements. From this we get $A(I_o) = (o.oo5\pm o.ooo2)s^{-1}$. P_o has been computed as a function of I by setting $dP/dt = o$. With $A(I) = (A(I)/A(I_o)) \, A(I_o) = (I/I_o) \, o.oo5 \; s^{-1}$ all parameters are given and P(I) can be calculated.

Because equation (1) predicts a linear dependence between $1/P$ and $1/I$ it is more instructive to show the function $1/P = f(1/I)$. The experimental data in Fig. 1 verify this prediction. Furthermore the absolute values of the experimental data are in good agreement with the theoretical curve. According to this agreement the theoretical assumptions which lead to the differential equation (1) seem to be right. Fig. 1 shows that one can get a higher equilibrium polarization P_o if the pumping light intensity I is increased. One possibility is to make use of a second ^{4}He lamp [5]. Because of the low absorption (about 2%) of the pumping light in the ^{3}He cell we have replaced the second ^{4}He lamp by a parabolic reflector which reflects the transmitted light once more to the ^{3}He cell. In this way we have got a maximum equilibrium nuclear polarization of 3o% instead of 25% without reflector. Even much more enhancement could be reached if R_+ which describes the fraction of the right circular polarized pumping light in the ^{3}He cell could be augmented.

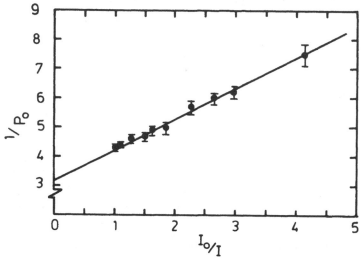

Fig. 1. The reciprocal equilibrium polarization $1/P_0$ versus the reciprocal light intensity I_0/I

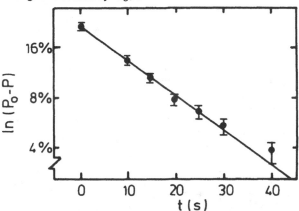

Fig. 2. The increase of polarization P during the "pumping up" process

References
1) F.D. Colegrove, L.D. Schearer, G.K. Walters, Phys. Rev.132(1963)2561
2) J.M. Daniels, R.S. Timsit, Can.J.Phys. 49(1971)525
3) R.S. Timsit, J.M. Daniels, Can.J.Phys. 49(1971)545
4) R. Beckmann, U. Holm, K. Lorenzen, Z.Physik 271(1974)89
5) Ch. Leemann et al., Helv.Phys.Acta 44(1967)141

THE "SUPERCUBE" SCATTERING CHAMBER FOR SPIN-1/2
AND SPIN-1 ANALYZING POWER MEASUREMENTS*

G. G. Ohlsen and P. A. Lovoi
Los Alamos Scientific Laboratory, University of California
Los Alamos, New Mexico 87545

We have constructed a scattering chamber, locally known as the
"supercube", specifically for making accurate measurements of spin-1/2
and spin-1 analyzing powers. The design aim is to make it routinely
possible to achieve accuracies of ± 0.001 on such measurements. This
level of performance has not yet been demonstrated except in one simple
case[1]. The cubical geometry, while complex, is believed to be capable
of implementing any imaginable spin-1/2 or spin-1 analyzing power measuring scheme.

The chamber is a scaled up version of an earlier "cube" chamber[1],
with an inside dimension of 61 cm. This large geometry is desirable
to minimize the errors induced by beam instability and by geometrical
imperfections. Cooled detector telescopes may view the target from left,
right, up, and down positions simultaneously. In fig. 1, the cutaway
view shows the basic geometry with a gas target and suitable collimator
assemblies in place. A solid target holder is also available. A second
set of independently movable arms is not yet fully implemented; this
will provide for good-geometry polarization monitoring or for doubling
the data acquisition rate. An auxiliary polarimeter chamber is attached
to the rear of the chamber. Beam may be admitted to the polarimeter by
remotely removing a plunger which forms the center collector of the faraday cup (see fig. 2).

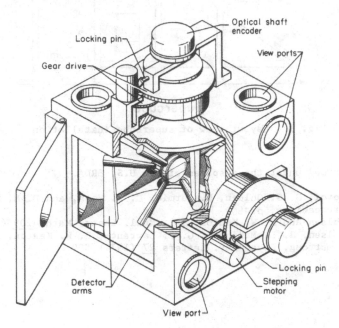

Fig. 1. Cutaway view of supercube

The chamber may be rotated about the beam direction. Four-way slits at the front of the chamber and as part of the faraday cup at the rear of the chamber are used to determine that a "proper flip"[2] is made when the rotation feature is employed. A beam steering feedback system maintains the centering on the entrance slits, which typically form a 2.5 mm x 2.5 mm aperture.

The rotation of the detector arms and the cube itself is under computer control. The arm positions are read out to an accuracy of ± 0.01 degrees via 14 bit optical shaft encoders. Taken together with the computer controlled spin direction, m-state selection, and beam polarization measurement (via the quench ratio[3]) that is available on the two LASL Lamb-shift sources, completely automated data-taking procedures are available.

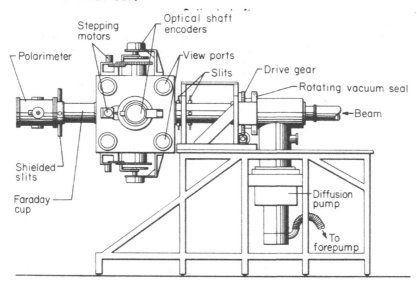

Supercube

Fig. 2. Layout view of supercube installation

References

*Work performed under the auspices of the U.S. ERDA.

1) P. A. Lovoi, G. G. Ohlsen, N. Jarmie, C. E. Moss, and D. M. Stupin, these proceedings, p.450
2) G. G. Ohlsen and P. W. Keaton, Jr., Nucl. Instr. Meth. 109 (1973) 41
3) G. G. Ohlsen, J. L. McKibben, G. P. Lawrence, P. W. Keaton, Jr., and D. D. Armstrong, Phys. Rev. Letters 27 (1971) 599

List of Participants

R. Abegg
University of Wisconsin
Madison, Wisconsin, USA

E.G. Adelberger
University of Washington
Seattle, Washington, USA

R. Albrecht
Max-Planck-Institut für Kernphysik
Heidelberg, Germany

K. Allenby
University of Birmingham
Birmingham, England

W. Arnold
University of Giessen
Giessen, Germany

J. Arvieux
Schweiz. Inst. f. Nuklearforschung
Villigen, Switzerland

D. Axen
University of British Columbia
Vancouver, Canada

A.D. Bacher
Indiana University
Bloomington, USA

B.L.G. Bakker
Vrije Universiteit
Amsterdam, Nederland

R. Balzer
Eidg. Techn. Hochschule
Zürich, Switzerland

F.C. Barker
The Australian National University
Canberra, Australia

H.H. Barschall
University of Wisconsin
Madison, Wisconsin, USA

A.K. Basak
University of Birmingham
Birmingham, England

E. Baumgartner
Universität Basel
Basel, Switzerland

V. Bechtold
Universität Karlsruhe
Karlsruhe, Germany

P.Bém
Czechoslovak Academy of Sciences
Rez u Prahy, Czechoslovakia

G. Bendiscioli
Università degle Studi di Pavia
Pavia, Italia

D. Besset
University of Geneva
Geneva, Switzerland

J. Beveridge
University of British Columbia
Vancouver, Canada

J. Bialy
Kernforschungszentrum
Karlsruhe, Germany

J. Birchall
Universität Basel
Basel, Switzerland

G. Bittner
Universität Erlangen
Erlangen, Germany

J.S. Blair
University of Washington
Seattle, USA

K. Bleuler
Inst. f. Theoretische Kernphysik
Bonn, Germany

D.O. Boerma
Rijksuniversiteit
Groningen, Nederland

I. Borbély
Joint Inst. f. Nucl. Research
Dubna; USSR

M. Bosman
Inst. de Physique Corpusculaire
Louvain-la Neuve, Belgium

D. Bovet
University of Neuchatel
Neuchatel, Switzerland

R.N. Boyd
University of Rochester
Rochester, USA

F.P. Brady
University of California
Davis, California

F.D. Brooks
University of Cape Town
Rondebosch, South Africa

L. Brown
Carnegie Inst. of Washington
Washington, USA

J. Bruisma
Vrije Universiteit
Amsterdam, Nederland

H.R. Bürgi
Eidg. Techn. Hochschule
Zürich, Switzerland

J. Bystricky
CEN Saclay
Gif-sur-Yvette, France

J.A. Cameron
MacMaster University
Hamilton, Canada

J.M. Cameron
University of Alberta
Alberta, Canada

R. Casparis
Universität Basel
Basel, Switzerland

R. Ceuleneer
Université de l'Etat à Mons
Mons, Belgium

J. Chauvin
Institut des Sciences Nucléaires
Grenoble, France

G. Clausnitzer
Universität Giessen
Giessen, Germany

T.B. Clegg
University of North Carolina
Chapel Hill, USA

H.E. Conzett
Lawrence Radiation Laboratory
Berkeley, USA

J.G. Cramer, Jr.
University of Washington
Seattle, Washington, USA

W. Dahme
Universität München
Garching, Germany

S.E. Darden
University of Notre Dame
Notre Dame, Ind., USA

R. De Leo
University di Bari
Bari, Italia

J.-C. Demijolla
University of Nancy
Nancy, France

R. de Swiniarski
Institut des Sciences Nucléaires
Grenoble, France

D. Dodder
Los Alamos Scientific Laboratory
Los Alamos, USA

A.D. Dohan
University of Manitoba
Winnipeg, Canada

P. Doleschall
Central Research Inst. f. Physics
Budapest, Hungaria

T.R. Donoghue
Ohio State University
Columbus, USA

W. Drenckhahn
Universität Erlangen
Erlangen, Germany

L. Drigo
University Padova
Padova, Italia

J.L. Durand
Institut des Sciences Nucléaires
Grenoble, France

H. Ebinghaus
Institut für Experimentalphysik
Hamburg, Germany

M. Eder
Fehrenstr. 23
Zürich, Switzerland

D. Ehrlich
Universität München
Garching, Germany

E. Elbaz
Institut des Physique Nucléaire
Villeurbanne, France

J.L. Escudié
CEN Saclay
Gif sur Yvette, France

E. Fabrici
Istituto Fisica Milano
Milano, Italia

F. Fernandez
Dept. Fisica Fundamental
Valladolid, Spain

D. Fick
Max-Planck-Institut f.Kernphysik
Heidelberg, Germany

E. Finckh
Universität Erlangen
Erlangen, Germany

F. Firk
Yale University
New Haven, Conn., USA

R. Fischer
Universität Hamburg
Hamburg, Germany

R. Fleischmann
Universität Erlangen
Erlangen, Germany

F. Foroughy
University of Neuchatel
Neuchatel, Switzerland

A. Fournier
University of Groningen
Groningen, Nederland

K. Frank
Universität Erlangen
Erlangen, Germany

R. Frick
Universität München
Garching, Germany

L. Friedrich
Kernforschungszentrum
Karlsruhe, Germany

T. Fujisawa
Inst. of Phys. and chem. research
Wako-Shi, Japan

R.B. Galloway
University of Edinburgh
Edinburgh, England

A. Garcia
University of Valencia
Valencia, Spain

R. Garrett
University of Auckland
Auckland, New Zealand

W.R. Gibbs
Los Alamos Scientific Laboratory
Los Alamos, USA

J. Gosset
CEN Saclay
Gif sur Yvette, France

C. Glashausser
Rutgers State University
New Brunswick, USA

H.F. Glavish
Stanford University
Stanford, USA

G. Graw
Universität München
Garching, Germany

G. Greeniaus
University of Geneva
Geneva, Switzerland

W. Grüebler
Eidg. Techn. Hochschule
Zürich, Switzerland

I. Gusdal
University of Manitoba
Winnipeg, Canada

H.H. Hackenbroich
Universität Köln
Köln, Germany

W. Haeberli
University of Wisconsin
Madison, Wisc., USA

G.M. Hale
Los Alamos Scientific Laboratory
Los Alamos, USA

S.S. Hanna
Stanford University
Stanford, California, USA

R.A. Hardekopf
Los Alamos Scientific Laboratory
Los Alamos, USA

E. Heer
University of Geneva
Geneva, Switzerland

G. Heidenreich
Universität Basel
Basel, Switzerland

G. Heil
Universität Erlangen
Erlangen, Germany

P. Heiss
Universität Köln
Köln, Germany

R. Henneck
Universität Erlangen
Erlangen, Germany

W.C. Hermans
Inst.for Nuclear Physics Research
Amsterdam, Nederland

R. Hess
University of Geneva
Geneva, Switzerland

A. Hofmann
Universität Erlangen
Erlangen, Germany

U. Holm
University of Hamburg
Hamburg, Germany

O. Huber
University of Fribourg
Fribourg, Switzerland

K. Imhof
University of Erlangen
Erlangen, Germany

M. Irshad
Université Laval
Québec, Canada

L. Jarczyk
Jagellonian University
Cracow, Poland

B. Jenny
Eidg. Techn. Hochschule
Zürich, Switzerland

H.P. Jochim
Max-Planck-Institut für Chemie
Mainz, Germany

R.C. Johnson
University of Surrey
Guildford, England

H. Jung
Eidg. Techn. Hochschule
Zürich, Switzerland

M. Junge
Inst. f. exp. Kernphysik
Karlsruhe, Germany

H. Kamitsubo
Inst. of Phys. and Chem. Research
Wako-Shi, Japan

I. Karban
University of Birmingham
Birmingham, England

P.W. Keaton, Jr.
Los Alamos Scientific Laboratory
Los Alamos, USA

H. King
Rutgers University
New Brunswick, N.Y., USA

H.-O. Klages
Universität Hamburg
Hamburg, Germany

S. Klein
Techn. Hogeschool Eindhoven
Eindhoven, Nederland

G.Klier
Universität Erlangen
Erlangen, Germany

L.D. Knutson
University of Washington
Seattle, Wash., USA

V. König
Eidg. Techn. Hochschule
Zürich, Switzerland

P. Krämer
Universität Erlangen
Erlangen, Germany

H.H. Krause
Universität Giessen,
Giessen, Germany

W. Kretschmer
Universität Erlangen
Erlangen, Germany

A.D. Krisch
University of Michigan
Ann Arbor, Michigan, USA

M. Krivopustov
Joint Institut for Nuclear Research
Moscow, UdSSR

D. Kröniger
Universität Erlangen
Erlangen, Germany

W. Kubischta
CERN
Geneva, Switzerland

J.A. Kuehner
McMaster University
Hamilton, Ontario, USA

E. Kuhlmann
Ruhr Universität Bochum
Bochum, Germany

K. Kutschera
Universität Erlangen
Erlangen, Germany

M. Lacombe
Physique Theorique et Particules
Elementaires
Paris, France

J. Lang
Eidg. Technische Hochschule
Zürich, Switzerland

C. Lechanoine
University of Geneva
Geneva, Switzerland

P. Leleux
Inst. de Physique Corpusculaire
Louvain-la-Neuve, Belgium

H. Lesiecki
Universität Tübingen
Tübingen, Germany

A. Lindner
Universität Hamburg
Hamburg, Germany

P. Lipnik
Institut de Physique Corpusculaire
Louvain-la-Neuve, Belgium

P.W. Lisowski
Triangle University
Durham, USA

B. Loiseau
University of Paris
Orsay, France

E.J. Ludwig
University of North Carolina
Chapel Hill, USA

J.E. Lynn
U.K.A.E.A.
Harwell, England

J.L. McKibben
Los Alamos Scientific Laboratory
Los Alamos, USA

G. Mack
University of Tübingen
Tübingen, Germany

J.S.C. McKee
University of Manitoba
Winnipeg, Canada

P. Macq
Unst. de Physique Corpusculaire
Louvain-la-Neuve, Belgium

R. Maschuw
Universität Hamburg
Hamburg, Germany

D.G. Mavis
Stanford University
Stanford, California, USA

B. Mayer
CEN Saclay
Gif sur Yvette, France

P.M.C. Melssen
Eindhoven University of Technology
Eindhoven, Nederland

G. Mertens
Universität Tübingen
Tübingen, Germany

J.P. Meulders
Institut de Physique Corpusculaire
Louvain-la-Neuve, Belgium

H. Meyer
Universität Basel
Basel, Switzerland

V. Meyer
Universität Zürich
Zürich, Switzerland

T. Minamisono
Stanford University
Stanford, USA

J.M. Nelson
University of Birmingham
Birmingham, England

G.J. Nijgh
Techn. Hogeschool
Eindhoven, Nederland

L.C. Northcliffe
Texas A & M University
College Station, Texas, USA

S. Oh
University of Manitoba
Winnipeg, Canada

J. Ohlert
Max Planck Institut für Chemie
Mainz, Germany

G.G. Ohlsen
Los Alamos Scientific Laboratory
Los Alamos, USA

N.T. Okumusoglu
University of Birmingham
Birmingham, England

Y. Onel
AERE Harwell
Didcot, Oxfordshire, England

H. Paetz gen. Schieck
Universität Köln
Köln, Germany

G. Pauletta
University of Neuchatel
Neuchatel, Switzerland

F. Pauss
Universität Graz
Graz, Austria

G. Perrin
Institut des Sciences Nucléaires
Grenoble, France

D.L. Pham
Institut des Sciences Nucléaires
Grenoble, France

M. Pignanelli
University of Milano
Milano, Italia

C. Pirart
Institut de Physique Corpusculaire
Louvain-la-Neuve, Belgium

G.R. Plattner
University of Basel
Basel, Switzerland

O.J. Poppema
Techn. Hogeschool Eindhoven
Eindhoven, Nederland

J.M. Potter
Los Alamos Scientific Laboratory
Los Alamos, USA

W.B. Powell
University of Birmingham
Birmingham, England

P.A. Quin
University of Wisconsin
Madison, Wisconsin, USA

J. Raynal
CEN Saclay
Gif sur Yvette, France

F. Resmini
University of Milano
Milano, Italia

R. Risler
Eidg. Techn. Hochschule
Zürich, Switzerland

E.L. Rizzini
University of Pavia
Pavia, Italia

A.B. Robbins
Rutgers State University
New Brunswick, New Jersey, USA

B.A. Robson
National University
Canberra, Australia

F. Rösel
University of Basel
Basel, Switzerland

P.M. Rolph
University of Birmingham
Birmingham, England

S. Roman
University of Birmingham
Birmingham, England

J. Rossel
Université de Neuchâtel
Neuchâtel, Switzerland

G. Roy
University of Alberta
Edmonton, Alberta, Canada

D. Rust
Indiana University
Bloomington, Ind.,USA

F.D. Santos
Faculdade de Ciencias
Lisboa, Portugal

F. Seiler
University of Basel
Basel, Switzerland

M. Sekiguchi
University of Tokyo
Midori-cho, Tanashi-shi, Tokyo, Japan

T. Seligman
Universität Köln
Köln, Germany

R.G. Seyler
Ohio State University
Columbus, Ohio, USA

H.S. Sherif
University of Alberta
Edmonton, Alberta, Canada

C.P. Sikkema
Rijksuniversiteit
Groningen, Nederland

M. Simonius
Eidg. Techn. Hochschule
Zürich, Switzerland

I. Sizov
Joint Institute for Nuclear Research
Moscow, UdSSR

R.J. Slobodrian
Université Laval
Québec, Canada

D. Spaargaren
Vrije Universiteit
Amsterdam, Nederland

P. Svenne
University of Manitoba
Winnipeg, Canada

G. Schatz
Universität Erlangen
Erlangen, Germany

U. Scheib
Universität Erlangen
Erlangen, Germany

P. Schiemenz
Universität München
Garching, Germany

P.A. Schmelzbach
Eidg. Techn. Hochschule
Zürich, Switzerland

F.K. Schmidt
Universität Karlsruhe
Karlsruhe, Germany

D. Schütte
Inst. für theoretische Kernphysik
Bonn, Germany

T.A.M. Schulte
Technische Hogeschool Eindhoven
Eindhoven, Nederland

J. Schwager
Universität Tübingen
Tübingen, Germany

P. Schwandt
University of Indiana
Bloomington, USA

W. Stach
Universität Erlangen
Erlangen, Germany

H. Staub
Universität Zürich
Zürich, Switzerland

E. Steffens
Institut für Experimentalphysik
Hamburg, Germany

H. Stöwe
Universität Köln
Köln, Germany

C. Stolk
Rijksuniversiteit
De Uithof, Utrecht, Nederland

G. Strassner
Kernforschungszentrum Karlsruhe
Karlsruhe, Germany

A. Strzałkowski
Jagellonian University
Cracow, Poland

G. Szaloki
Universität Basel
Basel, Switzerland

G. Tagliaferri
University of Milano
Milano, Italia

W.J. Thompson
University of North Carolina
Chapel Hill, USA

M. Thumm
Universität Tübingen
Tübingen, Germany

L.D. Tolsma
Eindhoven University of Technology
Eindhoven, Nederland

G. Tornielli
University of Padua
Padua, Italia

T.A. Trainor
University of Washington
Seattle, USA

O. Traudt
Max Planck Inst. für Chemie
Mainz, Germany

D. Trautmann
Universität Basel
Basel, Switzerland

P.B. Treacy
Australian National University
Canberra, Australia

R. van Dantzig
Inst. f. Nuclear Physics Research I.K.O.
Amsterdam, Nederland

P.J. van Hall
University of Technology
Eindhoven, Nederland

L. Vanneste
Katolieke Universiteit Leuven
Heverlee, Belgium

R. van Wageningen
Vrije Universiteit
Amsterdam, Nederland

J. Vavra
University of British Columbia
Vancouver, Canada

B.J. Verhaar
Technische Hogeschool
Eindhoven, Nederland

P. Viatte
Eidg. Techn. Hochschule
Zürich, Switzerland

F. Vogler
Universität Erlangen
Erlangen, Germany

H.V. von Geramb
Kernforschungsanlage Jülich
Jülich, Germany

H. Wäffler
Max Planck Institut für Chemie
Mainz, Germany

R. Wagner
Universität Basel
Basel, Switzerland

R.L. Walter
Duke University
Durham, North Carolina, USA

M. Wangler
Universität Erlangen
Erlangen, Germany

S.D. Wassenaar
Technische Hogeschool
Eindhoven, Nederland

J. Weber
University of Neuchatel
Neuchatel, Switzerland

Ch. Weddigen
Universität Karlsruhe
Karlsruhe, Germany

A. Weinig
Universität Bonn
Bonn, Germany

W. Weiss
Max Planck Inst.f.Kernphysik
Heidelberg, Germany

W.G. Weitkamp
University of Washington
Seattle, USA

H.R. Weller
Triangle Universities
Durham, North Carolina, USA

D. Werren
University of Geneva
Geneva, Switzerland

K. Wienhard
University of Giessen
Giessen, Germany

H.B. Willard
Case Western Reserve University
Cleveland, Ohio, USA

G. Zannoni
University of Padua
Padua, Italia

H. Zingl
Universität Graz
Graz, Austria

B. Zeitnitz
II. Inst.f.Experimentalphysik
Hamburg, Germany

Subject Index

Scattering and reactions on specific nuclei as well as the analyses are listed in paragraphs XIV to XVI. The page numbers of survey papers are underlined.

Author Index

Printed in the United States
By Bookmasters